Smith, Currie & Hancock's FEDERAL GOVERNMENT CONSTRUCTION CONTRACTS

A Practical Guide for the Industry Professional

SECOND EDITION

SMITH, CURRIE & HANCOCK EDITORS
Thomas J. Kelleher, Jr.
Thomas E. Abernathy, IV
Hubert J. Bell, Jr. Steven L. Reed

CHAPTER EDITORS
Thomas E. Abernathy, IV
Hubert J. Bell, Jr.
G. Fritz Hain
Douglas P. Hibshman
Y. Lisa Colon Herron Ramsey Kazem
Stephen J. Kelleher
Thomas J. Kelleher, Jr.
Lenny N. Ortiz
Steven L. Reed
Douglas L. Tabeling
Mark S. Wierman

Foreword by
Stephen E. Sandherr
Chief Executive Officer
The Associated General Contractors of America

WILEY

John Wiley & Sons, Inc.

Library of Congress Cataloging-in-Publication Data:

Smith, Currie & Hancock's Federal government construction contracts : a
    practical guide for the industry professional / editors, Thomas J Kelleher Jr. . . . et al.] ; contributors,
    Thomas E. Abernathy . . . [et al].
        p. cm.
    Includes index.
    ISBN 978-0-470-53976-7 (cloth/cd)
        1. Government purchasing—Law and legislation—United States. 2. Public contracts—United
States. 3. Letting of contracts—United States. 4. Construction contracts—United States. I. Kelleher,
Thomas J. II. Abernathy, Thomas E. III. Title: Smith, Currie and Hancock's Federal Government
Construction Contracts. IV. Title: Federal government construction contracts.
    KF865.S627 2010
    346.7302'3—dc22

                                                                                    2009045888

ISBN: 978-0-470-53976-7

Printed in the United States of America

V10008778_031819

Contents

9   Delays, Suspension, and Acceleration   **329**

12 Payment and Performance Bonds 463

13 Equitable Adjustments and Costs 483

FOREWORD

The Associated General Contractors of America (AGC) has remained the nation's leading construction trade association since it was founded in 1918 at the request of President Woodrow Wilson. With our membership of more than 33,000 firms, including 7,500 of America's leading general contractors, nearly 12,500 specialty contractors and more than 13,000 suppliers and service providers, the AGC is one of the most visible organizations in the construction industry.

AGC members construct all types of public, private and commercial projects including airports, flood control facilities and dams, courthouses, rail and transit facilities, hospitals, multi-family projects, as well as specialized military and defense related facilities. The federal government, as the single largest customer for construction projects in the United States, contracts for nearly every type of facility or building constructed by AGC members. The federal government's need for construction services extends worldwide. For that reason, the AGC has been a leader in developing strong relationships and partnering agreements with the various federal agencies. These efforts in the federal government construction market are not new for the AGC, but the current economic reality has greatly increased contractor opportunities on federal government projects which are important for all AGC members.

As part of our strategic mission, one of the AGC's core services for industry members is education, through a wide variety of online and in-person seminars and an extensive bookstore, all accessible at *www.AGC.org*. The primary goal of these publications is to provide industry professionals with materials that are timely, informative, and practical. To further that goal, the leadership of the Federal and Heavy Construction Division of AGC is pleased to join with John Wiley & Sons, Inc. to authorize the preparation of the Second Edition of Smith, Currie & Hancock's *Federal Government Construction Contracts: A Practical Guide for the Industry Professional*. Federal government construction contacting is an especially significant market segment for our industry at this time. More importantly, federal government contracts can appear to be a Byzantine complex of laws, regulations, and practices. The enactment of the American Recovery and Reinvestment Act created both new opportunities for contractors and added to the potential complexity of working for the federal government as an owner. This book provides industry professionals with a guide to understanding their obligations and rights, and it is a tool for orienting project management to the procedures and complexities of federal government construction contracting. I am confident this publication will help make the often challenging task of dealing with the federal government, easier to understand, and ultimately easier to manage productively.

Stephen E. Sandherr
Chief Executive Officer
The Associated General Contractors of America

PREFACE

Federal government contract law is essential to the American construction industry. It furnishes the theories and basic principles that contribute to the smooth running of the construction process and often provides the foundation for similar principles applicable to both private and state and local public construction contracts. When the construction process falters and a dispute arises involving a federal government construction contract, there are a variety of procedures available for the resolution of these differences. While these vary in their formality, an appreciation of these processes and the applicable principles should not be limited to lawyers. Individuals in management positions must be cognizant of what their contracts with the federal government and the law require of them. They also need to know what they can expect and require from others and ensure that their employees who are responsible for the direct management of these contracts understand their duties and obligations to the federal government and their own companies. Nor can lawyers focus solely on legal rules and procedures to the exclusion of the *business* of construction and expect to represent and assist their clients effectively. Federal government construction law and the business of construction are inextricably intertwined. We hope this book reflects this interrelationship in the topics that it covers and the various perspectives and approaches it employs.

Claims and disputes are necessarily addressed in any book on federal government construction law. They must be in any complete and competent analysis of that environment. However, that is only one aspect of *Federal Government Construction Contracts—A Practical Guide for the Industry Professional, Second Edition*. Our goal for this book is to help provide the kind of insight and understanding needed to *avoid* claims and disputes whenever possible. Reasonable recognition of the contractual allocation of the parties' rights, risks, and legal responsibilities, coupled with a spirit of communication and teamwork in the execution of the work, is far more likely to culminate in a successful project than an atmosphere rife with confrontation and dispute. Of course, the possibility of claims and disputes cannot be ignored. Careful attention and planning is required to avoid disputes and to deal with them effectively if they become inevitable.

Several developments in the last few years confirmed the need for a new edition of this book. These include, but are not limited to, comprehensive and demanding new requirements for contractor ethics and compliance programs, increased use by

the federal agencies of varying project delivery methods and the added requirements associated with the American Recovery and Reinvestment Act, etc. In addition, the hard economic reality of the last two years has prompted many firms to
look at opportunities with the federal government for the first time in many years.
Hence more emphasis is devoted in this Second Edition to materials that would be
useful to people who are being introduced to the federal government contracting for
the first time.

The primary purpose of this book is to assist contractors to understand the process of federal government contracts with its numerous "square corners." It may be
impossible to provide all the answers, but the book should assist in then identification of the key issues and questions. To make these materials more useful to you, we
have included checklists, sample forms, and summary "Lessons Learned and Issues
to Consider" for each chapter. These checklists are provided as a means to assist
the user of this book in applying the concepts and principles in a practical manner.
In addition, copies of these checklists may be electronically accessed at a support
Web site created by John Wiley, www.wiley.com/go/federalconstructionlaw, in an
MS Word format to permit the user to copy and adapt them as needed for a particular contractor's project and organization. Finally, *Federal Government Construction
Contracts: A Practical Guide for the Construction Industry, Second Edition* remains
a general teaching tool and is not a substitute for the advice of an attorney experienced in this field. Specific concerns and problems require the timely attention of
legal counsel familiar with federal government contracts.

We thank our clients, who have shared their insights and concerns and provided
the opportunities for the experiences that are shared in this book. We also owe substantial gratitude to the staff of The Associated General Contractors of America,
especially Marco A. Giamberardino, for that organization's assistance in organizing
an Industry Advisory Council of over two dozen general contractors and specialty
contractors. That council provided specific recommendations which are reflected in
the revised content and organization of this edition of our book We also owe our
gratitude to the construction industry as a whole for allowing us as construction
attorneys to participate in the challenges of the industry to avoid and resolve problems. We hope this work will contribute to the worthy goals of the industry. We also
hope that this book helps its readers understand their commitments to federal government construction contracting from concept to successful project completion.

The attorneys of Smith, Currie & Hancock LLP have practiced federal government contract law since the firm's establishment in 1965. During that time, we have
represented contractors on thousands of federal construction contract matters and
conducted hundreds of construction and government contract law seminars for clients, trade associations, colleges and universities, and professional groups. Our consistent goal has been to provide a practical, commonsense perspective on the legal
issues affecting our clients and the construction industry. In many respects this book
reflects a culmination and refinement of those educational endeavors, the practical
approach they entail, and the experience we have gained in our service to the federal
government contract bar. In that regard, we are especially proud of the fact that our
fellow practitioners have chosen three of our partners—Overton A. Currie, Thomas

E. Abernathy, IV, and Hubert J. Bell, Jr.—to serve as chair of the Section of Public Contract Law of the American Bar Association. Tom Abernathy and Hugh Bell are two of the four general editors for this book.

Thomas J. Kelleher, Jr. on behalf of
Smith, Currie & Hancock LLP
Atlanta, Georgia
January 2010

SMITH, CURRIE & HANCOCK LLP

A Firm Concentrating Its Practice on the Construction Industry

During the span of the last 45 years Smith, Currie & Hancock LLP, with offices in Atlanta, Georgia; Charlotte, North Carolina; Fort Lauderdale and Tallahassee, Florida; Las Vegas, Nevada; and Washington, D.C., has developed a nationally recognized practice focused on the construction industry and the variety of legal issues facing that industry. While the firm represents private and public clients working or located in all fifty states, as well as Mexico, Canada, Central and South America, Europe, Asia, and Africa, federal government construction contract law has been one of our principal practice areas since our founding.

After developing construction and employment law practices in the context of a general service firm, G. Maynard Smith, Overton A. Currie, and E. Reginald Hancock formed Smith, Currie & Hancock in 1965 to concentrate their practices in those areas in order to provide more effective, focused service to the firm's clients. Having trained and practiced law in the culture created by those three outstanding attorneys, the current members of the firm remain committed to a long-standing tradition of providing quality, cost-effective legal services to clients ranging from small, family-owned concerns to multibillion-dollar corporations.

In representing the many construction industry participants competing for and performing federal government construction contracts, we are necessarily involved in a wide variety of legal and business-related issues. The breadth of those issues is reflected by the spectrum of topics addressed in *Federal Government Construction Contracts—A Practical Guide for the Industry Professional, Second Edition.* The goal of this book, as well as its companion book, *Common Sense Construction Law, Fourth Edition,* is to provide an informative discussion of these topics for the construction professional without all of the specific details of a multivolume legal treatise. To accomplish that task in a practical and meaningful manner, this book reflects the collective efforts of many attorneys drawing on Smith, Currie's nearly 1000 years of collective experience in the areas of construction law and federal government contracts law.

Along with our extensive legal experience, Smith, Currie's attorneys understand the industry we serve. Many of the firm's attorneys have engineering degrees in addition to their legal education, and several worked in the construction industry prior to pursuing their law degrees. Others joined this firm after military service as government contracts legal counsel or have extensive training in public procurement. Three members of Smith, Currie & Hancock have served as chairs of the Section of Public Contract Law of the American Bar Association, and another partner has served as the chair of the American Bar Association Forum Committee on the Construction Industry.

Smith, Currie & Hancock has represented clients from the entire spectrum of the construction industry: contractors, subcontractors, construction managers, owners (public and private), architects, engineers, sureties, insurance companies, suppliers, lenders, real estate developers, and others. They include multinational and *Fortune* 500 companies and trade associations active in this multi-billion dollar industry as well as local and regional clients. While our attorneys have appeared in numerous reported court decisions and even more arbitrations, our primary goal has been to achieve resolution of differences by communication and agreement rather than formal litigation. Consequently, over the last forty-five years we have assisted in the amicable resolution of many more matters than these reported decisions. It is this vein of service to the construction industry that we authored this Second Edition of *Federal Government Construction Contracts—A Practical Guide for the Industry Professional*.

In addition to serving clients nationwide, Smith, Currie & Hancock attorneys have published numerous articles in trade magazines and other periodicals and have authored or coauthored dozens of books on construction and public contract law. Our lawyers maintain a heavy schedule of lectures and seminars sponsored by various trade associations, colleges, and universities, including The Associated General Contractors of America, the U.S. Army Corps of Engineers, Georgia Institute of Technology, Auburn University, the American Bar Association, the Practicing Law Institute, and the Associated Builders and Contractors, Inc.

BIOGRAPHICAL DATA

THOMAS E. ABERNATHY IV. Vanderbilt University, B.A. 1963 and J.D. 1967; Omicron Delta Kappa; Captain, U.S. Army, 1967–1971; taught government contract law at U.S. Army JAG School, 1969–71; co-editor: *Federal Government Construction Contracts*, John Wiley and Sons, Inc. (2008); contributing author: *Construction Business Handbook*, Aspen Publishers (2004); co-author: "Changed Conditions/Edition II," *Construction Briefings*, West Group, No. 2000-9 (September 2000); "Developments in Federal Construction Contracts," *Wiley Construction Law Update*, Wiley Law Publications (1992–1999); "Resolving Government Construction Claims Without Litigation," 93-10 *Briefing Papers*, Federal Publications, September 1993. Lecturer on construction and government contract matters for the American Bar Association,

Georgia Institute of Technology, University of Wisconsin Engineering School, Emory University and Florida State University Law Schools, McGraw-Hill/FW Dodge and government agencies including the U.S. Army Corps of Engineers and U.S. General Services Administration. Professional service memberships: American Bar Association, Section of Public Contract Law; Chair, Section of Public Contract Law (1983–1984); Fellow, American College of Construction Lawyers (since 1990); ABA, Forum on the Construction Industry; ABA Section of Litigation (Construction Committee) and Tort and Insurance Practice (Fidelity and Surety Committee); and American Arbitration Association Panel of Arbitrators. Member: State Bars of Georgia and Tennessee.

CLIFFORD F. ALTEKRUSE. B.A., Reed College (Meritorious Senior Thesis); J.D., University of South Carolina School of Law; Law Review, Order of the Coif. Lecturer on Alternative Dispute Resolution, Emory University School of Law, Georgia State University School of Law; Lecturer on Alternative Dispute Resolution and Construction Law for various CLE programs and professional groups. Professional service memberships: State Bar of Georgia (Past Chair and Member, Alternative Dispute Resolution Section); Atlanta Bar Association (Member, Construction Law and ADR Sections); American Bar Association (Member, Alternative Dispute Resolution Section and Forum on the Construction Industry); Lawyers Club of Atlanta; Consortium on Negotiation and Conflict Resolution. Member: State Bar of Georgia; State Bar of Texas (inactive).

MAURA ANDERSON. B.B.A., University of Notre Dame, 1991; M.A., Valparaiso University, 1992; J.D., Emory University, 1996. A LEED Accredited Professional. President, Greater Ft. Lauderdale Chapter of the National Association of Women in Construction; Member, the Associated Builders and Contractors Green Building Committee. Co-author: *The Hidden Legal Risks of Green Building*, 2009; speaker, The Risks of Green Building, Lorman Seminars, 2007–2009. Professional service memberships: American Bar Association, Atlanta Bar Association. Member: State Bars of Florida, Georgia and South Carolina.

WILLIAM L. BAGGETT, JR. A.B., Dartmouth College, 1984 (*magna cum laude*); Phi Beta Kappa; J.D., Vanderbilt University, 1987. Articles editor, *Vanderbilt Law Review*, 1986–87. Law clerk to the Honorable Albert J. Henderson, Senior Circuit Judge, United States Court of Appeals for the Eleventh Circuit, 1987–88. Professional service memberships: American Bar Association, Section of Litigation, Forum Committee on the Construction Industry; Lawyers Club of Atlanta. Member: State Bars of Georgia and Tennessee.

PHILIP E. BECK. B.S., University of Tennessee, 1978 (high honors); M.B.A., University of Tennessee, 1981; J.D., University of Tennessee, 1981; Order of the Coif; Moot Court Board; Omicron Delta Kappa; Beta Gamma Sigma; Phi Kappa Phi. Law Clerk to the Honorable Houston Goddard, Tennessee Court of Appeals, 1980–81. Co-author: "The Owner Contemplating Litigation and Its Alternatives," *Construction Litigation: Representing the Owner* (John Wiley and Sons, 1984; 2d ed. 1990); "The Contractor Contemplating Litigation and Its Alternatives," *Construction Litigation: Representing the Contractor* (John Wiley and Sons, 1986; 2d ed. 1992); "Construction Contracts," *Real Estate Transactions Handbook* (John Wiley and Sons, 1988; 2d ed. 1993; 3d ed. 2000). Professional service memberships: American Bar Association

Litigation and Public Contract Law Sections and ABA's Forum Committee on the Construction Industry; Member, Associated General Contractors of America, Inc's. Board of Directors. Member: State Bars of Georgia, Florida and Tennessee.

HUBERT J. BELL, JR. B.A., Davidson College, 1966; M.A., Pacific Lutheran University, 1974; J.D., University of Georgia, 1981 (*cum laude*); Order of the Coif. Author: "Inadvertent Disclosure of Privileged Material," *Georgia State Bar Journal,* 18(4) (May 1982); "The Economic Loss Rule; A Fair Balancing of Interests" (co-author), *Construction Lawyer* II(2) (April 1991); *Alternative Clauses to Standard Construction Contracts* (2d ed.) (contributing author) (Aspen Law and Business, 1998); "Expertise That Is 'Fausse' and Science That Is Junky: Challenging a Scheduling Expert" (co-author), *The Procurement Lawyer* 35(1) (Fall 1999). United States Army 1966–1978, USAR 1978–1997, Lieutenant Colonel, Retired, Federal Contracts Auditor, Department of the Army (Aviation Systems Procurement Officer). Professional service memberships: American Bar Association, Section of Litigation, Section of Public Contract Law (Council Member 1997–2000, Secretary, Vice-Chair, Chair-Elect, Chair 2003–2004), Forum Committee on the Construction Industry; Lawyers Club of Atlanta (President, 1998–1999). Member: State Bars of Georgia, Florida and the District of Columbia.

JAMES K. BIDGOOD, JR. B.S., Civil Engineering, Georgia Institute of Technology, 1974; M.B.A., Phillips University, 1980; J.D., Emory University, 1983 (with distinction); *Emory Law Journal*; Order of the Coif; USAF, 1974–1980. Frequent lecturer on topics of construction and environmental law, and "Green" construction. Co-author, "Cutting the Knot on Concurrent Delay," *Construction Briefings* (February 2008). Regional Coordinator, *State Public Construction Law Source Book* (CCH 2002). Professional service memberships: American Bar Association, Section of Public Contract Law; American Arbitration Association—Panel of Neutral Arbitrators. Member: State Bar of Georgia.

JAMES F. BUTLER, III: A.B., Duke University; J.D., University of Kentucky School of Law. He has extensive experience with design-build and EPC projects, representing EPC contractors, owners and designers, including development of contracts, RFP procedures and selection criteria for private and public owners. He has substantial EPC dispute resolution experience and has worked with the National Society of Professional Engineers Design-Build Task Force to develop a design-construct curriculum for the engineering community and construction industry and was awarded the Rich Allen Meritorious Service Award for his efforts. Adjunct Professor Georgia Institute of Technology's Building Construction Department; Mediator for the American Arbitration Association and a member of its Construction Industry National Panel of Arbitrators. Contributing author to the *Design-Build Contracting Handbook, Alternative Clauses to Standard Construction Contracts,* and *The AGC Environmental Risk Management Procedures Manual.* Member: State Bars of Georgia, Florida, Kentucky, Texas, Arkansas, Washington, New York and North Carolina.

MARK B. CARTER. B.S., University of Alabama, 1992; J.D., University of Georgia, 2003; 2006–2006 Georgia Super Lawyers' Rising Star; Design-Build Institute of America (Board Member 2006–2007); Practice concentrated in negotiation, arbitration, litigation, and mediation of public and private construction contracts; Experienced in federal and state bid protests as well as appellate practice before the United States Court of Appeals for the Federal Circuit. Member: State Bar of Georgia.

ROBERT C. CHAMBERS: Managing Partner of Smith, Currie & Hancock LLP. B.S.C.E., Georgia Institute of Technology, 1979 (high honors); M.S.C.E., 1980 (Study Emphasis Geotechnical Engineering); J.D., University of Georgia, 1985 (*magna cum laude*); Order of the Coif. Co-author: "Changed Conditions," *Construction Briefings,* No. 84-12, "Changed Conditions II," *Construction Briefings*, No. 2000-9, Federal Publications. Professional service memberships: American Bar Association Section of Public Contract Law; Section of Environmental Law; Forum Committee on the Construction Industry; State Bar of Georgia, Environmental Law Section. Member: State Bars of Georgia and The District of Columbia.

ROLLY L. CHAMBERS. B.A., University of North Carolina at Chapel Hill, 1975 (*with Honors*); J.D., Marshall-Wythe School of Law of the College of William and Mary, 1984. William and Mary Law Review, 1982–1984, Executive editor for student contributions, 1983–1984. N.C. Dispute Resolution Commission Certified Mediator—Superior Court, Estates and Guardianship. Professional service memberships: American Bar Association (Litigation Section and Construction Industry Forum), North Carolina Bar Association (Construction Law Section and Litigation Section), Mecklenburg County Bar Association. Member: State Bar of North Carolina.

IAN C. CLARKE. B.S. Electrical Engineering, Florida Agriculture and Mechanical University 1995 (with Honors); J.D. Emory University School of Law 2007; Appointed to steering committee for the ABA Forum on the Construction Industry, Division 1: Dispute Avoidance and Resolution; contributing author, *Common Sense Construction Law*, chapter 7: "Authority and Responsibility of the Design Professional"; author, "Specify 'Arbitration' with Clarity," *Construction Update* (July 2008); Member: State Bar of Georgia.

AUBREY L. COLEMAN, JR. Senior Partner of Smith, Currie & Hancock LLP. B.A., (*cum laude*) Tulane University, 1964, Phi Beta Kappa; LL.B., Vanderbilt University, 1967, Order of the Coif, Managing editor, *Vanderbilt Law Review*, 1966–1967. Captain, U.S. Army 1968–1969. Co-author: Georgia Construction Law, Professional Education Systems, 1984. Co-author: *Avoiding and Resolving Construction Claims* (Federal Publications, Inc., 2001); Author and lecturer on various aspects of construction law and construction claims preparation, presentation, and defense, including participation in programs presented by Georgia State University, Georgia Tech, University of Kentucky Law School, Emory Law School, Florida State Law School, Professional Education Systems, Inc., and Federal Publications, Inc.; Vice-Chair, ABA Fidelity and Surety Law Committee 1999–2001. Professional service membership: American Bar Association, Section of Public Contract Law. Member: State Bar of Georgia.

JAMES W. COPELAND. B.E., Vanderbilt University, 1988 (Dean's List, *Chi Epsilon*). J.D., Duke University, 1995 (Dean's Advisory Council). Admitted to the United States District Court for the Northern District of Georgia, the United States

Court of Claims, and the United States Court of Appeals for the Federal Circuit. Concentrated on litigation, arbitration, compliance, and transactional matters for prime contractors, subcontractors, and other participants in the construction industry. Professional service membership: American Bar Association, Section of Public Contract Law. Member: State Bar of Georgia.

MATTHEW E. COX. B.S., Brigham Young University, 1996; Mu Sigma Rho (Statistical National Honor Society); J.D., University of South Carolina, 1999; Law Clerk to the Honorable J. Derham Cole, 7th Judicial Circuit, State of South Carolina, 1999–2000; author of "Who is a Subcontractor Under the Miller Act?" *Common Sense Contracting* (Summer 2008); author of "Value Engineering Change Proposals—Who Gets What?" *Common Sense Contracting* (Summer 2007); represented multiple contractors regarding Miller Act claims. Member: State Bars of South Carolina, North Carolina and the District of Columbia.

F. ALAN CUMMINGS. B.S., Rhodes College and Auburn University, 1967. J.D., Florida State University, 1975 (with honors). Member, Florida State University Law Review, 1974–1975. Member: American Bar Association (Professional service memberships: American Bar Association, Section of Public Contract Law; Litigation Section; Forum Committee on the Construction Industry, 1982; Tort and Insurance Practice Section (Fidelity and Surety Committee). Member: State Bar of Florida.

JOSEPH J. DINARDO. B.S., Biology and Chemistry, Palm Beach Atlantic University, 1995 (*summa cum laude*); J.D., Cumberland School of Law, 1999; LL.M., International Business Transactions, Emory University School of Law, 2000. Author: "Who Is a Miller Act 'Subcontractor'?" *Common Sense Contracting*, Spring 2009; co-author Owners' Amending AIA A132-2009: *Standard Form of Agreement between Owner and Contractor, Construction Manager as Advisor Edition* and Contractors' Amending AIA A132-2009: *Standard Form of Agreement between Owner and Contractor, Construction Manager as Advisor Edition, Alternative Clauses to Standard Construction Contracts* (3d ed.) represented multiple contractors in filing and foreclosing claims of lien and prosecuting garnishment actions. Member: State Bar of Georgia; U.S. Patent and Trademark Bar.

KARL DIX, JR. B.S., Economics with major concentrations in Accounting and Decision Science from the Wharton School of Finance and Commerce at the University of Pennsylvania, 1980 (*magna cum laude*). J.D., Specialization in International Law from Cornell University, 1983 (*cum laude*). Captain, U.S. Army Corps of Engineers, Attorney Advisor, 1983–1987; Chief Counsel's Honors Program; Army Meritorious Service Medal with Oak Leaf Cluster. Professional service membership: American Bar Association, Section of Public Contract Law. Member: State Bars of New York, Georgia and Florida.

J. WILLIAM EBERT: B.S., Biology, University of Utah, 1976; J.D., University of Utah College of Law, 1980. Member: State Bars of Nevada and Utah (inactive).

ROBERT O. FLEMING, JR. Admitted to bar, 1977, Georgia. B.S. with honors from North Carolina State University in 1972. J.D., with distinction, from Duke University in 1977. Co-author: *Deposition Strategy, Law and Forms: Building Construction* (Matthew Bender, 1981). Adjunct Professor, Legal Aspects of Architecture, Engineering and the Construction Process, 1978, Construction Finance, 1979,

Georgia Institute of Technology. Approved Arbitrator for Construction Contract Disputes, American Arbitration Association. Member: State Bar of Georgia.

PHILIP L. FORTUNE. B.A., University of North Carolina at Chapel Hill, 1967, Gamma Beta Phi; J.D., University of Toledo, 1970. Member, *University of Toledo Law Review*; Instructor of Federal Procurement Law, Emory Law School, 1972. Author and lecturer on construction law, bonds, liens and preparation of claims, engineering practice and liability, and environmental protection laws including participation in programs presented by The Forum on the Construction Industry, Georgia Tech, Five County Builders and Contractors Association, Inc., Ft. Myers, Florida, the Carolinas AGC and Georgia AGC and various professional groups. Professional service membership: American Bar Association, Section of Public Contract Law. Member: State Bar of Georgia.

ROBERT J. GREENE. B.A., University of Utah (*cum laude*), 1973; J.D., University of Utah, 1978. Member; American Bar Association (Construction Industry Forum) and North Carolina Bar Association (Construction Law Section and Litigation Section), former member of National Panel of Arbitrators of the American Arbitration Association; Certified Mediator for North Carolina Superior Court; founding member, The Arbitration Group of the Carolinas. Member: State Bar of North Carolina.

C. DAMON GUNNELS: B.S., Accounting, University of Tennessee, 2002; J.D., University of Tennessee 2006, managing editor, *Transactions, Tennessee Journal of Business Law.* American Bar Association—Forum Committee on the Construction Industry. Georgia Branch, Associated General Contractors— Young Leadership Program; author, "Waiver of Right to Arbitrate," *Atlanta Bar Association Construction Law Section Newsletter;* co-author, "Florida Statutory Lien Rights and Arbitration," *Common Sense Contracting* (Fall 2009); represented numerous general contractors and subcontractors in complex construction disputes. Member: State Bar of Georgia.

G. FRITZ HAIN. B.S. Construction Management, Florida International University, 1997 (*magna cum laude*); J.D., Florida State University College of Law, 2008 (*magna cum laude*); Certified Building Contractor, State of Florida License No. CBC058197 since 1997. U.S. Green Building Council LEED Accredited Professional since 2003; represented multiple contractors and design-build entities on government contract bid protest and contract dispute matters. Professional service membership: American Bar Association; Member: State Bar of Georgia.

DIRK D. HAIRE. B.S., History and Political Science, Ball State University (*cum laude*); J.D., George Washington University Law School. Professional service memberships: ABA, Forum on the Construction Industry; Associated General Contractors of America, Board of Directors; Maryland Chapter, AGC of America, Board of Directors; Metro DC Chapter, AGC of America; Professional service membership: American Bar Association, Section of Public Contract Law. Member: State Bars of District of Columbia, Indiana, Maryland, and Minnesota.

EUGENE J. HEADY. B.S., Engineering, University of Hartford, 1981; Kappa Mu Honorary Engineering Society; J.D., Texas Tech University School of Law, 1996 (*cum laude*); Phi Delta Phi. Editor-in-chief, *Texas Tech Law Review*, 1995–1996.

Author: "Stuck Inside These Four Walls: Recognition of Sick Building Syndrome has Laid the Foundation to Raise Toxic Tort Litigation to New Heights," 26 *Tex. Tech. L. Rev.* 1041 (1995); "Contractors' Amending AIA A401-1997: Standard Form of Agreement between Contractor and Subcontractor;" "Subcontractors' Amending AIA A401-1997: Standard Form of Agreement between Contractor and Subcontractor," in *Alternative Clauses to Standard Construction Contracts* (Aspen Law and Business 1998). Co-author: Georgia chapter in *Fifty State Construction Lien and Bond Law*, Aspen Law and Business, 2000. Professional service memberships: Vice Chair Region IV-Georgia, American Bar Association, Section of Public Contract Law (1999-present); Forum on the Construction Industry. Member: State Bars of Georgia, Colorado, Florida and Texas.

Y. LISA COLON HERON. B.A. University of Miami 1996; J.D. University of Miami School of Law, 1999 (*cum laude*); Fellow, American Bar Association Forum Committee on Construction Industry; National Association of Minority Contractors, South Florida Chapter—board member. Lectures on key construction contract clauses—National Association of Minority Contractors; Turner School of Construction Management; represents contractors on small business contracting issues related to federal and local government contracting. Member: State Bar of Florida.

DOUGLAS P. HIBSHMAN. B.A., Bloomsburg University of Pennsylvania, 1993 (*magna cum laude*); J.D., University of Cincinnati College of Law, 1998; LL.M. Taxation, Georgetown University Law Center, 2006. Major, U.S. Marine Corps, Judge Advocate, 1999–2007. Law Clerk to the Honorable Mary Ellen Coster Williams, U.S. Court of Federal Claims, 2007–2008; Represented multiple contractors on GAO and U.S. Court of Federal Claims bid protest issues related to the award of federal government contracts, and on Small Business Administration size standard and NAICS Code protests. Professional service membership: American Bar Association, Section of Public Contract Law. Member: State Bars of Maryland, Ohio and The District of Columbia.

TIMOTHY W. JOHNSON. : A.B. (Government) Univ of So. Dakota 1974, with honors. J.D. Univ of Texas, 1977; Phi Beta Kappa; Legal Writing TQ—Texas Law School; presenter, Carolinas AGC Update, 2009; Vice-President and Corporate Officer for Labor Relations, Coca-Cola Enterprises 2001-2007; represented and advised multiple contractors on Davis-Bacon, Service Contract, FLSA, NLRB and related regulatory issues. Member: State Bar of Georgia.

KIRK D. JOHNSTON. B.S., Business Management and Marketing, Brigham Young University, 1996 (Dean's List); J.D., Santa Clara University School of Law, 2000; Represent government contractors in claims avoidance through assisting clients in every phase of the construction process, from the negotiation and drafting of federal contracts to educating clients through day-to-day review of projects, contracts, issues and project documents. Member: State Bars of California and Georgia.

REGINALD M. JONES. B.A, College of William and Mary, Williamsburg, Virginia, 1991; J.D., University of Georgia, 1999 (*cum laude*); Editorial Board, *Georgia Law Review*; Omicron Delta Kappa; captain, U.S. Army 1991–1996; author, "Lost

Productivity: Claims for the Cumulative Impact of Multiple Change Orders," 31 *Public Contract Law Journal* 1 (Fall 2001); author, "Update on Proving and Pricing Inefficiency Claims," 23 *The Construction Lawyer* 19 (Summer 2003), "Recovering Extended Home Office Overhead: What is the State of Eichleay?" 40 *The Procurement Lawyer* 8 (Fall 2004). Professional service memberships: American Bar Association, Public Contract Law Section (Chair, Construction Division 2009–2010) and Forum on the Construction Industry; Washington, DC and Virginia branches, Associated General Contractors of America, Inc.; Washington Building Council; Virginia Chapter, Associated Builders and Contractors, Inc. Member: State Bars of Georgia, Virginia and the District of Columbia.

S. GREGORY JOY. B.A. (with high distinction), University of Virginia, 1981, Phi Beta Kappa, Raven Honor Society; J.D., University of Virginia, 1984. Representative clients: owners, contractors, subcontractors and suppliers on numerous projects including public transit system, hospital, courthouse, school, university, office building, church, casino, steel mill, public park, public flood control, U.S. Navy wharf, federal and state military training facility, federal air traffic control facility, water and wastewater treatment facility, water, sewer and sanitary sewer line, federal warehouse, condominium and apartment complex, and other projects. Speaker at over 100 seminars on various construction and public contract law issues, as well as international contracting and environmental law issues. Co-editor and co-author, *Alternative Clauses to Standard Construction Contracts,* 3d ed. (Aspen Publishers 2009); second edition supplements (Aspen Publishers 2003–2008); co-author, *Common Sense Construction Law* (first, second, third and fourth editions) (John Wiley and Sons, 1997, 2001, 2005, 2009); co-author, *Federal Government Construction Contracts* (John Wiley and Sons, 2008); co-author, *Georgia Jurisprudence, Mechanic's and Materialmen's Liens; Other Liens on Real Property*, Property § 21B and 22, (Thomson West 2008); co-author, "Liens on Real Property," *Georgia Jurisprudence*, Property § 22 (Lawyers Cooperative Publishing 1995); author "Application of Selected American Laws to United States Companies: Foreign Corrupt Practices Act and Antiboycott Legislation," *Mercer Law Review*, (Winter 1992). Author of numerous manuals and articles regarding construction law, government contracts, international contracting, and environmental law. Professional service memberships: American Bar Association—Sections on Public Contract Law; Litigation and Environmental Law; Forum Committee on the Construction Industry. Member: State Bar of Georgia.

JASON D. KATZ. University of Virginia, 2001. J.D., Nova Southeastern University, 2005 (summa cum laude); *Law Review,* 2003–2005. Represented numerous contractors pursuing claims on public road and bridge projects. Member: State Bar of Florida.

RAMSEY KAZEM. B.S., Fairleigh Dickinson University, 1996 (*magna cum laude)*; M.B.A., Goizueta Business School, 2000; J.D., Emory University, 2000. Speaker at seminars on various topics including contract formation and interpretation, risk shifting clauses and project documentation. Represented government contractors on various contract claims and disputes including the successful challenge of the government's wrongful termination for default. Professional service membership: American Bar Association. Member: State Bar of Georgia.

STEPHEN J. KELLEHER. B.A., History, University of Notre Dame, 2001 (cum laude); M.Ed., University of Notre Dame, 2003; J.D., University of Virginia, 2008. Conducted ethics training for contractor in compliance with FAR Subpart 3.10; Represented contractors in multiple requests for equitable adjustments and claims relating to construction contracts as well as service contracts governed by the Service Contract Act; represented construction supplier in qui tam relator action under the Civil False Claims Act. Member: State Bar of Georgia.

THOMAS J. KELLEHER, JR. Senior Partner of Smith, Currie & Hancock LLP. A.B., Harvard University, 1965 (*cum laude*); J.D., University of Virginia, 1968; Co-author: "Inspection Under Fixed Price Construction Contracts," *Briefing Papers*, No. 766, Federal Publications, Inc., December 1976; "Preparing and Settling Construction Claims," *Construction Briefings*, No. 8312, Federal Publications, Inc., December 1983. *Construction Litigation: Practice Guide with Forms*, Aspen Law and Business; "Development in Federal Construction Contracts," *Wiley Construction Law Update*, Wiley Law Publications (1992–1999). The Judge Advocate General's School, U.S. Army, Procurement Law Division, 1970–1973. Creator and editor of the firm's construction law newsletter, *Common Sense Contracting*; co-editor of *Common Sense Construction Law* (4th ed.); co-editor of *Federal Government Construction Contracts* (2d ed). Professional service memberships: American College of Construction Lawyers; American Bar Association, Public Contract Law Section; Associated General Contractors of America, Inc.: Chair: Federal Acquisition Regulation Committee (1999–2003); Member of Corps of Engineers, Naval Facilities Engineering Command, Governmental Affairs, and Contract Documents Committees. Member: State Bars of Georgia and Virginia (inactive).

DAVID C. KING. B.A., Management, Maryville College, 1993 (*summa cum laude);* J.D., University of Georgia School of Law, 1996 *(cum laude*); President, Georgia Bar Association Tort and Insurance Practice Section; co-author, *Alternative Clauses to Standard Construction Contracts* (3d ed. 2009); successful representation of multiple contractors in protests of federal government contract awards. Professional service memberships: American Bar Association, Section of Public Contract Law, forum on the construction Industry. Member: State Bar of Georgia.

EVANGELIN LEE. A.B., University of Michigan, 2003; J.D., University of Nevada, William S. Boyd School of Law, 2006 (*cum laude*). Judicial Extern to Hon. Michael L. Douglas, Nevada Supreme Court, 2006. Executive Board, Asian Bar Association of Las Vegas. Professional service memberships: National Association of Women in Construction; American Bar Association Forum on Construction Industry,. Member: State Bar of Nevada.

S. ELYSHA LUKEN. B.A., University of Florida, 1996; J.D., Florida State University, 1999 (*magna cum laude).* Legislative editor and Editorial Board member, *Florida State University Law Review.* Professional service memberships: Florida Bar Appellate Practice Section; Florida Bar Trial Lawyers Section; Florida Bar Journal/News Editorial Board; American Bar Association, Construction Forum. Member: State Bar of Florida.

JOHN M. MASTIN, JR. B.S. Electrical Engineering, University of Alabama, 1976 (*cum laude*); J.D., University of Montana School of Law, 2005 (with honors). Member: American Bar Association, Forum on the Construction Industry; State Bars of Georgia and District of Columbia; *Common Sense Contracting*, 22(4) (Spring 2009); "Proving Acceleration," *Common Sense Contracting*, 22(1) (Spring 2008); "Contingent Payment Update," *Common Sense Contracting*, 21(1) (Spring 2007); co-author, "Construction Contracts," *Commercial Real Estate Transactions Handbook* (Aspen Publishers, 4th ed. 2009); represented owners, general contractors, and subcontractors in variety of construction matters in negotiation, mediation, arbitration, and litigation. Member: State Bar of Georgia.

ANDREW R. McBRIDE. B.A., History, Rutgers College, 1992; J.D., Emory University School of Law, 1996. Represented contractors on claims for variations in estimated quantities on disaster relief projects; represented sureties on Miller Act performance bond and payment bond claims. Member: State Bars of Georgia and New Jersey.

JOHN E. MENECHINO JR. B.A., Political Science and Economics, University of South Carolina, 1986 (*magna cum laude,* Phi Beta Kappa); J.D., University of South Carolina College of Law, 1989 (American Inns of Court Foundation). Academic fraternities: Gamma Beta Phi, Pi Sigma Alpha. Contributing author, *Construction Subcontracting: A Legal Guide* (John Wiley and Sons, 1991); *Alternative Clauses to Standard Construction Contracts* (John Wiley and Sons, 1991, supplements 1992, 1997; 2d ed. 1999). Professional service memberships: American Bar Association, Section of Public Contract Law; Forum on the Construction Industry; Tort and Insurance Practice Section; Fidelity and Surety Law Committee. Member: State Bar of Georgia.

GARRETT E. MILLER. B.A., University of North Carolina at Chapel Hill, 1996; J.D., University of Mississippi School of Law, 2001; Moot Court Board; Staff Member, *Journal of National Security Law*; Law School Honor Council Representative. Extensive representation of contractors on service and emergency disaster relief contracts for the U.S. Army Corps of Engineers, FEMA and other divisions of federal, state and local governments. Professional service memberships: American Bar Association, Section of Public Contract Law. Member: State Bar of Georgia.

GORAN MUSINOVIC. B.S., Business Administration, concentration in Accounting, collateral in International Business, minor in German, The University of Tennessee, 2006 (*summa cum laude*); J.D., The University of Tennessee College of Law 2009 (Concentration in Advocacy and Dispute Resolution, Order of Barristers, Moot Court Board). Member: State Bars of Georgia and Tennessee.

ERIC L. NELSON. B.A., California Polytechnic State University at San Luis Obispo; J.D., Washington and Lee University Law School. Co-author: "Delay Claims Against the Surety," *The Construction Lawyer*, 17(3) (July 1997); Co-author, "Trends in Construction Lost Productivity Claims," *Journal of Professional Issues in Engineering Education and Practice*, American Society of Civil Engineers (July 2004). Lecturer on construction claims for various professional organizations and seminars. Professional service memberships: American Bar Association,

Section of Public Contract Law (Chair Construction Division, Insurance and Bonding Subdivision 2001–present; Vice-chair Construction Division, Subcontracting Subdivision 2000–2001), Forum on the Construction Industry; Georgia Branch, Associated General Contractors of America, Inc. (Young Leadership Council 1998–2000). Member: State Bar of Georgia.

LENNY N. ORTIZ. B.S., Criminal Justice, Florida International University; J.D., Florida State University, 2001; Member of the Moot Court Team and Journal of Land Use and Environmental Law. Staff Attorney at the Florida Supreme Court, 2001–2002. Board Certified in Construction Law by the Florida Bar. Professional service memberships: American Bar Association; Broward County Bar Association; Broward County Bar—Construction Law Committee; Florida Bar—Construction Law Committee; ACE (Architecture, Construction and Engineering) Mentoring Program of America, board member and general counsel. Member: State Bar of Florida.

J. DANIEL PUCKETT. B.S. Mechanical Engineering, Georgia Institute of Technology, 2006; J.D., Florida State University College of Law, 2009. Member: State Bar of Georgia; U.S. Patent and Trademark Bar.

GENE F. RASH. B.S., North Carolina State University; J.D., Wake Forest University School of Law. Chi Epsilon National Civil Engineering Honors Society. Co-author: *Handbook of North Carolina Construction Law* (Carolinas AGC 1999). Professional service memberships American Society of Civil Engineers; North Carolina Bar Association; South Carolina Bar Association; American Bar Association; Member: State Bars of North Carolina and South Carolina.

STEVEN L. REED. B.S., Physical Geography (emphasis in mathematics), University of Georgia, 1972 (*magna cum laude*); J.D., University of Georgia School of Law, 1977. Administrative Judge/Mediator (ret.), Armed Services Board of Contract Appeals, 2000–2006; Administrative Judge/Mediator/Hearing Examiner, U.S. Army Corps of Engineers Board of Contract Appeals, 1988–2000; Division Trial Attorney, U.S. Army Engineer Division, Pacific Ocean, 1985–1988; Division Trial Attorney, U.S. Army Engineer Division, Europe, 1983–1985; Contracts Trial Attorney, U.S. Army Engineer Division, Pacific Ocean, 1981–1983; Litigator, U.S. Army Judge Advocate General's Corps, 1978–1981; State Prosecutor, Atlanta Metro Area, 1978. Colonel, U.S. Army Judge Advocate General's Corps (Reserve) (Ret.), 2002; Legion of Merit, 2002; Chief Senior Military Judge, Senior Military Judge, Military Judge, U.S. Army Reserve, 1996–2002; U.S. Army Field Artillery Officer, 1972–1975. Professional service memberships: American Bar Association, Section of Public Contract Law; State Bar of Georgia, Association for Conflict Resolution; Board of Contract Appeals Judges Association; Board of Contract Appeals Bar Association. Member: State Bars of Georgia and The District of Columbia.

RONALD G. ROBEY. B.A., Centre College and University of Kentucky, 1974 (with distinction in the Honors Program); J.D., University of Kentucky, 1977; Order of the Coif, lead articles editor, *Kentucky Law Journal*, 1976–1977. Author: "Construction Management—Avoid Being a Fiduciary," *Contractor Profit News* (October 1986); "A Telephone Sub-Bid Is Enforceable," *Contractor Profit News*, June 1985. Co-author: "Winning Strategies in Construction Negotiations,

Arbitration and Litigation," *Construction Contracts* (Practicing Law Institute, 1986). Chair of CLE construction seminars conducted in Atlanta, Orlando, and Las Vegas. Professional service memberships: American Bar Association, Section of Public Contract Law; Forum on the Construction Industry. Member: State Bars of Georgia, Florida, Kentucky, Michigan and Nevada.

CHARLES E. ROGERS. A.B., Duke University, 1989; J.D., Emory University, 1993. Lecturer on design, construction and handicap accessibility for various CLE programs and professional organizations. Professional service memberships: American Bar Association, Litigation Section and Construction Industry Forum; Atlanta Bar Association, Construction Section and Litigation Section. Member: State Bar of Georgia.

JESSICA L. SLATTEN. B.S., Conservation Biology, Tennessee Technological University, 2002 (*magna cum laude*); J.D., Florida State University College of Law, 2006 (*summa cum laude*); Rising Star designation in Construction/Surety Law by *Florida Super Lawyers*, 2009; author of "Avoid Losing a Bid Protest Before It Begins," *Common Sense Contracting*, 23(2) (Summer 2009). Member: State Bar of Florida.

KENT P. SMITH. B.A., Northwestern Oklahoma State University, 1963; JD, Washburn University School of Law, 1966 (*cum laude*); *Washburn Law Journal* Editorial board, alumni Fellow in Law, 1992. Law clerk to the U.S. Court of Appeals for the Tenth Circuit (1966–1967); attorney for the United States Atomic Energy Commission (1967–1968). Admitted to the U.S. Claims Court and the Fourth, Fifth, Tenth, Eleventh, and Federal U.S. Circuit Courts of Appeal. Author: "Contractor Default: How to Recognize It, What to Do About It, and How to Finish the Project," The Florida Bar (5th Annual Construction Contract Litigation Seminar), 1981; "Changed Conditions in Dredging Contracts," *World Dredging and Marine Construction*, 14(4) (1978); "Claims and the Construction Prime," *Federal Publications*. Co-author: "Differing Site [Changed] Conditions," *Federal Publications*, Briefing papers, 71-5. Lecturer on various construction law topics for the Federal Bar Association, the American Bar Association, the Practicing Law Institute, Emory University, Hamline University, the University of Wisconsin, and other educational institutions and organizations. Guest lecturer, Georgia Institute of Technology, Legal Aspects of Architecture, Engineering and Contracting. Professional service membership: American Bar Association, Section of Public Contract Law. Member: State Bars of Georgia and Kansas.

JOSEPH C. STAAK. B.S., Civil Engineering, Georgia Institute of Technology, 1974; J.D., University of Georgia, 1981 (*cum laude*); *Georgia Law Review*, 1979–1981; Research editor, 1980–1981; Design and Construction Engineer, USAF, 1974–1978. Co-author: *Georgia Construction Lien and Public Contract Bond Law* (Lorman Education Services, 1993); *Georgia Construction Law* (Lorman Education Services 1996); and *Government Construction Contracting Law: Federal and State Perspectives* (Lorman Education Services 2002). Lecturer on construction claims for various professional groups. Professional service memberships: American Bar Association, Section of Public Contract Law; American Society of Civil Engineers

(Associate Member). Certified in construction law by the Florida Bar. Member: State Bars of Georgia, Florida, Alabama and The District of Columbia.

CHARLES W. SURASKY. B.A., University of South Carolina, 1973 (*cum laude*); J.D., George Washington University, 1978 (with honors). Member, *George Washington Law Review* 1976–1978; Co-author "The Application of the U.C.C. to a Contractor's Relationship With Its Suppliers," *Construction Lawyer* (Fall 2009); Associate Director of the National Utility Contractors Association. Professional service memberships: American Bar Association, Section of Public Contract Law; forum on the Construction Industry. Member of the State Bars of Georgia, Florida, South Carolina and the District of Columbia.

DOUGLAS L. TABELING. B.A., University of Kentucky, 2000 (Honors Program); J.D., University of Kentucky, 2006. The Associated General Contractors, Young Leadership Program. Co-author, "Georgia Construction and Design Law," in *A State-by-State Guide to Construction and Design Law* (2d ed. 2009). Represented contractor in civil action under False Claims Act; advised contractors on compliance with public reporting requirements and Buy American provisions of 2009 American Reinvestment and Recovery Act; negotiated subcontract on federal government construction project. Professional service memberships: American Bar Association, Section of Public Contract Law; Forum on the Construction Industry. Member: State Bar of Georgia.

HELEN P. TAYLOR. B.S., Geography, University of South Carolina, 1997; M.S., Geography, University of Wisconsin at Madison, 2000; J.D., Tulane University Law School, 2005 (*cum laude).* Member: State Bar of Georgia.

JAMES B. TAYLOR. B.A., University of Tennessee, 2001; J.D., University of Tennessee School of Law, 2004 (*magna cum laude*). Co-author, "Cutting the Knot on Concurrent Delay," *Construction Briefings* 07-2; represented contractors in Miller Act claims regarding federal government contracts; advised clients regarding immigration, prevailing wage, union, and related labor issues in federal contracting. Member: State Bar of Georgia.

J. CLINT WALLACE. B.S., Real Estate, Risk Management/Insurance, Florida State University, 2005; J.D., Florida State University College of Law, 2008 (with honors). Author, "Contractors: Avoid Becoming a Banker," *Common Sense Contracting*; represented contractors on bid protest issues. Member: State Bar of Florida.

G. SCOTT WALTERS. B.C.E., Bachelor of Civil Engineering, The Georgia Institute of Technology, 1985, Chi Epsilon Honor Fraternity, 1984–1985; J.D., Widener University School of Law, 1996; External managing editor, *The Widener Law Symposium Journal,* 1995–1996. Professional service memberships: American Bar Association, Forum on the Construction Industry; Section of Environment, Energy and Resources; State Bar of Georgia: Environmental Law Section, Air and Waste Management Association-Georgia Chapter. Member: State Bar of Georgia.

GEORGE D. WENICK. B.S., University of Pittsburgh, 1972; J.D., University of Pittsburgh; Senior Claims Attorney, Office of Chief Counsel, Pennsylvania Department of Transportation, 1975–1984. Extensive national experience in public and private contracts, coupled with international experience primarily in China, Latin America, and Europe. Represented contractor in successful bid protest and eventual award of

$350 million contract for expansion of a major international airport; Represented contractor in bid protest concerning U.S. Government's award of a $100 million contract for satellite terminals; Represented constructor against construction manager in litigation concerning anthracite-burning power plant in Hanfeng, Shaanxi Province, PRC; Represented general contractor in bi-lingual arbitration against a Latin American Government, resulting in $51 million award in the contractor's favor; Represented EPC contractor in ICC arbitration in Paris against Italian developer concerning construction of two gas-fired power plants. Recognized by Chambers USA, Legal 500, and Georgia Super Lawyers. Languages: French, Spanish, and Italian. Professional service memberships: American Bar Association, Section of Public Contract Law; Forum on the Construction Industry. Member: State Bars of Georgia and Pennsylvania.

MARK S. WIERMAN. B.S., Naval Architecture, United States Naval Academy, 1988 (*cum laude*); M.B.A., Wake Forest, 1995 (with distinction); J.D., University of North Carolina, 2007. Member: State Bars of North Carolina, South Carolina and Virginia.

BRIAN A. WOLF. B.S.B.A. Real Estate and Urban Analysis, University of Florida, 1990; J.D. Stetson University College of Law, 1993; Stetson Law Review; intern, Honorable Judge Elizabeth A. Kovachevich, United States District Court, Middle District of Florida. Board Certified in Construction Law by The Florida Bar. Chairman, Florida Bar Construction Law Committee, RPPTL Section. Past Chairman, Construction Law Committee of the Broward County Bar Association. Author of Florida Lienlaw Online, Construction Publications, Inc. Chapter editor, *Common Sense Construction Law* (John Wiley and Sons, 3d ed., 2009). State Board of Directors, Associated Builders and Contractors Florida. Board of Directors, Associated Builders and Contractors Institute. Chairman, Legislative Committee Associated Builders and Contractors Florida East Coast Chapter. Florida Prestressed Concrete Association, Engineering Contractors Association and Underground Contractors Association. Member: State Bar of Florida.

JONATHAN Y. YI. B.A., University of Florida, 2001 (highest honors, University Scholar); J.D., University of Chicago Law School, 2004 (Rosensen Scholar). Represented service and construction in numerous government contracts matters, including bid preparation and compliance, local, state and federal level bid protests, and litigation before the Armed Services Board of Contract Appeals. Fluent in Korean. Member: State Bars of Florida, Georgia, and New York.

OF COUNSEL

ROBERT B. ANSLEY, JR. B.A., Vanderbilt University, 1962; J.D., Emory University, 1965; Bryan Honor Society. Co-editor-in-chief, *Journal of Public Law* and Emory Section, State Bar Journal of Georgia, 1965. Co-author: "Differing Site (Changed) Conditions," *Briefing Papers* No. 71-5 (Federal Publications, 1971). Member: State Bar of Georgia.

GLOWER W. JONES. A.B., Dartmouth College, 1958; LL.B., Emory University, 1963 (with distinction); Bryan Honor Society. Editor-in-chief of Emory Section of the *Georgia Bar Journal*, 1962–1963. Editorial Board of Georgia Bar Journal, 197579. Writer and lecturer for the American Bar Association National Institutes; "Punitive Damages as an Arbitration Remedy," *Journal of International Arbitrators* (June 1987); "Construction Contractors: The Right to Stop Work," *The International Construction Law Review* (July 1992 and October 1992) (Part II), Lloyds of London Press, Ltd.; "Who Bears the Risk of the Unexpected, Subsurface and Differing Site Conditions," *The International Construction Law Review* (1997)); editor and chapter author, *Alternative Clauses to Standard Construction Contracts* (Aspen Publishers, 2d ed., 1998); "Grounds for Confirming and Vacating Arbitration Awards," Salzburg, International Arbitration Superconference (June 2000). Member: Lawyers Club of Atlanta; Atlanta Bar Association; American Bar Association; (Chairman, International Public Construction Committee, Construction Division, Public Contract Law; 1992–1996); Council Member, International Bar Association, Section on Business Law; Arbitrator, American Arbitration Association; Vienna Arbitral Centre, Member, College of Fellows, The Chartered Institute of Arbitrators, 1998. Member: State Bar of Georgia.

PARTNERS EMERITUS

OVERTON A. CURRIE. (1926–2005).

LARRY E. FORRESTER. (1941–2006).

E. REGINALD HANCOCK. (1924–2004).

LUTHER P. HOUSE, JR. Phi Beta Kappa graduate of the University of Kentucky, 1955, and its College of Law. Assistant editor of the *Kentucky Law Journal*, 1957. Recipient of a Master of Laws degree from Yale Law School. Author and lecturer on all phases of construction contracting. Chair of Forum Committee of the Construction Industry of the ABA, 1991–1992; member: American Bar Association; American Trial Lawyers Association; and other professional societies.

GEORGE K. MCPHERSON, JR. B.S.B.A. and LL.B. Washington University. Omicron Delta Gamma. Assistant Solicitor General, Atlanta Judicial Circuit, Georgia 1964–1967. Member: State Bars of Georgia and Missouri.

BERT R. OASTLER. (1933–2002).

G. MAYNARD SMITH. (1907–1992).

JAMES ALLAN SMITH. B.A., Vanderbilt University, 1963; LL.B., University of Virginia, 1966. Member: State Bar of Georgia.

FEDERAL GOVERNMENT CONTRACTS AND COMMERCIAL CONTRACTS: A BRIEF COMPARISON

I. GOVERNMENT CONSTRUCTION CONTRACTING PROCESS: AN OVERVIEW

A. Introduction

Unlike commercial construction contracts, a government construction contract combines the expected statement of the scope of the work to be executed with terms and conditions that reflect the government's policies regarding contractual risk allocation, project management, and various social and economic objectives. While a description of the scope of the work, risk allocation terms, and project management requirements are common on all private and public construction projects, contractors need to recognize the significance of the various social and economic policies and their effect on all aspects of the project from contract award through execution of the work.

In addition to recognizing the multiple objectives related to the award and execution of a government construction project, potential government contractors need to appreciate that the federal government's departments and agencies awarding and administering contracts can have very different styles of management and organization. Understanding the federal government as a client and customer requires an investment of time to gain an appreciation of the differences between the various departments and even among the various offices within the same department that award and administer construction projects. Although extremely large in size, the federal government is not monolithic when it contracts for construction services.

Understanding the government's contracting process also requires an appreciation of the terminology or jargon commonly used by government contractors and agency personnel. Every business has its jargon, and federal construction contracting is no different. Finally, contractors must appreciate that the Internet is a major tool in government contracting from the initial steps in seeking to compete for an award to the final evaluation of the contractor's performance. Understanding and managing these tools is an essential step in becoming a successful government contractor.

B. Organization of This Book

With limited exceptions, the organization of this book follows the sequential steps of the government construction contracting process. **Chapter 1** provides an overview of the organization of several of the major federal agencies that award and administer construction contracts, the jargon or terminology used in the process, and a survey of many of the Internet sites involved in the contracting process. In addition, this chapter provides a brief comparison of commercial and government contract law, the sources of federal law affecting contractors and the performance of the contract work, and, last but not least, the federal government's comprehensive legal and regulatory scheme to promote the highest standards of business ethics and conduct.

Government contracts, whether for supply, service, or construction, illustrate the use of the procurement process to fill a perceived need. The initial steps in the contracting process are the authorization of funds, financing, and the delegation of authority to procure the work and administer the contract. (See **Chapter 2.**)

Once funding is in place, the procuring agency selects the project delivery method and contract type,[1] undertakes to solicit bids or proposals, and thereafter awards a contract for the work. This involves basic principles of contract law (offer, acceptance, authority to bind the government) and the selection of the actual procurement method (sealed bids or negotiated proposals) as well as the appropriate contract type and project delivery vehicle. **Chapter 3** discusses the contract formation process, relief for bid or proposal mistakes, and the resolution of bid protests. **Chapter 4** reviews various project delivery methods and contract types that the government may utilize in the procurement process.

For the past several decades, government contracts have been used to achieve social policies. These policies affect contractor selection (small business firms, service-disabled veteran-owned contractors, etc.) as well as performance of the work (labor laws, environmental laws, safety, etc.) and a preference for domestic (U.S.) products. These topics are addressed in **Chapter 5.**

Issues arising during performance of the work may include issues of contract interpretation, differing site conditions, delays, changes, inspection and acceptance, payment, bonding, and contract termination. These contract administration issues reflect the large majority of potential problems that a contractor may face during or after performance, and they are covered in **Chapters 6 to 13.**

Project documentation is important throughout contract performance and can affect the parties' rights and obligations under the various clauses. Consequently,

---

[1]*See* **Chapter 4** for an overview of the organization and contents of a typical government construction contract.

notice and documentation practices are addressed in **Chapter 14**. Regarding notice, government contractors need to consider that subcontracts and purchase orders are actually commercial (private) contracts being performed to satisfy the requirements of the prime contract with the government, and they should remember to flow down many of the federal government's terms and conditions into their subcontracts *and* purchase orders. These topics, which relate to the management of subcontracts, are beyond the scope of this book.[2]

Given the complexity of government projects and contracts, it is highly unlikely that all claims and disputes can be avoided. Even if a contractor is claim adverse, it needs to have an appreciation of the disputes process in the event a claim develops or appears likely. This topic is addressed in **Chapter 15**.

While not technically government contracts, projects funded by federal grants may have attributes similar to federal government contracts. Consequently, the role of the federal government and the effect of federal procurement principles on federally funded grant contracts are addressed in **Chapter 16**.

Finally, with the passage of the American Recovery and Reinvestment Act of 2009[3] (commonly called ARRA or the Recovery Act), Congress imposed several new requirements on contractors receiving Recovery Act funds in both a federal government contract and a grant-funded state/local construction project. Since these topics apply to a subset of federally funded projects, these requirements are addressed collectively in **Chapter 17**.

In an effort to provide a more practical perspective on the various topics addressed in this book, numerous **Checklists** are set forth throughout the book. These checklists are provided as a means to assist the user of this book in applying the concepts and principles in a practical manner. In addition, copies of these checklists are also included on the support Web site at www.wiley.com/go/federalconstructionlaw in a Word format to permit the user to copy and adapt them as needed for a particular contractor's project and organization.

C. Federal Agency Organization, Terminology, and Resources

The federal government procures construction services through multiple agencies. Screening the Federal Business Opportunities (FedBizOpps) Web site (*www.fbo.gov*) for notice of solicitations and awards posted in a 30-day period[4] reflects more than 3700 construction actions (solicitations and awards) involving 24 separate agencies of the federal government ranging from the Department of the Army—Corps of

---

[2]*See generally Common Sense Construction Law* (fourth edition) (ed. Thomas J. Kelleher, Jr. and G. Scott Walters [John Wiley & Sons 2009]) for a review of these issues and others (subcontract bidding, insurance, bankruptcy, purchase of goods under the UCC, etc.).

[3]Pub. L. 111-5.

[4]Search performed under the North American Industry Classification (NAICS) code "y"—Construction of Structures and Facilities (accessed August 21, 2009). Although it seems complex, this Web site is relatively easy to use.

Engineers (COE) with several hundred postings to the Architect of the Capitol and International Boundary and Water Commission with only a few postings each.

Regardless of the size of the project, nearly all contractors seek to gain an understanding of the client (owner) as part of the decision process on whether to compete for that contract. In the private sector, some call this activity "qualifying the owner." In reality, it is an effort by the contractor to understand the potential client and the anticipated project, and to conduct a self-evaluation of its capabilities for successful performance. In that context, the following questions or topics should be considered:

PROJECT QUALIFICATION CHECKLIST
- Has the contractor or its key project management personnel worked for that agency before? If so, what were the results and why?
- If the agency has multiple offices, what experience does the contractor have with the office that will administer the contract?
- Do the agency personnel that evaluate the proposal or bid remain responsible for the administration of the contract during performance?
- Has the agency or particular office previously built a project of similar type and complexity? If so, were there any problems? Potential subcontractors can be a useful source of information.
- Does the agency routinely change its project management staff as construction nears the punch list stage?
- Is the contracting officer located at the project site or in a relatively distant agency office? If so, does any government employee at the project site have contracting officer authority?
- Is the agency awarding and administering the contract also the eventual "owner" of the project, e.g., the Department of Veterans Affairs (VA) constructing a VA medical center, or is that agency essentially functioning as a construction manager? An example of the latter would be the Corps of Engineers constructing a project for use by the U.S. Air Force.
- What information is available regarding the experience and so on of the people the agency will place on the project site during the actual construction? (Again, potential subcontractors can be a useful source of information.)
- Does the agency routinely require project management, scheduling, or design coordination programs that require a significant new investment of contractor resources?

As a potential contractor evaluates these and similar questions, the contractor should recognize that federal agencies awarding and administering construction contracts are not organized uniformly. These structural differences may reflect differences in the agency's mission and, to some extent, historical practices. The next list illustrates some of the variety in agency organizational structure.

- **U.S. Army Corps of Engineers.** In 2009 the Corps (USACE) is geographically organized with one headquarters in Washington, D.C., and nine regional divisions including one in the Gulf Region of Southwest Asia overseeing 45 subordinate district offices in the United States and overseas and six specialized centers and laboratory facilities throughout the world.[5] In the United States, the Corps' District Offices are responsible for either civil works and/or military missions. USACE Civil Works District boundaries are set on the basis of watersheds. Military Districts are generally set within designated state boundaries. Given this organization and the fact that the Corps administers contracts for other agencies, the contracting officer may not be located near the project site. For example, it is not unusual to see a contract involving the location and disposal of munitions on former military training facilities awarded and administered by a contracting officer in Huntsville's Center of Expertise (CX) but performed in the Hawaiian Islands. Contractors need to consider whether there are potential issues created by the geographic remoteness of a project from the contracting officer.

- **Naval Facilities Engineering Command.** While not as complex as the USACE organization, NAVFAC has 10 Facilities Engineering Commands that report to two NAVFAC commands: NAVFAC Atlantic in Norfolk, Virginia, and NAVFAC Pacific in Pearl Harbor, Hawaii.[6] Similar to the COE, the Navy awards and administers projects for other Department of Defense branches, such as the U.S. Air Force, as well as for the Navy and Marine Corps. Consequently, there may be potential issues related to the distance between the project and that agency's contracting officer.

- **Department of Veterans Affairs.** The VA reflects a more centralized approach as it manages construction for the VA's health facilities (Veterans Health Administration) and approximately 80 national cemeteries in the National Cemetery Administration (NCA). The contracting officer for major VA projects is often located in that agency's Office of Construction and Facilities Management in the Washington, D.C., area.[7] Resident Engineers (RE) and Senior Resident Engineers (SRE) usually located at the project site are the primary point of contact with the contractor once a contract is awarded. These individuals may have limited contracting officer authority, as discussed in **Chapter 2.** Smaller construction projects (often minor renovations) may be awarded and administered by the staff at an individual VA facility.

With two dozen or more different agencies of the federal government awarding and administering contracts, contractors should anticipate that there will be differences in the administration of contracts among the agencies and even within the

---

[5]*www.usace.army.mil/about/Pages/Locations.aspx* (accessed November 3, 2009).
[6]*https://portal.navfac.navy.mil/portal/page/portal/navfac/* (accessed November 3, 2009).
[7]*www.cfm.va.gov/about/history.asp* (accessed November 3, 2009).

agencies. Just as a contractor typically performs a site investigation as part of its estimating process, a contractor should obtain as much information as possible regarding the agency's organization as it affects contract administration and the key agency personnel who will administer the contract on a day-to-day basis. Construction is very much a people business. Neither the federal government nor the various agencies are truly monolithic.

In addition to obtaining an appreciation of a particular agency's organization as it affects construction contract awards and project administration, a contractor should consider that government construction contracts often reference standards, design guides, and other technical publications used by various agencies (e.g., the Corps of Engineers, NAVFAC, the General Services Administration's Public Buildings Service, the VA, etc.). These standards or publications can provide critical information on the agency's expectations. For example, the expected level of detailed design development on a design-build project can vary substantially from agency to agency and from project to project. Acceptable practice on a private, commercial project may not be acceptable to a federal agency. If the agency's solicitation references a design guide or standard, a contractor's review of that document is an essential step in estimating the time and cost of performance.

Many agencies maintain virtual libraries on the Internet on which a contractor can access technical publications. For example, the VA's Technical Information Library at *www.cfm.va.gov/til/* contains materials on master specifications, design guides, and manuals. **Appendix A** to this book contains a listing of government contract–related Internet Web sites, including the reference libraries of the Corps of Engineers, NAVFAC, General Services Administration's Public Building Service, and the VA. This Web site data is also included on the support Web site at www.wiley.com/go/federalconstructionlaw.

D. Terminology and Jargon

Every industry and business uses jargon and acronyms, such as ERA (earned run average) in baseball. Federal government construction contracting is replete with both. For example, FAR is the acronym for the Federal Acquisition Regulation. In addition to providing information in the text of this book on acronyms and jargon commonly associated with government construction contracting, **Appendix B** to this book is a glossary of terms and acronyms often referenced or used in the award and administration of government construction contracts. A copy of this glossary is also found on the support Web site at www.wiley.com/go/federalconstructionlaw.

E. Internet-Based Resources

This book is intended to provide a construction professional with a reasonably comprehensive, basic resource and overview of the topics and issues that a government construction contractor may be required to address and needs to appreciate. To provide a single point of reference for all procurement-related information (potential contracting opportunities) and to reduce costs associated with the management of the procurement process, the government has created a number of Web sites that pertain to the construction contracting process, ranging from contractor registration, to access to technical manuals and standards. Attached to this book at **Appendix A** and included on the support Web site at www.wiley.com/go/federalconstructionlaw is a summary of the primary government contract–related Web sites along with a brief description of the purpose of each and the information that is available on each Web site.

II. RELATIONSHIP OF COMMERCIAL AND GOVERNMENT CONTRACT LAW

Since World War II, the federal government has consistently purchased or funded, directly or indirectly, a larger volume of construction services or work than any other single entity. While some agencies of the federal government, especially within the Department of Defense (DOD), have some capability to perform construction services with internal or agency resources, that capability is limited and often is used to support the military forces in their field operations rather than build substantial projects in the United States. Consequently, the government obtains nearly all of its needed construction work and services by contracting with private entities.

Basically, any reference to a *government construction contract* in this book means a contract directly with an agency of the federal government and does not include a contract awarded by a state or local public body or other entities using federal funds or financing.

The basic principles governing government construction contracts reflect the American common law of contracts, which evolved from the English common law. First, the parties to a contract must have the capacity to enter into that contract. Second, parties with capacity to contract generally may agree to whatever they wish, as long as their agreement does not run afoul of some legal authority or public policy. Thus, in private commercial contracts, an owner and a contractor may agree to a very risky undertaking in the context of a construction project, but they may not agree to do something illegal (e.g., gamble on the project's outcome). The former agreement reflects a policy of freedom of contract; the latter could violate a prohibition on gambling transactions. The law has long recognized that the government has the capacity to enter contracts.[8] Of greater importance is the issue of the contents of the contract and the parties' obligations under it.

A contract is traditionally defined as "a promise or set of promises, for the breach of which the law gives a remedy, or the performance of which the law in some way recognizes as a duty."[9] Thus, a contract is basically a set of promises made by one party to another party, and vice versa. In the United States, contract law reflects both the common law of contracts, as set forth in court decisions, and statutory law governing the terms of certain transactions.

---

[8] *United States v. Tingey,* 30 U.S. 115 (1831).
[9] Samuel Williston and Richard Lord, *Williston on Contracts* § 1:1 (Thomson/West 4th ed. 2007).

Similar to private contracts, government construction contracts contain or reflect both express and implied obligations or promises. Express contract obligations are those that are spelled out in the agreement or contract. Less obvious than the express duties under a contract, but just as important, are those obligations that are implied in every contract. Examples of these duties include the obligations of good faith and cooperation.

In the context of a construction project, one of the most important of these implied duties is the obligation that each contracting party cooperates regarding the other party's performance.[10] The fact that this obligation is implied rather than express is not reflective either of its importance or of the frequency with which it forms the basis for claims for compensation. Rather, the obligation to cooperate forms the very basis of the agreement between the parties.

The obligations to coordinate and cooperate are reciprocal and apply equally to all contracting parties. By way of illustration, an owner (public or private) owes a contractor an obligation to allow the contractor access to the site in order to perform its work; a prime contractor has a similar duty not to hinder or delay the work of its own subcontractors; and one prime contractor is obligated not to delay or disrupt the activities of other parallel prime contractors to the detriment of the government. Each example demonstrates that a contracting party owes an obligation of cooperation to the parties with which it has contracted. In addition, under certain circumstances, the duty to cooperate may extend to third parties with whom there is no direct contractual relationship.

In addition to the obligation of cooperation, the government, as the owner, and the contractor have other implied obligations, such as warranty responsibilities. The government's implied warranty of the adequacy of government-provided plans and specifications is of great importance to the contractor, and the breach of this warranty forms the basis of a large portion of contractor claims. The existence of an implied warranty in connection with government-furnished plans and specifications was recognized in *United States v. Spearin*.[11] The so-called *Spearin* doctrine has become well established in virtually every American jurisdiction that has considered the question of who must bear responsibility for the results of defective, inaccurate, or incomplete plans and specifications. In layman's language, the doctrine states that when an owner supplies the plans and specifications for the construction project, the contractor cannot be held liable for an unsatisfactory final result attributable solely to defects or inadequacies in the owner's plans and specifications. The key in this situation is the allocation of the risk of the inadequacies of the design to the contracting party that furnished the design or controlled the development of the design. Thus, in a design-build project, the design-build contractor, not the government, typically would bear the risk for a design error or deficiency.[12]

---

[10]*See* 13 Samuel Williston and Richard Lord, *Williston on Contracts* § 39:6 (Thomson/West 4th ed. 2000).

[11]248 U.S. 132 (1918).

[12]This risk allocation may be altered by the actions of the government. For example, in *M.A. Mortenson Co.,* ASBCA No. 39978, 93-3 BCA ¶ 26,189, the government furnished a conceptual structural design to the design-builder for estimating (bidding purposes). When it was determined that the conceptual structural design was inadequate, the government bore the risk of the cost of the additional steel and concrete to remedy the design problems, even though the contract was labeled as a "design-build" contract.

Similar to private contracts governed by the common law, the basic concept of breach of contract applies to government contracts. In private contractual relationships, a *breach of contract* results when one party fails in some respect to do what that party has agreed to do, without excuse or justification.[13] For example, a contractor's failure to use the specified trim paint color, or its failure to complete the work on time, constitutes a breach of contract. Public or private owners may likewise breach their contract obligations. Many contracts expressly provide, for example, that the owner will make periodic payments to the contractor as portions of the work are completed. If the owner unjustifiably fails to make these payments, this failure constitutes a breach of contract. Similarly, an owner may be held in breach for failing to meet other nonfinancial contractual obligations, such as the duty to timely review and return shop drawings and submittals. In short, any failure to live up to the promises that comprise the contract is a breach.

Whenever there is a breach of contract, the injured party has a legal right to seek and recover damages. In addition, if there has been a serious and *material* breach— that is, a breach that, in essence, destroys the basis of the parties' agreement—the injured party is justified in treating the contract as ended.[14]

Breach of contract actions are relatively rare in government contracting due to the fact that these contracts include remedy-granting clauses, such as the Changes clause,[15] the Default clause,[16] and the Suspension of Work clause.[17] These remedy-granting clauses, when combined with a very comprehensive disputes procedure that generally requires a contractor's continued performance pending claim resolution (see **Chapter 15**), effectively limit the application of traditional breach of contact theories and damages claims in government contracts. However, the concept of contractual terms limiting the scope of breach of contact liabilities and damages is not unique to government contracts, as illustrated by the provisions of the Uniform Commercial Code (UCC)[18] that provide for limitations on liabilities[19] and remedies.[20] All of these basic principles and concepts of contracting are reflected in both government contracting and private commercial contracts.

III. SOURCES OF FEDERAL LAWS AFFECTING GOVERNMENT CONSTRUCTION CONTRACTS

A. Contracts Awarded by Federal Agencies

The procurement and administration of government construction contracts, as well as the resolution of disputes on these projects, are governed by multiple statutes

---

[13]*See Restatement (Second) of Contracts* §§ 235, 236.

[14]*See generally* 17A Am. Jur. 2d *Contracts* § 528 (2007).

[15]FAR § 52.243-4.

[16]FAR § 52.249-10.

[17]FAR § 52.242-14.

[18]The UCC, which applies to the sale of goods and other commercial transactions, has been adopted in 49 states (Louisiana is the exception), the District of Columbia, and the Virgin Islands.

[19]UCC § 2-316.

[20]UCC §§ 2-718, 2-719.

and extensive regulations. Administrative boards of contract appeals and special courts have operated for decades for the sole purpose of resolving disputes on federal contracts.[21] Each year the boards and courts generate hundreds of decisions that collectively provide the single largest body of law in the area of construction disputes. Numerous fundamental principles of construction law have their genesis in the law of government construction contracts. It is impractical to speak of modern American construction law without the consideration of federal procurement law.

1. The Federal Acquisition Regulation and Its Supplements

Most government construction contracts reflect policies contained in statutes and in the Federal Acquisition Regulation.[22] Besides containing standard contract clauses, the FAR also sets forth extensive guidance to the federal agencies and their personnel regarding the award and administration of government construction contracts. In addition to the basic FAR, many of the federal agencies have their own supplements to the FAR. For example, the DFARS is the Defense Federal Acquisition Regulation Supplement. These supplements can substantially alter a contractor's rights, obligations, and remedies on a government contract with that agency. While possibly not as complex as the federal income tax regulations, the collective volume of these procurement regulations is extensive.[23]

Since the FAR contains in excess of 1,800 pages of materials, understanding its basic organization helps the user to navigate this procurement regulation. The FAR is subdivided into eight major subchapters containing 53 parts. Parts 1 to 51 contain substantive guidance and policy statements. Part 52 contains the clauses used in government contracts, and Part 53 contains examples of many of the standard forms used in contracting.

Each of these parts addresses a separate aspect of the acquisition process and contains policy guidance or direction, instructions on the use of contract provisions, and the text of the actual contract clauses. The eight major subchapters are:

---

[21]*See* **Chapter 15**.

[22]The United States Postal Service contracts under authority of the Postal Services Reorganization Act, 39 U.S.C. § 410(a), which exempts the Postal Service from the federal procurement laws and regulations governing traditional federal agencies. The Postal Service has its own regulations and policies contained in its Purchasing Manual. The Federal Aviation Administration is exempt from several procurement statutes pursuant to Pub. L. No. 104-50. Both agencies have boards and procedures to address claims and disputes.

[23]As published by the Government Printing Office, the FAR and its supplements are found in 48 C.F.R. (Web site: *http://ecfr.gpoaccess.gov*). Collectively, the FAR and its supplements total in excess of 4,100 pages of material. While the FAR contains separate parts or sections for particular types of contracts, those designations may be misleading. For example, FAR Part 36 is entitled "Construction and Architect-Engineer Contracts," but that part does not contain all of the provisions and regulatory guidance applicable to construction contracts. In addition, 41 C.F.R. Chapters 50, 51, 60 and 61 contain an additional 240 pages of regulations addressing wage and hour laws, affirmative action requirements, and other labor laws governing the performance of government contracts.

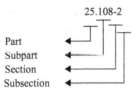

Figure 1.1   Makeup—FAR Number Citation

A. General (Parts 1-4)

B. Competition and Acquisition Planning (Parts 5-12)

C. Contracting Methods and Contract Types (Parts 13-18)

D. Socioeconomic Programs (Parts 19-26)

E. General Contracting Requirements (Parts 27-33)

F. Special Categories of Contracting (Parts 34-41)

G. Contract Management (Parts 42-51)

H. Clauses and Forms (Parts 52-53)

Within each of these parts are subparts, sections, and subsections. The FAR contains a numbering system that allows for discrete identification of every FAR paragraph. The digits to the left of the decimal point represent the part number. The numbers to the right of the decimal point and to the left of the dash represent, in order, the subpart (one or two digits), and the section (two digits). The number to the right of the dash represents the subsection. Subdivisions may be used at the section and subsection level to identify individual paragraphs. Figure 1.1 illustrates the structure of a typical FAR number citation (FAR 25.108-2).

Subdivisions in the text or a provision of the FAR below the section or subsection level consist of parenthetical alphanumerics using this sequence: (a)(1)(i)(A)(*1*)(*i*).

The various FAR contract clauses for all types of government contracts are found in Part 52. The numerical designation for each of these clauses contains a reference to the applicable substantive provision in the FAR, which provides guidance on its use. For example, Figure 1.2, the designation for the Changes clause for supply contracts (FAR 52.243-1), illustrates the makeup of the designation for a FAR clause.

Figure 1.2   Makeup—FAR Clause Citation

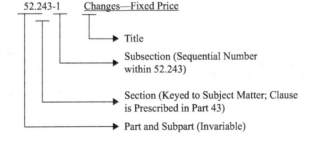

Although Part 36 of the FAR is entitled "Construction and Architect-Engineer Contracts," that part does not contain all of the provisions or policy guidance related to construction contracts. For example, the policy guidance for the Prompt Payment for Construction clause (FAR 52.232-27) is found in Part 32 of the FAR, which is entitled "Contract Financing." Many other key clauses similarly are found in various parts of the FAR; for example, the Suspension of Work clause (FAR 52.242-14), which obligates the government to compensate the contractor for certain government-caused delays, implements policy found in Part 42, "Contract Administration and Audit Services."

This organization can add some degree of potential confusion when determining whether a solicitation contains the correct FAR clause. Fortunately, FAR Subpart 52.3 contains a detailed matrix listing each of the clauses found in Part 52 together with a listing of principal types of contracts—for example, fixed-price construction contracts ("FP CON"). In the matrix column under FP CON are designations if a particular clause is generally authorized or required in that type of contract and a reference to the section of the FAR that prescribes the use of that clause. These designations are:

R = Required

A = Required when Applicable

O = Optional

By referring back to the substantive sections of the FAR (Parts 1-51), it is possible to review any policy guidance on the use of a particular clause or any variation of that clause.

Part 2 of the FAR contains definitions of many of the key words and terms used in the FAR. However, the listing of definitions is not comprehensive, as other definitions are found in other parts or subparts of the FAR. For example, very broad definitions of "subcontract" and "subcontractor" are set forth in FAR 44.101.[24]

In addition to the clauses required or authorized by the FAR, contractors also need to identify and review contract clauses that may be included in a contract pursuant to agency supplements to the FAR. As illustrated by the clauses in the Defense Federal Acquisition Regulation Supplement, DFARS 252.201-7000, Contracting Officer's Representative,[25] and the Department of Veterans Affairs Acquisition Regulation, VAAR 852.236-88, Contract changes—supplement,[26] these so-called supplements can substantially affect a contractor's obligations and limit its substantive rights.

---

[24]Those definitions are broader than the generally accepted understanding of these terms in the construction industry as they include vendors or materialmen providing supplies or equipment under purchase orders as well as firms performing work on the project site as subcontractors. This difference needs to be understood in the context of the administration of a government contract and the drafting of subcontracts and purchase orders for government contracts.

[25]*See* **Chapter 2.**

[26]*See* **Chapter 8.**

One major difference between most private commercial construction contracts and government contracts is the government's practice of *incorporating by reference* many key clauses from the FAR or the agency's FAR supplement into the construction contract. Upon reviewing a solicitation for a government project, a contractor may find a multipage listing of clauses with the FAR or the FAR supplement numerical designations. The significance of these clauses is not diminished by their listing on a multipage table of incorporated clauses. As part of its evaluation of the potential risks and obligations, a contractor should obtain and review each of those provisions. For a firm that is relatively new to government contracting, this review can be rather time intensive. Fortunately, most of the standard FAR clauses are revised on a relatively infrequent basis.[27] Once one becomes familiar with the FAR clauses and applicable agency supplements, the time needed to review subsequent solicitations from the same agency is substantially reduced.

Disputes arising out of or related to the performance of a government construction contract are governed by the Contract Disputes Act (CDA).[28] The CDA and its implementing regulations set forth a comprehensive approach to the resolution of contract claims by contractors and the government. (See **Chapter 15.**) The citations in this book are to the appropriate provisions of the CDA, other relevant statutes, and the applicable regulations, particularly the FAR, as well as to the various board and court decisions. The CDA and the other cited statutes are found in West Publishing Company's *United States Code Annotated.* Citations to the FAR and the agency supplements are found in Title 48 of the Code of Federal Regulations (CFR).

The current disputes procedure has its roots in practices that developed from the early part of the twentieth century. Over nearly 100 years, the process has evolved as efforts to remedy possible or actual procedural or substantive problems have been implemented. Understanding this history evolution provides a useful perspective on the current status of the disputes process. Consequently, **Appendix 1A** to this chapter provides an overview on the history and evolution.

2. Court, Board, and Bid Protest Decisions

Government contract case law is found in a variety of sources. Since 1921, selected bid protest decisions issued by the United States Government Accountability Office (GAO) have been published in the *Decisions of the Comptroller General of the United States.*[29] Beginning in 1974, Federal Publications, Inc., now part of the West Group, has published the *Comptroller General's Procurement Decisions (CPD)* service containing the full text of all of the GAO's bid protest decisions. Court decisions regarding bid protests have been issued by the federal district courts,[30] the

---

[27]For example, the standard Differing Site Conditions clause in government construction contracts, FAR 52.236-2, was last revised in April 1984.

[28]41 U.S.C. §§ 601 *et seq.*

[29]Formerly the General Accounting Office. Typically, about 10 percent of the GAO's decisions in a given year are published in that publication.

[30]The U.S. Federal District Courts' jurisdiction over bid protests ended as of January 2001.

various federal circuit courts of appeals, the United States Court of Federal Claims (and its predecessor courts), and the United States Court of Appeals for the Federal Circuit. The case law involving claims and disputes arising out of or related to the performance of a contract basically consists of the decisions of the various boards, United States Court of Claims, United States Claims Court, United States Court of Federal Claims (Court of Federal Claims), and the United States Court of Appeals for the Federal Circuit (Federal Circuit). On relatively rare occasions, the United States Supreme Court will consider and issue decisions directly addressing federal government contracts.[31]

The Court of Claims, which was abolished in 1982, had jurisdiction to entertain suits involving government contract claims, including claims under the CDA. When Congress abolished the Court of Claims, it created the Claims Court, now the Court of Federal Claims,[32] and granted to it all of the original jurisdiction of the Court of Claims.[33] At the same time, Congress also created a new United States Court of Appeals for the Federal Circuit.[34] The Federal Circuit reviews appeals of decisions from the boards and the Court of Federal Claims.[35] The Court of Federal Claims and the Federal Circuit view decisions of the old Court of Claims as binding precedent.[36]

B.   Contracts Funded by Federal Grants

Although not considered to be *government construction contracts*, many federal agencies provide grants to state, county, and municipal agencies to partially fund construction projects. These grant agreements may provide for the inclusion of clauses or application of federally mandated policies in the actual construction contracts. These grant agreements are addressed in **Chapter 16** of this book.

C.   Effect of Statutes and Regulations on Contractors

1.   Possible Conflicting Themes

When contracting with the federal government, contractors need to appreciate that there are two fundamental and potentially conflicting policies that may affect the parties' rights and obligations. One policy addresses the status of the United States when it enters into a contract in the commercial marketplace. This was summarized in *McQuagge v. United States*[37]:

---

[31]*See, e.g., S&E Contractors, Inc. v. United States,* 406 U.S.1 (1972).

[32]28 U.S.C. § 171.

[33]28 U.S.C. § 1491(a)(2). The United States Court of Federal Claims has the same basic jurisdiction but broadened to include nonmonetary claims.

[34]28 U.S.C. § 41.

[35]41 U.S.C. § 607(g)(1)(A); 28 U.S.C. § 1295(a)(10), (a)(3).

[36]*South Corp. v. United States,* 690 F.2d 1368, 1369 (Fed. Cir. 1982) (en banc); United States Court of Federal Claims Gen. Order No. 33, 27 Fed. Cl. xyv (1992).

[37]197 F. Supp. 460 (W.D. La. 1961).

In ordinary contractual relations with its citizens, the government enjoys the same privileges and assumes the same liabilities as does its citizens. This is distinguished from the situation where the sovereign is seeking to enforce a public right or protect a public interest, for example, eminent domain or an exercise of the taxing power. In the latter case the government is not bound by ordinary rules of private contract law or by doctrines of estoppel or waiver. When the government enters the market place, however, and puts itself in the position of one of its citizens seeking to enforce a contractual right (i.e., one which arises from express consent rather than sovereignty), it submits to the same rules which govern legal relations among its subjects.[38]

Many of the decisions that provide that the United States is bound by its contracts just as a private party involve questions of contract interpretation.[39] However, another theme in government contract cases reflects a statement by Justice Oliver Wendell Holmes, Jr., that "Men must turn square corners when they deal with the Government."[40] This statement would seem to imply that the government may have, in certain respects, a special status in its contractual relationships and that all of the rules governing contractual relationships may not apply in government contracts. While the standard contract provisions in a government construction generally reflect a balanced allocation of risks, there are many special requirements and legal principles, which every contractor must appreciate. These are the "square corners" of contracting with the government in the twenty-first century.

2. Authority and Public Policy Considerations

While the two themes just noted appear to conflict, the *McQuagge* decision referenced two conditions that are critical to understanding them. First, the government must be acting in a contractual capacity. Second, it must not be seeking to protect or enforce a public policy.

There is no question that the government has the capacity to enter into a contract.[41] However, a contract that is prohibited by statute or varies from mandatory procedures is not enforceable or binding on the government.[42] Similarly, the person or entity entering into a contract on behalf of the government must have the requisite authority to do so. If that person has the requisite authority to bind the government, the exercise of that authority usually involves a degree of discretion.[43] Consequently, if the contractual action by the government's representative reflects an error in judgment, the government usually is bound so long as the person was acting within the limits of that person's authority.[44]

---

[38]197 F. Supp. at 469; *see also Mann v. United States*, 3 Ct. Cl. 404, 411 (1867); *Hollerbach v. United States*, 233 U.S. 165 (1914).

[39]*See, e.g., Hollerbach v. United States*, 233 U.S. 165 (1914).

[40]*Rock Island, Ark. & La. R.R. v. United States*, 254 U.S. 141, 143 (1920).

[41]*United States v. Tingey*, 30 U.S. 115 (1831).

[42]*The Floyd Acceptances*, 74 U.S. 666 (1868).

[43]*Arizona v. California*, 373 U.S. 546 (1963); *United States v. MacDaniel*, 32 U.S. 1 (1833).

[44]*United States v. Winstar Corp.*, 518 U.S. 839 (1996); *Cooke v. United States*, 91 U.S. 389 (1875); *Liberty Coat Co.*, ASBCA No. 4119, 57-2 BCA ¶ 1576.

The key is ascertaining the limits of authority. This is one of those square corners for government contractors. The limits of authority question was addressed by the United States Supreme Court in *Federal Crop Insurance Corp. v. Merrill,*[45] which involved an issue of the ability of an unauthorized agent of a government agency to bind the United States. The Court rejected the application of the concept of apparent authority and ruled that the party dealing with the United States had the burden of ascertaining the *actual authority* of the government's representative. The Court, after reviewing the prior proceeding in the case, stated:

> That court [Supreme Court of Idaho] in effect adopted the theory of the trial judge, that since the knowledge of the agent of a private insurance company, under the circumstances of this case, would be attributed to, and thereby bind, a private insurance company, the Corporation [United States] is equally bound.

> The case no doubt presents phases of hardship. We take for granted that, on the basis of what they were told by the Corporation's local agent, the respondents reasonably believed that their entire crop was covered by petitioner's insurance. And so we assume that recovery could be had against a private insurance company. But the Corporation is not a private insurance company. It is too late in the day to urge that the Government is just another private litigant, for purposes of charging it with liability, whenever it takes over a business theretofore conducted by private enterprise or engages in competition with private ventures. Government is not partly public or partly private, depending upon the governmental pedigree of the type of a particular activity or the manner in which the Government conducts it. The Government may carry on its operations through conventional executive agencies or through corporate forms especially created for defined ends. *See Keifer & Keifer v. Reconstruction Finance Corp.,* 306 U.S. 381, 390. Whatever the form in which the Government functions, anyone entering into an arrangement with the Government takes the risk of having accurately ascertained that he who purports to act for the Government stays within the bounds of his authority. The scope of this authority may be explicitly defined by Congress or be limited by delegated legislation, properly exercised through the rule-making power. And this is so even though, as here, the agent himself may have been unaware of the limitations upon his authority.[46]

The *Federal Crop Insurance* decision reflects one of the two conditions expressed in *McQuagge*. The government must have entered into a valid contractual relationship. A valid contract can occur only if the government's representative is authorized to bind the United States. A similar condition applies to changes ordered by a representative of the government. Since the contractor bears the burden to ascertain the authority of the person with whom it is dealing, the often critical issue related to authority is addressed in several chapters of this book. (See **Chapters 2, 8, and 10.**)

---

[45] 332 U.S. 380 (1947).
[46] 332 U.S. at 383-384.

The second *McQuagge* exception to the general principle that the United States is bound to its contracts in the same manner as a private party referred to the enforcement of a *public right* or *public interest.* This exception is best illustrated by the decision of the United States Court of Claims in *G. L. Christian & Associates v. United States.*[47] The *Christian* decision involved a contractor's claim for its lost anticipated profits following the government's decision to terminate for convenience a large housing project. The contract did not contain a termination for convenience clause;[48] hence, there was no contractual preclusion on the recovery of lost anticipated profits. While acknowledging the basic principle that the government has the rights and ordinarily the liabilities of a private party when it enters into a contract, the Court of Claims held that the termination for convenience clause was incorporated into the contract by operation of law as it was a mandatory clause under the applicable procurement regulations.[49]

While the Christian doctrine appears to apply only to mandatory clauses that implement fundamental policy, a contractor generally is deemed to be on notice of these clauses. Contractual notice of the provisions to the contractor occurs following publication of the procurement regulation in the *Federal Register.*[50] If a regulation is not published in the *Federal Register,* the contractor still may be bound if it has actual notice or knowledge of it.[51] Given this doctrine, any government construction contractor needs to have a basic understanding of the key principles affecting the interpretation and enforcement of the standard mandatory clauses and the ability to obtain advice on these provisions. In addition, as the FAR provides guidance to the government's representatives on the award and administration of government contractors, a contractor should obtain or have access to the edition of the FAR and any agency supplements applicable to its contract.[52]

Another potential square corner for a government contractor follows from the principle that actions taken by a government official within the limits of that person's authority are presumed to be properly made unless contrary to law or regulation.[53] While this doctrine may operate to protect a contractor when the government's authorized representative makes what is later challenged as a bad business decision,[54] the same presumption that the contracting officer acted in good faith makes it very difficult to overturn actions such as termination for convenience on the basis that the action was an abuse of discretion, taken in bad faith, or motivated with malice toward the contractor.[55]

---

[47]312 F.2d 418 *rehearing denied* 320 F.2d 345 (Ct. Cl. 1965).

[48]*See* **Chapter 11** for a discussion of convenience terminations.

[49]312 F.2d at 427.

[50]41 U.S.C. § 4186.

[51]*Timber Access Indus. Co. v. United States,* 553 F.2d 1250 (Ct. Cl. 1977).

[52]Electronic versions of the FAR and its supplements can be accessed at the Government Printing Office's Web site, *http://ecfr.gpoaccess.gov.* (This Web site contains current versions of the FAR and its supplements and is updated on almost a daily basis.)

[53]*General Electric Co. v. United States,* 412 F.2d 1215 (Ct. Cl. 1969).

[54]*McQuagge v. United States,* 197 F. Supp. 460 (W.D. La. 1961); *Conrad Weihnacht Constr., Inc.,* ASBCA No. 20767, 76-2 BCA ¶ 11,963.

[55]*See Librach v. United States,* 147 Ct. Cl. 605, 612 (1959); *Kalvar Corp. v. United States,* 543 F.2d 1298 (Ct. Cl. 1976); *see also* **Chapters 2 and 11.**

IV.   PROCUREMENT INTEGRITY AND STANDARDS OF CONDUCT

Government contractors are expected to conduct business with a high degree of integrity and ethics. The consequences for failing to meet these expectations of integrity when dealing with the federal government can be extremely serious. The remedies that the government may utilize include contract cancellation, debarment, fines, damages, forfeiture of claims as well as criminal sanctions. Consequently, these expectations, as well as the related laws and regulations, must be understood and appreciated by contractors and subcontractors performing work for the government. The most common way that the government asserts its expectations of honesty and integrity is through the application of various anti-fraud and false claims statutes. The government effort to prevent fraud and false claims may become more focused on the construction industry in the near future. Past efforts, which have resulted in substantial payments to the government, primarily focused on the healthcare industry. However, allegations of abuses in Iraq and elsewhere have placed government construction and service contractors in the spotlight. An initial tool that the government employs to alert its contractors to the expectations related to integrity and ethics involve a broad variety of contractor certifications related to its actions and contract performance. Complete treatment of the details of the various laws and regulations and their interpretation is beyond the scope of this book and would justify, if not require, an entire separate book. However, contractors need to appreciate they will be held to a high standard of business ethics and conduct.[56]

A.   Importance of Certifications

Central to government contracting is the requirement that contractors and subcontractors must deal honestly with the government. This theme is reflected in the general standards of responsibility for a prospective contractor[57] and in the requirements for certification of cost or pricing data[58] and claims.[59]

Consistent with the expectation of a high standard of ethics and conduct, contractors are routinely required to provide certifications during all phases of the contracting process, from the initial solicitation to the resolution of claims. Often these certifications provide the initial foundation of the government's assertion of wrongdoing by a contractor. Consequently, no certification or affirmation of fact should be dismissed as just *another government form.*

---

[56]*See* FAR § 3.1002(a) (government contractors are expected to conduct themselves with the "highest degree of integrity and honesty").
[57]FAR § 9.104-1(d).
[58]FAR § 15.403.
[59]FAR § 33.207. *See also* **Chapter 15**.

While the subject matter and wording of contractor-provided certifications are subject to change, **Table 1.1** lists many of the certifications, affirmations, or representations currently required of a government construction contractor.[60]

While a few of these provisions reference potential liabilities associated with the various certifications, many are silent. In that regard, the FAR requires the inclusion of this clause in sealed bid procurements issued under FAR Part 14:

FAR § 52.214-4 FALSE STATEMENTS IN BIDS (APR 1984)

Bidders must provide full, accurate, and complete information as required by this solicitation and its attachments. The penalty for making false statements in bids is prescribed in 18 U.S.C. 1001.

This provision is not mandated for use in negotiated contracts awarded under FAR Part 15. However, 18 U.S.C. § 1001, the False Statements Act, which is referenced in the False Statements in Bids clause, is not limited in its application to sealed bid procurements.[61]

B. Overview of Federal Laws Related to Procurement Integrity/Standards of Conduct

Whether competing for or performing a government contract, every contractor needs to appreciate the broad scope of legislation intended to protect the government (the public) from a variety of prohibited activities. These laws prescribe a range of improper actions and the applicable civil and criminal sanctions. However, the government agencies perceive that the task of inspecting work and determining compliance with the contract requirements for billions of dollars in contracts every year is extremely difficult.

---

[60]Each contract, including those provisions incorporated by reference, should be screened to identify requirements for contractor certifications and representations, as the extent and scope of the requirements for certifications and representations are receiving extensive Congressional review and are becoming more detailed. For example, Section 872 of the Duncan Hunter National Defense Authorization Act of 2009 (Pub. L. 110-417) requires the Office of Management and Budget and the General Services Administration to establish, within one year of the effective date of that legislation (October 14, 2008), an information database on the integrity and performance of contractors that will be available to all federal agencies and grantees. That legislation also directs the adoption of regulations within one year of the effective date of the legislation requiring contractors with agency awards or grant contracts with a total value in excess of $10,000,000 to provide to the federal government detailed information similar to that currently found in FAR § 52.209-5 with certain critical differences. The new reporting requirements, representations, and database will cover a **five year** period, not three years, and will include disclosure of civil judgments "in connection with" the award or performance of a contract or grant with the federal government, default terminations, and the administrative resolution of suspension or debarment proceedings. The phrase "in connection with" is not defined in the Duncan Hunter Act.

[61]See also 15 U.S.C. § 645(d) (provides for criminal penalties for knowingly misrepresenting a firm's small business size status).

Table 1.1 Contractor Certifications and Representations

Title of Provision	FAR Reference	Basic Subject Matter
Certification of Independent Price Determination	FAR § 52.203-2	Price competition and actions to influence others in submitting offers in connection with a solicitation
Covenant Against Contingent Fees	FAR § 52.203-5	Agents engaged to solicit award
Taxpayer Identification	FAR § 52.204-3	Ownership and tax status of bidder/offeror
Certification Regarding Responsibility Matters	FAR § 52.209-5	Certification by offeror regarding whether it and/or any of its *principals* are or are not debarred, suspended, proposed for debarment, or declared ineligible for award of a federal agency contract; have or have not, within a three-year period preceding the offer on the contract, been convicted or had a civil judgment rendered against them for fraud or certain criminal offenses in connection any public contract or subcontract; are or are not presently indicted or civilly charged for fraud or certain criminal offenses in connection any public contract or subcontract; have or have not, within a three-year period preceding the offer on the contract, been notified of delinquent federal taxes in excess of $3,000 owed by the contractor or its *principals*
Small Business Program Representations	FAR § 52.219-1	Status of bidder/offeror under various SBA-related preference programs
Certificate of Current Cost or Pricing Data	FAR § 15.406-2	Applicable when contractor submits cost or pricing data for proposals or modifications (equitable adjustments)
Subcontractor Cost or Pricing Data—Modifications	FAR § 52.215-13	Applicable when subcontractor submits cost or pricing data for pricing of contract modifications (equitable adjustments)
Payrolls and Basic Records	FAR § 52.222-8	Certification that Davis Bacon wages fully paid and data on payroll records form (e.g., social security numbers) are accurate and complete

Affirmative Action Compliance	FAR § 52.222-25	Affirmative Action Program Status
Exemption from Application of Service Contract Act Provisions	FAR § 52.222-48	Contractor certification that services qualify as "commercial items" and priced based on catalog or market prices
Recovered Material Certification	FAR § 52.223-4	Applies if specifications required use of Environmental Protection Agency-designated products
NC State and Local Sales and Use Tax	FAR § 52.229-2	Certification and payment of North Carolina taxes
Disclosure Statement—Cost Accounting Practices And Certification	FAR § 52.230-1	Applicability of cost accounting standards to offeror
Payments under Fixed-Price Construction Contracts	FAR § 52.232-5	Amounts requested are only for performance in accordance with specifications, terms, and conditions of contract; payments to subcontractors have been made from previous payments; timely payments to subcontractors will be made; and payment request includes no amount that prime contractor intends to withhold (retain) from a subcontractor or supplier
Disputes	FAR § 52.233-1	Claims in excess of $100,000.00
Certification of Final Indirect Costs	FAR § 52.242-4	No unallowable costs are included in the costs used to establish indirect cost rates (Applies to cost reimbursement construction contracts)
Termination for Convenience Settlement Proposals (Total Cost Basis)	FAR § 53.301- SF 1436	Proposal reflects recognized commercial accounting practices and includes only those charges allocable to terminated contract and that are fair and reasonable
Termination for Convenience Schedule of Accounting Information	FAR § 53.301- SF 1439	Disclosure of contractor's accounting practices

Three different approaches have been adopted to address this difficulty. One approach involves the broad use of contractor furnished certifications and representations. Many of these are identified in **Table 1.1** of this chapter. These certifications and representations can serve at least three possible purposes.

(1) Alert the contractor signing the certification or representation to the significance of its signature.

(2) Simplify the government's proof in establishing a violation of an underlying statute.

(3) Create the basis for an action based solely on the false nature of the certification.

The final two purposes involve multiple civil and criminal statutes addressing prohibited conduct and the provision of economic incentives to those who report wrongdoing. In that regard, federal law provides substantial economic incentives or bounties for individuals to disclose fraudulent conduct by government contractors. In 1986 Congress amended the Civil False Claims Act, 31 U.S.C. §§ 3729-3733 to encourage third parties to identify and institute civil *qui tam* actions[62] involving allegations of fraudulent conduct and to share in the recovery of those actions. Coupled with this statute are requirements for self-reporting and/or hotlines as discussed in **Section IV.F** of this chapter.

The federal false claims and anti-fraud statutes are varied in terms of the subject matter of the prohibited conduct or activities. Some of the statutes provide for civil penalties or sanctions for prohibited activities while others provide for criminal sanctions. **Table 1.2** lists many of the statutes in the government's arsenal of remedies for contractor fraud and false claims.

C. Civil False Claims Act Actions

Although there are multiple statutes available to the government to combat improper conduct, fraud, and false claims, the civil False Claims Act[63] (FCA) is often invoked by the government as the preferred statutory basis for an action rather than the parallel criminal FCA statute or other anti-fraud statutes. Through 2008, an average of 457 new FCA matters were begun each year. There are several reasons for this preference: the *qui tam* provisions of the FCA, the proof of knowledge standard, and the way in which the statute deals with and assesses damages.

1. Qui Tam Provisions of the FCA

Included in the 1986 revisions of the False Claims Act, the *qui tam* provisions allow for private individuals to bring suits as "whistleblowers" against entities that may be liable under the FCA.[64] These suits are brought on behalf of the federal government.

---

[62]Essentially means private attorney general actions.
[63]31 U.S.C. §§ 3729 *et. seq.*
[64]31 U.S.C. § 3730(b)(1).

Table 1.2 Federal Anti-Fraud/False Claims Laws

Title	Statutory Reference	Subject Matter/Notes
Criminal Statutes		
Anti-Kickback Act	41 U.S.C. §§ 51–58	Prohibits payments by subcontractors at any tier to prime contractors or subcontractors to obtain a government contract
Conspiracy to Defraud	18 U.S.C. § 286; 18 U.S.C. § 371	Addresses claims and general conspiracy to defraud the government
False Claims Act, Criminal Liabilities	18 U.S.C. § 287	False claim need not have been paid by government to provide basis of liability
Theft from Federal Programs	18 U.S.C. § 666	Applies to theft from state and local public agencies receiving federal funds by "agents" of those agencies
False Statements Act	18 U.S.C. § 1001	Includes statements, false entries, oral and unsworn statements
Mail and Wire Fraud	18 U.S.C. §§ 1341–1350	Applies to use of mails and telecommunications to execute a scheme to defraud the United States
Major Fraud Act	18 U.S.C. § 1031	Applies to procurement fraud on a government contract or subcontracts there under valued at $1 million or more
Obstruction of Federal Audit	18 U.S.C. § 1516	Applies to any person employed on full-, part-time, or contractual basis to conduct an audit or a *quality assurance inspection* for or on behalf of the United States
Sarbanes–Oxley Act of 2002	18 U.S.C. § 1519	Applies to anyone who knowingly alters a document with intent to influence proper administration of any matter within jurisdiction of department or agency of the United States; violators subject to fines or imprisonment up to 20 years, or both
Civil Statutes		
Anti-Kickback Act	41 U.S.C. §§ 51–58	Prohibits kickback by subcontractors and suppliers
Contract Disputes Act of 1978	41 U.S.C. § 604	False or unsupported claims submitted to contracting officer; necessity for certification
False Claims Act	31 U.S.C. §§ 3729–3733	Applies to any request for money or property used on the government's behalf or to advance a government program or interest if the government provides any portion of the money or property or reimburses a third party for any portion of the money or property
Forfeiture of Claims Act	28 U.S.C. § 2515	Allows a special plea in United States Court of Federal Claims providing for forfeiture of entire claim if any part of it is tainted by fraud
Program Fraud Act	31 U.S.C. §§ 3801–3812	Administrative alternative to litigation in civil false statements and smaller false claims cases
Truth in Negotiations	10 U.S.C. § 2306a; 41 U.S.C. § 254	Cost or pricing data on negotiated contracts or subcontracts; modifications of contracts in excess of $650,000,[a] necessity for certification

[a] This amount is subject to adjustment for inflation every five years.

The individual bringing suit, whether a person or a corporate entity, called a relator in the litigation, is entitled to share in any recovery that results from the suit. Individuals who are in positions to know of potentially unlawful conduct have an incentive to bring the allegations to light primarily because of the prospect of sharing in any recovery from the contractor. The government benefits from the *qui tam* provisions because the individual bringing the FCA suit is often in a position to know a great deal more information concerning the activities of a contractor or subcontractor than the government agency administering the contract or the Department of Justice might know. The *qui tam* provisions in conjunction with other changes to the FCA made in 1986 have had an impact. According to Department of Justice statistics, the average number of FCA *qui tam* actions outnumbers non-*qui tam* actions by a ratio of nearly 2 to 1.[65] Since 1986, more than $21 billion has been recovered though FCA judgments and settlements. In fiscal year 2008 alone, recoveries amounted to at least $1.34 billion.[66] Of that recovery, the *qui tam* relator usually shares approximately 15% of the amount recovered.

2. *Proof of Knowledge Standard*

The criminal False Claims Act requires proof that the false statement was made with intent to deceive, was designed to induce a belief in the false statement or to mislead.[67] A "knowing" act means "[a]n act is done knowing if the defendant realized what he or she is doing, and did not act through ignorance, mistake or accident."[68] Intentional ignorance has been held to constitute constructive knowledge sufficient to satisfy this element of the offense.[69] In addition, the false statement need not be delivered to the government if it was relied on in the disbursement of funds provided by the government.[70]

In contrast, the civil FCA has a lower scienter, or knowledge, requirement. There is no requirement of specific intent to defraud. Unless an allegation of conspiracy is made, the level of "knowledge" for a civil FCA action is defined in 31 U.S.C. § 3729(b)(1) as:

(i) [having] actual knowledge of the information;

(ii) [acting] in deliberate ignorance of the truth or falsity of the information; or

(iii) [acting] in reckless disregard of the truth or falsity of the information.

In *United States ex rel Bettis v. Odebrecht Contractors of CA, Inc.,*[71] the court applied this standard to a civil false claims action related to a project for the construction of an embankment dam. During the bidding, the contractor Odebrecht underbid the contract so that the second-lowest bidder was approximately $30 million higher

[65]*See www.usdoj.gov/opa/pr/2008/November/fraud-statistics1986-2008.htm* (accessed July 14, 2009).

[66]*See www.usdoj.gov/opa/pr/2008/November/08-civ-992.html* (accessed July 14, 2009).

[67]*United States v. Lichenstein,* 610 F.2d 1272 (5th Cir. 1980), *cert. denied,* 447 U.S. 907 (1980).

[68]*United States v. Ibarra-Alcarez,* 830 F.2d 968 (9th Cir. 1987).

[69]*United States v. Petullo,* 709 F.2d 1178 (7th Cir. 1983).

[70]*Id.* at 1180.

[71]297 F.Supp. 2d 272, 277-8 (D.D.C. 2004).

and the estimate made by the Corps of Engineers was approximately $35 million higher than the contractor's bid. The plaintiff alleged in his FCA suit that the contractor underbid the project in order to seek adjustments to the contract price at a later date. The court rejected this argument as contrary to the reality of government contracting, which allows for flexibility through the process of equitable adjustments.[72] What the FCA requires is that a contractor knowingly make a claim for monies to which it would not otherwise be legitimately entitled.[73]

In cases asserting violations of the False Claims Act, courts will, when it is appropriate, impute the knowledge of employees to a contractor in order to impose direct liability upon the contractor. The imposition of direct liability on the contractor will depend on whether the employee at fault was acting within the scope of his or her employment with the intent to benefit the contractor. In *United States v. Dynamics Research Corp.*,[74] the district court faced the question of whether to impute the knowledge of the employees to the contractor. The employees used their position to influence the Air Force to purchase goods and services from third parties that in turn paid the employees when the employees provided the goods or services.[75] However, the court did not impute the knowledge of the employees to the contractor because the employees concealed their actions from their employer.[76] The court refrained from entering summary judgment for the government on the *direct liability* theory partly due to the lack of proof that the contractor enjoyed any benefit from the conduct of the employees. The court found that the contractor did not receive any payment and the government's argument that the contractor received the benefit of meeting contractual obligations as to minority-owned subcontractors was not persuasive.[77] Also, the court found that the employees did not hold positions in the company so that their actions were indistinguishable from the corporation.[78] The court ruled that the employees did not constitute the "apex of power" in the company. If the employees were the "apex of power," then questions of scope of employment and benefit would be irrelevant.[79]

The court then discussed whether it could impose *vicarious liability* on the contractor. Vicarious liability turns on whether the employee had **apparent authority**. If an employee occupies a position in which "according to the ordinary habits of persons in the locality . . . it is usual for such an agent to have a particular kind of authority, anyone dealing with him is justified in inferring that he has such authority."[80] In *Dyanmics Research*, the employee at fault occupied a position where he was to provide technical advice to the Air Force and to steer the Air Force to third-party contractors for needs relating to the prime contract. The court found that the

[72] *Id.* at 281.
[73] *Id.*
[74] 2008 WL 886035 (D.Mass. 2008).
[75] *Id.*
[76] *Id.* at *13.
[77] *Id.*
[78] *Id.*
[79] *Id.* (citing *United States. v. DiBona*, 614 F. Supp. 40, 44 (E.D. Pa. 1984).)
[80] *Id.* at *14 (quoting *Restatement (Second) of Agency 14* § 49(c).

employee had apparent authority and that the contractor could be held vicariously liable.[81] Under the theory of vicarious liability, there is no consideration of whether the employee was acting as to benefit the contractor, as under a direct liability theory.[82] If a court finds the employee had apparent authority for his actions, then it will not consider whether the company received a benefit. Instead, the court will hold the contractor vicariously liable. Ultimately the vicarious liability theory may be easier for the government to prove because it does not require proof of employee intent.

On a practical level, these two theories of liability under the FCA increase the benefit of "self-policing" by the contractor and the adoption of a meaningful ethics compliance program. A contractor should know what its employees are doing and what other business affairs they are involved in, and take steps to make sure that its employees are maintaining the same expectations of integrity that the contractor as a whole is required to maintain.

3. Fines and Damages

The monetary penalties for a contractor found liable under the False Claims Act may include fines, damages or both. Under the civil FCA, a court may assess a fine of between $5,000 to $10,000 fine for each false claim submitted to the government. In *United States v. United Technologies Corp.*,[83] the court found that the contractor had submitted 709 invoices to the Air Force for payment under its multi-year contract to furnish jet engines that were tainted by a false claim (misrepresentation) made at the time of the proposal on the contract. In finding that United Technologies was liable under the FCA, the court considered each individual invoice to be a separate claim and assessed fines in the amount of $7,090,000.[84] The basis for the false claim action was the contractor's explanation of the factual basis for the pricing of the engines over a multi-year period. Although the contractor stated to the government that it had utilized certain data in predicting future costs and prices, it did not utilize that data, and the court found the contractor was obliged to have followed through with its representation.[85] There were no damages because the court additionally found that the contractor reduced the price of the engines so that the government saved money over the term of the contract.[86] A contractor should recognize that even if the government is not harmed or even benefits from a false statement, as in *United Technologies*, the contractor still can be held liable for civil penalties under the False Claims Act.

However, the false claim in a case may not be each individual voucher submitted to the government. In *United States ex rel. Longhi v. Lithium Power Technologies, Inc.*,[87] the court disagreed with the government's contention that the defendant submitted 54 vouchers, each of which would incur the $10,000 penalty. Rather, the court

---

[81]*Id.* at *16.
[82]*Id.* at *15.
[83]2008 WL 3007997 (S.D. Ohio 2008).
[84]*Id.* at *12.
[85]*Id.* at *10.
[86]*Id.* at *12.
[87]530 F.Supp. 2d 888 (S.D. Tex. 2008).

looked at the "causative act" of the defendant and awarded the penalty only for each of the four contracts in the case.[88] What constitutes an individual false claim in a case appears to be a issue that may be given wide interpretation by a trial court.

In addition to the fines that may be assessed, treble damages may be awarded to the government.[89] The determination of the extent of the damages that the government suffered will vary depending on the type of contract and the circumstances of an individual case. In *Lithium Power*, the defendants misrepresented information about their history and qualifications in order to obtain a contract under the Small Business Innovation Research Program (SBIR). In assessing damages, the court disregarded the argument that the government received benefit from the research done by the defendants. Rather, it found that the purpose of the SBIR was to award contracts to foster entrepreneurship by and the development of eligible small businesses. Since the defendants were not eligible to participate in the program, as they had misrepresented their capabilities and resources,[90] the purpose of the program was not met and the government received no benefit. The court then assessed damages equal to the value of each of the four contracts that the defendants had been awarded.[91]

4. Potential Subcontractor Liabilities

The limit of potential liability of subcontractors under the False Claim Act centers on whether the FCA would apply if a subcontractor makes a false statement but that statement is not made directly to the government. In 2008, the Supreme Court of the United States addressed this issue in *Allison Engine Co. v. United States*.[92] In *Allison Engine*, the Navy contracted with two shipyards to build a fleet of destroyers. The shipyards then subcontracted with Allison Engine to construct generator sets. Allison Engine subcontracted with other firms to manufacture parts of the generator sets and to assemble the generator sets.[93] The subcontractors were required to submit a certificate of conformance to contract specifications along with delivering each unit. Employees of one of the subcontractors filed a False Claim Act suit alleging that invoices submitted by Allison Engine and the other subcontractors had not been done in accordance with contract requirements and that the subcontractors had falsified the certificates of conformance.[94]

The Supreme Court interpreted two sections of the FCA to determine whether liability under the act could be imposed when the false statement or claim is not made to the government itself but is made to a private entity. In the prime contract for the destroyers, the shipyards were paid sums of money in advance. The shipyards then received invoices and claims for payment by the subcontractors. The Supreme Court ruled that in a situation like this, the key issue is whether "a subcontractor . . . makes

---

[88]*Id.* at 901.
[89]31 U.S.C § 3729(a)(7).
[90]530 F.Supp. 2d at 898.
[91]*Id.* at 899.
[92]128 S.Ct. 2123 (2008).
[93]*Id.* at 2126.
[94]*Id.* at 2127.

a false statement to a private entity and does not intend the Government to rely on that false statement."[95] The subcontractors in *Allison Engine* were not held liable under the False Claims Act because the money used to pay the invoices, although originating from the government, had passed on to the shipyards, and the government no longer had any involvement in its disbursement nor any opportunity to rely on the false statements.

In 2009, Congress responded to the *Allison Engine* decision by passing amendments to the False Claims Act in the Fraud Enforcement and Recovery Act of 2009 (FERA).[96] These amendments were clearly intended to overrule the *Allison Engine* decision by eliminating the requirement that a false claim be made "to get" the claim paid by the government.[97] This amended provision is also retroactive to all claims pending on June 7, 2008, two days before the Court issued the *Allison* decision, demonstrating Congress's intent to substantially negate the effect of that decision.[98] The FERA also expands a contractor's potential liability under the FCA by changing the statute's definition of "claim." The new definition of claim now includes situations where a claim is made to a contractor where the "money is to be spent or used on the government's behalf or to advance a government program or interest."[99] Congress intentionally has closed the loophole allowed by *Allison Engine* by defining a claim to include situations where the government has already provided money to a higher-tier contractor.[100] Moreover, the expanded definition of "claim" may have an impact on contracts that do not involve the United States directly as a contracting party but instead are part of a federally funded grant to state, local, or private entities. For more on federal grants in general and the FCA's impact on those grants, see **Chapter 16**.

D. Other Remedies for Prohibited Conduct

In addition to assessing penalties and fines under the False Claims Act, the government has a number of other remedies that it can use to further accomplish the goal of preventing fraud in future contracts. These remedies are not used alone but will be employed in conjunction with the remedies of the FCA or other applicable fraud statutes, such as the Anti-Kickback Act, which calls for double damages.

1. Contract Cancellation Remedy

In addition to the specific remedies set forth in the various statutes, a government contractor faces the total cancellation of the underlying contract if it is tainted by

---

[95]*Id.* at 2130.
[96]PUB. L. 111-21.as codified 31 U.S.C. 3129(a)(1)(B) (2009).
[97]*Id.*
[98]In a subsequent proceeding in the *Allison* litigation, the retroactive provision of this statute was ruled to be unconstitutional under the Ex Post Facto Clause of the United States Constitution, Art. I, § 9, cl.3. *See United States v. Allison Engine Co., Inc.,* Nos.1:95-cv-970, 1:99-cv-923 (S.D. Ohio October 28, 2009).
[99]31 U.S.C. § 3129(b)(2)(ii).
[100]31 U.S.C. § 3129(b)(2)(ii)(1).

conduct that is considered to be a corrupt practice. In *United States v. Acme Process Equipment Co.,*[101] the contractor sued to recover breach of contract damages after the government canceled its contract. The cancellation was based on the fact that three of the contractor's employees accepted compensation for awarding subcontracts in violation of the Anti-Kickback Act. The contractor argued at the Court of Claims that contract cancellation was not an authorized remedy for a violation of the Anti-Kickback Act because both civil and criminal remedies were set forth in that statute. The Court of Claims accepted that argument on the grounds that Congress had intended to set forth the *entire* set of remedies available to the United States for a violation of that statute. The Supreme Court reversed that decision, holding that public policy requires that the United States be able to rid itself of a prime contract tainted by kickbacks. In such cases, the contractor would not be entitled to payment on a theory of *quantum merit* or otherwise, regardless of the incurrence of otherwise allowable performance costs.[102]

Applying this public policy, contract cancellation has been permitted when the contract was tainted by the making of false statements and false claims.[103] Similarly, in *Beech Gap, Inc.,*[104] the board upheld a termination for default following the conviction of the contractor's employees for submission of falsified test reports and pay estimates. The board refused to consider the contractor's argument that the government had superior knowledge of the alleged false test reports and pay estimates as an effort to relitigate an issue unsuccessfully litigated in the prior criminal action and dismissed the contractor's appeal.

Even if government insists on contract performance after becoming aware of the prohibited conduct, that action by the government does not operate to ratify the underlying contract. For example, in *Schuepferling Gmbh & Co., KG,*[105] the contract was tainted by bribery. Even though the government insisted on and accepted further performance by the contractor, those actions did not negate the government's right to void the contract *ab initio* (from the outset).

2. Other Remedies

Four decisions issued in 2006 and 2007 demonstrate the other remedies available to the government and give an indication of the heightened agency and court awareness of fraud and false claims as related to construction projects. These decisions illustrate the variety of remedial approaches and the ways in which a court will employ several statutes and remedies to protect the government's interest in eliminating fraud from government contracting.

The first decision, although concerning commercial banking rather than construction or procurement, is extremely important due to its holding that any claim

---

[101]385 U.S. 138 (1966).

[102]*United States v. Mississippi Valley Generating Co.,* 364 U.S. 520 (1961).

[103]*See Brown v. United States,* 524 F.2d 693 (Ct. Cl. 1973).

[104]ENGBCA Nos. 5585 et al., 95-2 BCA ¶ 27,879.

[105]ASBCA No. 46564, 98-1 BCA ¶ 29,659.

under a contract that has been tainted by fraud is forfeited under the Forfeiture of Fraudulent Claims Act, 28 U.S.C. § 2514, even if the wrongdoing is not directly related to the contract performance. In *Long Island Savings Bank, FSB v. United States*, the Federal Circuit specifically invoked the holding in a 50-year-old Court of Claims decision that stated when a contractor practices fraud against the government, the court does not have the right to divide the valid claims from the claims related to the fraud.[106] Therefore, even if a contractor has valid claims unconnected to the fraud, those claims will be forfeited, in effect increasing the penalty assessed against the contractor. However, the forfeiture remedy is appropriate only where the contract has been tainted from its inception.[107] In such a case, the government would have to prove that the contract itself was obtained by the contractor by means of a false statement.[108]

In *Veridyne Corp. v. United States*,[109] the government asserted a counterclaim against the contractor on the theory that it had committed fraud by underpricing modifications to its contract. Otherwise, the contract would have been rebid because the contractor was on the verge of "graduating" from eligibility in a § 8(a) program administered by the Small Business Administration.[110] The government argued the contract should be found to be void *ab initio* (i.e., void from its very inception) and that money paid to the contractor under modifications to the contract should be forfeited and returned to the government because the act of fraud.[111] However, the court stated that for such a remedy to be imposed, the government would have to prove that a bribe took place or that there was a violation of the conflict of interest laws.[112] The discussion in *Veridyne* does not diminish the potential liability of the contractor or restrict the remedies available to a great extent. In that case, the court declined to enter summary judgment, stating that there were still issues of fact to be resolved regarding the conduct of the contractor and government and the consequences of that conduct.[113] The contractor should note, as illustrated in *Long Island* and *Veridyne*, that fraudulent conduct to obtain a contract will likely result in the very harsh treatment (potentially total forfeiture) whereas fraud that occurs later may still permit the contractor to retain some of the money it earned on the contract. However, other remedies or penalties may be imposed against the contractor.

In *Morse Diesel Int'l, Inc. d/b/a AMEC Constr. Mgmt., Inc. v. United States*,[114] the contractor was found liable under the False Claims Act as well as the Anti-Kickback Act. The conduct that formed the basis for the government's claims included billing for the full amount of bond premiums when there was a discount or rebate agreement

---

[106]476 F.3d 917, 925 (Fed. Cir. 2007) citing *Little v. United States*, 138 Ct.Cl. 773, 152 F.Supp. 84 (1957).
[107]*Id.* at 926.
[108]*Id.*
[109]83 Fed. Cl. 575 (2008).
[110]*Id.* at 576.
[111]*Id.* at 581.
[112]*Id.* at 586.
[113]*Id.* at 589.
[114]79 Fed. Cl. 116 (2007).

with the bonding company; providing invoices from the bonding company marked "Paid" when payments had not been made; advance billings by reallocating $5.4 million in subcontractor line items, which were allegedly billed but not paid to the trade contractors, which was a violation of the progress payment certification provided for under the contract. These actions provided the basis for liability under the FCA. In addition, the government showed that the contractor had received kickbacks ("rebates" on the bond premium) from bonding companies that were in violation of the Anti-Kickback Act. The contractor argued during the trial that imposing penalties under the Anti-Kickback Act and the False Claims Act was duplicative or prohibitive.[115] The court rejected this argument on the grounds that Congress intended both statutes to be used to compensate the government and to heighten the degree of deterrence for future fraudulent conduct on the part of contractors. In addition to the penalties assessed under the fraud statutes, Morse Diesel's claims under the contract, in excess of $50 million, were forfeited.[116]

The fourth decision, *Daewoo Engineering and Construction, Ltd. v. United States*,[117] demonstrates the manner in which a court may find a contractor liable under the False Claims Act as well as the anti-fraud provisions of the Contract Disputes Act. In *Daewoo*, the contractor initially submitted a proposal to build a road around an island in Palau. This initial proposal was approximately $28 million less than the next lowest offer. Even after resubmitting its price prior to award, Daewoo was $13 million less than the next lowest offer and was awarded the contract by the Army Corps of Engineers.[118] The project was intended to take 1080 days but quickly met with problems associated with the soil conditions and specifications for the road. Daewoo submitted a certified claim based on delay for approximately $64 million.[119] The government responded claiming that Daewoo had violated the Contract Disputes Act and the civil FCA.[120] The Court of Federal Claims found that Daewoo had committed fraud largely based on testimony by one of its employees that the claim he certified contained about $50 million in claims that had not been incurred and were included in order to get the government's attention.[121] As a result of this finding, the court determined that Daewoo was liable to the government in the amount of the overstated claim—in excess of $50 million. In addition, the court found that Daewoo was liable under the civil FCA. Under this statute, the court only assessed a $10,000 fine because it did not have the evidence to find that the government had been damaged.[122] In addition to penalties under these two statutes, the court ruled that Daewoo would have had to forfeit any valid claim under 28 U.S.C § 2514 if it had proven liability on the part of the government.[123] Also, the court found that Daewoo had

---

[115]*Id.* at 122.
[116]*Morse Diesel Int'l, Inc. d/b/a AMEC Constr. Mgmt., Inc. v. United States*, 74 Fed. Cl. 601 (2007).
[117]73 Fed. Cl. 547 (2006) *aff'd* 557 F.3d 1332 (Fed. Cir. 2009).
[118]*Id.* at 550.
[119]*Id.* at 560.
[120]*Id.* at 581.
[121]*Id.* at 585.
[122]*Id.*
[123]*Id.* at 584.

committed fraud in the inducement by including in its bid statements that specific people and subcontractors would work on the project, although these people and companies never did, and also by submitting a schedule that it quickly abandoned and which the court doubted it had ever intended to keep.[124] The fraud in the inducement would have worked in conjunction with the forfeiture if Daewoo had been entitled to any recovery from the government.[125] The *Daewoo* case illustrates the wide range of theories that a court can employ in finding liability and imposing sanctions.

These recent cases are important illustrations of the wide range and severity of possible remedies if a contractor is found to have committed fraud against the government. The wide range of possibilities underscores the importance that a contractor ensures that it maintains the highest level of honesty and integrity in its dealings with the government.

E. Contractor Business Ethics and Conduct

1. *FAR Requirements Applicable to Contractors*

The FAR provisions detailing a contractor's obligations are found in FAR Subpart 3.10, Contractor Code of Business Ethics and Conduct; FAR Subpart 9.4, Debarment, Suspension and Ineligibility; and the implementing FAR clauses.[126] These provisions combine a *statement of expectations applicable* to all contractors; differing *mandatory requirements* depending on a company's size, the dollar value, and duration of a government contract; and *potential sanctions* applicable to *any contractor* if it fails to make certain required disclosures to the government. (*See* **Section IV.F** of this chapter.) The next sections provide an overview of these requirements and their potential effect on a government contractor.[127]

a. Contractor Standards of Conduct   The FAR addresses contractor codes of business ethics and conduct with a combination of a statement of expectations applicable to any contractor and mandatory requirements that vary with the size of the contractor and the value and duration of the contract.

STATEMENT OF EXPECTATIONS
- All contractors "must conduct themselves" with the "highest degree of integrity and honesty."
- All contractors "should have" a written code of business ethics and conduct.
- As part of an effort to promote compliance with the written code, all contractors "should have" an employee business ethics and training program suitable for the

---

[124]*Id.* at 587.
[125]*Id.* at 588.
[126]FAR §§ 52.203-13; 52.203-14.
[127]For a more comprehensive review of the background for these requirements and guidance on implementing an effective compliance program, *see Federal Government Contractor Ethics and Compliance Programs, Toolkit and Guidance* (Thomas J. Kelleher, Jr. [Associated General Contractors of America, 2009]).

size of the company that will "facilitate discovery and disclosure of improper conduct" and "ensure corrective measures" are carried out.

- If the contractor is aware that the government has overpaid on a contract financing or invoice payment, the contractor is expected to remit the overpayment amount to the government. A contractor may be suspended and/or debarred for knowing failure by a principal to "timely" disclose "credible evidence" of a significant overpayment.

The FAR does not define either "timely" or "credible evidence" in the regulations or the related contract clauses. However, the commentary that accompanied these provisions indicated that "timely disclosure" depended on the date that the contractor had "credible evidence" of a violation or the date of the contract award, whichever was later.[128]

MANDATORY REQUIREMENTS
Any contractor receiving a government contract in excess of $5 million[129] and with a contract duration of 120 days or more ("Covered Contract") shall:

- Have a written code of business ethics and conduct.
- Make a copy of that code "available" to each employee engaged in the performance of a Covered Contract.
- Exercise "due diligence" to prevent and detect criminal conduct.
- Promote an organizational culture that "encourages" ethical conduct and a commitment to compliance with the law.
- Make a "timely" written disclosure to the agency Inspector General, with a copy to the contracting officer, whenever the contractor has "credible evidence" of a violation of the civil False Claims Act (31 U.S.C. §§ 3729– 3733), or of federal criminal law involving fraud, conflict of interest, bribery, or gratuity violations.

b. Contractor Awareness Programs and Internal Control Systems   Except for small business concerns,[130] every contractor performing a Covered Contract must establish an ongoing business ethics and awareness program and an internal control system within 90 days of contract award.[131] The program shall include:

- Reasonable steps to communicate the contractor's compliance program standards and procedures.

---

[128]The FAR Councils stated that they adopted a "credible evidence" standard in lieu of "reasonable grounds to believe." The "credible evidence" standard is described as "a higher standard, implying that the contractor will have the opportunity to take some time for preliminary examination of the evidence to determine its credibility before deciding to disclose to the government. *See* 73 Fed. Reg. 667073. However, an opportunity to investigate seems to imply an obligation to investigate.

[129]FAR § 3.1004 authorizes agencies to reduce this monetary threshold.

[130]Based on the contractor's representation at the time of proposal or bid submission of its size in accordance with FAR Part 19, Small Business Programs, and 13 CFR Part 121. Commercial item suppliers are also excluded. *See* FAR § 52.203-13(c).

[131]Unless extended by the contracting officer.

- Effective training programs and dissemination of information appropriate to an employee's responsibilities and including training, "as appropriate," for the contractor's subcontractors.[132]

The internal control system shall:

- Establish procedures to facilitate timely discovery of improper conduct.
- Ensure corrective measures are promptly instituted and carried out.
- Assign responsibility for the internal control system at a "sufficiently high level" within the company and provide "adequate resources to ensure effectiveness" of the program and internal control system.
- Make reasonable efforts not to hire or engage as "principal"[133] of the contractor any person whom due diligence would have exposed as having engaged in conduct in conflict with the contractor's code of business ethics.
- Provide for periodic reviews to determine if the contractor's practices and procedures are in compliance with the contractor's code of business ethics and any special requirements of government contracting.
- Provide a monitoring and auditing process to detect criminal conduct.
- Make periodic evaluations of the effectiveness of the compliance program, particularly if criminal conduct has been discovered.
- Make periodic assessments of the risk of criminal conduct and modify the compliance program and internal control system to reduce that risk.
- Provide an internal reporting mechanism such as a hotline for employees to report improper conduct and instructions encouraging such reports.
- Provide for disciplinary action in the event of improper conduct or for failure to prevent or detect improper conduct.

When developing and implementing a business ethics and compliance program, contractors should expect that the Defense Contract Audit Agency (DCAA) will examine the contractor's compliance program during the course of an audit of a contractor's claim or proposal. Section 3 of Chapter 5 of DCAA's audit manual[134] states that the compliance program is an indicator of the "contractor's control environment" and provides detailed direction to auditors regarding the review of the contractor's compliance program for the purpose of assessing the effectiveness of the contractor's internal controls and risk of mischarging of costs to a contract.

---

[132]The FAR Councils did not further explain what is intended by the inclusion of subcontractors in the training requirement.

[133]FAR § 2.101 defines a "principal" as "an officer, director, owner, partner, or a person having primary management or supervisory responsibilities within a business entity (*e.g.*, general manager; plant manager; head of a subsidiary, division, or business segment; and similar positions)."

[134]*DCAA Contract Audit Manual* (*CAM*)2. Disclosure of Wrongdoing, Cooperation with Investigations, and Whistleblower Protection.

As noted above, the government now expects that a contractor will make a "timely disclosure" in writing to the agency Inspector General, with a copy to the contracting officer, whenever the contractor has "credible evidence" that a principal, employee, agent, or subcontractor of the contractor has committed a violation of the civil False Claims Act or a violation of federal criminal law involving fraud, conflict of interest, bribery, or gratuity violations in connection with the award, performance, or closeout of any government contract performed by that contractor or a subcontractor under that subcontract. The disclosure obligation related to a specific contract "continues until at least three years after final payment" on that contract.[135] Finally, the contractor is obligated to provide "full cooperation"[136] with any government agency responsible for audits, investigations, or corrective actions.

In addition to the requirements for disclosure and cooperation with the government, FAR Subpart 3.9, Whistleblower Protection for Contractor Employees, also addresses the potential for contractor efforts to discourage an employee from reporting perceived wrongdoing to the government. In addition to setting forth a detailed procedure in that subpart for investigating complaints about a contractor's alleged reprisal actions related to its employees, FAR § 3.903 policy provides:

> Government contractors shall not *discharge, demote or otherwise discriminate* against an employee as a reprisal for disclosing information to a Member of Congress, or an authorized official of an agency or of the Department of Justice, relating to a substantial violation of law related to a contract (including the competition for or negotiation of a contract). [Emphasis added.]

Consistent with that requirement, the FAR also expressly prohibits retaliation against an employee for making certain disclosures to the government. FAR § 3.907-2 policy details that prohibition:

> Non-Federal employers are prohibited from discharging, demoting, or otherwise discriminating against an employee as a reprisal for disclosing *covered information* to any of the following entities or their representatives:
> (1)The Board.[137]
> (2)An Inspector General.
> (3)The Comptroller General.
> (4)A member of Congress.

---

[135]These disclosure obligations are found in FAR § 3.1003(a)(2) and in two separate sections of FAR § 52.203-13. Subparagraph (b)(3) of FAR § 52.203-13 makes them applicable to all firms including small business concerns. Paragraph (c)(2)(F) restates them in the context of the elements of an internal control system. Therefore, even if a small business contractor is exempt from the requirement for an ethics awareness program and an internal control system, it *remains obligated* to comply with the disclosure provisions of FAR § 52.203-13. FAR § 3.1003(a)(2) states that any contractor's *"knowing"* failure to make a required disclosure provides grounds for suspension or debarment.

[136]"Full cooperation" is defined in FAR § 52.203-13(a).

[137]"Board" means the Recovery Accountability and Transparency Board established by Section 1521 of the American Reinvestment and Recovery Act of 2009 (ARRA). *See* **Chapter 17** for a review of ARRA.

(5) A State or Federal regulatory or law enforcement agency.

(6) A person with supervisory authority over the employee or such other person working for the employer who has the authority to investigate, discover, or terminate misconduct.

(7) A court or grand jury.

(8) The head of a Federal agency.[138]

"Covered information" is a defined term in that FAR Subpart. FAR § 3.907.1 defines that term in a rather broad manner:

> *Covered information* means information that the employee reasonably believes is evidence of gross mismanagement of the contract or subcontract related to covered funds, gross waste of covered funds, a substantial and specific danger to public health or safety related to the implementation or use of covered funds, an abuse of authority related to the implementation or use of covered funds, or a violation of law, rule, or regulation related to an agency contract (including the competition for or negotiation of a contract) awarded or issued relating to covered funds.

F.   Contractor Self-Reporting/Hotline Requirements

1.   Hotlines

Unless the contractor performing a Covered Contract has a business ethics and awareness program and an internal control system that includes a reporting mechanism with a company hotline and hotline posters, the contractor shall prominently display at all common work areas within business segments performing work on the contract and at the contract work sites any agency fraud hotline poster or the Department of Homeland Security fraud hotline poster as identified in the contract.[139] If the contractor uses a company Web site as a means of providing information to its employees, the contractor must also display an electronic version of the anti-fraud hotline posters on that Web site.

2.   Self-Reporting of Potential Violations

A contractor must make a "timely disclosure" in writing to the agency Inspector General, with a copy to the contracting officer, whenever the contractor has "credible

---

[138] Emphasis added.

[139] See FAR § 52.203-14, Display of Hotline Poster(s). A small business concern under 13 CFR Part 121 that elects not to adopt a business ethics and awareness program and an internal control system must post agency hotline posters, such as the DOD Hotline poster.. In accordance with FAR § 3.1004, the Display of Hotline Poster(s) clause does *not* apply to contracts performed entirely outside of the United States or for commercial items. The requirement for company hotlines as set forth in FAR § 52.203-13 does apply to contracts performed entirely outside of the United States if the dollar and duration thresholds are met. The Federal Highway Administration at 23 CFR § 635.119 requires that the anti- fraud notices be posted on each federally funded highway project.

evidence" that a principal, employee, agent, or subcontractor of the contractor has committed a violation of the civil False Claims Act or a violation of federal criminal law involving fraud, conflict of interest, bribery or gratuity violations in connection with the award, performance, or closeout of any government contract performed by that contractor or a subcontractor thereunder. The disclosure obligation related to a specific contract "continues until at least three years after final payment" on that contract.[140] Finally, the contractor is obligated to provide "full cooperation"[141] with any government agency responsible for audits, investigations, or corrective actions.

3. Flow-down Requirements

Contractors are required to flow down these requirements to subcontracts in excess of $5 million and 120 days in duration. Contractors are to verify their subcontractors' compliance with these requirements.[142]

G. Potential Nondisclosure Sanctions: Suspension and Debarment

While the Contractor Code of Business Ethics and Conduct clause at FAR § 52.203-13 contains explicit requirements for certain contractor disclosures, that clause does not contain an express statement of the consequences if the contractor fails to provide a disclosure as specified by that clause. The sanctions are addressed in a revised version of FAR § 3.1003(a)(2) and (3)[143] and are included in the revisions to the statement of the grounds for contractor suspension or debarment found in FAR §§ 9.407-2 and 9-406-2, respectively. These two provisions set forth as grounds for suspension or debarment of any contractor:

- "Knowing failure" by a "principal" of the contractor to make a "timely" written disclosure to the government in connection with a government contract awarded to that contractor when the contractor has "credible evidence" of:
  - Violation of federal criminal law involving fraud, conflict of interest, bribery or gratuity violations.

---

[140]These disclosure obligations are found in FAR § 3.1003(a)(2) and in two separate sections of FAR § 52.203-13. Subparagraph (b)(3) of FAR § 52.203-13 makes them applicable to all firms including small business concerns. Paragraph (c)(2)(F) restates them in the context of the elements of an internal control system. Therefore, even if a small business contractor is exempt from the requirement for an ethics awareness program and an internal control system, it *remains obligated* to comply with the disclosure provisions of FAR § 52.203-13. FAR § 3.1003(a)(2) states that any contractor's *"knowing"* failure to make a required disclosure provides grounds for suspension or debarment.

[141]"Full cooperation" is defined in FAR § 52.203-13(a).

[142]The FAR Councils' commentary to the regulation expressly states that there is *no requirement* for a contractor to "review or approve" a subcontractor's ethics code, compliance program, or internal control system. *See* 73 Fed. Reg. 67084.

[143]FAR § 3.1003(a)(2) extends the mandatory disclosure obligations related to wrongdoing or overpayments and the related sanctions to *all contractors*, not just those subject to the requirements of the clause at FAR § 52.213-13.

- Violation of the civil False Claims Act, (31 U.S.C. §§ 3729-3733).
- "Significant overpayments" on the contract other than those resulting from contract financing as defined in FAR § 32.001.[144]
- The disclosure sanctions applies to any government contract in existence as of the effective date of the new suspension/debarment regulations and reach back to closed contracts for a period of three years following final payment on that contract (i.e., December 12, 2005).[145]
- The disclosure obligation relates to the award, performance, or closeout of the contract or a subcontract under that contract.

The apparent purpose for the placement of the sanction provisions in parts 3 and 9 of the FAR was disconnect the sanctions of debarment or suspension from the presence or absence of the Contractor Code of Business Ethics and Conduct clause in a particular contract. Consequently, even if the clause is not in the contract or if the contract is less than the $5 million-120-day thresholds, or if the firm qualifies as a small business, every contractor performing a government contract remains subject to the potential sanctions of debarment or suspension for a failure to make one of these disclosures.

H. Defense Contract Audit Agency Fraud Indicators

Contractors should be mindful of the possibility of being audited either by the DCAA or by an agency's internal auditing service. The DCAA performs audit services for all DOD contracts as well as the majority of contract audit services for all other federal agencies.[146] The DCAA publishes the *DCAA Contract Audit Manual (CAM)*, which is the handbook for its auditors. The manual requires that "auditors should be familiar with specific fraud indicators" that are both listed in the *CAM* itself as well as a separate publication, *Handbook on Fraud Indicators for Contract Auditors*.[147] While the auditors are not responsible for proving fraud, they are required to find and report fraud indicators discovered during an audit to the appropriate law enforcement official. Additionally, if *qui tam* False Claims Act actions are filed against contractors, the Department of Justice attorneys often will seek information from DCAA audits to assist in the investigation of the claims.[148]

---

[144]FAR § 32.001 excludes payments made under the Payments under Fixed-Price Construction Contracts clause from this definition. An express requirement to notify the contracting officer of an overpayment on construction contracts is already set forth in FAR § 52.232-27(l). (*See also* FAR § 52.232-5(d) addressing refunds of "unearned amounts.")

[145]While these regulations are effective as of December 12, 2008, the FAR Councils expressed a clear intent that the sanctions for failure to make a disclosure apply to all existing contracts as of that date and to closed contracts up to three years following final payment. *See* 73 Fed. Reg. 67074.

[146]See *www.defense.gov/comptroller/defbudget/fy2007/budget_justification/pdfs/01_Operation_and_Maintenance/O_M_VOL_1_PARTS/DCAA.pdf* (accessed May 19, 2009).

[147]DCAA CAM 4-702.3147.DCAA CAM 4-709.

[148]DCAA CAM 4-709.

The *Handbook on Fraud Indicators* lists dozens of individual fraud indicators relating to labor, materials, and subcontractors among others. The type of contract and the work involved in performing the contract is central to determining which fraud indicators that an auditor will be especially on the lookout for. However, two main themes are apparent throughout the fraud indicators. First, the government has a major interest to protect in its procurement contracts, so auditors are instructed to investigate not just the government contracts that an individual contractor has been awarded but also the full scope of the contractor's operations including its business ethics and compliance program. For example, in evaluating patterns of labor costs, the *Handbook* contemplates that a DCAA auditor will gather information on the way labor is assigned to all of its contracts, both with the government and with private parties or internal divisions to determine whether the contractor is shifting labor costs to possibly defraud the government. The second theme is the highly subjective wording used in many of the fraud indicators. The indicators commonly contain words such as "significant," "weak," or "consistent" without defining how these words are to be used. The end result is that the auditor will use his or her own individual idea of what these indicators mean when deciding to refer the contractor to law enforcement.

➤ LESSONS LEARNED AND ISSUES TO CONSIDER

- Like private commercial contracts, government contracts are based on the concepts of an *exchange of promises* by the contracting parties and express and *implied obligations* binding on the parties.
- Construction projects are awarded by numerous agencies of the government. Each agency has a *unique organization* and mission. There are often key differences within the same agency from office to office.
- Similar to private work, construction for the government is a *people business,* and a contractor should gain an appreciation of the organization and operation of any agency for which it contemplates performing work.
- Government agencies maintain a *variety of Web sites* providing information on contracting opportunities, organization of a particular agency, and extensive online libraries listing standards and guides that often are incorporated into that agency's contracts by reference.
- Government construction contracts are replete with *jargon and acronyms.* These terms and abbreviations need to be understood by anyone doing business with a federal agency.
- While it is often stated that the United States submits to the same rules as private parties when it enters into a contract, there are important *exceptions* to that concept involving *authority* and fundamental *public policy* considerations.

(Continued)

- Federal government construction contract forms, policies, and procedures are devised from *multiple statutes* and a *comprehensive regulatory* system.
- Contractors must recognize that many key contract provisions are *incorporated by reference*. That practice by many government agencies does not diminish the significance of those provisions.
- To deter *fraud and false claims,* the federal government has available a broad spectrum of criminal and civil statutes that carry severe penalties for contractor wrongdoing.
- *Contractor certifications* are a key element of the government's effort to deter fraud and false claims. Such certifications should not be considered mere formalities.
- A growing trend in government contracting is a requirement that contractors employ programs encouraging employees to *report suspected fraud and wrongdoing.*
- Many key *resources* for government contractors are available on the Internet. Consistent with an effort to reduce reliance on paper, the federal government requires its contractors to report and post key information electronically.

APPENDIX 1A: BRIEF HISTORY OF THE DISPUTES PROCESS IN GOVERNMENT CONTRACTS

During World War I, the use of a board of contract appeals in the War Department became prevalent as the federal government sought to address issues arising out of the extraordinary increase in wartime procurement actions. Following that war, the use of boards to resolve contract claims diminished. However, with the increased volume of procurement during World War II, the nature of the boards and their practices became a concern to both industry and government procurement professionals. One decision, *Penker Construction Co. v. United States,*[149] highlighted the problems that could occur in the absence of established boards and fairly balanced rules of procedure. The saga of *Penker* was summarized by Joel F. Shedd Jr. in his excellent article, "Disputes and Appeals: The Armed Services Board of Contract Appeals."

> In *Penker* the contractor was refused permission to see the report on his claim that had been submitted by the constructing quartermaster, on the stated ground that it might be useful to him in prosecuting his claim against the government; and he was also told that no investigation would be made of the facts reported by the constructing quartermaster and that any doubts concerning interpretation of the specifications would be resolved in favor of the government. After congressional intervention, the Assistant Secretary of War told the contractor that he did not have time to hear appeals. The Assistant Secretary of War referred the appeal to a colonel, who referred it to a major, who referred it to a captain, who referred it back to the Quartermaster General, who referred it to the contracting officer, a brigadier general in his office, who referred it back to the captain who had prepared the contracting officer's decision from which the appeal has been taken. The Assistant Secretary of War's decision denying the appeal stated that he acted only in an administrative capacity, relying solely on the evidence and data presented to his office by the Office of the Quartermaster General. Under these facts, the Court of Claims held that the contractor had not received the kind of decision he was entitled to under the disputes clause and refused to accord any finality to such decision.[150]

The establishment of boards of contract appeals together with the further refinement of appeal rights to the United States Court of Claims resulted from the need to prevent the repetition of cases such as *Penker.* Following World War II, the contract claim disputes process essentially involved a three-step process. If the matter could not be resolved by agreement at the agency level, resolution required: (1) a decision by the contracting officer, (2) an appeal to the agency's board of contract appeals,

---

[149]96 Ct. Cl. 1 (1942).
[150]29 Law & Contemporary Problems at 50.

and (3) a limited right of appeal to the Court of Claims. A major problem with this process was that the boards were not authorized to decide "breach of contract" cases. This limitation on the board's jurisdiction precluded the resolution of "all disputes" at the board level because breach of contract claims had to be filed in the Court of Claims.

The modern disputes practices reflect the reforms and changes enacted with the Contract Disputes Act of 1978.[151] The details of the current process are set forth in **Chapter 15**. Many of the current procedures reflect an effort to address problems, enact reform, and provide a more efficient and credible process for all participants (federal government agencies and contractors).

Prior to the enactment of the CDA, the process for addressing contract disputes was a mixture of statutes, regulations. and interpretive case law. Federal government contracts contained a disputes clause, and every federal agency utilized a board of contract appeals. After a series of U.S. Supreme Court decisions,[152] the boards became the principal forum for the resolution of contractor claims, while the Court of Claims assumed the more limited role of an appellate court under the Wunderlich Act.[153] Except for the relatively unusual circumstance that could be characterized as a claim for breach of contract, nearly all claims arising under a contract had to be brought to the boards. However, the boards' jurisdiction was limited to "contract" claims, and any suit alleging breach of contract had to be filed in the Court of Claims.

Each agency board had its own rules and procedures, which had varying degrees of formality. In some agencies, board members or judges served only on a part-time basis. In those situations, the board judge often had other duties within the same agency that had awarded and administered the contract. In addition, due to the decision of the U.S. Supreme Court in *S&E Contractors, Inc. v. United States*,[154] the federal government had no right of appeal from an adverse decision by a board. The *S&E* decision also precluded efforts by an agency to obtain a review of an adverse board decision by the then General Accounting Office.[155]

Attempting to improve the overall disputes process, the CDA creates a comprehensive statutory basis for the disposition of contract disputes. The CDA applies to

---

[151]41 U.S.C. §§ 601-613.

[152]*United States v. Wunderlich*, 342 U.S. 98 (1951); *United States v. Moorman*, 338 U.S. 457 (1950); *United States v. Holpuch*, 328 U.S. 234 (1946).

[153]The reaction to the *Moorman* and *Wunderlich* decisions resulted in the passage of the Wunderlich Act, which limited the finality of board decisions. 41 U.S.C. §§ 321-322. This act was subsequently interpreted by the U.S. Supreme Court in *United States v. Bianchi*, 373 U.S. 709 (1963) and *United States v. Grace & Sons, Inc.*, 384 U.S. 424 (1966). *Bianchi* and *Grace* establish that a court reviewing a board decision was confined to the record created during the board proceeding and could not conduct an independent evidentiary hearing into issues not addressed by the board. Thus the boards became the primary fact-finding bodies, with significant emphasis placed on the development of a record that would support the board's findings with substantial evidence.

[154]406 U.S. 1 (1972).

[155]Now the Government Accountability Office.

any express or implied contract that is entered into by an "executive agency" of the federal government for the "procurement of [the] construction, alteration, repair, or maintenance of real property."[156] The act also applies to "the executive agency contracts for the procurement of property, other than real property, for the procurement of services and for the disposal of personal property, as well as for supplies.[157]

The term "executive agency" is defined in 41 U.S.C. § 601(2). It encompasses those entities that are commonly thought of as federal government agencies, such as the Department of Defense, the General Services Administration, the Department of Energy, the Department of Transportation, and the Department of Veterans Affairs. It also includes the U.S. Postal Service, the Postal Rate Commission, and various independent bodies and government corporations.[158]

The CDA's comprehensive statutory basis for resolution of disputes made significant changes to the old process. It makes the boards and their members more professional by requiring that all board members be full-time positions, and it is no longer possible for a board judge to function as an attorney for the agency on a part-time basis. Also, it gives the contractor a choice of a forum to appeal a contracting officer's final decision. Depending on the agency that awarded the contract, the contractor may elect to appeal to one of the two boards of contract appeals[159] or to file a suit on the contracting officer's final decision in the U.S. Court of Federal Claims (formerly the United States Claims Court). This concept is known as contractor's right of direct access. Furthermore, the CDA's provisions apply "notwithstanding any contract provision, regulation, or rules of law to the contrary."[160] As a result, it is not possible to agree by contract to limit the right of appeal to a particular forum.[161] In addition, the Act effectively reverses the *S&E* decision by giving the federal government the right to appeal an adverse board decision to the United States Court of Appeals for the Federal Circuit. Thus both parties are provided equal rights to appeal adverse board decisions.

---

[156]41 U.S.C. § 602(a).

[157]It is well established that the Contract Disputes Act applies to leases for real property. *See George Ungar,* PSBCA No. 935, 82-1 BCA ¶ 15,549; *Goodfellow Bros. Inc.,* AGBCA No. 80-189-3, 81-1 BCA ¶ 14,917; *Robert J. DiDomenico,* GSBCA No. 5539, 80-1 BCA ¶ 14,412. However, jurisdiction over a dispute outside of the terms of the lease, such as a decision to expand the area subject to the lease, has been rejected by a board. *See John Barrar & Marilyn Hunkler,* ENGBCA No. 5918, 92-3 BCA ¶ 25,074.

[158]The Act also contains provisions covering the Tennessee Valley Authority. *See* 41 U.S.C. § 602(b).

[159]Armed Services Board of Contract Appeals or the Civilian Board of Contract Appeals.

[160]41 U.S.C. § 609(b).

[161]*OSHCO-PAE-SOMC v. United States,* 16 Cl. Ct. 614 (1989).

2

AUTHORITY TO BIND THE GOVERNMENT, CONTRACT FINANCING, AND PAYMENT

I. AUTHORITY OF GOVERNMENT EMPLOYEES OR AGENTS

The federal government operates through delegated powers. Without a specific written delegation of power, government agencies, employees, or representatives have no authority to act on behalf of the government. Unauthorized acts of government employees or agents will not bind the federal government.[1]

In the twenty-first century, federal agencies administer and manage construction projects with a variety of representatives who may or may not be employees of that agency. Some are identified as contracting officer's representatives, technical specialists, inspectors, or a variety of other terms and labels. Some projects are managed by construction managers engaged under a separate contract with the government. Many of these individuals do not have the legal authority to bind the government or to change the contract price, schedule, requirements, terms, or conditions. However, these individuals may and do initiate actions that have the effect of changing the contract. This reality can create significant performance and financial risks for the contractor. When performing any government contract, a prudent contractor needs to anticipate these potential risks by:

(1) Establishing a procedure to ascertain the actual limits of authority of the government's representatives;

(2) Ensuring that the contractor's personnel clearly understand these authority limitations;

(3) Anticipating that a directive changing the contract may be issued by a government representative who may not have the requisite authority;

---

[1] *Wilber National Bank v. United States*, 294 U.S. 120 (1935).

(4) Establishing a procedure that balances the need to maintain a high level of cooperation with the representatives of the government while reasonably protecting the contractor's rights under the contract; and

(5) Providing appropriate written notice to those government representatives with the actual authority to bind the government.

A. Contracting Officer Authority

The person entering into and signing a contract on behalf of the federal government is called a *contracting officer*.[2] Contracting officers have authority to enter into, administer, or terminate contracts and make related determinations and findings.[3] Within the federal government, some high-level agency officials are designated contracting officers solely by virtue of their positions.[4] Other contracting officers are designated by name and must be appointed in writing, using a certificate[5] known as a warrant. The agency head appointing the contracting officer must provide clear written instructions regarding the scope and limits of that person's authority.[6]

The written information regarding the scope and limitations of a specific contracting officer's authority must be available for review by the public and other agency employees.[7] This is critical in determining whether the person purporting to bind the government has the necessary actual authority to do so. Anyone entering into an agreement with the government bears the risk of determining the actual limits of the authority of the government's representatives.[8] However, internal regulations purporting to limit a contracting officer's authority that have not been published or communicated to the contractor will not control.[9]

The term "contracting officer" also includes *authorized representatives* of the contracting officer acting within the limits of their authority as delegated by the contracting officer.[10] The Federal Acquisition Regulation (FAR) does not prescribe the scope of authority delegated by a contracting officer to an authorized representative. Each appointment letter sets out the scope and limits of authority delegated to the authorized representative for each project. Authorized representatives may be known by a variety of other titles, as explained below. This can result in confusion on a construction project regarding who has authority to make binding decisions. The contractor should obtain and read carefully any such appointment letter. When in

---

[2]FAR § 1.601(a).
[3]FAR § 1.602-1(a).
[4]FAR § 1.601(a).
[5]FAR § 1.603-3(a); Standard Form 1402, illustrated at FAR § 53.301-1402.
[6]FAR § 1.602-1(a).
[7]*Id.*
[8]*Federal Crop Ins. Corp. v. Merrill*, 332 U.S. 380 (1947) (in government contracting, the concept of "apparent authority" is not recognized).
[9]*Texas Instruments, Inc. v. United States*, 922 F.2d 810 (Fed. Cir. 1990) (nonpublic, internal directive did not divest contracting officer of authority to bind government).
[10]FAR § 2.101.

doubt, contact the contracting officer personally and in writing. It is always appropriate to request a copy of the contracting officer's warrant as well. Addressing and resolving potential authority questions should be standard pre-performance protocol for every project. If the government changes personnel during the project, this process should be repeated, as a similarity in title does not necessarily mean that the limits of authority are the same.

B. Authority of Authorized Representatives

Authorized representatives normally have less authority than a contracting officer and operate within a narrower scope of responsibility. Depending on the type of project and the agency involved, the representative could have a formal title, such as contracting officer representative (COR), resident officer in charge of construction (ROICC), area engineer (AR), senior resident engineer (SRE), resident engineer (RE), contracting officer technical representative (COTR), government technical representative (GTR), or government technical evaluator (GTE). An authorized representative could, however, lack a formal designation and simply function as a project representative, resident engineer, project engineer, project inspector, or the like.

It is crucial that contractors become familiar with the organization of the contracting agency and its jargon as it relates to contracting officials and operations officials. In some instances, a government employee may have dual responsibility for contracting and operational functions related to the agency's mission. Different agencies often have different titles for employees performing the same or similar role.

An *authorized representative,* whatever the title or lack of same, will have only the authority that was lawfully delegated, usually in writing, by the contracting officer. Actions of an authorized representative that are within that person's scope of authority are binding on the government.[11] However, a contractor who relies on the representations of a government official acting without authority does so at its own risk.[12]

It is important to remember that the title, duties, and scope of authority of an authorized representative will vary from agency to agency and from project to project. It is therefore imperative that contractors understand the full scope of an authorized representative's authority as soon as possible after receiving the contract award. Most, if not all, federal agency regulations require that the appointment of a contracting officer's authorized representative be in writing, that the writing specify the scope and any limitations on the representative's authority, and that a copy of the appointment document be provided to the contractor. Part of the contractor's premobilization checklist should include obtaining copies of any authorized representative's appointment letters issued by the contracting officer. Wherever any doubt

---

[11]*See Hudson Contracting, Inc.* ASBCA No. 41023, 94-1 BCA ¶ 26,466 (government bound when contracting officer allowed representative [Navy Project Manager/Project Engineer] to exercise broad authority as to contract administration "in his own name and position").

[12]*Niko Contracting Co., Inc. v. United States,* 39 Fed. Cl. 795 (1997); *see* **Chapter 8, Section III** for a discussion on an authorized representative's authority to issue changes to the contract requirements, and *see* **Chapter 10, Section VII** for a discussion on an authorized representative's authority to inspect and accept work.

exists regarding the scope of authority delegated to an authorized representative, a contractor should request written clarification from the contracting officer. A sample letter directed to the contracting officer requesting clarification on this issue can be found at **Appendix 2A**.

If another government employee gives direction to the contractor that causes or can cause extra work, delay, or expenses, the contractor needs to balance maintaining a high degree of cooperation on the project with the need to mitigate the potential financial risks of following the direction of an unauthorized government representative. **Appendix 2B** of this chapter is a sample letter to the contracting officer which seeks to achieve that balance.

These risks are clearly demonstrated from a review of both court and board decisions denying contractors' claims on the basis that the government's representatives did not have the authority to order a change to the contract. For example, in *Niko Contracting Co., Inc. v. United States*,[13] the Court of Federal Claims found that only the contracting officer had the express authority to approve modifications to Niko's contract. Consequently, the statements by the contracting officer's technical representative to the contractor that releases in change order requests could be ignored were ineffective because the COTR acted beyond the limits of the delegated authority.[14]

In *Winter v. Cath-dr/Balti Joint Venture*,[15] the Federal Circuit held that the contracting officer's primary representative in charge of managing the project and administering the contract (the Navy Resident Officer in Charge of Construction or Project Manager [PM]) did not have the authority to modify the terms of a contract and bind the government.[16] In this case, the contracting officer appointed the PM to be responsible for "construction management and contract administration," and notified the contractor of the authority provided to the PM.[17] The contractor performed extra work under the contract at the direction of the PM based on the belief that the PM appeared to have the authority to make changes to the contract. The Federal Circuit denied the contractor's claims for equitable adjustments to the contract price because it found that the contracting officer could not delegate express actual authority to the PM to make modifications that affected contract price as such a delegation of authority was expressly prohibited by the contract and by Department of Defense regulation.[18]

---

[13]39 Fed. Cl. 795 (1997).

[14]39 Fed. Cl. at 801.

[15]*Winter v. Cath-dr/Balti Joint Venture*, 497 F.3d 1339 (Fed. Cir. 2007).

[16]497 F.3d at 1344.

[17]497 F.3d at 1342.

[18]DFARS § 201.602-2 (Defense Federal Acquisition Regulation Supplement [DFARS]) applicable to contracts awarded by a branch of the Department of Defense, states that a contracting officer's representative "[h]as no authority to make any commitments or changes that affect price, quality, quantity, delivery, or other terms and conditions of the contract. . . ." The contract also included Navy prohibitions on contracting officer delegations, NAVFAC Clauses 5252.201-9300 and 5252.242-9300. *See* **Sections IV to VI of Chapter 8** for a discussion on the type of changes that a contracting officer or contracting officer's authorized representative can make to the contract requirements.

Recent decisions of the Armed Services Board of Contract Appeals (ASBCA) and the Civilian Agency Board of Contract Appeals (CBCA) reflect a strict adherence to the ruling in *Cath-dr/Balti*.[19] For example, in *States Roofing Corp.*,[20] the parties agreed that a full-time safety officer was not required. The contractor claimed it was directed by a Navy engineering supervisor to employ a full-time safety officer. Among other defenses, the government asserted that if the Navy supervisor did direct the contractor as alleged, he was not acting with authority on behalf of the contracting officer. The board agreed based on *Cath-dr/Balti*.

Under a janitorial services contract, *Corners and Edges, Inc.*,[21] a project officer directed extra work for trash-collecting procedures because a government incinerator was shut down. The contract provided that any guidance from the project officer that changed the terms and conditions of the contract was not valid. The contracting officer did not accept or ratify the project officer's directive. The claim was denied and the board conducted no further analysis into the nature of the directive.

Cases decided prior to *Cath-dr/Balti* sometimes took a less stringent approach to analyzing authority questions. In *Farr Bros., Inc.*,[22] under an Army Corps of Engineers contract, the ASBCA found that a contracting officer's representative had authority to direct a work stoppage. In that case, the contracting officer sent the contractor a letter designating a COR and stating that the COR "had full authority to act for the contracting officer, but was not empowered to take specifically listed actions."[23] The list of limitations included waiving contract requirements and approving change orders, but suspending work was not included in the list. The COR instructed the contractor to delay the start of excavation for seven days, and the contractor complied. The contractor later submitted a delay claim, which the government denied on the grounds that the COR did not have the authority to suspend the work. The board held for the contractor, finding the government's position was "not supported by the letter designating the project COR."[24] If DFARS § 201.602-2, which expressly prohibits the delegation of authority to issue change orders, had been effective at that time, the result might have been different.[25]

Project managers, engineers, inspectors, and other personnel can be the authorized representatives of the contracting officer for specific purposes without a formal delegation or title. For example, in *Walter Straga*,[26] the board found that a project

[19]*See Sinil Co., Ltd,* ASBCA No. 55819, 09-2 BCA ¶ 34,213; *States Roofing Corp.,* ASBCA Nos. 55500, 55503, 09-1 ASBCA ¶ 34,036; *American World Forwarders, Inc.,* CBCA No.888-Rate 09-1 BCA ¶ 34,26; *Delta Air Lines, Inc.,* CBCA No. 1306, 09-1 BCA ¶ 34,052.
[20]ASBCA No. 55500, 09-1 BCA ¶ 34,036. *See States Roofing Corp.,* ASBCA Nos. 55500, 55503, 09-1 BCA ¶ 34,036.
[21]ASBCA No. 55767, 09-1 BCA ¶ 34,019.
[22]ASBCA No. 42658, 92-2 BCA ¶ 24,991.
[23]*Id.*
[24]*Id.*
[25]497 F.3d 1345. DFARS § 201.602-2 restricts the authority that can be given to a COR, specifically stating that a COR "[h]as no authority to make any commitments or changes that affect price, quality, quantity, delivery, or other terms and conditions of the contract."
[26]ASBCA No. 26134, 83-2 BCA ¶ 16,611.

manager who responded to prebid technical inquiries was the proper official to do so and was therefore a representative of the contracting officer. The knowledge of the project manager was imputed to the contracting officer. The contracting officer could, of course, bind the government despite the absence of a formal title assigned to the project manager or a written delegation issued to the project manager naming the project manager as the contracting officer's authorized representative for technical responses.

In *Contractors Equip. Rental Co.*,[27] the contracting officer introduced an Air Force colonel (commander of the unit to be supported by contract performance) at the prebid conference as the "man to satisfy" regarding equipment. The board found the introduction and other facts and circumstances indicating that the contracting officer had yielded authority to the colonel tantamount to a de facto appointment as the contracting officer's authorized representative. The colonel's subsequent requests and instructions were found to be constructive changes that were binding on the government. The board's discussion implies that the contracting officer was aware of the colonel's directives.[28] This case may be best understood as an example of implied actual authority, discussed at **Section D.1** of this chapter.

A consistent theme in recent decisions by the courts and boards is that the delegation of authority by way of implied authority cannot vary from the express language of the contract. Therefore, among other reasons, the court in *S&M Management Inc. v. United States*[29] found no authority to modify a contract based on a government inspector's approval of certain work because the "contract expressly limited the authority to make changes . . . to the contracting officer." While many of these recent decisions regarding the absence of authority to order changes to a contract involved Department of Defense contracts, contractors working for other federal agencies or departments need to ascertain if those entities have similar limitations on the authority of contracting officer's representatives. When there is any uncertainty regarding the actual authority of a government representative to order a change to the contract, the better course of action is to send a letter to the contracting officer similar to that at **Appendix 2B**.

C. Actual versus Apparent Authority

Supreme Court Justice Holmes wrote: "Men must turn square corners when they deal with the Government."[30] This is especially true in government construction contracts with regard to the *actual* versus *apparent* authority of a representative of the government. *The doctrine of apparent authority does not apply to government contracts.* The onus is on the contractor to confirm that it is dealing with a government representative who has *actual* authority to bind the government: "[I]t is a well established

---

[27] ASBCA No. 13052, 70-1 BCA ¶ 8183.
[28] *Id.*
[29] 82 Fed. Cl. 240, 249 (2008).
[30] *Rock Island, Arkansas & Louisiana R.R. Co. v. United States*, 254 U.S. 141, 143 (1920).

principle that the party contracting with the government cannot rely upon apparent authority but instead has the burden of knowing the law and ascertaining whether the one purporting to contract for the government is staying within the bounds of his or her authority."[31]

In a private contract setting, the doctrine of *apparent authority* applies. Under this doctrine, a principal may be bound where conduct, words, or actions would lead a reasonable person to believe that the principal's agent had actual authority under the circumstances. The *Restatement (Second) of Agency* states:

> [A]pparent authority to do an act is created as to a third person by written or spoken words or any conduct of the principal which, reasonably interpreted, causes the third person to believe that the principal consents to have the act done on his behalf by the person purporting to act for him.[32]

In dealing with the government, the concept of apparent authority is not available as a means to bind the government. In an 1868 decision, *The Floyd Acceptances*,[33] the U.S. Supreme Court held that there were no officers in the government, "from the President down to the most subordinate agent, who [did] not hold office under the law, with prescribed duties and limited authority."[34] Since a government agent's authority is defined by law, a purported agent has no authority to bind the government without a lawful delegation of power. The government is not bound by the acts of its agents beyond the scope of their actual authority.[35] A prudent contractor dealing with the government should always obtain satisfactory proof, preferably in writing, that the government's agent has the actual authority for the transaction at hand.[36]

D. Alternate Theories to Bind the Government

In the absence of explicit actual authority, apparent authority, as discussed, is not sufficient to bind the government. There are, however, four alternate theories that *may* be used in an attempt to bind the government in cases involving representatives who lack express actual authority: (1) implied actual authority, (2) estoppel, (3) ratification, and (4) imputed knowledge.

Before reviewing these legal concepts, a few words of caution are appropriate. These concepts are described in this book to provide the construction professional a

---

[31]*Johnson v. United States,* 15 Cl. Ct. 169, 174 (1988), citing *Heckler v. Community Health Servs.,* 467 U.S. 51, 63 (1984); *Federal Crop Ins. Corp. v. Merrill,* 332 U.S. 380, 384 (1947); *Thanet Corp. v. United States,* 591 F.2d 629, 635 (Ct. Cl. 1979); *Hazeltine Corp. v. United States,* 10 Cl. Ct. 417, 440 (1986), *aff'd,* 820 F.2d 1190 (Fed. Cir. 1987).

[32]*Restatement (Second) of Agency* § 27 (1958).

[33]74 U.S. 666 (1868).

[34]74 U.S. at 677.

[35]*Leonardo v. United States,* 60 Fed. Cl. 126 (2004). *See* **Section IV.B** of this chapter for a discussion of the limitations on the personal liability of government employees for their actions.

[36]*See* 497 F.3d at 1346; *Dalaly v. United States,* 3 Cl. Ct. 203 (1983).

general overview of certain legal principles that *may* provide an avenue for relief if it is not possible to demonstrate actual authority. In government contracting, issues involving these concepts usually arise in the context of claims and the disputes resolution process related to constructive changes or constructive suspensions of the work.

Even when one or more of these concepts is invoked successfully by a contractor, claims and dispute resolution can be time consuming and expensive. Moreover, there is no certainty that a board or court will conclude that the facts warrant an application of one of these concepts. In short, these concepts should be viewed as possible lifelines or last-resort refuges when lack of authority is asserted by the government. The better practice is to provide written notice to the contracting officer describing the facts and circumstances that the contractor believes constitute the extra work, delay, and so on.[37] In that way the contractor's entitlement turns on the merits of the claim rather than the authority, or lack thereof, of the person acting as the government's representative.

1. Implied Actual Authority

The judge-made doctrine of implied actual authority has been developed through a series of federal court and board decisions. "Implied actual authority, like expressed actual authority, will suffice to hold the [g]overnment bound by the acts of its agents. 'Authority to bind the government is generally implied when such authority is considered to be an *integral part of the duties assigned to a government employee.*'"[38] Finding implied actual authority is a fact-driven inquiry in which the board or court must examine the nature of the duties assigned to determine, after considering all relevant facts and circumstances, whether the government's representative had the implied authority to bind the government.[39] Therefore, under certain facts and circumstances, actual authority can be implied and an express grant of authority is not necessary.[40]

The doctrine of implied actual authority can be used by a board or the Court of Federal Claims as a gap-filler. In order to find implied actual authority, the board or court must conclude that the government intended to grant such authority but failed to do so explicitly due to some oversight or that such authority was *inherent* in a particular position. If an internal regulation, contract, or some form of communication to the contractor expressly states that a government representative has no actual authority, a board or court generally will not find implied actual authority.[41]

In addition, in order to find implied authority, the representative must have had at least some modicum of actual authority. In *California Sand & Gravel, Inc. v. United States,*[42] the U.S. Claims Court stated:

---

[37]*See* **Chapter 14** concerning documentation and notice.

[38]*Perri v. United States,* 53 Fed. Cl. 381, 398 (2002) (quoting J. Cibinic & R. Nash, *Formation of Government Contracts* 43 (1982) (emphasis in original).

[39]*Leonardo v. United States,* 60 Fed. Cl. 126 (2004).

[40]*Cruz-Pagan v. United States,* 35 Fed. Cl. 59 (1996).

[41]*Aero-Abre, Inc. v. United States,* 39 Fed. Cl. 654 (1997).

[42]22 Cl. Ct. 19 (1990), *aff'd,* 937 F.2d 625 (Fed. Cir. 1991).

[A] person with some limited actual authority impliedly may have broader authority. However, a person with no actual authority may not gain actual authority through the court-made rule of implied actual authority. Specifically, the court may not substitute itself unconditionally for the executive agency in granting authority to an unauthorized person. The most a court can do is interpret the limited authority of an authorized person in a broader manner than ordinarily would be the case. As a predicate to a finding of implied actual authority, there must be, at the least, some limited, related authority upon which the court can "administer" the law so as not to ignore the policies and decisions of those persons charged with managing government programs.[43]

The Claims Court held that an agent of the government without contracting officer authority could not be granted that authority, expressly or impliedly, by other agents who did not have the authority to make the grant.[44]

2. Estoppel

Estoppel is an equitable doctrine of fairness, which is invoked to avoid injustice in particular factual circumstances.[45] Estoppel is used to prevent a party from escaping liability for statements, actions, or inactions relied on by another contracting party and therefore prevent undue hardship to a contractor who detrimentally relied on an earlier inconsistent position of the other party.[46] Due to the various contexts in which it may be applied, estoppel can be a confusing concept.

In private construction contracting, the two dominant estoppel claims are *promissory estoppel* and *equitable estoppel*. "Promissory estoppel is used to create a cause of action, whereas equitable estoppel is used to bar a party from raising a defense or objection it otherwise would have, or from instituting an action that it would be otherwise entitled to bring."[47] "Promissory estoppel is a court-developed doctrine for enforcing promises that reasonably induce action or inaction and that are binding to the extent necessary to avoid injustice."[48]

Every contract requires *consideration* to be enforceable. In other words, there must be an exchange of value. Promissory estoppel is a substitute for consideration where the parties have not bargained for an exchange of value, but one party acts on the promise of the other and confers value or incurs liability as a result. The principle is that a promise made without consideration may be enforceable and create a contractual relationship where the promise induces reasonably expected detrimental reliance. Promissory estoppel is defined as:

---

[43]22 Cl. Ct. at 27.
[44]*Id.*
[45]*Heckler v. Community Health Servs. of Crawford Co., Inc.,* 467 U.S. 51 (1984).
[46]*Miller Elevator Co., Inc. v. United States,* 30 Fed. Cl. 662 (1994).
[47]*Knaub v. United States,* 22 Cl. Ct. 268 (1991) quoting *Jablon v. United States,* 657 F.2d 1064 (9th Cir. 1981).
[48]*Id.* quoting *Durant v. United States,* 16 Cl. Ct. 447 (1988).

(1) A promise which the promisor should reasonably expect to induce action or forbearance on the part of the promisee or a third person and which does induce such action or forbearance is binding if injustice can be avoided only by enforcement of the promise. The remedy granted for breach may be limited as justice requires.[49]

Promissory estoppel is rarely used to establish the existence of a government construction contract, and the Court of Federal Claims does not have jurisdiction to consider claims based on promissory estoppel.[50]

It is well established that equitable estoppel *cannot* be used as a basis to create authority to bind the government:[51] "It is essential to a holding of estoppel against the United States that the course of conduct or representation be made by officers or agents of the United States who are acting within the scope of their [actual] authority."[52] Authority by estoppel is a misnomer in dealing with the federal government. Therefore, authority must first be found through implication or another recognized alternate theory before equitable estoppel may be invoked to prevent a government representative from denying authority.

Once the necessary authority has been established, equitable estoppel may be available against the government if some form of affirmative misconduct is shown in addition to the traditional elements of estoppel.[53] The common-law elements of equitable estoppel are:

(1) misleading conduct, which may include not only statements and action but silence and inaction, leading another to reasonably infer that rights will not be asserted against it; (2) reliance upon this conduct; and (3) due to this reliance, material prejudice if the delayed assertion of such rights is permitted.[54]

In summary, to invoke the doctrine of equitable estoppel against the government, it is necessary to show that the government's representative has authority and that the government's representative engaged in some form of affirmative misconduct. In addition, the claimant must have reasonably relied on the conduct and must show that it will suffer material prejudice if the government is not bound to the acts of its representative.

---

[49]*Restatement (Second) of Contracts* § 90 (1981). Promissory estoppel is often invoked by contractors to enforce a bid or quote from a prospective subcontractor or vendor that was used by the contractor in its bid/proposal to the owner. *See Drennan v. Star Paving Co.,* 333 P.2d 757 (Cal. 1958); *Crook v. Mortenson-Neal,* 727 P.2d 297 (Alaska 1986).

[50]*Gregory v. United States,* 37 Fed. Cl. 388 (1997).

[51]*Conner Brothers Constr. Co., Inc. v. United States,* 65 Fed. Cl. 657, 692 (2005).

[52]65 Fed. Cl. at 693, quoting *Emeco Indus., Inc. v. United States,* 485 F.2d 652, 657 (Ct. Cl. 1973).

[53]65 Fed. Cl. at 693.

[54]*Id.*

3. Ratification

The FAR defines ratification as "the act of approving an unauthorized commitment by an official who has the authority to do so."[55] The government will be bound when the acts of an unauthorized agent are expressly or impliedly adopted by a properly authorized government representative. For a valid ratification, the ratifying official must have the authority to bind the government, must have actual or constructive knowledge of the unauthorized agreement, and must have expressly or impliedly adopted the agreement.[56] The authorized representative must be sufficiently aware of the material facts underlying the unauthorized action in order to ratify and bind the government. In terms of a government construction project, notice by the contractor to the contracting officer may be critical to establish that the contracting officer ratified the directions or actions of the on-site government representative. (*See* **Appendix 2B** as well as **Chapters 3, 8, and 14.**)

In *California Sand and Gravel, Inc. v. United States,*[57] the Claims Court found that a simple discussion in which the government's authorized representative told the contractor to "work something out" was not enough to ratify an unauthorized modification of a contract.[58] In *Williams v. United States,*[59] however, the Court of Claims found it "incredible" that the contracting officer would not know about an unauthorized agreement for the contractor to do extra work at the base where the contracting officer was located. Consequently, the court held that the contracting officer, by taking no action, ratified the otherwise unauthorized agreement.

Since FAR § 1.602-3 lists specific requirements for ratification, it is doubtful that a contracting officer will agree to ratify an unauthorized commitment absent satisfaction of all specified requirements. In particular, note that "[u]nauthorized commitments that would involve claims subject to resolution under the Contract Disputes Act of 1978[60] should be processed in accordance with [FAR] 33.2, Disputes and Appeals."[61]

4. Imputed Knowledge

Closely related to ratification is the concept of imputed knowledge. Under the common law, the liability of a principal is affected by the knowledge of its agent.[62] In government contracting, the knowledge of an unauthorized representative has been imputed to the authorized representative where the nature of the relationship creates a presumption that the authorized representative would be informed. Imputation of knowledge can be critical where a contract requires actual notice of events such as constructive changes, delays, and so on.

[55]FAR § 1.602-3(a).
[56]*Aero-Abre v. United States,* 39 Fed. Cl. 654 (1997).
[57]22 Cl. Ct. 19 (1990).
[58]22 Cl. Ct. at 27–28.
[59]127 F. Supp. 617 (Ct. Cl. 1955).
[60]41 U.S.C. §§ 601–13.
[61]FAR §1.602-3(b)(5).
[62]*Restatement (Second) of Agency* § 272 (1958).

Courts and boards often have found various representatives to be the "eyes and ears" of the contracting officer and imputed the knowledge of the representative to the contracting officer. For example, in *U.S. Fed. Eng'g & Mfg., Inc.,*[63] the ASBCA imputed the knowledge of the government's project engineer to the contracting officer binding the government even though the contracting officer had no actual knowledge of the required additional work.

In *Sociometrics, Inc.,*[64] the knowledge of the contracting officer's day-to-day contract administrator was imputed to the contracting officer, thereby binding the government and entitling the contractor to payment for extra work even though a contractually required option to order the work had not been formally exercised. The representative was the eyes and ears of the contracting officer, justifying imputation of knowledge and a conclusion that the government was obligated to compensate the contractor for the extra work.[65]

On construction projects, issues of ratification and imputed knowledge often relate to the actions of the government's inspectors in rejecting work. In that context, a field inspector commonly has been found to be the government's representative on the job and the inspector's knowledge has been imputed to the contracting officer.[66] However, written notice to the contracting officer describing the inspector's direction and the basis why the contractor believes that the direction is a change is always preferable. The sample letter at **Appendix 2B** can be tailored to the specific factual situation in order to provide written notice.

E.   Presumption of Government Good Faith—Immunity of Government Employees/Agents

A contractor aggrieved by the action of a contracting officer or contracting officer's representative or other government representative often wants to recover damages against the employee as well as the government. In that regard, contractors must recognize there are significant obstacles to recovery.

1.   Presumption of Good Faith

The *Restatement (Second) of Contracts* provides: "Every contract imposes upon each party a duty of good faith and fair dealing in its performance and its enforcement."[67] The Uniform Commercial Code (UCC), which applies to the sale of goods and materials, states: "Every contract or duty within this Act imposes an obligation of good faith in its performance or enforcement."[68] "Good faith" is defined by the UCC as

---

[63]ASBCA No. 19909, 75-2 BCA ¶ 11,578.
[64]ASBCA No. 51620, 00-1 BCA ¶ 30,620.
[65]*Id.*
[66]*See, e.g., Raby Hillside Drilling, Inc.,* AGBCA No. 75-101, 78-1 BCA ¶ 13,026 (knowledge of inspectors with day-to-day contract administration responsibilities imputed to CO). *See* **Chapter 10** for a further discussion of inspection related changes.
[67]*Restatement (Second) of Contracts* § 205 (1981); *see also Centex Corp. v. United States,* 395 F.3d 1283 (Fed. Cir. 2005) (implicit in every contract are the duties of good faith and fair dealing between the parties).
[68]UCC § 1-203.

"honesty in fact and the observance of reasonable commercial standards of fair dealing."[69] Among the obligations imposed by the implied covenant of good faith and fair dealing are not to interfere with the other party's performance and not to act so as to destroy the other party's reasonable expectations regarding the benefits of the contract.[70] Government contractors requesting a price adjustment, time extension, or attempting to respond to a threatened default often assert claims that involve issues of the government's alleged breach of the duty of good faith and fair dealing.

There is a strong presumption in U.S. jurisprudence that government representatives exercise their duties in good faith.[71] Clear and convincing evidence to the contrary generally is required to overcome this presumption of good faith in favor of the government.[72] The Federal Circuit in *Am-Pro Protective Agency, Inc. v. United States* (a case alleging bad faith) described the clear and convincing burden in these words:

> A requirement of proof by clear and convincing evidence imposes a heavier burden upon a litigant than that imposed by requiring proof by preponderant evidence but a somewhat lighter burden than that imposed by requiring proof beyond a reasonable doubt. Clear and convincing evidence has been described as evidence which produces in the mind of the trier of fact an abiding conviction that the truth of a factual contention is *"highly probable."*[73]

This is a heavy burden for the contractor to satisfy. Recent decisions of the Court of Federal Claims have reached conflicting results regarding the presumption of good faith and *Am-Pro*'s clear and convincing evidentiary standard—with some cases simply adopting the standard with little or no discussion and some narrowing the holding to include only those cases in which a government official is accused of fraud or quasi-criminal wrongdoing in the exercise of official duties.[74]

There is a long line of Supreme Court cases holding that when the federal government enters into a contract, it should have the same rights and obligations as private parties and likewise be governed by general contract law—including the implied

---

[69]UCC § 1-201(19).

[70]*Centex Corp. v. United States,* 395 F.3d 1283 (Fed. Cir. 2005).

[71]*Am-Pro Protective Agency, Inc. v. United States,* 281 F.3d 1234, 1239 (Fed. Cir. 2002), citing *Knotts v. United States,* 128 Ct. Cl. 489, 492 (1954) (stating "we start out with the presumption that the official acted in good faith").

[72]281 F.3d at 1239.

[73]281 F.3d at 1240 (emphasis in original), citing *Price v. Symsek,* 988 F.2d 1187, 1191 (Fed. Cir. 1993).

[74]*See, e.g., Boston Edison Co. v. United States,* 64 Fed. Cl. 167 (2005) (adopted "clear and convincing" standard in case where government allegedly delayed contractor's performance); *J. Cooper & Assocs. v. United States,* 53 Fed. Cl. 8 (2002) (adopted "clear and convincing" standard in case where government allegedly acted in bad faith by refusing to issue additional work orders); *Tecom, Inc. v. United States,* 66 Fed. Cl. 736 (2005) ("clear and convincing" standard applies only to cases in which "a government official is accused of fraud or quasi-criminal wrongdoing in the exercise of his official duties"); *Helix Elec., Inc. v. United States,* 68 Fed. Cl. 571 (2005) (refused to adopt "clear and convincing" standard in case where government allegedly impeded contractor's efforts to obtain its rights under the contract).

duty of good faith and fair dealing.[75] Following the reasoning in these cases, the presumption of good faith would seem to have no relevance when government officials act in a contractual context.[76] There is a movement in federal procurement circles advocating this position, which would level the playing field by requiring good faith and fair dealing from both the government and its construction contractors, treating both with the same presumption as to good faith action.

In summary, the strong presumption of good faith conduct of government officials, and the need for its rebuttal by clear and convincing evidence appears to be endorsed by the decisions in the Federal Circuit, while some decisions in the subordinate Court of Federal Claims attempt to limit the presumption to only those cases involving fraud or quasi-criminal wrongdoing. However, these courts sometimes issue opinions that are very fact dependent, confuse the government's different roles, and import the good faith presumption from the sovereign to the commercial context without rigorous analysis. Commentators on federal procurement have argued that the presumption of good faith should be abolished when the government is acting as a private contracting party.[77] Contractors need to appreciate that demonstrating *bad faith* is a difficult challenge regardless of the standard or test followed by a board or the Court of Federal Claims. Finally, where a contractor asserts that a government official not only failed to comply with the duty of good faith and fair dealing but affirmatively engaged in bad faith, the contractor will have the burden of showing, by clear and convincing evidence, that there was "some specific intent to injure" the contractor.[78]

2. Limitations on Liability

Under the Federal Employees Liability Reform and Tort Compensation Act,[79] a government employee, representative, or agent acting within the scope of his or her employment generally has absolute immunity from civil actions or proceedings seeking monetary damages. In such actions seeking monetary compensation, only the government can be sued—not the individual.[80]

---

[75]*Cooke v. United States*, 91 U.S. 389 (1875); *United States v. Boswick*, 94 U.S. 53 (1877); *Lynch v. United States*, 292 U.S. 571 (1934); *United States v. Winstar Corp.*, 518 U.S. 839 (1996); *Mobil Oil Exploration & Producing Southeast, Inc. v. United States*, 530 U.S. 604 (2000); *Franconia Assocs. v. United States*, 536 U.S. 129 (2002).

[76]*See Tecom v. United States*, 66 Fed. Cl. 736 (2005); *Centex Corp. v. United States*, 395 F.3d 1283 (Fed. Cir. 2005).

[77]*See* Stuart B. Nibley, *Unraveling the Mixed Message Government Procurement Personnel Receive: Message 1: Act Absolutely in the Government's "Best Interests" Message 2: Act "Ethically"* 36 Pub. Con. L. J. 23 (2006); Marshall Doke, Proposal for Public Comment, Commercial Practices Legislation, Acquisition Advisory Panel (May 5, 2005); W. Stanfield Johnson, *Needed: A Government Ethics Code and Culture Requiring Its Officials to Turn "Square Corners" When Dealing with Contractors*, 19 Nash & Cibinic Rep. ¶ 47(Oct. 2005); *see also Centex Corp. v. United States*, 395 F.3d 1283, 1304-08 (Fed. Cir. 2005).

[78]*Am-Pro Protective Agency, Inc.* 281 F.3d at 1240, quoting *Kalvar Corp. v. United States*, 543 F.2d 1298, 1302 (Ct. Cl. 1976).

[79]Federal Employees Liability Reform and Tort Compensation Act of 1988, Pub. L. No. 100-694.

[80]*See* 28 U.S.C. § 2679(b).

In Section 5 of the Act, Congress enumerated two exceptions to this absolute immunity rule. A government employee acting within the scope of that person's employment may be sued when an injured plaintiff brings: (1) a *Bivens* action,[81] which seeks damages for a constitutional violation by a government employee; or (2) an action under a federal statute that authorizes recovery against a government employee.[82] These exceptions will rarely be applicable in federal government construction contracts, and an injured contractor's sole remedy, almost always, will be against the government under the Federal Tort Claims Act (FTCA).[83] Under the FTCA, the government "shall be liable . . . in the same manner and to the same extent as a private individual under like circumstances."[84]

In *Aversa v. United States,*[85] the U.S. Court of Appeals for the First Circuit found that while federal law determines whether a person is a federal employee and the extent of that person's federal responsibilities, state law governs whether the person was acting within the scope of that employment and those responsibilities.[86] The federal court will therefore look to the local state's tort law to determine whether the government representative's alleged acts are within the scope of employment.[87] However, an independent contractor doing work for the government is not considered to be a federal employee and does not have the benefit of the immunity afforded to government employees.[88]

II. AVAILABILITY OF FUNDING

In addition to authority to bind the government, a contracting officer must have the funds available to pay for the goods, services, or construction work that are the subject of the contract. The next discussion addresses the source and administration of government contract funds.

A. Federal Budget Process and Contract Financing

The federal government's process for authorizing projects and funds, appropriating funds, and managing the financing of government construction projects is a complex topic. It involves legislative activities at the congressional level and financial management activities from the White House (Executive Office of the President, Office

---

[81]*See Bivens v. Six Unknown Fed. Narcotics Agents,* 403 U.S. 388 (1971).
[82]*United States v. Smith,* 499 U.S. 160 (1991), citing 28 USC § 2679(b)(2).
[83]Federal Tort Claims Act, 28 U.S.C. § 1346(b) and § 2671 *et seq.*
[84]*Knowles v. United States,* 91 F.3d 1147, 1150 (8th Cir. 1996).
[85]99 F.3d 1200 (1st Cir. 1996).
[86]*Id.*
[87]*See, e.g., Wilson v. Drake,* 87 F.3d 1073 (9th Cir. 1996).
[88]*See Linkous v. United States,* 142 F.3d 271 (5th Cir. 1998) (case dismissed when physician was found to be an independent contractor and not an employee of the government); *see also Rodrigues v. Sarabyn,* 129 F.3d 760 (5th Cir. 1997).

of Management and Budget [OMB]) down to and including the contracting agency or military department. Although a contractor may not need to master the details of the legislative and agency process to fund projects, it does need to have an appreciation of the basic principles and those issues that can affect a contractor working for the government.

As the initial step in the project funding process, federal agencies formulate their spending needs or budget and the administration submits this budget to Congress on an annual basis. The government's fiscal year runs from October 1 of a calendar year to September 30 of the next calendar year, and the budget cycle is, in theory, scheduled to have funding in place for the planned operations at the start of a given fiscal year. The reality of the budget process seldom matches the theory. The result can present uncertainties and risks for contractors.[89]

Each federal agency develops its budget by working with the OMB. The OMB staff reviews these budgets in consultation with the executive branch and assists the president in developing an overall budget, officially known as the Budget of the United States Government. The president's budget must be submitted to Congress no later than the first Monday in February.

The House and Senate Budget Committees take the president's budget, and each chamber approves what is called a budget resolution. Leading Budget Committee members from the House of Representatives and Senate review and negotiate changes to the separate budget resolutions to develop a consensus agreement, called a conference report. In theory, Congress adopts the conference report in April or May and passes the budget resolution, which is technically a concurrent resolution, not a formal law requiring the president's signature.

The congressional budget resolution triggers legislative activity—appropriations bills—that are submitted to the president for signing into law. These appropriations bills fund the various federal agencies, which in turn fund government construction contracts.

B. Appropriations Process

In government contracting, there are two types of relevant statutes: authorizing statutes and appropriating statutes. A federal agency's project, plan, or program generally must first be authorized by Congress, but the authorizing statutes do not actually fund them. Funding for a particular program, plan, or project comes with the passage of specific appropriation legislation. Each project will not necessarily require authorization but may be included in an overall authorization bill for project types (such as maintenance and repair contracts for Army facilities).

An example of specific project authorization can be found in federal highway projects. One function of the Federal Highway Administration is to provide assistance to states by funding state highway improvement projects. The *authorization*

---

[89]The uncertainties can range from having a proposal available for acceptance for an extended period of time pending receipt of funds to the risk of administering a project and subcontracts on an incrementally funded project. See **Section II.E** of this chapter.

for these activities is codified under Title 23 of the Code of Federal Regulations. However, the *funding* for a specific authorized highway improvement project or program comes from appropriation acts that fund the Federal Highway Administration's activities. As such, the authorizing and appropriation acts are generally, but not always, separate and distinct.

Appropriation statutes provide guidance and place limitations as to how these appropriated funds may be used for specific projects. Federal contracting officers and finance and accounting officials, who have the statutory authority to bind the government and spend federal appropriated funds, must be responsive to these key appropriation statutes, which may contain provisions affecting contract awards and management of the contracting process. Understanding these laws is crucial because these statutes define how federal funds may be spent. In addition, federal law provides civil and criminal sanctions for federal officials who violate those appropriation laws, whether done willfully or inadvertently.

The principles of appropriation law are found in specific statutes and in the Constitution of the United States. The Constitution and the governing statutes both enforce the overriding principle that government funds may not be expended unless those funds have been appropriated for a certain purpose or project to be committed within a specified time period. In other words, no money can be paid out of public funds unless Congress has made an appropriation for that purpose and time. Outside of the Constitution, the federal statutes that further define and limit how federal funds may be expended are generally known as federal appropriation and anti-deficiency statutes. These federal laws addressing the expenditure of federally appropriated funds are found at 31 U.S.C. § 1301 through 31 U.S.C. § 9703.

C. Administering Appropriated Funds

Government agencies are subject to certain limitations in administering appropriated funds. These limitations generally are broken into three categories: (1) purpose, (2) time, and (3) amount. Under 31 U.S.C. § 1301 (the purpose statute), appropriations shall be applied only to projects for which the appropriations were made. Under 31 U.S.C. § 1552 (the time statute), where an appropriation is made available for an obligation for a definite period of time, that appropriation must be committed for use during this period of time or the authority to use the appropriation expires. Finally, pursuant to the Anti-Deficiency Act,[90] agencies are strictly forbidden to obligate more than the amount of funds that were appropriated.

Once an agency has access to appropriated funds, the next step is to *obligate* those funds to the purpose for which they were appropriated.[91] Following the appropriation of funds, the agency's administrative process includes the subdivision of funds to, and sometimes within, an agency. Depending on the nature of the fund division and subdivision process, limits may be established on fund availability for a particular project or group of projects.

---

[90]The Anti-Deficiency Act is comprised of more than 20 related statutes scattered among several sections of Title 31. The key provision of the Act is 31 U.S.C. § 1341(a)(1).

[91]*See* 31 U.S.C. § 1301(a).

Part of the fund administration process includes the use of codes reflecting the accounting classification of funds: construction, services, operation, maintenance, and the like. The codes are often referred to as *fund cites*. Fund cites are found in the *Accounting and Appropriation* data blocks of the four forms commonly used for contract awards and modifications.[92]

Agencies may award contracts subject to the *availability of funds*. If a contract is awarded on this basis, it is required to include an Availability of Funds clause,[93] which conditions the government's liability under the contract on funds being made available for the contract and the contractor receiving notice that such funds are available. This funding vehicle may be used where the lead time for the particular contract requires the contract to be awarded prior to the agency's receipt of the funds for the new fiscal year. Under the regulations governing the use of this clause in a contract, the government may not accept services or supplies until the funds are available.[94]

In construction, several factors control how appropriated funds may be obligated. A cardinal rule in this area is that the agency may not split funds over separate projects. Also, an agency may not treat clearly interrelated construction activities as separate projects. In this regard, a project includes all work necessary to produce a complete and usable facility or a complete and usable improvement to an existing facility.

D. Anti-Deficiency Act

The United States Constitution gives Congress the power of the purse, expressly stating that "[n]o money shall be drawn from the Treasury, but in Consequence of Appropriations made by Law. . . ."[95] Since 1870, Congress has passed various legislative acts to prevent the executive branch from entering contracts that call for spending money in excess of amounts appropriated. These enactments are collectively known as the Anti-Deficiency Act.[96] The Anti-Deficiency Act prohibits the government's obligation of funds in advance of an appropriation. Simply put, the government is required to pay as it goes.

The Act covers these areas:

---

[92]*See* Block 14 of Standard Form (SF) 26, Award/Contract, Block 21 of SF 33, Solicitation, Offer, and Award, Block 23 of SF 1442, Solicitation, Offer and Award (Construction, Alteration, or Repair), and Block 12 of Standard Form 30, Amendment of Solicitation/Modification of Contract, illustrated at FAR §§ 53.301-26, 33, 1442, and 30, respectively.

[93]FAR §§ 32.703-2, 52.232-18. Typically, this clause is used for operations and maintenance contracts and continuing service-type contracts, but it is authorized by FAR § 52.301 for inclusion in fixed-price construction contracts.

[94]FAR § 32.702.

[95]U.S. CONST. art. I, § 9, cl. 7.

[96]31 U.S.C. § 1341(a)(1).

(1) The making or authorizing an expenditure from, or creating or authorizing an obligation under, any appropriation or fund in excess of the amount available in the appropriation or fund unless authorized by law, 31 U.S.C. § 1341(a)(1)(A)

(2) Involving the government in any obligation to pay money before funds have been appropriated for that purpose, unless otherwise allowed by law, 31 U.S.C. § 1341(a)(1)(B)

(3) Accepting voluntary services for the United States, or employing personal services not authorized by law, except in cases of emergency involving the safety of human life or the protection of property, 31 U.S.C. § 1342

(4) Making obligations or expenditures in excess of an apportionment or reapportionment, or in excess of the amount permitted by agency regulations, 31 U.S.C. § 1517(a)

A government officer or employee who violates 31 U.S.C. § 1341(a) (obligate or expend funds in excess or advance of appropriation),[97] § 1342 (voluntary services prohibition), or § 1517(a) (obligate or expend funds in excess of an apportionment or administrative subdivision of appropriated funds as specified in an agency's regulation) "shall be subject to appropriate administrative discipline including, when circumstances warrant, suspension from duty without pay or removal from office."[98] In addition, an officer or employee who knowingly and willfully violates any of the three sections cited "shall be fined not more than $5,000, imprisoned for not more than 2 years, or both."[99] These sanctions obviously are intended to make contracting officers mindful of the need not to obligate the government beyond the available funds.

In summary, while fiscal law *may not normally* be of concern to a contractor, it is crucial to understanding the legal pressures imposed on federal disbursing officers and other officials who have the statutory authority to pay for goods and services from federally appropriated funds.

Although the sanctions of the Anti-Deficiency Act essentially target government employees, a contractor needs to be sensitive to the fact that the available funds for a project may affect the timing, negotiation, and final pricing of changes outside of the resolution and payment of claims under the Contract Disputes Act.[100]

---

[97]*Hercules, Inc. v. United States,* 516 U.S. 417 (1996).

[98]31 U.S.C. §§ 1349(a), 1518.

[99]31 U.S.C. §§ 1350, 1519. A contracting officer's exposure for an unauthorized payment includes personal liability to reimburse the government for the improper expenditure. The government may also seek to recover the funds from the entity that received them.

[100]*See* **Chapter 15** for a discussion of claims under the Contract Disputes Act. Although not usually necessary, a contractor does have the right to request confirmation that funds are available for the scope of work required by a contract modification.

E. Multiyear or Incrementally Funded Contracts

A multiyear contract or continuing contract is a contract that extends for more than one fiscal or budget year. An agency using a multiyear contract must comply with the Anti-Deficiency Act, 31 U.S.C. § 1341, which prohibits contracts that obligate the government beyond the appropriations period (**see Section II.D**). While multi-year contracts are allowable under the FAR, the traditional practice in government contracting has been to fully fund construction contracts, even though the construction of those projects may extend over several successive fiscal years.[101] If an agency proposes to award a project that is incrementally funded or funded over several years, the contractor needs to evaluate the risks presented by this type of work.

Government contracts that are incrementally funded, or funded over successive years prior to actual appropriations (e.g., a contract containing a continuing con-tracts clause), are executed by the contracting officer with the words *subject to the availability of funds* or other limitations on the government's obligation under the contract.[102] The purpose of this language is to avoid a violation of the federal anti-deficiency laws and define the parties' obligations and rights under that contract.

Historically, the Corps of Engineers used continuing contracts on its large civil works construction projects (e.g., locks, dams, extensive flood prevention projects, and related facilities). The Corps was authorized under the River and Harbor Act of 1922 to use contracts with a continuing contracts clause for certain projects.[103] As a result of congressional concerns that the Corps was misusing continuing contracts, the Energy and Water Development Appropriations Act of 2006[104] severely restricted the Corps' use of continuing contracts.

The Corps now must obtain the approval of the Assistant Secretary of the Army (Civil Works) before using a continuing contract. In addition, the Corps' primary acquisition focus must be on fully funded or incrementally funded projects. Under certain limited circumstances, the Corps may use a continuing contract when it has approval and a "studied, deliberate, and credible plan" in place to limit the contractor to the amount appropriated and to properly fund future contract obligations.[105] The Corps can no longer include a continuing contracts clause in its contracts as standard

---

[101]See DFARS § 232.703-1. Historically, the primary exception to that practice has been certain civil works construction projects. However, some military construction projects are incrementally funded. In general, DFARS § 232.703-1 provides that a fixed-price contract can be incrementally funded only when the contract is for severable services, the contract does not exceed one year in length, and the contract is funded with available (unexpired) funds. Additionally, incrementally funded contracts are permitted when directly authorized by Congress or when the contract is funded with research and development appropriations.

[102]This language is also included in negotiated contract modifications and negotiated contract settlements where an agreement has been reached on the contractor's entitlement, but at the time the document is executed there are insufficient funds available to fund the modification or negotiated settlement.

[103]42 Stat. 1043. See Continuing Contracts clause prescribed at EFARS § 32.705-100 and found at EFARS § 52.232-5001 (this is the so-called True Continuing Contracts clause).

[104]Pub. L. No. 109-103.

[105]See U.S. Army Corps of Engineers, Engineer Circular 11-2-189, "Execution of the Annual Civil Works Program" (Dec. 31, 2005). *www.usace.army.mil/CECW/Documents/cecwm/exe_guide/FY06_CW_Impl_EC_contract.pdf*.

operating practice but can consider continuing contracts only as a last resort for civil works construction.

An example of an availability of funds clause as used by the Department of Interior, Bureau of Reclamation, was addressed in detail by the Court of Federal Claims in *PCL Construction Services, Inc. v. United States.*[106] The court described the availability of funds provision in this way:

> The contract as signed also contained the "subject to the availability of funds" clause . . ., which provided that the government's obligation under the contract "is contingent upon the availability of appropriated funds." That clause provided that no legal liability on the part of the government arises "until funds are made available to the Contracting Officer" and "until the Contractor receives notice of availability [of funds], to be confirmed in writing by the Contracting Officer." 48 C.F.R. § 52.232-19 (1990)[107]

The availability of funds clause in the *PCL* case contained these key elements:

- Made the government's liability under the contract contingent on the availability of funds.
- Required the contracting officer to notify the contractor in writing of the availability of funds or that no additional funds would be made available.
- Provided an indication of the anticipated funding over subsequent years but did not warrant that information.
- Stated that no payment would be made for work performed after funds were exhausted, unless and until sufficient additional funds were made available to the contracting officer.
- Advised the contractor that prosecution of the work at a rate that exhausted funds before the end of a fiscal year would be at the contractor's risk.
- Permitted, *but did not require,* the contractor to continue with the work even though the funds were exhausted.
- Stated that no interest would be payable to the contractor resulting from the contractor's election to continue working after funds were exhausted.
- Provided for a time extension due to an exhaustion of funds but precluded any recovery under the Suspension of Work clause.

An availability of funds clause puts prospective contractors on notice of funding limitations and some of the risks and obligations related to that type of provision. Although traditionally associated with large civil works projects, this type of provision and incremental funding is now being used for more traditional government construction projects, such as military barracks. In that context, the contracting activity—for example, a Corps of Engineers District—may issue a request for

---

[106]41 Fed. Cl. 242 (1998).
[107]41 Fed. Cl. at 248.

proposal (RFP) seeking design-build proposals on a $60 million project. If the project is not fully funded, the RFP will include a schedule of the anticipated funding stream. The RFP will also include the Limitation of Government's Obligation clause[108] found in the DFARS, and the funding stream schedule might indicate that 10 percent of the award amount would be funded in the first year.

To an extent, the parties' rights and obligations under the DFARS clause are similar to those in the *PCL* decision. However, there are some significant differences. These include:

- Agreement by the contractor to perform up to a point at which the total amount payable by the government *including* reimbursement in the event of a termination for convenience "approximates" the amount allotted to the contract.
- Requires the contractor to advise the government 90 days prior to the date, in the contractor's best judgment, that the total amount payable by the government, including any cost for a termination for convenience, will *approximate* 85 percent of the total amount of monies that allotted to the contract.[109]
- Expressly states that the contractor is *not authorized* to continue working on the project beyond that funding allotment.
- Expressly states that the government is *not obligated* to reimburse the contractor in excess of the amount of funds allotted to the contract regardless of anything to the contrary in the Termination for Convenience clause.
- Expressly states that the total amount payable to the contractor includes the estimated termination settlement expenses and costs.[110]

Contractors reviewing this type of clause and the prospect of a fixed-price contract with a stated duration of several years need to consider these issues:

- Award of subcontracts and purchase orders and management of the actions taken by subcontractors in awarding lower-tier purchase orders and sub-subcontractors.
- Terms and conditions of subcontracts and purchase orders.
- How are prices for work, materials, and equipment fixed in a period of possible sharp price escalation within the funds allotted at the time of award?
- Risk of price escalation if key subcontracts and purchase orders are not fully priced at the time of award of the initial contract by the government.
- Risk of exposure to Miller Act claims by lower-tier suppliers and sub-subcontractors since a prospective waiver of a firm's Miller Act rights is void under 40 U.S.C. § 3133(c).

---

[108]DFARS § 252.232-7007. See **Appendix 2C** for the full text of this clause.
[109]Notice requirements under comparable Limitation of Cost clauses have been enforced. *See Int'l Tech Corp.*, ASBCA No. 54136, 06-2 BCA ¶ 33,348.
[110]*See* FAR § 49.206.

Good contracting practices and common sense dictate that any and all contract provisions included by the government that limit or make incremental the availability of funds be included in requests for quotations issued to prospective subcontractors and suppliers. Subcontracts or purchase orders awarded should include the provisions so that subcontractors and vendors are on notice of the limitations. Coordination and information sharing between the contractor and its subcontractors can work to avoid exceeding funding limitations and disputes. It is prudent to require subcontractors to provide the same type of notices to the general contractor that are required to be submitted to the government under the prime contract.

Due to the long-term nature of multiyear contracts, the contracting officer may include an economic price adjustment clause in that contract.[111] Such clauses often state that an official labor or material price index be used as the basis for any price adjustment. However, it is very rare to see economic price adjustment clauses used in domestic construction contracts. If that type of clause is absent from a fixed-price contract, the contractor *normally* bears the risk of unexpected price escalation.[112] All of these issues reflect risks that must be considered in developing a fixed-price proposal and any contingency included in that price.

III. CONTRACT PAYMENT PROCEDURES

Although it is important for government contractors to be generally familiar with the authorization and appropriation process, it is crucial for contractors to understand the invoicing, payment, and payment assurance areas of government contracting. These topics govern the transfer of funds from the government to the contractor and are vitally important to the contractor's ability to continue its work and to succeed in the government contract field.

A. Invoicing

The contractor's responsibility for submitting invoices to the government will control whether and when the contractor is paid and also whether the contractor may be entitled to interest on late payments. Therefore, it is important that invoices are submitted with the appropriate information and in the correct format.

FAR § 52.232-5, *Payments Under Fixed-Price Construction Contracts (Sept. 2002)*, sets forth the prime contractor's responsibilities for submitting proper invoices. This clause sets forth specific items of information that must be included with the invoice, including:

---

[111]*See* FAR § 17.109(b); *Kings Point Mfg. Co.,* Comp. Gen. Dec. B-220224, 85-2 CPD ¶ 680.
[112]*Spindler Constr. Corp.,* ASBCA No. 55007, 06-2 BCA ¶ 33,376.

(i) An itemization of the amounts requested, related to the various elements of work required by the contract covered by the payments requested.

(ii) A listing of the amount included for work performed by each subcontractor under the contract.

(iii) A listing of the total amount of each subcontract under the contract.

(iv) A listing of the amounts previously paid to each such subcontractor under the contract.

(v) Additional supporting data in a form and detail required by the Contracting Officer.

Along with each progress payment request, the contractor also must *certify* that the request for payment includes only work performed in accordance with the specification and the terms and conditions of the contract. In addition, the contractor must certify that it has *not* billed for any amounts that it intends to withhold or retain from a subcontractor or supplier in accordance with the terms and conditions of the subcontract.[113] That certification provides:

I hereby certify, to the best of my knowledge and belief, that—

(1) The amounts requested are only for performance in accordance with the specifications, terms, and conditions of the contract;

(2) All payments due to subcontractors and suppliers from previous payments received under the contract have been made, and timely payments will be made from the proceeds of the payment covered by this certification, in accordance with subcontract agreements and the requirements of Chapter 39 of Title 31, United States Code;

(3) This request for progress payments does not include any amounts which the prime contractor intends to withhold or retain from a subcontractor or supplier in accordance with the terms and conditions of the subcontract; and

(4) This certification is not to be construed as final acceptance of a subcontractor's performance.

(Name)
(Title)
(Date)

B. Payment

FAR § 52.232-5 also governs payments for fixed-price construction contracts. It states that the government shall make progress payments on a monthly basis based on estimates of accomplished work that meets the standards of quality established under the contract, as approved by the contracting officer.

---

[113]*See* **Chapter 1** for a further discussion of certifications.

As to retainage, if the contracting officer finds that satisfactory progress was achieved during any period for which a progress payment is to be made, the contracting officer must authorize payment to be made in full. However, if satisfactory progress has not been made, the contracting officer may retain a maximum of 10 percent of the amount of the payment as retainage until satisfactory progress is achieved.

In addition to work performed, the government shall, upon request, reimburse the contractor for the amount of premiums paid for performance and payment bonds (including coinsurance and reinsurance agreements, when applicable) after the contractor has furnished evidence of full payment to its surety.

Final payment on a fixed-price construction contract requires three elements to be satisfied: (1) the completion and acceptance of all work; (2) the presentation of a properly executed voucher; and (3) the execution of a release of all claims against the government arising by virtue of the contract, other than claims, in stated amounts, that the contractor has specifically excepted from the operation of the release.[114]

C. Prompt Payment Act

The federal Prompt Payment Act[115] was passed in 1982 in an effort to address the problem of slow payment by the government to its contractors. It sought to resolve issues related to slow payment by requiring the government to pay invoices within a certain number of days after receipt or pay interest on past-due sums and, potentially, a penalty if interest is not paid. For construction contracts, the Prompt Payment Act sets up two separate schemes to determine when an agency must make payments of money due to a contractor. First, the Act requires that an agency make progress payments for work done within 14 days of receipt of a proper invoice.[116] Second, for final payments, an agency must make the payment within 30 days of receipt of a proper invoice from the contractor.[117] Additionally, the FAR prohibits an agency from making a final payment earlier than 7 days before the due date.[118] This means, in practice, that a contractor can expect that an agency will make final payment sometime between 23 to 30 days from the date the agency receives the invoice.

1. Flow-Down Requirements

The Prompt Payment Act requires that certain payment provisions relating to construction contracts be incorporated into the subcontracts entered into by prime contractors. Section 3905(b) of the Act requires prime contractors who are awarded

---

[114]*See* **Chapter 15** for a discussion of final payment releases.
[115]31 U.S.C. §§ 3901 *et seq.*
[116]31 U.S.C. § 3903(6); FAR § 32.904(d)(1)(i).
[117]31 U.S.C. § 3903(a)(1)(B); FAR § 52.232-27(a)(1)(ii).
[118]FAR § 32.906(a).

federal projects to include prompt payment terms in their subcontracts, thereby obligating the general contractor to pay subcontractors within *seven* days of payment by the government.[119] The Act also requires prime contractors to direct their subcontractors and suppliers to incorporate similar prompt payment provisions in the agreements with their lower-tier contractors.[120] These requirements are included in the standard Prompt Payment for Construction Contracts clause.[121] Contractors should review both their standard subcontract and purchase orders to determine if the terms of payment are consistent with the requirements of this clause.

2.   Retainage

Under the Prompt Payment Act, government contractors are allowed to include retainage provisions in subcontracts[122] and to withhold payments "in accordance with the subcontract agreement," where the general contractor notifies the contracting officer of its actions.[123] Additionally, if the government contractor seeks to withhold funds from a subcontractor for improper work, the contractor is not allowed to request payment for these amounts from the government.[124]

3.   Interest Penalties

If the prime contractor fails to comply with the Prompt Payment Act, it may be liable to the government for an interest penalty. That is, the prime contractor may be liable to the federal government for an interest penalty and also be exposed to its subcontractors for a similar penalty in the event that it fails to pay its subcontractors on time.[125] However, despite these penalties, the prime contractor is itself entitled to interest in the event a payment request is not timely paid by the government.

The contractor does not have to make a separate demand or claim for interest under the Prompt Payment Act because interest is due automatically and must be included and noted in the government's payment. The applicable interest rates for Prompt Payment Act claims are found in the *Federal Register.* For prime contractors, the interest on late government payments begins to accrue on the day after the payment due date and is compounded every 30 days for up to 12 months.

If the federal agency fails to include the interest penalty with a late payment, the contractor must, within 40 days of the late payment, submit a written notice to the contracting officer explaining that the contractor was entitled to interest for the late payment and that the agency has failed to correct its failure within a 10-day period from the date of the late payment.

---

[119]31 U.S.C. § 3905(b).

[120]31 U.S.C. § 3905(c).

[121]FAR §§ 52.232-27(c)(1), 52.232-27(c)(3).

[122]*But see* FAR § 52.232-5 (Payments Under Fixed-Price Construction Contracts). Paragraph (4) of the certification reflects a policy that the government effectively holds the retainage on the subcontract until the prime contractor is ready to release it.

[123]FAR § 52.232-27(d)(3).

[124]FAR § 52.232-27(h).

[125]FAR § 52.232-27(e)(6).

4. *Subcontractor Prompt Payment Claims*

To support the goals of the Prompt Payment Act, the Act provides subcontractors with assistance from the government in pursuing prompt payment claims and also provides the subcontractor the information needed to evaluate whether the Act has been violated. Specifically, the contracting officer has the discretion to undertake an investigation into the subcontractor's allegations. If an investigation is undertaken and it is determined that the prime contractor's payment certifications are inaccurate, then an administrative or other remedial action must be initiated.[126] In addition, the subcontractor may also request that the government furnish information regarding the prime contractor's payment bond.[127] Last, if a subcontractor believes that it has not been timely paid, it may demand the documentation needed to determine whether payments have been made in accordance with the Act.[128]

5. *Government Overpayments to Contractors*

If the contractor discovers that a portion or all of a previously pay application constituted payment for performance that fails to "conform to the specifications, terms, and conditions" of the contract, FAR § 52.232-5 imposes certain affirmative obligations on the contractor related to this "unearned amount."[129] These include:

- Notification to the contracting officer of the performance deficiency
- Obligation to pay the government interest on the unearned amount until the contractor notifies the contracting officer that the deficiency has been corrected; or the contractor reduces a subsequent pay request by an amount equal to the unearned amount

The FAR also requires contractors who receive a duplicate contract financing or invoice payment from the government for work not performed to immediately notify the contracting officer of the overpayment and remit the amount overpaid.[130] Along with the returned payment, the contractor must inform the contracting officer of: (1) the circumstances surrounding the overpayment (e.g., duplicate payment, erroneous payment, liquidation errors; (2) date(s) of overpayment); (3) the affected contract number and delivery order number, if applicable; (4) the affected contract line item or subline item, if applicable; and (5) the contractor's point of contact; and provide a copy of the remittance and supporting documentation.[131]

The failure of a contractor to disclose to the contracting officer the receipt of a "significant overpayment" by the contractor is grounds for suspension or disbarment

---

[126]FAR § 32.112-1(c).

[127]FAR § 28.106-6(c).

[128]FAR § 32.112-2.

[129]FAR § 52.232-5(d). *See also* FAR § 9-406-2(b)(1)(vi) providing that a contractor may be debarred if a preponderance of the evidence shows that the contractor failed to disclose that it received significant overpayments on a contract.

[130]Prompt Payment for Construction Contracts (Oct. 2008), FAR § 52.232-27(l).

[131]FAR § 52.232-27(l).

of the contractor from future contracts with the government.[132] The discovery of "credible evidence" of a "significant overpayment" triggers a contractor's duty to disclose the overpayment to the contracting officer, and the disclosure of the overpayment must be "timely."[133] Although the FAR does not define what constitutes a "significant overpayment" to a contractor or the period of time in which a contractor must notify the government of a significant overpayment to be considered "timely," a commonsense definition of these terms should be applied by contractors when analyzing their duty to report overpayments to the contracting officer. The suspension or debarment of a contractor can be a fatal blow to a company's reputation and its financial stability. Therefore, the mandatory nature of reporting government overpayments should not be taken lightly.

IV. GOVERNMENT SETOFF/DEBT COLLECTION RIGHTS

Under FAR § 32.602, there are several situations in which the government obtains a right to pursue the contractor for funds related to the contract, including:

(i) Damages or excess costs related to defaults in performance

(ii) Breach of contract obligations concerning progress payments, advance payments, or government-furnished property or material

(iii) Government expense of correcting defects

(iv) Overpayments related to errors in quantity or billing or deficiencies in quality

(v) Retroactive price reductions resulting from contract terms for price redetermination or for determination of prices under incentive-type contracts

(vi) Overpayments disclosed by quarterly statements required under price redetermination or incentive contracts

(vii) Delinquency in contractor payments due under agreements or arrangements for deferral or postponement of collections

(viii) Reimbursement of costs paid by the government where a postaward protest is sustained as a result of an awardee's misstatement, misrepresentation, or miscertification

These situations give rise to the government's capacity as a debt collector in government contracting.

FAR § 32.610 states that the government's action to collect these debts is initiated by a demand for payment, which shall be made as soon as the responsible official has computed the amount of refund due. This demand must include a description of the

---

[132]FAR §§ 9.406-2(b)(1)(vi), 9.407-2(a)(8).
[133]*Id.*

debt and a notice advising the contractor of its right to seek a deferment of the debt under FAR § 32.613.

A. Government Setoff

The government's ability to collect contract debts is facilitated by the ability to set off amounts owed to it by a contractor from amounts the government owes to the contractor, including amounts owed under more than one contract. Under FAR §§ 32.611 and 32.612, the government has the right to set off debts owed by the contractor (related to the categories set forth earlier) against funds the government owes to the contractor. The government's withholding may begin 30 days after the demand for payment is issued.

However, the government's set-off right is not unlimited. For example, if the contract is properly assigned under the Assignment of Claims Act of 1940,[134] the government must scrupulously respect the rights of the assignee in any withholding of payments.

B. Federal Claims Collection Act

Government contract debts are governed by the Federal Claims Collection Act (Collection Act),[135] in addition to Part 32 of the FAR. The Collection Act specifies certain procedural and notice requirements for the federal government to recoup overpayments on government contracts. Section 3716 of the Collection Act authorizes the government to recoup overpayments through administrative offsets.[136]

V. ASSIGNMENTS/NOVATIONS

A. Assignments

In order to assist a contractor in financing ongoing contract work, 31 U.S.C. § 3727 and Part 32 of the FAR provide that a contractor may make an assignment of claims against the government to a third party *only* under specified conditions and/or circumstances. The specified limited conditions where a contractor may assign its rights to a third party for monies to which it is due from the government for contract performance are set forth in FAR § 32.802 (Conditions):

Under the Assignment of Claims Act, a contractor may assign moneys due or to become due under a contract if all the following conditions are met:

(a) The contract specifies payments aggregating $1,000 or more.
(b) The assignment is made to a bank, trust company, or other financing institution, including any Federal lending agency.

---

[134]31 U.S.C. § 3727 and 41 U.S.C. § 15.
[135]31 U.S.C. § 3701 *et seq.*
[136]31 U.S.C. § 3716.

   (c) The contract does not prohibit the assignment.

   (d) Unless otherwise expressly permitted in the contract, the assignment—

      (1) Covers all unpaid amounts payable under the contract;

      (2) Is made only to one party, except that any assignment may be made to one party as agent or trustee for two or more parties participating in the financing of the contract; and

      (3) Is not subject to further assignment.

   (e) The assignee sends a written notice of assignment together with a true copy of the assignment instrument to the—

      (1) Contracting officer or the agency head;

      (2) Surety on any bond applicable to the contract; and

      (3) Disbursing officer designated in the contract to make payment.

An assignment of claims against the government cannot be made to anyone other than a financial institution. If an assignment of a claim against the government is made to someone other than a financial entity, the assignment is invalid as the party to whom the assignment is made has no privity with the government. "Privity" means that there is a contractual relationship between the government and the other party. For example, a prime contractor that has a contract with the government has contractual relations (i.e., privity) with the government. A subcontractor typically has no privity of contract with the government. Accordingly, only a prime contractor may take legal action against the government for issues that arise under the contract.[137]

Privity of contract must be shown for all parties attempting to maintain a claim as a contractor against the government under the Contract Disputes Act.[138] In short, a basic tenet of government contract law—that the government consents to be sued only by those (i.e., a contractor) with whom it has privity of contract, although exceptions exist within very limited circumstances—is emphatically embodied in the Contract Disputes Act.[139]

B. Novations

After a contract is executed with the government, a contractor may contemplate either selling the company to another firm or making a decision to reorganize the company. In this instance, the contractor must provide a novation request to the contracting officer.

Basically, a novation agreement is the substitution of one party in a contract with another party. A novation agreement is a legal instrument executed by (a) the contractor (transferor), (b) the successor in interest (transferee), and (c) a third party (in this case, the government). Under a novation agreement, the transferor guarantees

---

[137]See **Chapter 15** for a discussion of subcontractor claims and the disputes process.
[138]41 U.S.C. §§ 601 *et seq.*
[139]See *Oakland Steel Corp. v. United States,* 33 Fed. Cl. 611 (1995).

performance of the contract, and the transferee assumes all obligations under the contract. As provided by FAR § 42.1204:

(a) 41 U.S.C. 15 prohibits transfer of Government contracts from the contractor to a third party. The Government may, when in its interest, recognize a third party as the successor in interest to a Government contract when the third party's interest in the contract arises out of the transfer of—

(1) All the contractor's assets; or

(2) The entire portion of the assets involved in performing the contract.

However, if a novation agreement is proposed, the government has the right not to accept the novation if the contracting officer does not believe it is in the government's overall best interests. As provided by FAR § 42.1204(c):

When it is in the Government's interest not to concur in the transfer of a contract from one company to another company, the original contractor remains under contractual obligation to the Government, and the contract may be terminated for reasons of default, should the original contractor not perform.

➤ LESSONS LEARNED AND ISSUES TO CONSIDER

- *Unauthorized* acts of government employees or agents will not bind the government.
- The burden is on the contractor to establish that it is dealing with a government employee or agent having *actual authority* to bind the government.
- The doctrine of *apparent authority* does not apply to bind the government.
- Some agency supplements to the FAR expressly limit the extent of the authority that may be delegated to a *contracting officer's representative*. An individual's title may not accurately reflect that person's actual authority.
- Read the contract with careful attention to provisions that *limit* delegation of authority from the contracting officer.
- Obtain and read delegation letters from the contracting officer, *literally* and *strictly construing* the terms of any delegation.
- Establish a *procedure to ascertain* the documentation setting forth the extent of the *actual authority* of any representative of the government.
- If a government representative or employee other than the contracting officer issues an instruction that may involve extra cost or time or otherwise modify the terms of the contract, provide *appropriate written* notice of that instruction to the contracting officer.

(Continued)

- Authority to bind the government generally is *implied* when such authority is considered to be an integral part of the duties assigned to a government employee.
- Actual authority *cannot be created* through estoppel.
- The government will be bound when the acts of an unauthorized agent are expressly *or impliedly ratified* by a properly authorized government representative.
- The knowledge of an unauthorized representative *will be imputed* to the authorized representative where the nature of the relationship creates a presumption that the authorized representative would be informed.
- A government employee, representative, or agent acting within the scope of his or her employment generally has absolute *immunity* from being sued.
- There is a *strong presumption* that government employees discharge their duties lawfully and in good faith. A rebuttal of this presumption generally requires clear and convincing evidence to the contrary.
- Contractors must be *familiar with* the basic principle of federal appropriation law: The government *cannot spend* what has not been appropriated.
- The general rule in looking at the use of appropriated funds is whether the use relates to the *purpose* of the appropriation, is made within the *time period* of the appropriation, and whether the use is within the *amount* of the funds appropriated.
- The Anti-Deficiency Act provides for administrative and *criminal penalties* where an agency spends more funds than were allocated; this Act operates as a harsh limitation on what the government may and may not do in relation to contract funding.
- A contract may be *funded incrementally* if it is made "subject to the availability of funds"; however, the contractor must be made aware of such a limitation and its consequences on the proposed project.
- It is crucial for contractors to be aware of and closely follow the FAR *invoicing procedures*, as those procedures control whether and how much the contractor is paid.
- Prime contractors generally have a right to be paid by the government within 30 days of *receipt of a proper invoice* by an authorized government employee.
- A contractor is required to *timely notify the contracting officer if it receives a significant overpayment* for work not performed. Failure to do so can lead to suspension or debarment of the contractor.
- Subcontractors generally have a right to be paid by their prime contractors within *seven days* of the prime contractor's receipt of payment from the government.
- Government contracts may be assigned only in very specific situations; typically, they may be *assigned only to a financial institution.*

APPENDIX 2A: SAMPLE LETTER TO CONTRACTING OFFICER REQUESTING CLARIFICATION ON THE SCOPE OF AN AUTHORIZED REPRESENTATIVE'S AUTHORITY

To be sent via U.S. Mail (if no response, send a follow up via Certified Mail-Return Receipt Requested) (if e-mailed, follow up with a hard copy letter).

[Name],   Contracting Officer

[Address]

    ATTN: [Office Name or Mail Code]

    SUBJECT: Request for Clarification Regarding the Scope of Authority Delegated to [Authorized Representative's Name and Title] by the Contracting Officer

              Contract No. [Number]

              [Project Name or Other Identifier]

Dear [Name of Contracting Officer]:

We hereby request clarification regarding the scope of authority delegated to [Authorized Representative's Name and Title] in connection with the above-referenced contract. In general, we request clarification regarding [Authorized Representative's Name and Title]'s authority to make changes to the contract requirements, to inspect and to accept or reject the contract work performed [or other subject of inquiry]. Specifically, we request clarification on [Authorized Representative's Name and Title]'s authority to [add any specific examples of the Authorized Representative's exercise of authority here].

Please provide a full explanation of the authority delegated to [Authorized Representative's Name and Title], any limitations placed on the exercise of that authority, and the period of time covered by the delegation. In that regard, please include with your explanation a copy of the letter of appointment delegating such authority to [Authorized Representative's Name and Title].

Our firm is committed to achieving a high degree of cooperation on this project, and we will perform as directed by [Authorized Representative's Name and Title] unless and until you direct otherwise. However, in order to avoid needless disputes regarding questions of authority, we intend to provide you or your successor, if any, appropriate written communications if we believe that a directive from the [Authorized Representative's Title] may affect the contract price, time or terms, and conditions.

Sincerely,

APPENDIX 2B: NOTIFICATION OF POTENTIAL CHANGES BY A CONTRACTING OFFICER'S REPRESENTATIVE OR OTHER GOVERNMENT REPRESENTATIVE

To be sent via e-mail with PDF on letterhead with follow-up hard copy via U.S. Mail (send letter both ways to ensure that a copy is received).

[Name],   Contracting Officer

[Address]

    ATTN: [Office Name or Mail Code]

    Subject: Possible Change Order Directive

              Contract No. [Number]

              [Project Name or Other Identifier]

Dear [Name of Contracting Officer]:

The purpose of this letter is to confirm the direction provided to our firm on [insert date] by [insert name of government representative] who is the [insert title such as Contracting Officer's Representative] on this project. On the date noted above, [insert name] instructed our firm's [insert title such as Superintendent] to [provide summary of direction of change or delay; testing requirement; sequence of activities; etc., with sufficient detail to enable the contracting officer to appreciate the effect of the directive]. It is our conclusion that this direction is a change to the contract's requirements that will entitle [insert contractor's name] to an adjustment in the contract price and time [modify as appropriate].

In order to avoid disrupting the progress of the work, we will need to start implementation of this instruction by [insert date—provide a date with sufficient lead time to allow for delivery of communications and evaluation by the contracting officer]. Therefore, if the government does not want us to perform in accordance with the instructions summarized above, please advise us in writing or via e-mail by that date.

If there are any questions or if additional information is needed, please do not hesitate to contact [me or insert name of appropriate contractor's representative].

Sincerely,

Cc: [name of government's representative issuing instruction]

APPENDIX 2C: LIMITATION OF GOVERNMENT'S OBLIGATION (MAY 2006)

(MAY 2006)

(a) Contract line item(s) ____* through ____* are incrementally funded. For these item(s), the sum of $____* of the total price is presently available for payment and allotted to this contract. An allotment schedule is set forth in paragraph (j) of this clause.

(b) For item(s) identified in paragraph (a) of this clause, the Contractor agrees to perform up to the point at which the total amount payable by the Government, including reimbursement in the event of termination of those item(s) for the Government's convenience, approximates the total amount currently allotted to the contract. The Contractor is not authorized to continue work on those item(s) beyond that point. The Government will not be obligated in any event to reimburse the Contractor in excess of the amount allotted to the contract for those item(s) regardless of anything to the contrary in the clause entitled "Termination for Convenience of the Government." As used in this clause, the total amount payable by the Government in the event of termination of applicable contract line item(s) for convenience includes costs, profit, and estimated termination settlement costs for those item(s).

(c) Notwithstanding the dates specified in the allotment schedule in paragraph (j) of this clause, the Contractor will notify the Contracting Officer in writing at least ninety days prior to the date when, in the Contractor's best judgment, the work will reach the point at which the total amount payable by the Government, including any cost for termination for convenience, will approximate 85 percent of the total amount then allotted to the contract for performance of the applicable item(s). The notification will state (1) the estimated date when that point will be reached and (2) an estimate of additional funding, if any, needed to continue performance of applicable line items up to the next scheduled date for allotment of funds identified in paragraph (j) of this clause, or to a mutually agreed upon substitute date. The notification will also advise the Contracting Officer of the estimated amount of additional funds that will be required for the timely performance of the item(s) funded pursuant to this clause, for a subsequent period as may be specified in the allotment schedule in paragraph (j) of this clause, or otherwise agreed to by the parties. If after such notification additional funds are not allotted by the date identified in the Contractor's notification, or by an agreed substitute date, the Contracting Officer will terminate any item(s) for which additional funds have not been allotted, pursuant to the clause of this contract entitled "Termination for Convenience of the Government."

(d) When additional funds are allotted for continued performance of the contract line item(s) identified in paragraph (a) of this clause, the parties will agree as to the period of contract performance which will be covered by the funds. The provisions of paragraph (b) through (d) of this clause will apply in like manner to the additional allotted funds and agreed substitute date, and the contract will be modified accordingly.

(e) If, solely by reason of failure of the Government to allot additional funds, by the dates indicated below, in amounts sufficient for timely performance of the contract line item(s) identified in paragraph (a) of this clause, the Contractor incurs additional costs or is delayed in the performance of the work under this contract and if additional funds are allotted, an equitable adjustment will be made in the price or prices (including appropriate target, billing, and ceiling prices where applicable) of the item(s), or in the time of delivery, or both. Failure to agree to any such equitable adjustment hereunder will be a dispute concerning a question of fact within the meaning of the clause entitled "Disputes."

(f) The Government may at any time prior to termination allot additional funds for the performance of the contract line item(s) identified in paragraph (a) of this clause.

(g) The termination provisions of this clause do not limit the rights of the Government under the clause entitled "Default." The provisions of this clause are limited to the work and allotment of funds for the contract line item(s) set forth in paragraph (a) of this clause. This clause no longer applies once the contract is fully funded except with regard to the rights or obligations of the parties concerning equitable adjustments negotiated under paragraphs (d) or (e) of this clause.

(h) Nothing in this clause affects the right of the Government to terminate this contract pursuant to the clause of this contract entitled "Termination for Convenience of the Government."

(i) Nothing in this clause shall be construed as authorization of voluntary services whose acceptance is otherwise prohibited under 31 U.S.C. 1342.

(j) The parties contemplate that the Government will allot funds to this contract in accordance with the following schedule:

On execution of contract $____

(month) (day), (year) $____

(month) (day), (year) $____

(month) (day), (year) $____

3

CONTRACT FORMATION

I. BASIC PRINCIPLES OF CONTRACT FORMATION

A contract is a legally enforceable promise or set of promises. The *Restatement (Second) of Contracts* specifically defines a contract as "a promise or a set of promises for the breach of which the law gives a remedy, or the performance of which the law in some way recognizes as a duty."[1] In a commercial context, the parties may generally agree to whatever they wish, as long as the agreement does not run afoul of applicable law or an overriding public policy. A government contract, however, is subject to numerous statutory and regulatory limitations and restrictions regarding what the parties can bargain for and how the contract must be structured and executed. Although there are significant differences between the parameters and scope of private and government contracts, the principles of contract formation are the same.

As a threshold matter, the formation of any contract requires an objective manifestation of voluntary, mutual assent to an exchange of promises. Simply put, the parties must agree to the same contractual terms and conditions. This mutuality of intent may be shown by demonstrating the existence of an offer and a reciprocal acceptance of that offer.[2] In addition to offer and acceptance, the common law requires consideration, something of value bargained for and exchanged by the contracting parties.

A valid and enforceable government construction contract must meet these requirements: "mutual intent to contract including offer and acceptance, consideration, and a Government representative who had actual authority to bind the Government."[3] Authority to bind the government is discussed in **Chapter 2**. The requirements of offer, acceptance, and consideration are briefly explained next.

A. Offer—Acceptance—Consideration

1. Offer

The existence of an offer may be established by showing the offeror's (the party making the offer) manifestation of willingness to enter into a bargain. The offer must be made in a manner that justifies another person's understanding that assent to that

[1]*Restatement (Second) of Contracts* § 1 (1981).
[2]*Anderson v. United States,* 344 F.3d 1343, 1353 (Fed. Cir. 2003).
[3]*Trauma Serv. Group v. United States,* 104 F.3d 1321, 1326 (Fed. Cir. 1997).

bargain is invited.[4] An offer gives the offeree (the party receiving the offer) the power to create a contract by accepting the offer.[5]

Even though the contractor's offer is based on the agency's solicitation, the government generally requires that the construction contractor make the offer, thereby allowing the government to maintain the preferred position of the offeree with the power to create the contract upon acceptance. The government typically invites offers by issuing a request for proposals (RFP), an invitation for bids (IFB), or a request for quotations (RFQ). These government invitations for offers are commonly referred to as *solicitations*. The offer/bid must be submitted in accordance with the terms of the government's solicitation for it to be a valid and acceptable offer.

Generally, in sealed bidding, a bid may be modified or withdrawn before the time set for the bid opening.[6] Thereafter, under the "Firm Bid Rule," the bid may not be withdrawn absent a mistake in the bid.[7] In negotiated procurement, offers (proposals) normally may be withdrawn at any time before the award of the contract.[8]

2. Acceptance

Acceptance is a manifestation of assent by the offeree to the terms made or invited by the offer.[9] If the legal effect of the offeree's action makes the offeror's promise enforceable, then acceptance has occurred.[10] The government's acceptance may be manifested by communication of the acceptance by an authorized person, typically the contracting officer or an authorized representative of the contracting officer, with the actual authority to accept the offer.

In *D&S Universal Mining, Inc.,* the Comptroller General explained acceptance in these terms:

> In order for a binding contract to result, the contracting officer must unequivocally express an intent to accept an offer. Also, the acceptance of a contractor's offer by the Government must be clear and unconditional; it must appear that both parties intended to make a binding agreement at the time of the acceptance of the contractor's offer.[11]

Merely notifying a contractor that it is the apparent low bidder does not constitute acceptance.[12] The government must manifest a "clear-cut expression" of its intention to accept the offer or bid.[13]

---

[4]*Anderson v. United States,* 344 F.3d 1343, 1353 (Fed. Cir. 2003).
[5]Joseph M. Perillo, *Calamari and Perillo on Contracts* § 2.5, 32 (5th ed., West 2003).
[6]FAR § 14.303.
[7]*Refining Assocs. v. United States,* 109 F. Supp. 259 (Ct. Cl. 1953); *see also* Block 12 of Standard Form 33 and Block 11 of Standard Form 1447 in FAR § 53.301.
[8]FAR § 15.208(e).
[9]*Restatement (Second) of Contracts* § 50(1) (1981).
[10]E. Allan Farnsworth, *Contracts* § 3.3, 113 (3d ed., Aspen L. & Bus. 1999).
[11]Comp. Gen. Dec. B-200815, 81-2 CPD ¶ 186 (1981).
[12]*See Goldberger Foods, Inc. v. United States,* 23 Cl. Ct. 295 (1991); *see also DeMatteo Constr. Co. v. United States,* 600 F.2d 1384 (Ct. Cl. 1979).
[13]*Greenwood Co.,* ASBCA No. 12232, 67-2 BCA ¶ 6650.

3. Consideration

Consideration is bargained-for legal detriment. "Consideration can be described as the price bargained and paid for a promise."[14]

In government construction contracts, consideration usually is the exchange of promises to perform. The construction contractor's bid or proposal is a promise to perform construction services for a certain amount of money. The government, in exchange for the promise to perform, promises to pay the contractor the agreed-upon contract price: a bargained-for set of promises legally recognized as a *contract*. The existence of consideration is rarely challenged or at issue in the government contracts arena.

B. Express Contracts

When the parties manifest their mutuality of intent—offer and acceptance, supported by consideration—by words, their contract is said to be express.[15]

An express contract may be written or oral (i.e., an express written contract or an express oral contract). The key is that the parties' mutual assent to an agreement is expressed by words.

There is no federal Statute of Frauds,[16] as there is in private contract law, that requires certain classes of contracts to be written to be enforceable. There are, however, numerous federal statutory and regulatory requirements for certain documents to be in writing.[17] A prudent government contractor should ensure that all agreements are in writing. Generally, oral agreements provide a very unreliable basis for granting a contractor any relief.[18]

C. Implied Contracts

In addition to express contracts, which comprise the vast majority of government construction contracts, a contract may exist by implication. Under the common law of contracts, there are two kinds of implied contracts, one implied-in-fact and the other implied-in-law.[19] Government contract law recognizes implied-in-fact contracts but not implied-in-law contracts.

1. Implied-in-Fact Contracts

An implied-in-fact contract may be created by the conduct of the parties. The elements of an implied-in-fact contract are the same as those of an express contract (i.e.,

---

[14]John Cibinic, Jr. & Ralph C. Nash, Jr., *Formation of Government Contracts* 247 (3d ed. [George Washington University 1998]).

[15]Joseph M. Perillo, *Calamari and Perillo on Contracts* § 1.8, 21 (5th ed., West 2003).

[16]*See Restatement (Second) of Contracts* § 110 (1981).

[17]*See, e.g.,* 10 U.S.C. § 2305(b)(3); 31 U.S.C. § 1501(a)(1); 41 U.S.C. § 253b(c); FAR §§ 2.101; 14.408-1; 15.504.

[18]*Edwards v. United States,* 22 Cl. Ct. 411 (1991).

[19]This discussion is provided for background information in order to place some of the basic principles of government contract law in context and to illustrate some of the differences between government contracts and commercial contracts.

mutual intent to contract, offer, acceptance, and consideration). The difference is that the manifestation of assent in an implied-in-fact contract is inferred from the parties' actions or course of conduct rather than found in expressed words.

An implied-in-fact contract is based on a meeting of the minds, which, although not found in writing, is inferred as fact from the conduct of the parties showing their mutual understanding under the circumstances.[20] The parties' meeting of minds must be evidenced by some definitive conduct, act, or sign.[21]

In the government contracting arena, implied-in-fact contracts are recognized as valid contracts as long as they comply with all statutory and regulatory requirements of express contracts. Agency procedures must be followed before a binding implied-in-fact contract can be formed.[22] In addition, the government's course of conduct, upon which the contractor relied, must be the conduct of a government representative with actual authority. (*See* **Chapter 2**.) Implied-in-fact contracts are rare in government contracting.

2. Implied-in-Law Contracts

An implied-in-law contract, sometimes referred to as a quasi-contract or constructive contract, is not a contract in the true sense but is imposed by operation of law on the grounds of equity and justice. A contract implied-in-law is one in which no actual agreement between the parties occurred but where contractual obligations are imposed on equitable principles to prevent injustice.[23] An implied-in-law contract presupposes the unjust enrichment of one party at the expense of another. Implied-in-law contract claims are not actionable under the Tucker Act[24] or the Contract Disputes Act[25] because the federal government has not waived its sovereign immunity from suits based on a contract implied-in-law.[26] As such, implied-in-law contracts are not recognized or actionable in the government contracts realm.

D. Ratification of Unauthorized Agreements

In the absence of an express or implied-in-fact contract, a binding agreement may still be found by *ratification*. Under common law, ratification is the "affirmance by a person of a prior act which did not bind him but which was done or professedly done on his account, whereby the act, as to some or all persons, is given effect as if originally authorized by him."[27] Federal courts recognize two types of ratification: (1) individual ratification and (2) institutional ratification.[28]

---

[20]*Baltimore & Ohio R.R. Co. v. United States*, 261 U.S. 592, 597 (1923).
[21]261 U.S. at 598.
[22]*Harbert/Lummus Agrifuels Projects v. United States*, 142 F.3d 1429 (Fed. Cir. 1998).
[23]*Contel of California, Inc. v. United States*, 37 Fed. Cl. 68 (1996).
[24]28 U.S.C. §§ 346(a)(2) and 1491(a)(1).
[25]41 U.S.C. §§ 601 *et seq.*
[26]*See Merritt v. United States*, 267 U.S. 338 (1925); *United States v. Mitchell*, 463 U.S. 206 (1983); *Russell Corp. v. United States*, 537 F.2d 474 (Ct. Cl. 1976).
[27]*Schism v. United States*, 316 F.3d 1259, 1289 (Fed. Cir. 2002).
[28]*See Gary v. United States*, 67 Fed. Cl. 202 (2005).

1. Individual Ratification

Individual ratification, in the government construction contracts setting, is the act of approving an unauthorized commitment by a government official who has the authority to do so.[29] An *unauthorized commitment* is an agreement that is not binding solely because the government representative who initially made it lacked the authority to enter into that agreement on behalf of the government.[30] "Under the doctrine of individual ratification, a superior [government] official must (1) possess authority to contract, (2) possess full knowledge of the material facts surrounding the unauthorized action, and (3) knowingly confirm, adopt, or acquiesce in the unauthorized action."[31]

2. Institutional Ratification

Institutional ratification occurs when the government seeks and receives the benefits from an otherwise unauthorized contract.[32] However, officials with ratifying authority must know of the promise, as such knowledge is a key element of an institutional ratification claim.[33] An official with the power to ratify must know the material facts relating to the acceptance of the benefits and must agree to accept them for an unauthorized promise to bind the government under the doctrine of institutional ratification.[34]

Federal government policy discourages the use of ratification authority.[35] Contractors should view ratification as a *last resort* remedy and avoid becoming involved in informal agreements to perform government construction work for an agency. Nonpayment by the government is a clear risk of informal agreements, and litigation over a ratification claim is almost certain.

II. THE GOVERNMENT PROCUREMENT PROCESS

The federal government spends billions of dollars each year procuring goods and services of all kinds. Federal procurements range from multibillion-dollar federal construction projects and the acquisition of major Department of Defense weapons systems to the procurement of several hundred dollars' worth of office supplies by an agency. In order to bring organization and fairness to this procurement process, the federal government has implemented numerous statutes, regulations, and policies to instruct and direct agencies on how to conduct procurements from beginning

---

[29]67 Fed. Cl. at 215.

[30]*Id.*

[31]*Id.*

[32]*Janowsky v. United States,* 133 F.3d 888, 891–92 (Fed. Cir. 1998); *see also Digicon Corp. v. United States,* 56 Fed. Cl. 425, 426 (2003); *Silverman v. United States,* 679 F.2d 865 (Ct. Cl. 1982).

[33]*Gary v. United States,* 67 Fed. Cl. 202, 216 (2005).

[34]67 Fed. Cl. at 217.

[35]FAR § 1.602-3(b)(1); *see* **Chapter 2, Section I** for further discussion of ratification.

to end. All contractors who intend to provide goods or services to the federal government must have a fundamental understanding of how the procurement process works in order to successfully compete for and perform procurement contracts. A hypothetical construction project sponsored by the U.S. Army Corps of Engineers (Corps) is used next to explain the sequence and parameters of the federal procurement process.

The procurement process begins when the Corps identifies a need for construction services. Once that need is identified, the Corps assigns a contracting officer to set up and manage the procurement process. An initial determination by the procuring activity (contracting officer) addresses the kind of procurement method that will be used to award the contract. Generally, federal construction contracts are awarded through either sealed bid procurements or negotiated procurements. Sealed bid procurements focus primarily on obtaining the lowest price from a responsive and responsible bidder; negotiated procurements focus on obtaining the *best value* for the job. Negotiated procurements have become more and more popular over the last 20 years and will be used for this example.

Following the selection of the negotiated procurement method, a request for proposals (RFP) is drafted and publicly issued by the Corps on FedBizOpps (the Federal Business Opportunities Web site) at *www.fbo.gov*. While contractors should routinely monitor this site for upcoming government projects, once a contractor becomes interested in and starts to develop a proposal for a specific project, it is imperative to monitor the Web site for any amendments to the requirements of the RFPs. The RFP instructs offerors on how to submit a proposal for the construction project, explains what criteria will be used to evaluate proposals, and establishes a time frame for receiving proposals and selecting a contract awardee.

After all proposals have been received by the Corps from interested contractors, the Corps begins the evaluation phase of the procurement process. Negotiated procurements seek to award the contract to the contractor providing the *best value* to the government. For this reason, the Corps will look not only at a contractor's proposed price but also at such factors as the contractor's key personnel, management, and past performance on other projects. The evaluation of proposals can take many months and may require contractors to submit several rounds of revised proposals. A negotiated procurement allows the Corps to have detailed discussions with offerors, which enable offerors to refine their proposals to best suit the needs of the Corps. Once the Corps has completed the evaluation process, it selects the *best value* contractor for the contract award.

Unless the procurement is a multiple award task order (MATOC) solicitation, only one contractor can receive the contract award. Even if there are multiple awardees, one or more disappointed offerors who thought their proposals offered the *best value* for the government to may question the Corps' evaluation process and award decision. Those disappointed contractors may seek to stop the Corps from allowing performance to begin on the project through a *bid protest*, which can be filed with the Corps, the Government Accountability Office (GAO), or the U.S. Court of Federal Claims. Bid protests can be filed before the award of a contract to correct a defect in the RFP or after the award of a contract to correct a defect in the evaluation process.

The GAO and the Court of Federal Claims have significant influence and authority to correct any inequities or mistakes that occur during the procurement process.

After the contract is executed by the contractor and performance begins, the contractor is bound to follow the contract and the direction of the Corps' authorized representatives in constructing the project. The performance of the contract, like the procurement process itself, is regulated by numerous government statutes, regulations, and policies. Contractors must have a fundamental understanding of their responsibilities before performance begins. The ultimate goal of all contractors performing work for the government should be to perform the project through a spirit of partnering with the contracting agency. Contractors that perform contracts in a competent and efficient manner are much more likely to receive positive performance evaluations than contractors that encounter major problems during performance. Outstanding performance evaluations are a key factor in receiving future work from the government.

In sum, the procurement process varies from agency to agency and project to project. The procurement principles followed by the individual federal agencies are, for the most part, identical. The keys to success for government contractors is to have a fundamental working knowledge of the federal procurement process and each agency's approach to the negotiated procurement process, submit the best drafted proposals[36] possible, and perform contracts with the goal of receiving the highest possible past performance evaluation.

III. COMPETING FOR THE AWARD

Federal government construction contracts are generally awarded through either sealed bid procurement or negotiated procurement. Sealed bid procurements generally focus on selecting the lowest-priced bid for contract award. Negotiated procurements, however, have multiple levels of evaluation and generally focus on selecting the *best value* proposal for contract award. The similarities and differences between these two procurement methods are discussed in more detail next.

A. Sealed Bids

Competitive sealed bidding, previously known as formal advertising, is a process in which the government solicits bids by issuing an invitation for bids (IFB). Typically, bids are delivered in a sealed envelope to be opened at the time and place specified in the IFB. At the bid opening, the bids are read aloud to the public in attendance. Generally, the contract award is based on the lowest responsive bid from a responsible bidder.

---

[36]On a sealed bid procurement, the goal must be to submit a responsive, low bid from a contractor that is considered to be responsible and otherwise eligible for the award.

Sealed bidding gives all qualified contractors the opportunity to compete for government contracts and gives the government the benefits of competition while also avoiding favoritism, collusion, or fraud.[37] The Federal Acquisition Regulation (FAR) provides that sealed bidding shall be used for domestic construction projects if:

(1) Time permits the solicitation, submission, and evaluation of sealed bids;
(2) The award will be made on the basis of price and other price-related factors;
(3) It is not necessary to conduct discussions with the responding offerors about their bids; and
(4) There is reasonable expectation of receiving more than one sealed bid.[38]

The required elements of sealed bidding are:

(a) *Preparation of invitations for bids.* Invitations must describe the requirements of the Government clearly, accurately, and completely. Unnecessarily restrictive specifications or requirements that might unduly limit the number of bidders are prohibited. The invitation includes all documents (whether attached or incorporated by reference) furnished prospective bidders for the purpose of bidding.

(b) *Publicizing the invitation for bids.* Invitations must be publicized through distribution to prospective bidders, posting in public places, and such other means as may be appropriate. Publicizing must occur a sufficient time before public opening of bids to enable prospective bidders to prepare and submit bids.

(c) *Submission of bids.* Bidders must submit sealed bids to be opened at the time and place stated in the solicitation for the public opening of bids.

(d) *Evaluation of bids.* Bids shall be evaluated without discussions.

(e) *Contract award.* After bids are publicly opened, an award will be made with reasonable promptness to that responsible bidder whose bid, conforming to the invitation for bids, will be most advantageous to the Government, considering only price and the price-related factors included in the invitation.[39]

Under sealed bidding, the government does not have the option of evaluating the technical merits of a bid. If a contract award is made, the contract must go to the responsible bidder with the lowest responsive bid. Therefore, it is important that the bid documents fully describe the project in exacting detail to ensure that the government receives what it pays for and that the low bidder is bound by contract to perform quality work in a satisfactory manner.

B. Competitive Negotiation

Historically, the preferred method for construction contracting in government procurement was through formal advertising (sealed bidding), as discussed earlier.

---

[37]*See United States v. Brookridge Farm, Inc.,* 111 F.2d 461 (10th Cir. 1940).
[38]FAR § 6.401; *see also* FAR § 36.103.
[39]FAR § 14.101.

Negotiated procurement was another method of contracting but was used only under very specific circumstances set forth in the applicable statutes and the FAR. Under negotiated procurement procedures, the federal government issued a request for proposals (RFP). Generally, proposals were received by the government, and the government entered into negotiations with the offerors within the competitive range based on the terms of the solicitation. After submission of best and final offers, award was made to the offeror whose offer was deemed to be the most favorable to the government under the provisions of the RFP. Often this was the lowest evaluated price.

After the Federal Acquisition Streamlining Act of 1994 (FASA),[40] the Clinger-Cohen Act of 1996,[41] and the rewrite of FAR Part 15 in 1997, more and more federal government construction contracts are being awarded under a competitive negotiation technique known as *best value*. Best value is the expected outcome of an acquisition that provides the government the greatest overall benefit in response to the requirement.[42] FAR § 15.101 describes the best-value selection process in these terms:

> An agency can obtain best value in negotiated acquisitions by using any one or a combination of source selection approaches. In different types of acquisitions, the relative importance of cost or price may vary. For example, in acquisitions where the requirement is clearly definable and the risk of unsuccessful contract performance is minimal, cost or price may play a dominant role in source selection. The less definitive the requirement, the more development work required, or the greater the performance risk, the more technical or past performance considerations may play a dominant role in source selection.[43]

An agency may consider many different evaluation factors when conducting a best-value procurement to determine which proposal will be the most advantageous for the government. The evaluation factors used by an agency to conduct a best-value procurement are based on the particular needs of the agency and the goals of the procurement being conducted. Each agency has broad discretion in selecting the factors to be used to evaluate proposals and in determining the relative importance of each evaluation factor in relation to other factors. In general, an agency may consider some or all of these evaluation factors or subfactors during a best-value procurement: cost or price (must always be considered during a best-value procurement), past performance, experience, management and key personnel, and technical expertise.

The best-value process permits trade-offs among cost or price and noncost factors and allows the government to accept other than the lowest-priced proposal.[44] The trade-off decisions must be based on a reasoned explanation by the agency, but the agency's selection official has very broad discretion.[45]

---

[40]Pub. L. No. 103-355, 108 Stat. 3243.
[41]Pub. L. No. 104-106, § 4001, 110 Stat. 186,642.
[42]FAR § 2.101.
[43]FAR § 15.101.
[44]FAR § 15.101-1.
[45]*See Widnall v. B3H Corp.,* 75 F.3d 1577 (Fed. Cir. 1996); *see also* FAR § 15.308.

FAR § 15.101-1 provides:

(a) A tradeoff process is appropriate when it may be in the best interest of the Government to consider award to other than the lowest priced offeror or other than the highest technically rated offeror.

(b) When using a tradeoff process, the following apply:

    (1) All evaluation factors and significant subfactors that will affect contract award and their relative importance shall be clearly stated in the solicitation; and

    (2) The solicitation shall state whether all evaluation factors other than cost or price, when combined, are significantly more important than, approximately equal to, or significantly less important than cost or price.

(c) This process permits tradeoffs among cost or price and non-cost factors and allows the Government to accept other than the lowest priced proposal. The perceived benefits of the higher priced proposal shall merit the additional cost, and the rationale for tradeoffs must be documented in the file in accordance with [FAR] 15.406.

The best-value procurement process has been the subject of many protests to the procuring agencies and the GAO as well as actions in the United States Court of Federal Claims and the United States Court of Appeals for the Federal Circuit. The GAO and federal courts give substantial deference to the discretion of the government in determining which offer is the most advantageous to the government. However, the ultimate standard or test applied by both the GAO and the federal courts is whether the government complied with (a) the FAR requirements for negotiated best-value procurements and (b) whether the agency followed the selection and evaluation procedures set forth in the RFP.[46] An agency's evaluation determinations and award decision will not be disturbed by the federal courts or the GAO unless the agency has acted arbitrarily or violated applicable procurement statutes or regulations in reaching its conclusions.[47]

FAR Part 15 provides agencies with substantial flexibility in conducting a negotiated procurement.[48] For example, following an evaluation of a proposal, agencies may take one or more of these actions:

- Award without *discussions* if the RFP advised the offerors of that possible action.

- If an award without discussions is contemplated, the agency may engage in communications with an offeror to obtain *clarifications* of the offeror's proposal.[49]

---

[46]FAR § 15.304; 75 F.3d 1577 (Fed. Cir. 1996); *see also Grumman Data Systems Corp. v. Dalton*, 88 F.3d 990 (Fed. Cir. 1996); *TRW v. Unisys Corp.*, 98 F.3d 1325 (Fed. Cir. 1996).

[47]*Alion Science and Tech. Corp. v. United States*, 74 Fed. Cl. 372, 374 (2006) ("Generally, a reviewing court does not disturb an agency award unless it is arbitrary, capricious, an abuse of discretion, or otherwise violates applicable procurement law"); *see also Matrix Int'l Logistics, Inc.*, Comp. Gen. Dec. B-272388.2, 97-2 CPD ¶ 89 (the GAO will not disturb an agency evaluation of proposals or award decision where the agency acted reasonably and consistently with regards to the procurement evaluation factors).

[48]*See* FAR §§ 15.202, 15.306.

[49]*Discussions* and *clarifications* are almost terms of art as used in FAR Part 15. Clarifications are described as "limited exchanges" to address specific topics such as past performance evaluations. *See* FAR § 15.306(a)(2).

- If the government plans on establishing a *competitive range,* certain communications are authorized under FAR § 15.306(b) to facilitate that process and decision.
- Once a competitive range is established, certain proposals may be eliminated from consideration. Thereafter, *exchanges* with those offerors within the competitive range are termed *discussions.* FAR §§ 15.306(d) and (e) describe the topics for discussion and the limits on those discussions during exchanges with the offerors in the competitive range.
- Following that process, offerors are given an opportunity to submit final *proposal revisions,* and the contracting officer is required to establish a *common cut-off date* for receipt of such revisions.[50]

The submission of competitive proposals on a complex design-build or construction management project can be very costly. Consequently, it is not unusual for agencies to use multistep processes to determine whether potential offerors appear to have a potential for success.[51] A multistep evaluation process saves prospective contractors from the cost of developing a complete technical and price proposal for every solicitation and allows the agency essentially to *short list* the viable competitors. On a few occasions, an agency may authorize the payment of *stipends* to those firms on the short list that do not receive the contract award.

C. Effect of Government Estimate on Authority to Award Contract

Regardless of the procurement methodology, contracting officers must purchase construction services from responsible sources at *fair and reasonable* prices.[52] One of the analytical techniques the government may use in evaluating whether a proposal or bid is fair and reasonable is to compare the offered price with an independent government cost estimate.[53]

The government is required to prepare an estimate of the construction costs and time for performance of construction contracts and construction contract modifications anticipated to exceed the "simplified acquisition threshold" (generally, $100,000).[54] A government estimate should represent the agency's best projection of the reasonable costs for the construction services being procured,[55] and the estimate should be based on reliable, accurate, and current information.[56]

Government estimates are used for internal purposes in helping to determine the reasonableness of proposals or bids. Although the government estimate generally

---

[50]*See* FAR § 15.307.
[51]*See* FAR § 15.202.
[52]FAR § 15.402(a).
[53]*See* FAR §§ 15.404-1; 36.214(b)(2).
[54]FAR § 36.203(a) mandates the preparation of government estimate for construction contracts and modifications exceeding the simplified acquisition threshold; FAR § 2.101 defines the simplified acquisition threshold, which is generally $100,000.
[55]*See Process Control Technologies, a Div. of GMC Enters., Inc. v. United States,* 53 Fed. Cl. 71, 77 (2002).
[56]*See Blue Dot Energy Co.,* Comp. Gen. Dec. B-253390, 93-2 CPD ¶ 145.

is disclosed when sealed bids are publicly opened, there is no public opening of proposals received in response to an RFP, and the government estimate is not made available to offerors.[57] A bid is not unreasonable or unacceptable simply because it exceeds the government estimate unless a particular statute or regulation provides to the contrary.[58] For example, the Corps of Engineers is required to reject bids for river and harbor improvements (e.g., dredging contracts) if the bids exceed the government estimate by more than 25 percent.[59] The Corps is not, however, required to award a contract to the low bidder simply because its bid is within 25 percent of the estimate.[60] Formerly, construction bids involving small-business set-asides that exceeded the government's estimate of the fair market price by more than 10 percent were required to be rejected. This is no longer the law, and the 10 percent threshold has been removed.[61]

Assuming the funds are available, a contracting officer may have the authority to award a contract even where the lowest bid or proposal exceeds the government estimate. There is no statute or regulation that circumscribes the authority of the contracting officer to make such an award, as long as the offered price is determined to be fair and reasonable.

However, an agency or contracting officer may cancel a solicitation and reject all bids or proposals on the basis that the offered prices were unreasonable because they exceeded the government estimate. In making such a determination, the GAO requires some meaningful margin of difference between the low responsible bid and the government estimate. A meaningful margin does not necessarily require a double-digit differential. For example, in *Building Maintenance Specialists, Inc.,* the contracting officer's cancellation of the solicitation and rejection of all bids was upheld where the low bid exceeded the government estimate by only 7.2 percent.[62] An agency's procurement decisions, including the rejection of all proposals or bids, will be upheld unless shown to be arbitrary, capricious, an abuse of discretion, or otherwise not in accordance with law.[63]

Contracting officers have broad discretion in the use of government estimates, but it is not without limit. For example, if the original government estimate is unreasonably low, a contracting officer's decision to reject all bids after opening and cancel the invitation may be found to be arbitrary and capricious unless there is a compelling reason to reject the bids.[64] In negotiated construction procurements, FAR § 36.214(b)(2) addresses the process for another use of a government estimate. If the

---

[57]*See Overstreet Elec. Co., Inc. v. United States,* 47 Fed. Cl. 728, 732 (2000) (cost estimate used for determining price reasonableness).

[58]*2 Government Contract Awards: Negotiation and Sealed Bidding* § 11:5; *see also* FAR § 15.405(a) (price can be reasonable even when it exceeds the agency's initial negotiating position).

[59]*See* 33 U.S.C. § 624 (2); *see also Atkinson Dredging Co., Inc.—Reconsideration,* Comp. Gen. Dec. B-250965.2, 93-2 CPD ¶ 31 (1993).

[60]*Atkinson Dredging Co., Inc.—Reconsideration,* Comp. Gen. Dec. B-250965.2, 93-2 CPD ¶ 31.

[61]FAR § 19.506.

[62]Comp. Gen. B-186441, 76-2 CPD ¶ 233 (1976).

[63]5 U.S.C. § 706(2)(A); 28 U.S.C. § 1491(b)(4).

[64]FAR § 14.404-1; *see also Great Lakes Dredge & Dock Co. v. United States,* 60 Fed. Cl. 350 (2004).

contractor's proposed price is "significantly lower" than the government's estimate, the contracting officer is directed to make sure that both the prospective contractor and the government understand the scope of the work. If errors in the government's estimate are revealed by this negotiation process, the government's estimate must be adjusted.[65]

D. Prebid/Preproposal Clarifications

In preparing a proposal or bid, a prospective offeror or bidder may identify possible ambiguities or have questions about the project. Most agencies conduct *prebid* conferences to provide additional information and receive questions. In addition, many solicitations provide information regarding the submission of questions including a *point of contact* and a deadline for receipt of questions. Contractors should follow these procedures carefully and not substitute e-mails or oral communications for the procedures set forth in the RFP/IFB. The risk of failing to follow the proscribed procedure is a determination that the government's response to the inquiry is not binding on the government.[66] If the government's reply to a question is "bid it as you see it," the offeror/bidder assumes a substantial risk if it submits a proposal knowing of a defect or ambiguity in the solicitation documents without placing the government on notice of the contractor's understanding of the requirement in a written communication separate from the bid or proposal submission.[67] Alternatively, if the government's response to the inquiry is consistent with the offeror/bidder's interpretation and that interpretation is relied on in pricing the work, the government may be precluded from advancing a different interpretation during performance.[68] A contractor should consider these guidelines when seeking a prebid or preproposal clarification:

- Seek *clarifications* of questions or possible ambiguities in accordance with the procedure set forth in the RFP/IFB.
- *Avoid reliance* on communications from government officials that are not written and issued by the contracting officer.
- If necessary, document *in writing* any information provided by the government *prior* to submitting the proposal or bid.
- Do *not qualify* a sealed bid or include a letter of *clarification* with the bid, as that is very likely to render the bid nonresponsive.
- Be prepared to demonstrate *actual reliance* on the information provided by the government in response to a question.

---

[65]FAR § 36.214(b)(2). Obviously, this process should enable a contractor to determine whether there are any errors or omissions in its estimate.

[66]*Diamond Aircraft Indus., Inc.,* Comp. Gen. B-289309, 2002 CPD ¶ 35 (offeror could not rely on informal advice, oral or via e-mail, from the government).

[67]*Robins Maint., Inc. v. United States,* 265 F.3d 1254 (Fed. Cir. 2001).

[68]*P.J. Dick, Inc. v. General Services Administration,* GSBCA No. 12151, 96-1 BCA ¶ 27,955.

- If there is no response to a question or the response is not meaningful, consider *filing a protest* with the agency or the GAO before the due date for submission of proposals or bids. If a protest is not asserted before that date, the government will argue that the contractor's right to later assert a protest on that grounds is waived.

E. Electronic Bids and Reverse Bid Auctions

1. Overview

The government has procedures in place for electronic procurement, and FAR Part 15, which governs negotiated procurement, has been rewritten to allow federal agencies to use online auction technology. Conduct that favors one offeror over another, reveals an offeror's technical solution, or knowingly furnishes source selection information contrary to regulatory or statutory requirements is still prohibited.[69] In electronic bidding (e-bidding), contractors submit their bids electronically rather than actually delivering a sealed bid in an envelope. In a reverse bid auction, prospective contractors compete in a real-time, online auction by bidding down the price to provide the products or services sought by the government.

2. Risks of Reliance on the Internet

The use of the Internet and e-mail are common tools for conducting business and government contracting. However, reliance on these electronic means of communication can be risky for government contractors. Several decisions from the Comptroller General indicate the problems e-bidding presents for federal contractors. To a large degree, a contractor appears to bear nearly all of the risk associated with the use of the Internet and e-mail.

In *PMTech, Inc.*,[70] a contractor bidding online attempted to timely submit its bid but failed to get anything other than the cover sheet for its proposal in by the time responses to the solicitation were due. It was the next day before PMTech's bid was completely submitted, and it was thereafter rejected as untimely. PMTech protested to the Comptroller General, and the Comptroller General rejected PMTech's arguments. The Comptroller General's decision stated, in part: "We view it as an offeror's responsibility, when transmitting its proposal electronically, to insure the proposal's timely delivery by transmitting the proposal sufficiently in advance of the time set for receipt of proposals to allow for timely receipt by the agency."[71] In a decision similar to those involving late delivery of bids or late delivery of faxes, the Comptroller General concluded that PMTech took a certain amount of risk when it waited until only 13 minutes before the deadline for receipt of proposals

---

[69]FAR § 15.306(e).
[70]Comp. Gen. Dec. B-291082, 2002 CPD ¶ 172.
[71]*Id.*

to begin to attempt to transmit its proposal. Similarly, in *Sea Box, Inc.*,[72] the Comptroller General held that while the bid was received by the federal government's e-mail system server before the deadline, it was untimely because it failed to reach the specified e-mail box by the stated deadline. In the *Sea Box* decision, the Comptroller General pointed out that the protester sent its bid, consisting of seven e-mails, approximately 11 minutes before the stated deadline. The e-mails reached the agency's initial point of entry for e-mail before the deadline and were held for a period of time, then were sent to a virus-scanning server and subsequently arrived 7 to 24 minutes late at the e-mail box. The agency rejected the bid as late, and the Comptroller General agreed.

Even if the communication or e-mail is somehow lost in the government's e-mail system, the offeror appears to bear the risk of nondelivery and the rejection of its proposal by the agency as untimely. For example, in *Lakeshore Engineering Services,*[73] an amendment to the RFP by the Navy required the offerors to submit their offers/price proposals by e-mail. Lakeshore submitted its price proposal to the designated e-mail address provided in the amendment and requested an electronic delivery receipt. Even though Lakeshore's server generated a confirmed delivery status notification, the e-mailed price proposal was never received by the agency. Consequently, the Navy eliminated Lakeshore's proposal from the competitive range for being untimely. Based on the information provided by the Navy, the GAO concluded that the "delivery receipt" only confirmed that the message was successfully relayed from Lakeshore's system, not that it was received by the agency's e-mail system. There was no explanation on how or why the e-mail was lost in the government's system. In that context, the GAO ruled that the offeror bore the risk that its e-mail communication was delivered at the designated place at the proper time and denied Lakeshore's protest.

In *Tishman Construction Corp.,*[74] the prospective contractor submitted identical proposals electronically and on paper, as was required by the RFP for construction management services. The paper version of the proposal was received by the contracting activity approximately 73 minutes after the stated deadline while the electronic version was received approximately 50 minutes prior to the time for receipt of proposals. The agency rejected the proposal as untimely. The Comptroller General disagreed and sustained Tishman's protest, relying on a prior decision in which a bidder submitted its bid at two separate locations, as required by the solicitation, but only one was timely.[75] In that earlier decision, the Comptroller General found that the contractor did not obtain a competitive advantage as the agency had a complete copy of the bid in a timely manner, and no competitive advantage was obtained. Under those circumstances, the Comptroller General considered the late bid at the second location to be a minor informality.

---

[72]Comp. Gen. Dec. B-291056, 2002 CPD 181; *see also Symetrics Indus., LLC,* Comp. Gen. Dec. B-298759, 2006 CPD ¶ 154 (e-mailed proposal reached contracting officer's inbox one minute late although it was in the agency's server three minutes before the deadline. Protest against rejection of the proposal as untimely was denied).

[73]Comp. Gen. B-401434, 2009 CPD ¶ 155.

[74]Comp. Gen. Dec. B-292097, 2003 CPD ¶ 94.

[75]*ABT Assocs., Inc.,* Comp. Gen. Dec. B-226063, 87-1 CPD ¶ 513.

In *USA Information Systems Inc.*,[76] the Comptroller General denied a protest where the solicitation materials were available only on the Internet. The procuring agency had posted an amendment to the solicitation with a short response time and did not specifically advise the protester of the amendment. The Comptroller General found that the protester had not taken reasonable steps to be made aware of the amendment, such as registering for e-mail notification or checking the Internet site, and it was for those reasons that the protester had insufficient time to protest the terms of the solicitation.

3. Reverse Bid Auctions

Reverse bid auctions are a relatively new procurement technique for some government agencies. Reverse bid auctions are like bidding for items on eBay except that the bidder is, in effect, bidding against itself and the competition by offering lower bids in response to information on the Web site regarding its competitors' prices. In part, reverse bid auctions are an effort to take advantage of auction psychology. While the solicitation has a cutoff date and time for *bid submission,* the process allows for an extension of that time when a revised *bid price* is received. The process can be extremely risky for a contractor that makes winning the contract at all costs the highest priority. At the same time, the agency has no assurance that it will receive a bidder's most competitive price.

In *MTB Group, Inc.*,[77] the Comptroller General determined that an online reverse auction was a proper method of procurement by the Department of Housing and Urban Development. The contractor argued that the reverse auction was prohibited because it forced vendors to disclose their price information. However, the Comptroller General ruled that nothing in the FAR prohibited contractors from voluntarily disclosing their price information and that reverse bid auctions were permissible.[78]

Firms submitting bids or proposals electronically must adhere to the basic and long-standing rules regarding timeliness of the submission of bids or proposals, just as if they were submitting those same bids in a paper format.

Reverse bid auctions for procurement of construction services have been strongly criticized by construction trade associations. For example, the Associated General Contractors of America contend that reverse auctions do not guarantee the lowest price, may encourage imprudent bidding, and may contravene federal procurement laws.[79]

IV. BID GUARANTEES (BONDS)

A. Requirement for Bid Guarantee

In government contracting, the FAR refers to a bid bond as a *bid guarantee* and directs the contracting officer to require a bid guarantee whenever the agency requires

---

[76]Comp. Gen. Dec. B-291488, 2002 CPD ¶ 205.

[77]Comp. Gen. Dec. B-295463, 2005 CPD ¶ 40.

[78]Even if an agency indicates a reluctance to use reverse bid auctions for a construction procurement as a matter of basic policy, a procuring activity may seek to issue the solicitation as a *commercial item* procurement under FAR Part 12 and use a reverse bid auction.

[79]*See* Associated General Contractors of America White Paper on Reverse Auctions for Procurement of Construction, *www.agc.org/cs/advocacy/legislative_activity/Reverse_Auctions* (accessed November 4, 2009).

the contractor to provide payment and performance bonds.[80] Miller Act payment and performance bonds are required on virtually all construction contracts in excess of $100,000.[81] FAR § 28.101-2(a) provides that the Bid Guarantee clause at FAR § 52.228-1 shall be included in all solicitations (IFBs or RFPs). That clause provides:

(a) Failure to furnish a bid guarantee in the proper form and amount, by the time set for opening of bids, may be cause for rejection of the bid.

(b) The bidder shall furnish a bid guarantee in the form of a firm commitment, e.g., bid bond supported by good and sufficient surety or sureties acceptable to the Government, postal money order, certified check, cashier's check, irrevocable letter of credit, or, under Treasury Department regulations, certain bonds or notes of the United States. The Contracting Officer will return bid guarantees, other than bid bonds, (1) to unsuccessful bidders as soon as practicable after the opening of bids, and (2) to the successful bidder upon execution of contractual documents and bonds (including any necessary coinsurance or reinsurance agreements), as required by the bid as accepted.

A valid bid bond is a material requirement of the solicitation and furnishing a defective bid bond is grounds for rejection of a bid as nonresponsive.[82] Although this section of the FAR and the clause refers to *bid* and *bidder*, the requirement applies to negotiated contracts as long as there is a requirement for payment and performance bonds. Under this provision, the guarantee bond covers the offeror/bidder until the point the offer/bid is rejected or until the firm that is awarded the contract executes the contract and provides the required Miller Act payment and performance bonds.

The amount of the bid guarantee is set by the government agency either as a percentage of the proposal or bid price or as a stated dollar amount. Typically, the percentage is set at 10 percent; however, other percentages may be used in different circumstances and by different agencies.

B. Purpose for Guarantee and Surety's Liability

In sealed bid procurements, one practical purpose for the bid bond is to keep the bidder from withdrawing its bid during the bid acceptance period specified by the government in Section 17 of Standard Form 1442, In a sealed bid procurement, the bid acceptance period is a material requirement affecting bid responsiveness.[83] The surety that underwrites the bid bond will be liable to the government if the bidder withdraws its bid before the expiration of the bid acceptance period, fails to enter into the contract, or fails to provide the required payment or performance bonds.[84] The bid bond also protects the government's bidding process by ensuring that bidders have a financial stake in the process and are not able to submit bids without consequence.

---

[80]FAR § 28.101-1.

[81]FAR § 28.102-1. (*See* **Chapter 12** for a discussion of Miller Act payment and performance bonds.)

[82]*See Sundt Corp.*, Comp. Gen. Dec. B-274203, 96-2 CPD ¶ 171; *General Elevator Co., Inc.*, Comp. Gen. Dec. B-226976, 87-1 CPD ¶ 385.

[83]*See Perkin Elmer Corp.*, Comp. Gen. Dec. B-236175, 89-2 CPD ¶ 352.

[84]*See Aeroplate Corp. v. United States*, 67 Fed. Cl. 4, 12 (2005).

The fundamental purpose of the bid guarantee is explained by FAR § 52.228-1, which states:

> (d) If the successful bidder, upon acceptance of its bid by the Government within the period specified for acceptance, fails to execute all contractual documents or furnish executed bond(s) within 10 days after receipt of the forms by the bidder, the Contracting Officer may terminate the contract for default.

After this 10-day period has expired, the government may proceed against the bid bond surety to recover the costs associated with the offeror's/bidder's nonperformance (failure to execute the contract and/or provide the required Miller Act bonds). The surety will be liable to the government to the extent that the government suffered damage resulting from the contractor's failure to execute the contract and/or furnish the payment and performance bonds. This damage is generally the additional cost associated with having to make an award to another firm at a higher price. Basically, the surety's obligation is limited to the difference between its contractor principal's offer/bid and the next lowest acceptable price subject to a cap of the total amount of the bond.[85]

In negotiated procurements, FAR § 15.208(e) provides that an offeror is free to withdraw its offer by written notice at any time prior to award. Consequently, the purpose of the bond or guarantee in negotiated procurements is directly related to the offeror's fundamental obligations to execute the contract and furnish the required Miller Act bonds following receipt of a notice of award.

C.   Surety Defenses

A surety has several possible defenses to the government's bid bond claim when a bidder fails or refuses to execute the government contract and/or provide the bonds required by the bid solicitation: (1) a material change in the scope of the contract has occurred after bid submission;[86] (2) a delay in awarding the contract beyond the time set forth in the solicitation has occurred;[87] (3) a material bid mistake has occurred and it would be unconscionable to enforce the bid because the mistake did not occur as a result of negligence, the mistake was one of fact as opposed to judgment, and the government will not be prejudiced;[88] and (4) a failure by the government to mitigate its damages.[89] However, a failure to mitigate the damages defense requires a showing that the government was arbitrary and capricious in its reprocurement actions.[90]

---

[85] *See Peerless Ins. Co. v. United States,* 674 F. Supp. 1202 (E.D. Va. 1987). (In theory, the bidder/offeror's liability extends to the full amount of the excess cost of reprocurement. *See* FAR § 52.228-1(e).)

[86] *See City of Devils Lake v. St. Paul Fire & Marine Ins. Co.,* 497 F. Supp. 595, 597-8 (D. ND 1980).

[87] *See Hennepin Water Dist. v. Petersen Const. Co.,* 297 N.E.2d 131, 134-5 (Ill. 1973).

[88] *See James Cape & Sons Co. v. Mulcahy,* 700 N.W. 2d 243, 256 (Wis. 2005).

[89] *See Peerless Ins. Co. v. United States,* 674 F. Supp. 1202 (E.D. Va. 1987).

[90] *Id.*

V.   RESPONSIVE BIDS AND PROPOSALS

A sealed bid on a government contract usually will not be considered by an agency unless the bid is "responsive," which means that it complies with all material requirements of the solicitation. Responsiveness differs from responsibility because responsiveness focuses on whether the bid, as submitted, is an offer to perform the exact tasks spelled out in the IFB and whether acceptance of that bid will bind the contractor to perform in strict conformance with the invitation.[91]

Failure of a contractor to comply carefully with all the requirements for competitive bidding established by the solicitation may result in the bid being declared "nonresponsive" or, if an award has been made, may render the contract void or prevent the contractor from receiving full payment for work performed. In determining the responsiveness of bids, the bidder's intent to be bound must be clearly ascertainable from the face of the bid.[92] However, a bid's deviation from the terms of the solicitation must be *material* to render a bid nonresponsive.[93]

A deviation is considered *material* if it gives a bidder a substantial competitive advantage that prevents other bidders from competing equally. A deviation is also material if it goes to the substance of the bid or prejudices other bidders. The deviation goes to the substance of the bid if it affects price, quantity, quality, or delivery of the items offered.[94] A contractor submitting a sealed bid for a government project must take the contract as presented. Thus, any qualification of a bid that limits or changes one or more of the terms of the proposed contract subjects the contractor to the risk of being deemed nonresponsive. For example, in *Lift Power, Inc.*,[95] a contractor was found to be nonresponsive where it reserved the right in the bid to change its price if costs should increase.[96] Obviously, such a qualification, if accepted by the government, could have given the contractor an unfair price advantage over other bidders.[97]

Including reservations or conditions in a bid generally renders the bid nonresponsive. According to the FAR, a bid is nonresponsive if the bidder includes a condition that:

(1)   Protects against future changes in conditions, such as increased costs, if the total possible cost to the government cannot be determined;

(2)   Fails to state a price and indicates that price shall be price in effect at time of delivery;

(3)   States a price but qualifies it as being subject to price in effect at time of delivery;

(4)   When not authorized by the invitation, conditions or qualifies a bid by stipulating that it is to be considered only if, before date of award, the bidder receives (or does not receive) an award under a separate solicitation;

[91]*Prestex, Inc. v. United States*, 320 F.2d 367 (Ct. Cl. 1963).
[92]*Jarke Corp.*, Comp. Gen. Dec. B-231858, 88-2 CPD ¶ 82.
[93]A true *minor* informality may be cured or waived. *See* FAR § 14.405; *Bilt-Rite Contractors, Inc.*, Comp. Gen. B-259106.2, 95-1 CPD ¶ 220.
[94]FAR § 14.404-2(d).
[95]Comp. Gen. Dec. B-182604, 75-1 CPD ¶ 13.
[96]*See Kipp Constr. Co.*, Comp. Gen. Dec. B-181588, 75-1 CPD ¶ 20.
[97]*See Chemtech Indus., Inc.*, B-186652, 76-2 CPD ¶ 274.

(5)   Requires that the government determine that the bidder's proposed product or service meets applicable government specifications; or

(6)   Limits rights of the government under any contract clause.[98]

(7)   Fails to unambiguously express a commitment to the minimum bid acceptance period.[99]

The majority of the rules concerning bid responsiveness are aimed at preventing a contractor from having "two bites at the apple." In other words, the concept of bid responsiveness is used to guard against the low bidder having the opportunity, after bids are opened and all prices are revealed, to accept or reject an award based on some contingency that the bidder created for itself and that only applies to, and works to the advantage of, that bidder.

The prohibition against a bidder having two bites at the apple also applies when a defect in the bid or an ambiguity in a solicitation subjects the intended bid to differing interpretations. For example, in *Caprock Vermeer Equipment, Inc.*,[100] a bidder for an equipment supply contract included descriptive literature in its bid upon which the contractor wrote "optional." The literature indicated that the product to be supplied by the bidder deviated from the requirements in the solicitation. The Comptroller General found that there were two possible interpretations of the bid, at least one of which rendered the bid nonresponsive. The Comptroller General, therefore, upheld the government's rejection of the bid.

A bid is nonresponsive if the bidder attempts to make the bid contingent upon some act or event. In *Hewlett Packard*,[101] the Comptroller General found a bid to be nonresponsive where the bidder sent a transmittal letter stating that the bid was contingent upon the removal of a contract clause. The Comptroller General found that the contingency rendered the bid nonresponsive because the bidder sought to change the terms of the contract to the sole advantage of the bidder.

A bid will also be considered nonresponsive if the bidder deviates from the bidding requirements by failing to acknowledge addenda, particularly where the addenda contain a statutorily required provision.[102] Also, an oral, rather than written, acknowledgment of an amendment is unacceptable.[103] However, failure to acknowledge an addendum that has only a negligible effect on contract performance and does not affect contract price may be waived.[104]

According to the Comptroller General, nonresponsive bids also include bids that fail to acknowledge an amendment that would impose a new legal obligation (even if it would have no effect on price)[105] and that fail to certify that a small business product will be provided.[106]

---

[98]FAR § 14.404-2(d).
[99]*Sundt Corp.*, Comp. Gen. Dec. B-274203, 96-2 CPD ¶ 171.
[100]Comp. Gen. Dec. B-217088, 85-2 CPD ¶ 259.
[101]Comp. Gen. Dec. B-216530, 85-1 CPD ¶ 193.
[102]*Grade-Way Constr. Co. v. United States*, 7 Cl. Ct. 263 (1985).
[103]*Alcon, Inc.*, Comp. Gen. Dec. B-228409, 88-1 CPD ¶ 114.
[104]FAR § 14.405(d)(2).
[105]*American Sein-Pro*, Comp. Gen. Dec. B-231823, 88-2 CPD ¶ 209.
[106]*Delta Concepts, Inc.*, Comp. Gen. Dec. B-230632, 88-2 CPD ¶ 43.

A determination that a bid was nonresponsive because it was materially unbalanced was upheld by the Comptroller General even though the bidder contended the "unbalancing" resulted from allocated technical evaluation and preproduction costs of the first articles produced under the contract, which were approximately 15 times greater than the unit prices of the remaining production quantity.[107] An "unbalanced bid" is one that allocates a disproportionate part of the contract price to a line item for which payment will be made early in contract performance—thus, in effect, constituting an advance payment to the contractor. The Comptroller General said that where the costs incurred to produce the first articles are a necessary investment in the production quantity, the costs should be amortized over the total contract rather than allocated solely to the first articles. The reason for rejecting "front-loaded" bids is that the greatly enhanced first article prices provide funds to a firm in the early period of contract performance and are, in essence, an interest-free loan to which a contractor is not entitled.[108]

Rejection of a bid for nonresponsiveness may also be proper when the principal on the bid bond submitted by the bidder is not the same legal entity as the offeror on the bid form. Generally, a surety can be obligated on a bid bond only if the principal named in the bond fails to execute the contract. The refusal of a nonbidding entity to contract with the awarding authority does not result in a forfeiture of the bid bond. Defective bid bonds constitute a substantial deviation from the solicitation and ordinarily require rejection of the bid as nonresponsive because such bonds do not protect the public from contractors that refuse to enter into contracts with the government.[109]

Many deviations such as those just discussed may be considered "material"; however, minor (or *de minimus*) irregularities may be waived by the awarding authority.[110] This long-established policy permitting waiver of minor irregularities or informalities preserves the focus of competitive bidding on lowest price by discouraging questions over matters not affecting the substance of the bid.[111]

The basic rule observed in connection with minor irregularities is that the defect or variation in the bid must have negligible significance when contrasted with the total cost or scope of the invitation for bids. Deviations affecting price, quantity, quality, delivery, or completion are generally material and merit especially stringent standards to protect against any bidder obtaining a competitive advantage.[112] For an irregularity in a bid to be waived, it must be so inconsequential or immaterial that the bidder does not gain a competitive advantage after all bids have been examined. Thus, a minor irregularity may be found where the bidder fails to initial a price change in its bid before bid opening;[113] fails to mark its bid envelope with the

---

[107]*M.C. General, Inc.,* Comp. Gen. Dec. B-228334, 87-2 CPD ¶ 572.

[108]*Fidelity Technologies Corp.,* Comp. Gen. Dec. B-232340, 88-2 CPD ¶ 511.

[109]*See Yank Waste Co.,* Comp. Gen. Dec. B-180418, 74-1 CPD ¶ 190; *but see* Comp. Gen. Dec. B-178684 (Dec. 26, 1973).

[110]51 Comp. Gen. 62 (1971).

[111]41 Comp. Gen. 721 (1962).

[112]*Prestex, Inc. v. United States,* 320 F.2d 367 (Ct. Cl. 1963); FAR § 14.405.

[113]Comp. Gen. Dec. B-211870, 83-2 CPD ¶ 243.

solicitation number, date, and time of bid opening;[114] or fails to provide incidental information requested by the invitation.[115]

The determination of what constitutes a minor irregularity generally is left to the discretion of the contracting officer,[116] and courts allow contracting officers substantial discretion in determining what constitutes a minor irregularity.

A number of Comptroller General decisions have helped to define what are minor irregularites, and they include:

(1) The omission of unit prices under circumstances where they could be calculated by dividing total prices by estimated quantities.[117]

(2) The insertion of the wrong solicitation number on a bid bond.[118]

(3) The omission of a principal's signature on a bid bond when the bond is submitted with a signed bid.[119]

(4) An ambiguous bid price if the bid is low under all reasonable interpretations.[120]

(5) A failure to include required information on affiliates.[121]

(6) A failure to acknowledge an amendment to the solicitation that would not have a material impact on price[122] or only a trivial impact on price.[123]

(7) A failure to acknowledge an amendment reducing the quantity of items to be ordered where the amendment imposed no obligations not already in the original invitation and had no impact on the bid price.[124]

(8) A failure to provide equipment description information when the solicitation did not make it clear a failure would result in bid rejection.[125]

It is not uncommon for government representatives and construction industry professionals to use the term "responsiveness" in negotiated procurements. However, in a strict sense, the concept of *responsiveness* generally does not apply to negotiated procurements.[126] The give-and-take of negotiated procurements allows agencies to use flexible procedures not available in sealed bids. The government is not required to reject an offer that varies from the requirements of the RFP, but it may take various steps to bring the proposal into compliance while preserving fair competition

---

[114]Comp. Gen. Dec. B-210251, 83-1 CPD ¶ 87.

[115]Comp. Gen. Dec. B-215162, 84-2 CPD ¶ 413.

[116]*Excavation Constr., Inc. v. United States*, 494 F.2d 1289 (Ct. Cl. 1974).

[117]*GEM Eng'g Co.*, Comp. Gen. Dec. B-231605.2, 88-2 CPD ¶ 252.

[118]*Kirila Contractors, Inc.*, Comp. Gen. Dec. B-230731, 88-1 CPD ¶ 554.

[119]*P-B Eng'g Co.*, Comp. Gen. Dec. B-229739, 88-1 CPD ¶ 71.

[120]*NJS Dev. Corp.*, Comp. Gen. Dec. B-230871, 88-2 CPD ¶ 62.

[121]*A & C Bldg. & Indus. Maint. Corp.*, Comp. Gen. Dec. B-229931, 88-1 CPD ¶ 309.

[122]*Adak Communications Sys., Inc.*, Comp. Gen. Dec. B-228341, 88-1 CPD ¶ 74.

[123]*Star Brite Constr. Co.*, Comp. Gen. Dec. B-228522, 88-1 CPD ¶ 17 ($2,000 out of a $118,000 difference between low and second low bid).

[124]*Automated Datatron, Inc.*, Comp. Gen. Dec. B-231411, 88-2 CPD ¶ 137.

[125]*Houston Helicopters, Inc.*, Comp. Gen. Dec. B-231122, 88-2 CPD ¶ 149.

[126]*ManTech Telecomms. and Info. Sys. Corp. v. United States*, 49 Fed. Cl. 57 (2001).

among all offerors.[127] The FAR's flexible negotiation procedures can allow a firm in the competitive range to cure any RFP exceptions or deviations through discussions and the submission of best and final offers.[128] However, if the proposal substantially deviates from the requirements of the RFP, the offeror risks a determination that its proposal is not in the competitive range[129] or that the agency will make an award without discussion to a competitor whose proposal is deemed by the agency to be more *responsive* to the agency's requirements.[130]

VI.   THE EVALUATION PROCESS

Once the government receives all proposals or bids in response to a RFP or an IFB, the government evaluates the proposals or responsive bids in accordance with the evaluation criteria established by the RFP or the IFB. In general, the evaluation process requires the government to evaluate each responsive bid or proposal received against the stated evaluation factor(s) and then rank each proposal or bid in relation to those of all other responsive offerors. At the conclusion of this analysis and ranking process, the government selects the best-value proposal or the lowest evaluated bid for the contract award.

The evaluation process is the culmination of a significant amount of planning and effort on the part of both the offerors and the government, and the process concludes with the ultimate contract award decision to one of the offerors. As such, the evaluation process is subject to a great deal of scrutiny by offerors and the government alike and generates many disputes and protests from disappointed offerors. It is therefore imperative that contractors have a fundamental understanding of the evaluation process in order to protect their interests during the procurement process and to ensure that they are in the best possible position to compete for a contract award.

A.   Sealed Bid Evaluation Process

The bid evaluation process in a sealed bidding procurement must be performed based solely on the factors identified in the IFB to ensure that bidders are able to compete for the contract award on a fair and equal basis.[131] Only price and *other price-related factors* may be considered by the government, and the award of the contract must be made without discussions to the responsible bidder whose bid is most advantageous to the government.[132] This evaluation process can vary from the simple to the more complex based on the price-related evaluation factors set forth in the IFB.

Under a sealed bid procurement, the basic bid evaluation process for the government is generally quite simple since a bidder must submit the lowest responsive bid price based on the work to be performed under the terms of the IFB to be eligible

---

[127]*Id.*
[128]FAR § 15.306(d); *see also Hollingsead Int'l,* Comp. Gen. Dec. B-227853, 87-2 CPD ¶ 372.
[129]*See* FAR §§ 15.305, 15.306(b).
[130]*See* FAR § 15.306(a).
[131]*Alliance Properties, LLC,* 61 Comp. Gen. 48, 81-2 CPD ¶ 357.
[132]10 U.S.C. § 2305(b)(3); 41 U.S.C. § 253b(c).

for an award.[133] However, the evaluation process can become more complex depending on the other price-related factors considered by the agency. If identified in the IFB,[134] the FAR authorizes consideration of these factors: foreseeable costs or delays to the government resulting from transportation costs to deliver goods or supplies to the government, the location of goods and supplies, differences in inspection of goods or supplies, life cycle costs, federal, state or local tax costs, and applicable laws or regulations such as the Buy American Act.[135]

Contractors must carefully analyze the bid evaluation method and the other price-related factors to be evaluated by the government during a sealed bidding evaluation process and structure their bids accordingly. Contractors have no opportunity to clarify any deficiencies in their bids in a sealed bidding procurement and must submit the best and most accurate bid possible to ensure that they maximize their chances of receiving the contract award.

B. Competitive Negotiation Evaluation Process

The evaluation of proposals during a negotiated procurement is a multitiered process that begins with the submission of proposals by the offerors, continues through negotiations between the government and the offerors, and concludes with the government's evaluation of best and final offers and the selection of a contract awardee. This process can take as little as a few days for simple small-value procurements or it may extend over many months, or even longer, for complex and costly procurements. The purpose of this lengthy evaluation process is to enable the government to evaluate the relevant strengths and weaknesses of each proposal and to select the contractor capable of providing the best value to the agency.

All agency evaluations must be conducted in accordance with the criteria designated by the RFP. Although an agency has significant discretion in establishing the scope and parameters of the evaluation process, once the evaluation process is set in the RFP, the agency must strictly follow it.[136] Evaluations that do not comply with the material terms of the RFP will be set aside by the federal courts and the GAO. Likewise, the federal courts and the GAO must find unlawful and set aside a procurement conducted in a manner that is "arbitrary, capricious, an abuse of discretion, or otherwise not in accordance with law."[137]

The FAR does not prescribe any specific methodology for agencies to use to evaluate proposals during a negotiated procurement, only that the evaluation process be conducted fairly and impartially.[138] As such, agencies have a tremendous amount of discretion in determining the evaluation factors to be considered and the methodology to be used in evaluating offerors with regard to those evaluation factors.[139] An agency

---

[133]*Alliance Properties, LLC*, 61 Comp. Gen. 48, 81-2 CPD ¶ 357.

[134]FAR § 14.201-5(c).

[135]FAR § 14.201-8.

[136]10 U.S.C. § 2305(b); 41 U.S.C. § 253(b); FAR § 15.305(a).

[137]*PGBA, LLC v. United States*, 389 F.3d 1219, 1226 (Fed. Cir. 2004) (citing 5 U.S.C. § 706(2)(A)).

[138]FAR § 15.306(a).

[139]*Alion Science and Tech. Corp. v. United States*, 74 Fed. Cl. 372, 374 (2006); *see also Matrix Int'l Logistics, Inc.*, Comp. Gen. Dec. B-272388.2, 97-2 CPD ¶ 89.

may consider many different evaluation factors when conducting a best-value procurement to determine which proposal will be the most advantageous for the government. Evaluation factors must represent the key areas of importance to be considered in the source selection decision and allow for a meaningful comparison of competing proposals.[140] An agency may consider cost or price data during a best-value procurement as well as a contractor's past performance, experience, management and key personnel, and technical expertise to determine the best value to the government. Therefore, a prospective offeror needs to carefully study the stated evaluation criteria and structure its proposal to meet the agency's requirements and the criteria for the evaluation of the proposals. As a general rule, it is not prudent to ignore any requirement or criteria as unnecessary or unimportant.

The analysis of the relevant strengths and weaknesses of the proposals submitted to the government is the heart of the evaluation process and is the foundation that supports the contract award decision. Agencies must rate or rank proposals in accordance with a methodology established by the RFP. Agencies are authorized to use "any rating method or combination of methods, including color or adjectival ratings, numerical weights, and ordinal rankings" to evaluate proposals.[141] Agencies are required to document in writing the relative strengths, weaknesses, and risks of each proposal and maintain this documentation in the contract file.[142] This documentation generally is prepared in a narrative form by the agency's personnel as they simultaneously rate or rank each proposal.

The two most common evaluation factors to be considered in a negotiated procurement are price/cost to the government and the past performance of the offeror. Just as in sealed bidding, price or cost to the government must be considered by an agency in all negotiated procurement evaluations.[143] The government must ensure that the prices of supplies and services proposed by offerors are both "fair and reasonable" based on the cost evaluation methods established by the RFP.[144] It is important to note that the lowest-priced proposal does not necessarily guarantee an offeror the contract award. Negotiated procurements generally seek to obtain the best value for the government, and the lowest-priced proposal does not always equate with the best value due to differences in respective contractors' past performance, proposed plan of performance, technical expertise, key personnel, and so on.

C. Evaluation of Past Performance

In recent years, past performance has become a major evaluation factor commonly used in negotiated procurements to evaluate contractors competing for a new award. The past performance determinations made by agencies during the proposal evaluation process have generated numerous challenges and protests by contractors that generally

---

[140]FAR § 15.304(b).
[141]FAR § 15.305(a).
[142]*Id.*
[143]FAR § 15.304(b).
[144]FAR § 15.402(a).

claim that the government has either erroneously devalued their past performance or erroneously upgraded the past performance of a competing contractor.

The evolution of past performance as a major evaluation factor began taking shape in 1994 when Congress passed the Federal Acquisition Streamlining Act.[145] Congress acknowledged in FASA that a contractor's past performance should be evaluated during a procurement to ascertain whether that contractor should receive future work. Section 1091 of FASA provides that "past contract performance of an offeror is one of the relevant factors that a contracting official of an executive agency should consider in awarding a contract." The government believes that the use of performance evaluations will sufficiently motivate contractors to perform at the highest level or, to the extent they are not, to improve their performance before they are rated by a procuring agency or department.

1. Purpose of Past Performance Evaluations

A contractor's past performance evaluations are a significant factor in an agency's determination of a best-value award. Procuring agencies believe that they are better able to predict the quality of a contractor's performance and the associated customer's satisfaction by analyzing the contractor's performance on previous projects. The theory behind the evaluation of contractor's performance and its use in future procurements is twofold. First, Congress believes that an active dialogue between the contractor and the government, while a contract is being performed, will result in better performance by the contractor. This active dialogue is encouraged by the knowledge of both the contractor and the government that a contractor's performance during the project will be rated by the government for use in future procurements. Second, contractors will presumably provide their best efforts in the performance of a contract if they are aware that their performance evaluations will be used by other procuring activities on future procurements. This incentive for contractors to perform up to the highest standards theoretically creates a larger pool of high-quality contractors that will be available to procuring activities for new contracts.

The government must be careful to limit the scope and content of performance evaluations to matters related to a contractor's actual contract performance. In May 2000, the Office of Federal Procurement Policy issued a government-wide memorandum providing that contractors could not be given "downgraded" past performance evaluations for filing protests and claims or for deciding not to use alternative dispute resolution (ADR). That same memorandum provided that contractors could not be given inflated performance evaluations for refraining from filing protests and claims or agreeing to use ADR.[146]

In *Nova Group Inc.*,[147] the contractor's past performance rating was downgraded because it had pursued contract claims against the government on nine occasions over 15 years. The GAO found no evidence suggesting the claims were indicative

---

[145]Pub. L. 103-355 (Oct. 13, 1994).
[146]*Best Practices for Collecting and Using Current and Past Performance Information*, Office of Federal Procurement Policy, Office of Management & Budget, Executive Office of the President, May 2000.
[147]B-282947, 99-2 CPD ¶ 56.

of poor performance, frivolous, or filed in bad faith, and therefore held that the contracting officer could not use legitimate contractor claims to downgrade a contractor's proposal with regard to past performance.

Several decisions from the Comptroller General have addressed the relevance and use of past performance information in current procurements. In *C. Lawrence Construction Co., Inc.*,[148] the GAO found that a Corps of Engineers' requirement to provide at least five past performance references for relevant contracts, defined in the RFP as similar construction contracts in the $5 to $10 million range, was not unduly restrictive and was consistent with the Army's position that evaluating at least five projects provided a "comfort zone" with respect to a prospective contractor's overall performance trends.

The Comptroller General has also ruled that it is reasonable for an agency to consider the *specific* past performance of a contractor to which award was being made more favorably than the *general* past performance of the protesting contractor.[149] Similarly, agencies have the right to look at the contractor's performance of all past contracts, regardless of size or scope, because that information enables the agency to predict whether the contractor will satisfactorily perform the requirements of the new contract.[150] The smaller dollar value or scope of work of a previous contract does not render the contract performance irrelevant for past performance evaluation purposes.[151] Agencies may also consider negative past performance comments even though those comments had not been documented contemporaneously with the past performance.[152]

2. *Reviewing Past Performance Evaluations during Source Selection*

As part of the overall evaluation process for a procurement, the procuring agency should confirm the accuracy of any prospective contractor's past contract information and assign a performance risk rating. Final past performance ratings may be reflected in a color, adjectival, number, or some other rating system, depending on the particular agency policy for ranking offerors or the rating scheme established by the RFP. The presence and severity of problems in a prospective contractor's overall work record should be considered as part of the performance risk assessment, as should the demonstrated effectiveness of any corrective action taken by the contractor. When a prospective contractor's past performance reflects problems, the procuring agency must evaluate the extent to which the government played a part in that poor performance. Naturally, the procuring agency should look for the areas of performance that are most critical to the procurement being sought.

---

[148]*C. Lawrence Constr. Co., Inc.*, Comp. Gen. Dec. B-289341, 2002 CPD ¶ 17.
[149]*M&W Constr. Corp.*, Comp. Gen. Dec. B-288649.2, 2002 CPD ¶ 30.
[150]*The Standard Register Co.*, Comp. Gen. Dec. B-289579, 2002 CPD ¶ 54.
[151]*Dan River, Inc.*, Comp. Gen. Dec. B-289613, 2002 CPD ¶ 80.
[152]*Kathpal Technologies, Inc.*, Comp. Gen. Dec. B-291637.2, 2003 CPD ¶ 6.

The Office of Federal Procurement Policy has provided this guidance to procuring agencies drafting past performance evaluation criteria:

(1) *Use Past Performance as a Distinct Factor.* The past performance factor should be distinct and identifiable in order to reduce the chances of its impact being lost within other factors and to ease the evaluation process. However, if integrating past performance with other non-cost/price factors provides a more meaningful picture, each agency should use its own discretion. The key is to not dilute the importance or impact of past performance when determining the best value contractor.

(2) *Choose Past Performance Subfactors Wisely.* Tailor the subfactors to match the requirement and to capture the key performance criteria in the statement of work. Carefully consider whether subfactors add value to the overall assessment, warrant the additional time to evaluate and enhance the discrimination among the competing proposals.

(a) *Quality of Product or Service.* The offeror will be evaluated on compliance with previous contract requirements, accuracy of reports, and technical excellence to include quality awards/certificates.

(b) *Timeliness of Performance.* The offeror will be evaluated on meeting milestones, reliability, responsiveness to technical direction, deliverables completed on-time, and adherence to contract schedules including contract administration.

(c) *Cost Control.* The offeror will be evaluated on the ability to perform within or below budget, use of cost efficiencies, relationship of negotiated costs to actuals, submission of reasonably priced change proposals, and providing current, accurate, and complete billing timely.

(d) *Business Relations.* The offeror will be evaluated on the ability to provide effective management, meet subcontractor and SDB [Small Disadvantaged Business] goals, cooperative and proactive behavior with the technical representative(s) and Contracting Officer, flexibility, responsiveness to inquiries, problem resolution and customer satisfaction. The offeror will be evaluated on satisfaction of the technical monitors with the overall performance, and final product and services. Evaluation of past performance will be based on consideration of all relevant facts and circumstances. It will include a determination of the offeror's commitment to customer satisfaction and will include conclusions of informed judgment. However, the basis for the conclusions of judgment should be substantially documented.

(3) *Subcontractor, and Teaming, and Joint Venture Partner's Past Performance.* For the purpose of evaluation of past performance information, offerors shall be defined as business arrangements and relationships such as joint ventures, teaming partners, and major subcontractors. Each firm in the business arrangement will be evaluated on its performance under existing and prior contracts for similar products or services.[153]

---

[153]*Best Practices for Collecting and Using Current and Past Performance Information*, Office of Federal Procurement Policy, Office of Management & Budget, Executive Office of the President, May 2000, pp. 19-20 (available at *www.whitehouse.gov/omb/rewrite/procurement/contract_perf/best_practice_re_past_perf.html* (accessed July 27, 2009)).

Procuring agencies using the guidance just set forth should first determine how well a prospective contractor has performed on past contracts and how relevant that performance is to the new procurement. That past performance rating is then used along with other rated evaluation factors, such as cost or price and technical expertise, in a comparative assessment to ascertain the most highly rated offeror for contract award.

In general, a procuring agency is given the "greatest deference possible" with regard to its evaluation and rating of an offeror's past performance during the source selection process.[154] Agencies are able to revise its past performance rating of an offeror numerous times during the source selection process before it comes up with a final rating.[155] These past performance ratings generally will not be disturbed by the courts or the GAO in a bid protest unless the ratings are unreasonable or inconsistent with the terms of a solicitation.[156]

It is not unusual for a government agency to request that the offeror identify key anticipated subcontractors in its proposal. That request may indicate that the agency will consider past performance evaluation information on these subcontractors. In that regard, FAR § 15.305(a)(2)(iii) provides, in part:

> The [proposal] evaluation should take into account past performance information regarding . . . subcontractors that will perform major or critical aspects of the requirement when such information is relevant to the instant acquisition.

The RFP should advise the prospective offerors of the extent to which the agency plans to consider past performance information on anticipated subcontractors. In *Singleton Enterprises*,[157] the GAO held that a solicitation contained a latent ambiguity when it stated that the past performance of the "offeror" would be considered. Given the language in FAR § 15.305(a)(2)(iii), the GAO ruled that the agency's refusal to consider subcontractor past performance was improper absent a clear statement in the RFP of the agency's intent.

When reviewing an RFP, prospective offerors should attempt to determine whether the past performance of its prospective subcontractors[158] will be considered by the agency. If there is a request for identification of anticipated subcontractors or a design firm in a design-build proposal, an offeror should obtain clarification on that issue *before* submitting the proposal. If the agency indicates that subcontractor or design firm past performance information will be considered, it is imperative that the contractor have all of the key members of its anticipated proposal team share their respective past performance evaluations before the contractor finally sets its team.[159]

---

[154]*Fort Carson Support Svcs. v. United States*, 71 Fed. Cl. 571, 598 (2006).
[155]*Id.*
[156]*Consolidated Eng'g Servs. v. United States,* 64 Fed. Cl. 617, 637 (2005).
[157]B-298576, 2006 CPD ¶ 157.
[158]If the RFP seeks a design-build proposal, a design firm's past performance evaluations are likely to be very important, if not critical. The government agencies rate and maintain a past performance database on design firms as well as contractors.
[159]A firm's past performance evaluations are available to it and to government agencies. Consequently, a contractor cannot obtain independent access to the reports and evaluations on potential subcontractors and/or design firms.

When the government communicates with contractors to establish a competitive range after the receipt of proposals, it must hold discussions with contractors whose past performance evaluation prevents the contractor from being placed in the competitive range if the contractor has not had a prior opportunity to respond to the adverse information.[160] In a negotiated procurement where the contract award will be made without discussions, a contractor *may* be given the opportunity to address any deficiencies in its past performance record if it has not previously had the opportunity to respond.[161] Regardless of whether a contractor has the opportunity to negotiate with the government regarding its past performance record or not, contractors should make efforts to preemptively explain to the government any deficiencies in their past performance record before the evaluation process begins. Hoping the government will not discover any adverse past performance information is not the prudent path to follow because the government will likely discover any deficient past performance of an offeror during the evaluation process. Preemptively explaining adverse performance prior to the evaluation allows a contractor to mitigate any negative inference the contracting officer or source selection team will draw from the adverse past performance information.

3. Preparing and Maintaining Contractor Performance Ratings

Agencies generally obtain past performance information on offerors in two ways. First, agencies obtain past performance information directly from the offerors or from references submitted by the offerors. This past performance information is generally required to be provided under the terms of the RFP and comes in the form of performance evaluation forms or surveys, narrative descriptions of the contractor's performance, or letters from a contractor's references describing the contractor's past performance.

Second, agencies obtain past performance information from data collection systems that are regulated by FAR Subpart 42.15. In general, FAR § 42.1502(a) requires agencies to prepare performance evaluations on all contractors that perform contracts in excess of the simplified acquisition threshold as defined by FAR § 2.101 (contracts over $100,000 in value). The content and format of these evaluations are left up to the discretion of the agency but must be tailored to the size, content, and complexity of the contract being performed.[162] Agencies are required to prepare a contractor performance evaluation when the contract is complete and also are encouraged to prepare interim performance evaluations for contracts that exceed one year in length, including option periods.[163]

Agencies must share these performance evaluations with other agencies as needed to support future award decisions.[164] Agencies are authorized to use automated or web-based past performance information systems to maintain contractors'

---

[160]FAR § 15.306(b)(1)(i).
[161]FAR § 15.306(a)(2).
[162]FAR § 42.1502(a).
[163]*Id.*
[164]FAR § 42.1503(c).

performance information and evaluations as long as those systems have adequate safeguards to protect the confidential nature of that information.[165] Two of the widely used agency databases in government construction contracting is the Construction Contractor Appraisal Support System (CCASS) used by the Department of Defense and the Contractor Performance System (CPS) maintained by the National Institute of Health (NIH).[166] Both systems are part of the Past Performance Information Retrieval System (PPIRS) available to all federal agencies.

CCASS is an example of a web-based past performance information system that is used to maintain contractors' performance information and evaluations. CCASS is used by the Department of Defense (DOD) to track and store the performance evaluations of construction contractors providing services to DOD agencies. DOD regulations require that all construction contractor evaluations be completed on DD Form 2626, entitled "Performance Evaluation (Construction)."[167] DD Form 2626 enables DOD agencies to rate construction contractors in five major performance categories: quality control, effectiveness of management, timely performance, compliance with labor standards, and compliance with safety standards. The form also enables agencies to provide a narrative description of a contractor's performance.

DOD also uses a similar system known as ACASS (Architect-Engineer Contractor Appraisal Support System) to track and store the performance evaluations of architects and engineers. The Army Corps of Engineers maintains and serves as the executive agent for both the CCASS and ACASS systems. All DOD agencies are required to use CCASS and ACASS to maintain the performance evaluations of construction contractors and architects and engineers.[168] The data maintained in the CCASS and ACASS systems is also shared with the DOD Contractor Performance Assessment Reporting System (CPARS) and the PPIRS.[169] The PPIRS system is a government-wide web-based repository that stores contractor past performance evaluations for all federal government agencies and enables these agencies to access and use the past performance evaluations of other agencies for source selection purposes.

A contractor's past performance information on a specific project will be maintained by the contracting agency for up to three years after the completion of the

---

[165]FAR § 42.1503(d).

[166]Department of Veterans Affairs contracts using the NIH past performance system include a requirement that the contractor register on the NIH Contractor Performance System (CPS) Web site at *https//epccontractor.nih.gov*.

[167]DFARS § 236.201.

[168]DFARS §§ 236.201; 236.604.

[169]As reflected in a July 29, 2009, memorandum, "Improving the Use of Contractor Performance Information," from the Deputy Administrator of the Office of Management and Budget (OMB), that, as of July 1, 2009, all federal agencies are required to submit an electronic record of contractor past performance in the government's Past Performance Information Retrieval System. PPIRS is a web-enabled, enterprise application that provides timely and pertinent contractor past performance information to the DOD and federal acquisition community for use in making source selection decisions. PPIRS assists acquisition officials by serving as the single source for contractor past performance data. *www.ppirs.gov*. The July 29, 2009, OMB memorandum instructs agencies to consider a contractor's achievement of small business subcontracting goals in performance evaluations when the contract includes a Small Business Subcontracting Plan. *See* **Chapter 5**.

contract and will be able to be accessed by other agencies via the PPIRS and the other web-based performance evaluation systems during that period of time.[170] Contractors are entitled to access their own performance records, but not those of other contractors, from these performance evaluation systems.

In addition to the PPIRS system described above, the requirements for detailed information regarding a contractor's past performance continue to expand in scope and duration. For example, Section 872 of the Duncan Hunter National Defense Authorization Act of 2009[171] requires the Office of Management and Budget and the General Services Administration to establish, within one year of the effective date of that legislation (October 14, 2008), an information database on the integrity and performance of contractors that will be available to all federal agencies and public bodies receiving federal grants. That legislation also mandates the adoption of regulations within one year of the effective date of the legislation requiring contractors with agency awards or grant contracts with a total value in excess of $10,000,000 to provide to the federal government detailed information similar to that currently found in FAR § 52.209-5 (**See Chapter 1**) with certain critical changes and additional topics. The new reporting requirements and database will cover a **five year** period, not three years, and will include disclosure of civil judgments "in connection with" the award or performance of a contract or grant with the federal government, default terminations, debarments and suspensions, and the administrative resolution of suspension or debarment proceedings. Unfortunately, the phrase "in connection with" is not defined in this legislation. The clear purpose of this legislation is to provide all federal agencies with access to information that might bear on the determination of a contractor's responsibility under FAR Subpart 9.1 and to make similar information available to grant recipients.

As a practical matter, contractors should obtain and understand the agency policy on evaluating performance prior to commencing work on any government project.[172] The evaluation criteria and process used by the agency should be a high priority during partnering meetings as well as during periodic meetings with representatives of the government as the work progresses. No contractor wants to be surprised with an adverse performance rating at the end of a contract. Understanding the performance evaluation process and methods used by the agency during the performance period will help contractors avoid this scenario.

4. Challenges to an Agency's Performance Rating

A contractor's past performance history is critical to its ability to successfully compete in competitive best-value procurements. One bad performance evaluation from a government agency can adversely affect a contractor's ability to compete for and obtain government contracts for years to come. Many past performance disputes between a

---

[170]*See D.F. Zee's Fire Fighter Catering,* Comp. Gen. Dec. B-280767.4, 99-2 CPD ¶ 62; *see also* FAR § 42.1503(e); 58 Fed. Reg. 3575 (1993).

[171](Pub. L. 110-417).

[172]For example, DOD agencies such as the Navy and the Corps of Engineers utilize DD Form 2626, Performance Evaluation (Construction). *See Federal Government Contractor Past Performance Evaluations - Toolkit and Guidance* (Marco A. Giamberardino et al. [Associated General Contractors of America, 2010]).

contractor and an agency originate as the result of the government's terminating a contract for default based on what is perceived to be a contractor's unacceptable performance of the contract requirements.[173] Consequently, any contractor performing work on a federal government project needs to appreciate the past performance evaluation process and develop a plan to avoid receiving a disappointing evaluation. In the event a contractor receives an unfavorable past performance evaluation, that contractor needs to be aware of the procedures for challenging and correcting that adverse evaluation.

Contractors initially may challenge an unfavorable past performance evaluation at the agency level. FAR § 42.1503 provides that contractors have the opportunity to respond to performance evaluations after the contracting officer has issued and provided a copy of the final performance assessment to the contractor. Agencies must provide completed performance evaluations to contractors as soon as practicable, and contractors have 30 days after receipt of that evaluation to review, comment on, or rebut the conclusions made in the performance evaluation.[174]

If the contractor provides a rebuttal for a part or all of the performance ratings, the contracting officer and lead assessor must work with the contractor to see if a mutual agreement can be reached on the contractor's ratings. If no agreement can be reached, the contractor may seek review of its rating at least one level above the contracting officer. In the event that no mutual agreement can be reached between the contractor and the agency regarding the performance evaluation, the ultimate conclusion regarding the performance evaluation rests with the agency.[175] Any rebuttal statements provided by the contractor must be attached to the performance evaluation and provided to any acquisition officials evaluating the contractor for future contracts. No performance evaluation, regardless of whether it is positive or negative, is to be retained in the PPIRS/CPARS/CCASS/ACASS systems longer than three years after completion of contract performance.

In the event a contractor is not successful in changing an unfavorable performance evaluation at the agency level, it can pursue a challenge to the agency's performance evaluation at the Court of Federal Claims and possibly at the Armed Services Board of Contract Appeals (ASBCA). In *Todd Construction. LLP v. United States*,[176] the Court of Federal Claims held that contractors can challenge an unfavorable performance evaluation under the Contract Disputes Act (CDA) as long as the challenge satisfies the requirements of a viable CDA claim.[177] The *Todd* case involved a roofing contractor that received an unsatisfactory performance evaluation from the U.S. Army Corps of Engineers. The contractor was unsuccessful in changing the performance evaluation at the agency level, and the unfavorable performance evaluation was posted in the CCASS system.

The contractor filed suit at the Court of Federal Claims in an attempt to remove or amend the evaluation. Over the government's objections, the court found that a

---

[173]*See Widnall v. B3H Corp.*, 75 F.3d 1577 (Fed. Cir. 1996); *Aerospace Design & Fabrication, Inc.*, Comp. Gen. Dec. B-278896.3, 98-1 CPD ¶ 139.

[174]FAR § 42.1503(b).

[175]*Id.*

[176]85 Fed. Cl. 32 (2008).

[177]*See* **Chapter 15.** An action under the Administrative Procedures Act, 5 U.S.C. § 702, may be a possible alternative challenge.

contractor's challenge to the unfavorable performance evaluation constituted a viable claim under the CDA and that the Court of Federal Claims had jurisdiction over such claims. The Court of Federal Claims' ruling in *Todd*, along with the previous rulings of that court in *Record Steel and Construction, Inc. v. United States*[178] and *BLR Group of America, Inc. v. United States*[179] signify an important change in the court's view of contractor past performance challenges. Although these decisions make clear that contractors can challenge performance evaluations at the Court of Federal Claims, the court has yet to define the remedy to which a contractor is entitled if the court finds a performance evaluation is erroneous or unjustified.[180]

In limited circumstances, contractors may challenge an agency's performance evaluation at the ASBCA. Historically, the ASBCA has declined to exercise CDA jurisdiction over a contractor's challenge of an unfavorable performance evaluation.[181] However, the board will consider such challenges when they involve the contractor's rights and obligations under the terms of the contract. In *Sundt Construction, Inc.*,[182] a contractor received an unfavorable performance evaluation from the U.S. Air Force at the conclusion of the contract even though the contractor and the Air Force had an oral agreement that guaranteed the contractor a satisfactory performance evaluation if the contractor withdrew its claims for equitable adjustments. The contractor filed a CDA claim at the ASBCA requesting that the board direct the Air Force to amend the performance evaluation to reflect the agreement between the parties. The ASBCA exercised jurisdiction over the case not because it involved a claim to amend a contractor's performance evaluation but because it involved the breach of a contract agreement between the parties by the Air Force regarding the performance evaluation rating the contractor would receive. The holding of the ASBCA in *Sundt* indicates that while the board will not exercise jurisdiction over a direct challenge to a contractor's performance evaluation, it will exercise jurisdiction over such challenges when the contractor's performance evaluation is wrapped up in the parties' rights and obligations under the contract.

Although asserting a challenge to an unjust performance evaluation at the Court of Federal Claims or one of the boards may provide some measure of relief for a contractor, using the disputes process as a remedy should be viewed as the last option. Rather, contractors should make the achievement of an outstanding performance evaluation as one of its priorities from the day it receives the contract award. Contractors can improve their chances of receiving a favorable performance evaluation by doing following these steps:

(1)   If there is an initial project kick-off partnering meeting, the contractor should advise all participants that it intends to strive to achieve an outstanding rating. Make that achievement one of the contractor's stated goals from the very

---

[178]62 Fed. Cl. 508 (2004).

[179]84 Fed. Cl. 634 (2008).

[180]*Todd Construction. LLP v. United States*, 88 Fed. Cl. 235 (2009).

[181]*See, e.g., TLT Constr. Corp.*, ASBCA No. 53769, 02-2 BCA ¶ 31,969; *CardioMetrix*, ASBCA No. 50897, 97-2 BCA ¶ 29,319.

[182]ASBCA No. 56293, 09-1 BCA ¶ 34,084.

beginning. If there is no initial partnering session, place achievement of an *Outstanding* past performance rating on the preconstruction meeting agenda.

(2)   Engage the government's representative in a discussion of the various categories of performance in order to gain insight regarding the practical application of the concepts evaluated in the performance evaluation report. Understand the government's standards of evaluating performance and its expectations regarding the contractor's performance.

(3)   Periodically address the performance evaluation categories at project meetings with the government's representatives. Ask "How are we doing? And "What can we do better?"

(4)   If there are deficiencies or criticisms of its performance, the contractor should take action to address these and improve on them. Ask the government's representatives if improvement has been achieved. Remember, the key is to avoid being unpleasantly surprised at the end of the project.

By following some or all of these steps during the performance of a contract, contractors not only improve their chances of receiving a favorable performance evaluation but make it easier to challenge an unfavorable performance evaluation in the event one is received. It is much easier to show that a performance evaluation is unjust or erroneous when a contractor obtained positive feedback from the government's representatives all throughout its performance only to receive an unfavorable evaluation at the conclusion of the contract.

D. Standards for Responsibility Determinations

Responsibility determinations focus on whether the contractor has the necessary technical expertise, managerial structure and assets, financial resources, and integrity to perform the contract work. A *responsible* bidder/offeror is a contractor capable of undertaking and completing the work in a satisfactory fashion. The contracting officer has the obligation to make an affirmative responsibility determination before awarding a contract.[183] Responsibility determinations must be based on the most recent and reliable information available.[184] The determination of responsibility is based on the contractor's ability to perform the specified work at the time work is to commence, not at the time of bidding. Consequently, a contractor is not required to have the ability to perform at the time the bid or proposal is submitted. A contractor will be deemed responsible if it has or can obtain the apparent ability to perform the work as of the date work is to start.[185] In certain circumstances, the bidder may demonstrate its ability to provide the contract work through a subcontract arrangement.[186] This determination is based on the best information available at the time of

---

[183]FAR § 9.103(b).
[184]FAR § 9.105-1(b)(1); *see also Hayes Int'l Corp. v. United States,* 7 Cl. Ct. 681 (1985).
[185]Comp. Gen. Dec. B-176227, 52 Comp. Gen. 240 (1972), 1972 CPD ¶ 97.
[186]*Id.*

award.[187] Requirements related to responsibility may be satisfied at any time prior to award,[188] but an agency is under no obligation to wait indefinitely while a bidder attempts to satisfy the responsibility criteria.[189]

Federal agencies will consider a number of factors in determining whether a prospective contractor is responsible. These factors fall into two general categories, ability and reliability/ethics. First, a contractor must have the ability to perform the work required by the solicitation. In determining the ability of a contractor to perform, government agencies will consider and evaluate the contractor's financial resources, facilities and equipment, technical knowledge, management structure and assets, relevant experience, and licenses and permits. The second general category of responsibility standards addresses the contractor's demonstrated reliability and ethics. In this category, procurement officials will consider the integrity of the contractor and its performance on previous projects.

FAR § 9.104-1 specifies that a contractor must demonstrate these qualifications in order to be considered responsible:

(1)   Have adequate financial resources to perform the contract, or the ability to obtain them.

(2)   Be able to comply with the required or proposed delivery or performance schedule, taking into account all existing commercial and governmental business commitments.

(3)   Have a satisfactory performance record.

(4)   Have a satisfactory record of integrity and business ethics.

(5)   Have the necessary organization, experience, accounting and operational controls, and technical skills, or the ability to obtain them.

(6)   Have the necessary production, construction, and technical equipment and facilities, or be able to obtain them.

(7)   Be otherwise qualified and eligible for award under applicable laws and regulations.

The procuring activity usually will evaluate the past performance of the contractor's predecessor companies, key personnel, and major subcontractors.[190] A contractor may be deemed to lack responsibility by a federal agency due to its relationship with other business entities. In *OSG Product Tankers LLC v. United States*,[191] the Court of Federal Claims upheld a defense agency's decision which found that an offeror lacked responsibility to perform a contract due to the federal conviction of the offeror's parent company for misconduct unrelated to the offeror's procurement activities. The procuring agency believed that the offeror would be unable to obtain

---

[187]*See Roada, Inc.*, Comp. Gen. Dec. B-204524.5, 82-1 CPD ¶ 438; *B&W Stat Laboratory, Inc.*; *QUAL-MED Assocs., Inc.*, Comp. Gen. B-188627, 77-2 CPD ¶ 151.

[188]*Acquest Dev. LLC*, Comp. Gen. B-287439, 2001 CPD ¶ 101.

[189]*See Roada, Inc.*, Comp. Gen. Dec. B-204524.5, 82-1 CPD ¶ 438.

[190]FAR § 15.305(a)(2)(iii).

[191]82 Fed. Cl. 570 (2008).

the required security clearances to perform the contract due to its parent company's convictions and found the offeror to lack responsibility. The Court of Federal Claims dismissed the offeror's challenge to the agency's responsibility determination and found the agency's decision to be rational and fully supported by the facts.

All available information, whether part of the bid or proposal or not, should be submitted by the contractor and considered by the government to resolve responsibility questions. Since the contracting officer is making a quasi-judicial decision when determining whether a contractor is *responsible* within the meaning of the governing statute, the contractor is entitled to due process. Therefore, a finding by a procuring activity that a contractor is not responsible should be supported by a record establishing (1) the facts on which the decision was based, (2) details of the investigation that disclosed these facts, and (3) the opportunity that was offered to the contractor to present its qualifications.

The GAO has refused to entertain challenges to responsibility determinations absent an allegation of fraud on the part of the procuring agency. The result has been the tacit acceptance of agency determinations in the vast majority of cases.[192] However, where an agency determination of nonresponsibility is so unreasonable as to be arbitrary and capricious, the affected contractor may be given an opportunity to be reevaluated by the agency or permitted to recover its bid or proposal preparation costs.[193]

If the contracting officer reaches an initial determination that a small business concern is not responsible, the agency is obligated to refer that determination to the Small Business Administration (SBA) as set forth in FAR Subpart 19.6. As discussed in **Chapter 5**, the SBA has the authority to make a final determination on the responsibility of a prospective small business contractor.

E. Debarment and Suspension

Debarment and suspension are government-imposed sanctions that prohibit companies and individuals from contracting with the government. Debarments exclude contractors from government contracting and government-approved subcontracting for a reasonable, specified period.[194] Debarments usually are limited to a maximum of three years.[195] A contractor proposed for debarment must be given notice and an opportunity to be heard.[196] Suspensions are temporary disqualifications used when the government determines that immediate action is necessary to protect the government's interest pending the completion of an investigation or legal proceeding.[197] Suspensions are, in a sense, temporary debarments in which the investigatory period could possibly exceed the standard three-year debarment.[198] Debarment and suspension

---

[192]*Central Metal Prods.*, 54 Comp. Gen. 66 (1974); 4 C.F.R. § 21.5(c).

[193]*L. A. Anderson Constr. Co.*, GSBCA No. 6235, 82-1 BCA ¶ 15,507.

[194]FAR § 9.403.

[195]FAR § 2.101.

[196]FAR § 9.406-3.

[197]FAR § 9.407-1(b)(1).

[198]*See Frequency Elecs., Inc. v. U.S. Dept. of the Air Force*, 151 F.3d 1029 (4th Cir. 1998) ("temporary" suspension of almost five years upheld).

are meant to protect the government and taxpayers from dishonest or illegal contractor conduct and are not to be employed for purposes of punishing contractors.[199]

There are two generally recognized types of debarment: procurement and nonprocurement (sometimes called inducement). The grounds for procurement debarment are commission of fraud or a criminal offense showing a lack of business integrity, a serious violation of contract terms, or other causes showing a lack of present responsibility.[200]

Specifically, contractors can be debarred or suspended for a host of dishonest and illicit behavior related to the performance of a contract, such as fraud, bribery, tax evasion, false statements, willful failure to perform or violation of the terms of a federal contract, conflict of interest, lack of business integrity, or receiving *significant* overpayments from the government.[201] Contractors are expressly required to timely disclose to the government any of its or its subcontractors' known violations of (1) federal law involving fraud, conflict of interest, bribery, or gratuity, (2) the False Claims Act, or (3) significant overpayments resulting from the performance of a contract.[202] *See* **Chapter 2, Section III.C.5.** for a discussion of overpayments by the government.

This duty to disclose applies to any contractor that has received a contract award. Not only is the contractor required to disclose the violations of federal law of its own employees, it is also required to disclose the known violations of any of its subcontractors. This duty to disclose is triggered when the contractor obtains *credible evidence* of its or its subcontractors' wrongdoing, and this duty to disclose lasts for three years after final payment on any government contract awarded to the contractor.[203]

The timeliness of a contractor's disclosure is measured from the date it obtains credible evidence of wrongdoing. As such, a contractor that sits on or consciously ignores evidence of wrongdoing that it has a duty to disclose does so at its own peril.

The grounds for nonprocurement debarment are spelled out in the Nonprocurement Common Rule (NCR) that provides for government nonprocurement suspension and debarment.[204] The NCR is codified in the *Code of Federal Regulations* at 2 C.F.R. Part 180 and requires that procurement and nonprocurement debarments and suspensions be treated reciprocally and recognized throughout the government.[205] The NCR grounds for debarment generally incorporate the causes in the FAR but introduce four new grounds: (1) knowingly doing business with an ineligible person, (2) failing to pay debts to any federal agency (except the Internal Revenue Service), (3) willful violation of a statute or regulation applicable to a public agreement, and (4) violation of a voluntary exclusion agreement or a debarment or suspension settlement agreement.[206]

Debarment and suspension determinations are discretionary and not automatically triggered. These decisions are not made by contracting officers but by agency

---

[199]FAR § 9.402(b).
[200]FAR § 9.406-2.
[201]FAR §§ 9.406-2, 9.407-2.
[202]*Id.*
[203]*Id.*
[204]53 Fed. Reg. 19,160 (May 26, 1988).
[205]*See* 70 Fed. Reg. 51,863 (Aug. 31, 2005); 68 Fed. Reg. 66,534 (Nov. 26, 2003).
[206]2 C.F.R. § 180.800(c)(2)-(4).

heads or designees at levels above the contracting officer. Even where grounds for disbarment or suspension have been clearly established, the debarring or suspending official must determine whether the contractor is *presently* responsible by examining the seriousness of the contractor's acts, remedial measures taken, or mitigating circumstances.[207]

If a contractor is debarred, suspended, or proposed for debarment, FAR § 9.404 requires that name of that firm, detailed information regarding the contractor such as its address, DUNS Number, Taxpayer Identification Number, grounds for the agency action, and the agency point of contact be provided to the General Services Administration for inclusion in the web-based Excluded Parties List System (EPLS). If a firm is listed on the EPLS, agencies are precluded from awarding new contracts to that firm unless the head of the agency determines that there is a compelling reason to do so.[208] Existing contracts *may* be completed or terminated in the discretion of the agency.[209] However, adding work by change order, the exercise of option, or actions to extend the duration of the contract are not permitted unless the head of the agency makes a written determination that there are compelling reasons to do so.[210]

Contractors should not take the obligations to report the wrongdoing of its employees and its subcontractors lightly because the suspension or debarment of a contractor can be a fatal blow to a company's reputation and its financial stability. When in doubt about whether an event or incident should be reported to the government, contractors should seek legal advice in order to become fully aware of their rights and obligations.

VII. LATE BIDS AND PROPOSALS

As a general rule, "late is late," and any bid or proposal received after the exact time specified in the solicitation will not be considered.[211] Moreover, the bidder/offeror usually bears the risk of assuring timely submission of the bid/offer or modification to its bid or proposal, regardless of the mode of submission. This applies whether the bid/offer is mailed, hand delivered, faxed, sent via e-mail, or submitted through an agency's web-based electronic bid/offer collection platform.[212]

The FAR does, however, permit acceptance of late bids or proposals under certain limited circumstances. The FAR permits acceptance of a late *bid* where (1) the bid is received before the award is made, (2) accepting the late bid would not unduly delay the acquisition, (3) the bid was transmitted by an authorized electronic means and received by the government no later than 5:00 PM one working day prior to the bid

[207]2 C.F.R. § 180.125; FAR §§ 9.406-1(a), 9.407-1(b).

[208]FAR § 9.405(a).

[209]FAR § 9.405-1(a).

[210]FAR § 9.405-1(b).

[211]*See* FAR §§ 14.302(a), 14.304(b)(1), 15.208(b)(1), 52.214-7(b)(1), 52.215-1(c)(3)(ii)(A).

[212]*See* FAR §§ 14.304, 15.208, 52.214-31(g); *Lakeshore Eng'g Servs*, Comp. Gen. B-401434, 2009 CPD ¶ 155; *Amigo-JT, Joint Venture,* Comp. Gen. Dec. B-292830, 2003 CPD ¶ 224; *PMTech, Inc.,* Comp. Gen. Dec. B- 291082, 2002 CPD ¶ 172.

date, *and* (4) there is acceptable evidence that the bid was received at the designated government installation and was under the government's *control prior to the designated time.*[213]

A late *proposal* may be accepted where (1) the proposal is received before the award is made, (2) accepting the late proposal would not unduly delay the acquisition, *and* (3) there is acceptable evidence that the proposal was received at the designated government installation and was under the government's control prior to the designated time.[214]

In addition to these regulatory conditions, an offer that is received late may be accepted where improper government action is the sole or paramount cause for the tardy delivery and the integrity of the procurement process would not be adversely affected by acceptance.[215] If a bidder has done all it could and should do to fulfill its responsibility, it should not be predjudiced if the bid did not arrive as required because the government failed in its own responsibility.[216]

A late bid may be considered for award if the government's affirmative misdirection made timely delivery impossible or government mishandling after timely receipt was the sole or main cause for the bid's late receipt at the designated location.[217] In *Palomar Grading and Paving,*[218] the GAO found that the government was the paramount cause of the late receipt of a bid because the solicitation instructions specified the wrong zip code for bid delivery, making it impossible for UPS to deliver the bid on time.

In *Hospital Klean of Texas,*[219] a FedEx representative found the doors of the government facility locked when it attempted bid delivery at 8:47 AM on the bid closing date. After knocking and getting no response, the FedEx representative left and subsequently delivered the bid on the next business day. The Court of Federal Claims held that the bidder, through its agent FedEx, had not "done all it could" to ensure timely delivery and did not allow the late acceptance.[220] The court found that the bidder should have made an attempt to redeliver later that same day or call the contracting officials identified in the solicitation materials.[221] Delays in gaining access to a government building are not unusual and should be expected.[222]

The bid opening officer has the authority to determine when the time set for opening the bids has arrived and must announce that decision to those present. The bid opening officer's time declaration is binding unless it is shown to be unreasonable under the circumstances.[223] In *General Engineering Corp.,*[224] the bid opening officer, noting

---

[213]FAR § 14.304(b)(1).

[214]FAR § 15.208(b)(1).

[215]*See Hospital Klean of Texas, Inc. v. United States,* 65 Fed. Cl. 618 (2005); *St. Charles Travel,* Comp. Gen. Dec. B-226567, 87-1 CPD ¶ 575.

[216]*Palomar Grading & Paving, Inc.,* Comp. Gen. Dec. B-274885, 97-1 CPD ¶ 16.

[217]*Hospital Klean of Texas, Inc. v. United States,* 65 Fed. Cl. 618, 623 (2005).

[218]*Palomar Grading & Paving, Inc.,* Comp. Gen. Dec. B-274885, 97-1 CPD ¶ 16.

[219]65 Fed. Cl. 618 (2005).

[220]*Id.* at 623.

[221]*Id.* at 624.

[222]*Econ, Inc.,* Comp. Gen. Dec. B-222577, 86-2 CPD ¶ 119.

[223]*See* FAR § 14.402-1(a); *see also General Eng'g Corp.,* Comp. Gen. Dec. B-245476, 92-1 CPD ¶ 45; *Swinerton & Walberg Co.,* Comp. Gen. Dec. B-242077.3, 91-1 CPD ¶ 318.

[224]Comp. Gen. Dec. B-245476, 92-1 CPD ¶ 45.

the time on the clock in the bid opening room, announced that no more bids would be accepted. Moments later, before any bids had been announced, another bidder tendered its bid at what was later determined to be at least 1 minute and 29 seconds before the exact time for the bid opening. The Comptroller General held that the bid was properly rejected because the fact that the wall clock was approximately a minute and a half fast did not render the bid opening official's time determination unreasonable.

Bids and proposals may be considered late even if the procuring agency issues amendments to the solicitation allowing for the later acceptance of the bid/proposal. In *Geo-Seis Helicopters, Inc. v. United States,* [225] the Court of Federal Claims determined that an offeror's revised proposals were both considered "late" and could not be accepted by the procuring agency. On the day both revised proposals were due, and before the specified deadline for receipt of the revised proposals, the offeror notified the procuring agency that its revised proposals would be late due to weather. The agency issued amendments to the solicitation prior to the deadlines to receive the revised proposals, which gave offerors more time to submit revised proposals. The offeror submitted its revised proposals after the initial time deadlines for receipt but before the revised deadlines implemented by the agency's amendments to the solicitation. The Court of Federal Claims found that the agency's *post-hoc* amendments allowing the late receipt of revised proposals was contrary to the *late is late* provisions of the FAR and therefore unreasonable.

VIII.   RELIEF FOR CONTRACTOR BID/PROPOSAL MISTAKES

Mistakes are not uncommon in the rush and inherent complexity involved in submitting construction bids and proposals. Construction contractors do not work under ideal conditions in the hurry to meet the deadline for submitting bids and proposals. Most courts recognize that honest, sincere people, even in the exercise of ordinary care, can make mistakes of such a fundamental character that it would be unfair for the government to take advantage of a clearly erroneous proposed price.[226]

The law recognizes two types of mistakes: "mutual" and "unilateral." A mutual mistake occurs when the contractor and the government are both under the same erroneous belief as to the same fact or facts. A very old case providing an example of a mutual mistake involved two parties who both thought a cow was barren when in fact she was with calf and agreed to the sale of the cow at a price based on the fact she was barren. [227] If, however, one party's erroneous assumption is not shared by the other side, the mistake is unilateral,[228] and relief usually is not available to

---

[225]77 Fed. Cl. 633 (2007).

[226]*See Hunt Constr. Group, Inc. v. United States,* 281 F.3d 1369 (Fed. Cir. 2002); *Chernick v. United States,* 372 F.2d 492 (Ct. Cl. 1967); *Pavco, Inc.* ASBCA No. 23783, 80-1 BCA ¶ 14,407; *Black Diamond Energies, Inc.,* Comp. Gen. Dec. B-241370, 91-1 CPD ¶ 119.

[227]*Sherwood v. Walker,* 33 N.W. 919 (Mich. 1887).

[228]E. Allan Farnsworth, *Contracts* § 9.3, 623 (3d ed., Aspen L. & Bus. 1999).

the mistaken party. The contractor's relief can depend on whether the mistake is mutual or unilateral and whether the mistake is discovered before or after the contract award.

A. Preaward

In negotiated procurement, a proposal usually may be withdrawn or a mistake corrected by the contractor at any time prior to offer acceptance.[229] Sealed bids, however, are subject to the "firm bid rule," which generally prohibits modification or withdrawal of bids for a reasonable time after the bid opening while the bids are analyzed to determine if an award can be made.[230] Nevertheless, the doctrine of mistake allows withdrawal or modification of a bid preaward under certain circumstances.[231]

A contractor may obtain relief for only certain types of mistakes. As a threshold matter, the mistake must be a clear-cut clerical or arithmetic error, or a misreading of the specifications.[232] An error in business judgment is not a "mistake" for which relief is available.[233]

Simple negligence by the bidder is not a bar to relief as the reasonableness of the mistake is not relevant under the doctrine of mistake.[234] However, as stated, the mistake must meet the gateway matter of qualifying for relief (clear-cut clerical or arithmetic error, or a misreading of the specifications) before rescission, reformation, or other affirmative contract relief is available.

In *Liebherr Crane Corp.*,[235] the court held that a contractor's failure to read the bid specifications is not the same as misreading the specifications; the failure to read was a mistake in judgment rather than a compensable mistake, and the contractor was not entitled to reformation of the contract. When a contractor bids a project, it makes many business judgments and assumes the risk that its judgments may be wrong.[236]

The contracting officer has an affirmative duty to examine all bids for mistakes.[237] Where the contracting officer has reason to believe that a mistake may have been made, the contracting officer shall notify the bidder of the suspected mistake and request a verification of the bid.[238] When the bidder alleges a mistake, the issue is processed in accordance with FAR § 14.407.

---

[229]*See* FAR §§ 52.215-1, 15.001, 15.208.

[230]*See W.A. Scott v. United States,* 44 Ct. Cl. 524 (1909); *Refining Assocs. v. United States,* 124 Ct. Cl. 115, (1953).

[231]*See* FAR §§ 14.407 *et seq.*

[232]*Will H. Hall and Son, Inc. v. United States,* 54 Fed. Cl. 436, 440 (2002).

[233]*Liebherr Crane Corp. v. United States,* 810 F.2d 1153, 1157 (Fed. Cir. 1987).

[234]*See BCM Corp. v. United States,* 2 Cl. Ct. 602 (1983); *PHT Supply Corp. v. United States,* 71 Fed. Cl. 1 (2006)).

[235]810 F.2d 1153, 1157 (Fed. Cir. 1987).

[236]*Id.*

[237]FAR §§ 14.407-1; 14.407-3. The contracting officer's notification should alert the bidder to the basis for the suspected mistake in bid, e.g., bid price much lower than other bids or the government estimate.

[238]*Id.*

An apparent clerical mistake may be corrected by the contracting officer before the award upon verification of the intended bid.[239] For other mistakes, withdrawals or corrections are often allowed preaward where the contractor meets the evidentiary standards of the FAR.[240]

The evidentiary standards for mistakes disclosed before award are summarized in this way:

> If the bid is responsive, and the bidder provides "clear and convincing evidence" both of the existence of the mistake and the bid actually intended, an agency head may make a determination "permitting the bidder to correct the mistake." If this correction would result in displacing one or more lower bids, such a determination shall not be made unless the existence of the mistake and the bid actually intended are "ascertainable substantially from the invitation and the bid itself." If the alleged mistake evidence is not "clear and convincing, a higher official may permit a withdrawal of the bid, or the agency head may determine that the bid be neither withdrawn nor corrected."[241]

B. Postaward

There is no relief available for a contractor's unilateral mistake discovered after award unless the contractor can show that the contracting officer knew or should have known of the contractor's mistake at the time the bid was accepted.[242] A government construction contract may be reformed or rescinded after award if the contracting officer accepts a bid with actual or constructive knowledge that the bid contains a material mistake.[243] The contractor's unilateral mistake must be apparent to charge the contracting officer with notice of the mistake.[244] If the contractor discovers a mistake following award, it must move promptly to request relief or run the risk that the board or court may let the chips lie where they fall due to the contractor's inaction.[245]

In order to succeed on a claim for mutual mistake after the award of contract, the party seeking relief must prove these four elements:

(1) The parties to the contract were mistaken in their belief regarding a fact.
(2) That mistaken belief constituted a basic assumption underlying the contract.
(3) The mistake had a material effect on the bargain.
(4) The contract did not put the risk of the mistake on the party seeking reformation.[246]

---

[239]FAR § 14.407-2.

[240]*See* FAR § 14.407-3.

[241]W. Noel Keyes, *Government Contracts Under the Federal Acquisition Regulation* § 14.40 (3d ed., West 1986).

[242]*Bromley Contracting Co. v. United States,* 794 F.2d 669, 672 (Fed. Cir. 1986).

[243]*Will H. Hall and Son, Inc. v. United States,* 54 Fed. Cl. 436, 440 (2002); *see also Turner-MAK (JV),* ASBCA No. 37711, 96-1 BCA ¶ 28,208.

[244]FAR § 14.407-4(c)(2).

[245]*See Turner-MAK (JV),* ASBCA No. 37711, 96-1 BCA ¶ 28,208; *United States v. Hamilton Enters., Inc.,* 711 F.2d 1038 (Fed. Cir. 1983).

[246]*C.W. Over & Sons, Inc. v. United States,* 54 Fed. Cl. 514, 525 (2002).

Postaward mistakes in sealed bids and negotiated procurements under FAR Part 15 are processed in accordance with the procedures established in FAR § 14.407-4.[247] The FAR states in pertinent part:

(a) When a mistake in a contractor's bid is not discovered until after award, the mistake may be corrected by contract modification if correcting the mistake would be favorable to the Government without changing the essential requirements of the specifications.

(b) In addition to the cases contemplated in paragraph (a) of this section or as otherwise authorized by law, agencies are authorized to make a determination—

(1) To rescind a contract;

(2) To reform a contract-

(i) to delete the items involved in the mistake; or

(ii) to increase the price if the contract price, as corrected, does not exceed that of the next lowest acceptable bid under the original invitation for bids; or

(3) That no change shall be made in the contract as awarded, if the evidence does not warrant a determination under subparagraph (1) or (2) above.

(c) Determinations under paragraph (b)(1) and (2) above may be made only on the basis of clear and convincing evidence that a mistake in bid was made. In addition, it must be clear that the mistake was—

(1) mutual; or

(2) if unilaterally made by the contractor, so apparent as to have charged the contracting officer with notice of the probability of the mistake.

(d) Each proposed determination shall be coordinated with legal counsel in accordance with agency procedures.[248]

Since negotiated procurements allow for withdrawal of the offer prior to award, mistake issues are far more common on sealed bid procurements. Consequently, Part 15 of the FAR does not address preaward mistake issues. However, FAR § 15.508 states that mistakes in a contractor's proposal that are disclosed after award will be processed substantially in accordance with FAR § 14.407-4.

IX. BID PROTESTS

A. Evolution of Bid Protest Rights and Procedures

Whether the procurement is a sealed bid or a negotiated procurement, challenges to a federal agency's conduct on a particular procurement action are labeled *bid protests*. The procedures and principles for resolving bid protests have evolved over many

---

[247]For discovery of mistakes in proposals after award, *see* FAR § 15.508; for discovery of mistakes in bids after award, *see* FAR § 14.407-4.
[248]FAR § 14.407-4.

decades. The first agency to formally address bid protests and questions involving alleged bid mistakes was the General Accounting Office (GAO; now the Government Accountability Office). Acting on its authority under 31 U.S.C. §§ 71, 74, and 82 to settle and adjust claims by or against the United States, the GAO adopted the practice of providing advance opinions to the federal agencies on questions such as the legality of proposed awards. Beginning in 1921, the GAO published selected decisions on such award questions.[249]

The federal agencies acquiesced and followed these GAO advisory decisions. At least as early as 1926, the GAO published a decision recommending the addition of the phrase "or equal" when a specification called for a particular brand name product to be provided to the government.[250] The term "protest" by a bidder appears in a GAO published decision at least as early as 1930.[251]

For many years, the GAO was the only entity to which a party disappointed with an agency's handling of a solicitation and award could challenge (protest) that action. This limitation on avenues for relief was reinforced by the 1940 decision of the United States Supreme Court in *Perkins v. Lukens Steel Co.*[252] in which the Court held that the then current procurement act did not bestow "litigable rights" on those who wanted to contract with the government.

The *Lukens Steel* decision has never been reversed or distinguished by the Supreme Court. However, over the next three decades, federal courts developed alternative doctrines or theories that provided disappointed bidders a measure of relief. In 1956, the Court of Claims held that an aggrieved bidder could recover its bid preparation costs as damages if its bid had not been evaluated in good faith.[253] In 1970, the U. S. Court of Appeals for the District of Columbia ruled in *Scanwell Lab., Inc. v. Shaffer* that a disappointed bidder could bring an action under the Administrative Procedures Act, 5 U.S.C. § 702, to challenge a procurement action and that a federal district court could issue an injunction barring an award by a federal government agency.[254] This decision was followed by other federal courts with the result that *Scanwell* actions in federal district courts became an alternative forum to the GAO for bid protests in the latter part of the twentieth century.

In 1996, the Administrative Disputes Resolution Act established concurrent jurisdiction in both the federal district courts and the Court of Federal Claims to render judgment on an interested party's (bid protester's) objection to (1) a federal agency solicitation, (2) a proposed award or the award of a contract, or (3) an alleged statutory violation in connection with a procurement. This concurrent jurisdiction extended to both preaward and postaward protests. Although injunctive and declaratory relief were permitted, monetary damages were limited to bid or proposal preparation costs. Finally, as of January 1, 2001, Congress rescinded the bid protest jurisdiction of the

---

[249]1 Comp. Gen. 21 (1921) and 1 Comp. Gen. 304 (1921).
[250]5 Comp. Gen. 835 (1926).
[251]*See* 9 Comp. Gen. 25 (1930).
[252]310 U.S. 113 (1940).
[253]*Heyer Products Co. v. United States,* 140 F. Supp. 409 (Ct. Cl. 1956).
[254]424 F.2d 859 (D.C. Cir. 1970).

federal district courts, leaving the Court of Federal Claims as the only judicial forum available to challenge a government contract procurement or award.

The GAO remains an alternative forum for the resolution of challenges to a federal agency's procurement actions. One apparent consequence of the *Scanwell* line of decisions was the adoption by the GAO of more comprehensive bid protest procedures.[255] These procedures reflect an effort to provide an expeditious resolution of bid protests with reasonable due process safeguards for the interested parties.

The current choice of dual forums to challenge agency procurement actions and decisions clearly reflect concepts and principles that have evolved over the past century. The procedures and principles governing these actions and the rights of parties have also evolved as the Congress and the courts have sought to address apparent problems in the procurement process that may be prejudicial to certain offerors or bidders.

B. Multiple Forums—Basic Principles

When submitting a bid or proposal for a government construction project, contractors must rely not only on their own evaluation of the IFBs or the RFPs issued by the agency but also on a proper application of the procurement procedures that are applicable to that agency. Increased competition for contracts combined with economic cycles that reduce the total volume of contracts available for award have forced both contractors and their attorneys to learn the rules governing competitive procurements, the enforcement of rules governing sealed bids or competitive proposals, and related bid protest rules and regulations.

A guiding principle in government bid protests is quick action. Whether the protest is filed with an agency, the GAO, or the Court of Federal Claims, the disappointed bidder/offeror must act within a very limited time frame. Some protests must be filed prior to the deadline for the submission of sealed bids or receipt of proposals or the right to file a protest may be lost. Moreover, under some circumstances, the contracting agency may award the contract even though a protest is still pending. Once the award is made and performance has started, the reviewing body will be reluctant to reverse the decision of the awarding authority. Thus, obtaining a favorable ruling on the merits of a bid protest may prove to be an empty victory if the contract has already been awarded to another contractor and performance of the contract has started.

The protesting contractor often bases its protest on grounds such as a defect in the solicitation, the failure of the apparent successful bidder/offeror to meet one or more of the requirements in the solicitation such as lowest evaluated price, responsiveness to the solicitation, or the agency's failure to evaluate the bid or offer consistent with the evaluation criteria in the solicitation. A protester has several options for filing a bid protest. A bid protest may be filed with the:

(1) Contracting agency responsible for the procurement; or
(2) Comptroller General of the Government Accountability Office (GAO); or
(3) Court of Federal Claims

[255]4 C.F.R. Part 21.

It is also possible to file protests with more than one of these three entities, depending on the nature and timing of the protest.[256]

C. Who May Protest: Standing

In order to be able to file a protest, a party must have standing to protest, that is, be an "interested party." An "interested party" is an actual or prospective bidder or offeror whose direct economic interest would be affected by the award of a contract or by the failure to award a contract.[257] Any "interested party" may file a protest with the GAO alleging an irregularity in the solicitation or the award of a federal government contract. In several cases, protesters that did not even submit offers in response to the solicitation were nevertheless held to be interested parties. For example, protesters that were precluded from submitting a proposal because of an unduly short response time or a restrictive specification,[258] or who otherwise were denied the opportunity to compete,[259] have been found to have standing to protest.

The Competition in Contracting Act defines an *interested party* for purposes of a protest action filed in the Court of Federal Claims in this way:

> The term "interested party", with respect to a contract or solicitation or other request for offers . . . means an actual or prospective bidder or offeror whose direct economic interest would be affected by the award of the contract or by failure to award the contract.[260]

The Federal Circuit has ruled that Congress intended to adopt this definition for purposes of bid protest actions filed in the Court of Federal Claims under the Tucker Act, 28 U.S.C. § 1491.[261] To qualify as an "actual or prospective bidder," a protesting party must have submitted a timely bid or proposal or must be expecting to submit a bid or proposal before the closing date of the solicitation, and have the capability to perform the contract at the time the proposal period ends.[262] Thus, a firm that objects to the bundling of multiyear contracts into one procurement, but does not have the capability to compete for or perform the contracts until the latter years of the multiyear performance, does not have standing as an interested party.[263] In contrast, an incumbent contractor that objected to an agency decision to perform

---

[256]Until August 1996, the General Services Administration Board of Contract Appeals (GSBCA) had jurisdiction along with the GAO to hear and decide protests involving procurements under the Brooks Act (40 U.S.C. § 759) for automatic data processing equipment software, maintenance services, and supplies. The GSBCA's jurisdiction for such protests was eliminated in 1996, and the GAO now has jurisdiction over such protests. 40 U.S.C. § 759 *et seq.*

[257]4 C.F.R. § 21.0(a); FAR § 33.103(d)(2).

[258]*Vicksburg Fed. Bldg. Ltd. Partnership,* Comp. Gen. Dec. B-230660, 88-1 CPD ¶ 515.

[259]*Afftrex, Ltd.,* Comp. Gen. Dec. B-231033, 88-2 CPD ¶ 143; *REL,* Comp. Gen. Dec. B-228155, 88-1 CPD ¶ 125.

[260]31 U.S.C. § 3551 (2)(A).

[261]*Am. Fed'n of Gov't Employees v. United States,* 258 F.3d 1294, 1302 (Fed. Cir. 2001).

[262]*Rex Serv. Corp. v. United States,* 448 F.3d 1305 (Fed. Cir. 2006).

[263]*Space Exploration Technologies Corp. v. United States,* 68 Fed. Cl. 1 (2005).

the contract work in-house without first resoliciting the work did have standing to protest because the federal government's action effectively deprived it of an opportunity to compete for the work.[264]

D. Federal Agency Protests

Protests to the federal agency must be addressed to the contracting officer or other official designated to receive protests.[265] Many solicitations disclose the name of the person to whom a protest should be addressed. If it is feasible under the agency's rules, a protester should submit the protest at a level above the contracting officer, because the actions or decisions of the contracting officer and its representatives may be the grounds for the protest. The agency is required to provide a procedure for the protester to request an independent review above the contracting officer level, either as an initial review of the protest or as an appeal from the contracting officer's decision.[266]

The protest needs to be logically presented, and must include: (1) the name, address, and fax and telephone numbers of the protester; (2) the solicitation or contract number; (3) a detailed statement of the legal and factual grounds for the protest, including a description of the resulting prejudice to the protester; (4) copies of relevant documents; (5) a request for a ruling by the agency; (6) a statement of the form of relief requested; (7) all information establishing that the protester is an "interested" party; and (8) all information establishing the timeliness of the protest.[267]

Upon receipt of a protest before award, the contract may not be awarded pending agency resolution of the protest unless the award is justified in writing for urgent and compelling reasons or determined in writing to be in the government's best interest.[268] The justification must be approved at a level above the contracting officer.[269] If the award is withheld pending agency resolution of the protest, the contracting officer is to notify the other bidders/offerors whose bids or offers may become eligible for award and, if appropriate, request that those bidders or offerors extend the time for acceptance.[270] If an extension cannot be obtained, the agency is authorized to consider proceeding with the award.[271]

A disappointed offeror/bidder usually has the right to request a debriefing concerning the reasons its proposal was not selected for award. When a protest issue arises on a negotiated procurement under FAR Part 15, the date of the protesting party's debriefing request determines whether the agency is required to suspend performance pending resolution of the protest by the agency. The first step is the submission of a "timely" request for a debriefing. Whether the request is for a preaward debriefing[272] or postaward debriefing,[273] the request for the debriefing must be in

---

[264]*Labat-Anderson, Inc. v. United States,* 65 Fed. Cl. 570, 575 (2005).
[265]FAR § 33.103(d)(3).
[266]FAR § 33.103(d)(4).
[267]FAR § 33.103(d)(2).
[268]FAR § 33.103(f)(1).
[269]*Id.*
[270]FAR § 33.103(f)(2).
[271]*Id.*
[272]FAR § 15.505.
[273]FAR § 15.506.

writing and submitted to the contracting officer within *three days* after the date when the protesting contractor was notified of the agency's adverse action.[274] If the request for a debriefing is made *verbally*, the contractor has not complied with the applicable FAR provisions defining a "timely debriefing" request. Just as the agency is required to provide written notification of an offeror's exclusion from the competition or award to a competitor, the contractor must make a written request for a debriefing. As a practical matter, disappointed offerors should always request a postaward debriefing to determine why they did not receive the contract award and how they were rated compared to other offerors. Valuable information can be obtained during an agency debriefing that may provide the offeror the factual basis that can serve as the basis of a protest of the agency's actions.

Upon receipt of a protest within 10 days after contract award or within 5 days after a debriefing date offered to the protester under a "timely" debriefing request,[275] the contracting officer shall immediately suspend performance pending resolution of the protest by the agency, including any review at a higher agency level. Performance will not be suspended if the agency justifies, in writing, continued performance for urgent and compelling reasons or if continued performance is determined, in writing, to be in the government's best interests.[276] That justification or determination must be approved at a level above the contracting officer, or by another official, pursuant to agency procedures.[277]

An agency protest will not extend the time for obtaining a performance stay from the GAO. However, federal agencies may voluntarily suspend performance of the contract after the denial of the agency protest if the protester subsequently protests to the GAO.[278] Under the applicable regulations, agencies are to use their best efforts to resolve protests within 35 days after the protest is filed.[279]

Bidders and offerors are often reluctant to protest to the contracting agency because the agency is being asked to judge the actions of its own employees. Protests to the GAO or the Court of Federal Claims provide more of an opportunity for a "neutral, third-party review" of the agency's actions and positions.[280] Agency protests may be less successful than protests to the GAO if the basis for the protest involves an unusual issue or one that is not a clear and direct violation of applicable

---

[274]Under FAR § 15.505(a)(1), an offeror must *submit* a written request for a debriefing within three days after receipt of notice that it has been excluded from the competition. Under FAR § 15.506(a)(1), the offeror's written request for a debriefing must be *received* within three days after the date on which the offeror received written notice of the award to another firm. These two FAR provisions contemplate that a written request made on the third day must be received on the same day. As a practical matter, this means that a written request for a debriefing on the third day should be faxed and e-mailed to increase the potential for receipt by the agency on that day.

[275]If the contractor postpones the debriefing from the offered date, it risks losing the benefit of this performance suspension provision.

[276]FAR § 33.103(f)(3).

[277]*Id.*

[278]FAR § 33.103(f)(4).

[279]FAR § 33.103(g).

[280]Even if there may be a "neutral third-party review," the Court of Federal Claims has clearly stated that a disappointed bidder faces a heavy burden when seeking to challenge "best value" procurements. *See Park Tower Mgmt., Ltd., v. United States*, 67 Fed. Cl. 548 (2005).

laws or regulations. The agency may be more likely to ratify the actions of its employees if there is any basis for such actions.

However, there are circumstances where an agency protest may be advantageous to a bidder. An agency protest will, in almost all circumstances, be less costly and time consuming than a GAO or Court of Federal Claims protest. Moreover, an agency protest generally offers the protester the quickest means of obtaining relief either preaward or postaward because the agency is capable of correcting its own error much faster than the GAO or the Court of Federal Claims could bring about the correction.

In a preaward protest, because the agency has a vested interest to expedite the procurement process, the protester may have the opportunity to obtain a quick decision from the contracting agency. If that decision is adverse to the protester's position, the protester may then seek relief in another forum, such as the GAO. However, the protester needs to be mindful that if the agency protest is unsuccessful, the protester has 10 days from the actual or constructive knowledge of the agency decision to file a protest to the GAO.[281] In a postaward protest, the protester will likely obtain an expedited decision from the agency due to the agency's desire to begin contract performance.

A disappointed offeror/bidder initially may submit a bid protest to the federal agency that is sponsoring the procurement. Federal regulations do not discuss in detail the procedures involved in submitting an agency protest, but some general guidelines are provided at FAR Subpart 33.1.

Agency protests may be filed before or after contract award by a bidder/offeror or a prospective bidder/offeror if its direct economic interest is affected by the award of the contract or by the failure of the agency to award the contract.[282] The protester must submit the protest in writing.[283] The agency receiving the protest must respond using a method that provides evidence of receipt of the protest.[284]

An interested party may object to any of these issues or agency actions in a protest:

(1) A solicitation or other request by an agency for offers for a contract for the procurement of property or services.
(2) The cancellation of the solicitation or other request.
(3) An award or proposed award of the contract.
(4) A termination or cancellation of an award of the contract if the written objection contains an allegation that the termination or cancellation is based in whole or in part on improprieties concerning the award of the contract.[285]

The interested party and the federal agency are to use their best efforts through open and frank discussion to try to resolve concerns raised by an interested party

---

[281]4 C.F.R. § 21.4(a)(3); FAR § 33.103(d)(4).
[282]FAR § 33.101.
[283]*Id.*
[284]FAR § 33.103(h).
[285]FAR § 33.101(ii)(1-4).

before the submission of a protest.[286] However, there are time limits for agency protests. Protests based on alleged apparent improprieties in the solicitation must be filed before bid opening or the closing date for the receipt of proposals.[287] In all other cases, protests are to be filed not later than 10 days after the basis for the protest is known or should have been known, whichever is earlier.[288] Two such dates that trigger the start of the 10-day time frame are the offeror's receipt of the notification that it has been excluded from the competition and the offeror's receipt of the notification that award has been made to another firm. An agency can consider an untimely protest if the agency determines that the protest raises issues significant to the agency's acquisition system.[289] The GAO has similar rules on the timing of protests.[290] (In general, the protester should submit a protest as soon as it becomes aware of the basis for the protest, because the likelihood of success declines significantly once an award has been made.)

E. GAO Protests

This is a summary of the important factors to consider in filing a protest with the GAO. Keep in mind that these regulations are subject to change (including the initial deadline for filing a protest). The current bid protest regulations, as published in the *Federal Register* and the *Code of Federal Regulations*, must be followed to ensure a valid and timely protest.[291]

Protests based on alleged improprieties in a solicitation must be filed prior to bid opening or the time set for receipt of initial proposals.[292] As a result, the Comptroller General has held that protests regarding improprieties that were apparent prior to bid opening are untimely even if submitted with a bid[293] or proposal.[294] Generally, all other protests must be filed no later than 10 calendar days after the basis of the protest is known or should have been known, whichever is earlier. Protests challenging a procurement conducted on the basis of competitive proposals under which a debriefing is requested and, once requested, is then required must be filed within 10 calendar days after the debriefing.[295] GAO protests following an agency protest must be filed within 10 days after the agency's initial adverse action on the agency protest.[296]

Offerors need to be aware of the time limits for requesting and securing preaward or postaward debriefings because the date of a debriefing directly impacts the amount

---

[286]FAR § 33.103(b).
[287]FAR § 33.103(e).
[288]*Id.*
[289]*Id.*
[290]4 C.F.R. § 21.2.
[291]*See* 4 C.F.R. § 21.0, *et seq.*; *see also* 31 U.S.C. § 3553.
[292]4 C.F.R § 21.2 (a)(1).
[293]*Fredrico Enter., Inc.,* Comp. Gen. Dec. B-230724.3, 88-1 CPD ¶ 450.
[294]*Darome Connection,* Comp. Gen. Dec. B-230629, 88-1 CPD ¶ 461.
[295]4 C.F.R. §§ 21.0(e) and 21.2(a)(2).
[296]*Id.*

of time an offeror has to file a protest with the GAO. Offerors must request preaward debriefings in writing to the contracting officer within three days after receiving notice that the offeror was excluded from the competition.[297] Likewise, offerors have three days after receiving notice of contract award to submit a written request for a postaward debriefing to the contracting officer.[298] Although offerors are entitled to receive postaward debriefings from the procuring agency, requests for preaward debriefings may be denied by the agency, for compelling reasons, if it is not in the best interest of the government to conduct a debriefing at that time.[299] Offerors that fail to request a debriefing within the required period of time waive their rights to a debriefing.[300] As a practical matter, offerors should always ask for a preaward or postaward debriefing upon learning that they were excluded from the award competition or did not receive the contract award. Debriefings often provide offerors valuable information regarding the agency evaluation process, the information considered by the agency, and the perceived strengths and weaknesses of their own proposals and those of other offerors. The information obtained during a debriefing often can be used to challenge the agency's actions in a bid protest proceeding.

After receiving a bid protest, the GAO has one working day to notify the procuring agency of the protest.[301] Under most circumstances, a contracting agency cannot award a contract after the agency has received notice of the protest and while the protest is pending.[302] The head of an agency may authorize award of a contract notwithstanding a protest (1) upon a finding that urgent and compelling circumstances that significantly affect the interests of the United States will not permit waiting for the Comptroller General's decision (as long as the award was otherwise likely to occur within 30 days after the making of the finding), and (2) the Comptroller General is advised of that agency's finding.[303]

Similar suspension of performance rules apply if the contract has already been awarded. If a protest is filed within 10 calendar days after the award (or 5 calendar days after the date offered for a required debriefing) and the agency is notified of the protest, the contracting officer may not authorize contract performance to begin while the protest is pending. Where contract performance has already begun, the contracting officer shall immediately direct the contractor to cease performance and suspend related activities.[304] The head of the contracting agency can authorize performance of the contract notwithstanding the protest (1) upon a written finding that the performance is in the best interest of the United States, or urgent and compelling circumstances that significantly affect the United States will not permit waiting for the Comptroller General's decision, and (2) after the Comptroller General has been notified of that finding.[305]

---

[297] FAR § 15.505(a)(1).
[298] FAR § 15.506(a)(1).
[299] FAR §§ 15.505(b); 15.506(a)(1).
[300] FAR §§ 15.505(a)(3); 15.506(a)(3).
[301] 4 C.F.R. § 21.3(a).
[302] 4 C.F.R. § 21.6 and 31 U.S.C. § 3553(c)(1).
[303] 31 U.S.C. § 3553(c)(2) and (3).
[304] 31 U.S.C. § 3553(d)(3)(A).
[305] 31 U.S.C. § 3553(d)(3)(C).

Under 4 C.F.R. § 21.1(b), a protest must be in writing and addressed to General Counsel, Government Accountability Office, 441 G Street, NW, Washington, D.C. 20548, Attention: Procurement Law Control Group. The protest must include the name, street address, electronic mail address and telephone and facsimile numbers of the protester and be signed by the protester or its representative. It must also identify the contracting activity and the solicitation and/or contract number and include a detailed statement of the legal and factual grounds of the protest, including copies of all relevant documents. Finally, the protest must set out all information establishing that the protester is an interested party, set out information establishing the timeliness of the protest, specifically request a ruling by the Comptroller General, and state the form of relief requested.[306]

A protest shall not be deemed filed unless it is actually received by the GAO within the time for filing and is accompanied by a certificate that a copy of the protest, together with relevant documents not issued by the contracting agency, was concurrently served on the contracting agency that is designated in the solicitation or, if there is no designated entity, to the contracting officer.[307] No formal briefs or other technical forms of pleadings are required.[308]

The GAO's regulations permit consideration of untimely protests raising significant issues. For example, the Comptroller General invoked its discretion to consider an untimely protest under the "significant issue" exception in *Reliable Trash Service Co.,*[309] where the agency's record clearly indicated that bids could not have been evaluated on a common basis. In *Associated Professional Enterprises, Inc.,*[310] however, the Comptroller General held that the "good cause" exception to the timeliness rule will be limited in future cases to circumstances in which a compelling reason beyond the protester's control prevents timely filing.

In most cases, the agency that is the object of the protest is required to issue an agency report commenting on the bases for the protest within 30 days of receiving notice of the protest from the GAO.[311] The protester must submit comments to the GAO and the other participating parties on the agency report within 10 calendar days of receipt of the report or the protest will be automatically dismissed.[312]

The Comptroller General is required to render a decision within 100 days after the protest is filed.[313] If the Comptroller General finds that the protested solicitation, proposed award, or actual award did not comply with statute or regulation, it will likely recommend that the agency implement corrective action to amend its error(s). Corrective action can take many forms, including a recommendation from the GAO that the agency refrain from awarding the contract or exercising options under the

---

[306]4 C.F.R. § 21.1(c).
[307]4 C.F.R. §§ 21.0(g) and 21.1(e).
[308]4 C.F.R. § 21.1(f).
[309]*Reliable Trash Service Co. of MD., Inc.,* Comp. Gen. Dec. B-234367, 89-1 CPD ¶ 535.
[310]*Associated Prof'l Enters., Inc.,* Comp. Gen. Dec. B-235066.2, 89-1 CPD ¶ 480.
[311]4 C.F.R. § 21.3(c).
[312]4 C.F.R. § 21.3(i). The GAO reserves the right to extend or reduce the comment period.
[313]4 C.F.R. § 21.9(a).

contract, terminate the contract, recompete the contract, issue a new solicitation, award a contract consistent with the law, or take other actions as deemed appropriate.[314] If the Comptroller General determines that the agency has not followed applicable statutes or regulations, it may find the protester is entitled to its bid or proposal preparation costs, protest costs, and reasonable attorneys' fees.[315]

Comptroller General decisions are advisory only, and agencies are not required to follow them.[316] However, the Competition in Contracting Act requires an agency to provide a full report to Congress explaining any refusal to follow a GAO decision.[317] As such, GAO decisions are rarely, if ever, not fully implemented by a federal agency.

F. Court of Federal Claims Protest Actions

The Court of Federal Claims does not have time deadlines comparable to those of GAO for filing a protest. The court will accept jurisdiction over a protest so long as it is filed within the general (six-year) statute of limitations on actions against the United States. The court has specifically rejected assertions that it should adopt the GAO's rules on the timeliness of filing protests.[318] However, the court, like the GAO, requires protests alleging improprieties, ambiguities, or defects in a solicitation to be filed before bids or proposals are due to be received by the procuring agency.[319] In *Blue and Gold Fleet*,[320] the Court of Appeals for the Federal Circuit held that when a contractor has the opportunity to object to the terms of a solicitation containing a *patent*, or obvious, error or ambiguity and fails to do so before the close of the bidding process, the contractor waives its ability to raise the patent error or ambiguity in a bid protest at the Court of Federal Claims.

This rule exists to prevent contractors from waiting until after the evaluation process and after contract award to raise challenges to the terms of the solicitation because such a tactic leads to wasted effort on the part of the procuring agency and time-consuming and costly litigation for the government and the contract awardee. It is therefore imperative for contractors to thoroughly review and analyze all pertinent portions of a solicitation and to raise any potential concerns regarding the solicitation with the agency well in advance of the date offers are due. The contractor should consider filing a preaward bid protest at the GAO or the Court of Federal Claims should the agency fail to address those concerns when the terms of the solicitation prejudice the contractor or put it at a competitive disadvantage compared to other offerors.

---

[314] 4 C.F.R. § 21.8(a).

[315] 4 C.F.R. § 21.8(d).

[316] *Ameron, Inc. v. U.S. Army Corps of Eng'rs*, 809 F.2d 979, 995 (3d Cir. 1986), *cert. granted*, 485 U.S. 958, *cert. dismissed*, 488 U.S. 918 (1988). (The GAO is an arm of Congress, whereas the federal agencies are part of the Executive Branch.)

[317] 31 U.S.C. § 3554(e)(1).

[318] *Software Testing Solutions, Inc. v. United States*, 58 Fed. Cl. 533, 535-6 (2003).

[319] *Blue & Gold Fleet, L.P. v. United States*, 492 F.3d 1308 (Fed. Cir. 2007).

[320] *Id.*

In addition to the patent error/ambiguity rule, the Court of Federal Claims will also invoke the doctrine of laches if it concludes that a protest action is "stale" and the protester (plaintiff) was not diligent in seeking relief.[321] Therefore, one of the most important aspects of bid protests at the Court of Federal Claims is quick action. The earlier a protest is filed, the more willing the Court of Federal Claims will be to implement injunctive relief prohibiting the evaluation of proposals, the award of a contract, or the performance of the contract.

The bid protest jurisdiction at the Court of Federal Claims is founded on the Tucker Act, 28 U.S.C. § 1491. The Court of Federal Claims is substantially similar to the GAO bid protest process in that a protester:

(1)   must be an *interested party*;

(2)   must challenge improprieties on the face of the solicitation or apparent to the bidder prior to the time proposals are due;

(3)   is capable of requesting a stay of the contract award or a stay of contract performance under the Court's injunctive relief powers; and

(4)   may obtain its bid and proposal preparation costs in the event its protest claims are successful.

The Court of Federal Claims, unlike the GAO, has the ability to grant a protester declaratory or injunctive relief under its bid protest jurisdiction with which an agency must comply. The nonmonetary relief available to protesters takes several forms, which include:

(1)   a temporary restraining order;

(2)   a preliminary injunction;

(3)   declaratory relief; or

(4)   a permanent injunction.

The court may use one or a combination of these forms of relief to direct an agency to refrain from awarding the contract or exercising options under the contract, terminate the contract, recompete the contract, issue a new solicitation, award a contract consistent with the law, or take other actions.

Often a contractor will want to obtain a temporary restraining order or an injunction to prevent the award of a contract or commencement of contract performance by another bidder. This is particularly true on contracts where the agency has indicated that it intends to award the contract or allow contract performance despite the previous implementation of an automatic stay by the GAO.

A temporary restraining order and an injunction suspend any further activity on the contact, whether award or performance, while the court or appropriate agency has the opportunity to decide the merits of the protest. The court can grant a temporary restraining order for no more than 10 days but can tailor the grant of injunctive relief for as long as deemed necessary by the court to provide relief to the protester.

---

[321]*EDP Enters., Inc. v. United States,* 56 Fed. Cl. 498, 501 (2003) (delay in seeking enforcement of rights is evidence of a lack of irreparable harm).

A protester can be required to post a bond for security if a restraining order is issued.[322] The most a contractor can expect in the way of monetary relief in a Court of Federal Claims protest is the cost of bid or proposal preparation. Protesters are not capable of recovering lost or anticipated profits.[323] The protester may also recover certain legal fees and costs under the Equal Access to Justice Act if it satisfies the requirements of that Act.[324]

G. Protests Related to Socioeconomic Preference Programs

1. Overview

As discussed in **Chapter 5**, government contracting has long been a vehicle to promote social and economic change. Examples of the policies and programs designed to benefit particular social and economic sectors include various programs administered by the Small Business Administration (SBA), such as:

- Certificates of Competency affecting responsibility determinations of small business concerns;
- Small business set-asides;
- SBA Section 8(a) contracts;
- HUBZone contracts;
- Service Disabled Veteran Owned Set-Asides; and
- Mentor-Protégé Agreements

A basic requirement to benefit from one or more of these programs is that the contractor be a *small business concern* under the SBA's regulations.[325] In order to obtain the benefits of that status, a bidder or offeror represents on a particular procurement that it is a small business concern under the size standard applicable to that solicitation and that it has *not* been determined by the SBA to be other than a small business.[326] The contracting officer is required to accept this representation unless (1) another bidder/offeror or interested party challenges that status or (2) the contracting officer has a reason to question the representation.[327]

Given the importance of a concern's status as a small business and the fact that small business status involves a self-representation, there is a potential for challenges regarding the size status or eligibility of a firm under one or more of the SBA's programs. Generally, these issues or questions fall under the jurisdiction of the SBA. A common element in these SBA programs is a requirement that the bidder or offeror

---

[322]Fed. R. Civ. P. 65(c).
[323]*Rockwell Int'l Corp. v. U.S.,* 8 Cl. Ct. 662 (1985); *Heyer Products Co. v. U.S.,* 140 F. Supp. 409 (Ct. Cl. 1956).
[324]28 U.S.C. § 2412(d)(1)(A); *Crux Computer Corp. v. U.S.,* 24 Ct. Cl. 223 (1991). *See* **Section H** of this chapter and **Chapter 15.**
[325]*See* **Chapter 5**; 13 C.F.R. Part 121.
[326]FAR § 19.301(a).
[327]FAR § 19.301(b).

represent (self-certify) its small business status.[328] Once a firm affirmatively represents its small business status during a procurement, it is subject to having its status protested by other bidders/offerors (*interested parties*). As detailed in **Section G.2**, the SBA protest procedures are substantially similar from SBA program to program. Small business status protests are addressed by the SBA's Office of Hearings and Appeals (OHA). The OHA decisions may be researched at this Web site: *www.sba .gov/aboutsba/sbaprograms/oha/ohadecisions/index.html*

2. Summary of SBA Protest Procedures

Table 3.1 summarizes the SBA protest procedures and deadlines for challenging a firm's claimed small business status.[329] In all cases, a protest by a bidder or offeror must be submitted to the contracting officer.

Protests submitted after the five-day deadline to the contracting officer still must be forwarded by the SBA but must be dismissed by the SBA as untimely.[330]

H. Recovery of Attorney's Fees and Expenses in Bid Protests

Since many bid protests involve complex issues of procurement law, it is not unusual for a protesting party to involve legal counsel. Often the fees and expenses in pursuing a bid protest at the GAO or the Court of Federal Claims can be significant. Both the GAO and the Court of Federal Claims provide procedures that may entitle a successful protesting party to recover some or all of its legal fees and expenses. The basis and standards for relief in each forum are discussed next.

1. GAO Protests

Under the Competition in Contracting Act (CICA) provisions set forth in 31 U.S.C. § 3554(c)(1) and FAR § 33.104(h), the GAO is authorized to recommend that an agency pay bid and proposal preparation costs and certain fees and expenses associated with filing and pursuing a bid protest where the GAO determines that a solicitation, proposed contract award, or an actual contract award does not comply with applicable law or regulation. These costs may include (1) bid and proposal preparation costs and (2) reasonable attorney, consultant, and expert fees. These limits are placed on these fees:

---

[328]The SBA's regulations detail the penalties for misrepresentation and false statements regarding a firm's status. These are contained in 13 C.F.R. § 121.108 for small business, 13 C.F.R § 124.501 for 8(a) small business, 13 C.F.R. § 124.1011 for small disadvantaged business, 13 C.F.R. § 125.29 for SDVO small business, and 13 C.F.R. § 126.900 for HUBZone small business.

[329]These procedures and time periods are subject to change in the *Federal Register,* the Code of Federal Regulations, and the FAR. Check the current regulations for the applicable procedures and time periods for action.

[330]13 C.F.R. § 121.1004. FAR § 19.302(j) implies that the information in the untimely protest to the SBA may be considered by the SBA for future procurements. SBA's regulations do not include a similar representation of future consideration. If a new protest arises, the prudent course is to resubmit the information within the required time frame rather than expect the SBA to locate that information.

Table 3.1 SBA Protest Procedures

Program	Protest Initiated By	Timing Requirements	Writing Requirements	Information Required
Small Business Size Status[a]	Offeror or other interested party	*Bids:* Protest to contracting officer within 5 business days of bid opening	Verbal, confirmed in writing[b]	Specific detailed factual basis for protest
		Negotiated Contracts: Protest to contracting officer within 5 business days after contracting officer's notification of identity of prospective awardee	Verbal, confirmed in writing[b]	Specific detailed factual basis for protest
	Contracting officer	Anytime	Yes	Protester's submission plus information required by 13 C.F.R. § 121.1006
Small Disadvantaged Business Status[c]	Other offerer who is responsive or within the competitive range	Same as size status protests	Yes	Specific facts and allegations
	Contracting officer	Anytime	Yes	Protester's submission plus information required by FAR § 19.305(e)
HUBZone Status[d]	Other offeror or interested party on competitive procurements	Same as size protests[d]	Yes	Specific grounds for protest
	Contracting officer	Anytime	Yes	Specific grounds for protest
SDVO Status[e]	Other offeror or interested party on competitive procurements	Same as size protests[f]	Yes	Specific grounds for protest
	Contracting officer	Anytime	Yes	All specific grounds for protest

[a]FAR §19.302; 13 C.F.R. § 121.1004.

[b]A protest may be initiated orally to the contracting officer, but a confirming letter must be received by the contracting officer within the five-day period or postmarked no later than one day after the verbal protest. *See* FAR § 19.302(d)(1); 13 C.F.R. § 121.1005.

[c]FAR §§ 19.304, 19.305; 13 C.F.R. § 124, Subpart B.

[d]FAR § 19.306.

[e]FAR § 19.307.

[f]*See* 13 C.F.R. § 125.25(d).

- *Attorney's fees.* These may not exceed $150 per hour unless the federal agency determines, on the recommendation of the GAO, that a cost of living or a special factor justifies a higher fee.[331]
- *Consultants and expert witness fees.* These may not exceed the highest rate of compensation for expert witnesses paid by the federal government pursuant to 5 U.S.C. § 3109 and 5 C.F.R. § 304.105.[332]

Operating within those restrictions, the GAO has applied local cost of living factors to increase the allowed attorney's fees rate[333] and has determined that hourly rates for attorneys as high as $475 were reasonable, given the issues in the case and the customary rates charged by other attorneys with comparable experience for similar work.[334] Rates as high as $80 per hour have been accepted for the work of legal assistants.[335] However, the GAO closely reviews the actual time entries (descriptions and amount of time) for reasonableness[336] and excludes time that does not appear to have been spent pursuing[337] the bid protest before the GAO. If the federal agency unreasonably delays consideration of the protesting firm's claim for reimbursement of the fees and expenses in pursuing a bid protest, the GAO has recommended payment of the additional costs incurred by the protester in presenting its claim for reimbursement of the bid protest fees and expenses before the GAO.[338]

Certain costs and expenses that a contractor might feel are related to a protest are not recoverable. For example, the time spent exploring settlement is not considered as time spent in pursuit of the protest.[339] Although the rates of the protesting firm's employees may include actual rates of compensation, plus reasonable overhead and fringe benefits,[340] reimbursement may not include profit or be based on so-called market rates.[341] Reimbursement for consultant fees have been limited to the highest rate of pay for a federal government employee, even where the consultant billed at a higher rate.[342] Finally, the Comptroller General, in *Princeton Gamma-Tech, Inc.*,[343] has held that costs incurred in connection with agency-level protest cannot be reimbursed under the GAO's rules permitting reimbursement of costs for a prevailing protester.

The GAO will include any recommendation that the agency pay the protester's costs of filing and pursuing the protest before the GAO and/or bid or proposal

---

[331]FAR § 33.104(h)(5)(ii) provides that the $150 per hour cap on attorneys' fees "constitutes a benchmark as to a 'reasonable' level for attorneys' fees for small businesses."

[332]FAR § 33.104(h)(5)(i).

[333]*Department of State-Costs*, B-295352.5, 2005 CPD ¶ 145.

[334]*Courtsmart Digital Sys., Inc.-Costs*, B-292995.7, 2005 CPD ¶ 47. The CICA statutory cap was not applied since the protesting party established that it was a small business. The GAO's decision made no reference to FAR § 33.104(h)(5)(ii).

[335]*Id.*

[336]*Blue Rock Structures, Inc.-Costs*, B-293134.2, 2005 CPD ¶ 190.

[337]*Id.*

[338]*Galen Med. Assocs., Inc.-Costs*, B-288661.6, 2002 CPD ¶ 114.

[339]*Id.*

[340]*SKJ Assocs., Inc.-Costs*, B-291533.3, 2003 CPD ¶ 130.

[341]*Id.*

[342]*ITT Fed. Servs. Int'l Corp.-Costs*, B-296783.4 *et al.*, 2006 CPD ¶ 72.

[343]*Princeton Gamma-Tech, Inc.-Costs*, Comp. Gen. Dec. B-228052.5, 89-1 CPD ¶ 401.

preparation costs in its decision. Under the GAO's bid protest rules, the protester is required to file its claim for costs and expenses "detailing and certifying the time expended and costs incurred" with the agency within 60 days after receipt of the GAO's decision. Failure to file the claim within the required time frame may result in forfeiture of the right to recover those costs and expenses.[344]

On occasion, the federal government agency may elect to initiate corrective action under 4 C.F.R. § 21.8 after a protest has been filed with the GAO but before the GAO issues a decision on the matter. The GAO's bid protest rules allow the GAO to recommend that the agency pay the costs of filing and pursuing the protest if the protesting party files a request with the GAO within 15 days of the date on which the protester learned (or should have learned) that the GAO had closed the protest based on the agency's decision to take corrective action.[345]

Although not stated in its bid protest regulations, the GAO has consistently rejected requests that the agency pay a protester's costs for filing and pursuing a bid protest before it if the agency takes corrective action, except in rare circumstances where the GAO concludes that the agency "unduly delayed" taking corrective action in the face of a clearly meritorious protest.[346] The "unduly delayed" standard basically refers to delay after the protest is filed with the GAO.[347] Furthermore, the GAO has recommended that agencies pay protester's costs if the federal agency unduly delays implementing the promised or GAO recommended corrective action that led to the dismissal of the earlier protest.[348]

2. Court of Federal Claims Protest Actions

Under the Tucker Act, the court's jurisdiction is limited to awarding bid preparation and proposal costs to a successful protester if that protester had a substantial chance of receiving the contract award.[349] The Court of Federal Claims has rejected claims by protesters seeking to recover bid protest costs incurred pursuing protests at the agency or GAO level on the grounds that the court lacked jurisdiction over such claims.[350]

Notwithstanding that the Court of Federal Claims does not have jurisdiction over claims for bid protest costs beyond bid preparation and proposal costs, parties that qualify for recovery of legal fees under the Equal Access to Justice Act have received awards of legal fees from the court pursuant to that statute in the context of bid protests.[351]

---

[344]4 C.F.R. § 21.8(f)(1).

[345]4 C.F.R. § 21.8(e).

[346]*Pemco Aeroplex, Inc.-Recon. and Costs*, B-275587.5, 97-2 CPD ¶ 102; *T Square Logistics Serv. Corp.-Costs*, B-297790.4, 2006 CPD ¶ 78.

[347]*J&J/BMAR-Joint Venture, LLP-Costs*, B-290316.7, 2003 CPD ¶ 129.

[348]*Commercial Energies, Inc.-Recon. and Declaration of Entitlement to Costs*, B-243718.2, 91-2 ¶ CPD 499.

[349]28 U.S.C. § 1491(b)(2).

[350]*S.K.J. Assocs., Inc. v. United States*, 67 Fed. Cl. 218 (2005).

[351]28 U.S.C. § 2412; *Chapman Law Firm Co. v. United States*, 65 Fed. Cl. 422 (2005); *Filtration Dev. Co., LLC v. United States*, 63 Fed. Cl. 612 (2005). *See* **Chapter 15** for a detailed discussion of the elements needed to recover legal fees under the Equal Access to Justice Act.

➤ LESSONS LEARNED AND ISSUES TO CONSIDER

- A contract is a legally enforceable *promise* or set of promises.
- A valid government contract must have mutual intent to contract including offer and acceptance, consideration, and a government representative who had *actual authority* to bind the government.
- In an *express contract,* the parties manifest their mutuality of intent by words.
- An *implied-in-fact contract* is inferred from the conduct of the parties.
- An *implied-in-law contract* is imposed by operation of law to prevent injustice. However, such *contracts are not actionable* in government construction contracting.
- An unauthorized contract may be *ratified* by a government official with the authority to do so.
- Government construction contracts are awarded through either *sealed bid* procurement or *negotiated* procurement. Negotiated *best value* procurements are the more common procurement process for government construction contracts.
- *Best-value* negotiated procurement permits trade-offs among cost or price and noncost factors and allows the government to accept other than the lowest-priced proposal.
- One of the analytical techniques the government may use in evaluating whether a bid or proposal is *fair and reasonable* is to compare the offered price with an independent government cost estimate.
- If the original government estimate is unreasonably low, a contracting officer's decision to reject all bids and cancel the invitation may be found to be arbitrary and capricious unless there is a *compelling reason* to reject the bids.
- *Bid bonds* are required to be submitted with bids and proposals to ensure that the bidder selected for contract award executes the contract and provides the required Miller Act bonds.
- *Responsiveness* focuses on whether the bid, as submitted, is an offer to perform the exact tasks spelled out in the invitation for bids and whether acceptance will bind the contractor to perform in strict conformance with the invitation.
- Contractors must be intimately familiar with the process designed for *evaluating bids/proposals* in the IFB/RFP to ensure they are in the best possible position to receive the contract award.
- A contractor's *past performance* is generally evaluated during a current acquisition process to ascertain whether that contractor should receive future work.
- Performance evaluations are completed and stored by government agencies in electronic databases, such as PPIRS, CPARS, CCASS, and ACASS. Contractors must be familiar with how performance evaluations are conducted and how they can be *challenged* if an adverse evaluation is received.
- *Responsibility* focuses on whether the offeror has the necessary technical, managerial, and financial capability and integrity to perform the work.

(Continued)

- Initial determinations that a *small business contractor* is not responsible must be referred to the SBA, which is the final decision maker on that issue.
- *Debarments* exclude a contractor from government contracting and government-approved subcontracting for a reasonable, specified period.
- *Suspensions* are temporary disqualifications of a contractor pending the completion of investigation or legal proceedings, when it has been determined that immediate action is necessary to protect the government's interest.
- Congress has enacted legislation mandating the establishment of a new database containing detailed information related to a contractor's *lack of responsibility* over a 5 year span of time. This database will be available to all federal agencies and to public bodies receiving federal grants.
- Contractors have an *affirmative duty* to inform the government of any dishonest or illicit behavior in connection with the performance of a contract by itself or any of its subcontractors.
- A late bid generally will be *rejected* unless improper government action is the sole or paramount cause of the tardy delivery.
- A contractor may obtain *relief* for a bid/proposal mistake that is a clerical or arithmetic error or a misreading of the specifications.
- An error in *business judgment* is not a mistake for which equitable relief is available.
- The contracting officer has an *affirmative duty* to examine all bids and proposals for mistakes.
- *Bid protests* related to a procurement may be filed with the agency, the GAO, or the Court of Federal Claims.
- A protestor must have *standing* to file a bid protest at the GAO or the Court of Federal Claims (i.e., it must be an *interested party* with a direct economic interest in the procurement).
- Protests related to matters involving eligibility as a small business are within the *jurisdiction* of the Small Business Administration.
- Regardless of the forum for the consideration of the protest, *timeliness* is critical.
- Protests regarding *defects or ambiguities* in a solicitation must be filed before the due date for receipt of the bid/proposal. If that deadline is missed, the right to challenge that defect is waived in many cases.
- Under limited circumstances, the GAO or the Court of Federal Claims *may award* a successful protester its costs and expenses, including an allowance for legal fees in pursuing a protest action.

4

CONTRACT TYPES

I. INTRODUCTION

A wide variety of contract types have been developed to govern the relationship between the government and its contractors due to the broad scale and complexity of government contracting. The government has created multiple contract forms with the goal of obtaining quality workmanship and timely performance at a reasonable price[1]—this concept is used frequently referred to as *best value*. In seeking the *best value*, one government technique is the identification and allocation of risk. The risk is allocated by contract terms related to the pricing arrangements as well as the methods and types of project delivery. The concept of a *fair and reasonable* price must be understood in the context of the allocation of risk and responsibilities.

In selecting contract types the government considers such factors as the level of anticipated competition, urgency of the work required, and the type and complexity of the requirement.[2] Analysis of these factors allows the government to assess the degree of risk in contract performance resulting from its novelty, complexity, stability of design, quantity, duration, market conditions, and other features. As uncertainties increase, contract types that assign less risk to contractors are more likely to be used. Although the government strives to place as much risk as reasonably possible on the contractor, it recognizes that at some point it must pay too high a price for contingencies involved with novel or unique projects.

Government contractors should become familiar with the different contract types and their underlying risk allocation principles when considering whether to compete for government projects. The various project delivery systems and contract types used by federal government agencies are examined in the next sections.

---

[1]Contracting at *fair and reasonable* prices is a basic concept, which is restated in multiple provisions of the FAR. *See,* e.g., FAR § 2.101 definition of "pricing"; and FAR §§ 15.402, 15.403-3, 15.404-1, 15.405. Even in a competitive sealed bid procurement, contracting officers are admonished to determine that sealed bid prices are "reasonable" before making an award. *See* FAR § 14.408-2. In negotiated construction procurements, FAR § 36.214(b)(2) directs contracting officers to "make sure both the offeror and the Government estimator completely understand the scope of the work" when the contractor's proposed price is "significantly lower" than the government's estimate.
[2]FAR § 16.104.

II. ORGANIZATION OF A TYPICAL CONSTRUCTION CONTRACT

Chapter 3, Contract Formation, discusses the procedures used by the government to solicit bids or proposals for award of construction contracts. Once the award is made, the resulting contract includes the solicitation provisions and sets forth the duties and responsibilities of both government and the contractor. The typical construction contract contains the clauses required by statute, executive order, the Federal Acquisition Regulation (FAR), and other clauses prescribed in the federal agency's acquisition regulations. The table of contents for a federal government construction contract typically has these items:

- Standard Form 1442: Solicitation, Offer and Award (Construction, Alteration or Repair)
- Standard Form 24: Bid Bond
- Standard Form 25: Performance Bond
- Standard Form 25A: Payment Bond
- Instructions to Bidders or Proposers
- Description of Work/Supplies/Services to be provided
- Bid Form or Proposal Pricing Form
- Evaluation Factors for Award of the Contract
- Representations and Certifications
- Contract Clauses Expressly Set Forth in the Solicitation
- Contract Clauses Incorporated by Reference into the Solicitation
- Special Conditions or Contract Requirements
- General Requirements
- Contract Specifications
- Contract Drawings

Prior to submitting a proposal or bid for a government construction contract, the contractor should review all of the clauses, terms, and conditions contained in or incorporated by reference in the solicitation and recognize that all of them are important regardless of whether the full text of the clause is set forth in the solicitation.

III. PROJECT DELIVERY CATEGORIES

The phrase "contract type" has several different connotations. One way to group government contracts involves the methodology by which projects are delivered to the government. This section examines the principal methodologies used for project delivery including traditional design-bid-build, design-build, construction management, and the early contractor involvement type of delivery system.

A. Design-Bid-Build Contracts

Government construction contracts traditionally were awarded through competitive sealed bidding using the design-bid-build project delivery method. Even in today's *best-value* selection process, design-bid-build remains as a viable delivery method for certain projects. This process involves the prebid preparation and review of design information. In design-bid-build work, design and construction proceed sequentially. Construction begins only after the design is complete. The process is initiated by the government's recognition of a need and the formation of the project's general concept. This concept ultimately is reflected in a complete set of plans and specifications for the entire project—from the initial site work to the final interior finishes. The plans and specifications are then used to solicit bids or proposals from general contractors, which rely on the scope of work defined by the plans and specifications as the basis for their pricing and to solicit subcontractor bids.

The traditional design-bid-build approach is reflected in and reinforced by industry customs and practices. Because of its long use in government contracts, the problems that arise are fairly predicable and can be resolved through established procedures and remedies. Under the *Spearin* doctrine, it is well established that the government generally will be liable to the contractor for additional costs associated with defects in the project plans and design specifications.[3] Similarly, subcontractors are responsible to the general contractor for their work, and they look to the general contractor, or through it to the government, for resolution of design-related problems.

The traditional design-bid-build model affords certain advantages to both the contractor and the government. It provides both with a complete design and the best opportunity to obtain a fixed price through competitive sealed bidding or competitive proposals before performance begins. Also, the government maintains exclusive control over the design professional and the contractor throughout performance of their contract work. This approach generally has been proven reliable and satisfactory.

However, there are certain disadvantages to the design-bid-build concept. The sequence of completing the design before beginning performance or construction is arguably not the most effective use of time and money. Waiting for a complete design before obtaining pricing or commencing performance exposes the government to inflation and delay in use of the construction project. Also, the use of complete plans and specifications to solicit competitive bids or proposals and make award on a fixed-price basis may encourage contractors to use the lowest-priced acceptable standards, which may lead to disputes regarding the interpretation of that standard. The distinction between "design" and "performance" specifications is not always clear, leaving unsuspecting contractors with unanticipated design responsibility associated with so-called *diagrammatic* designs and *performance* specifications. However, in many cases the contractor is normally not responsible for verifying that a detailed design is

---

[3] *United States v. Spearin,* 248 U.S. 132 (1918).

adequate to meet the contract's performance criteria.[4] Responsibility for design may also shift to the contractor in connection with contractor-proposed changes in equipment or installation of the work.[5] New approaches to project delivery and clearer allocations of risk and responsibility have been used by the government to avoid the disadvantages of the traditional design-bid-build approach.

B. Design-Build Contracts

In the mid-1990s the FAR was amended to authorize the use of design-build selection procedures for government construction contracting.[6] The critical difference between design-build contracting and the traditional design-bid-build approach is that the contractor, or *design builder,* is generally responsible for both the design and construction of the project. As a result, the design builder is liable for both design problems and construction defects to the extent that the government did not control the design or provide a conceptual design.[7]

Federal government design-build construction contracts typically are awarded through a two-phase procedure.[8] The procedure can be fairly straightforward, although the clarity of the evaluation criteria and their application are often a concern within the construction industry. The government develops, either in house or by separate contract, a general scope of work statement that defines the project and provides prospective offerors with sufficient information to enable them to submit proposals in response to a request for proposals (RFP). This work statement may or may not include partial design information. In phase one, the government solicits qualifications from design-build firms.

Design-build firms are not required to submit detailed design or price information during the first procurement phase.[9] The government is required to select candidates based on their technical qualifications rather than price. In addition to the scope of work, the phase one solicitation must identify the phase one evaluation factors, phase two evaluation factors, and the maximum number of offerors that will be selected to submit phase two proposals. At the end of the first phase, the government generally selects between three and five offerors to participate in phase two.

---

[4]*J.E. Dunn Constr. Co. v. General Services Administration,* GSBCA No. 14477, 00-1 BCA 30,806; *see also Santa Fe Eng'rs, Inc.,* ASBCA No. 24469, 92-1 BCA ¶ 24,665 *aff'd, Santa Fe Eng'rs v. Kelso,* 19 F.3d 39 (Fed. Cir. 1994); *SAE/American-Mid-Atlantic, Inc.,* GSBCA No. 12294 *et al.,* 98-2 BCA ¶ 30,084; *Morrison-Knudsen Co., Inc.,* ASBCA No. 32476, 90-3 BCA ¶ 23,208.

[5]*See Trescon Corp.,* ENGBCA No. 5253, 88-3 BCA ¶ 21,163 (contractor assumed design responsibility associated with contractor-provided design revisions).

[6]*See* FAR Subpart 36.3 (implementing 10 U.S.C. § 2305a and 41 U.S.C. § 253m).

[7]Many design-build contracts are essentially "bridging" designs in which the "design builder" completes the design provided by the government ("connects the dots"). In those cases, the determination of the party responsible for the design and the risk of application of the *Spearin* doctrine can present complex factual issues. *See M.A. Mortenson Co.,* ASBCA No. 39978, 93-3 BCA ¶ 26,189 (contractor entitled to equitable adjustment for concrete and steel quantities in excess of those represented in conceptual design); *Donahue Elec., Inc.,* VABCA No. 6618, 03-1 BCA ¶ 32,129 (contractor entitled to equitable adjustment for increased costs associated with boiler when government provided 50 percent design drawings to be used to "complete" design).

[8]*See* FAR Subpart 36.3 (implementing 10 U.S.C. § 2305a and 41 U.S.C. § 253m).

[9]FAR § 36.303-1.

After the government selects the most highly qualified offerors, or *short-listed* firms, these firms are requested to submit phase two competitive proposals.[10] Phase two proposals must include technical proposals and cost or price information.[11] Following review of phase two proposals, the government awards a contract for the design and construction of the project to the design-build firm whose proposal is the "most advantageous to the United States."[12] In making this determination, the contracting agency's consideration is limited to cost or price and the specific evaluation factors set forth in the solicitation.[13]

The resulting contract represents the government's acceptance of the offeror's proposal, which becomes part of the contract. Under *traditional rules* of contract formation, the terms of the proposal generally would control in the event of a discrepancy between the contractor's proposal and the RFP. However, boards of contract appeals typically have not applied this traditional concept when faced with conflicts between proposals and RFP documentation.[14] In addition, some agencies, including the Corps of Engineers, have developed order of precedence clauses specifically designed to reverse the traditional concept of offer and acceptance in the design-build context.

In traditional design-bid-build construction, bidders bid to perform the government contract requirements set forth in the solicitation, and the successful bidder is bound by those requirements. Because of the differences in design-build procurement procedures, one might expect that these rules would not be applied in the design-build context. However, in *United Excel Corp.,* a leading case involving a conflict between the contractor's proposal and RFP specifications, the Department of Veterans Affairs Board of Contract Appeals (VABCA) declined to apply a different rule of contract interpretation than that applicable to traditional design-bid-build contracts.[15]

In *United Excel,* the specifications in a design-build RFP were held to control over a conflicting proposal despite the fact that it was a design-build contract. Specifically, RFP material specifications provided for the use of "aluminum or steel" diffusers. Another specification in the RFP documents required steel diffusers in particular rooms. The design-builder's proposal was based on aluminum, which was the lower-cost alternative.

The contractor argued that the two incongruent specifications created an ambiguity. Although an ambiguity in a contract generally is construed against the drafter, here the government, when the ambiguity is "patent," or obvious, on its face, the ambiguity will be construed against the contractor that failed to seek resolution of the ambiguity. In this case, the contractor argued that patent ambiguity rule of construction is unduly harsh on a contractor that is forced to bid on incomplete plans inherent in the design-build context. The contractor specifically sought a new rule that would allocate fault through the principles of comparative negligence.

---

[10]FAR § 36.303-1(b).
[11]FAR § 36.303-2.
[12]41 U.S.C. § 253b(d)(3).
[13]*Id.*
[14]*See United Excel Corp.,* VABCA No. 6937, 04-1 BCA ¶ 32,485.
[15]*Id.*

The board refused the contractor's request for an equitable adjustment for the additional costs associated with using the more expensive steel diffusers because the contractor did not make a preproposal inquiry. The board did not discuss the traditional rules of offer and acceptance or the legal effect of making the offeror's design-build proposal a part of the resulting contract. Instead, it found nothing in the law or the contract to establish a new rule of law, based on comparative negligence, for allocating the risk of patent ambiguities in the specifications of a design-build RFP.[16] Thus, design-builders should not assume that ambiguities in an RFP will be resolved in favor of their proposals.

Federal agencies have also attempted to preclude a conclusion that the contractor's proposal is the final offer, the terms of which would control after acceptance. For example, the Corps of Engineers utilizes a special order of precedence clause in design-build solicitations that is applied when elements of the contractor's proposal differ from the solicitation. That provision provides:

In the event of conflict or inconsistency between any of the provisions of this contract, precedence shall be given in the following order:

(1) Betterments: Any portions of the accepted proposal which both conform to and exceed the provisions of the solicitation.

(2) The provisions of the solicitations.

(3) All other provisions of the accepted proposal.

(4) Any design products including, but not limited to, plans, specifications, engineering studies and analyses, shop drawings, equipment installation drawings, etc. These are "deliverables" under the contract and are not part of the contract itself. Design products must conform with all provisions of the contract, in the order of precedence herein.

The term "betterment" has not been defined, but typically means an improvement that increases value.[17] Design-build firms submitting proposals to the Corps of Engineers or other agencies employing similar clauses should be mindful of this type of provision and carefully assess their proposals and RFP documents for consistency. If the proposal varies from the RFP or offers a system or installed product that is arguably of a different standard or quality than that *specified* in the RFP, the offeror should clearly identify and clarify that difference prior to submitting its initial technical and price proposal to the agency.[18]

---

[16]*Id.*

[17]*Black's Law Dictionary* (8th ed. 2004).

[18]While many negotiated awards follow an opportunity for discussions with the government's representatives, most RFPs contain a provision providing that the government may award a contract without dicussions. *See* FAR § 52.215-1(f)(4).

C. Construction Management Contracts

Another construction contracting technique that federal agencies have used is construction management (CM). In this procurement method, the government employs a professional "construction manager" to administer the design and construction processes and control construction costs. The construction management firm is employed because of its professional knowledge of the construction process, including cost and schedule control. Typically it is responsible for selection and on-site management of specialty or trade contractors and quality control during construction. In addition, construction managers often are required to evaluate design criteria against current market conditions to ensure constructability within the project budget.

The term "construction management" describes a broad range of services. Generally, the role of the construction manager depends on whether that firm most closely resembles the architect or general contractor (GC) in traditional construction. The key distinction is whether the risk of completing the project on time and within budget is the responsibility of the construction manager. When these risks have been contractually shifted to the construction manager, the delivery system is often known as *CM at risk* or *CM/GC*. When some or all of the risk of completion or cost is retained by the government, the term "CM agency" is generally used.

However, these terms often are used in a less than consistent manner by the federal agencies and the contractors providing CM services to the government. The CM label may be misleading, and it is essential that any CM-type contract be carefully reviewed during the proposal stage to evaluate issues and questions such as:

- CM discretion in the trade contractor (subcontractor) selection process. Must the CM justify to the government using other than the low-bonded firm?
- Is there a guaranteed maximum price (GMP)? If so, are there conditions or limits on its application?
- Risk of cost overruns once the trade contractors are engaged.
- Risk of cost overruns if the final design scope cannot be procured at the estimated price or GMP.
- Responsibility for trade contractor coordination. Who bears the risk of the delay or disruption?
- Who actually controls the trade contractors in terms of authority to direct performance, withhold progress payments, and so on?
- Risk of time overruns. The absence of a liquidated damages clause may or may not be beneficial to the CM firm.
- Risk of quality control issues, long-term warranties, and so on.
- Risk of gaps or overlaps in the buyout of the trade contractors.
- Availability of a contingency. Who controls it? What is it used for?
- Who bonds the project?
- Does the contract have unusual requirements for contractor-prepared composite coordination drawings?

Currently there is no standard-form government contract for CM services used by the various federal agencies. Agencies such as the General Services Administration (GSA) and the Corps of Engineers allow their offices certain latitude in determining how to use CM services and assign risks and responsibilities to a firm functioning as a CM. This variation in terms and conditions, as well as financial risks, must be evaluated on a project-by-project basis.

D. Early Contractor Involvement Contracts

The CM at risk or CM/GC delivery system is a common description for project delivery in the private sector. In government contacting, some agencies label a similar delivery system as *early contractor involvement* (ECI) or *integrated design construct* (IDc). The GSA refers to the ECI delivery system as an "at risk Construction Manager as constructor" (CMc) delivery system. The VA's description of its IDc project delivery method also states that it is a similar to a CM at risk contract.[19] Although there are variations in terminology and labels, the ECI form of contracting provides a general overview of this type of CM at risk project delivery vehicle in government construction contracting.

ECI, formerly known as the integrated design bid build (IDBB) delivery system, was first utilized in 2007 by the Corps of Engineers (Corps). Rather than issue a design-build solicitation with certain performance criteria, the Corps decided that it would separately engage both the design professional and the CM/GC or constructor at the initial stage of the project. Each party (design professional and constructor) provides estimates for the cost of construction as the design is developed. The constructor's proposal and estimate will include an initial target cost, an initial target profit with incentive provisions, and a ceiling price that is the maximum to be paid to the contractor absent an adjustment under a contract clause providing for an equitable adjustment.[20] The intent of this process is to align the designer's "design to" target cost and the constructor's target price over successive stages and to have both parties engage in a collaborative process ultimately to achieve a mutually acceptable design and final price.

Under the ECI delivery system, the agency solicits a contractor for preconstruction services, which are to be performed contemporaneously with the services of the design professional. Once the preconstruction services and design are nearing completion, the agency may utilize an option in the ECI contract to award the construction phase of the project to that same contractor. In essence, a project delivered using ECI involves a contract to provide services prior to construction and an option for the agency to award construction of the project to the service provider at a firm fixed price.

The contractor may be required to perform a variety of preconstruction services, depending on the specific needs of the project and the scope of work to be performed

---

[19]For the purposes of this chapter, the term "ECI" will be used to describe both it and the IDc project delivery method used by the VA.

[20]*See* FAR § 16.403-2.

concurrently by the design team. Typical preconstruction services include providing input on the constructability of the proposed design, value engineering analysis, and construction phasing and sequencing. Involving the contractor in the early stages of the project gives the agency the opportunity to identify potential issues with the project and to increase the likelihood of an on-time and on-budget completion of the project.

To an extent, the constructor (contractor) appears to provide many of the same preconstruction services as a more traditional CM firm. However, that firm may become the contractor (constructor) on a fixed-price basis. The intent of the process is to allow the government user a greater degree of control over the design than might be achieved in a design-build delivery method while engaging both the constructor and design professional in a process that facilitates contemporaneous consideration of design and construction cost issues. Since both entities are under contract with the government, agencies such as the Corps and the VA reserve the right to adjust the target or ceiling price as desired, revise the scope to maintain a previously established price, and resolve differences between the constructor and the design professional.[21]

On relatively large ECI, IDc, or CM at-risk projects, the agency may elect to initially employ a fixed-price incentive contract (FAR § 16.403) with a provision to later enter into a guaranteed maximum price or firm fixed-price contract.[22] One potentially complicating aspect of this approach in terms of management and expense is the likely requirement that this type contract will specify the use of the Earned Value Management System (EVMS or EVM) for the management of the project and for reporting progress and cost expenditures. The Defense Acquisition University describes the use of EVMS, which is mandated by Office of Management and Budget (OMB) in its Circular A-11, as a technique to relate resource planning, schedule, and cost and to encourage contractors to use effective internal cost and management systems.[23]

---

[21]Although not specifically addressed in the information describing the ECI delivery vehicle, contractors should seek clarification from the Corps of Engineers regarding the application of the cost accounting standards (FAR 30) and 48 C.F.R. Chapter 99 (FAR Appendix) to the contract. Compliance with these standards can be a costly process.

[22]*See* **Section V.C.1** of this chapter for a discussion of fixed-price incentive contracts. In reviewing a RFP with this type of project delivery system and contract, a contractor should anticipate that the modification to a firm fixed-price contract generally will trigger the application of the requirements for the submission of cost or pricing data by the contractor and its subcontractors under FAR Subpart 15.4 if the resulting contract (subcontract) price is not based on adequate price competition. In addition, even though the final contract price is a firm fixed-price, the contractor's costs are subject to audit and review under the Cost Principles found in FAR Part 31. This audit review may include items such as **executive compensation** (FAR § 31.205-5), **insurance costs** (FAR § 31.205.19), **employee morale expenses** (FAR § 31.205-13), etc. In addition, if the contract price exceeds the thresholds set forth in FAR Subpart 9903.2, that contractor may become subject to the Cost Accounting Standards. *See* FAR Part 31 and Chapter 99. The appropriate time to determine if a contractor's accounting system, which may have been developed for a competitive bid firm fixed-price environment, is sufficient for these additional requirements is prior to the submission of a proposal to a federal agency.

[23]*See https://acc.dau.mil.evm* (accessed August 5, 2009).

The FAR mandates the use of EVMS for major acquisition systems and other acquisitions in accordance with agency procedures. FAR § 234.203(c) requires the use of the earned value management clause at FAR § 52.234-4. That clause provides:

EARNED VALUE MANAGEMENT SYSTEM (JULY 2006)

(a) The Contractor shall use an earned value management system (EVMS) that has been determined by the Cognizant Federal Agency (CFA) to be compliant with the guidelines in ANSI/EIA Standard-748 (current version at the time of award) to manage this contract. If the Contractor's current EVMS has not been determined compliant at the time of award, see paragraph (b) of this clause. The Contractor shall submit reports in accordance with the requirements of this contract.

(b) If, at the time of award, the Contractor's EVM System has not been determined by the CFA as complying with EVMS guidelines or the Contractor does not have an existing cost/schedule control system that is compliant with the guidelines in ANSI/EIA Standard-748 (current version at time of award), the Contractor shall—

    (1) Apply the current system to the contract; and

    (2) Take necessary actions to meet the milestones in the Contractor's EVMS plan approved by the Contracting Officer.

(c) The Government will conduct an Integrated Baseline Review (IBR). If a pre-award IBR has not been conducted, a post award IBR shall be conducted as early as practicable after contract award.

(d) The Contracting Officer may require an IBR at—

    (1) Exercise of significant options; or

    (2) Incorporation of major modifications.

(e) Unless a waiver is granted by the CFA, Contractor proposed EVMS changes require approval of the CFA prior to implementation. The CFA will advise the Contractor of the acceptability of such changes within 30 calendar days after receipt of the notice of proposed changes from the Contractor. If the advance approval requirements are waived by the CFA, the Contractor shall disclose EVMS changes to the CFA at least 14 calendar days prior to the effective date of implementation.

(f) The Contractor shall provide access to all pertinent records and data requested by the Contracting Officer or a duly authorized representative as necessary to permit Government surveillance to ensure that the EVMS conforms, and continues to conform, with the performance criteria referenced in paragraph (a) of this clause.

(g) The Contractor shall require the subcontractors specified below to comply with the requirements of this clause: [*Insert list of applicable subcontractors.*]

_____

_____

_____

Department of Defense policy[24] requires the use of EVMS on all cost or incentive contracts of $20 million or more and specifies that the EVMS shall comply with the guidelines in the American National Standards Institute/Electronic Industries Alliance Standard 748, Earned Value Management Systems (ANSI/EIA-748).[25] EVMS focuses on the contractor's cost of performance, schedule objectives, and schedule achievement.

In addition to effectively requiring a contractor to evaluate the adequacy of its cost documentation system, a contractor also needs to determine during the proposal phase if the solicitation documents include a detailed critical path method (CPM) network specification to monitor progress and payments (cash flow). If the RFP contains a requirement for a detailed CPM, the CPM's cost loading of cash flow schedule of values should be evaluated to identify any inconsistencies with the EVMS standards and any inconsistent or duplicative requirements should be reconciled or clarified to avoid unnecessary effort or costs.[26] In its acquisition manual, GSA sets forth this guidance to that agency regarding the use of EVM and its application to a contractor's schedule for performance.[27]

(3) *Performance Schedule.* To ensure that the management control system [EVM] is integrated, the program manager is required to define requirements in the work statement for a schedule showing the sequence of events and the critical path for program milestones or deliverables. *Offerors should be required to use this schedule in preparing their proposals,* and the performance schedule will ultimately result in an Integrated Master Schedule after completion of the IBR [Integrated Baseline Review]. Sample work statement language follows:

"The Contractor must establish a performance schedule that describes in sufficient detail the sequence of events needed to accomplish the requirements of the contract. The performance schedule must also reflect congruent CWBS elements. The Contractor must ensure the performance schedule portrays an integrated schedule plan to meet the milestones and delivery requirements of the

---

[24]*See* DFARS § 234.201. The requirement for the use of the EVMS also creates another basis for a detailed review and audit of the contractor's cost accounting and management systems by representatives of the Defense Contract Management Agency (DCMA) and the Defense Contract Audit Agency (DCAA). *See* Section 11-202.6e of the *DCAA Audit Manual* that discusses DCAA's right to an initial audit of all data affecting contract costs including direct and indirect costs, budgets, and operating forecasts.

[25]A resource for determining the intent of this standard is the "Earned Value Management System Intent Guide" (August 2006) published by the National Defense Industrial Association, Program Management System Committee. A copy of the standard can be obtained at *www.geia.org*; click on the online store. The National Defense Industrial Association's Web site is *www.ndia.org*. Search under "earned value management."[accessed November 6, 2009].

[26]Many *cost* values in a CPM schedule of values may reflect a distribution of line items in the proposal/bid, or subcontractors' values (prices) with the possible addition of the general contractor's overhead and profit.

[27]*See* General Services Administration Acquisition Manual (GSAM) Subpart 534.2, Earned Value Management Systems. The GSAM is available on the Web at *www.acquisition.gov/gsam/gsam.html* [accessed November 6, 2009].

contract. The performance schedule also must identify the program's critical path. The performance schedule is to be constructed using a software tool compatible with standard scheduling software. *The Contractor must submit the performance schedule at the post-award conference* and an updated version monthly in program status reviews." [Emphasis added]

The references to developing a detailed schedule at the time of proposal and to submit a performance schedule at the postaward conference may require an investment of time and expense in the proposal phase that exceeds normal practice. Similarly, if the GSA Acquisition Manual contemplates a submission of a complete performance schedule at the postaward conference, this may entail detailed schedule planning with limited input from the subcontractors or trade contractors. When reviewing a solicitation that contemplates the use of EVM, a potential offeror should seek clarification of these requirements and expectations prior to submitting its price proposal.

Similarly, the ECI contract provisions may also include the standard Payments under Fixed-Price Construction Contract clause found at FAR § 52.232-5. This clause typically contemplates payment from an approved schedule of values, possibly as contained in a schedule developed in accordance with the contract's CPM specification. In that context, "cost" is often the *price* being paid to a subcontractor for some work activity with the possible addition of some portion of the contractor's overhead and profit. A contractor's schedule of values and any related CPM activity values may not reflect or report the subcontractor's cost. However, the FAR EVMS clause, as set forth earlier, contemplates that designated subcontractors shall be required to comply with the ANSI/EIA-748 Standard.

In addition to obtaining a clear understanding from the agency on the relationship of the EVMS requirements and any requirements for the use of a CPM schedule and payment under FAR § 52.232-5, a contractor, proposing on a project subject to the EVMS, needs to evaluate its own budget and cost accounting system in light of a likely audit by DCAA. Finally, the contractor should consider the possible application of these requirements to its prospective subcontractors at the proposal stage as many subcontractors may be unable or unwilling to provide the level of cost reporting contemplated by the standard and clause.

IV. INDEFINITE DELIVERY CONTRACTS

Indefinite delivery contracts are used by the government to acquire supplies (by issuance of a delivery order) or services (by issuance of a task order) when the time or quantities of services or supplies are not fully understood at the time of award.[28] There are three types of indefinite delivery contracts: (1) definite quantity contracts,

---

[28]FAR §§ 16.501-1 and 16.501-2(a).
[29]FAR § 16.501-2.

(2) requirements contracts, and (3) indefinite quantity contracts.[29] These contracts may provide for any appropriate cost or pricing arrangement for an estimated quantity of supplies or services.[30] The primary motivation underlying this contract type is the government's desire to limit its obligation to the minimum quantity of goods or services specified in the contract while having a prearranged contractual arrangement for the acquisition of additional goods or services.[31]

A. Definite Quantity Contracts—FAR § 16.502

A definite quantity contract provides for delivery of a definite quantity of specific supplies or services for a fixed period, with deliveries or performance to be scheduled at designated locations upon order. "Task" or "delivery" orders are issued to the contractor when the government determines its desired locale and/or timing for delivery or performance. Definite quantity contracts are used frequently when it can be determined in advance that a definite quantity of supplies or services will be required during the contract period and the supplies or services are readily available or available within a short lead time.

B. Requirements Contracts—FAR § 16.503

Requirements contracts provide for filling all actual requirements at designated government activities for certain supplies or services during a specific contract period, with deliveries or performance scheduled by placing orders with the contractor. This type of contract obligates the government to order all of its requirements, if any, from the contractor, and the contractor promises to fill all requirements.[32] The government may be held in breach of contract if it performs the contracted for work internally.[33] Before entering into a requirements contract, the government is required to provide a realistic estimated total quantity of goods or services. The estimate may be obtained from records of previous requirements or consumption, or by other means, and should be based on the most current information available.[34] However, the only limitation on the government's freedom to vary its requirements after contract award is that it be done in good faith.[35] Contractors typically have not been successful in recovering damages due to the government's revised requirements.[36]

---

[30]*Id.*

[31]*See id.*

[32]The government breaches the contract when it purchases its requirements from any other source. *See Satellite Servs., Inc.,* B-280945, 98-2 CPD ¶ 125 (solicitation for requirements contract that contained a disclaimer clause purporting to allow government to order services from another contractor rendered contract illusory).

[33]*C&S Park Serv., Inc.,* ENGBCA No. 3624, 78-1 BCA ¶ 13,134.

[34]*AGS-Genesys Corp.,* ASBCA No. 35302, 89-2 BCA ¶ 21,702.

[35]*Id.*

[36]*See, e.g., L&C Europe Contracting Co., Inc.,* ASBCA No. 53270, 04-2 BCA ¶ 32,748.

C.   Indefinite Delivery/Indefinite Quantity Contracts—FAR § 16.504

When the government wishes to purchase services or supplies but is unable to determine the exact amounts required for a project, the agency may consider using an indefinite delivery/indefinite quantity contract. An indefinite delivery/indefinite quantity contract (commonly referred to as ID/IQ contract) provides for an indefinite quantity, within stated limits, of supplies or services during a fixed contractual period. The government places delivery orders (for supplies) or task orders (for services) for individual requirements. Contracts of this type must require the government to order and the contractor to furnish at least a minimum quantity of goods or services, and, if ordered, the contractor must furnish any additional quantities, which may not exceed the stated maximum. Similarly, minimum and/or maximum quantities also may be required for individual task or delivery orders. Indefinite quantity or delivery contracts frequently are used when it is deemed unadvisable for the government to commit itself for more than a minimum quantity. These contracts are used only when recurring need is anticipated. ID/IQ contracts are used commonly by some agencies to procure construction services.

The government's use of ID/IQ contracts for construction projects was recently upheld in *Tyler Construction Group v. United States*.[37] The general contractor in *Tyler* contended that the FAR's authorization of ID/IQ contracts for "services" does not include construction projects. The court disagreed with the contractor, noting that the FAR grants procurement officials authority to use innovative approaches to satisfy the government's procurement needs. The only relevant limit to this permissive exercise of authority would be a statutory or regulatory provision that precludes such authority. Since no prohibition exists in the FAR or elsewhere, the court upheld the use of ID/IQ contracts for construction.

When considering whether to compete for an ID/IQ contract, contractors should consider the minimum order as well as the potential long-term commitment of resources, such as bond capacity and personnel, because task orders are issued and performed at later dates. In addition, the contractor should recognize that the agency's obligation to purchase supplies or services does not extend beyond the specified minimum. Thus, once the agency has purchased the minimum from the contractor, the agency is free to purchase the same supplies or services from other suppliers or even negotiate lower prices with the same contractor. For example, an asbestos abatement contractor entered into a Naval Facilities Engineering Command (NAVFAC) ID/IQ contract to encapsulate asbestos for $5.00 per square foot.[38] Once the government had met the specified $50,000 minimum order, it renegotiated the contract price down to $0.23 per square foot. The contractor argued that the government implicitly threatened the contractor, by forcing the contractor to perform work without

---

[37]*Tyler Constr. Group v. United States*, 570 F.3d 1329 (Fed. Cir. 2009). This ruling reflects the current FAR policy that an agency's acquisition team members exercising personal initiative and sound business judgment may adopt any strategy, policy, or procedure that is in the best interests of the government and is not addressed in the FAR or prohibited by law. *See* FAR § 1.102(d). This policy statement is often advanced as providing a broad grant of discretion to contracting officers.

[38]*Abatement Contracting Corp. v. United States*, 58 Fed. Cl. 594 (2003).

compensation, and improperly classified the remaining abatement work to be performed in an attempt to lower the contract price. The Court of Federal Claims held for the government, stating that once the government had purchased the minimum work at the specified price, it was free to look to another party or to purchase the services at a different price from the same contractor. The contractor's allegations of threats and unfair dealing did not amount to duress, and the court noted that the assertion of the legitimate ID/IQ contract right to purchase the bare minimum does not violate the duties of good faith and fair dealing.

Agency errors, negligence, and misstatements do not obligate the agency to purchase more than the minimum amount stated in an ID/IQ contract. For example, in one case the Defense Industrial Supply Center (DISC) awarded an ID/IQ contract for aluminum sheets, and the specified minimum quantity was 10 percent of the solicitation's estimated annual value of aluminum. The contractor later complained that the DISC negligently prepared the annual estimates.[39] The contractor argued that the DISC's stated annual estimates were significantly overstated due to the agency's errors. The ASBCA ruled in favor of the DISC, stating that the agency had purchased the minimum quantity and that allegations of the agency's superior knowledge, negligent misrepresentation, and misstatement of the annual quantity estimates in the solicitation were immaterial. The board noted that "less than ideal contracting tactics" do not constitute a breach of contract as long as the agency purchases the specified minimum.

On occasion, an agency will award a contract containing estimates of the services or work to be performed and some version of a variation in quantity clause. That clause may appear to trigger the right to an equitable adjustment if there is a substantial variation from the estimated quantity in the actual work ordered by the government. If there is a failure to order the estimated quantity, the question is whether the estimates are just that, as opposed to some form of guarantee. Alternatively, does the variation in quantity clause take precedence and entitle the contractor to an equitable adjustment in the unit price? In *Brink's/Hermes Joint Venture v. Department of State*,[40] the Civilian Board of Contract Appeals rejected the government's assertion that "estimates are just estimates" and ruled that the terms of the contract's variation in quantity clause controlled the determination of the parties' rights. As the actual quantities of some work items amounted to only 3 percent of the estimated quantity in the contract, the board held that the contractor was entitled to reprice the unit rates to account for the variation (underrun).[41]

D. Multiple Task Order Contracts/Single Task Order Contracts

A review of construction solicitations on FedBizOpps (Federal Business Opportunities) quickly demonstrates the degree to which individual sealed bid procurements

---

[39]*Transtar Metals, Inc.*, ASBCA No. 55039, 07-1 BCA ¶ 33,482.
[40]CBCA No. 1188, 09-2 BCA ¶ 34,209.
[41]While this appeal involved a security guard services contract, the board's analysis drew on prior construction cases interpreting variation in estimated quantities clauses.

have been surpassed by best value requests for proposals under FAR Part 15. Many of these RFPs contemplate that the agency will award indefinite duration, indefinite quantity, multiple award task order contracts (ID/IQ MATOCs). In this project delivery system, the agency subsequently issues task orders, which are competed among only those contractors selected during the MATOC award phase. Often the stated rational for this approach is that the agency can select its contractors on factors other than low price, as the agency can consider factors such as: (1) technical merit, (2) past performance, (3) price, and (4) use of small businesses as subcontractors. In addition, the agencies report that the use of a MATOC delivery vehicle reduces the acquisition lead time and allows better coordination of the timing of solicitations in a given regional area. In some MATOCs, the eventual award of a task order may be made on a low-price, technically acceptable basis.

For contractors that are accustomed to competing on a sealed bid basis (award to the low, responsive, and responsible bidder), the use of the best-value MATOC delivery system can result in substantially reduced opportunities to compete for awards because all work of a certain type or in a given geographic area will be essentially restricted to award to those firms previously selected as the MATOC contractors. In *Weeks Marine, Inc. v. United States,* [42] a dredging contractor filed a bid protest in the U.S. Court of Federal Claims challenging a decision by the South Atlantic Division of the Corps of Engineers to include all dredging work in the South Atlantic Division area in a five-year, best-value MATOC procurement. The Court of Federal Claims upheld the challenge and concluded that the Corps' asserted justification for using a MATOC rather than the traditionally employed, sealed bid procurement process lacked any factual support and had no rational basis.[43] On appeal, the Federal Circuit reversed that decision on the basis that the Court of Federal Claims could not substitute its judgment for that of the agency.[44] If the agency put forth a rational basis and consideration of the relevant factors, the court should apply a "highly deferential" rational basis standard of review. As a practical matter, this decision indicates that a challenge to an agency decision to select a particular delivery system such as a MATOC will be extremely difficult regardless of prior practices.

ID/IQ contracts for service tasks or deliveries of supplies may be awarded to a single contractor (SATOCS) or to multiple contractors (MATOCS). Due to the competition that develops between awardees in a multiple award system, agencies are required, to the maximum extent practicable and except in limited circumstances, to make multiple awards of ID/IQ contracts under a single solicitation for the same or similar supplies or services.[45]

Awarding an ID/IQ contract to multiple sources typically lowers the prices of the goods and provides the agency with improved flexibility and speed in the procurement process. Although multiple awards benefit the agency and are statutorily preferred, the contracting officer still must determine whether multiple awards are

---

[42]79 Fed. Cl. 22 (2007).
[43]79 Fed. Cl. at 33.
[44]*Weeks Marine, Inc. v. United States,* 575 F.3d 1352 (2009).
[45]*See Nations, Inc.,* B-272455, 96-2 CPD ¶ 170.

appropriate. In general, the contracting officer avoids situations in which one contractor maintains an exclusive specialty. This may occur when one aspect of the contract's scope of work may realistically be performed by only one awardee. Thus, the contracting officer should consider the nature of the contract's requirements; the expected duration and frequency of the orders; the skills and materials necessary to perform; and the ability to maintain competition among multiple awardees throughout the contract's duration.[46]

A sole-source award of an ID/IQ contract is appropriate when: only one contractor is capable of performing the work due to the unique nature of the requirements; more favorable terms may be had; the administration costs of multiple awards would exceed the expected benefits of multiple awards; the projected orders are so integrally related that only one contractor can perform them; the contract value does not exceed the simplified acquisition threshold; or the best interests of the agency would not be served with multiple awards.[47]

V. PRICING CATEGORIES

Another way to categorize contract types is the method by which the price is determined and the contractor is paid. The various pricing arrangements used by the federal government to compensate contractors are described next.

A. Fixed-Price Contracts—FAR Subpart 16.2

The most frequently used government construction contract is the fixed-price type. Fixed-price contracts provide for a firm price or, in appropriate cases, an adjustable price based on factors other than cost.[48] Fixed-price contracts that provide for an adjustable price may include a ceiling price, a target price, or both. Within these parameters, adjustable fixed-price contracts provide for an upward or downward revision of the contract price upon the occurrence of specified contingencies.[49] The predictability of performance costs and time is the primary factor in determining whether a firm-fixed-price or adjustable fixed-price contract will be used.

1. Firm-Fixed-Price Contracts—FAR § 16.202

Firm-fixed-price contracts provide for a price that is not subject to any adjustment on the basis of the contractor's cost experience in performing the project.[50] This contract type places maximum risk and full responsibility for all costs and resulting profit or

---

[46]FAR § 16.504 (c).
[47]*Id.*
[48]FAR § 16.201.
[49]FAR § 16.203-1.
[50]FAR § 16.202-1.

loss on the contractor.[51] It also provides the maximum incentive for the contractor to control costs and perform efficiently and imposes the least administrative burden on the parties. To the extent the contractor is able to perform at a cost below the firm fixed price, it is able to increase its profit.

The FAR states that the government's selection of the contract type requires the exercise of sound judgment by the contracting officer. The objective is to select[52] a contract type and price that will result in reasonable contractor risk and provide the contractor with the greatest incentive for efficient and economical performance. A firm-fixed-priced contract utilizes the basic profit motive of business enterprise and typically is used when the risk involved is minimal or can be predicted with an acceptable degree of certainty.[53]

For example, firm-fixed-price contracts are appropriate when the government acquires construction services, commercial items, or other supplies or services on the basis of reasonably definite functional or detailed specifications. These factors typically exist when: (1) there is sufficient price competition (whether sealed bids or competitive proposals); (2) there are reasonable price comparisons with prior procurements of the same or similar supplies or services made on a competitive basis or supported by valid cost and pricing data; (3) the government is capable of making a realistic estimate of the probable costs of performance; or (4) performance uncertainties can be identified and reasonable estimates of their potential cost impact can be made.[54] In the latter situation, contractors frequently are required to accept a firm fixed price in connection with a specific contractual assumption of identified risks.[55]

Firm-fixed-price contracts frequently are used for procurement of construction services, particularly those subject to design specifications. The use of firm-fixed-price contracts normally is not appropriate for more indefinite procurements, such as those involving the development of major systems.[56] The use of firm-fixed-price contracts for research and development has been specifically limited.[57]

2. Fixed-Price Contracts with Economic Price Adjustment— FAR § 16.203

In appropriate cases, fixed-price contracts may contain provisions allowing adjustment to the fixed price.[58] These contracts are referred to as fixed-price contracts with economic price adjustment. They may include a ceiling price or target price that may be revised only by equitable adjustment under stated circumstances.[59] Within

---

[51]*Id.*

[52]FAR § 16.103(a) provides that selection of the contract type is "generally a matter for negotiation." As a practical matter, there is little, if any, negotiation with federal agencies regarding the selection of the type of contract used for government construction contracts.

[53]FAR § 16.103(b).

[54]FAR § 16.202-2.

[55]*Id.*

[56]*See United Pac. Ins. Co. v. United States,* 464 F.3d 1325 (Fed. Cir. 2006).

[57]FAR § 35.006(c) (the use of cost-reimbursement-type contracts usually is appropriate).

[58]FAR § 16.201.

[59]*Id.*

these parameters, fixed-price contracts with economic price adjustment provide for an upward or downward revision of the contract price upon the occurrence of specified circumstances.

There are three general types of economic price adjustments: (1) adjustments based on established prices; (2) adjustments based on actual costs of labor or material; and (3) adjustments based on cost indexes of labor or material.[60] Adjustments based on established prices are based on increases or decreases in the agreed-on published or otherwise established prices of specific items. Adjustments due to the actual costs of labor or materials are based on measurable increases or decreases in the specified costs of labor and/or materials that the contractor experiences during performance. Adjustments using cost indexes are based on increases or decreases in labor or material cost indexes that are identified in the contract. Unlike cost-reimbursement contracts, adjustments to fixed-price contracts are not conditioned on the contractor's unique cost experience.[61]

Fixed-price contracts with economic price adjustment protect both parties from identifiable price fluctuations by eliminating the need for the inclusion of contingencies in the bid or proposal. Typically they are used when there is serious doubt concerning the stability of market or labor conditions that will exist during contract performance and contingencies that otherwise would be included in a firm-fixed-price contract can be identified and covered separately in the contract.[62] Price adjustments based on established prices normally should be restricted by industry-wide contingencies. Adjustments based on labor and material costs are expected to be limited to contingencies beyond the contractor's control.

Typically, contracts with economic price adjustment contain detailed procedures for establishing baseline cost or pricing data. When they do not, the contracting officer is responsible for obtaining adequate information to establish the base level from which adjustment will be made.[63] The contractor is responsible for demonstrating the variance from baseline data to support an upward adjustment of price. It is important to note that a contractor may waive its entitlement to adjustment by failing to submit its request and supporting data within the time required and procedures set forth in the contract.[64]

Prior to entering into fixed-price contracts with economic price adjustment, contracting officers must make specific findings that this type of contract is necessary either to protect the contractor and the government from significant fluctuations in labor or material costs or to provide for contract price adjustments in the event of changes in the contractor's established prices.[65] If these specific findings are not made, fixed-price contracts with economic price adjustment are prohibited.[66] As a practical

---

[60]FAR § 16.203-1.

[61]See **Section V.B** of this chapter.

[62]FAR § 16.203-2.

[63]FAR § 16.203-2(b).

[64]*Riggs Nat'l Bank of Washington, D.C. v. General Services Administration,* GSBCA No. 14061, 97-1 BCA ¶ 28,920 (contract provision clearly stating that tenant would lose rights if time restrictions not met is strictly construed).

[65]FAR § 16.203-3.

[66]*Id.*

matter, fixed-price contracts with economic price adjustments are not common in domestic (performed within the United States) government construction projects.

Absent the inclusion of an economic price adjustment provision in a fixed-price contract, the contractor *normally* bears the risk of unexpected price inflation, as illustrated by the decision in *Spindler Construction Corp.*[67] In *Spindler*, the contractor's structural steel subcontractor experienced a 23 percent price increase ($199,008.29) in the cost of prefabricated steel components that occurred over a period of several months. The contractor sought relief based on the legal principle of "commercial impracticability."

The Armed Services Board of Contract Appeals did not reject the doctrine of commercial impracticability. Rather, it held that a 23 percent increase was *not sufficient* to obtain relief. The board noted that in prior decisions, it had held that price increases of 57 and 70 percent were not sufficiently great to justify relief. Finally, the board stated that under a firm-fixed-price contract with no provision for economic price adjustment, the contractor *normally assumes* the risk of price increases.

Spindler reflects the traditional risk allocation in a firm-fixed-price government contract. Some contracting officers may attempt to categorically reject price escalation claims on the basis of the *Spindler* decision and the absence of an economic price adjustment clause in fixed-price construction contracts. This is an overly broad reading of *Spindler* and the decisions cited in it. In *Spindler*, the ASBCA used the phrases "normally assigns" and "general risk" when discussing the allocation of the price escalation risk to the contractor. However, the board did not state, as an absolute rule, that there were *no circumstances* under which the risk of price escalation would be borne by the government.

In *Spindler*, there was no assertion that the government caused delays to the work that affected the procurement of the steel. Similarly, there was no discussion of changes delaying the procurement of the steel. Those omissions are significant because it is well established that a contractor whose work is delayed by the government is entitled to recover resulting wage and material price escalation even though the contract does not contain a price escalation clause.[68] Consequently, the normal or general assignment of risk of price escalation in a government contract is subject to being shifted to the government if the contractor can link the escalation to an event (change, differing site condition, or delay, etc.) for which the government is responsible.

3. *Fixed-Price Contracts with Price Redetermination—FAR §§ 16.205 - 16.206*

It is possible to negotiate a firm-fixed-priced price for goods or services to be provided only within certain limited durations. However, costs during extended performance are less certain, and fixed-price contracts with price redetermination may

---

[67]ASBCA No. 55007, 06-2 BCA ¶ 33,376.
[68]*See Sydney Constr. Co., Inc.*, ASBCA No. 21377, 77-2 BCA 12,719 *and Triple "A" South*, ASBCA No. 43684, 94-2 BCA ¶ 26,609.

be used to address such cost increases. This contract type allows for partial perform-ance, followed by a redetermination of price based on experience and other exter-nal factors. Fixed-price contracts with price redetermination are divided into two subtypes: (1) fixed-price contracts with prospective price redetermination, in which subsequent performance costs are redetermined after an initial period of per-formance; and (2) fixed-price contracts with retroactive price redetermination, in which the price for work performed is subject to redetermination based on the con-tractor's actual cost experience.[69] Retroactive redetermination is permitted only for research and development contacts with an estimated price of $100,000 or less and a reasonable firm-fixed price cannot be negotiated.[70]

B.   Cost-Reimbursement Contracts—FAR Subpart 16.3

Cost-reimbursement contracts provide for payment of allowable incurred costs through various methods described in the contract. Prior to entering into contracts of this type, the government establishes an estimate of the total cost for purposes of appropriating funds and fixes a ceiling that the contractor may not exceed (except at its own risk) without approval by equitable adjustment. The decision to use cost-reimbursement contracts is within the contracting officer's discretion, subject to the governmental policy to use it in circumstances when firm pricing cannot be achieved with reasonable certainty.[71] Cost-plus-a-percentage-of-cost contracts are not permit-ted in government contracting.[72]

In order to use a cost-reimbursable contract, the government must contract with a contractor that has an adequate internal cost accounting system.[73] Similarly, the gov-ernment must exercise appropriate oversight to ensure the use of efficient methods, cost controls, and cost accounting practices. To be allowable, costs must be reason-able, allocable, properly accounted for, and not specifically disallowed.[74] Further, absent an equitable adjustment, costs must be below any limitation of costs or *ceiling* imposed by contract.[75] Cost-type contracts are not permissible for the acquisition of commercial items.[76] See **Chapter** 13 for a discussion of certified cost or pricing data that is often required in the context of an award of a cost-type contract.

---

[69]FAR §§ 16.205, 16.206. (However, it is rare to see a federal construction contract using this type of contract.)

[70]FAR § 16.206-2.

[71]*See Fluor Enters. Inc. v. United States*, 64 Fed. Cl. 461 (2005) (decision to use cost-type contract upheld considering uncertainty of volume of work required); *Surface Tech. Corp.*, B-288317, 01 CPD ¶ 147 (se-lection of cost-type contract found reasonable considering unpredictable nature of the requirements).

[72]FAR § 16.102; *see also* 10 U.S.C. 2306(a) and 41 U.S.C. 254(b).

[73]*See CrystaComm, Inc.*, ASBCA No. 37177, 90-2 BCA ¶ 22,692 (contractor failed to establish required cost accounting system).

[74]*See* FAR § 31.201-2; these cost principles are addressed in **Chapter** 13. In addition, certain negotiated cost-reimbursement-type contracts may result in the application of the Cost Accounting Standards (CAS). *See* FAR Subpart 30.2 and the FAR Appendix at 48 C.F.R. § 9903.201-1. CAS compliance may be an administrative burden and costly if a contractor has to revise its traditional accounting practices.

[75]FAR § 52.232-20 (fully funded); FAR § 52.232-22 (incrementally funded).

[76]FAR § 16.301-3(b).

1. Cost-Type Contracts—Subtypes

Cost-type contracts are further categorized based on the methodology used to determine contract price. All cost-type contracts derive their contract price by reference to the contractor's cost experience on particular projects. The most frequently used cost-type contracts are described next.

a. "Pure" Cost Contracts—FAR § 16.302 A *pure* cost contract, referred to simply as a cost contract, is a cost-reimbursement contract in which the contractor receives no fee. This type of contract typically is reserved for research and development work, particularly with nonprofit educational institutions or for contracts between the federal government and other nonprofit organizations.

b. Cost-Sharing Contracts—FAR § 16.303 A cost-sharing contract is a cost-reimbursement contract in which the contractor receives no fee and is reimbursed only for a predetermined portion of its allowable costs. This type of contract may be used when the contractor agrees to absorb a portion of the costs, with the expectation of receiving some benefit from the completed project other than direct monetary compensation.

c. Cost-Plus-Fixed-Fee Contracts—FAR § 16.306 Cost-plus-fixed-fee contracts are the most commonly used cost-reimbursement contract. They provide for payment to the contractor on the basis of a negotiated fee above cost. The fixed fee is set at the time of contract formation and does not vary in relation to actual costs. The fixed fee, however, may be adjusted as a result of changes in the work performed under the contract. This contract type permits contracting for work that might otherwise present too great a risk to the contractor and too great a price contingency for the government. However, this contract type provides the contractor with only minimum financial incentive to control costs.

Cost-plus-fixed-fee contracts are used frequently when the contract is for performance of research or preliminary exploration or study and the level of effort required is unknown.[77] This type of contract normally is not used in development of major systems once preliminary exploration, studies, and risk reduction have indicated a high degree of probability that the project is achievable and firm performance objectives and pricing can be developed.[78] Cost-plus-fixed-fee contracts have been used in conjunction with novel or unique construction projects. In some instances, a cost-plus-award-fee contract has been used on such projects to provide an incentive for superior performance and effective cost management.

C. Incentive-Type Contracts—FAR Subpart 16.4

Because most cost-type contracts provide contractors little economic incentive to decrease or efficiently manage costs, the government developed cost and fee incentives

---

[77]FAR § 16.306(b)(1)(i).
[78]FAR § 16.306(b)(2).

to discourage inefficiency and waste. Incentive-type contracts typically offer predetermined incentives for certain levels of technical performance or delivery.[79]

Incentive contracts (except for award-fee contracts, treated in **Section V.C.2**) include a target cost, a target profit or fee, and a profit or fee adjustment formula. Within the constraints of any contractual price ceiling or minimum or maximum fee, the contractor's profit or fee is adjusted in relation to the target cost. Generally, when actual costs meet the target, the contractor will be paid the target profit or fee. When actual costs exceed the target, profit or fee will be reduced. Similarly, when actual costs are below target, the contractor is entitled to an upward adjustment of profit or fee in accordance with the contractual formula.[80]

In lieu of target costs, incentive-type contracts may target performance criteria or characteristics for a specific product. Performance targets may include, for example, objective factors such as strength, efficiency, or ability to withstand ranges of temperature. Like cost-incentive contracts, a contractor's profit or fee is adjusted based on achievement of predetermined criteria.[81]

Incentives in the form of increased or decreased profit also may be based on compliance with delivery schedules. The FAR specifically provides for the use of multiple forms of incentives within single contracts.[82] Because various incentives are often interdependent, providing multiple incentives in a single government contract frequently will require contractors to consider trade-offs among incentives. For example, an incentive based only on performance criteria does not encourage minimizing cost while incentives based on both cost and performance encourage both efficient cost and excellent performance.

1. Fixed-Price Incentive Contracts—FAR § 16.403

A fixed-price incentive contract is a fixed-price contract that provides for adjustment of profit and establishment of the final contract price by application of a formula based on the relationship of total final negotiated cost to total target cost. The final price is typically subject to a cost ceiling that is negotiated at the inception of the contract.[83]

Fixed-price incentive contracts often are used when a firm-fixed-price contract is deemed unsuitable, the nature of the supplies or services being acquired are such that the contractor's assumption of some cost responsibility will provide an incentive for effective cost control and performance, and the contract also includes technical performance incentives.[84]

Another variation of the fixed-price incentive contract is the successive target incentive contract. This type of contract sets out initial targets and adjustment formulas and identifies a production point at which a firm target, adjustment formula, and ceilings are negotiated by the parties.[85]

[79]FAR § 16.402-2.
[80]*See, e.g.,* FAR § 16.402-1(b).
[81]*See, e.g.,* FAR § 16.402-2.
[82]*See* FAR § 16.402-4.
[83]*See CTA, Inc.,* ASBCA No. 47062, 00-2 BCA ¶ 30,946.
[84]FAR § 16.403.
[85]FAR § 16.403-2.

Traditionally, fixed-price incentive contracts have been used more commonly on agency supply contracts rather than construction projects. However, the Corps of Engineers, the GSA, and the VA recently have begun to utilize a new project delivery technique called early contractor involvement (ECI) or integrated design and construct (IDc) on several construction projects. These project delivery methods combine features of construction management-type preconstruction services and fixed-price negotiated construction contracts with the use of a separate design firm under contract with the government agency.[86]

At the beginning of the project, the agency utilizes a fixed-price incentive contract with successive target costs, a target profit and a ceiling price. That is, the agency sets a ceiling price that is to be paid to the contractor, excluding amounts paid pursuant to equitable adjustments or other special circumstances, and also specifies an initial target cost for the project. Once the design has reached or is approaching completion, the agency and the contractor negotiate a firm target cost. The parties may negotiate either a firm fixed price, which includes a firm target profit, or a formula for establishing the final price. If the latter option is chosen, the final cost is negotiated at completion and the profit is calculated using the formula. The new project delivery system and the possible use of an Earned Value Management System (EVMS) as well as the application of the FAR Part 31 Cost Principles with detailed audits by the Defense Contract Audit Agency or other government auditors is further discussed in **Section III.D** of this chapter.

2. Fixed-Price Contracts with Award Fees—FAR § 16.404

Award fee provisions may be used in fixed-price contracts when the government wishes to motivate the contractor and other incentives cannot be used because there is not a method by which contractor performance may be measured objectively. These contracts establish a fixed price (including reasonable profit) that the contractor will be paid for satisfactory performance and establish criteria through which the contractor may be entitled to additional compensation for exemplary performance. Contracts with award fees are authorized only when procedures have been established for conducting award fee evaluations and an award fee board has been established.[87]

3. Cost-Reimbursement Incentive Contracts—FAR § 16.304

A cost-plus-incentive-fee contract is a cost-reimbursement contract that provides for an initially negotiated fee to be adjusted later by a formula based on the relationship of all actual allowable costs to total target costs. The same basic principles applicable to other incentive-based contracts apply to cost-plus-incentive-fee contracts.

---

[86]The ECI process is further described in **Section III.D** of this chapter.
[87]FAR § 16.404(b).

VI. OPTIONS

Another procurement method utilized by federal government agencies involves the inclusion of options within contracts. An option allows the government maximum flexibility because the government is provided with a unilateral right within a stated period to purchase additional supplies or services or to extend a contract.[88] As a result, options often are used when the government's requirements beyond a minimum threshold are not certain and full funding has not yet been secured.

The government's use of options is restricted by regulation.[89] Agencies are precluded from using options in circumstances where the contractor will incur undue risks, as would be the case when: the price or availability of labor or materials is not reasonably foreseeable; market prices are likely to change substantially; or the option addresses known government requirements for which funds have been appropriated.[90] Agencies are further discouraged from using options when procurement of option items can be accomplished through subsequent competitive acquisition procedures.[91] Solicitations containing options will state the basis of evaluation, either inclusive or exclusive of the option; and, when appropriate, must inform bidders that the government anticipates that it will exercise the option upon award. Options may be priced by the contractor using any of the pricing mechanisms addressed earlier or by unit price.[92] Unpriced options and other "agreements to agree" are enforceable between the government and contractors if conditioned on an obligation to bargain in good faith.[93] Otherwise, such agreements are unenforceable.[94]

When it is anticipated that the government will exercise one or more options contained in a solicitation, the contracting officer typically is required to evaluate the options prior to awarding a contract.[95] In certain circumstances, such as when it is reasonably certain that funding will not be available for the option, the government need not evaluate options prior to awarding a contract.[96]

A. Exercising Options

The government must comply with applicable statutes and regulations in exercising an option.[97] Agencies may exercise options only after determining that: (1) funds are available; (2) the requirement covered by the option fulfills an existing government need; (3) exercise of the option is the most advantageous method of fulfilling the

---

[88]See FAR § 17.202.

[89]Id.

[90]Id.

[91]FAR § 17.202(b)(1)(ii).

[92]FAR § 17.203.

[93]Aviation Contractor Employees, Inc. v. United States, 945 F.2d 1568 (Fed. Cir. 1991).

[94]Restatement (Second) Contracts § 33.

[95]FAR § 17.206(a).

[96]FAR § 17.206(b).

[97]Golden West Refining Co., EBCA No. C-9208134, 94-3 BCA ¶ 27,184; New England Tank Indus. of N.H., Inc., ASBCA No. 26474, 90-2 BCA ¶ 22,892.

government's need, price, and other factors considered; and, unless exempted, (4) proper notice of the proposed exercise is published in accordance with FAR Part 5.[98] In determining whether to exercise an option, contracting officers must consider whether a new solicitation would produce a better price or more advantageous offer, an informal analysis of the market indicates the option is advantageous, and whether the time between contract award and exercise of the option is sufficiently short so that exercise of the option is most advantageous.[99] These items are to be considered in addition to price.[100]

The government must exercise an option according to the terms of the option.[101] An agency's improper exercise of an option may constitute a breach of contract. In *White Sands Constr., Inc.*, for example, the Corps of Engineers awarded an ID/IQ contract requiring at least 60 days notice prior to executing an option to extend the contract.[102] When the contractor received the required notice 7 days after the 60-day deadline had passed, the contractor brought an appeal based on breach of contract. The ASBCA held for the contractor, finding that preliminary notice is an integral component of the option execution process and, therefore, the Corps had not properly executed the option.

If a contractor contends that an option was exercised improperly and performs, it may be entitled to an equitable adjustment.[103] The government has substantial discretion in determining whether to exercise an option. The decision not to exercise an option is generally not protestable.[104] Conversely, the determination to exercise an option is subject to protest by other potential bidders/offerors.[105]

B. Total Contract Duration with Options

Generally, a contract containing options may not exceed five years in duration.[106] Further, contracts with options must state the period within which the option(s) may be exercised.[107] This period must be set so as to provide the contractor adequate lead time to allow continuous production.[108] Option periods within service contracts may be extended beyond the contract completion date. Typically, extensions of the option

---

[98]FAR §17.207(c); *see also* FAR § 5.201 (agencies required to provide notice of proposed contract actions).

[99]FAR § 17.207(d).

[100]*Id.*

[101]*Lockheed Martin Corp. v. Walker,* 149 F.3d 1377 (Fed. Cir. 1998).

[102]ASBCA No. 51875, 04-1 BCA ¶ 32,598.

[103]*See id.; Lockheed Martin Corp.,* ASBCA No. 45719, 00-2 BCA ¶ 31,025.

[104]*See Young Robinson Assoc., Inc.,* B-242229, 91-1 CPD ¶ 319 (contractor cannot protest failure to exercise option because it is a matter of contract administration).

[105]*See Alice Roofing & Sheet Metal Works, Inc.,* B-283153, 99-2 CPD ¶ 70 (protest denied when agency reasonably determined that option exercise was most advantageous means of satisfying needs).

[106]FAR § 17.204(e); *see Gen. Dynamics C4 Systems, Inc.,* ASBCA No. 454988, 08-1 BCA ¶ 33,779 (recognizing exception for information technology contracts).

[107]FAR § 17.204(b).

[108]*Id.*

period are accomplished when funding is not available for the option quantity within the fiscal year of contract completion.[109]

VII.   PROJECT DELIVERY AND CONTRACT-TYPE RISK ANALYSIS

In analyzing government construction contract risks, the contractor should review the project delivery system and the contract types and clauses. Since the primary contract clauses are reviewed in separate chapters of this book, the risks involved in the delivery system and type of contract are recapped below.

The project delivery system selected for each contract will determine the range of risks involved. The design-bid-build category separates the risks with the government being responsible for the design and the contractor being responsible for construction within the time and price set forth in the contract. The design-build project delivery system can shift much of the risk for design and construction to the design-build contractor. The construction management category has varied risks ranging from very little risk where the CM only provides services as an agent for the government to a high-risk contract where the CM is responsible for completing the project on time and within budget. Finally, the ECI category is a hybrid of these previous delivery systems in that the government has separate contracts with the designer and the builder, but the designer and builder interact during the design process as in design-build contracts and ultimately the contractor is responsible for completing the project on time and within budget.

In addition to the risks in the various project delivery categories, other risks involve quantity of contract work and the price or calculation of the amount to be paid for the work. The indefinite delivery contract category places the quantity of work risk on the contractor with the work ranging from only the amount needed by the government in a requirements contract, to a minimum order or an order within a stated minimum or maximum amount as in the definite and indefinite quantity contracts. The payment amount or basis for calculating the amount to be paid for the work is determined by the pricing category selected by the government, which ranges from a firm-fixed-price contract (with or without price adjustment or redetermination) to cost-reimbursement contracts (with costs only, shared costs, or cost plus fixed fee). In addition, these fixed-price and cost-reimbursement contracts also may have incentives or award fees.

In particular, a contractor should recognize that there can be a fundamental difference in the level of audit and cost oversight of a fixed-price contract depending on the type of solicitation. If the final firm fixed-price was not based on adequate price competition, the contractor's entire cost of performance, as well as its accounting system, may be subject to examination in accordance with the Cost Principles of FAR Part 31. Depending upon the value of the contract, the Cost Accounting Standards may apply. These factors should not deter a firm from competing for an

---

[109]FAR § 17.204(d).

award of a contract. However, they may affect how the contractor evaluates the cost and expense associated with performance. If the requirements are appreciated at the proposal stage, there is less risk of a problem and dispute as the project nears completion and close-out.

VIII. BUILDING INFORMATION MODELING

Building information modeling (BIM) is not a contract type or project delivery vehicle in a strict sense. Rather, it is a computer-based technology using software to convert a traditional two-dimensional design into three dimensions. It allows the project participants to engage in virtual design and construction of the project. However, BIM may substantially alter how the parties engage in the construction process and may change the allocation of the parties' risks and responsibilities in contrast to the more traditional contract types and project delivery vehicles.

Many federal agencies are considering the use of BIM as a means to facilitate the identification of design problems, develop space take-offs, integrate building systems or components into the project design, and provide a more effective long-term building maintenance/renovation tool than the traditional as-built drawings. Beginning in 2007, the GSA became the first federal agency to require the use of BIM on all newly funded major construction and renovation projects. The GSA currently requires a spatial program validation—an application of BIM technology used to quickly and precisely assess a proposed design's performance with respect to spatial requirements. The use of BIM in validating spatial design programs is intended to help eliminate the need to overdesign a building in order to compensate for unreliable data that are often inherent in non-BIM technologies. Beyond using BIM for spatial validation, the GSA is also encouraging the use of BIM technology for applications such as four-dimensional phasing, which allows project participants to better understand the consequences of construction sequencing. The GSA also uses laser scanning to efficiently and accurately capture a building's spatial data.

The Corps of Engineers is also a strong proponent of BIM technology. In October 2006, the Engineer Research and Development Center of the Corps of Engineers issued its BIM study as ERDC-TR-06-10. That document sets forth a comprehensive plan for implementation of BIM on civil works and military construction over the next decade. In this study, the Corps described Building Information Modeling in this manner.

1.1 BACKGROUND

Building Information Modeling is an emerging technology with the potential to enable significant improvement in the speed, cost, and quality of facility planning, design, construction, operations, and maintenance. According to the National Institute of Building Sciences:

A Building Information Model (BIM) is a digital representation of physical and functional characteristics of a facility. As such it serves as a shared knowledge resource for information about a facility forming a reliable basis for decisions during its lifecycle from inception onward.

The potential of BIM stems both from its value as an open interchange mechanism between the tools used to perform the various functions of the AEC industry (standards) and the ability of computational tools to manipulate the model directly, with or without human intervention (computability). In a typical BIM-enabled process, the data model serves as the principal means for communication between activities conducted by professionals. When fully implemented, BIM will increase reuse of design work (decreasing redesign effort); improve the speed and accuracy of transmitted information used in e-commerce; avoid costs of inadequate interoperability; enable automation of design, cost estimating, submittal checking, and construction work; and support operation and maintenance activities.[110]

Recognizing that there are costs and risks associated with BIM, especially as the technology matures, the Corps announced a phased approach to BIM beginning in 2008.[111] The Corps' roadmap calls for its eight Centers of Standardization to achieve an initial level of operating capability for BIM in 2008, 90 percent compliance with the National BIM Standard (NBIMS) by the end of 2010, and full operational capability with NBIMS by the end of 2012. The Corps' contract requirements typically specify the use of BIM during the design phase of a project or mandate the use of BIM to produce construction documents or other project deliverables.

BIM is a technology and, as with many other modern technologies, the technology behind BIM must be implemented with software to produce practical benefits for the construction industry. Numerous software providers offer BIM software packages. Although each provider's BIM software offers unique benefits and applications, all BIM software must achieve certain technological standards in order for the software to exchange information seamlessly with other BIM applications. This concept, known as interoperability, recently has become a concern as various government agencies and private parties have specified the use of different and, at times, incompatible requirements for BIM applications.

Theoretically, all BIM software is developed so that any participant on a BIM-enabled project is able to provide model data that can be read and seamlessly interpreted by users of any BIM software. This means that businesses should be able to purchase BIM software from any vendor, secure in the knowledge that any one BIM application will be compatible with all other applications. In practice, however, the various BIM software providers have implemented the BIM standards differently. The Corps of

---

[110]D. K. Smith (2006), Presentation, "Building Information Models: A Revolution in the Construction Industry." Accessible through URL: *www.nibs.org/BIM/BIM_Revolution.pdf*.
[111]ERDC TR-06-10, p. 9.

Engineers has required the use of BIM software from specific vendors in many of its contracts. A copy of an example of a Corps BIM specification is included on the support Web site at www.wiley.com/go/federalconstructionlaw. The Corps' specification of a single format may lead project participants to partially abandon other formats, thus possibly delaying the adoption of BIM throughout the industry. Interoperability is an evolving issue, and the contractual requirements, technology, and software applications are rapidly changing. The importance of this issue to the success of BIM will motivate the involved parties to work on quickly providing potential solutions.

Recognizing that BIM presents significant opportunities and risks for the construction industry, the Associated General Contractors of America published "The Contractor's Guide to BIM,"[112] which addresses how this software-based three-dimensional process works, the collaboration opportunities and issues, the technical issues, especially the ability of the available software to interoperate, and cost issues for contractors considering BIM. Additionally, ConsensusDOCS LLC has a BIM addendum, ConsensusDOCS No. 301, to address the contractual relationships between the parties on a BIM project. A copy of the addendum is included on the support Web site. Also, several books further address key issues with BIM.[113]

As many government agencies announce plans to utilize BIM over the next five to ten years and expect their contractors to become proficient in the use of this technology, contractors need to consider these potential issues:

BIM CHECKLIST FOR CONTRACTORS

- Determine if BIM is required or desired by the agency. Is the use of BIM an evaluation factor in the award of the contract?
- How does the agency expect the contractor to use BIM on the specific project? Federal agencies may have very different goals and requirements.
- What software is required for the project? Will the current software be fully interoperable with other BIM software used by other contributors on the project?
- Does the agency require the BIM documents to be produced in a certain format? Will the software be fully compatible?
- Determine the resources (equipment and personnel) needed to implement BIM for a particular project or government agency.
- Consider the legal responsibilities for a party's contribution to the model as well a party's access to the model.
- What is the standard of care for each party's contribution to or use of the model?

---

[112]This guide is available for purchase as a hard copy or as a free download of a pdf document at: *www .agc.org/bookstore*. Enter document number 2926.

[113]Dana K. Smith and Michael Tardiff, *Building Information Modeling, A Strategic Implementation Guide* (Hoboken, NJ: John Wiley & Sons, 2009); Chuck Eastman et al., *BIM Handbook, A Guide to Building Information Modeling* (Hoboken, NJ: John Wiley & Sons, 2008).

- Consider the consequences (cost, time, and responsibility) if the BIM process flags a design conflict.
- Determine the procedures and protocols for designating projections derived from a BIM model.
- What legal representations are made as to the dimensional accuracy of the models?
- Who is responsible for any cost, time, and liability related to any design revisions made during a collaborative BIM design process?
- Determine the storage and retrieval requirements for electronic files and data.
- Understand data security issues and various access levels to the BIM models.
- Learn about design rights (*intellectual property rights*) for certain data generated during the BIM design process.
- Understand insurance and bonding risks as BIM is used to facilitate providing preconstruction services.
- Will BIM be utilized for the RFI process?
- Understand subcontract/purchase order terms and conditions.

➤ LESSONS LEARNED AND ISSUES TO CONSIDER

- Different contract types are used by government agencies to *allocate risks.* Prior to submitting a proposal or bid, every contractor should thoroughly analyze the contract risks placed on the contractor, especially any specific contractual assumption of risk.
- *Contract types* are categorized by project delivery, indefinite delivery, pricing, and the nature or purpose of the contract. Communicate with the contracting officer for clarification of the contract type before entering into the contract.
- All clauses in a government contract are significant. Recognize that many *critical clauses* and *specification requirements* may not be shown in the full text of the solicitation documents.
- The type of contract entered into is not controlled by labels or the title of the contract, and government construction contracts must be examined beyond the first page to assess the *risks, rights,* and *obligations* of the parties.
- The traditional approach to construction using *design-bid-build* has many strengths but also weaknesses that have prompted pursuit of other approaches to the construction process.
- *Design-build contracting* represents the most radical departure from the traditional design-bid-build approach to construction by vesting nearly all design and construction responsibilities and resulting liabilities in one party. The dramatic alteration of the traditional roles of the parties in design-build requires special attention to make certain the contract sets out the mutually understood and specific rights and responsibilities of each party.

(Continued)

- Even in the traditional build-to-design approach, contractors can assume *discrete design liability* as the result of performance specifications, value engineering proposals, the shop drawing process, and where secondary design review responsibility is imposed by standard contract clauses.

- The manner in which *alternative contracting methods* differ from clearly defined and accepted practices and roles requires careful attention to avoid unanticipated problems and disputes.

- *Construction management* generally involves an entity with diverse expertise in design, construction, and management in the design and construction process. The precise role of a construction manager on any project, however, can be determined only by reference to specific contract language. CM contracts must be reviewed carefully for specific *risk allocation* provisions.

- At least three agencies are experimenting with a design-to-cost and parallel construct-to-cost process described as *construction manager as constructor, early contractor involvement,* or *integrated design and construct* contracts in lieu of traditional construction management or design-build project delivery systems. These forms of contracting and project delivery are essentially comparable to construction manager at-risk contracts in the private sector.

- Earned value management is a standard for measuring performance and cost on major acquisitions, generally those in excess of $20 million using a cost-reimbursement or incentive-type contract. Construction contractors should seek preproposal clarification of the relationship of the EVM requirements, the related ANSI/EIA-748-A standard, and the agency's contract provisions addressing scheduling and the FAR payment clauses.

- Federal government construction contractors must understand the fundamental differences between *fixed-price* and *cost-reimbursement* contracts.

- Effective and reasonable *accounting procedures* are critical to demonstrate allowable costs when contracting on a cost-reimbursement basis.

- Provisions allowing for upward or downward adjustment in *adjustable fixed-price contracts* should be clearly understood and any ambiguities resolved prior to contract execution.

- Fixed-price contracts subject to economic adjustment are not *cost-reimbursable* contracts, and criteria allowing for adjustments should be understood before entering into the contract.

- All *notice* provisions contained in contracts should be examined carefully. A contractor may waive entitlement to price adjustment by failing to submit a request and supporting data as required and within the times set forth in the contract.

- *Incentive-type* contracts should be examined carefully for potential trade-offs between multiple incentives. Contractors should strive to negotiate the maximum upward pricing adjustment based on the most achievable incentive criteria.

- If the final firm fixed-price of a contract is not based on *adequate price competition*, it is probable that the contractor may subject to more comprehensive and detailed audits by the government, as well as obligations to satisfy the FAR Part 31 Cost Principles for all aspects of its cost of contract performance rather than just the costs associated with requests for equitable adjustments or claims. Compliance with these requirements may present an additional overhead expense, risk, and burden for a contractor accustomed to performing traditional fixed-price contracts where the price was based on adequate price competition.

• Federal agencies are adopting innovative technology, such as *Building Information Modeling.* It is unlikely that this technology will be limited only to very large projects.

5

SOCIOECONOMIC
POLICIES

I. INTRODUCTION

The government often chooses to pursue its social and economic policy goals by exercising its considerable financial clout. For example, the tax code, with its credits and deductions for favored behaviors such as homeownership and charitable contributions is one example of that practice. Federal procurement is another. The government promotes its public policies by targeting its spending, by altering its typical acquisition procedures through noncompetitive preferences for particular groups of contractors and goods, and by attaching mandates to its funds in the form of required compliance with various labor, employment, and environmental standards.

The use of government contracting to effect social and economic change can be traced back to 1932 with the formation of the Reconstruction Finance Corporation (RFC), created by President Herbert Hoover to alleviate the financial crisis of the Great Depression. During World War II, other government agencies were formed to assist small businesses in competing for wartime contracts. In 1942 the Smaller War Plants Corporation promoted effective utilization of small businesses producing war materiel and essential civilian supplies. During the Korean War, Congress created the Small Defense Plants Administration (SDPA). The SDPA certified small businesses to the RFC. The RFC was succeeded by the Small Business Administration (SBA) in 1953.

This chapter addresses government policies and programs aimed at achieving certain social or economic benefits through the award and the administration of government contracts. These policies include set-asides and preferences in procurement for different categories of small or disadvantaged businesses, equal opportunity and affirmative action employment requirements, domestic product purchasing preferences, workplace safety and labor standards, and environmental regulations.

II. SMALL BUSINESS CONTRACTING PROGRAMS

A. Overview of Policies and Programs

Small business is a major element of government contracting, especially in government construction contracting. Congress has authorized a variety of programs to implement its policy[1] of advancing the interests of small businesses through the procurement process. These programs grant advantages and preferences to specific categories of small business concerns, such as those located in historically underutilized business zones (HUBZones) as well as those owned by service-disabled veterans (SDVO), by women (WOSB), or by other socially and economically disadvantaged individuals (including small disadvantaged businesses [SDBs[2]]and Section 8(a) program participants).[3] The policy also permits—and in practice, requires—agencies to create preferential set-asides for qualifying small business concerns (SBCs) of any kind. (All references hereafter to "small business programs" include both set-asides and various programs devoted to particular types of small businesses.) Reflecting these policies and programs, the federal government obligated no less than $83.2 billion in contract awards in fiscal year (FY) 2007 to businesses qualifying as small business concerns under the various small business programs.[4] That amount represented approximately 22 percent of the total value of all government contracts awarded pursuant to similar procurement actions during that fiscal year.

---

[1]"It is the declared policy of the Congress that the Government should aid, counsel, assist, and protect, insofar as is possible, the interests of small-business concerns in order to preserve free competitive enterprise, to insure that a fair proportion of the total purchases and contracts or subcontracts for property and services for the Government (including but not limited to contracts or subcontracts for maintenance, repair, and construction) be placed with small-business enterprises, ... and to maintain and strengthen the overall economy of the Nation." 15 U.S.C. § 631(a). *See also* 41 U.S.C. § 252(b); FAR § 19.201(a).

[2]The SBD program shared the goal of the Section 8(a) program to assist socially and economically disadvantaged (i.e., minority-owned) small business concerns. The SDB program was challenged and found to violate the right to equal protection. *Rothe Dev. Corp. v. Department of Defense*, 545 F.3d 1023 (Fed. Cir. 2008) The court held that when Congress re-enacted 10 U.S.C. § 2323 (Section 1207) in 2006, it did not have a "strong basis in evidence" to conclude that the Department of Defense was a passive participant in racial discrimination in relevant markets across the country and that race-conscious remedial measures were necessary. On February 29, 2009, the district court issued a broad injunction effectively eliminating Section 1207 and all race-based preferences it established for SBD and 8(a) firms. *Rothe Dev.*, 606 F. Supp. 2d 648 (W.D. Tex. 2009). On March 10, 2009, the Department of Defense issued a memorandum stating all contracts relying exclusively on the authority of Section 1207 program must cease. The memorandum further stated that non–race-based set-asides under the Small Business Act, including Section 8(a) and HUBZone set-asides, remain valid.

[3]*See* 15 U.S.C. § 644(g)(1).

[4]Federal Procurement Data System, Small Business Goaling Report FY 2007, available at: *www.sba .gov/idc/groups/public/documents/sba_homepage/fy2007sbgr.html* (accessed June 5, 2009). The true figure is almost certainly higher. This amount does not reflect a number of exclusions, including contracts awarded and performed outside of the United States and contracts awarded by agencies funded predominately by agency-generated sources, such as the U.S. Postal Service and the U.S. Mint. The amount, however, does include supply and service contracts in addition to construction contracts. Goals for each agency vary. Goals for each fiscal year are available at *www.sba.gov/aboutsba/sbaprograms/ goals/inden.html*.

Congress has also enacted statutory goals for procurement from small business concerns on an overall basis and for specific categories of SBC. The current government-wide goals for small business participation in procurement contracts as a percentage of the total value of all prime contract awards are:[5]

Category	Not Less than Goal
Small Business Concerns	23%
HUBZones	3%
SDVOs	3%
WOSBs	5%
Section 8(a)/SDBs	5%

Congress also authorized each agency to establish annual goals that present, for that agency, the maximum practicable goals for prime contract awards in each of the various categories of SBCs. As a result, the agency goals for specific categories may vary. For example, for FY 2009 the goal for prime contract awards to SDVO concerns for the Department of Defense (DOD) small business concerns (SBCs) [6] was 3 percent, while the Department of Veterans Affairs (VA) goal[7] for the same SBC category was 7 percent.

In addition to goals for prime contract awards to SBCs, the agencies also establish goals for small business subcontracting by the agency's prime contractors. These can have a substantial effect on a general contractor's buyout of subcontracts and management of a construction contract. For example, the FY 2009 subcontracting goals for DOD were:

Small Business Concerns	37.2%
HUBZones	3%
SDVOs	3%
WOSBs	5%
Section 8(a)/SDBs	5%

The figures for prime and subcontract awards to SBCs are overall goals for awards of all types of contracts. The agency goals for construction contracts (prime and subcontract) are often substantially higher as many agencies use construction contracts as a means to offset shortfalls in SBC awards in other types of procurements, such as weapons systems. For example, in FY 2007, DOD awarded $269.3 billion in prime contracts of which $19.9 billion (7.3 percent) were construction prime contracts. DOD also reported that for FY 2007, it awarded $55 billion in prime contracts to all SBCs of which $7.8 billion, or 14.2 percent, were construction prime contract awards to SBCs.[8] In the same report to the SBA, DOD stated that its FY 2007 goal for small business awards of construction contracts was 40 percent of all prime

---

[5] 15 U.S.C. § 644(g)(1).

[6] DOD Office of Small Business Programs, *www.acq.osd.mil/osbp/statistics/goal.htm* (accessed July 31, 2009).

[7] Jnauary 28, 2008, Memorandum for Under Secretaries, Assistant Secretaries, Other Key Officials, Deputy Assistant Secretaries, and Field Facility Directors from the Secretary of Veterans Affairs.

[8] January 31, 2008, letter from the Deputy Director, DOD Office of Small Business Programs, to the Associate Administrator, SBA's Government Contracting Business Development.

contract construction awards and that the $7.8 billion in actual SBC contract awards represented 39.3 percent of the total ($19 billion) of all construction contract awards.[9] As illustrated by these statistics, construction contracts awards have been and will likely remain a primary tool used by federal agencies to meet their small business goals, resulting in more opportunities for SBCs in the construction market than in other segments of federal procurement.

B. Eligibility and Size Determinations

The SBA administers and supervises the various small business programs. That agency establishes eligibility requirements for the programs and makes eligibility determinations regarding individual contractors. In order to be eligible to participate in small business programs, a business must first meet certain qualifying criteria regarding its organization, its activities, and its size.

1. Defining "Small Business Concerns"

A contractor must qualify as a small business concern to be eligible for small business programs. The Federal Acquisition Regulation (FAR) defines a "concern" as any business entity: (1) organized for profit (even if its ownership is in the hands of a nonprofit entity), (2) with a place of business located in the United States or its outlying areas, and (3) that makes a significant contribution to the U.S. economy through payment of taxes and/or use of American products, material, and/or labor. The permissible organizational structures of a concern as a business entity include, but are not limited to, individual proprietorships, partnerships, limited liability companies, corporations, joint ventures, associations, and cooperatives.[10] Nearly any commonly understood form of business is permissible.

For a concern, as generally defined, to qualify as a *small business concern,* it must be: (1) independently owned and operated, (2) not dominant in the field of operation in which it is bidding on contracts, and (3) qualified under the criteria and size standards established by the SBA. "Not dominant in its field of operation" means not exercising a controlling or major influence on a national basis in a kind of business activity in which a number of business concerns are primarily engaged. In order to determine whether a concern is "dominant" under that definition, the SBA considers business volume, the number of employees, financial resources, competitive status or position, ownership or control of materials, processes, patents, license agreements, facilities, sales territory, and the nature of the business's activity.[11]

---

[9]*Id.*

[10]FAR § 19.001. This definition mirrors the definition found in the SBA's regulations at 13 C.F.R. § 121.105(b).

[11]*Id. See also* 15 U.S.C. § 632(a)(1).

2. *Size Standards*

The SBA establishes "criteria and size standards" by industry.[12] Most, but not all, of these size standards are generally defined by a concern's average annual receipts[13] and/or its number of employees.[14] The standards can be based on either, both, or neither of these two factors. The standards differ among industries because of variances in the number and the size of firms in each industry generally. The maximum amount of average annual receipts ranges from $750,000 in many of the agricultural industries to $33.5 million for certain construction-related businesses. If the size standard is based upon the number of employees, the maximum number of employees ranges from 100 to 1,500 but typically falls between 500 and 1,500.

In the construction industry, size standards are defined by annual receipts rather than by the number of employees. The subsectors for "Construction of Buildings" and for "Heavy and Civil Engineering Construction" put the limit at $33.5 million. "Dredging and Surface Cleanup Activities" are limited to $20 million in average receipts, and "Land Subdivision" is $7 million. The subsector for "Specialty Trade Contractors" limits SBCs to those under $14 million in annual receipts.[15]

The contracting officer determines which industry classification is appropriate for any particular solicitation. The solicitation should identify the size standard that applies by listing the applicable North American Industry Classification System (NAICS) group number found in 13 C.F.R. § 121.201. If products or services from different industry classifications are required in the same solicitation, the solicitation must identify the appropriate size standard for each.[16] If such a solicitation calls for two or more items with different size standards, and if a contractor may submit an offer on individual items, then the contractor must meet the standard for each item for which it submits an offer. However, if such a solicitation requires a contractor to submit an offer on all items, then the contractor need only meet the size standard of the item comprising the largest percentage of the total value of the contract.[17]

---

[12]13 C.F.R. § 121.201.

[13]"Receipts" means total income plus the cost of goods sold. 13 C.F.R. § 121.104(a). "Average annual receipts" are calculated using the last three completed fiscal years, based on the fiscal year of the business concern, for concerns that have been in business at least that long. 13 C.F.R. § 121.104(c)(1). If a concern has not been in business for three complete fiscal years, then average annual receipts are calculated by dividing its total receipts for the entire period that the concern has been in business by the number of weeks in that period and then multiplying by 52. 13 C.F.R. § 121.104(c)(2).

[14]Given the relatively large number of employees typically allowed as a maximum, this "limitation" is rarely an issue for a small construction firm.

[15]13 C.F.R. § 121.201 Sector 23. Size standards are changed periodically by the SBA. The most current size standards are available on the Government Printing Office Web site where the current C.F.R. is posted. Go to *ecfr.gpoaccess.gov* and click on Title 13 and follow links to 13 C.F.R. Part 121 [accessed Nov. 9, 2009].

[16]FAR § 19.303(a).

[17]FAR § 19.102(e); 13 C.F.R. § 121.407.

3. Affiliates

Size determinations are made not only by considering the size of the individual entity offering to contract but also by including the size of its *affiliates*. A business concern that otherwise would meet the applicable size standard to qualify as a small business on its own can become ineligible if the combined size of that concern and its affiliate(s) exceeds the maximum threshold for the solicitation.[18] Business concerns are affiliates of each other if, either directly or indirectly, one *controls or has the power to control* the other, or if, either directly or indirectly, another concern controls or has the power to control both.[19] Whether the power to control is actually exercised is irrelevant.[20]

In making affiliation determinations, the SBA will consider all appropriate factors and the totality of the circumstances. Particular factors that the SBA will consider are common ownership, common management, previous ties, and contractual relationships.[21] Affiliation can be specifically determined through common ownership of stock. An individual or an entity is an affiliate of a concern if the individual or entity owns, controls, or has the power to control either not less than 50 percent of the concern's voting stock or a block of stock that affords control because it is large relative to any other blocks of voting stock.[22]

Common management as an indicator of affiliation might take the form of interlocking management, where the officers, directors, employees, or principal stockholders of one concern form a working majority of the directors or officers of another.[23] The SBA also will consider whether the concerns have common facilities or office space.[24] Note that the SBA is authorized to determine that concerns are affiliates even when no single factor is sufficient to constitute affiliation.[25]

4. SBC Subcontracting to Large Businesses and Affiliation

Small businesses are permitted to subcontract portions of set-aside contracts to large businesses[26] unless prohibited by statute, by regulation, or by the terms of the solicitation.[27] Although the existence of a subcontract relationship is not itself dispositive

[18] *Texas-Capital Contractors, Inc. v. Abdnor,* 933 F.2d 261 (5th Cir. 1990); *Aloha Dredging & Constr. Co. v. Heatherly,* 661 F. Supp. 738 (D.D.C. 1987).
[19] *Cytel Software, Inc.,* SBA No. SIZ-2006-10-12-60 (Nov. 20, 2006). The SBA's Office of Hearings and Appeals decides appeals involving size status. *See* **Chapter 3.**
[20] 13 C.F.R. § 121.103; FAR § 19.101(1). *DMS Facility Servs., LLC,* SBA No. SIZ-4913 (Mar. 12, 2008).
[21] 13 C.F.R. § 121.103; FAR § 19.101.
[22] 13 C.F.R. § 121.103(c)(1); FAR § 19.101(3); *Cytel Software, Inc.,* SBA No. SIZ-4822 (Nov. 20, 2006).
[23] FAR § 19.101(6)(i).
[24] FAR § 19.101(6)(ii). *See also* 61 Fed. Reg. 3281, Jan. 31, 1996 ("common facilities" as a factor is not listed at 13 C.F.R. § 121.103(a)(2) as an affiliation indicator; however, the list at that provision is not intended by the SBA to be exhaustive, and the SBA continues to maintain that it has the flexibility to make an appropriate affiliation determination based on the totality of the circumstances. 13 C.F.R. § 121.103(a)(5)).5
[25] FAR § 19.101; 13 C.F.R. § 121.103.
[26] A "large business" is a business concern that does not qualify as a "small business concern" for purposes of the procurement at issue.
[27] 13 C.F.R. § 125.6; *O.V. Campbell & Sons Indus., Inc.,* Comp. Gen. Dec. B-216585, 85-1 CPD ¶ 385.

of affiliation, contractors should be aware that businesses will be deemed to be affiliates if the nature of the subcontract relationship is such that the large-business subcontractor has the power to control the small-business concern.[28]

There are also limits on the amount of work that may be subcontracted by small business prime contractors. For example, in general construction contracts, the prime contractor must perform at least 15 percent of the cost of the contract (not including the cost of materials) with its own employees; for construction by special trade contractors, at least 25 percent.[29] On occasion, an agency may incorrectly state the amount of permissible subcontracting in a solicitation (either an invitation for bids (IFB) for sealed bidding or a request for proposals (RFP) for competitive proposals). The Armed Services Board of Contract Appeals has held that such an error did not relieve the small business contractor from complying with the correct figure. The board concluded that the contractor should have been aware of the applicable regulation and the government's error at the time it submitted its bid.[30]

A small business that subcontracts with a large business can be found to be an affiliate of the large business under the "ostensible subcontractor" rule. An ostensible subcontractor is one that performs primary and vital requirements of a contract or of an order under a multiple award schedule contract or a subcontractor upon which the prime contractor is unusually reliant. A contractor and its ostensible subcontractor are treated as joint venturers and therefore affiliates for size determination purposes.[31] The purpose of the ostensible subcontractor rule is to prevent large firms from forming relationships with small firms to evade SBA size requirements.[32] The SBA can also find a protégé (small business) and its large-business mentor affiliated as long as it does not base its determination solely on the mentor-protégé relationship.[33] **Appendix 5A** to this chapter is a chart reflecting a summary of the SBA's regulations on subcontracting and joint ventures involving contracts awarded to SBCs on a restricted (set-aside) basis.

C. SBC Restricted Procurements (Set-asides)

Federal agencies are authorized to restrict procurements, either in whole or in part, for award only to small business concerns.[34] These reserved contracts are commonly labeled *set-asides*. Contracting officers, either on their own initiative or working with the SBA, initially decide which solicitations to set aside. The contracting officers must ensure that a *fair proportion* of government contracts are set aside in each industry for which the FAR has identified a small business concern size standard.[35]

---

[28]*See* 13 C.F.R. § 121.103(f); *Aloha Dredging & Constr. Co. v. Heatherly,* 661 F. Supp. 738 (D.D.C. 1987).

[29]15 U.S.C. § 644(o); 13 C.F.R. § 125.6. Note that repetitive subcontracts with the same large business can be seen by the SBA as an indicator of affiliation.

[30]*Sarang-National Joint Venture,* ASBCA No. 54992, 06-2 BCA ¶ 33,347.

[31]13 C.F.R. § 121.103(h)(4).

[32]*TKTM Corp.,* SBA No. SIZ-4885 (Jan. 31, 2008).

[33]*Id.*

[34]15 U.S.C. § 644(i); FAR Subpart 19.5.

[35]*See* 13 C.F.R. § 121.201.

Although the FAR authorizes both *total set-asides*[36] and *partial set-asides*,[37] acquisitions of construction services are set aside in total, as the FAR does not authorize partial set-asides for construction.[38] In addition to set-asides for SBCs, federal agencies also restrict (set aside) contracting opportunities for specific categories of SBCs in an effort to meet the goals set by Congress. The second group of set-asides includes contracting opportunities reserved for Section 8(a) contractors, HUBZone, SDVO, and WOSB concerns.

1. Total Set-asides

Total set-asides are selected under the *rule of two*. That is, if an agency determines that there is a *reasonable expectation* that offers will be submitted by at least two responsible small business concerns offering different products at a fair market price, then the entire amount of any individual acquisition or class of acquisitions exceeding $100,000 must be set aside for SBCs.[39] Such contracts may not be awarded as a total set-aside for SBCs if the cost of the contract would exceed the *fair market price*,[40] which is determined by a proposal analysis conducted by the contracting officer using reasonable price guidelines.[41] However, every contract with a value anticipated at more than $3,000, but less than $100,000,[42] must be reserved for SBCs when the contracting officer reasonably determines there is a reasonable expectation (1) of receiving two or more competitive offers from responsible small business concerns and (2) award will be made at fair market prices.[43]

Broad discretion is given to contracting officers in making set-aside determinations.[44] An agency should make reasonable efforts to explore the market to determine whether a *reasonable expectation* small business interest exists, but it need not perform an in-depth survey.[45] However, it also may not rely solely on past procurement history when faced with evidence that qualified SBCs will submit offers.[46]

2. SBA Section 8(a) Program

The SBA Section 8(a) program is named after Section 8(a) of the Small Business Act of 1953 and is designed to assist qualifying minority-owned businesses. Under

---

[36]*See* FAR § 19.502-2.

[37]*See* FAR § 19.502-3.

[38]FAR § 19.502-3(a) ("The contracting officer shall set aside a portion of an acquisition, except for construction. . . ."). "Construction" is defined at FAR § 2.101.

[39]FAR § 19.502-2(b).

[40]FAR § 19.501(i).

[41]FAR § 19.202-6(a). *See also* FAR § 15.404-1 (addressing reasonable price guidelines).

[42]The anticipated value thresholds increase to more than $15,000, but less than $250,000, for acquisitions of supplies or services that are to be used to support a contingency operation or to facilitate defense against or recovery from nuclear, biological, chemical, or radiological attack.

[43]FAR § 19.502-2(b).

[44]*Int'l Filter Mfg., Inc.*, Comp. Gen. Dec. B-299368, 2007 CPD ¶ 71.

[45]*McGhee Constr., Inc.*, Comp. Gen. Dec. B-249235, 92-2 CPD ¶ 318; *JT Constr. Co.*, Comp. Gen. Dec. B-254257, 93-2 CPD ¶ 302.

[46]*FKW Inc.*, Comp. Gen. Dec. B-249189, 92-2 CPD ¶ 270.

Section 8(a), the SBA directly enters into contracts with federal agencies and, in turn, subcontracts the work to certain SBCs.[47] More than $13.4 billion of government contracts of all types were awarded under the 8(a) program during fiscal year 2007.[48]

a. Eligibility In order to participate in the Section 8(a) program, a concern must qualify as a "socially and economically disadvantaged"[49] SBC. Those terms describe a small business concern[50] as both owned and controlled by socially and economically disadvantaged individuals. That is, at least 51 percent of the concern must be unconditionally owned by such individuals, and they must manage the daily business operations on a full-time basis, including setting strategic policy.[51] SBA's regulations further state, in pertinent part:

> Disadvantaged individuals managing the concern must have managerial experience of the extent and complexity needed to run the concern. A disadvantaged individual need not have the technical expertise or possess a required license to be found to control an applicant or Participant if he or she can demonstrate that he or she has ultimate managerial and supervisory control over those who possess the required licenses or technical expertise. However, where a critical license is held by a non-disadvantaged individual having an equity interest in the applicant or Participant firm, the non-disadvantaged individual may be found to control the firm.[52]

Social disadvantage and economic disadvantage are unique characteristics, each of which must be shown. Social disadvantage is established either by membership in a designated group (including Black Americans, Hispanic Americans, Native Americans, and Asian Americans)[53] or by evidence of (1) one objective distinguishing feature that has contributed to social disadvantage,[54] (2) personal experiences of substantial and chronic social disadvantage in American society,[55] and (3) negative impact on entry into or advancement in the business world because of the disadvantage.[56]

---

[47]15 U.S.C. § 637; 13 C.F.R. § 124.501(a) (also permitting the SBA to delegate contract execution to agencies).

[48]Federal Procurement Data System, Small Business Goaling Report FY 2007, available at: *www.sba .gov/idc/groups/public/documents/sba_homepage/fy2007sbgr.html* (accessed June 5, 2009). This amount includes supply and service contract awards in addition to construction contracts.

[49]13 C.F.R. § 124.101.

[50]Section 8(a) program applicants and participants must meet the general definition and eligibility requirements associated with all "small business concerns." *See* **Section I.A** of this chapter.

[51]15 U.S.C. § 637(a)(4). *See also* 13 C.F.R. § 124.105 (defining "unconditionally owned" for purposes of the 8(a) program); 13 C.F.R. § 124.106 (defining "control" for purposes of the 8(a) program).

[52]13 C.F.R. § 124.106.

[53]13 C.F.R. § 124.103(b).

[54]13 C.F.R. § 124.103(c)(2)(i) (including features such as race, ethnic origin, gender, physical handicap, long-term residence in an environment isolated from the mainstream of American society, or other similar causes not common to individuals who are not socially disadvantaged).

[55]13 C.F.R. § 124.103(c)(2)(ii).

[56]13 C.F.R. § 124.103(c)(2)(iii).

Economic disadvantage is established by demonstrating that one's "ability to compete in the free enterprise system has been impaired due to diminished capital and credit opportunities as compared to others in the same or similar line of business that are not socially disadvantaged."[57] The SBA's regulations further describe the relevant considerations:

> In considering diminished capital and credit opportunities, SBA will examine factors relating to the personal financial condition of any individual claiming disadvantaged status, including personal income for the past two years (including bonuses and the value of company stock given in lieu of cash), personal net worth, and the fair market value of all assets, whether encumbered or not. SBA will also consider the financial condition of the applicant compared to the financial profiles of small businesses in the same primary industry classification, or, if not available, in similar lines of business, which are not owned and controlled by socially and economically disadvantaged individuals in evaluating the individual's access to credit and capital. The financial profiles that SBA compares include total assets, net sales, pre tax profit, sales/working capital ratio, and net worth.[58]

Individuals must submit a narrative describing their economic disadvantage and personal financial information to the SBA. Married individuals, unless legally separated, also must submit the personal financial information of their spouse.[59]

For initial eligibility for admission to the 8(a) program, an individual's net worth must be less than $250,000. For continued eligibility after admission, the individual's net worth must remain less than $750,000. The ownership interest in the concern and one's equity in his or her primary personal residence are excluded for net worth purposes; however, they are not excluded for asset valuation or access to capital and credit purposes.[60]

In order to participate in the Section 8(a) program, a small business concern also must establish that it possesses reasonable prospects for success in competing in the private sector. To do so, it must have been in business in its primary industry classification for at least two full years immediately prior to the date of its application to the program, unless a waiver is granted. Other factors considered include access to credit and capital, technical and managerial experience, operating history, and record of performance.[61]

A disadvantaged individual may acquire eligibility only once;[62] therefore, only one concern owned and controlled by that individual may enroll in the program.[63] Program participants remain in the program for a maximum of nine years,[64] and

[57]13 C.F.R. § 124.104(a).
[58]13 C.F.R. § 124.104(c).
[59]13 C.F.R. § 124.104(b).
[60]13 C.F.R. § 124.104(c)(2).
[61]13 C.F.R. § 124.107.
[62]13 C.F.R. § 124.108(b).
[63]13 C.F.R. § 124.108(b)(3).
[64]13 C.F.R. § 124.2.

they are required each year to demonstrate that they remain eligible.[65] As part of this annual review, program participants must provide information regarding their efforts to obtain contracts outside the program. The failure to meet specified targets for non-8(a) contracts during the final five years of eligibility (years five through nine in the program, known as the transitional stage) may lead to restrictions on future awards through the program or termination from the program.[66]

Potential penalties for misrepresentation of disadvantaged status for qualification for the 8(a) program include ineligibility for future participation in any small business programs, suspension or debarment from government contracting, fines up to $500,000, and imprisonment for up to 10 years.[67] A contractor may also be liable under the civil False Claims Act[68] if it submits vouchers for payment under a contract awarded pursuant to the 8(a) program while deliberately withholding information which would establish that it was no longer eligible for participation in the 8(a) program.[69]

b. Contract Awards   The SBA, qualified 8(a) contractors, and procuring agencies all can identify procurements suitable for inclusion in the program. Contractors either may request that the SBA contact the agency or may contact the agency directly to request that the contract be considered for inclusion.[70] Although no particular procurement may be forced into the program, the prime contract participation goal of 5 percent for small disadvantaged businesses[71] encourages agencies to consider using the program when appropriate.

The SBA awards 8(a) program contracts through both sole-source awards and limited competitions.[72] In sole-source awards, either the agency has identified and has proposed a particular participating contractor[73] or the SBA will identify a contractor.[74] If the SBA is selecting the contractor for a possible award from among two or more program participants, then "the selection will be based upon relevant factors, including business development needs, compliance with competitive business mix requirements (if applicable), financial condition, management ability, technical capability, and whether award will promote the equitable distribution of 8(a) contracts."[75] A program participant becomes ineligible to receive sole-source awards if the total value of all 8(a) contracts (both sole-source and competitive) it has been awarded exceeds a certain amount.[76] This does not restrict the concern's eligibility for future competitive awards through the 8(a) program.

---

[65]13 C.F.R. § 124.509(c).
[66]13 C.F.R. § 124.509(d).
[67]15 U.S.C. § 645(d).
[68]31 U.S.C. §§ 3729-3733.
[69]*Ab-Tech Constr., Inc. v. United States,* 31 Fed. Cl. 429 (1994).
[70]13 C.F.R. § 124.501.
[71]15 U.S.C. § 644(g)(1).
[72]*See* 13 C.F.R. § 124.503.
[73]13 C.F.R. § 124.503(c).
[74]13 C.F.R. § 124.503(d).
[75]13 C.F.R. § 124.503(d)(3).
[76]13 C.F.R. § 124.519. For firms having revenue-based size standards (such as construction firms), the limit is five times that size standard or $100 million, whichever is less. For firms having employee-based size standards, the limit is $100 million.

Competition among 8(a) contractors is required for procurements offered and accepted into the program if: (1) there is a reasonable expectation that at least two eligible participants will submit offers at a fair market price; (2) the anticipated award price of the construction contract, including options, will exceed $3.5 million; and (3) the requirement has not been accepted for award as a sole source 8(a) procurement on behalf of a tribally owned or Alaska Native Corporation-owned concern.[77] Competition is permissible for procurements below the threshold amount if the SBA approves an agency's request for competition, particularly when technical competitions are appropriate or when a large number of potential awardees exist.[78] Competitions are conducted by contracting agencies in accordance with the FAR.[79] The SBA, at its discretion, may restrict competitions to concerns in the developmental stage of the 8(a) program (years one through four)[80] or, in the case of construction procurements, to concerns within a geographic boundary.[81]

The SBA and contracting agencies have very broad discretion in determining which contracts are placed into the 8(a) program. Protests challenging a decision to procure under the 8(a) program require a showing either of possible bad faith on the part of the government or of violation of acquisition regulations.[82] The FAR requires the SBA to consider whether selecting a particular contract for inclusion would have an *adverse impact* on an individual small business, a group of small businesses located in a specific geographical location, or other small business programs.[83] However, the SBA's adverse impact analysis is aimed narrowly at protecting incumbent small businesses that are performing an offered requirement outside the 8(a) program.[84] Unlike facilities maintenance or other services, all construction contracts, by definition, are deemed new requirements that are not subject to adverse impact determinations.[85] Even if a protesting contractor meets the presumption of adverse impact under the regulations, the SBA maintains the discretion to accept a requirement into the 8(a) program.[86]

3. Historically Underutilized Business Zones

The Historically Underutilized Business Zone (HUBZone) program provides contracting assistance in the form of set-asides and price evaluation preferences for

---

[77]FAR § 19.805-1; *cf.* 13 C.F.R. § 124.506(a) (indicating an anticipated award price threshold of $3 million [$5 million for manufacturing contracts]).

[78]13 C.F.R. § 124.506(c).

[79]13 C.F.R. § 124.507(a).

[80]13 C.F.R. § 124.507(c)(1).

[81]13 C.F.R. § 124.507(c)(2).

[82]4 C.F.R. § 21.5(b)(3); *Designer Assocs., Inc.,* Comp. Gen. Dec. B-293226, 2004 CPD ¶ 114. Bad-faith allegations usually are not successful since they must be proved by "virtually irrefutable evidence that the contracting agency directed its actions with the specific and malicious intent to injure the protestor. *Information Res., Inc.,* Comp. Gen. Dec. B-271767, 96-2 CPD ¶ 38 at 2. Government officials are presumed to act in good faith. *Superior Landscaping Co.,* Comp. Gen. Dec. B-310617, 2008 CPD ¶ 33 at 4.

[83]13 C.F.R. § 124.504(c).

[84]*Id.; see also Korean Maint. Co.,* Comp. Gen. Dec. B-243957, 91-2 CPD ¶ 246.

[85]13 C.F.R. § 124.504(c)(1)(ii)(B).

[86]*Catapult Tech. Ltd.,* Comp. Gen. Dec. B-294936, B-294936.2, 2005 CPD ¶ 14.

qualifying small business concerns in an effort to increase employment opportunities, investment, and economic development in HUBZone areas. The program was established as part of the Small Business Reauthorization Act of 1997.[87] The government-wide goal for participation by HUBZone SBCs in procurement contracts is 3 percent of the total value of all prime contract awards for each fiscal year.[88] More than $8.4 billion in contract awards were granted to HUBZone contractors in fiscal year 2007.[89] The program applies to all federal agencies. The SBA's Web site at *www.sba.gov/hubzone/* provides basic information on this program and a link to a map that helps a firm determine if it is located in a HUBZone. Just click on the link: "Are You in a HUBZone?"

a. Qualifying for the HUBZone Program   A HUBZone is an area located within (1) a census tract in which 50 percent or more of the households have an income that is less than 60 percent of the median gross income for the area; (2) a nonmetropolitan county[90] with an unemployment rate not less than 140 percent of the statewide or the national average, whichever is less, or with a median household income less than 80 percent of the statewide median; (3) an Indian reservation; (4) a base closure area (for a period of five years either from December 8, 2004, or from the date of the base closure, whichever is later); or (5) a redesignated area.[91] In addition to qualifying as a small business,[92] a concern must meet three additional criteria to qualify for the HUBZone program. It must (1) have its principal office located in a HUBZone, (2) be at least 51 percent unconditionally and directly owned and controlled by U.S. citizens, and (3) 35 percent of its employees must reside in a HUBZone.[93] "Control" means both day-to-day management and long-term decision-making authority.[94] Although a percentage of the employees must reside in a HUBZone, the owner(s) need not. A concern must attempt to maintain the qualifying percentage of employees during contract performance.[95] There is no time limit on participation in the HUBZone program,[96] but a contractor must recertify every three years.[97]

b. Contracting Assistance   HUBZone small business concerns are eligible for three types of contracting assistance: set-asides, sole-source awards, and price evaluation preferences.[98] Agencies must set aside acquisitions exceeding the simplified

---

[87]Pub. L. 105-135.

[88]15 U.S.C. § 644(g)(1).

[89]Federal Procurement Data System, Small Business Goaling Report FY 2007, available at: *www.sba .gov/idc/groups/public/documents/sba_homepage/fy2007sbgr.html* (accessed June 5, 2009).

[90]A nonmetropolitan county is a county that was not located in a metropolitan statistical area at the time of the most recent census. 13 C.F.R. § 126.103.

[91]13 C.F.R. § 126.103.

[92]*See* **Section I.A** of this chapter.

[93]13 C.F.R. § 126.200(b). *See also* 13 C.F.R. § 126.201 (defining ownership for purposes of the HUBZone program); and 13 C.F.R. § 126.202 (defining control for purposes of the HUBZone program).

[94]13 C.F.R. § 126.202.

[95]13 C.F.R. § 126.200(b)(5).

[96]13 C.F.R. § 126.502.

[97]13 C.F.R. § 126.500.

[98]13 C.F.R. § 126.600. *See also* FAR §§ 19.1305–19.1307.

acquisition threshold (currently $100,000 in most cases)[99] for competition restricted to HUBZone contractors if there is a reasonable expectation that offers at a *fair market price* will be received from two or more HUBZone contractors.[100] The fair market price[101] is determined by a proposal analysis conducted by the contracting officer using reasonable price guidelines.[102] A contracting officer also has the discretion to set aside acquisitions below the simplified acquisition threshold under the same circumstances.[103] Contracting officers must consider HUBZone set-asides before considering other HUBZone assistance programs or small business set-asides.[104] This was recently affirmed by the Government Accountability Office (GAO) in *Mission Critical Solution*,[105] when the GAO decided that the Army improperly awarded a contract to an 8(a) firm before determining whether the acquisition should be set aside for a HUBZone small business. The ruling signifies the GAO's position that set-asides under the HUBZone program take precedence over all other set-asides. If a procurement qualifies for competition restricted to HUBZone contractors, then it must be so restricted even if it also qualifies for other set-aside programs. Even if only one acceptable offer for the set-aside acquisition is received from a HUBZone contractor, the contracting officer should award the contract to that concern. If no acceptable offers are received, then the acquisition can be set aside under other small business programs.[106]

A contracting officer may award contracts to HUBZone contractors on a sole-source basis without considering whether to set aside the procurement under the general small business program if: (1) only one HUBZone contractor can satisfy the requirement, (2) the value of the procurement is more than the simplified acquisition threshold and less than $3.5 million, and (3) the award can be made at a *fair and reasonable price*.[107] In full and open competitions for procurements exceeding the simplified acquisition threshold, HUBZone contractors receive a price evaluation preference of 10 percent. The factor is added to all offers except those from other SBCs.[108]

Similar to the SBA's broad discretion regarding the Section 8(a) program, contracting officers are given broad discretion to determine whether HUBZone contractors are *responsible offerors* for purposes of their qualifying status regarding HUBZone set-asides. A contracting officer must not unreasonably fail to consider available

---

[99]41 U.S.C. § 403(11); 10 U.S.C. § 2302(7); FAR § 2.101.

[100]FAR § 19.1305(b).

[101]*See* FAR § 19.501(i).

[102]FAR § 19.202-6(a); FAR § 15.404-1.

[103]FAR § 19.1305(c). *Saturn Landscape Plus, Inc.*, Comp. Gen. Dec. B-297450-3, 2006 CPD ¶ 70.

[104]FAR § 19.1305(a).

[105]Comp. Gen. Dec. B-401057, 2009 CPD ¶ 93, citing 15 U.S.C. § 657a and FAR § 19.1301(b). In a July 10, 2009, memorandum, the director of the Office of Management and Budget (OMB) instructed federal agencies not to follow this GAO ruling in determining a selection of a procurement for a particular set-aside. *See www.whitehouse.gov/omb/assets/memoranda_fy2009/m09-23.pdf*.

[106]*See* FAR Subpart 19.5.

[107]FAR § 19.1306 (except $5.5 million for manufacturing contracts).

[108]FAR § 19.1307.

relevant information and must not otherwise violate a statute or regulation.[109] A contracting officer, for example, may not ignore information that would be expected to have a strong bearing on whether an offeror should be found responsible.[110] In *Wild Building Contractors, Inc.*,[111] the GAO denied a protest against a HUBZone set-aside contract award based on allegations that the HUBZone contractor that received the contract award was closely affiliated with another contractor—allegedly jeopardizing the awardee's HUBZone status—because the contracting officer was aware of and gave reasonable consideration to the information and because the contractor awarded the contract was indeed certified by the SBA as qualified for the HUBZone program, even though the GAO did, in fact, find a close affiliation.[112]

4. Service-Disabled Veteran-Owned Business

The Veterans Benefit Act of 2003[113] created a procurement program for small business concerns owned and controlled by service-disabled veterans.[114] Like HUBZone SBCs, service-disabled veteran-owned small businesses must be at least 51 percent unconditionally owned and controlled by one or more qualifying individuals.[115]

Agencies may set aside, but are not required to set aside, acquisitions exceeding the micropurchase threshold (typically $2,000 to $3,000)[116] for competition restricted to SDVO SBCs if there is a reasonable expectation that offers at a *fair market price* will be received from two or more SDVO SBCs.[117] As with HUBZone set-asides, even if only one acceptable offer for the set-aside acquisition is received from an SDVO offeror, the contracting officer should award the contract to that concern. If no acceptable offers are received, then the acquisition can be set aside under the program for all SBCs.[118] SDVO SBCs are eligible for sole-source awards under the same circumstances as HUBZone SBCs: (1) only one qualifying concern

---

[109]4 C.F.R. § 21.5(c).

[110]67 Fed. Reg. 79,834; *see also Verestar Gov't Servs. Group,* Comp. Gen. Dec. B-291854, B-291854.2, 2003 CPD ¶ 68 (during comment period for promulgation of 4 C.F.R. § 21.5(c), GAO considered a protest argument of the type described).

[111]Comp. Gen. Dec. B-293829, 2004 CPD ¶ 131, 2005 CPD ¶ 48.

[112]Note, however, that the bid protest before the GAO in this example is not the same as a direct challenge before the SBA of an offeror's HUBZone status qualification.

[113]15 U.S.C. § 657f.

[114]A "service-disabled veteran" is a veteran with a disability that was incurred or aggravated in line of duty in the active military, naval, or air service. 13 C.F.R. § 125.8; 38 U.S.C. § 101(16).

[115]13 C.F.R. § 125.8(g). *See also* 13 C.F.R. § 125.9 (defining ownership for purposes of the SDVO small business concern program); 13 C.F.R. § 125.10 (defining "control" for purposes of the SDVO SBC program).

[116]"Micropurchase threshold" means $3,000, except for acquisitions of construction subject to the Davis-Bacon Act ($2,000), acquisitions of services subject to the Service Contract Act ($2,500), and acquisitions of supplies or services that are to be used to support a contingency operation or to facilitate defense against or recovery from nuclear, biological, chemical, or radiological attack ($15,000 in the case of any contract to be awarded and performed, or purchase to be made, inside the United States; $25,000 in the case of any contract to be awarded and performed, or purchase to be made, outside the United States). FAR § 2.101.

[117]FAR § 19.1405; 15 U.S.C. § 657f(b). The "fair market price"—*see* FAR § 19.501(i)—is determined by a proposal analysis conducted by the contracting officer using reasonable price guidelines, FAR § 19.202-6(a); FAR § 15.404-1.

[118]FAR § 19.1405(c).

can satisfy the requirement, (2) the value of the procurement is more than the simplified acquisition threshold (currently $100,000)[119] and less than $3.5 million, and (3) the award can be made at a fair and reasonable price.[120] The government-wide goal for participation by SDVO SBCs in procurement contracts is 3 percent of the total value of all prime contract awards for each fiscal year.[121] Before proceeding with a SDVO set-aside, the contracting agency must first reasonably consider whether the conditions for a HUBZone set-aside exist. If so, the contracting agency must proceed with the HUBZone set-aside.[122]

D. Small Business Subcontracting Plans

Many very large government construction contracts are beyond the prime contracting capacity (especially bonding capacity) of most small businesses. However, the government has devised a program by which small businesses are assisted in gaining subcontracts from large contractors. Most procurement contracts exceeding the simplified acquisition threshold (currently $100,000)[123] must include a clause obligating the contractor to give small businesses the maximum practicable opportunity to participate in the performance of the contract to the fullest extent consistent with efficient contract performance.[124]

Additional requirements are placed on large businesses awarded contracts with values of more than $1 million for construction contracts (or more than $550,000 in all other contracts).[125] The contractor must provide a subcontracting plan that is approved by the contracting officer and included in the contract. The plan must include assurances that it will include the clause entitled "Utilization of Small Business Concerns" in all subcontracts that offer further subcontracting opportunities and that all subcontractors, except SBCs, will adopt plans reflecting the requirements of FAR § 52.219-9.[126] These goals can present a challenge on large, complicated

---

[119]41 U.S.C. § 403(11); 10 U.S.C. § 2302(7); FAR § 2.101.

[120]FAR § 19.1406 (except $5.5 million for manufacturing contracts). *Cf.* 15 U.S.C. § 657f(a) (indicating an anticipated award price threshold of $3 million [$5 million for manufacturing contracts]).

[121]15 U.S.C. § 644(g)(1).

[122]*Int'l Program Group, Inc.*, Comp. Gen. Dec. B-400278; B-400308, 2008 CPD ¶ 172.

[123]41 U.S.C. § 403(11); 10 U.S.C. § 2302(7) (doubling that amount in the case of any contract to be awarded and performed, or purchase to be made, outside the United States in support of a contingency operation or a humanitarian or peacekeeping operation for DOD, NASA, or the Department of Homeland Security (which includes the U.S. Coast Guard); FAR § 2.101 (increasing the amount to $250,000 and $1 million for acquisitions of supplies or services that are to be used to support a contingency operation or to facilitate defense against or recovery from nuclear, biological, chemical, or radiological attack inside or outside the United States, respectively; pursuant to 41 U.S.C. § 428a).

[124]15 U.S.C. § 637(d)(1); FAR §§ 19.702, 52.219-8.

[125]15 U.S.C. § 637(d)(4); FAR § 19.702.

[126]FAR § 19.704; FAR § 52.219-9. (Standard Forms (SFs) 294 and 295 currently restrict the reporting by prime contractors to only first-tier subcontracts.) On June 1, 2007 the Electronic Subcontracting Reporting System (eSRS) was implemented. Its purpose is to streamline the process of reporting on subcontracting plans and provide agencies with access to analytical data on first-tier subcontractor performance under the subcontracting plans applicable to those firms. The eSRS eliminates the need for paper submissions and processing of the SF 294 and SF 295. FAR § 52.219-9(b). The eSRS Web site is *www.esrs.gov.*

design-build projects. Failure to perform in *good faith* regarding the plan is a material breach of the contract.[127] Such a breach entitles the government, in addition to any other available remedies, to liquidated damages in an amount equal to the actual dollar amount by which the contractor failed to achieve its subcontracting goal.[128]

A key to demonstrating a good faith effort to achieve the goal is a contractor's documentation of its efforts. If no documentation is maintained and the goals are not achieved, the contractor's position is very difficult to explain. In sealed bid acquisitions, the subcontracting plans may not be negotiated, but communications between an agency and a contractor are allowed. In such cases, subcontracting plans may be submitted after bids are received and opened[129] so long as they are submitted within the time limit prescribed by the contracting officer.[130] In competitively negotiated acquisitions, the proposed subcontracting plan can be a technical factor by which the government can make award determinations. Further, agencies may include an incentive clause increasing the contractor's profit based on the extent to which subcontract awards to small businesses exceed the negotiated percentage goal.[131]

E. SBC Self-Certification of Status

1. Small Business Concerns

Small business concerns initially self-certify. Self-certification is a required part of a contractor's offer in response to a solicitation.[132] A contractor's failure to certify its status as a small business concern at the time of bidding in a sealed bid procurement is treated as a minor informality if the bidder's size status is not necessary to determine whether the bid meets the IFB's material requirements and does not affect the responsiveness of the bid.[133] However, when solicitations require contractual commitments related to the set-aside (e.g., commitments to use materials obtained from a small business or not to subcontract certain amounts of the work), failure to self-certify could render a bid nonresponsive, which would not be curable after bid opening.[134]

Most issues of improper certifications are addressed at the preaward stage by protests to the SBA or referral to the SBA by the procuring activity.[135] If an incorrect certification is timely challenged by the contracting officer or by one or more other offerors in the procurement, then the agency generally is expected to withhold the award pending receipt of a decision from the SBA.[136] If an award has been made

---

[127]FAR § 52.219-9(k).

[128]FAR § 19.705-7.

[129]*CH2M Hill, Ltd.,* Comp. Gen. Dec. B-259511, 95-1 CPD ¶ 203; *Devcon Sys. Corp.,* Comp. Gen. Dec. B-197935, 80-2 CPD ¶ 46, 59 Comp. Gen. 614 (1980).

[130]FAR § 19.702(a)(2).

[131]FAR § 19.708(c)(1); FAR § 52.219-10.

[132]13 C.F.R. § 121.405(a). Size as of date price is provided. 13 C.F.R. §121.404(a).

[133]*Gracon Corp.,* Comp. Gen. Dec. B-224344, 86-2 CPD ¶ 41.

[134]*Wright Assocs., Inc.,* Comp. Gen. Dec. B-238756, 90-1 CPD ¶ 549.

[135]FAR Subpart 19.3.

[136]FAR § 19.302.

before a timely protest is filed, the protest decision generally does not apply to that acquisition.[137] However, if the SBA's decision upholds the size protest, the agency has the discretion to terminate the previous award. Termination can happen immediately if the protestor is next in line for the contract and is ready and able to step in and perform.[138] Since protests regarding a contractor's size status often arise in the context of traditional bid protests, the time deadlines and procedures for these protests are addressed in **Chapter 3, Section IV.G.**

The SBA has exclusive and conclusive authority to determine whether a business concern is indeed small for the purpose of any given solicitation. If the certification was knowingly false, the contractor may face suspension or debarment from federal contracting[139] or criminal liability under the federal false statements statute.[140] Punishments under the Small Business Act for misrepresentations of one's status as a small business concern in order to obtain a contract include ineligibility for future participation in any small business programs for up to three years, suspension or debarment from government contracting, fines up to $500,000, and imprisonment for up to 10 years.[141] A contactor might also be in jeopardy of civil fraud liability for making a knowingly false certification that results in award of a contract and requests for payment.[142]

2. HUBZone

A HUBZone SBC must be certified by the SBA as a HUBZone SBC. There is no HUBZone self-certification process. If the SBA determines that a concern is qualified as a HUBZone then it will issue a certification and add the concern to the list of Qualified HUBZone SBCs on its Internet Web site at *http://sba.gov/hubzone*.[143] The HUBZone concern must have an updated business profile in the Central Contractor Registration (CCR). A contracting agency cannot rely on the CCR database to verify a certification when a HUBZone concern attempts to self-certify. The contracting agency must refer the question to the SBA.[144]

---

[137]FAR § 19.302(i).

[138]*Landmark Constr. Corp.,* Comp. Gen. Dec. B-281957.3, 99-2 CPD ¶ 75.

[139]FAR § 9.406-2.

[140]18 U.S.C. § 1001. *See also* FAR § 52.214-4 (False Statements in Bids (Apr. 1984)).

[141]15 U.S.C. § 645(d)(2). The SBA's regulations setting forth penalties for misrepresentations and false statements in connection with small business programs are located at 13 C.F.R. §§ 121.108 (small businesses generally), 124.501 (Section 8(a) small businesses), 124.1011 (small disadvantaged businesses), 125.29 (veteran-owned or service-disabled veteran-owned small businesses), 126.900 (HUBZone small businesses).

[142]31 U.S.C. §§ 3729-3733.

[143]FAR § 19.1303 (b)

[144]*AMI Constr.,* Comp. Gen. Dec. B-286351, 2000 CPD ¶ 211. On January 1, 2004 the SBA, OMB, the General Services Administration, and DOD integrated the PRO-Net and DOD's CCR databases, which created one portal for entering and searching small business sources.

3. SDVO

There is no federal SDVO business certification program. The SDVO business owner self-certifies his or her service-disabled status and small business status. To be eligible for the SDVO business program, a veteran must be able to produce an adjudication letter from the U.S. Department of Veterans Affairs or a Department of Defense Form 214, Certificate of Release or Discharge from Active Duty stating that he/she has a service-connected disability.

F. Mentor-Protégé Agreements

Mentor-protégé programs are organized by the SBA and by a number of government agencies,[145] most prominently the Department of Defense, to further the goals of small business programs. The SBA mentor-protégé program is designed to encourage approved mentors to provide various forms of assistance to eligible 8(a) participants as protégés through a mentor-protégé agreement. The purpose of the mentor-protégé relationship is to enhance the capabilities of the protégé and improve its ability to compete successfully for federal contracts. The SBA's mentor-protégé program assists Section 8(a) program participants with technical and managerial support. Relationships with more established businesses provide participants unique joint venture and subcontracting opportunities. Mentors can also take up to a 40 percent equity interest in their protégés to help them raise capital. No determination of affiliation or control will be found between a protégé firm and its mentor based on the mentor-protégé agreement, assistance provided pursuant to the mentor-protégé agreement, or ownership up to 40 percent of equity.

To qualify as a mentor, a concern must demonstrate that it: possesses favorable financial health (including profitability for at least two years), good character, does not appear on the federal list of debarred or suspended contractors, and can impart value to a protégé firm due to lessons learned and practical experience gained through the 8(a) program or from its general knowledge of government contracting.[146] To be a protégé, a firm must be in the developmental stage of the 8(a) program, have not yet received an 8(a) contract, have a size that is less than half the size standard corresponding to its primary NAICS code, and be in good standing in the 8(a) program.[147]

The mentor and protégé firms must enter into a written agreement setting forth the protégé's needs and describing the assistance the mentor is committed to providing that will address those needs. The agreement also must specify that the mentor will provide such assistance to the protégé for at least one year. The SBA must

---

[145]Agencies with mentor-protégé programs include NASA, the Federal Aviation Administration, and the Departments of State, Defense, and Energy. In August 2009, the GSA adopted its own regulations for mentor-protégé programs. The regulations went into effect on September 14, 2009. *See* GSAR Subpart 519.70.

[146]13 C.F.R. § 124.520(b).

[147]13 C.F.R. § 124.520(c).

approve the agreement. The agreement must set forth the assistance to be provided by the mentor for promoting real and sufficient gains significant to the protégé. The agreement must not be merely a vehicle to enable a non-Section 8(a) mentor to receive 8(a) contracts. Federal agencies operating a mentor-protégé program require the mentor and protégé to identify themselves. The agency will not assist in pairing. The mentor must have an active and approved subcontracting plan, and the protégé must be a certified SBC under one of the SBA plans.

Under the DOD Mentor-Protégé program, protégé firms can include small disadvantaged business concerns; business entities owned and controlled by an Indian tribe; business entities owned and controlled by a Native Hawaiian Organization; qualified organizations employing the severely disabled; women-owned small business concerns; service-disabled veteran-owned small business concerns; or HUBZone small business concerns.[148] In the DOD program, there generally are two types of Mentor-Protégé agreements.[149] *Direct reimbursement agreements* are those in which the mentor receives reimbursement for allowable costs of developmental assistance provided to the protégé. *Credit agreements* are those in which the mentor receives a credit toward SDB subcontracting goals based on the cost of developmental assistance provided to the protégé.

G. Limitations on Subcontracting and Joint Ventures

As summarized for comparison purposes in **Appendix 5A** to this chapter, the federal government has placed limitations on subcontracting involving SBCs, 8(a) set-asides, HUBZones, SDVO small business concerns, as well as joint ventures with large businesses. Each of these of limitations is addressed in the next sections.

1. Subcontracting to a Large Business

An 8(a) set-aside SBC acting as a prime contractor must perform at least 15 percent of the cost of the work (excluding materials) with its own employees.[150] A specialty trade SBC must perform at least 25 percent of the cost of work (excluding materials) with its own employees.[151] Repetitive subcontracts with the same firm are viewed as indicia of affiliation. A HUBZone SBC must perform at least 50 percent of the cost of labor with its own employees or the employees of other qualified HUBZone SBCs. A SDVO SBC operating as a prime must perform at least 15 percent of the cost of the work (excluding material) with its own employees. A specialty trade SDVO must perform at least 25 percent of the cost of the work (excluding materials) with its own employees.[152]

---

[148]*See* DFARS Subpart 219.71.
[149]DFARS § 219.7102.
[150]13 C.F.R. § 125.6.
[151]*Id.*
[152]13 C.F.R. § 124.10.

A large business may provide bonding assistance to an SBC without the finding of an affiliation.[153] The SBA may view repetitive subcontracts with the same firm as indicia of affiliation even if percentage limits are followed.

2. Joint Ventures

A large business may not joint venture on an 8(a) set-aside absent a mentor-protégé agreement. An 8(a) firm may enter into one or more joint ventures with other SBCs to perform an 8(a) set-aside contract provided the SBA approves the joint venture agreement.[154] Members of a joint venture are deemed affiliates for the purpose of each acquisition in which they participate with limited exceptions. A large business may not joint venture with a HUBZone concern on a HUBZone set-aside. A HUBZone SBC may joint venture with another HUBZone SBC on a HUBZone set-aside. A large business may not joint venture with a SDVO concern on a SDVO set-aside. SDVOB concerns may joint venture with other SBCs provided the joint venture meets certain conditions.[155]

H. Certificates of Competency and Responsibility Determinations

As discussed in **Chapter 3,** contracting officers are required to make an affirmative determination that a prospective contractor is responsible prior to awarding a contract to that firm.[156] However, if the prospective contractor is an SBC, the SBA plays a major role in the final determination of responsibility. If a contracting officer determines that a small business contractor appears to lack certain elements of responsibility including, but not limited to, capability, competency, capacity, credit, integrity, perseverance, tenacity, and limitations on subcontracting, the contracting officer is directed to withhold award and refer the matter along with detailed information relating to the proposed award to the SBA Government Contract Area Office serving the area in which the prospective contractor's business headquarters is located.[157]

The contracting officer is required to advise the SBA of the agency's determination that the offeror or bidder is not responsible and specify those elements of responsibility that the contracting officer found lacking.[158] Upon receipt of this notification, the SBA has 15 business days (or such longer period as agreed to by the agency and the SBA) to inform the prospective contractor of the agency's determination and advise that firm of its right to seek a certificate of competency (COC) from the SBA, review the agency's determination, and either issue or deny a COC. The SBA may

---

[153]*David Boland, Inc.*, SBA No. SIZ-4965 (June 12, 2008).

[154]13 C.F.R. § 124.513. (Each joint venture entity must comply with the SBA's rule setting specific limits on the number of business opportunities that a specific joint venture entity may pursue in a two year period, the *three in two* rule. *See* 13 C.F.R. § 121.103(h)).

[155]13. C.F.R. § 125.15(b)(2).

[156]*See* FAR § 9.103.

[157]FAR § 19.602-1. There are a few exceptions on the referral requirement. For example, if the contractor is on the list of suspended or debarred contractors, no referral to the SBA is required.

[158]FAR § 19.602-1(c)(1).

limit its review to those elements of nonresponsibility identified by the contracting officer, but it has the discretion to examine other factors affecting responsibility of the prospective contractor.

Basically, the SBA's determination to issue or deny a COC is final and binding on the agency and the contractor. However, the goal of the process is minimize disagreements between the agency and the SBA. Accordingly, there are various procedures to allow the agency to "appeal" a decision of the SBA to issue a COC effectively reversing an agency determination of nonresponsibility.[159] For example, on contracts in excess of $25 million, the agency appeal is submitted directly to the SBA's headquarters for review and action. Regardless of the contract size, the SBA's determination is final;[160] however, the SBA may reconsider a decision if the contractor seeking the COC submitted false information or omitted material information.[161]

III. EQUAL EMPLOYMENT OPPORTUNITY AND AFFIRMATIVE ACTION PROGRAMS

Federal statutes and executive orders regulate the employment practices of government contractors. Collectively, Executive Orders No. 11246 and No. 11141, the Rehabilitation Act of 1973, and the Vietnam Era Veterans Readjustment Assistance Act (VEVRAA) of 1974 have detailed record-keeping and affirmative action requirements that federal contractors must satisfy. Possible penalties for failure to comply with these laws and regulations include contract revocation (default termination) and debarment.

A. Office of Federal Contract Compliance Programs

The Office of Federal Contract Compliance Programs (OFCCP)[162] has been delegated the responsibility to ensure that federal contractors do not discriminate against individuals based on certain protected classifications and that federal contractors abide by the requirements that they create, maintain, and implement affirmative action plans. The OFCCP also has the authority to initiate compliance evaluations to ensure that covered entities are in compliance with their nondiscrimination and affirmative action obligations. The OFCCP can conduct compliance evaluations by any one or by any combination of methods:

(1) A desk audit of the contractor's written affirmative action plan
(2) An on-site review to investigate unresolved problem areas identified in the contractor's written affirmative action plan (including examination of personnel and employment policies)

---

[159]*See* FAR § 19.602-3(b).
[160]*See* FAR § 19.602-4.
[161]FAR § 19.602-3(c).
[162]41 C.F.R. Part 60.

(3) An off-site review of records

(4) A compliance check (i.e., a visit to the facility to determine whether information submitted is complete and accurate)

(5) An on-site "focused review" (restricted to analysis of one or more components of the contractor's organization or employment practices)

The OFCCP has the authority to implement enforcement proceedings and regulations pursuant to Executive Order No. 11246 and the VEVRAA.[163] Although the Department of Labor has enforcement authority with regard to the Rehabilitation Act of 1973, the regulations it has adopted parallel those implemented under Executive Order No. 11246 and the VEVRAA.

B. Executive Order No. 11246

Executive Order No. 11246[164] prohibits employment discrimination based on race, color, religion, sex, or national origin by contractors and subcontractors operating under federal service, supply, use, or construction contracts and by contractors and subcontractors performing under federally assisted construction contracts. All contracts and subcontracts subject to Executive Order No. 11246 must include a clause pledging not to discriminate because of race, color, religion, sex, or national origin and to take affirmative action to ensure that applicants are employed and that employees are treated during employment without regard to those protected classifications. These dual obligations are included in an Equal Employment Opportunity clause that all contracting federal agencies are required to include in their contracts with private employers.[165] Executive Order No. 11246 applies to all contractors doing business with the federal government under contracts or subcontracts that exceed $10,000. Moreover, employers that have 50 or more employees and that have federal contracts worth at least $50,000 are required to prepare and to maintain written affirmative action plans.

C. Vietnam Era Veterans Readjustment Assistance Act of 1974

The VEVRAA requires government contractors and subcontractors to take affirmative action to employ and to advance in employment qualified disabled veterans and veterans of the Vietnam era. The Act's coverage is triggered when a contractor or a subcontractor holds contracts of $10,000 or more. Employers that have 50 or more employees and contracts of $50,000 or more must also prepare written affirmative action plans. Coverage is not triggered, however, by federally assisted contracts or employment agreements.

Part of an employer's affirmative action obligation under the VEVRAA is inclusion in each covered contract and subcontract—and in any contractual modifications,

---

[163]38 U.S.C. § 4212.
[164]30 Fed. Reg. 12,319 (1965).
[165]FAR § 52.222-26.

renewals, or extensions—a clause declaring that the employer will not discriminate against an employee or an applicant with regard to any position for which the employee or the applicant is qualified because he or she is a disabled veteran or veteran of the Vietnam era. The VEVRAA requires that all covered employers list job openings with the local employment service office. (This is also referred to as the "mandatory listing" requirement.)

D. Rehabilitation Act of 1973

The Rehabilitation Act of 1973 requires certain federal contractors and subcontractors that enter into contracts in excess of $10,000 to take affirmative action to employ and to advance in employment individuals with disabilities and not to discriminate against qualified individuals with disabilities.[166] All employers to which the Act applies must include an equal employment opportunity clause in their federal contracts. In addition, all employers with at least 50 employees and with a federal contract of at least $50,000 must prepare and maintain an affirmative action plan that must be updated annually.

E. Executive Order No. 11141

Contractors and subcontractors are prohibited from discriminating against employees or employment applicants based on age unless it can be shown that age is a bona fide occupational qualification or a statutory requirement.[167] Agencies must bring this policy to the attention of contractors; however, no implementing clause need be included in the contract.[168]

IV. BUY AMERICAN ACT

The Buy American Act (BAA)[169] generally restricts government agencies to purchasing only domestic goods for their own use or for use in construction by the agency's contractors. In an era of global trade and diminishing domestic manufacturing, compliance with this restriction is complicated and challenging. The BAA requires construction contractors to use domestic *construction materials* (and, in supply contracts, requires contractors to supply domestic *end products*). A contracting officer may acquire nondomestic (or "foreign") construction materials or end products only if the circumstances justify an exception to the general requirement.

Although the basic BAA has been in effect for over 50 years, it is not the only version of the Buy American Act that affects construction contractors. When Congress

[166] 29 U.S.C. § 793.
[167] 29 Fed. Reg. 2477 (1964).
[168] FAR § 22.901(c).
[169] 41 U.S.C. §§ 10a–10d.

enacted the American Recovery & Reinvestment Act of 2009 (ARRA or Recovery Act), Public Law 111-5, in 2009, it included legislation setting forth new and different BAA provisions affecting construction projects funded by ARRA. This statute was implemented with interim FAR provisions issued on March 31, 2009.[170] The AARA Buy American requirements and the implementing regulations are complex and set forth different standards from the basic BAA, which may present challenges for a contractor performing a project that is funded by both Recovery Act and non-Recovery Act appropriations. See **Chapter** 17 for an overview of the various Recovery Act requirements and regulations affecting government construction contracts.

A. Exceptions to Domestic Purchasing Requirements

Contractors, subcontractors, and suppliers may use nondomestic construction materials during performance of a construction contract (and a contracting officer may purchase nondomestic end products) if: (1) the head of the agency determines the cost of domestic products or materials to be *unreasonable*;[171] (2) the head of the agency grants a waiver based on a determination that the purchase of such products or materials is inconsistent with the public interest;[172] (3) the end product is not mined, produced, or manufactured in the United States in sufficient and reasonably available commercial quantities of a satisfactory quality;[173] (4) the award value of the contract is less than the micropurchase threshold[174] (currently $2,500 for purposes of the BAA);[175] or (5) the products or materials are not for public use within the United States.[176] In a construction procurement, if an exception to the BAA mandate is granted by the head of the agency prior to accepting offers based on impracticability or an unreasonable increase in the cost of the contract, then that exception must be noted in the specifications.[177] An exception to the BAA regarding specific construction materials can be granted prior to award upon a request made to the contracting officer by an offeror.[178]

1. Unreasonable Costs

The reasonableness of the cost of domestic products and materials might not be determined until after offers have been submitted. When an offer proposing to use

[170]74 Fed. Reg. 14623-14633.

[171]41 U.S.C. §§ 10a, 10b(a), 10d; FAR §§ 25.103, 25.202.

[172]*Id.*

[173]*Id. See also* FAR § 25.104 (listing specific articles that have been determined to be nonavailable for purposes of supply procurements); *Midwest Dynamometer & Eng'g Co.,* Comp. Gen. Dec. B-252168, 93-1 CPD 408.

[174]*Id.*

[175]41 U.S.C. § 428.

[176]41 U.S.C. §§ 10a, 10b(a), 10d; FAR §§ 25.103, 25.202.

[177]41 U.S.C. § 10b(a).

[178]FAR § 25.203. For DOD military construction projects using steel, see requirements in DFARS § 236.274.

any foreign materials has the lowest total cost, that offer and the lowest-cost offer proposing the use of only domestic materials are compared under a regulatory scheme requiring the addition of a percentage of the cost of any proposed foreign products or materials to the total cost of that offer. That additional percentage, called an "evaluation factor," creates a preference for domestic products and materials—and for contractors offering to use them—in the evaluation of offers by increasing the price of an offer proposing to use foreign products and materials for purposes of the evaluation. After adding the evaluation factor to the offer proposing foreign products or materials, the total cost of that offer is compared to the lowest responsive offer proposing only domestic products and materials. Note that the evaluation factor is not used to provide a preference for one offer proposing foreign supplies or materials over another.[179] In a construction procurement, the contracting officer must add to the offered price 6 percent of the cost of any foreign construction material offered in the proposal for exception from the BAA. Offerors may submit alternative proposals using domestic construction materials in order to avoid rejection of their offer based on the evaluation preferences of the BAA.[180]

2. Trade Agreements

The Trade Agreements Act of 1979 (TAA)[181] supersedes the BAA restrictions by authorizing the president to waive any otherwise applicable law, regulation, procedure, or practice regarding government procurement. Thus, pursuant to a number of varying trade agreements and acts, the U.S. Trade Representative establishes detailed regulations[182] that apply the TAA for procurements exceeding certain thresholds. The threshold for construction acquisitions is no less than $7,443,000.[183] The regulations also put limits on the types of products, countries of origin, and types of procurements involved.

B. Differentiating between End Products and Construction Materials

Construction materials and end products that qualify as *domestic* and comply with the BAA are not subject to offer-evaluation cost increases. The base definitions pertaining to a product or material's domesticity are the same. A *domestic construction material* is either unmanufactured construction material mined or produced in the United States or construction material manufactured in the United States, more than 50 percent of the cost of the components of which is comprised of components

---

[179]FAR § 25.101(c).

[180]FAR § 25.204. In a procurement of supplies, if an offer proposing foreign end products represents the lowest offer before applying this provision of the BAA, then the contracting officer increases the price of that offer by 12 percent of the cost of the foreign end products if the lowest offer proposing domestic end products is from a small business concern and by 6 percent if the lowest offer is from a large business. FAR § 25.105. The DOD applies a 50 percent increase to the cost of foreign end products in a procurement of supplies. DOD FAR Supplement (DFARS) § 225.502.

[181]19 U.S.C. §§ 2501 *et seq.*

[182]FAR Part 25, § 25.402.

[183]FAR §§ 25.202, 25.402.

mined, produced, or manufactured in the United States. Similarly, a *domestic end product* is either an unmanufactured end product mined or produced in the United States or an end product manufactured in the United States, more than 50 percent of the cost of the components of which is comprised of components mined, produced, or manufactured in the United States.[184]

In practice, however, the BAA functions very differently when applied to end products or to construction materials because of the fundamental difference in the nature of the two. With regard to end products, determinations are made on the final items produced by the contractor and purchased by the government. An end product is an item in its final form, and its subparts are *not* considered individually (except to the extent that the subparts are the components of the end product). In contrast, construction materials are considered individually even though they are merely pieces of a many-layered puzzle being integrated or incorporated into a larger project: a bridge, a building, a dam, and so on.

For example, a supply contract for end products could be for trucks, truck engines, or truck tires. In any case, the test is applied only to the end item being acquired: the truck, the engine, or the tire. In a contract for supplying trucks, the issue simply is whether 50 percent of the cost of the components of the truck as a whole were manufactured or produced in the United States. The BAA is applied differently, however, to a construction contract for an Army barracks. It is not the end item—the building—to which the test is applied. BAA requirements apply to the materials used to build the barracks. The questions, then, become: What is a construction material and what is a component? Where is the line drawn? Is a fuse box a singular construction material? Or are the wiring, the switches, and the casing distinct construction materials? According to the FAR:

> Construction material means an article, material, or supply brought to the construction site by a contractor or subcontractor for incorporation into the building or work. The term also includes an item brought to the site preassembled from articles, materials, or supplies. However, emergency life safety systems, such as emergency lighting, fire alarm, and audio evacuation systems, that are discrete systems incorporated into a public building or work and that are produced as complete systems, are evaluated as a single and distinct construction material regardless of when or how the individual parts or components of those systems are delivered to the construction site. Materials purchased directly by the Government are supplies, not construction material.[185]

The key is distinguishing between construction materials and components. For example, when do steel beams that are fabricated in Japan but modified, as needed, in the United States become construction material instead of components? In *S.J. Amoroso Construction Co. v. United States*,[186] the steel beams were considered construction

---

[184]FAR § 25.003.
[185]FAR § 25.003.
[186]12 F.3d 1072 (Fed. Cir. 1993).

material before additional domestic components were added. Importantly, the steel beams were delivered to the site before they were modified. As a consequence, the contractor that used the beams was held to be in violation of the BAA.

C. Government Remedies

Violations of the BAA can result in significant penalties, including termination for default.[187] More often the government may seek an equitable deduction in the contract price[188] or simply require the contractor to pay the additional cost to obtain compliant construction material.[189] Willful violations might also subject a contractor to penalties under the False Claims Act.[190]

V. LABOR STANDARDS

A number of statutes mandate labor standards that must be observed by government contractors. These statutes regulate minimum wages, overtime wages, and working conditions. Compliance with these statutes is critical because penalties for violations can include contract termination and even debarment for up to three years.

A. Basic Labor Laws

1. Fair Labor Standards Act

The Fair Labor Standards Act (FLSA) is a federal law requiring employers covered by the Act to pay employees a minimum wage and overtime pay. The activities of the employee and not the activities of the employer determine whether the FLSA covers the employees in question. Most construction trade workers perform duties covered by the minimum wage and overtime provisions of the FLSA. However, work performed by executive, administrative, or professional employees as defined in the FLSA are exempt from the minimum wage and overtime requirements of the Act even if performed in conjunction with the work of other employees on a construction site. At times, an employee may be subject to coverage for some work and not covered for other work performed during the same week. In that case, the employee is entitled to coverage for the entire week, as long as the covered work was not isolated and sporadic.[191]

Coverage also depends on the existence of an employer-employee relationship. Although individual employees are covered by the FLSA, independent contractors

---

[187]*Two State Constr. Co.,* DOTCAB No. 78-31, 81-1 BCA ¶ 15,149.
[188]*See LaCoste Builders, Inc.,* ASBCA Nos. 29884 et al., 88-1 BCA ¶ 20,360 (contractor declined to offer equitable deduction and was eventually terminated for default).
[189]*Worcester Bros. Co.,* ASBCA No. 49014, 99-2 BCA ¶ 30,519.
[190]*United States v. Rule Indus., Inc.,* 878 F.2d 535 (1st Cir. 1989).
[191]29 C.F.R. § 776, Subpart B.

hired to perform a service for the employer are not covered. This distinction can be misleading, however, because of the broad definition of the employer-employee relationship in the FLSA. An employee is defined by the FLSA as "any individual employed by an employer."[192] An employer is defined as "any person acting directly or indirectly in the interest of an employer in relation to an employee."[193] An independent contractor must be truly independent.

The regular rate of pay is based on the number of hours worked during a standard workweek. A workweek is defined as seven consecutive days, or 168 hours. An employer must calculate the wage by considering each workweek separately and may not average weeks in which less than the statutory minimum wage was earned. The employee is entitled to regular pay for the first 40 hours worked during a workweek and one and a half times the regular rate of pay for each hour worked in excess of 40 hours.

The FLSA does not require that employees be paid by the hour. Compensation systems involving weekly, monthly, or yearly salaries or piecework rates are perfectly acceptable as long as the total straight-time compensation divided by the straight-time hours worked equals the minimum wage. Dividing straight-time compensation by the number of straight-time hours results in the *regular rate of pay*. The regular rate of pay determines the amount of overtime due to a particular employee.

The federal minimum wage for nonexempt employees is currently $7.25 per hour for the first 40 hours worked.[194] In most cases, employees must receive payment free and clear in cash or negotiable instrument. The only exception: Employers may credit against the minimum wage the reasonable costs incurred in paying for room, board, or lodging customarily provided to employees.[195]

Other expenses may not be credited against the minimum wage. For example, although the FLSA does not prevent employers from requiring employees to wear uniforms, it does prevent employers from forcing employees to pay for the uniforms or the cleaning of the uniforms if doing so would push the standard rate of pay below the required minimum wage. Employers also may not deduct expenses for tools of the trade, breakage, or suspected theft if the deduction will send the weekly wage below the statutory minimum. Deductions for theft resulting in a weekly wage below the minimum standard may be applied only after the guilt of an employee has been determined in a criminal proceeding.

There are certain situations where it is difficult for the employer to discern whether time spent by the employee constitutes compensable or noncompensable work time. Any time considered *work time* will affect the regular rate of pay as well as the overtime calculation for each employee. Determining what constitutes compensable work time is vitally important to complying with the provisions of the FLSA. Employers face a challenge when determining whether time spent by the employee preparing for the day's work should be compensated. The key question is whether the

---

[192]29 U.S.C. § 203(e)(1).
[193]29 U.S.C. § 203(d).
[194]29 U.S.C. § 206(a)(1).
[195]29 U.S.C. § 203(m).

time spent by the employee outside of scheduled work time predominantly benefits the employer. If so, then the employee should be compensated.

The Portal-to-Portal Act[196] was enacted by Congress to clarify this situation. The Act allows employers to exclude activities that occur either prior to the time on any given workday at which an employee begins working ("preliminary time") or after the time on any given workday at which he or she stops working ("postliminary time").[197] The Portal-to-Portal Act eliminates from compensable time activities such as travel and walking time before and after work. However, preliminary and postliminary time is compensable if it is considered an integral part of the principal job. If integral, then the activity is characterized as "preparatory" and is compensable work time. For example, although time spent washing hands and changing clothes at the end of a workday usually is not compensable, time spent filling up the fuel tanks of delivery vehicles most likely is compensable.

The distinction between preparatory and preliminary may be difficult to determine. As in most instances of wage and hour law, the key inquiry is whether the main beneficiary of the time in question is the employer or the employee. If the questioned activity primarily benefits the employee, then the time is most likely not considered work time. If, however, the employer is the prime beneficiary, then wages for the time spent during the activity must be paid.

Employers must pay employees for all time spent *on duty.* In many situations it is fairly simple to determine when a particular employee is on or off duty. Problems develop, however, when employers attempt to determine whether to compensate employees who are *on-call,* or waiting to be called to work. When considering whether to compensate such employees, employers must pay close attention to all of the factors in order to ensure compliance with the FLSA.

Employees who are waiting for materials to arrive or waiting to work while on duty generally must be compensated. It is fairly clear that if the employee is completely under the control of the employer and is unable to pursue his or her own interests, then compensation is required. For example, time spent waiting because of machinery breakdowns or waiting for deliveries is compensable.

Employees who are completely relieved of duty are not entitled to compensation for idle hours. An employee is considered completely relieved of duties if told "in advance that he may leave the job and that he will not have to commence work until a definitely specified hour has arrived."[198] In short, employees are off duty if they are able to spend the idle time pursuing their own interests.

2. Contract Work Hours and Safety Standards Act

The Contract Work Hours and Safety Standards Act (CWHSSA)[199] covers laborers and mechanics on contracts exceeding $100,000 for public works of the United

[196]29 U.S.C. §§ 251-262.
[197]29 U.S.C. § 254(a).
[198]29 C.F.R. § 785.16(a).
[199]40 U.S.C. §§ 3701-3708.

States or the District of Columbia. The law requires overtime wages beyond a 40-hour week[200] and specifies certain health and safety requirements. Violations of the overtime-wages provisions are punishable by a withholding of underpaid wages by the contracting agency, plus liquidated damages of $10 per day per person for each day of violation. Prime contractors are liable for their subcontractor's nonpayment of overtime wages. The Act also permits debarment of up to three years for violations of the health and safety provisions, which require employers to provide working conditions and surroundings that are sanitary and free from hazards or dangers. Debarment is also a potential remedy for willful violation of the wage provisions.[201] These construction health and safety standards are established by the Secretary of Labor in consultation with the Department of Labor Advisory Committee on Construction Safety and Health. Intentional violations of any part of the Act may be punished by fines or by imprisonment of not more than six months.

As with the Davis-Bacon Act, most disputes or issues related to this labor standard must be submitted to the Department of Labor (DOL) for resolution. However, in *Myers Investigative & Security Service, Inc. v. Environmental Protection Agency* (EPA),[202] the board ruled that a contractor could assert a Contract Disputes Act claim that EPA's failure to incorporate a revised wage standard into the contract was a breach of contract, which negated the contract clause allowing for the imposition of liquidated damages for a violation of the CWHSSA. On that basis, the board concluded that it could take jurisdiction over that claim.

B. Davis-Bacon Act

The Davis-Bacon Act (DBA)[203] requires contractors to pay mechanics and laborers a "prevailing wage"[204] on federal construction projects performed in the United States that exceed $2,000. A violation of the DBA can justify debarment of the contractor if the Comptroller General finds that the contractor "disregarded their obligations to employees and subcontractors."[205]

1. Applicability

The DBA traditionally has been the most frequent basis for DOL or contracting agency employment-related actions involving contractors or their subcontractors on construction projects. The DBA applies to construction activity performed on *the site of the work*. Construction activity generally does not include manufacturing,

---

[200]It is the policy of the United States that overtime not be utilized, whenever practicable. FAR § 22.103-2.

[201]*Janik Paving & Constr., Inc. v. Brock,* 828 F.2d 84 (2nd Cir. 1987).

[202]GSBCA No. 16587-EPA, 05-2 BCA ¶ 32,989.

[203]40 U.S.C. §§ 3141-3148; FAR Subpart 22.4; 29 C.F.R. Part 5.

[204]40 U.S.C. § 3142(b). DOL "prevailing wage" determinations for the area in which the project is being performed are seen by many as reflective of the local union wage agreements and job classifications.

[205]40 U.S.C. § 3144(b)(1). This standard is more liberal to the contractor than the equivalent debarment provision of the Service Contract Act. Private causes of action also exist under the DBA. *Hartt v. United Constr. Co.,* 655 F. Supp. 937 (W.D. Mo. 1987), *aff'd without opinion,* 909 F.2d 508 (8th Cir. 1990).

supplying materials, or performing service/maintenance work.[206] The *site of the work* usually is limited to the geographical confines of the construction job site.[207] Transportation of materials to and from the project site is not considered to be construction for the purposes of the DBA.[208] The DBA also may apply to construction work performed under a nonconstruction contract—for example, an installation support contract. If the contract requires a substantial and segregable amount of construction, the DBA applies.[209]

The prevailing wage is the key to Davis-Bacon labor standards. "Wages" includes both basic hourly rates for various classifications of labor needed for the project and fringe benefits. DOL wage determinations are not subject to review by the GAO, agency boards of contract appeals, or the United States Court of Federal Claims. However, the effect of a DOL decision on the contractual relationship between the government and its contractors may be the subject of a contract claim that can be adjudicated by the court or the boards.[210] Laborers and mechanics employed by the prime contractor and all subcontractors at every tier are covered. Working foremen who devote more than 20 percent of their time during a workweek to performing duties as a laborer or mechanic are also covered.[211]

Many DBA disputes involve issues regarding the proper classification of work to a particular craft (wage rate) and accurate record keeping. Employees who *work with the tools* part of the time and also perform work as laborers can lead to alleged violations and enforcement questions.

2. Enforcement

Enforcement of the DBA may begin with either the DOL or the contracting agency. The contracting officer must withhold contract payments if the contracting officer believes that a violation of the DBA exists or if requested to do so by the DOL. If an alleged violation of the DBA is not resolved at the local level, the DOL resolves the dispute. Disputes related to the interpretation and enforcement of the DBA are not subject to the Disputes clause, even though the contracting officer makes the initial

---

[206]FAR § 22.402.

[207]*Ball, Ball & Brosamer, Inc. v. Reich,* 24 F.3d 1447 (D.C. Cir. 1994).

[208]*Building & Constr. Trades Dep't, AFL-CIO v. Dep't of Labor Wage Appeals Bd,* 932 F.2d 985 (D.C. Cir. 1991). *Cf.* 29 C.F.R. § 3.2(b); 29 C.F.R. § 5.2(j).

[209]DFARS § 222.402-70. The DFARS includes specific tests to assist in the determination of whether the DBA (repair) or SCA (maintenance) applies.

[210]*Clevenger Roofing and Sheet Metal Co. v. United States,* 8 Ct. Cl. 346, 351-54 (1985); *Chem-Care Co.,* ASBCA No. 53614, 04-1 BCA ¶ 32,593; *Joe E. Woods, Inc.,* DOTCAB No. 2777, 96-2 BCA 28,551 (holdings of the predecessor boards of contract appeals, including the former DOTCAB, are binding as precedent in the Civilian Board of Contract Appeals, *Business Mgmt. Research Assocs., Inc. v. General Services Administration,* CBCA No. 464, 07-1 BCA ¶ 33,486); *Gerald Moving & Warehousing Co.,* Comp. Gen. Dec. B-225618, 87-1 CPD ¶ 59. *But see Burnside-Ott Aviation Training Ctr., Inc. v. United States,* 985 F.2d 1574, 1579-81 (Fed. Cir. 1993) (addressing whether the contractual effect of wage decisions by the DOL may be adjudicated before the U.S. Court of Federal Claims).

[211]FAR § 22.401; 29 C.F.R. § 5.2(m).

withholding of funds.[212] However, if the dispute is based on contractual rights and obligations of the parties, there exists a basis to submit the claim or dispute to the board or to the Court of Federal Claims.[213]

Violations of the DBA can lead to termination and even debarment. For example, consider *Glaser Construction Co., Inc. v. United States*.[214] In August 1996, a contractor and the Department of Veterans Affairs entered into a construction contract for the renovation and alteration of two wings of a building at a VA medical center. The contract included the standard default clause,[215] providing that the government may terminate the contract if the contractor refuses or fails to prosecute the work with the diligence that will ensure completion within the time specified in the contract, including any extension. The contract also included a standard clause entitled "Contract Termination—Debarment,"[216] which provided that a breach of any of a number of the labor-related clauses in the contract (including the DBA and the CWHSSA) could be grounds for termination of the contract and for debarment of contractor. Further, the contract provided that the contractor could be terminated for default if it committed "sufficiently serious"[217] violations of the Buy American Act.

The notice to proceed was issued to the contractor on September 26, 1996, with contract completion required by November 1, 1997. As typically happens in renovation and alteration contracts, several change orders were issued by the VA. Midway through the project, seven contract modifications had been issued, which granted the contractor time extensions totaling 37 days and changing contract completion to December 8, 1997. In early September 1997, the contractor submitted a revised construction schedule to the VA that proposed a project completion date of May 2, 1998. The VA's contracting officer responded to the contractor, rejecting the proposed revised project completion date, and told the contractor that the contractual completion date remained December 8, 1997. In early October 1997, the project completion date was extended by contract modification to December 12, 1997.

The VA issued a cure notice to the contractor in late October 1997 stating that the contractor's most recent progress schedule was unacceptable, set out the VA's belief that the contractor's president's statement that the contractor would not be able to meet the December 12, 1997, completion date constituted an anticipatory breach, and required that the contractor cure its progress-related problems within 15 days after receipt of the notice. The cure notice also advised the contractor that it could be terminated for default if it failed to cure the conditions that the VA believed were endangering performance. The contractor responded in late October 1997 and asserted that problems created by the VA had delayed the project and impacted its ability to complete performance by December 12, 1997.

---

[212]*Emerald Maint., Inc. v. United States*, 925 F.2d 1425 (Fed. Cir. 1991). Federal district courts can entertain appeals from DOL decisions. *See Building & Constr. Trades Dep't AFL-CIO v. Sec'y of Labor*, 747 F. Supp. 26 (D.D.C. 1990).

[213]*See Central Paving, Inc.*, ASBCA No. 38658, 90-1 BCA ¶ 22,305.

[214]52 Fed. Cl. 513 (2002).

[215]FAR § 52.249-10 Default (Fixed-Price Construction) (Apr. 1984).

[216]FAR § 52.222-12.

[217]FAR § 25.206(c)(4).

On December 12, 1997, 16 percent of the contract work remained. The contracting officer notified the contractor that day that the completion time would expire that day at midnight and rejected the contractor's claims that the VA had delayed the contractor's work. However, the letter also indicated that it was in the government's best interest to allow the contractor to continue performing under the contract in a default status and that the VA expected the contractor to be complete with the project by January 21, 1998, so that the VA could take beneficial occupancy by February 1, 1998. The contractor failed to complete performance by January 21, 1998, and by a final decision of that same date the VA terminated for default the contractor's right to proceed under the contract.

Subsequent to the issuance of the final decision, violations of both the DBA and the BAA by the contractor came to the attention of the VA. Subsequently, the contractor and its president were debarred for these violations. However, neither the DBA nor the BAA violations had been considered by the contracting officer when deciding whether to terminate the contract for default.

The contractor appealed the termination for default to the Court of Federal Claims. Part of the evidence presented by the VA to support the termination for default was the contractor's violations of the DBA and the BAA and the subsequent disbarment of the contractor and its president. The contractor argued that because the VA had not relied on those violations as a basis for the termination and did not reference the violations in the final decision, they could not be matters relied on by the VA to now support the earlier termination. The contractor also argued that without a contracting officer's final decision that addressed the DBA and BAA violations, the court was without jurisdiction even to consider the VA's argument.

The court first addressed the question of whether it had jurisdiction to sustain the termination on the grounds of violations of the DBA and the BAA absent a contracting officer's final decision on those issues. It concluded that notwithstanding the language of the Contract Disputes Act, which states that "[a]ll claims by the government against a contractor relating to a contract shall be the subject of a decision by the contracting officer,"[218] there were sufficient prior decisions to uphold a termination for default based on a post-hoc justification even though the new justification had not been the subject of a contracting officer's final decision.[219] The court thus concluded that it did have jurisdiction to consider the VA's claim that the decision to terminate the contractor for default was justified or could be justified by the violations of the BAA and/or the DBA. As the DOL had given the contractor an opportunity to challenge the DBA violations, and as the contractor had failed to avail itself of that opportunity, the court concluded that the violations of the DBA alone were sufficient to justify the termination for default and that it was unnecessary to consider the question of whether the contractor also had violated the BAA.

---

[218]41 U.S.C. § 605(a).

[219]*See* **Chapter 11, Section II.C.** for a discussion of the government's ability to sustain a termination for default on alternative or new grounds.

C. Service Contract Act

The Service Contract Act of 1965 (SCA)[220] provides that contractors performing any "service contract"[221] shall pay their employees not less than the FLSA minimum wage. Contracts in excess of $2,500 are subject to prevailing wage and fringe benefit determinations. These wage determinations are either set by the DOL[222] or established by a preceding contractor's collective bargaining agreement.[223] Although the SCA does not apply directly to construction contracts, the SCA covers facility support services, such as grounds maintenance and landscaping, as well as the operation, maintenance, or logistical support of a federal facility.[224] These types of projects can include work often performed by a construction contractor which may, therefore, find itself subject to the SCA for that project. In addition to its compensation provisions, the SCA also prohibits employment in hazardous or unsanitary working conditions. Remedies for noncompliance include debarment for up to three years,[225] contract termination,[226] and withholding of payment.[227]

D. E-Verify

Executive Order No. 12989, as amended by President George W. Bush on June 9, 2008, requires federal contractors and subcontractors to use the Department of Homeland Security (DHS) E-Verify system to check whether their employees are eligible for lawful employment in the United States. The amended Executive Order provides:

> Executive departments and agencies that enter into contracts shall require, as a condition of each contract, that the contractor agree to use an electronic employment eligibility verification system designated by the Secretary of Homeland Security to verify the employment eligibility of: (i) all persons hired during the contract term by the contractor to perform employment duties within the United States; and (ii) all persons assigned by the contractor to perform work within the United States on the Federal contract.

[220]41 U.S.C. §§ 351-358. Violations of the SCA provide for the debarment of the contractor absent unusual circumstances (41 U.S.C. § 354(a)) and contract cancellation (41 U.S.C. § 352 (c)). *Universities Research Assocs., Inc. v. Coutu,* 450 U.S. 754 (1981).

[221]The definition of a "service contract" excludes construction, alteration, or repair of public works of the United States as well as painting or decorating. FAR § 22.1003-3.

[222]Such wage determinations by DOL are known as the "prevailing wage" determination. FAR § 22.1002-2; 29 C.F.R. § 4.143. This wage determination typically includes multiple classifications of workers and varying rates. A major area of risk for the contractor involves the classification of certain activities and wage rates.

[223]FAR § 22.1008-2(b); 29 C.F.R. § 4.163; *Klate Holt Co. v. Int'l Bhd. of Elec. Workers,* 868 F.2d 671 (4th Cir. 1989); *Professional Servs. Unified, Inc.,* ASBCA No. 45799, 94-1 BCA ¶ 26,580.

[224]FAR § 22.1003-5; 29 C.F.R. § 4.130. These types of contracts may include activities normally considered to be "construction."

[225]41 U.S.C. § 354.

[226]41 U.S.C. § 352(c).

[227]41 U.S.C. § 352(a).

E-Verify is an Internet-based employment eligibility verification system operated by the U.S. Citizenship and Immigration Services (USCIS) in partnership with the Social Security Administration (SSA). Before an employer can use the E-Verify system, the employer must enroll in the program and agree to the E-Verify Memorandum of Understanding (MOU) required for program participants. The terms of the MOU are established by USCIS and are not negotiated with each participant.

Although the FAR E-Verify clause[228] initially was issued with an effective date in January 2009, mandatory implementation of E-Verify was delayed repeatedly by administrative action and by litigation. However, following resolution of many of the judicial challenges to the rule, the Secretary of Homeland Security issued instructions making E-Verify applicable as of September 8, 2009.

Employers with federal contracts or subcontracts that contain the FAR E-Verify clause required to use E-Verify to determine the employment eligibility of:

(1) Employees performing direct, substantial work under the federal contract or subcontract;

(2) Newly hired employees regardless of whether they are working on a federal contract or subcontract.

As of its effective date, this is a mandatory clause for all construction contracts in excess of the simplified acquisition threshold ($100,000). Excluded from this mandatory coverage are contracts that are *only* for work that will be performed outside the United States, or contracts having a period of performance of 120 days or less.

VI. ENVIRONMENTAL LAWS

The government uses the procurement and the administration of contracts to implement a number of environmental policies.

A. Energy Efficiency and Sustainable Design

Under regulations promulgated pursuant to the Energy Policy Act of 2005,[229] it is the government's policy to acquire services and supplies that promote energy and water efficiency, that advance the use of renewable energy products, and that help foster markets for emerging technologies. That policy extends to all acquisitions, regardless of amount.[230] If cost effective and available, agencies are required to purchase energy-efficient items. This requirement explicitly applies to contracts for design, construction, renovation, or maintenance of a public building, and the specifications for such a contract must include the requirement.[231]

---

[228]FAR § 52.222.54. (This clause includes a mandatory flow-down requirement to all subcontracts involving construction with a value in excess of $ 3,000 and including work performed in the United States.)
[229]Pub. L. 109-58, 119 Stat. 594.
[230]FAR § 23.202.
[231]FAR § 23.203.

Section 109 of the Energy Policy Act requires new federal buildings to be designed 30 percent below American Society of Heating, Refrigerating, and Air-Conditioning Engineers standards or the International Energy Code, to the extent that the technologies employed are life cycle cost effective. All agencies must identify new buildings in their budget requests and identify those that meet or exceed the standard. In addition, the Act requires sustainable design principles to be applied to new and replacement buildings.

Executive Order No. 13423[232] sets specific performance standards for energy efficiency in building design. The objective is to pursue design of high-performance buildings. Each agency head is required to ensure that new construction and major renovation of buildings comply with the Guiding Principles set forth in the Federal Leadership in High Performance and Sustainable Buildings Memorandum of Understanding (2006) and that 15 percent of the existing federal capital asset building inventory of the agency as of the end of fiscal year (September 30) 2015 incorporates the sustainable practices in the Guiding Principles. The Guiding Principles are: (1) employ integrated design principles, (2) optimize energy performance, (3) protect and conserve water, (4) enhance indoor environmental quality, and (5) reduce the environmental impact of materials.

B. Recovered Materials

The Resource Conservation and Recovery Act of 1976 (RCRA)[233] requires agencies to specify the use of recovered materials (rather than new) for certain EPA-designated products to the maximum extent practicable without jeopardizing the intended use of the item.[234] Exceptions to this requirement may be based on agency determinations that the items are not reasonably available in a reasonable time, that the items will not meet applicable performance standards, or that the items are available only at an unreasonable price.

C. National Environmental Policy Act

The National Environmental Policy Act (NEPA) requires that the government must consider environmental issues involved in any major proposal or action. A formal Environmental Impact Statement (EIS) must be prepared in connection with all "major Federal actions significantly affecting the quality of the human environment."[235] Prior to determining whether an EIS is necessary, a more concise document called an Environmental Assesment (EA) typically is prepared. Preparation of the EA can result in a finding of no significant impact (FONSI). Construction projects

---

[232]72 Fed. Reg. 3919 (Jan. 24, 2007).
[233]42 U.S.C. §§ 6901-6992k.
[234]FAR § 23.402; 42 U.S.C. § 6962.
[235]42 U.S.C. § 4332(2)(C).

are particularly likely to implicate NEPA, requiring the preparation of an EA-FONSI or, where necessary, an EIS. Failure by the government or a contractor properly to navigate these waters can lead to schedule delays, changes, and extra costs.[236]

D. Leadership in Energy and Environmental Design

Leadership in Energy and Environmental Design (LEED) is a voluntary rating system for sustainable building and design. LEED provides building owners and operators a framework for identifying and implementing practical and measurable green building design, construction operations, and maintenance solutions. LEED can be applied to any type of building. Federal agencies with a significant construction program have adopted policies requiring LEED certification for all agency construction. The Department of Agriculture, GSA, NASA, and the U.S. Navy require that all new construction and major renovation obtain LEED Silver certification. The U.S. Navy was the first federal agency to certify a LEED project.[237]

VII. EMPLOYEE SAFETY AND THE OCCUPATIONAL SAFETY AND HEALTH ACT

A. Occupational Safety and Health Act

The Occupational Safety and Health Act ("the Act")[238] was enacted in 1970 with the intent to "regulate commerce among the several States and with foreign nations and to provide for the general welfare, to assure so far as possible every working man and woman in the Nation safe and healthful working conditions and to preserve our human resources. . . ."[239] In light of the potential hazards often present on construction sites, the Act has particular significance to the construction industry. Accordingly, in order to avoid civil and criminal liability, construction industry employers must be cognizant of their responsibilities under the applicable provisions of the Act.

The Act requires employers to comply with certain safety standards and furnish a work environment for employees that is "free from recognized hazards that are causing or are likely to cause death or serious physical harm" to employees. Employer liability for violations of the Act can include injunctions, civil and criminal fines ranging from $5,000 to $70,000 per violation, depending on the severity, and imprisonment.[240]

---

[236]*Curry Contracting Co.,* ASBCA No. 53716, 06-1 BCA ¶ 33,242.

[237]The Navy obtained LEED Silver certification for a 58,000-square-foot drill hall at the Great Lakes Training Center. The project was completed in 2008.

[238]29 U.S.C. §§ 651-678.

[239]29 U.S.C. § 651(b).

[240]29 U.S.C. § 666.

Employers have dual responsibilities under the Act. Employers are principally required to follow codified regulations regarding unique aspects of their respective work environments.[241] In addition to the regulatory guidelines, however, the Act also imposes a "general duty" on employers to maintain a safe and healthful work environment by eliminating otherwise unregulated working conditions that may be hazardous to the health or safety of employees.[242]

The Act vests the Secretary of Labor with the responsibility of implementing safety standards through rule-making proceedings.[243] The Secretary of Labor is also responsible for conducting on-site inspections to ensure employer compliance with the requirements of the Act.[244] Compliance checks can be initiated as a result of routine inspections or employee complaints. The Secretary has the authority to obtain a warrant for inspection if the employer refuses to allow inspectors access to the facility. If the employer receives a citation for alleged workplace hazards, it can challenge the citation by seeking review before the Occupational Safety and Health Review Commission.[245] If the employer is unsuccessful in its challenge before the commission, it can seek redress in the federal court system.[246]

B. Agency Safety Requirements

In addition to Occupational Safety and Health Act, project safety is a major priority for every government agency. FAR Part 23 includes detailed requirements addressing hazardous material identification, material safety data, and toxic chemical release reporting. Like many major private owners, the government agencies emphasize safety both during the preaward evaluation of potential contractors and during performance of the work. Safety is a major element in the government's past performance evaluation of its contractors.[247] From a contractor's perspective, safety is essential in terms of the savings in human cost and the impact of safety problems or incidents on the progress of the work, past performance evaluations, insurance costs, and so on.

Many agencies adopt written safety manuals to provide specific standards for safe performance of the work. For example, the *Safety and Health Requirements Manual*,[248] adopted by the U.S. Army Corps of Engineers and used by multiple

---

[241] 29 U.S.C. § 654(a)(2).
[242] 29 U.S.C. § 654(a)(1).
[243] 29 U.S.C. § 655.
[244] 29 U.S.C. § 657.
[245] 29 U.S.C. § 661.
[246] 29 U.S.C. § 660.
[247] *See* DD Form 2626, Performance Evaluation (Construction). *See also Federal Government Contractor Past Performance Evaluations - Toolkit and Guidance* (Marco A. Giamberardino et al. [Associated General Contractors of America, 2010]).
[248] U.S. Army Corps of Engineers, *Safety and Health Requirements Manual*, EM 385-1-1 (2008), available at: *www.usace.army.mil/CESO/Pages/EM385-1-1,2008new!.aspx* (available in English and in Spanish). (This manual is revised periodically by the Corps of Engineers.)

agencies, contains detailed and specific guidance and requirements covering safe construction practices. The introductory sections of this manual provide these general instructions to contractors:

SECTION 1

PROGRAM MANAGEMENT

01.A GENERAL

01.A.01 No person shall be required or instructed to work in surroundings or under conditions that are unsafe or dangerous to his or her health.

01.A.02 The employer shall be responsible for initiating and maintaining a safety and health program that complies with the US Army Corps of Engineers (USACE) safety and health requirements.

01.A.03 Each employee is responsible for complying with applicable safety *and occupational health* requirements, wearing prescribed safety *and health* equipment, *reporting unsafe conditions/activities,* preventing avoidable accidents, *and working in a safe manner.*

01.A.04 Safety and health programs, documents, signs, and tags shall be communicated to employees in a language that they understand.

01.A.05 *Worksites with non-English speaking workers shall have a person(s), fluent in the language(s) spoken and English, on site when work is being performed, to translate as needed.*[249]

Required compliance with this manual or similar publications is routine on government construction projects. A prudent contractor should obtain and review the applicable manual to ascertain its effect on the means and methods of performance in order to be able to cover the cost thereof in its bid or proposal. In many instances, a safety plan will be a required preconstruction submittal, disapproval of which could delay project work. If there are questions regarding the application or interpretation of a particular provision, they should be addressed no later than the start of the specific work activity in order to avoid the disruption caused by a government-directed work suspension due to a perceived or a real safety problem.

In addition to requiring adherence to safety manuals, such as the Corps of Engineers' EM 385-1-1, many contracts also contain requirements for contractor-provided on-site safety personnel. The number of personnel and their necessary level of training and experience need to be considered at the prebid/proposal stage so that the cost is included in the price for the work. Specific qualifications and requirements, such as letters of intent for named personnel, may be a preaward evaluation factor and/or will be a required submittal item after award. A contractor should evaluate the special conditions, the specifications, and any safety manual referenced in the solicitation and should consider the following Site Safety Checklist.

---

[249]*Id.* at 1 (emphasis original).

SITE SAFETY STAFF REQUIREMENTS CHECKLIST
- Is there a requirement for a separate safety staff distinct from the project management team?
- How many individuals are required in the safety organization?
- What are the required education, language skills, and/or experience for each member of the safety organization?
- May experience be substituted for formal education?
- What documentation is required to obtain approval of safety personnel and a safety plan?
- To whom must each member of the safety staff report?
- May any member of the safety staff be provided by or be employed by a subcontractor or a vendor?
- When must each member of the safety staff be physically present at the job site?
- Does the contract require *full-time* presence or presence when certain work is *ongoing*?
- Are there restrictions on the duties that can be assigned to the safety staff members in addition to their safety functions?
- Do the contractor's subcontracts and/or purchase orders clearly bind those working on-site to the safety program?

No one desires an unsafe work environment; the cost is too great. Careful analyses of the applicable required safety procedures can help avert job disruptions and cost overruns as well as contribute to a safer work environment.

➤ LESSONS LEARNED AND ISSUES TO CONSIDER

- Contract amounts totaling approximately *one-fourth* of all federal prime-contract procurement dollars are awarded to *small business concerns*.
- Construction contract awards typically comprise a *relatively large* proportion of the awards to small business concerns.
- A major element of the preference programs for small business construction contractors includes various types of *set-asides*.
- In questions of a prospective contractor's *responsibility*, nearly all agency determinations that a small business contractor is not responsibility must be referred to the SBA for review under the Certificate of Competency process. If the SBA acts within the applicable time frames, the SBA's determination is final and binding on the agency.

(Continued)

- Construction contracts awarded to large business contractors will contain detailed and, at times, challenging requirements for *goals* related to subcontracting to small business firms.
- Failure to demonstrate a *good faith* effort to achieve those goals may result in the imposition of liquidated damages. The availability of *specific and detailed documentation* of the prime contractor's efforts to achieve the goals set forth in the contract is essential to demonstrating a good faith effort.
- *Eligibility for any small business program* requires meeting both the general criteria for small business concerns and the specific criteria of each small business program.
- *Misrepresentation* of status as an SBC can lead to exclusion from small business programs, debarment from government contracting altogether, and even criminal fines and imprisonment.
- Significant limitations exist in connection with small business *affiliations*, especially when the affiliated business is not an SBC or when a joint venture agreement is involved.
- Contractors awarded federal contracts or employed on federally assisted projects are subject to equal employment opportunity, affirmative action, and antidiscrimination laws and regulations. Violations under these laws or regulations can cause the imposition of severe consequences, including *default termination or debarment* or both.
- Contractors are required to use domestic construction materials, absent an exception. Generally it is difficult to demonstrate that any item delivered to the construction site will be considered a *component* rather than a *construction material* no matter what the item later becomes *during* the course of the construction of the project.
- Labor laws, particularly the Davis-Bacon Act, require *strict compliance* with wage, benefit, and overtime regulations as well as significant record-keeping burdens and the attendant administrative costs. Failure to comply can result in a contractor being terminated for default or even debarred.
- As of September 8, 2009 use of the E-Verify system became a mandatory requirement for nearly every government construction contract performed within the United States. The flow-down requirement to subcontracts has an *extremely low threshold* of $3,000.
- Environmental concerns are becoming more prominent features of federal contract work in the form of *green requirements* for enhanced energy efficiency, sustainable design, and LEED design, construction, and certification.
- Safety is a major requirement on government construction projects. Contractors must evaluate applicable *safety manuals* and contract requirements for *on-site* safety personnel, safety procedures, and safety plans. These matters may be considered by the government as a preaward evaluation factor. Further, safety-related submittals often must be approved before the government will allow work at the job site.

SBA Program Requirements, Joint Ventures and Subcontracting

Teaming Arrangements with a Small Business Concern (SBC)	8(a) Set-Aside	HubZone Set-Aside	SDVO Set-Aside
Teaming with Disadvantaged SBC as prime, Large GC as Subcontractor	For general construction, SBC prime must perform at least 15% of the cost of the work (excluding "materials") with its own employees. **For specialty trade contractors, the SBC prime must perform at least 25 percent of the cost of the work (excluding "materials") with its own employees.** *See* 13 C.F.R. 125.6. Repetitive subcontracts with the same firm could be seen as evidence or an indication of affiliation.	In addition to the requirements in 13 C.F.R. 125.6, **HUBZone SBC must perform at least 50 percent of the cost of personnel with its own employees or the employees of other qualified HUBZone SBCs.**	For general construction, SDVO SBC prime must perform at least 15% of the cost of the work (excluding "materials") with its own employees. **For specialty trade contractors, the SDVO SBC prime must perform at least 25 percent of the cost of the work (excluding "materials") with its own employees.** *See* 13 C.F.R. 124.10 and 125.6. Repetitive subcontracts with the same firm could be seen as evidence or an indication of affiliation.
Open Joint Venture (JV)	Large business may not JV with 8(a) on an 8(a) Set-Aside absent a **Mentor-Protégé agreement.** If approved by the SBA, 8(a) SBC may enter into one or more JVs with other SBCs to perform 8(a) Set-Asides. *See* 13 C.F.R. 124.513. The SBA must approve the JV agreement and any revisions to that agreement. **See Mentor-Protégé below.**	Large business may not JV with a HUBZone firm on a HUBZone Set-aside. HUBZone SBC may JV with another HUBZone SBC on a HUBZone Set-aside.	Large business may not JV with a SDVO SBC on a SDVO Set-aside. SDVO SBC may JV with one or more other SBCs. SBA specifies certain provisions of the JV agreement, *See* 13 C.F.R. 125.15(b)(2).
Teaming with Disadvantaged SBC as prime, Large GC as Subcontractor. (Large GC is indemnifying disadvantage's bond.)	Bonding assistance standing alone is not evidence or an indication of affiliation. *See Size Appeal of David Boland, Inc.,* SBA No. SIZ-4965 (2008). SBA may view repetitive subcontracts with the same firm as evidence or an indication of affiliation even if percentage limits are followed.	Bonding assistance standing alone is not evidence or an indication of affiliation. *See Size Appeal of David Boland, Inc.,* SBA No. SIZ-4965 (2008). SBA may view repetitive subcontracts with the same firm as evidence or an indication of affiliation even if percentage limits are followed.	Bonding assistance standing alone is not evidence or an indication of affiliation. *See Size Appeal of David Boland, Inc.,* SBA No. SIZ-4965 (2008). SBA may view repetitive subcontracts with the same firm as evidence or an indication of affiliation even if percentage limits are followed.
Mentor-Protégé	Large business may act as a "mentor" to a protégé and own up to 40% of the protégé firm. Protégé must qualify under the SBA's 8(a) program. SBA may approve a large business having more than one protégé. *See* 13 C.F.R. 124.520(b)(2). SBA must approve the terms of the mentor-protégé agreement. *See* 13 C.F.R. 124.520(e). Once approved, the mentor-protégé may joint venture for certain contracts. *See* 13 C.F.R. 124.520(d). The SBA must approve the terms of the JV agreement. *See* 13 C.F.R. 124.513.	May also qualify as an 8(a) SBC. There is no separate mentor-protégé program for this category of SBC.	May also qualify as an 8(a) SBC. There is no separate mentor-protégé program for this category of SBC.

(Continued)

Teaming Arrangements with a Small Business Concern (SBC)	8(a) Set-Aside	HubZone Set-Aside	SDVO Set-Aside
Teaming Arrangements with a Small Business Concern	**WOSB**	**Unrestricted Small Business Set-aside (SBC Set-aside)**	**SDB Program Ruled Unconstitutional Nov. 2008**
Teaming with Disadvantaged SBC as Prime, Large GC as Subcontractor	For general construction, SBC prime must perform at least 15% of the cost of the work (excluding "materials") with its own employees. **For specialty trade contractors, SBC prime must perform at least 25% of the cost of the work (excluding "materials") with its own employees.** *See* 13 C.F.R. 125.6. Repetitive subcontracts with the same firm could be seen as evidence or an indication of affiliation.	For general construction, SBC prime must perform at least 15% of the cost of the work (excluding "materials") with its own employees. **For specialty trade contractors, SBC prime must perform at least 25% of the cost of the work (excluding "materials") with its own employees.** *See* 13 C.F.R. 125.6. Repetitive subcontracts with the same firm could be seen as evidence or an indication of affiliation.	
Open JV	Large business may not JV with a WBE on a WBE set-aside. No separate rules for Women-Owned SBCs. See rules governing unrestricted SBCs and set-asides for that category of firm.	Large business may not JV with SBC on a contract set-aside for a SBC. Members of a JV are deemed affiliates for the purpose of that procurement with limited exceptions. The SBA's rules on the number of permitted proposals by JVs seem unrealistic. *See* 13 C.F.R. 121.103(h), which limits the number or permissible JV offers to three over a two-year period.	
Teaming with Disadvantaged SBC as Prime, Large GC as Subcontractor (Large GC Indemnifies disadvantaged's bond.)	Bonding assistance standing alone is not evidence or an indication of affiliation. *See Size Appeal of David Boland, Inc.,* SBA No. SIZ-4965 (2008). SBA may view repetitive subcontracts with the same firm as evidence or an indication of affiliation even if percentage limits are followed.	Bonding assistance standing alone is not evidence or an indication of affiliation. *See Size Appeal of David Boland, Inc.,* SBA No. SIZ-4965 (2008). SBA may view repetitive subcontracts with the same firm as evidence or an indication of affiliation even if percentage limits are followed.	
Mentor-Protégé	May also qualify as an 8(a) SBC. There is no separate mentor-protégé program for this category of SBC.	May also qualify as an 8(a) SBC. There is no separate mentor-protégé program for this category of SBC.	

6

GOVERNMENT CONTRACT INTERPRETATION

I. IMPORTANCE OF CONTRACT INTERPRETATION

The contract is the foundation of virtually every relationship in the construction industry. This is particularly the case in the context of federal government contracting, where written contracts provide the primary evidence of the parties' agreement and obligations. Contract interpretation is the process of determining the precise meaning of the terms embodied within a written contract. It involves deciding the meaning of words, resolving conflicts among provisions of the documents, and evaluating the parties' likely intent in the context of ambiguous language or unforeseen events and circumstances. Familiarity with these basic rules of contract interpretation can help avoid the problems and disputes that may arise in the context of the award and performance of government contracts.

II. GOAL OF CONTRACT INTERPRETATION

The written word is by nature potentially imprecise. Writing for the U.S. Supreme Court, Mr. Justice Holmes once described the interpretation challenge in this manner:

> A word is not a crystal, transparent and unchanged, it is the skin of a living thought and may vary greatly in color and content according to the circumstances and the time in which it is used.[1]

The challenge for parties to a contract, boards, and courts remains unchanged today. It is the need to determine the reasonable meaning of contract language based on the circumstances surrounding its use. Although few construction contractors were English majors or minors in school, a basic appreciation of the principles of

[1] *Towne v. Eisner*, 245 U.S. 418, 425 (1918).

contract interpretation is essential. For a contractor, the skin of a living thought is embodied in the project's plans, specifications, as well as the general and special provisions of the contract.

The goal of contract interpretation is to ascertain and enforce parties' intended meaning of contract terms. Because it is rarely possible to determine what was in the minds of the parties at the time of contracting, federal courts and boards rely on the objective expressions of the parties. This entails interpretation through the eyes of a "reasonable contractor."[2] A party's own subjective intent, not expressed at the time of contracting, is usually not relevant and will not be considered for purposes of contract interpretation.[3]

In interpreting government contracts, the courts and boards rely on various rules. The principal interpretive rules are summarized in *Restatement (Second) of Contracts* § 202, "Rules in Aid of Interpretation." The application of these rules to government contracts is examined in the sections that follow.

A. Reasonable and Logical Meaning

The primary contract interpretation rule is that the *reasonable, logical meaning* of the contract language is presumed to be the meaning intended by the parties. This rule overrides all other rules of contract interpretation.[4]

According to this rule, contract language is interpreted as it would be understood by a reasonably intelligent and logical contractor familiar with the facts and circumstances surrounding the contract. Courts and boards use two primary sources of information to determine this objective intent: (1) the language used by the parties in the contract, and (2) the facts and circumstances surrounding contract formation. If the principal purpose of the parties is ascertainable, that purpose is given great weight.[5]

B. Contract Interpreted as a Whole

Another fundamental principal of contract interpretation is that a contract must be considered as a whole, giving effect to all of its parts and words.[6] In interpreting a contract, no part of the document should be rendered meaningless or viewed as mere surplusage.[7] Each part of the agreement should be examined with reference to all other parts, because one clause may modify, limit, or give meaning to another.[8] Similarly, when several documents form an integral part of one transaction, a court

---

[2]*Corbetta Constr. Co. v. United States*, 461 F.2d 1330 (Ct. Cl. 1972).
[3]*Hughes Comm. Galaxy, Inc. v. United States*, 26 Cl. Ct. 123 (1992), *rev'd on other grounds*, 998 F.2d 953 (Fed. Cir. 1993).
[4]*Alvin Ltd. v. United States Postal Serv.*, 816 F.2d 1562 (Fed. Cir. 1987).
[5]*Restatement (Second) of Contracts* § 202(1) (2007).
[6]*New Valley Corp. v. United States*, 119 F.3d 1576 (Fed. Cir. 1997); *McDevitt Mech. Contractors, Inc. v. United States*, 21 Cl. Ct. 616 (1990).
[7]*Id.*; *Monster Gov't Solutions, Inc. v. U.S. Department of Homeland Security*, DOTCAB No. 4532, 06-2 BCA ¶ 33,312.
[8]*T. Brown Constructors, Inc., v. Pena*, 132 F.3d 724 (Fed. Cir. 1997).

or board may read these together with reference to one another even where the documents involved do not specifically refer to one or the other. A similar rule applies to documents attached to the contract or clauses, standards or manuals that are incorporated by reference. Therefore, an interpretation that leaves portions of the contract meaningless generally will be rejected.

C. Conflicts Avoided

When reading the contract as a whole, its provisions must, if possible, be harmonized.[9] For example, when performance of a contract requires removal of certain items and specifies that the contractor must furnish *all work* necessary for performance of the contract, a contract clause entitled *selective demolition* will not be interpreted to alleviate the contractor's obligation to remove items necessary to complete the work.[10] When a more harmonious interpretation is reasonably available, that interpretation will control. Likewise, because the contract must be considered as a whole, ambiguity may not be created by viewing a single term in isolation.

D. Normal and Ordinary Meaning

Words used in a contract will be given their normal and ordinary meaning unless there is evidence that a contrary meaning was intended by the parties.[11] An intended meaning other than normal and ordinary may be evidenced by the context in which terms are used or by the circumstances surrounding formation of the contract.[12]

1. Special Meaning

Certain terms and phrases used repeatedly in government contracts have evolved into terms of art. These special terms, when used in the context of government contracts, will be given their special meaning.[13] One example is the phrase "equitable adjustment." Under an ordinary meaning interpretation, an equitable adjustment might be reasonably construed as permitting a contractual adjustment for recovery of unearned but reasonably anticipated profit. However, the consistent practice in the field of government contracts has allowed an equitable adjustment to cover an allowance for profit on work actually done but has not included unearned but anticipated profits.[14] An ordinary meaning interpretation cannot be used to override the historical interpretation of common terms and phrases used in the context of government contracts.

---

[9]*Data Enters. of the Northwest v. General Services Administration,* GSBCA No. 15,607, 04-1 BCA ¶ 32,539.

[10]*See Coker Corp.,* GSBCA No. 6918, 84-1 BCA ¶ 17,007.

[11]*Slingsly Aviation, Ltd.,* ASBCA No. 50473, 03-1 BCA ¶ 32,252; *The Master Builders,* ASBCA No. 26129, 82-2 BCA ¶ 15,842.

[12]*See VION Corp. v. General Services Administration,* GSBCA No. 12565-P, 94-1 BCA ¶ 26,555.

[13]*See Gen. Builders Supply Co. v. United States,* 187 Ct. Cl. 477 (1969).

[14]*Id.*

2. Technical Terms

Terms may acquire nonstandard or technical meanings in certain industries or trades. Industry-specific meanings may differ substantially from the normal and ordinary meaning assigned to contract terms. When interpreting contracts, technical meanings will override the normal and ordinary meanings when circumstances indicate that the parties intended to use the technical meaning of the term.[15] Technical terms are prevalent within the construction industry and will be interpreted by courts and boards in accordance with their industry-accepted meaning unless a contrary intention is expressed. Likewise, terms used in government contract forms and regulations will be interpreted in the context of federal government contracting.

E. Party-Defined Terms

It is common practice for parties to define the terms they use in a contract. The definition of terms within government contracts is commonplace. These agreed-upon definitions are the clearest manifestation of the parties' intent. Therefore, courts and boards will abide by the parties' chosen definitions.[16] Contractors should also recognize that many terms are defined in the Federal Acquisition Regulation (FAR), such as "contract,"[17] "subcontract"[18] and "subcontractor."[19] Often, these definitions are not set forth in the contract but will be used to interpret FAR clauses. Some of these definitions—that is, "subcontract" and "subcontractor"—may have meanings that differ from those commonly understood in commercial transactions or industry conventions.

F. Status of Mandatory FAR Clauses

Unlike many commercial and state/local public construction contracts, it is a routine practice for many of the federal agencies to include several pages of FAR or FAR supplement clauses by reference. As part of the process of understanding its obligations and the intent of the contract, a contractor should adopt a process to obtain a basic appreciation of those referenced clauses and to identify any newly added or revised versions.[20] The fact that a clause is *only* incorporated by reference does not affect its importance on the parties' rights and obligations under the contract. The FAR also requires that certain contract clauses be included in government contracts of particular types. If certain mandatory clauses are not included in a contract, whether in full text or incorporation by reference, they are nevertheless deemed included by operation of law.[21] Courts and boards have adopted a standardized test to determine if a

---

[15]*P.J. Dick Contracting, Inc.,* PSBCA No. 1097, 84-1 BCA ¶ 17,149.

[16]*Guy F. Atkinson Co., Inc.,* ENGBCA No. 4891, 86-1 BCA 18,555.

[17]*See* FAR § 2.101.

[18]*See* FAR § 44.101.

[19]*Id.* (Those FAR definitions include vendors supplying materials under purchase orders.)

[20]Most lists of incorporated FAR clauses provide the clause's FAR number, title, and date of the adoption of that version, e.g., FAR § 52.211-13, Time Extensions (Sep. 2000).

[21]*G.L. Christian & Assocs. v. United States,* 312 F.2d 418 (Ct. Cl. 1963).

mandatory FAR clause is deemed to be included within a government contract.[22] If the applicable FAR clause is (1) mandatory and (2) expresses a significant or deeply ingrained strand of public procurement policy, it will be deemed controlling even if omitted from a government contract.[23]

III.   CIRCUMSTANCES SURROUNDING CONTRACT INTERPRETATION

Courts and boards frequently interpret contracts based on the facts and circumstances surrounding the contract's formation. This evidence comes in three forms: (1) evidence of discussions and conduct; (2) evidence of the parties' prior dealings; and (3) evidence of custom or usage within the applicable industry. The next sections examine the application of these facts and circumstances.

A.   Discussions and Conduct

Discussions and conduct of contractor and government representatives can be persuasive when a subsequent interpretation of a contract becomes necessary. For example, a contractor may become aware of a possible ambiguity at a prebid conference and request clarification.[24] The government's clarification may serve as proof that the parties agreed on the resolution of a possible ambiguity and on a common interpretation of the contract. At a minimum, sufficient clarification from the government is an expression of its intended meaning. Absent a preaward objection by the contractor, the government's expressed intention will control.

Prebid or preproposal conduct may also indicate that one party should be held to the other party's interpretation, when that interpretation was communicated to the other party. This can be done expressly through discussions or implied by conduct.[25] If the other party, knowing this interpretation, remains silent or does not object or offer a contrary interpretation, this interpretation will be binding on the parties.[26] Further, evidence of discussions or conduct occurring after a contract is awarded, but prior to controversy, may indicate the reasonableness of one party's interpretation.

1.   Use of Parol Evidence

Evidence of the parties' discussions and conduct before and at the time a written contract is signed ("parol evidence") generally is not admissible as evidence to resolve

---

[22]*Parcel 49C Ltd. P'ship v. General Services Administration,* GSBCA No. 16377, 05-2 BCA ¶ 33,098; *S.J. Amoroso Constr. Co. v. United States,* 12 F.3d 1072 (Fed. Cir. 1993).
[23]*Id.*
[24]*Engineered Demolition, Inc. v. United States,* 70 Fed. Cl. 580 (2006).
[25]*L. P. Fleming, Inc.,* PSBCA No. 5197, 06-1 BCA ¶ 33,193.
[26]*Fessel, Siegfriedt, & Moeller Advertising,* HUDBCA No. 90-5360-C10, 06-1 BCA ¶ 33,128; *Amerifab Indus., Inc.,* ENGBCA No. 4981, 87-1 BCA ¶ 19,400; *Shadrick Contracting Co., Inc.,* ASBCA No. 14613, 71-1 BCA ¶ 8647.

a dispute regarding contract interpretation. Reducing a contract to writing has legal consequences. Traditionally, the law has imposed rules that limit the use of external or parol evidence to vary or contradict the unambiguous written terms used in a contract. Courts and boards refer to this concept as the "parol evidence" rule.[27]

Evidence of discussions and conduct prior to execution of a contract is, however, not automatically barred by the parol evidence rule. The first question that must be answered to determine whether parol evidence will be considered is whether the contract is a final and complete expression of the parties' agreement (a *fully integrated* contract). If the contract is final and complete, normally parol evidence cannot be used to vary or contradict its unambiguous terms.[28] For example, the government has been prohibited from relying on extrinsic evidence of its intent to include compensation for a suspension of work due to delay in the government's execution of a modification when the language of the modification provided it was entered into to compensate the contractor for "the foregoing changes."[29] Relying on this unambiguous language, it was held that the modification provided compensation only for the delay to the contractor in completing the changed work and not for the premodification suspension.[30] Conversely, if a term is ambiguous, courts and boards frequently allow extrinsic evidence concerning the parties' negotiations to ascertain the intended meaning of contractual language.[31]

2. Parol Evidence and Authority to Bind the Government

Discussions between a contractor and government representatives may be used to interpret the terms or specifications in a contract but, as illustrated by one decision, for these discussions to be heard and potentially used to interpret the contract, the government representative must have the authority to bind the government. In *P.R. Burke Corp. v. United States*, the contractor repeatedly sought clarification of the directive in a contract for a sewage treatment plant to "remain in operation" while the contractor performed repairs.[32] At one point during the discussions over this term, an assistant plant manager, a government representative, told the contractor that the plant could go offline for "not more than 60 days."[33] The court found that the assistant plant manager was not the contracting officer and the contractor failed to show that the assistant plant manager was given the authority to bind the government.[34]

---

[27]*Fluor Daniel, Inc. v. Regents of Univ. of California*, EBCA No. C-9909296, 02-2 BCA ¶ 32,017; *Insulation Specialties, Inc.*, ASBCA No. 52090, 03-2 BCA ¶ 32,361.

[28]*Lockheed Martin Tactical Defense Systems, Inc.*, ASBCA No. 46797, 00-2 BCA ¶ 30,919; *SCM Corp. v. United States*, 675 F.2d 280, 230 Ct. Cl. 199 (1982).

[29]*Algernon Blair, Inc.*, ASBCA No. 25825, 87-1 BCA ¶ 19,602; *Insulation Specialties, Inc.*, ASBCA No. 52090, 03-2 BCA ¶ 32,361.

[30]*Algernon Blair, Inc.*, ASBCA No. 25825, 87-1 BCA ¶ 19,602; *but see Bell BCI Co. v. United States*, 570 F.3d 1337 (Fed. Cir. 2009) (change order releases interpreted to preclude later submission of a cumulative impact claim).

[31]*Bannum, Inc. v. Department of Justice, Federal Bureau of Prisons*, DOTCAB No. 4452, 06-1 BCA ¶ 33,228; *Turner Constr. Co. v. General Services Administration*, GSBCA No. 15502, 05-1 BCA ¶ 32,924.

[32]*P.R. Burke Corp. v. United States*, 277 F.3d 1346, 1350 (Fed. Cir. 2002).

[33]*Id.* at 1354.

[34]*Id.* at 1355.

Based on this case, if a court or board determines that parol evidence of discussions between a contractor and a government representative is needed to interpret a contract term or specification, the individual with whom the contractor has the discussions would likely have to possess the authority to bind the government. For further discussion on who has the authority to bind the government and in what circumstances, see **Chapter 2, Section I.**

B. Parties' Prior Dealings

Many government contracts are awarded to contractors with past experience with government construction contracts. A contractor may have a prior history of dealings with a particular agency or with an individual contracting officer. When parties have dealt with each other previously, courts and boards may look at their earlier behavior and practices to help interpret their current contract. Similarly, a prior course of dealing may be used to demonstrate that an explicit requirement of the contract is not binding because that requirement was not enforced in the past.[35] Typically, however, evidence of an established pattern of prior dealings may be offered to aid a court or board in contract interpretation but cannot be used to vary or modify the clear express terms of a written contract. The "parol evidence" rule prevents such use of such extrinsic evidence.

The admission of prior dealings serves the purpose of showing what the parties intended by the language in a contract. For example, the Armed Services Board of Contract Appeals (ASBCA) interpreted a services contract to require the contractor to furnish specific personnel even though the contract did not expressly call for them. The contractor was obligated to provide these people based on the fact that it had furnished such personnel under prior similar contracts.[36] Because the personnel had been furnished under prior contracts, the contractor's undisclosed intention not to be bound to continue this practice was not considered.[37]

Another application of the principle of predispute interpretations of contract requirement is illustrated by the decision of the Veterans Affairs Board of Contract Appeals (VABCA) in *Centex Bateson Construction Co., Inc.*,[38] involving the contractor's claim for the work required to wet seal a window wall as a result of the VA's insistence on a 15-pound-per-square-inch (psf) field water penetration test. The board held that although the specifications did not state a field water test pressure for the window wall system, the contract contemplated that the window wall would resist water penetration greater than a standard commercial installation because the contract specified a high (15 psf) laboratory water penetration test pressure with no stated leakage tolerance. During performance, the government followed a lesser field test pressure of 12 psf (no leakage) and 15 psf (minimum leakage). The board

---

[35]*Gresham & Co., Inc. v. United States,* 200 Ct. Cl. 97 (1972) (government barred from enforcing specification based on unconditional acceptance of nonconforming items on prior 21 contracts).
[36]*Benning Aviation Corp.,* ASBCA No. 19850, 75-2 BCA 11,355.
[37]*Id.*
[38]VABCA No. 4802, 97-2 BCA ¶ 29,194.

held that a reasonable reading of the contract requirements as a whole was that the field-test standard was the same as the specified laboratory standard of 15 psf, within reasonable tolerances, for field installation conditions. Thus, considering the application of reasonable tolerances, the VA had not changed the specifications.

During performance of the work, the window wall subcontractor had not questioned or objected to the 15-psf pressure standard—even during the height of the on-site problems with field-test failures. Shortly after completion of the window wall, the subcontractor submitted its initial claim to the prime contractor. Even then the subcontractor failed to identify changed testing requirements as a basis for its claim. In addition, during the actual performance, the subcontractor had not undertaken to track the costs of complying with increased pressure standards. It was nearly two years after the window wall was installed before the subcontractor, for the first time, argued that it was entitled to a price adjustment for the wet sealing and the 15-psf field-test pressure.

The VABCA held that the subcontractor recognized, prior to the initiation of its claim, that it had to meet the 15-psf field-test standard. The board went on to hold that the subcontractor's position regarding the field-testing standard and wet sealing at the time the issues arose during performance was a more credible basis on which to interpret the contract than were the subcontractor's later contentions.

The VABCA also denied the subcontractor's claim for a price adjustment for the additional work required to wet seal the window wall system based on the subcontractor's contention that a window wall meeting the field-test performance standard of 15 psf could not be built under the specifications. The contract documents left it to the subcontractor's discretion to design a system meeting the *performance requirements* within the specified architectural parameters providing basic dimensions, profiles, and sight lines of members. Although it was clear that the subcontractor's window wall system, as designed, could not meet the contract performance requirements without the application of exterior sealant, the sealant application was held to be simply another change in the subcontractor's design that enabled the window wall system to perform as required and for which the subcontractor, not the government, was responsible.

As previously stated, extrinsic evidence is generally not admissible to show an intent entirely different from what is clearly stated in the contract. A prior course of dealing, however, may show that the contract is not the final and complete agreement of the parties. If this showing is made, parol evidence is then properly considered to determine the intent of the contracting parties.

Prior conduct may also support an argument of waiver or estoppel. A party may be prevented from enforcing an explicit contract requirement if in its prior dealings it did not require compliance with the requirement.[39] Therefore, the intent to vary from a prior course of dealing should be expressly stated in subsequent contracts.

---

[39]*North Star Alaska Housing Corp. v. United States,* 76 Fed. Cl. 158, 194 (2007) citing *Sperry Flight Systems v. United States,* 548 F.2d 915 (Ct. Cl. 1977). *See also T&M Distributors, Inc.,* ASBCA No. 51405, 00-1 BCA ¶ 30,677.

C. Custom and Usage

Evidence of customs within a particular industry may be used to show that the parties intended for an ordinary word to have a specialized meaning. However, courts and boards are divided on the role of such evidence.[40] One line of decisions holds that evidence of trade practice and custom may be admitted to show the meaning of an ambiguous contract term but not to override a seemingly unambiguous term.[41] The second line of cases maintain that evidence of trade practice and custom may be introduced to show that a term which appears on its face to be unambiguous has, in fact, a specialized meaning other than its normal and ordinary meaning.[42]

A party seeking to assert a trade custom or practice must present evidence that the custom is well established.[43] One method of establishing trade custom is to show the interpretations of other offerors or bidders on that contract.[44]

Similarly, a technical word will be given its ordinary meaning in the industry unless it is shown that the parties intended to use it in a different context. The appropriate meaning of ambiguous technical terms may also be clarified by the introduction of extrinsic evidence. For example, in a classic case, a Texas court allowed the introduction of evidence of custom to establish the intended meaning of the contract term "working days" as it related to the owner's right to assess liquidated damages for delay in completion.[45] More recently, the United States Claims Court relied on trade practice to interpret patently ambiguous pipe-wrapping requirements in a federal government contract.[46]

In *Metric Constructors v. National Aeronautics and Space Administration*, the Federal Circuit attempted to provide clear guidance on the use of evidence of custom and trade practice in the context of contract interpretation.[47] The court stated that if a contracting party seeks to prove that a contract term is ambiguous because of trade usage or custom, the court will first hear evidence of the trade usage and custom to determine the context in which the parties entered the contract.[48] Within that context, the court will use the trade usage and custom evidence to decide whether the term was reasonably susceptible to different interpretations at the time of contracting.[49] Then the contracting party must show that it reasonably relied on a different interpretation

---

[40]*Metric Constructors v. National Aeronautical and Space Administration*, 169 F.3d 747 (Fed. Cir. 1999).
[41]*R. B. Wright Constr. Co. v. United States*, 919 F.2d 1569 (Fed. Cir. 1990); *George Hyman Constr. Co. v. United States*, 564 F.2d 939 (Ct. Cl. 1977); *WRB Corp. v. United States*, 183 Ct. Cl. 409 (1968).
[42]*See Restatement (Second) of Contracts* § 222; *Hensel Phelps Constr. Co. v. General Services Administration*, GSBCA No. 14744, 01-1 BCA ¶ 31,249; *Hoffman Constr. Co.*, DOTBCA No. 2150, 93-2 BCA ¶ 25,803; *Gracon Corp.*, IBCA No. 2271, 89-1 BCA ¶ 21,232; *but see Nielsen—Dillingham Builders JV v. United States*, 43 Fed. Cl. 5 (1999) (court rejected use of trade practice to resolve an ambiguity created by clearly conflicting contract provisions).
[43]W.G. *Cornell Co. v. United States*, 376 F.2d 299 (Ct. Cl. 1967).
[44]*See Eagle Paving*, AGBCA 75-156, 78-1 BCA ¶ 13,107.
[45]*Lewis v. Jones*, 251 S.W. 2d 942 (Tex. Ct. App. 1952).
[46]*Western States Constr. Co., Inc. v. United States*, 26 Cl. Ct. 818 (1992). *See also Metric Constructors, Inc. v. National Aeronautical and Space Administration*, 169 F.3d 747 (Fed. Cir. 1999).
[47]*Metric Constructors v. National Aeronautical and Space Administration*, 169 F.3d 747 (Fed. Cir. 1999)
[48]*Id*. at 752.
[49]*Id*.

of the term.[50] The court bases this analysis on the need to hear evidence that reflects the contracting parties' true intent and to avoid hearing "post hoc explanations of [their] conduct."[51] The contractor should bear in mind that evidence of trade usage and custom cannot be used to create an ambiguity when the contract is not susceptible of differing interpretations. Trade practice and custom "does not trump other canons of interpretation." [52]

IV. RESOLVING AMBIGUITIES

The rules of contract interpretation just discussed may not resolve every ambiguity in a contract. Many government contracts contain an *order of precedence* clause that will control which terms apply in the case of conflicting contract provisions. Common law rules of precedence may be used in the absence of a controlling contract clause. If an ambiguity remains, or cannot be resolved by an order of precedence rule or clause, courts and boards will apply one of these risk-allocation principles to resolve the conflict: (1) the ambiguity should be construed against the party that failed to request a clarification of the ambiguity; or (2) the ambiguity should be construed against the drafter. Each of these principles is discussed in the sections that follow.

A. Order of Precedence

It may be impossible to interpret a contract without resolving a direct conflict between different terms. Government contracts are complex and frequently contain numerous sections drafted by different people or agencies. When two or more conflicting provisions cannot be harmonized, the rules of contract interpretation establish an order of precedence that may resolve the conflict.

Many government contracts include an *order of precedence* clause expressly stating which provisions control in the case of a conflict.[53] For example, the order of precedence clause may state that the specifications take precedence over the drawings, the special conditions take precedence over general conditions, and so on.

The order of precedence may be modified to reflect the project delivery method used by the agency on a particular procurement. For example, the following is an order of precedence clause used by the Corps of Engineers in a design-build contract.

SCR-41 DESIGN BUILD CONTRACT - ORDER OF PRECEDENCE

(a) The contract includes the standard contract clauses and schedules current at the time of the contract award. It entails (1) the solicitation in its entirety,

---

[50]*Id.*
[51]*Id.*
[52]*Id.*
[53]*See, e.g., General Eng'g & Machine Works v. O'Keefe,* 991 F.2d 775 (Fed. Cir. 1993), citing *Hensel Phelps Constr. v. United States,* 886 F.2d 1296 (Fed. Cir. 1989).

including all drawings, cuts, illustrations, and any amendments, and (2) the successful offeror's accepted proposal. The contract constitutes and defines the entire agreement between the Contractor and the Government. No documentation shall be omitted which in any way bears upon the terms of that agreement.

(b) In the event of conflict or inconsistency between any of the provisions of this contract, precedence shall be given in the following order:

1) Betterments: Any portions of the accepted proposal which both conform to and exceed the provisions of the solicitation.

2) The provisions of the solicitation. (See also Contract Clause: SPECIFICATIONS AND DRAWINGS FOR CONSTRUCTION.)

3) All other provisions of the accepted proposal.

4) Any design products including, but not limited to, plans, specifications, engineering studies and analyses, shop drawings, equipment installation drawings, etc. These are "deliverables" under the contract and are not part of the contract itself. Design products must conform with all provisions of the contract, in the order of precedence herein.[54]

As noted, reference is generally made to an order of precedence clause to resolve an unavoidable conflict. For example, in *Manhattan Construction Co. v. United States,* the contract included an order of precedence clause that stated in the event of discrepancies between specifications and drawings, the specifications controlled.[55] The contractor sought to advance a claim for the installation of additional steam traps on the basis that the specifications controlled over the drawing details. The court ruled against the contractor because it found a reasonable contractor would have read the drawings in conjunction with the specifications, thus avoiding any ambiguity and need to resort to an order of precedence provision. In government contracts, mandatory FAR clauses generally cannot be altered or overridden by a conflicting provision and the operation of an order of precedence clause absent compliance by the government with the *deviation procedures* set forth in FAR Subpart 1.4.

In the absence of an order of precedence clause, general common law rules of precedence will apply. For example, it is a basic rule of contract interpretation that general terms and provisions in a contract yield to specific ones.[56] It is also a general rule of contract interpretation that when specific requirements or definitions are itemized and spelled out, anything not expressly included is deemed to be excluded.[57] Additionally, handwritten terms take precedence over typewritten terms, and typewritten terms take precedence over printed or form terms.[58]

---

[54]*Strand Hunt Contr., Inc.,* ASBCA No. 55671, 55813, 08-2 BCA ¶ 33,868.

[55]*Manhattan Constr. Co. v. United States,* 2008 WL 355519 (Fed. Cl. 2008).

[56]This rule is generally known as *ejusdem generis. See In Matter of Sellers (Wayne C.) and Sellers and Co.,* HUDBCA No. 89-4260-D8, 1989 WL 87567 (1989), citing *United States v. Alpers,* 338 U.S. 680 (1950).

[57]This rule is generally known as *expressio unius est exclusio alterius* (the express mention of the one is the exclusion of the other). *See* 17A CJS Contracts § 312.

[58]*See, e.g., Authentic Architecture Millworks, Inc. v. SCM Group,* 586 S.E. 2d 726 (Ga. Ct. App. 2003); *Wood River Pipeline Co. v. Willbros Energy Services Co.,* 738 P.2d 866 (Kan. 1987).

B. Duty to Seek Clarification

Obvious ambiguities in government contracts may give rise to a duty of the nondrafting party, typically the contractor, to request clarification. In addition, the government frequently includes clauses in solicitation documents that require requests for clarification or interpretation be submitted within a specified number of days prior to bid opening or the closing date for receipt of proposals. As will be discussed, if the government fails to provide the clarification requested, the contractor's *expressed interpretation* may become binding on the government.

1. Patent versus Latent Ambiguities

Ambiguities are either obvious (patent) or may arise only in certain circumstances (latent). An obvious or patent ambiguity is one that is readily apparent from the wording of the contract.[59] By contrast, language containing a latent ambiguity initially appears to be clear and unambiguous but actually contains an underlying ambiguity that becomes apparent only after close examination or presentation of extrinsic facts.[60]

A bidder or offeror has an obligation to seek clarification of patent ambiguities or inconsistencies that appear in the solicitation documents.[61] As stated, government construction solicitations typically contain an express provision imposing an affirmative duty on a contractor to seek clarification of patent ambiguities.[62] However, the lack of such a provision does not relieve the contractor of its duty to request clarification of obvious ambiguities. For example, the Court of Federal Claims has held that, when a provision in the solicitation package conflicts directly and openly with a provision in a referenced handbook, a contractor has an affirmative obligation to seek clarification of such an obvious ambiguity.[63] Because the contractor in that case did not alert the contracting officer to the patent discrepancy, the court barred the contractor from recovering any compensation caused by the conflicting provisions within the solicitation and handbook.[64]

The difficulty arises in determining whether an ambiguity was obvious prior to bidding. One factor used to make this determination is whether other bidders/offerors requested clarification prior to bidding or submitting a proposal.[65] Ultimately, the question of whether the ambiguity was obvious, giving rise to the duty to seek clarification, will depend on "what a reasonable man would find to be patent and glaring."[66]

---

[59]*See Big Chief Drilling Co. v. United States,* 15 Cl. Ct. 295 (1988).

[60]*See AWC, Inc.,* PSBCA No. 1747, 88-2 BCA ¶ 20,637.

[61]*White v. Edsall Constr. Co.,* 296 F.3d 1081 (Fed. Cir. 2002); *Newsome v. United States,* 676 F. 2d 647 (Ct. Cl. 1982); *Westar Revivor, Inc.,* ASBCA Nos. 52837, 53171, 06-1 BCA ¶ 33,288.

[62]*Blount Brothers Constr. Co. v. United States,* 346 F.2d 962 (Ct. Cl. 1965).

[63]*Nielsen-Dillingham Builders, JV v. United States,* 43 Fed. Cl. 5 (1999) (government contract specifications commonly incorporate multiple standards, guide specifications, and handbooks).

[64]*Id.; see also Big Chief Drilling Co.,* 15 Cl. Ct. 295 (1988).

[65]*See W.M. Schlosser Co.,* VABCA No. 1802, 83-2 BCA ¶ 16,630.

[66]*Max Drill, Inc. v. United States,* 427 F.2d 1233 (Ct. Cl. 1970).

2. Agency Failure to Clarify

The rule requiring contractors to seek clarification of patently ambiguous terms pre-supposes that such inquiry will yield a response that will clarify the government's intent.[67] For example, in *Engineered Demolition, Inc. v. United* States,[68] the scope of work included the removal of radiologically contaminated waste on a unit price basis. The contract drawings indicated that the quantity of waste was 6,600 cubic yards. However, the specifications stated that the estimated quantity of waste was 8,080 cubic yards. In response to the contractor's prebid inquiry, the government affirmed that the larger quantity was correct and the contractor based its price on that information. The actual quantity was 6,677 cubic yards, and the contractor submitted a claim based on that underrun in the estimated quantity. The court rejected the government's argument that the quantity was merely an estimate because it had provided a prebid confirmation that the higher number was correct when there was no basis or justification for that action.[69]

A government failure to timely respond to a contractor's request for clarification may also operate to bind the government to the contractor's expressed interpretation.[70] However, if the request for clarification receives no meaningful response, it may be necessary for the contractor to send a follow-up confirmation of its position before submitting its bid or proposal.[71] In addition, even if the contractor does not express its interpretation of the ambiguous clause, a term that is not patently (obviously) ambiguous may be construed against the government under the rule of construing ambiguous terms against the drafter.

In some cases, contractors have attempted to be excused from the duty of seeking clarification based on an argument that the request would have been futile or the government would not have properly responded.[72] These attempts have largely failed due to the difficulty of demonstrating that requests for clarification would not have been properly answered.[73]

C. Construction against the Drafter

The risk of ambiguous contract language generally belongs to the party responsible for drafting the ambiguity unless the nondrafting party knew, or should have known, of the ambiguity.[74] Three requirements must be met for this principle to apply.

---

[67]*See Engineered Demolition, Inc. v. United States,* 70 Fed. Cl. 580 (2006); *Peter Kiewit Sons' Co.,* ASBCA No. 17709, 74-1 BCA ¶ 10,430.

[68]70 Fed. Cl. 580 (2006).

[69]70 Fed. Cl. at 592.

[70]*BMY, Division of Harsco Corp.,* ASBCA No. 36805, 93-2 BCA ¶ 25,684 (prebid inquiry regarding inclusion of tax in bid price unanswered and contractor reconfirmed its interpretation prior to submitting its bid).

[71]*Community Heating & Plumbing v. Kelso,* 987 F.2d 1575 (Fed. Cir. 1993) (no follow-up by bidder after receiving a nonresponsive answer to initial inquiry). *See also Constr. Service Co.,* ASBCA No. 4998, 59-1 BCA ¶ 2077; *Southside Plumbing Co.,* ASBCA No. 8120, 1964 BCA ¶ 4314.

[72]*Id.; S. Head Painting Contractor, Inc.,* ASBCA No. 26249, 82-2 BCA ¶ 15,886.

[73]*S. Head Painting Contractor, Inc.,* ASBCA No. 26249, 82-2 BCA ¶ 15,886; *see also NBM Constr. Co.,* ASBCA No. 37095, 89-3 BCA ¶ 22,252.

[74]The technical name for this interpretative rule is *contra proferentem.* See *Freeman & Co. v. Bolt,* 968 P.2d 247 (Idaho App. 1998); *United States v. Turner Constr. Co.,* 819 F.2d 283 (Fed. Cir. 1987).

(1)   There must truly be an ambiguity—that is, the contract must have at least two reasonable interpretations. A nondrafting party's interpretation need not be the only reasonable interpretation for this principle to apply.[75]

(2)   One of the two parties must have drafted or chosen the ambiguous contract language.

(3)   The nondrafting party must demonstrate that it relied on its reasonable interpretation.[76]

Typically the rule of interpretation against the drafter is applied against the government, which commonly drafts most government contracts. However, in a limited number of cases, the rule has been applied against the contractor. This typically occurs when a contractor's proposed modification or additional terms is accepted by the government and incorporated into the contract verbatim.[77]

V. ALLOCATION OF RISKS AND OBLIGATIONS

The primary purpose of all contracts is to allocate risks and obligations among the contracting parties. Typically, this allocation is by express contact terminology. However, several implied contractual obligations are read into all government contracts.

A. Implied Duties

The most fundamental duty implied in all contracts is the duty not to interfere with the performance of the other contracting party. Nearly all implied contractual duties stem from this basic concept. Additionally, the government is deemed to provide adequate and accurate information concerning contract performance and is expected not to withhold pertinent knowledge.

1. Warranty of Plans and Specifications

Perhaps the most important implied contract obligation is that the party responsible for furnishing completed design information impliedly warrants its adequacy and sufficiency.[78] In the context of a government construction contract, this rule was set forth by the United States Supreme Court in the seminal case of *United States v. Spearin*:

---

[75]*Fry Communications, Inc. v. United States*, 22 Cl. Ct. 497 (1991); *Bennett v. United States*, 371 F.2d 859 (Ct. Cl. 1967); *Gall Landau Young Constr. Co.*, ASBCA No. 25801, 83-1 BCA ¶ 16,359.

[76]*Turner Const. Co., Inc. v. United States*, 367 F.3d 1319 (Fed. Cir. 2004); *Interstate Gen. Govt. Contractors v. Stone*, 980 F.2d 1433 (Fed. Cir. 1992); *Fruin-Colnon Corp. v. United States*, 912 F.2d 1426 (Fed. Cir. 1990).

[77]*See Canadian Commercial Corp. v. United States*, 202 Ct. Cl. 65 (1973); *S.S. Mullen, Inc.*, ASBCA No. 8808, 1964 BCA ¶ 4449.

[78]*Big Chief Drilling Co. v. United States*, 26 Cl. Ct. 1276 (1992); *Ordnance Research, Inc. v. United States*, 609 F.2d 462 (Ct. Cl. 1979).

[I]f the contractor is bound to build according to plans and specifications prepared by the owner, the contractor will not be responsible for the consequences of defects in the plans and specifications.[79]

Often referred to as the *Spearin* doctrine, this implied duty has been recognized by the federal courts, boards of contract appeals, and the courts of nearly every state. The government impliedly warrants the adequacy and sufficiency of the completed plans and specifications to the contractor, even when the design is prepared by the government's independent architect or engineer.[80] Generally in federal construction contracts, all delays caused by defective design specifications are compensable.[81]

a. Design versus Performance Specifications   *Design specifications* provide a detailed design and the precise manner or method of performing the contract.[82] *Performance specifications,* however, specify the performance characteristics that are to be obtained by the contractor and leave the details of the design to the contractor's own ingenuity.[83] The *Spearin* doctrine does not apply to *performance specifications.* However, the government still can be liable for a contractor's unanticipated difficulties under a performance specification if the contractor shows that the government-furnished performance specification was impossible or commercially impracticable to achieve.[84] A performance specification is commercially impracticable if it can be met only at an excessive and unreasonable cost.[85]

In any project, the plans and specifications are intended to direct the contractor in building the structure to meet the government's needs and requirements. The specifications describe these requirements using a design specification or a performance specification, and sometimes both. Unfortunately, there are times when a project incorporates both design and performance specifications that create conflict in meeting the government's goals and questions regarding the allocation of risk.

The case of *J.E. Dunn Construction Co. v. General Services Administration*[86] illustrates these issues. The drawings for the north elevation of the courthouse project showed a circular plaza with six columns that were approximately four stories tall. On top of these columns was a semicircular glass curtain wall that extended to the seventh floor, and above the seventh floor another semicircular glass curtain wall rose to the penthouse of the building. Precast columns continued the semicircular shape of the structure from the ground level to the penthouse behind the curtain wall assembly, and the columns framed the north elevation plaza and the lower and upper north curtain wall assemblies.

[79]248 U.S. 132, 136 (1918).
[80]*Greenhut Constr. Co.,* ASBCA No. 15192, 71-1 BCA ¶ 8845.
[81]*Daly Constr., Inc. v. Garrett,* 5 F.3d 520 (Fed. Cir. 1993); *American Line Builders, Inc. v. United States,* 26 Cl. Ct. 1155 (1992); *Chaney & James Constr. Co. v. United States,* 421 F.2d 728 (Ct. Cl. 1970).
[82]*Weststar Revivor, Inc.,* ASBCA No. 53171, 06-1 BCA ¶ 33,288.
[83]*See Precision Dynamics, Inc.,* ASBCA No. 50519, 05-2 BCA ¶ 33,071.
[84]*Int'l Elec. Corp. v. United States,* 646 F.2d 496 (Ct. Cl. 1981).
[85]*Oak Adec, Inc. v. United States,* 24 Cl. Ct. 502 (1991); *W.F. Magann Corp. v. Diamond Manufacturing Co.,* 775 F.2d 1202 (4th Cir. 1985).
[86]GSBCA No. 14477, 00-1 BCA ¶ 30,806.

The curtain wall subcontractor estimated material and engineering costs on the assumption that the drawing considered the structural design requirements and that any building movement had been factored into the design. However, when the curtain wall shop drawings were submitted, the architectural cladding consultant reported that the north curtain wall would not accommodate the contract's deflection criteria. It did not provide for the specified vertical movement, and the approved correction involving changes in glass size, anchor design, and horizontal rail configuration. That solution required additional engineering, new dies and extrusions, new aluminum for the redesigned horizontal mullions, and additional costs for fabrication, notching, and coping.

While acknowledging the deflection criteria were performance specifications, the curtain wall subcontractor argued that the government had to ensure the specified design accommodated the deflection criteria because the contract imposed design requirements on the sizes, shapes, and profiles of the curtain walls. The General Services Administration (GSA) asserted that determining the means and methods of accommodating building movement criteria during the design and engineering of the curtain walls was the subcontractor's responsibility. According to GSA, the drawings were only to be considered *diagrammatic,* and it was therefore unreasonable for the subcontractor to compromise performance in favor of architectural detailing. Again referencing the specifications, the government noted that the drawings were just a starting point and were expected to be modified at the subcontractor's discretion to meet performance criteria.

The board recognized that many specifications are a mixture of performance and design, and the extent to which one supersedes the other dictates the degree of discretion allotted the subcontractor to meet specified requirements. The board referenced several other cases to illustrate the issue.

In *Santa Fe Engineers, Inc.,*[87] the drawings did not indicate the exact location of duct openings in the floor slabs, so the board concluded that the duct chase and slab penetration drawings were performance specifications because no dimensions were given; nor was the structural steel framing needed to support the concrete surrounding the nondimensioned openings shown.

In *SAE/American-Mid Atlantic, Inc.,*[88] however, the contractor was entitled to rely on the contract drawings for a metal stud backup wall system that showed studs spaced every 16 inches, even though that spacing did not meet wind load specifications. The board rejected the government's argument that designing for wind load resistance was the contractor's responsibility because the specifications did not state the metal studs at the windows needed to be spaced significantly closer than the 16 inches shown uniformly throughout the drawings. The contractor was allowed to recover its additional costs.

A third decision, *Morrison-Knudsen Co., Inc.,*[89] involved drawings for a fuel system that powered radar sites in Alaska. The drawings were held to be design

---

[87]ASBCA No. 24469, 92-1 BCA ¶ 24,665, *aff'd*; *Santa Fe Engineers, Inc. v. Kelso,* 19 F.3d 39 (Fed. Cir. 1994).
[88]GSBCA No. 12294, *et al.,* 98-2 BCA ¶ 30,084.
[89]ASBCA No. 32476, 90-3 BCA ¶ 23,208.

specifications because they contained exact dimensions for the fuel system and the configuration of the piping. When the contractor built in accordance with those requirements and had to overcome a defective design for the fuel system, the board decided the contractor was entitled to a contract adjustment. Finally, in the *Leslie-Elliott Constructors, Inc.*,[90] the drawings for an automatic sprinkler system were found to be primarily design specifications, even though the pipe diameters were to be developed by the contractor. This was because the drawings were definitive in describing the number and location of mains, branch lines, and sprinklers as well as the length of the piping. The board concluded the drawings defined the type of system to be used by the contractor and were not, as the government argued, merely showing only the overall picture or general scheme of the sprinkler system.

Based on these decisions, the board in *J.E. Dunn* decided the specifications required the curtain wall subcontractor to follow the drawings in building the project. For example, Paragraph 1.02A of the specifications for the courthouse project stated, "The requirements shown by the details are intended to establish the basic dimensions of the module and sightlines, joint and profiles of members." It went on to state that the drawing details are "requirements" and "within these parameters, the contractor is responsible for the design and engineering of the window system, including whatever modifications or additions may be required to meet the specified requirements and maintain the visual design concept for the entire project."

The board noted the subcontractor's discretion was confined by the requirements shown on the drawing details. Although the contractor could make modifications, any alteration had to conform to the details and maintain the visual design concept of the building. Other specification provisions emphasized the binding nature of these drawing details. Paragraph 2.02A required the subcontractor to "provide shapes and profiles, as shown" for aluminum members.

Paragraph 1.02B described the subcontractor's responsibility for engineering systems and when to engineer systems by stating: "It is however intended that conditions not detailed shall be developed through the contractor's shop drawings to the same level of aesthetics and in compliance with the performance criteria as indicated for the detailed areas and stipulated in the specifications." The board noted that the contract drawings depicted the vertical sections with specific-sized details.

The board acknowledged that the performance specifications were important. The contract required the subcontractor to comply with the design requirements and the drawing details and to use its ingenuity and skill in the performance specifications. However, after reviewing these possibly conflicting contract specification provisions, the board could not support the government's position that the drawing details were merely diagrammatic or that the written specifications subordinated the drawing details to the performance requirements.

b. Liability for Errors, Conflicts, and Omissions   The *Spearin* doctrine can serve as both a shield and a sword for a party that is not responsible for furnishing design data. When the government furnishes design specifications containing errors,

---

[90]ASBCA No. 20507, 77-1 BCA ¶ 12,354.

conflicts, and/or omissions, the contractor will not be liable for an unsatisfactory final result if the contractor performs in accordance with the government-furnished plans and specifications. If an inadequacy in a government-furnished design results in delay, disruption, or additional costs to the contractor, the contractor may use the *Spearin* doctrine as the basis for claims for additional time and compensable delay.[91]

For example, in a recent case arising from the construction of two helicopter hangers, the Federal Circuit affirmed a board decision awarding a contractor additional costs under the *Spearin* doctrine, where the government's design called for the hangar doors to be constructed and rigged with three "pick" points, which would not work.[92] In another case, the Seventh Circuit enforced the *Spearin* doctrine under Illinois law, holding that a city impliedly warranted the suitability of a specified quarry to produce adequate armor rock.[93]

c. Designated Materials or Supplier   The government typically is not deemed to impliedly warrant that specified materials will be commercially available.[94] Courts and boards generally hold that it may be reasonable to assume that by listing approved sources of supply, the government may warrant the general ability of those sources to perform. However, suppliers' willingness to perform and ability to perform within any specific time period is not impliedly warranted by the government.[95]

An exception to this rule is when the government specifies the use of *standard products*. In this case, the contractor is entitled to an implied warranty of commercial availability.[96] Similarly, the government is required to disclose any knowledge of a specified product's lack of commercial availability.[97] Two decisions by the ASBCA illustrate the facts that can affect the resolution of issues related to proprietary specifications.

In *C&D Construction, Inc.*,[98] the board found that the contractor, by failing to file a prebid protest, assumed the risk of meeting an allegedly proprietary specification. However, in *Logics, Inc.*,[99] the board found that the government's knowledge of the unavailability of a proprietary item shifted the risk of unavailability from the contractor to the government.

The government is generally entitled to strict compliance with its specifications and is not obligated to accept substitutes.[100] However, most government construction

---

[91]*R. L. Hamm & Assoc., Inc. v. England,* 379 F.3d 1334 (Fed. Cir. 2004); *USA Petroleum Corp. v. United States,* 821 F.2d 622 (Fed. Cir. 1987); *Felton Constr. Co.,* AGBCA No. 406-9, 81-1 BCA ¶ 14,932; *R.M. Hollingshead v. United States,* 111 F. Supp. 285 (Ct. Cl. 1953).
[92]*White v. Edsall Constr. Corp.,* 296 F.3d 1081 (Fed. Cir. 2002).
[93]*Edward E. Gillen Co. v. City of Lake Forest,* 3 F.3d 192 (7th Cir. 1993).
[94]*Franklin E. Penny Co. v. United States,* 207 Ct. Cl. 842 (1975).
[95]*Id.*
[96]*J.W. Bateson Co.,* ASBCA No. 19823, 76-2 BCA ¶ 12,032.
[97]*Haas & Haynie Corp.,* GSBCA No. 5530, 84-2 BCA ¶ 17,446.
[98]ASBCA No. 48,590, 97-2 BCA ¶ 29,283.
[99]ASBCA Nos. 46914, 49364, 97-2 BCA ¶ 29,125.
[100]*Carothers Constr. Co. Inc.,* ASBCA No. 41268, 93-2 BCA ¶ 25,628; *J.L. Malone & Assocs.,* VABCA 2335, 88-3 BCA ¶ 20,894, *aff'd,* 879 F.2d 841 (Fed. Cir. 1989).

contracts include an important exception to this rule. The *brand name or equal* provision of the Material and Workmanship clause[101] provides that the specification of a brand name or proprietary product shall be construed to allow contractors to use any product that is equal to the brand name or proprietary product.

As a general rule, there is no implied warranty that the specified materials are commercially available.[102] However, the risk of unavailability may shift to the government if there are special circumstances, such as where the government has a prior relationship with the supplier of the unavailable material and there are no specifications for that material which the contractor can use to either make the material itself or have it made.[103]

In *C&D Construction,* the invitation for bids for backup generator sets specified the use of four-stroke engines. During the bid preparation period, only Cummins and Caterpillar produced four-stroke engines that complied with the specifications. Since the Cummins engine was not manufactured domestically, C&D believed that use of the Cummins engine would violate the Buy American Act. (See **Chapter 5** for a discussion of the Buy American Act.)

Relying on the brand name or equal clause, C&D included in its bid a two-stroke engine produced by Detroit Diesel. C&D believed the two-stroke engine was the functional equivalent of a four-stroke engine. However, C&D did not challenge the allegedly proprietary specification prior to bid submission.

After receiving award of the contract, C&D submitted the two-stroke engine for approval by the contracting officer. The contracting officer rejected the submittal and insisted that C&D provide a four-stroke engine. Because the Caterpillar engine was significantly more expensive than the Cummins engine, the contracting officer did grant C&D a Buy American Act waiver for the Cummins engine. C&D submitted a claim for the price differential between the Cummins engine and the Detroit Diesel engine.

The ASBCA denied C&D's appeal. The board found that C&D could not prove that the contract specifications were written around the "proprietary characteristics" of one manufacturer. Since four-stroke engines were available from two sources, Caterpillar and Cummins, the board concluded that the specification was not proprietary and the *brand name or equal* section of the Material and Workmanship clause did not apply. The board noted that C&D's proper remedy would have been to file a bid protest with the General Accounting Office[104] on the grounds that specification of a four-stroke engine unduly restricted competition.

In *Logics, Inc.,*[105] the government awarded Logics a contract for 64 low-voltage rectifier filters. These filters were a subcomponent in a radar system designed to detect enemy fire and manufactured exclusively for the government by Hughes Aircraft Co. (Hughes). The contract required Logics to test all of the rectifier filters. The test apparatus for the rectifier filters was depicted schematically in the contract

---

[101]FAR § 52.236-5.
[102]*Franklin E. Penny Co. v. United States,* 524 F.2d 668 (Ct. Cl. 1975).
[103]*Aerodex, Inc. v. United States,* 417 F.2d 1361 (Ct. Cl. 1969).
[104]Now the General Accountability Office.
[105]ASBCA Nos. 46914, 49364, 97-2 BCA 29,125.

specifications as a stock, commercially available transformer. In fact, the test transformer depicted in the contract was a unique design specially developed by Hughes, which had never been manufactured by any other contractor. There was not sufficient data included in the contract either to build or to purchase the required test apparatus. In addition, Hughes had changed the design of the test apparatus it furnished to the government. However, Logics included a *certification* with its bid that it had assured itself that all needed parts and materials were available. Logics further agreed that nonavailability of parts or materials would not be an excusable cause of late delivery.

After beginning performance, Logics found that it could not buy, build, borrow, or lease the necessary test apparatus. The government subsequently agreed to provide the test apparatus but was not able to do so until more than two and a half years after contract award. However, Logics was never able to agree on a revised testing and delivery schedule with the government. Four and a half years after the contract was awarded, the contracting officer terminated Logic's contract for default.

On appeal, Logics argued that the contract specifications were defective in specifying what appeared to be a stock, commercially available test apparatus when, in fact, the product was unique, its specifications were known only to Hughes, and Hughes would sell transformers only to the government. The government admitted that the transformer was unique but argued that Logics assumed the risk of its unavailability because the invitation for bids warned Logics that the government would not furnish the transformer. The government also argued that Logics' certification shifted the risk of material nonavailability to the contractor.

The board rejected the government's arguments finding that contract specifications were misleading when they depicted an item necessary for contract performance as being a standard commercial item when it is, in fact, a unique item. The board further found that specifications were also misleading when an erroneous representation was made that adequate technical data are available to either make or purchase any component needed for performance. Reversing the default termination, the board then concluded that these misrepresentations overcame the contractor's certification that it had ascertained the availability of all needed materials.

2. Defenses and Prerequisites

A contractor cannot recover under the *Spearin* doctrine if it knew or, through the exercise of reasonable care, should have known of the defective nature of the design prior to submitting its bid.[106] Thus, when solicitation documents direct bidders to conduct a prebid site inspection, the contractor will be deemed to have prior knowledge of any inaccuracies in the design that could have been discovered through a reasonable site inspection, even if the contractor never inspects the site.[107]

---

[106] *Blount Bros. Constr. Co. v. United States,* 346 F.2d 962 (Ct. Cl. 1965).

[107] *Stuyvesant Dredging Co. v. United States,* 834 F.2d 1576 (Fed. Cir. 1987); *Johnson Controls, Inc. v. United States,* 671 F.2d 1312 (Ct. Cl. 1982); *Allied Contractors, Inc. v. United States,* 381 F.2d 995 (Ct. Cl. 1967).

In addition to the lack of actual or constructive knowledge of design errors, contractors seeking the benefit of the *Spearin* doctrine must prove that the defective specification was the cause of the performance difficulties. The contractor has the burden of proving reasonable reliance on the subject design specification.[108]

B. Impracticability/Impossibility of Performance

The law will not require an impossible or impracticable act. Thus, contractors may obtain relief if performance is found to be impossible or greatly different from what was expected. These risk allocation devices vary from those discussed earlier because they are not premised on government breach or fault. Instead, contractors are excused from performing impossible or impracticable actions based on the assumption that both the contractor and the government were mistaken as to the ability to perform.[109]

The classic application of the impossibility doctrine is when some supervening event, such as destruction of a building or death of a necessary party to the contract, occurs. Courts and boards have long recognized that when performance of a contract is impossible, performance will be excused.[110] As this doctrine evolved, the courts and boards recognized that there are also circumstances where performance, though not impossible, is commercially infeasible.

A contractor may obtain relief under the doctrine of impracticality when it can demonstrate that performance of the contract is substantially more difficult or expensive than the parties expected. Specifically, a contractor must demonstrate: (1) the occurrence of a contingency or something unexpected, (2) the risk of the contingency was not allocated by agreement or custom, and (3) the occurrence or contingency rendered performance of the contract commercially impracticable.[111] Whether performance of a specified task is commercially impracticable is assessed on an *objective basis*.[112] In other words, if another contractor could or has successfully performed under the specification, it is likely that a claim of commercial impracticability will be rejected.[113]

---

[108]*R. L. Hamm & Assoc., Inc. v. England*, 379 F.3d 1334 (Fed. Cir. 2004) (contractor entitled to rely on quantities set forth in the solicitation by the government and was not required to reverse engineer those quantities during the proposal stage); *see also Gulf & Western Precision Eng'g Co. v. United States*, 211 Ct. Cl. 207 (1976); *Ball, Ball & Brosamer, Inc.*, IBCA No. 2103-N, 93-1 BCA ¶ 25,287.

[109]The common law doctrine of mutual mistake is addressed in *SMC Info. Systems, Inc. v. General Services Administration*, GSBCA No. 9371, 93-1 BCA ¶ 25,485.

[110]*See MMI Capital, LLC v. General Services Administration*, GSBCA No. 16739, 2006 WL 2170507 (2006), citing *ESB, Inc.*, ASBCA No. 22914, 81-1 BCA ¶ 15,012.

[111]*See Transatlantic Financing Corp. v. United States*, 363 F.2d 312 (D.C. Cir. 1966); *see also Restatement (Second) of Contracts* § 266(1).

[112]*ASC Systems Corp.*, DOTCAB No. 73-37, 78-1 BCA ¶ 13,119.

[113]*Id.*; *see also Koppers & Co. v. United States*, 186 Ct. Cl. 142 (1968).

C. Superior Knowledge

Offerors or bidders for federal government contracts are entitled to all information possessed by the contracting agency and pertinent to contract performance. As such, contractors are entitled to an implied warranty that all information that is vital for preparation of estimates and for performance of the contract has been disclosed. In the leading case on the topic, *Helene Curtis Indus. v. United States,* the Court of Claims described the concept in this way:

> [T]he Government, possessing vital information which it was aware the bidders needed but would not have, could not properly let them flounder on their own. [T]he Government—where the balance of knowledge is so clearly on its side—can no more betray a contractor into a ruinous course of action by silence than by the written or spoken word.[114]

The Court of Federal Claims has subsequently refined the elements underlying a "superior knowledge" claim. The government can be held liable for breach of contract for nondisclosure of superior knowledge when: (1) the contractor undertakes to perform without vital knowledge of facts that would affect performance costs or direction; (2) the government was aware the contractor had no knowledge of and had no reason to know of the facts; (3) any contract specification supplied misled the contractor, or did not put it on notice to inquire; and (4) the government failed to provide the relevant information.[115] In these circumstances, the contractor may recover damages proximately caused by the government's nondisclosure of vital information.

D. Assumption of Risk

When a contractor has either expressly or impliedly assumed the risk of events impacting contract performance, it may not rely on risk allocation principles to avoid the consequences of those events. The most common means of establishing an assumption of risk is by reference to exculpatory contract provisions. For example, a contract clause that is clearly worded to indicate to the contractor that the government does not expressly or impliedly warrant the adequacy and accuracy of furnished information will generally be enforced.[116] Further, in the absence of some other risk-allocation principle, contractors typically are deemed to assume all risks related to the level of effort required to perform a contract. For example, under normal circumstances, contractors have been held to have assumed the risk of labor shortages, price escalations, and availability of materials.[117] When risks are assumed either expressly or impliedly, contractors are expected to include a contingency in their estimates. Therefore, a subsequent equitable adjustment is not permitted.

---

[114]*Helene Curtis Indus., Inc. v. United States,* 160 Ct. Cl. 437, 444 (1963).

[115]*Northrop Grumman Corp. v. United States,* 63 Fed. Cl. 12 (2004), citing *GAF Corp. v. United States,* 932 F.2d 947 (Fed. Cir. 1991).

[116]*See Wunderlich Contracting Co. v. United States,* 173 Ct. Cl. 180 (1965).

[117]*See, e.g., DK's Precision Machining & Manufacturing,* ASBCA No. 39616, 90-2 BCA ¶ 22,830; *Bescast, Inc.,* ASBCA No. 38149, 90-3 BCA ¶ 23,244; *Spindler Constr. Corp.,* ASBCA No. 55007, 06-2 BCA ¶ 33,376.

E. Risk Allocation in Design-Build Contracts

Certain risk allocation principles and implied duties may apply differently in the design-build context. For example, in design-build contracts, the design-builder and not the government has the implied duty to furnish adequate and sufficient design information. In the design-build context, the contractor generally bears the risk of creating a design that meets the government's performance requirements and specifications. In *Gee & Johnson v. United States*,[118] the contractor was responsible for designing a building for the Navy. After construction, the building suffered damage from water, which was found to be the result of the lack of "flashing" in precast concrete sills. The contractor denied liability because it was contractually obligated to design a building in accordance with the "applicable building code in effect at the time." The applicable building code did not require flashing. However, the Court of Federal Claims ruled that the contractor was liable for the damages because the contract incorporated by reference a guide specification that included a requirement that walls should include flashing.

Similar to the practice of incorporating numerous FAR clauses by reference, government construction contracts often incorporate by reference multiple guide specifications or manuals. Although these referenced documents may affect the interpretation of any government construction contract, a contractor performing under a design-build or performance specification often bears the risk that its design satisfies the requirements or standards set forth in the referenced manuals or guides.

However, because at least conceptual design information typically is provided to design-builders by the government, when the design-builder contracts to "complete" the design, the government has been held responsible for affirmative errors in the furnished preliminary design information.[119] In *Donahue Electric Inc.*, the VA specified that a certain type of boiler be installed in the building. The board stated that a "properly written and administered design-build contract transfers the risk of design insufficiency" to the contractor.[120] However, by including a detailed specification on the type of boiler and eliminating the contractor's discretion, the government effectively shifted the risk of the adequacy of the boiler to itself.[121] In the construction setting, it is rare that a design-build contract is written to shift all risk to the contractor. However, it may not be easy to determine whether a particular specification is a design or performance specification.

As illustrated by the clause in **Section IV.A** of this chapter, a recent trend in design-build contracts is to include a "Betterments" clause within the order of precedence. A typical Betterments clause will state that any portion of an accepted proposal which conforms to and exceeds the provisions of the solicitations will control over conflicting language in other parts of the contract.[122] The intended purpose of such clauses is that any design beyond what is indicated in the solicitation will be

---

[118]*Gee & Jenson Engineers, Architects and Planners v. United States*, 2008 WL 4997488 (Fed. Cl. 2008).

[119]*Donahue Electric, Inc.*, VABCA No. 6618, 03-1 BCA ¶ 32,129; *see also M.A. Mortenson Co.*, ASBCA No. 39978, 93-3 BCA ¶ 26,189.

[120]*Id.*

[121]*Donahue Electric, Inc.*, VABCA No. 6618, 03-1 BCA ¶ 32,129.

[122]*Strand Hunt Constr., Inc.*, ASBCA No. 55671, 55813, 08-2 BCA ¶ 33,868 (2008).

incorporated into the contract and the final agreed-on contract price and will not be the basis of any valid claim.

VI. STANDARD FAR CLAUSES AFFECTING ALLOCATION OF RISKS AND CONTRACT INTERPRETATION

Standard clauses mandated by the FAR may affect the allocation of risk in federal government contracts. Three notable standard FAR clauses are the Permits and Responsibilities clause,[123] the Material and Workmanship clause,[124] and the Specifications and Drawings clause.[125]

A. FAR § 52.236-7 Permits and Responsibilities Clause

The Permits and Responsibilities clause, a mandatory clause in all fixed-price or cost-reimbursement government construction contracts, provides:

> The Contractor shall, without additional expense to the Government, be responsible for obtaining any necessary licenses and permits, and for complying with any Federal, State, and municipal laws, codes, and regulations applicable to the performance of the work. The Contractor shall also be responsible for all damages to persons or property that occur as a result of the Contractor's fault or negligence. The Contractor shall also be responsible for all materials delivered and work performed until completion and acceptance of the entire work, except for any completed unit of work which may have been accepted under the contract.

The Permits and Responsibilities clause obligates the contractor to obtain any and all necessary licenses and permits without additional expense to the government. In addition, the clause mandates compliance with all applicable laws and regulations, makes the contractor generally responsible for all damage caused to persons or property due to the contractor's fault or negligence, and makes the contractor generally responsible for the work and materials on the project prior to the government's acceptance. Absent the application of one or more of the risk allocation principles discussed earlier, courts and boards typically hold that the contractor has assumed the risk of excess costs, damages, or delay associated with matters addressed in this standard clause.[126]

---

[123] FAR § 52.236-7.
[124] FAR § 52.236-5.
[125] FAR § 52.236-21.
[126] *Management Resource Assocs., Inc.*, ASBCA No. 49457, 03-1 BCA ¶ 32,141; *see, e.g., Taisei Rotec Corp.*, ASBCA No. 50669, 02-1 BCA ¶ 31,739 (contractor held responsible for damages to helicopter when scaffolding, which the contractor failed to properly erect in a Navy hangar, collapsed); *Hills Materials Co.*, ASBCA Nos. 42410, 42411, 92-1 BCA ¶ 24,636 (no recovery allowed for added cost due to revised Occupational Safety and Health Administration standard issued after award). *See also Green v. Tecom, Inc.*, 566 F.3d 1037 (Fed. Cir. 2009).

B. FAR § 52.236-5 Material and Workmanship Clause

The Material and Workmanship clause is another mandatory clause in government construction contracts. It provides:

(a) All equipment, material, and articles incorporated into the work covered by this contract shall be new and of the most suitable grade for the purpose intended, unless otherwise specifically provided in this contract. References in the specifications to equipment, material, articles, or patented processes by trade name, make, or catalog number, shall be regarded as establishing a standard of quality and shall not be construed as limiting competition. The Contractor may, at its option, use any equipment, material, article, or process that, in the judgment of the Contracting Officer, is equal to that named in the specifications, unless otherwise specifically provided in this contract.

(b) The Contractor shall obtain the Contracting Officer's approval of the machinery and mechanical and other equipment to be incorporated into the work. When requesting approval, the Contractor shall furnish to the Contracting Officer the name of the manufacturer, the model number, and other information concerning the performance, capacity, nature, and rating of the machinery and mechanical and other equipment. When required by this contract or by the Contracting Officer, the Contractor shall also obtain the Contracting Officer's approval of the material or articles which the Contractor contemplates incorporating into the work. When requesting approval, the Contractor shall provide full information concerning the material or articles. When directed to do so, the Contractor shall submit samples for approval at the Contractor's expense, with all shipping charges prepaid. Machinery, equipment, material, and articles that do not have the required approval, shall be installed or used at the risk of subsequent rejection.

(c) All work under this contact shall be performed in a skilful and workman-like manner. The Contracting Officer may require, in writing, that the Contractor remove from the work any employee the Contracting Officer deems incompetent, careless, or otherwise objectionable.

The Material and Workmanship clause specifies that all equipment and materials incorporated into the work be "new and of the most suitable grade for the purpose intended . . ." unless provided otherwise in the contract. Further, the clause specifies that reference to specific items or materials shall be construed as requiring an *or equal* product; again, unless specifically provided otherwise. Finally, the clause provides generally that all work shall be performed in a skillful and workmanlike manner.

The language in the Material and Workmanship clause related to the reference to products or materials by trade name or catalog number is commonly referred to as the *brand name or equal* provision of that clause. In *Jack Stone Co. v. United*

States,[127] the court interpreted an earlier version of the clause as requiring the contracting officer to consider an alternative product to a brand name product if the proposed alternative meets the essential requirements set forth in the specifications, functions in the same manner as the brand name product, and provides the same standard of quality.[128] The proposed substitute need not be identical to the specified product.[129] Subsequently, the *or equal* concept was applied to proprietary specifications in *William R. Sherwin v. United States*.[130]

In general, for a contractor to invoke this clause and to recover under the Changes clause for a contracting officer's refusal to allow the submittal and use of an *or equal* product, the contractor must show:

(1) That the specification was proprietary;
(2) That it submitted information to the contracting officer showing an equal alternative product; and
(3) That the proposed substitute was the same standard of quality as the proprietary item.[131]

This clause does not preclude the government from writing specifications around the design specifications of particular products. Consequently, if the bidder/offeror was aware that the specification required a proprietary product when it was preparing its bid/proposal, it must submit a protest. Failure to protest will likely preclude any future relief.[132] However, the government must call out the salient or essential characteristics in order that contractors can intelligently prepare bids or proposals and select vendors.[133] Similarly, if the government chooses a brand name specification, the *salient characteristics* must be described.[134] Although the government can reject a product, the bidder/offeror is not expected to guess at the essential or salient characteristics.[135]

C. FAR § 52.236-21 Specifications and Drawings for Construction Clause

The Specifications and Drawings clause is another mandatory clause in all fixed-price government construction contracts. Subsection (a) provides of that clause:

---

[127]344 F.2d 370 (Ct. Cl. 1965).

[128]*Id.*

[129]*Urban Plumbing & Heating Co. v. United States*, 408 F.2d 382 (Ct. Cl. 1969).

[130]436 F.2d 992 (Ct. Cl. 1971).

[131]*Harvey Constr. Co., Inc.*, ASBCA No. 39310, 92-3 BCA ¶ 25,162; *see also Hook Constr., Inc. v. General Services Administration*, GSBCA No. 16470, 05-1 BCA ¶ 32,862.

[132]*See American Renovations & Constr. Co., v. United States*, 45 Fed. Cl. 44 (1999).

[133]*S&D Constr. Co.*, VABCA No. 3885, 95-2 BCA ¶ 27,609.

[134]*North American Constr. Corp.*, ASBCA No. 47941, 96-2 BCA ¶ 28,496; *Blount Brothers Corp.*, ASBCA No. 31203, 88-3 BCA ¶ 20,878.

[135]*Northrop Grumman Corp.*, ASBCA No. 52178 *et al.*, 05-2 BCA ¶ 32,992.

(a) The Contractor shall keep on the work site a copy of the drawings and specifications and shall at all times give the Contracting Officer access thereto. Anything mentioned in the specifications and not shown on the drawings, or shown on the drawings and not mentioned in the specifications, shall be of like effect as if shown or mentioned in both. In case of difference between drawings and specifications, the specifications shall govern. In case of discrepancy in the figures, in the drawings, or in the specifications, the matter shall be promptly submitted to the Contracting Officer, who shall promptly make a determination in writing. Any adjustment by the Contractor without such a determination shall be at its own risk and expense. The Contracting Officer shall furnish from time to time such detailed drawings and other information as considered necessary, unless otherwise provided.

In many contracts, the specifications may set forth or describe more than one material or component, each of which appears to be acceptable, while the drawings depict only one. In a 2006 decision, the Federal Circuit addressed the "like effect" provision of the Specifications and Drawings clause in connection with contract specifications that gave the contractor the option to use one of two materials (precast concrete or polystyrene) as concrete void retainers.[136] Both were described in the specifications. However, the contract drawings contained notes describing in detail and sizing only a precast concrete void retainer.

The contractor planned on using the less expensive polystyrene product. The government insisted that the drawings operated to narrow the contractor's choices to the precast retainer. The ASBCA agreed with the government and denied the contractor's claim for excess costs associated with using the precast product.[137] Relying on the *like effect* language in the clause, the Federal Circuit reversed, holding that the drawings, which depicted only a precast concrete retainer, operated to provide specific information on the concrete product. The drawings did not eliminate the choices set forth provided in the specifications.[138]

VII. INTERPRETATION OF SUBCONTRACTS UNDER FEDERAL CONTRACTS

Nearly all government contracts of significant size or complexity involve multiple subcontracts and, often, numerous tiers of contracts among the various parties that supply goods and services for a project. Frequently, clauses contained in government contracts are repeated or incorporated by reference into subcontracts awarded by the general contractor. Often incorporated contract clauses originate from a Federal Acquisition Regulation which mandates that the clause, in whole or in part, be flowed

---

[136]*Medlin Constr. Group, Ltd. v. Harvey,* 449 F.3d 1195 (Fed. Cir. 2006).
[137]*Medlin Constr. Group, Ltd.,* ASBCA No. 54772, 05-1 BCA ¶ 32,939.
[138]*Medlin Constr. Group, Ltd. v. Harvey,* 449 F.3d 1195 (Fed. Cir. 2006).

down to subcontractors and suppliers. These clauses have a unique history and are the subject of years of interpretation by the boards and courts.

A review of those clauses with flow-down requirements reveals that the FAR does not provide uniform direction to contractors on the method or manner by which a contractor is to flow down the provision. Some clauses appear to stipulate that the contractor should incorporate the clause verbatim, others call for a substantive flow-down, and a third group direct the contractor to address the subject matter in the subcontract or purchase order.

Of the third group, the Termination for Convenience of the Government (Fixed-Price)[139] clause can present the most difficulty for a contractor. Nothing in the clause addresses a flow-down requirement. However, FAR § 49.108-5 expressly conditions a contractor's ability to include in its termination for convenience settlement proposal any amount awarded to a subcontractor in a final judgment for the subcontractor's lost profits on the contractor's reasonable effort to include in the subcontract a termination clause similar to the FAR clauses.[140] Failure to address that potential issue during the negotiation or award of a subcontract could have costly consequences.

Addressing flow-down requirements can be a time-consuming process. **Appendix 6A** is a table listing many of the FAR clauses authorized or required for use in government construction contracts together with a legend indicating the FAR's guidance on the method and manner of incorporating the clause in subcontracts or purchase orders. In reviewing this table, contractors and subcontractors should recognize that many mandatory clauses, such as the Changes clause, the Differing Site Conditions clause, the Buy American Act clause, and the Permits and Responsibilities clause, do not contain flow-down requirements. However, prudence and experience dictate that the subcontract provisions need to address these topics and should be coordinated with the applicable FAR provisions. (See **Chapter 14** for an additional discussion of subcontract terms and conditions.) Finally, the table at **Appendix 6A** is provided as an example and is subject to change based on FAR revisions and the agency's determination regarding which clauses to include in any government contract. Each contract needs to be reviewed to identify the specific applicable flow-down requirements.

Government contractors must carefully consider the implication and effect of standard government contract clauses in their subcontracts because these clauses will not necessarily be interpreted by a federal court or board (or under federal law) in the event a dispute with a subcontractor arises. Frequently, disputes between federal government contractors and their subcontractors are resolved through arbitration or by litigation in a state or federal court. Absent the presence of certain limited circumstances, such as a subcontract provision stating that the contract will be interpreted

---

[139]FAR § 52.249-2 (Alternate 1).
[140]*See* FAR § 49.502(e).

in accordance with the law of federal government contracts, arbitrators or judges often will apply the law of the applicable state to interpret subcontracts. That law is not always in line with the federal interpretation.

Federal law, including federal interpretive principles, will apply in disputes between general contractors and subcontractors when (1) the federal government possesses a sufficient and direct interest in the outcome of the litigation, *and* (2) the application of state law would conflict with a federal policy or interest, or frustrate a specific objective of federal legislation.[141] Both federal and state courts are reluctant to identify a sufficiently direct and immediate federal interest to warrant the application of federal law to interpretation of subcontracts.[142] Instances in which a sufficient federal interest has been found have been generally limited to those involving a direct impact on national security.[143] Thus, it is often the case that state law, even if conflicting with federal interpretation, will apply to the interpretation of subcontracts entered into by federal government contractors.

State courts, as well as federal courts applying state law in diversity actions, frequently rely on federal interpretation of federal government contract clauses as persuasive authority. *Persuasive* authority, however, is just that; it is not binding on a particular court. Common clauses used in government contracts and subcontracts that may be interpreted differently when state law is applied include disputes clauses, clauses regarding equitable adjustments, and termination for convenience clauses.

The requirement to initially submit disputes to the government as required by a disputes clause incorporated into a subcontract has been the subject of controversy.[144] Similarly, subcontractors have argued that, contrary to long-standing federal interpretation, all expected profits (including expected profit for work not performed) may be awarded upon a termination for convenience pursuant to an incorporated contract clause.[145] The remedies available under an incorporated equitable adjustment clause have also resulted in differing interpretations when state law is applied.[146] Thus, federal government contractors must carefully assess and prepare for such risks when negotiating subcontracts involving federal projects. Assumption that federal interpretation will be adhered to can lead to unexpected results.

---

[141]*Northrop Corp. v. AIL Systems, Inc.*, 959 F.2d 1424, 1428-29 (7th Cir. 1992); *see also Boyle v. United Technologies Corp.*, 487 U.S. 500, 507 (1988).

[142]*Id.*; *see also Woodward Governor Co. v. Curtiss-Wright Flight Systems, Inc.*, 164 F.3d 123 (2nd Cir. 1999); *Linan-Faye Constr. Co., Inc. v. Housing Authority of the City of Camden*, 49 F.3d 915, 920-21 (3rd Cir. 1995).

[143]*See, e.g., United States v. R.H. Taylor*, 333 F.2d 633 (5th Cir. 1964); *American Pipe & Steel Corp. v. Firestone Tire & Rubber Co.*, 292 F.2d 640 (9th Cir. 1961).

[144]*United States v. R.H. Taylor*, 333 F.2d 633 (5th Cir. 1964).

[145]*Linan-Faye Constr. Co., Inc. v. Housing Authority of the City of Camden*, 49 F.3d 915, 920-21 (3rd Cir. 1995).

[146]*See, e.g., American Pipe & Steel Corp. v. Firestone Tire & Rubber Co.*, 292 F.2d 640 (9th Cir. 1961).

➤ LESSONS LEARNED AND ISSUES TO CONSIDER

- In government contracts, the written contract almost always *incorporates by reference* numerous contract clauses, guide specifications, and manuals. These must be reviewed and understood to obtain a complete understanding of the contract. The fact that a provision is incorporated only by reference *does not reduce* its importance.

- The *written* contract entered into between contractor and the government provides the primary evidence of the relationship between the parties.

- When contracts are not *ambiguous,* the written terms will control to the exclusion of all oral representations or documents not identified as contract documents—extraneous evidence cannot be used to render a written contract ambiguous.

- The meaning of contract terms will be assessed on an *objective basis*; subjective intent as to meaning should be expressed and resolved at the time of contracting.

- Contracts will be *interpreted as a whole,* with reference to all sections and incorporated contract documents. Courts and boards will attempt to harmonize all provisions.

- If conflicting provisions cannot be harmonized, *order of precedence* rules will govern.

- Special meaning within industries and technical terms will be given their *special or technical* meaning when the parties' status indicates such meaning was intended. Otherwise, the rule of normal and ordinary meaning applies.

- Conduct under *previous contracts* will be considered as evidence of intent on subsequent contracts. If a different course of action is intended, it should be expressed at inception of subsequent contracts.

- Contractors should address all *identified ambiguities* in prospective contracts with the contracting agency *prior to submitting* a proposal or bid. As a general rule, *patent ambiguities* that are not questioned by the contractor prior to award will not be construed against the government.

- Contractors must carefully assess the potential assumption of *design responsibility* before suggesting remedies to potential design problems.

- The distinction between *design* and *performance* specifications should be carefully analyzed and addressed prior to contract execution.

- The potential for *varying interpretations* of government contract provisions in subcontracts should be considered when drafting and awarding subcontracts for federal government project.

APPENDIX 6A: SAMPLE TABLE OF FAR CLAUSES WITH FLOW-DOWN REQUIREMENTS

FAR Reference	Title of FAR Clause	Additional Flow-down Requirements (see legend below)
52.203-7	Anti-Kickback Procedures	S-FD
52.203-11	Certification and Disclosure Regarding Payments to Influence Certain Federal Transactions	V-FD
52-203-12	Limitation on Payments to Influence Certain Federal Transactions	S-FD(A)
52-203-13	Contractor Code of Business Ethics and Conduct	S-FD(A)
52-203-14	Display of Hotline Posters	S-FD(A)
52.203-15	Whistleblower Protections Under the American Recovery and Reinvestment Act of 2009	S-FD
52.204-2	Security Requirements	S-FD
52.204-9	Personal Identity Verification of Contractor Personnel	V-FD
52.214-26	Audit and Records—Sealed Bidding	V-FD(A)
52.214-28	Subcontractor Cost or Pricing Data—Modifications—Sealed Bidding	S-FD
52.215-2	Audit and Records—Negotiation	V-FD(A)
52.215-12	Subcontractor Cost or Pricing Data	S-FD
52.215-13	Subcontractor Cost or Pricing Data—Modifications	S-FD
52.215-14	Integrity of Unit Prices	S-FD
52.215-15	Pension Adjustments and Asset Reversions	S-FD
52.215-18	Reversion or Adjustment of Plans for Postretirement Benefits (PRB) Other Than Pensions	S-FD
52.215-19	Notification of Ownership Change	S-FD
52.219-9	Small Business Subcontracting Plan	S-FD (subcontractors (except small business concerns) that receive sub-contracts in excess of $1,000,000 for construction must adopt a plan that complies with the requirements of the clause at 52.219-9)
52.222-4	Contract Work Hours and Safety Standards Act—Overtime Compensation	V-FD
52.222-11	Subcontracts (Labor Standards)	S-FD
52.222-21	Prohibition of Segregated Facilities	V-FD
52.222-26	Equal Opportunity	V-FD
52.222-27	Affirmative Action Compliance Requirements for Construction	V-FD
52.222-35	Equal Opportunity for Special Disabled Veterans, Veterans of the Vietnam Era, and Other Eligible Veterans	V-FD
52.222-36	Affirmative Action for Workers with Disabilities	V-FD

(Continued)

FAR Reference	Title of FAR Clause	Additional Flow-down Requirements (see legend below)
52.222-37	Employment Reports on Special Disabled Veterans, Veterans of the Vietnam Era, and Other Eligible Veterans	V-FD
52.222-50	Combating Trafficking in Persons	S-FD
52.222-54	Employment Eligibility Verification	S-FD(A)
52.223-7	Notice of Radioactive Materials	V-FD
52.223-14	Toxic Chemical Release Reporting	S-FD (with addition of FAR § 52.223-13)
52.223-15	Energy Efficiency in Energy-Consuming Products	V-FD
52.224-2	Privacy Act	V-FD
52.225-13	Restrictions on Certain Foreign Purchases	S-FD
52.225-19	Contractor Personnel in a Designated Operational Area or Supporting a Diplomatic or Consular Mission Outside the United States	S-FD
52.227-1	Authorization and Consent	S-FD(A)
52.227-9	Refund of Royalties	S-FD
52.227-10	Filing of Patent Applications—Classified Subject Matters	S-FD
52.227-11	Patent Rights—Ownership by the Contractor (Short Form)	S-FD(A)
52.227-13	Patent Rights—Ownership by the Government	S-FD(A)
52.228-3	Worker's Compensation Insurance (Defense Base Act)	S-FD
52.228-4	Worker's Compensation and War Hazard Insurance Overseas	S-FD
52.228-5	Insurance—Work on a Government Installation	S-FD
52.230-2	Cost Accounting Standards	S-FD (with exceptions)
52.230-3	Disclosure and Consistency of Cost Accounting Practices	S-FD (with exceptions)
52.230-6	Administration of Cost Accounting Standards	S-FD (consistent with FAR §§ 52.230-2 and 52.230-3
52.232-27	Prompt Payments for Construction Contracts	SM-FD
52.236-13	Accident Prevention	V-FD(A)
52.244-6	Subcontracts for Commercial Items	V-FD
52.247-63	Preferences for U.S. Flag Air Carriers	SM-FD
52.247-64	Preference for Privately Owned U.S. Flag Commercial Vessels	SM-FD
52.248-3	Value Engineering—Construction	SM* **(See Chapter 8)**
52.249-2	Termination for Convenience of the Government	SM* (see FAR §§ 49.108-5 and 49.502) **(See Chapter 11)**
52.250-1	Indemnification Under Public Law 85-804	SM*

The legend provides general information regarding the nature of the flow-down directions contained in the various FAR clauses. There is no single uniform approach to the flow-down requirements in the FAR.

LEGEND

S-FD Substance with Flow Down to Lower Tiers

S-FD(A) Substance with further Flow Down to Lower Tiers (Alter Parties)

V-FD Verbatim with further Flow Down to Lower Tiers

V-FD(A) Verbatim with further Flow Down to Lower Tiers (Alter Parties)

SM Subject Matter

SM* Subject Matter—Special Treatment or Attention

Not every one of these provisions will be found in the typical fixed-price construction contract. However, FAR § 52.301 contains a matrix of clauses that are *Required, Required when Applicable,* or *Optional* clauses for use in fixed-price construction contracts. All of these clauses fall into one of those three categories.

7

DIFFERING SITE CONDITIONS

I. HISTORICAL OVERVIEW

A. Introduction

Site conditions that are materially different from those contemplated at the time the contract price was estimated are a source of many disputes between owners and contractors. Unanticipated conditions often result in extra costs, and they also can substantially delay and disrupt job progress. These delays usually occur at the beginning of a job, which can have a greater overall impact on the project.

The history of the government contracting process reflects attempts to reduce some of the contingencies that contractors face in performing their work. A contingency reflects a risk that a prospective contractor can treat in one of two undesirable ways: (1) raise the contract price to cover the possibility that the contingency will occur, which can needlessly increase the federal government's costs and give the contractor a *windfall* if it *does not* occur; or (2) take a chance that the contingency will *not* occur, which can cause the contractor harmful *losses* (and discourage reasonable bids or proposals on future projects) if it *does* occur.

Major contingencies in construction work can involve *unexpected* conditions at the *site*. These conditions may be unexpected either because they (1) *varied* from conditions *indicated* in the contract documents, or (2) even though not in conflict with the contract documents, the conditions could not *reasonably* have been *anticipated*.

The federal government began dealing with this problem in 1927 by using a Changed Conditions clause in its fixed-price construction contracts,[1] and that type of clause has continued in use to the present day. The purpose of the clause is to place the risk of certain reasonably unexpected site conditions on the *federal government*

---

[1] See Report to the President of the United States by the Director of the Bureau of the Budget (1923), Report of Chairman of Interdepartmental Board of Contracts & Adjustments, pp. 140-141; Kendall, Changed Conditions as Misrepresentations in Government Construction Contracts, 35 Geo Wash. L. Rev. 978, 982 (1967), 4 YPA 187; Report to the President of the United States by the Director of the Bureau of the Budget (1927), Interdepartmental Board of Contracts & Adjustments, p. 93.

by granting a price increase and a time extension to a contractor that encounters such differing conditions. The effect of the clause is thus to lower bid or proposal prices.

B. Differing Site Condition Defined

A *differing site condition*—or *changed condition,* as it is sometimes called—is a physical condition encountered in performing the work that was not visible or otherwise known to the contractor when it submitted its proposal or bid and that is materially different from the condition believed to exist at the time of pricing the contract. Often this condition could not have been discovered by a reasonable site investigation. Examples of a *changed condition* or *differing site condition* problems include soil with inadequate bearing capacity to support the building being constructed; soil that cannot be reused as structural fill; unanticipated groundwater (static or percolating); quicksand; muck; rock formations (or excessive or insufficient quantities of rock); artificial (man-made) subsurface obstructions; and concealed (latent) conditions in an existing structure.

C. Responsibility for Differing Site Conditions

Under a traditional contract risk-allocation analysis, the contractor generally bears the risk of unforeseen site conditions and is expected to protect itself against unforeseen conditions by including a contingency factor in its proposed price or bid. This traditional risk analysis was articulated by the U.S. Supreme Court in *United States v. Spearin*[2] in the following manner:

> Where one agrees to do, for a fixed sum, a thing possible to be performed, he will not be excused or become entitled to additional compensation, because unforeseen difficulties are encountered. Thus one who undertakes to erect a structure upon a particular site, assumes ordinarily the risk of subsidence of the soil. [Citations omitted]

This basic principle remains in effect in the twenty-first century, absent special factual circumstances or a contract provision reallocating the risk of unforeseen site conditions.

The basic flaw in this approach is that a contractor cannot accurately value a true unknown. Even if included, the price contingency may end up being totally inadequate or, alternatively, grossly conservative. The one constant is that including any contingency increases contract prices and thus works to the detriment of the owner if adverse conditions are not encountered. Since the government enters into thousands of separate construction contracts in any given year, the potential cumulative effect of these contingencies is so substantial that reallocation of the risk of differing site conditions to the government makes economic sense.

---

[2]248 U.S. 132 (1918).

To alleviate some of the problems associated with unexpected subsurface conditions, versions of differing site condition clauses commonly are included in many construction contracts. The underlying reason for the presence of this widely used provision has been explained by many courts, including the United States Court of Claims in *Foster Construction C.A. & Williams Brothers Co. v. United States*:[3]

> The purpose of the changed conditions clause is thus to take at least some of the gamble on subsurface conditions out of bidding. Bidders need not weigh the cost and ease of making their own borings against the risk of encountering an adverse subsurface, and they need not consider how large a contingency should be added to the bid to cover the risk. *They will have no windfalls and no disasters.* The Government benefits from more accurate bidding, without inflation for risks which may not eventuate. It pays for difficult subsurface work only when it is encountered and was not indicated in the logs. [Emphasis added]

II.   FEDERAL GOVERNMENT CONTRACT CLAUSE

The text of the current Differing Site Conditions clause used in government contracts was adopted in 1984 and is set forth in the Federal Acquisition Regulation (FAR) at FAR § 52.236-2. It provides:

(a) The Contractor shall promptly, and before the conditions are disturbed, give a written notice to the Contracting Officer of (1) subsurface or latent physical conditions at the site which differ materially from those indicated in this contract, or (2) unknown physical conditions at the site, of an unusual nature, which differ materially from those ordinarily encountered and generally recognized as inhering in work of the character provided for in the contract.

(b) The Contracting Officer shall investigate the site conditions promptly after receiving the notice. If the conditions do materially so differ and cause an increase or decrease in the Contractor's cost of, or the time required for, performing any part of the work under this contract, whether or not changed as a result of the conditions, an equitable adjustment shall be made under this clause and the contract modified in writing accordingly.

(c) No request by the Contractor for an equitable adjustment to the contract under this clause shall be allowed, unless the Contractor has given the written notice required; *provided,* that the time prescribed in (a) above for giving written notice may be extended by the Contracting Officer.

---

[3] 435 F.2d 873 (Ct. Cl. 1970); *see generally* Abernathy and Shull, *Construction Business Handbook,* Chapter 15: Differing Site (Changed) Conditions (Aspen Publishers 2004); Currie, Abernathy, and Chambers, *Changed Conditions,* Construction Briefings No. 84-12 (Federal Publications, 1984).

(d) No request by the Contractor for an equitable adjustment to the contract for differing site conditions shall be allowed if made after final payment under this contract.

A. Types of Conditions Covered

The clause defines differing site conditions as "subsurface" or "subsurface or latent physical conditions." A review of some of these situations illustrates the importance of *reviewing the exact language* of any differing site conditions clause and the related contract provisions in any contract to determine which conditions are covered. For example, where actual grade elevations turn out to be lower than those shown on the contract drawings (requiring additional fill to meet the grade requirements), a contractor should be able to recover under the standard FAR clause's "latent physical conditions" language.[4]

B. Type I and Type II Changed Conditions

The FAR clause provisions identify two distinct types of unanticipated conditions that are compensable. These are usually designated as Type I and Type II changed conditions.

A Type I changed condition is described as a physical condition "differing materially from those indicated in the contract."

A Type II changed condition is described as unknown physical conditions at the site, of an *unusual* nature, which differ materially from those ordinarily encountered and generally recognized as inhering in work of the character required by the contract.

C. Notice Requirements

The FAR clause requires that the contractor stop work and give written notice upon encountering an unexpected condition, *before* disturbing it, so that the contracting officer or a representative of the contracting officer will have an opportunity to inspect and evaluate the condition. The contracting officer or other authorized representative of the government, such as the resident engineer, should be *notified immediately* when materially different conditions are encountered. If verbal notice is provided, that should be confirmed *in writing* as soon as possible. By giving the government the option of investigating the condition and, if appropriate, determining how best to proceed, the contractor greatly increases the likelihood of resolving any resulting claim in an expeditious and mutually acceptable manner.

---

[4]*See Ace Constructors, Inc., v. United States,* 70 Fed. Cl. 253 (2006); *Anthony P. Miller, Inc. v. United States,* 422 F.2d 1344 (Ct. Cl. 1970).

D. Operation of the Differing Site Conditions Clause

The Differing Site Conditions clause provides a mechanism for dealing with an adverse, changed site condition. However, a contract adjustment is not automatic. To obtain an adjustment under the clause, the contractor must first establish that the situation is covered by the clause. Before examining what typically must be proven, it is important to remember what the contractor is *not* required to prove.

A contractor providing notification of a suspected differing site condition does not need and should not attempt to assert or prove fault, bad faith, or defective design by the government or its representatives. There are simply some situations where differing, unanticipated conditions are encountered. This is especially true when dealing with subsurface work or with older structures where only sketchy construction history is available.

The Differing Site Conditions clause allows the contractor to be reimbursed for its reasonable additional costs, regardless of whether the government's representatives knew or were unaware of the actual conditions. By including a Differing Site Condition clause, the government has assumed a portion of the risk for such conditions in exchange for the contractor not including a contingency in its bid.

The converse is also true: The government may be entitled to a cost reduction if conditions prove less onerous than expected. Although downward adjustments are not common, they do occur. Those credits are consistent with the central purpose of the clause, which is to more closely base the contract price on the reasonable value of the work actually performed, thereby eliminating unnecessary risks to each party.

III. RECOVERY FOR A TYPE I CHANGED CONDITION

A. Type I Defined

In order to recover for a Type I changed condition—where the actual conditions are at variance with the conditions "indicated" by the contract documents—the contractor must demonstrate: (1) that certain conditions are *indicated* by the plans, specifications, and other contract documents; (2) that it *relied* on the physical conditions *indicated in the contract;* (3) the nature of the *actual* conditions encountered; (4) the existence of a *material variation* between the conditions indicated and the conditions actually encountered; (5) that *notice* was given; and (6) that the change condition resulted in *additional* performance *costs* and/or *time.*

The initial emphasis in Type I changed condition situations is on what conditions were "indicated" in the contract. Some statement or representation must be contained in the contract documents regarding what conditions could be expected. The actual conditions must differ from that statement or representation.

What is meant by "indicated in the contract" has been defined by numerous board and court decisions. Materials or data that are referenced in the bid or proposal documents may be considered to be indicated in the contract documents and the contractor's rights to recovery under the Differing Site Conditions clause may be affected by that information.[5] It is not required that the indications (upon which the contractor is entitled to reasonably rely) be affirmatively expressed on the plans or in specific contract provisions. Instead, such indications may be a reasonable inference based on reading the contract and any referenced information as a whole. Thus, the contractor may be able to compare actual conditions, not only with the express representations in the contract documents but also with all reasonable inferences and implications that can be drawn from those documents.

B. Type I Examples

Examples of situations where *express* representations of conditions in the contract documents were found to have differed materially from the actual conditions encountered include such items as:

(1) *Variance from anticipated blow counts.* Soil conditions with actual blow counts that were one-third to one-half the strengths indicated by the contract borings constituted a changed condition. The contractor that encountered this condition during the construction of two underground garages was entitled to additional compensation.[6]

(2) *Excavated materials not suitable as fill.* The contract specifications required that soil materials located on site be excavated and reused as fill. However, the specified excavation and recompaction was prevented by the physical properties of the soil, which differed materially from the contract indications. In this case the contractor was entitled to an equitable adjustment for a Type I changed condition.[7]

Examples of nonexpress, or *implied,* contract indications include:

(1) *Hidden roof system not disclosed.* A roofing contractor that was required to demolish and remove an existing roofing system in addition to the roof indicated in the contract specifications and drawings was determined to have a valid Type I Differing Site Condition claim. Nothing contained in the contract

[5]*See Ace Constructors, Inc. v. United States,* 70 Fed. Cl. 253 (2006) (site elevation data); *Billington Contracting, Inc.,* ASBCA No. 54147, 05-1 BCA ¶ 32,900 (data referenced in bid documents but located 760 miles away from project site); *Bean Stuyvesant, L.L.C.,* ASBCA No. 53882, 06-2 BCA ¶ 33,420 (drilling logs referenced in the bid documents).

[6]*Baltimore Contractors, Inc. v. United States,* 12 Cl. Ct. 328 (1987). *See also Kilgallon Constr. Co. Inc.,* ASBCA No. 52583, 03-2 BCA ¶ 32,380, where boring logs indicating that the top six inches of excavation would be "hard crust" were reasonably interpreted to mean the material could be broken up by construction equipment was a Type I differing site condition when the hard crust turned out to have two to three times the compressive strength of concrete; *Comtrol, Inc. v. United States,* 294 F.3d 1357 (Fed. Cir. 2002), claim for quicksand as a Type I Differing Site Condition denied where solicitation stated hard material may be encountered and defined hard material but also stated that bidders should include in bid the disposal of surface and subsurface water.

[7]*Southern Paving Corp.,* ASBCA No. 74-103, 77-2 BCA ¶ 12,813; *see also W.R. Henderson Constr., Inc.,* ASBCA No. 52938, 02-1 BCA ¶ 31,741.

specifications or drawings indicated the existence of the additional roof system. Additionally, an inspection of the roof revealed no evidence that any additional roofing work had been performed after the as-built drawings had been prepared.[8]

(2) *Suitable equipment for work.* The Armed Services Board of Contract Appeals upheld a Differing Site Condition claim and stated that the "compaction, and clearing and grubbing" requirements were sufficient contract indications. The ASBCA concluded that, while the contract documents made no express representation regarding subsurface conditions, the compaction, clearing, and grubbing requirements led the contractor to reasonably believe it could utilize heavy equipment to perform its work. The board stated that "where, as here, design requirements cannot be met and procedures and equipment reasonably anticipated cannot be used, the situation represents a classic example of a Type I differing site condition."[9] Similarly, boards have found a Type I differing site condition when the existing on site material referenced in the contract as available for use was unsuitable as controlled fill because of high plasticity,[10] that a Type I condition existed where ongoing and pervasive seepage kept the subgrade wet, preventing or impairing compaction and grading,[11] and that the contractor had established a Type I differing site condition because the soil encountered at the site differed materially from the conditions indicated in the contract.[12]

(3) *Unanticipated sloughing of soils.* A tunneling contractor that encountered "running" ground conditions that were not disclosed by the contract soils information was granted relief under the Differing Site Conditions clause for encountering a Type I changed condition. The contractor was required to grout in order to stop the sloughing.[13]

(4) *Dry conditions implied by specified construction procedures.* When the construction procedures and design requirements set forth in the contract documents, read as a whole, indicated subsurface conditions permitting excavation "in the dry," but actual conditions made it impossible or impracticable to excavate in this manner, the court concluded that a changed condition had been encountered.[14]

---

[8] *Southern Cal. Roofing Co.,* PSBCA No. 1737 *et al.,* 88-2 BCA ¶ 20,803.
[9] *Kinetic Builders, Inc.,* ASBCA No. 32627, 88-2 BCA ¶ 20,657.
[10] *PK Contractors, Inc.,* ASBCA No. 53576. 04-2 BCA ¶ 32,661.
[11] *CEMS, Inc. v. United States,* 59 Fed. Cl. 168 (2003).
[12] *B.W. Farrell, Inc.,* ASBCA No. 53311, 06-2 BCA ¶ 33,322.
[13] *Shank-Artukovich v. United States,* 13 Cl. Ct. 346 (1987).
[14] *See Foster Constr., C.A. v. United States,* 193 Ct. Cl. 587 (1970). *But see Tricon-Triangle Contractors,* ENGBCA No. 5113, 88-1 BCA ¶ 20,317 (denying a Type I differing site condition claim where the presence of groundwater could be implied from the contract provision requiring the contractor to maintain a dewatering system).

(5) *Implied thickness of concrete floor.* The comparison of a 6-inch drain connection shown on the drawings with a cross section of concrete floors on the same drawings indicated the concrete floors were about 6 inches thick. When concrete floors up to 24 inches thick were encountered, a differing site conditions claim was allowed.[15]

IV. RECOVERY FOR A TYPE II CHANGED CONDITION

A. Type II Defined

Type II changed conditions differ significantly from those just discussed because it is possible to recover even where the contract is *silent* about the nature of the condition. To establish a Type II changed condition, the contractor must show that the conditions encountered are *unusual* and *differ materially* from those reasonably anticipated, given the nature of the work and the locale.

To qualify as sufficiently *unknown and unusual,* the condition encountered does not have to be in the nature of a geological freak—for example, permafrost in the tropics.[16] Instead, all that is generally required is that the unknown physical condition be one that was reasonably unanticipated, based on an examination of the contract documents and the site.

The key to recovery for a Type II changed condition is the comparison of actual conditions with what was reasonably expected at the time of the bid or proposal submission. This inquiry into reasonable expectations will raise questions of the contractor's *actual and constructive* knowledge of working conditions in the particular area. For example, awareness of a condition at the site that is common knowledge to other contractors working in the area, and thus reasonably ascertainable by inquiry, may be attributed to the contractor. Consequently, when a contractor failed to visit the work site and the plans and specifications noted potential problems, the resulting failure to discover obvious physical conditions caused the General Services Administration Board of Contract Appeals to conclude that the bidder's judgment was simply a "guess . . . premised in error" that formed no basis for recovery as a Type II changed condition.[17]

B. Type II Examples

Examples of Type II changed conditions include:

(1) *Subsurface water.* A water table found to be much higher than reasonably could have been anticipated has been held to be a changed condition,

---

[15]*J.E. Robertson Co. v. United States,* 437 F.2d 1360 (Ct. Cl. 1971).
[16]*See Ruff v. United States,* 96 Ct. Cl. 148 (1942); *Western Well Drilling Co. v. United States,* 96 F. Supp. 377 (D. Cal. 1951).
[17]*See L.B. Samford, Inc.,* GSBCA No. 1233, 1964 BCA ¶ 4309.

where dry and stable subsurface conditions were reasonably anticipated (but not indicated).[18]

(2) *Buried timber/rubble.* In leveling land that had been cleared, the board found that a contractor had no notice of buried timbers, although the contract required the disposal of surface stumps, roots, and other trash encountered. The buried trees thus constituted a Type II changed condition.[19] Similarly, submerged piling in a dredge-filled area warranted changed conditions relief.[20] However, another case reached the opposite result and held that the presence of buried stumps should have been anticipated because the site was in a fill area that contained some protruding stumps, new sprouts, and new branches—indicating growth from buried stumps.[21]

(3) *Undersized floor joists.* During performance of a contract to renovate certain family housing units, the contractor did not encounter 2" × 8" floor joists, as would normally be encountered. Rather, the contractor found that over 80 percent of the joists were much closer to 7 inches in height. This required substantial shimming and other modifications, which resulted in extra costs. The contractor did not have a Type I differing site condition claim, because the contract plans did not give an exact representation as to the size of the floor joists. However, since the actual dimensions of the joists differed significantly from the conditions an experienced contractor would reasonably expect to encounter on a project of this type, the undersized joists did constitute a compensable Type II differing site condition.[22]

(4) *Oversized walls.* In performing a contract to renovate an existing hospital, the contractor encountered a four-course-thick brick and masonry wall on the interior of the hospital. Such a massive wall was unusual for an interior partition; therefore, the contractor recovered the extra costs associated with the removal of the wall as a Type II differing site condition.[23]

(5) *Utilities.* When a contractor discovered that a third party had performed previous wiring in such a way that the phasing and wiring required by its contract could not be accomplished without extra work, this unknown condition warranted payment.[24] An undisclosed sewage line encountered in attempting

---

[18]*Loftis v. United States*, 110 Ct. Cl. 551, 76 F. Supp. 816 (1948); *see also Blount Bros. Corp.*, ENGBCA No. 2803, 70-1 BCA ¶ 8256.

[19]*Morgan Constr. Co.*, IBCA No. 299, 1963 BCA ¶ 3855; *see also Virginia Beach Air Conditioning Corp.*, ASBCA No. 42638, 92-1 BCA ¶ 24,432; *Josun, Inc.*, AGBCA No. 80-113-4, 88-2 BCA ¶ 20,590.

[20]*Caribbean Constr. Corp.*, IBCA No. 90, 57-1 BCA ¶ 1315.

[21]*Gillioz Constr. Co.*, 2 CCF 1211, W.D. BCA ¶ 826 (1944).

[22]*Kos Kam, Inc.*, ASBCA No. 34684, 88-1 BCA ¶ 20,246; *see also Minter Roofing Co., Inc.*, ASBCA No. 29837, 90-1 BCA ¶ 22,279 (extra studs around door openings).

[23]*Hercules Constr. Co.*, VABCA No. 2508, 88-2 BCA ¶ 20,527; *see also Penn Envtl.l Control, Inc.*, VA-BCA No. 3726, 94-2 BCA ¶ 26,790.

[24]*Dodson Elec. Co.*, ASBCA No. 5280, 59-2 BCA ¶ 2342.

to dig a manhole was found to be a changed condition.[25] Similarly, a differing site condition was found to exist when a contractor installing conduit pipe under an airfield perimeter road encountered a sewer line that was not indicated on the contract documents and was not a condition that would generally be expected.[26] However, in a different setting, a contractor encountered sewers, gas lines, water lines, and coaxial cables that were not shown on the plans, but a changed conditions claim was denied because the site was in a heavily built-up area and manholes were shown on the plans.[27]

(6) *Peculiar structural conditions.* A dock painting contractor was entitled to an equitable adjustment for extra work due to unusual conditions that reasonably could not have been anticipated at the time of contracting. The additional work was caused by peculiar structural features of the dock to be painted that, in combination with the air pressure from incoming tides, caused a continuous water seepage or mist over the dock. The contract documents, the contractor's prebid site inspection, and the contractor's experience as a painting contractor were not sufficient to provide notice of this unusual condition.[28]

(7) *Unusually tough soil.* The contractor's difficulties compelled a finding that the soil was unusually tough and materially different from conditions ordinarily encountered or generally recognized as inhering in earth and dam construction.[29]

(8) *"Double-poured" roof.* Where an existing roof system was found to be "double-poured," and therefore much thicker and expensive to replace, the court concluded that a Type II changed condition might be found.[30]

(9) *Ceiling tiles.* Where ceiling tiles had to be replaced during the installation of a fire alarm and sprinkler system, a Type II differing site condition was found because the tiles were extremely dirty and many were peeling away from their vinyl backs or were held in place by heavy objects.[31]

(10) *Miscellaneous items.* Where the contractor encountered beer cans, live ammunition, and ladies' underwear in cleaning a duct system in a military barracks, the contractor was granted relief for a Type II changed condition.[32]

---

[25]*Neale Constr. Co.*, ASBCA No. 2753, 58-1 BCA ¶ 1710; *see also Baltimore Contractors, Inc. v. United States*, 12 Cl. Ct. 328 (1987); *R. A. Glancy & Sons, Inc.*, VABCA No. 2327, 89-3 BCA ¶ 20,068.

[26]*Unitec, Inc.*, ASBCA No. 22025, 79-2 BCA ¶ 13,923.

[27]*H. Walter Schweigert*, ASBCA No. 4059, 57-2 BCA ¶ 1433; *see also Cee Tee Co.*, DOTCAB No, 1183, 82-1 BCA ¶ 15,467 (buried telephone cable adjacent to existing telephone building held not to be a Type II condition). *But see Green Constr. Co.*, ASBCA No. 46157, 94-1 BCA ¶ 26,572 (oversized manhole bottom held to be a Type II condition).

[28]*Warren Painting Co.*, ASBCA No. 18456, 74-2 BCA ¶ 10,834.

[29]*Servidone Constr. Corp. v. United States*, 19 Cl. Ct. 346 (1990).

[30]*Lathan Co., Inc. v. United States*, 20 Cl. Ct. 122 (1990); *see also Vega Roofing Co., v. Int'l Boundary and Water Comm'n*, GSBCA No. 13576-IBWC, 97-2 BCA ¶ 28,990 (Type II claim denied when two layers of roofing material were readily apparent from a reasonable site inspection).

[31]*Fire Security Systems, Inc. v. General Services Administration*, GSBCA No. 12120, 97-2 BCA ¶ 28,994.

[32]*Community Power Suction Furnace Cleaning Co.*, ASBCA No. 13803, 69-2 BCA ¶ 7963.

A Type II differing site condition may result not only from a variance in the type or quantity of a material encountered, but also from the unusual performance of an expected material. Thus, even though clay was expected to be encountered, when, as a result of percolating water, the clay behaved in an unusual, erratic fashion, with an unexpected tendency to slide, there was a changed condition.[33] Similarly, the unexpected shrinkage of soil, which materially increased the number of cubic yards of earth in a dam, was an unexpected property of the soil that constituted a changed condition.[34] A contractor was also allowed to recover for the additional cost of handling a subsurface water condition, although subsurface water was to be expected, but the place where it was encountered and the rate of its flow were unusual and unforeseeable.[35] As can be seen from a comparison of these decisions, the resolution of Type II Differing Site Conditions claims are controlled by the specific facts related to the project.

V. OTHER CONDITIONS: WEATHER AND QUANTITY VARIATIONS

A. Weather and Site Conditions

As a basic rule, weather and the effect of weather on the work site or project is not a basis for recovery under the Differing Site Conditions clause. This principle was established by the Court of Claims in *Arundel Corp. v. United States.*[36] In the *Arundel* decision, the contract basically involved unit price dredging work. After the bid was accepted, a hurricane struck the project site and scoured out a substantial quantity of material from the work site. The remaining material was more difficult to dredge. Although the contract contained a version of a Differing Site Condition clause, the court rejected the contractor's contention that it applied to a changed site condition created by weather. In the court's view, the Differing Site Condition clause applied to latent conditions extant at the time of award and did not apply to weather-created conditions, which are typically described as acts of God.[37] The analysis in *Arundel* continues to be applied by the boards and the Court of Federal Claims.[38]

For example, in *John Massman,* the contract documents for scour protection work at a lock and dam site on the Mississippi River contained certain river flow data entitled "Approximate Average Monthly Flow," which the contract stated was provided as a guide for scheduling purposes. The actual river flows were much higher, but the

---

[33]*Paccon, Inc.,* ASBCA No. 7643, 1962 BCA ¶ 3546. *See also J. Lawson Jones Constr. Co., Inc.,* ENG-BCA No. 4363, 86-1 BCA ¶ 18,719; *Ballenger Corp.,* DOTCAB 74-32, 84-1 BCA ¶ 16,973; *but see Kilgallon Constr. Co., Inc.,* ASBCA No. 51601, 01-2 BCA ¶ 31,621.
[34]*Guy F. Atkinson,* IBCA No. 385, 65-1 BCA ¶ 4642.
[35]*Norair Eng'g Corp.,* ENGBCA No. 3568, 77-1 BCA ¶ 12,225.
[36]103 Ct. Cl. 688 (1945), *cert. denied* 326 U.S. 752 (1945).
[37]103 Ct. Cl. at 710.
[38]*John Massman Contracting Co. v. United States,* 23 Cl. Ct. 24, 31 (1991); *Turnkey Enters. v. United States,* 597 F.2d 750, 759 (Ct. Cl. 1979); *Commercial Contractors Equip., Inc.,* ASBCA No. 52930, 03-2 BCA ¶ 32,381.

court rejected the argument that the variance between the historical averages and the actual river flow constituted a differing site condition. Similarly, in *Dennis T. Hardy Electric, Inc.*,[39] the ASBCA rejected an argument that higher-than-expected water tables caused by heavy rainfalls entitled the contractor to recover the costs for dewatering. As the contract documents contained no representations on the level of the water table, the claim was rejected.

Severe weather may, in certain situations, be a factor giving rise to a Differing Site Condition claim. In *D.H. Dave and Gerben Contracting Co.*,[40] the contract drawings did not include any reference to groundwater at the construction site. However, during performance, an excessive amount of rain fell on the site, resulting in delays due to inadequate drainage at the site. The board ruled that the contractor encountered a changed condition at the site that was the result of the excessive rains, inadequate drainage, and a fluctuating water table. Since the combination of these factors made performance impossible, the board ruled that there was a differing site condition. The modern standard is neatly summarized in *Kilgallon Construction Co., Inc.*,[41] where the board stated that the contractor must "prove that interaction of the rain with the pre-existing and unknown site condition produced unforeseeable consequences."

Although it is clearly established law that severe weather alone is not grounds for a differing site condition claim, severe weather may be a factor in obtaining relief as part of a defective specification claim. In *D.F.K. Enterprises, Inc. v. United States*, the contractor won a contract to paint water storage tanks at the White Sands Missile Range.[42] The contract included a chart indicating the anticipated number of adverse weather days at the site for each calendar month. Although the contract stated elsewhere that wind should be taken into account as a source of adverse weather, and accordingly incorporated into a bid, the chart of days only accounted for precipitation and did not account for wind at all.[43] The contractor developed its bid based on the weather chart with the assumption that it accounted for all sources of adverse weather. Severe winds later delayed performance. After the contractor filed a claim stemming from defective specifications, the court ruled in favor of the contractor, finding that the chart was an affirmative representation of past weather conditions at the site and that the contractor had relied on the chart in preparing its bid.[44] Similarly, in *P.K. Contractors, Inc.*, the government provided data on the flow of a river that proved to be erroneous.[45] After the government instructed the contractor to continue working despite water flow higher than expected, the board held there to be a compensable change to the contract.

Although weather may be a factor either in conjunction with a Type II Differing Site Condition claim or as an element in a defective specification or change claim,

[39]ASBCA No. 47770, 97-1 BCA ¶ 28,840; *but see United Contractors v. United States,* 368 F.2d 585 (Ct. Cl. 1966) (contract contained affirmative representation that water would not be encountered).
[40]ASBCA No. 6257, 1962 BCA ¶ 3493.
[41]ASBCA No. 516101, 01-2 BCA ¶ 31,621.
[42]*D.F.K. Enters., Inc. d/b/a Am. Coatings v. United States,* 45 Fed. Cl. 280 (1999).
[43]*Id.* at 283.
[44]*Id.* at 288.
[45]ENGBCA Nos. 4901, 5584, 92-1 BCA ¶ 24,583.

these theories are extremely fact specific. Both *D.F.K.* and *P.K. Contractors* were situations where the government provided information concerning weather or natural conditions that was then used by the contractor to prepare the bid and perform. It would be unlikely for these theories to be applied to any situation where the contractor accepted the duty and the risks of ascertaining weather data in preparation of its proposal.

B. Quantity Variations and Site Conditions

Construction contracts involving extensive earth work, rock removal, dewatering, and so on often contain unit prices and estimated quantities for those work items. In *Perini Corp. v. United States*,[46] the contract documents for a dam construction project contained an estimated quantity for pumping water out of coffer dam areas. The contractor's bid for the applicable pay items greatly exceeded the other bids, but its overall price was low and it received the award. Ultimately the contractor pumped over 25 times the estimated quantity for one of the pay items.[47] The contract did not contain any type of a Variation in Estimated Quantities clause,[48] and the government sought to adjust the unit price on the basis of the Differing Site Conditions clause contained in the contract. Reversing a board decision in favor of the government, the Court of Claims concluded that there were sufficient indications in the contract to alert the parties to the fact that the relevant estimated quantities for pumping water were inaccurate and that the contractor had concluded from its site investigation that large quantities of water would be encountered.[49]

In contrast, in *Gregg, Gibson & Gregg, Inc.*,[50] the board did allow recovery under the Differing Site Conditions clause for a quantity variation of 6 percent in acreage to be cleared when the quantity was set forth as a precise figure. Consequently, whether a variation from a quantity provided in the contract documents is a basis for recovery under the Differing Site Conditions clause may depend on the preciseness of the data provided, the quantity variance, and the contractor's reliance on the estimated quantity in preparing its proposal or bid.[51]

Currently, most government construction contracts with work or pay items based on estimated quantities contain a Variation in Estimated Quantity clause[52] as well as the Differing Site Conditions clause. As a general rule, if the quantity variation is the result of a differing site condition or a change to the work, the Variation in Estimated Quantity clause does *not* control the calculation of the equitable adjustment in the contract price.[53]

---

[46]*Perini Corp. v. United States*, 381 F.2d 403 (Ct. Cl. 1967).

[47]It pumped 7,564,610 units of water as compared to the estimated quantity of 302,000 units set forth in the bid documents.

[48]*See* **Section II.C of Chapter 8** of this book for a discussion of that clause.

[49]381 F.2d 403, 412.

[50]ENGBCA No. 3041, 71-1 BCA ¶ 8677.

[51]*See AFGO Eng'g Corp.*, VACAB No. 1236, 79-2 BCA ¶ 13,900.

[52]FAR § 52.211-18.

[53]*See United Contractors v. United States*, 368 F.2d 585 (Ct. Cl. 1966); *Met-Pro Corp.*, ASBCA No. 49694, 98-2 BCA ¶ 29,776 (Differing Site Conditions clause controlled); *Morrison-Knudsen Co. v. United States*, 397 F.2d 826 (Ct. Cl. 1968) (Changes clause controlled).

VI. FACTORS AFFECTING RECOVERY

Although the federal government includes a standard Differing Site Conditions clause in its contracts, some contracts also include other provisions that attempt to minimize or reduce claims under them. These include clauses relating to site inspection, notice, and various other provisions that seek to limit the ability of contractors to rely on information provided to them during the proposal or bidding process. Whether these additional contract clauses or provisions will bar or limit recovery under a Differing Site Conditions clause usually depends on a variety of other factors that are described next.

A. Site Investigations

Requests for proposals and invitations for bids commonly require contractors to visit the site prior to submitting their proposals or bids. Government construction contracts routinely require the contractor to warrant that it has made a site inspection and include site investigation clauses. For example, FAR § 52.236-3, Site Investigation and Conditions Affecting the Work, states:

> (a) The Contractor acknowledges that it has taken steps reasonably necessary to ascertain the nature and location of the work, and that it has investigated and satisfied itself as to the general and local conditions which can affect the work or its cost, including but not limited to . . . The Contractor also acknowledges that it has satisfied itself as to the character, quality, and quantity of surface and subsurface materials or obstacles to be encountered insofar as this information is reasonably ascertainable from an inspection of the site, including all exploratory work done by the Government, as well as from the drawings and specifications made a part of this contract. . . .
>
> (b) The Government assumes no responsibility for any conclusion or interpretations made by the Contractor based on the information made available by the Government. . . .

Such a requirement does not automatically nullify the effect of a Differing Site Conditions clause if one is present and does not necessarily obligate the contractor to discover hidden conditions at its peril.[54] A contractual requirement that the contractor make a site investigation does not obligate a prospective contractor to discover hidden subsurface conditions that would not be revealed by a reasonable preaward inspection.[55] The adequacy of the site investigation is measured by what a reasonable, intelligent contractor, experienced in the particular field of work involved, could be expected to discover, not what a highly trained expert might have found.[56]

---

[54]*Farnsworth & Chambers Co. v. United States,* 171 Ct. Cl. 30 (1965).
[55]*Warren Painting Co., Inc.,* ASBCA No. 18456, 74-2 BCA ¶ 10,834; *Maintenance Eng'rs,* ASBCA No. 17474, 74-2 BCA ¶ 10,760; *John G. Vann v. United States,* 190 Ct. Cl. 546, 573 (1970).
[56]*Stock & Grove, Inc. v. United States,* 493 F.2d 629 (Ct. Cl. 1974); *Commercial Mech. Contractors, Inc.,* ASBCA No. 25695, 83-2 BCA ¶ 16,768.

The term "site investigation" is often interpreted to mean, essentially, a sight investigation, and not to extend to making independent subsurface, geotechnical investigations.[57] However, this is not always the case. The contractor will be responsible for being aware of all information reasonably made available to it as well as all information that could be gained by a "reasonable" site inspection under the circumstances. For example, in *Bean Stuyvesant LLC*,[58] the ASBCA denied a dredging contractor's Type I Differing Site Conditions claim, concluding that the contractor had failed to prove that: (1) the conditions indicated in the contract documents differed materially from those conditions actually encountered; (2) the conditions encountered were reasonably unforeseeable based on all the information available to the contractor at the time of bidding; and (3) the contractor reasonably relied on its interpretation of all contract and contract-related documents.

Likewise, the ASBCA held in *Tri-Ad Constructors*[59] that a contractor was not entitled to an equitable adjustment for installing more electrical cable than anticipated because the contractor failed to conduct a prebid site inspection as required by the contract. From its reading of an electrical wiring diagram, the contractor believed that seven electrical substations were located immediately above the main bank of underground ducts running between two switching stations. If the contractor had made an inspection of the site, it would have seen that the electrical substations, each the size of an automobile, were offset some 300 feet from the main line, requiring loops between this duct bank and each substation. The contractor was charged with the knowledge it would have obtained from a reasonable site inspection, and the ASBCA concluded that even the most cursory inspection would have revealed the need for additional cable between each substation and the main ductline.[60]

Failure to consider and compare information set forth on the contract drawings with that obtained from a site visit can result in the denial of a Differing Site Conditions claim on the basis that the contractor's site inspection was not adequate

---

[57]*Gulf Constr. Group, Inc.*, ENGBCA No. 5850, 93-1 BCA ¶ 25,229, stating a site investigation provision does not override the Differing Site Conditions clause and contractor is not required to discover hidden conditions.

[58]ASBCA No. 53882, 06-2 BCA ¶ 33,420. *See also Larry D. Barnes, Inc. v. United States*, 45 Fed. Appx. 907 (Fed. Cir. 2002) (claim denied because contractor failed to conduct any prebid site inspection and because underground obstructions were foreseeable); *Conner Bros. Constr. Co., Inc. v. United States*, 65 Fed. Cl. 657 (2005) (reasonable site inspection would have revealed the differing site conditions, which were foreseeable); *Southern Comfort Builders, Inc. v. United States*, 67 Fed. Cl. 124 (2005) (contractor's failure to attend a prebid site visit did not relieve it from the knowledge it would have gained by attending); *Vega Roofing Co. v. Int'l Boundary and Water Comm'n*, GSBCA No. 13576-1BWC, 97-2 BCA ¶ 28,990 (reasonable site investigation would have revealed actual roof condition in a roof replacement project).

[59]ASBCA No. 34732, 89-1 BCA ¶ 21,250; *cf. O.K. Johnson Elec. Co., Inc.*, VABCA No. 3464, 94-1 BCA ¶ 26,505 (contractor failed to make an above-the-ceiling examination), which was found to readily feasible, *with H. T. Lyons, Inc.*, ASBCA No. 36901, 89-1 BCA ¶ 21,426 (obstructions found above the ceiling, site investigation impossible).

[60]*See also McCormick Constr. Co. v. United States*, 18 Cl. Ct. 259 (1989) (denying a contractor's Differing Site Condition claim because a reasonable site investigation would have revealed the possible subsurface condition).

or reasonable. For example, in *Conner Brothers Construction Co., Inc. v. United States*,[61] a hospital renovation project required the removal of sections of heating, ventilation, and air-conditioning (HVAC) ductwork; grilles; and diffusers; and the installation of new sections with grilles and diffusers. Representatives of both the subcontractor and the general contractor made site inspections. However, neither individual compared the locations of the existing grilles and diffusers to locations for the replacement grilles and diffusers shown on the drawings depicting the new work. In addition, while making their above-ceiling site inspections, neither could observe the type of connections (flexible or hard-piped) between the grilles and diffusers and the main duct in the existing HVAC system due to a lack of light and the above-ceiling congestion. Rather than submitting an inquiry to the contracting officer, the estimate and bid price anticipated that existing connections were flexible and that the new grilles and diffusers would be placed in the same ceiling locations as the existing ones. When those assumptions proved incorrect, the contractor submitted a claim for additional compensation on behalf of itself and its subcontractor. The court denied recovery based on its conclusion that the prebid site investigation was inadequate due to the failure to compare the existing locations of grilles and diffusers to the new locations and the failure to submit a prebid inquiry regarding the above-ceiling conditions that could not be observed.

However, a finding that a contractor's preproposal or prebid site investigation was inadequate does not necessarily result in the total denial of a Differing Site Conditions claim. This result is illustrated by the court's extensive analysis of the contractor's differing site conditions claim arising out of the construction of a dam in North Texas.[62] In that case, the court determined that the contractor, Servidone, encountered a Type II differing site condition, largely due to the unexpected high concentration of a type of montmorillonite clay in the materials to be used as fill. This concentration of clay particles made working with the soil extremely difficult. Notwithstanding that finding, the court reduced the contractor's recovery on that claim due to its conclusion that Servidone's site inspection was not adequate for a contractor that had not previously performed similar work in that general location. In the court's opinion, a reasonable contractor with no prior experience with the local materials or conditions should have consulted knowledgeable local firms regarding the expected nature of the soil in that area and, if permissible, taken soil samples for evaluation. Based on those conclusions, the court reduced Servidone's recovery to reflect the conditions which a contractor that was knowledgeable and experienced with typically expected local soils would anticipate and price.[63]

Contractors and subcontractors should develop a site investigation checklist for use when preparing an estimate for any project and orienting its project management to the provisions and requirements of the Differing Site Conditions clause. **Appendix 7A** to this chapter is a sample site investigation checklist. In order to facilitate its use, an

---

[61]65 Fed. Cl. 657 (2005).

[62]*Servidone Constr. Corp. v. United States*, 19 Cl. Ct. 346 (1990).

[63]The court record included extensive testimony and the estimate of a local, experienced contractor.

electronic version of the same document is included on the support Web site at www
.wiley.com/go/federalconstructionlaw. In addition to requiring that a contractor conduct a reasonable site investigation, some solicitations also require a contractor to
review documents concerning the site conditions that are made available for inspection prior to bidding but are not provided to the contractor as part of the bid package. If
the contractor fails to review the available documents before submitting its bid, it may
later be precluded from recovering for conditions that are different from expected
but that could have been determined from a review of the documents made available.[64] For example, a contractor's differing site conditions claim was denied when
the contractor failed to review records of previous dredgings that contained information regarding the nature of the materials to be dredged and that were available to the
contractor prior to bidding.[65]

B. Exculpatory and Risk Shifting Provisions

Contracts may contain broad exculpatory clauses that purport to disclaim any government liability for the accuracy of plans, specifications, borings, and other subsurface
data. For example, FAR § 52.236-4, Physical Data, instructs contracting officers to:

> . . . [I]nsert the following clause in solicitations and contracts when a fixed-
> price construction contract is contemplated and physical data (e.g., test borings,
> hydrographic, weather conditions data) will be furnished or made available to
> offerors. All information to be furnished or made available to offerors before
> award that pertains to the performance of the work should be identified in the
> clause. When subparagraphs are not applicable they may be deleted.

PHYSICAL DATA (APR 1984)

> Data and information furnished or referred to below is for the Contractor's
> information. The Government shall not be responsible for any interpretation of
> or conclusion drawn from the data or information by the Contractor.
>
> (a) The indications of physical conditions on the drawings and in the specifi-
> cations are the result of site investigations by_____ *[insert a description
> of investigational methods used, such as surveys, auger borings, core borings,
> test pits, probings, test tunnels].*

---

[64]*Id. See also Stuyvesant Dredging Co. v. United States,* 834 F.2d 1576 (Fed. Cir. 1987); *G&P Constr. Co.,
Inc.,* ASBCA No. 49524, 98-1 BCA ¶ 29,457.
[65]*Stuyvesant Dredging Co.,* 834 F.2d at 1581; *see also Billington Contracting, Inc.,* ASBCA Nos. 54147,
54149, 05-1 BCA ¶ 32,900 (denying a changed conditions claim where specification notified bidders that
previous dredging records were available and contractor did not review those records that would have
disclosed the same conditions which were in the claim); *see also Randa/Madison Joint Venture III v. Dahlberg,*
239 F.3d 1264 (Fed. Cir. 2001) regarding contractor's review of subsurface data that is referenced in the
contract documents; *Thomas J. Young, Jr.,* PSBCA Nos. 3885, 3983, 98-2 BCA ¶ 29,772 regarding reference to a full soils testing report alerted bidders to the availability of that information.

Many decisions have held that these clauses do not have the sweeping effect the broad language of the clause may indicate, and as a result they normally will not allow such clauses to eliminate the relief provided by the Differing Site Conditions clause.[66] For example, in *Woodcrest Construction Co. v. United States*,[67] the Court of Claims allowed the contractor to recover under the Changed Conditions clause despite the extremely broad exculpatory provisions in the contract. The court stated:

> The effect of an actual representation is to make the statements of the Government binding upon it, despite exculpatory clauses which do not guarantee the accuracy of a description. . . . Here, although there is no [express] statement which can be made binding upon the Government, there was in effect a description of the site, upon which plaintiff had a right to rely, and by which it was misled. Nor does the exculpatory clause in the instant case absolve the Government, since broad exculpatory clauses . . . cannot be given their full literal reach, and, "do not relieve the defendant of liability for changed conditions as the broad language thereof would seem to indicate."[68] [G]eneral portions of the specifications should not lightly be read to override the Changed Conditions Clause. . . . [69]

Similarly, contract clauses that seek to disclaim particular standard clauses often will not be enforced. In *Syblon-Reid Co.*, a contract for the removal of sediment from a canal included a clause that required all bidders to be present for a single group site inspection and placed the responsibility of estimating the amount of material to be removed solely on the contractor.[70] More material was removed than expected, and the contractor filed a Differing Site Conditions claim. The board ruled that the disclaimer clause in the contract did not operate to eliminate the Differing Site Conditions clause because "it was not possible to ascertain the quantity of work from the required site inspection."

Even when a contract lacks a Differing Site Conditions clause *and* contains extensive exculpatory language, it still may be possible for the contractor to recover *if* it can show, for example, that an independent subsurface investigation was not feasible and that it was thus forced to rely on information provided by the owner.[71] Another factor affecting the enforceability of disclaimer clauses is the time allowed for the investigation.[72] In government contracting, a disclaimer that conflicts with the

---

[66]*See* Currie, Abernathy & Chambers, Changed Conditions, Construction Briefings No. 84-12 (Dec. 1984). *See also Travelers Cas. and Sur. Co. of Am. v. United States,* 75 Fed. Cl. 696 (2007) (stating that exculpatory language in a site investigations clause does not relieve the government from responsibility for its contractual indications).

[67]408 F.2d 406 (Ct. Cl. 1969).

[68]*Fehlhaber Corp. v. United States,* 151 F. Supp. 817, 825 (Ct. Cl.), *cert. denied,* 355 U.S. 877 (1957).

[69]*United Contractors v. United States,* 368 F.2d 585, 598 (Ct. Cl. 1966).

[70]IBCA No. 1313-11-79, 82-2 BCA 16,105.

[71]*North Slope Technical Ltd. v. United States,* 14 Cl. Ct. 242 (1988), which involved frozen ground during the winter bid period.

[72]*Farnswoth & Chambers Co. v. Uinted States,* 171 Ct. Cl. 30 (1960) (government allowed one month for bid preparation).

Differing Site Conditions clause generally will not be interpreted in a manner to negate the relief provided by the standard Differing Site Conditions clause.[73]

However, this basic principle does not mean that all exculpatory clauses are ineffective. The Court of Claims, in *United Contractors v. United States*, stated that clear and unambiguous language could operate to override a standard contract provision.[74] However, the court in this case did not find that the Changed Conditions clause was overridden by an exculpatory clause in the contract. In *VECA Electric Co.*, the board ruled that warnings in the specifications about the "general nature of the drawings" reinforced the conclusion that the contractor's interpretation of the drawings was unreasonable because the drawings were diagrammatic and not specific as to the exact nature of the electrical conduits involved in the contract.[75] In *Hardwick Bros. Co. II v. United States*,[76] the contract included disclaimers that data and information included in the contract was "for information only." The court ruled against the contractor's differing site conditions claim using the exculpatory clause in conjunction with evidence of a laundry list of things the contractor should have done in performing the contract.[77] These two cases appear to show that a court or board will enforce a disclaimer particularly when the contractor has acted unreasonably in other aspects of performance.

In addition to exculpatory clauses, a contractor needs to consider other risk-shifting provisions in a contract that might affect its rights under a Differing Site Conditions clause. Many design-bid-build construction contracts contain requirements that reflect performance specifications rather than detailed design specifications. Design-build projects provide the contractor even greater latitude to develop a design solution meeting the government's requirements. Even though the Differing Site Conditions clause is a mandatory provision in design-build or other contracts containing performance specifications,[78] the government may attempt to assert that these performance or design-build duties alter the contractor's obligations and rights under that clause. Although performance specifications have been included in government construction contracts for many years and no board or court has interpreted those requirements as nullifying the relief promised by the Differing Site Conditions clause, it is plausible that the government could attempt to hold a design-build contractor's site inspection to a higher standard. While this would not nullify the remedy available under the Differing Site Conditions clause, it could have the effect of making it more difficult for a design-builder to recover under this clause. Therefore,

---

[73]*See* Report to the President of the United States by the Director of the Bureau of the Budget (1923), Report of Chairman of Interdepartmental Board of Contracts & Adjustments, pp. 140–141; Kendall, Changed Conditions as Misrepresentations in Govt. Construction Contracts, 35 Geo Wash. L. Rev. 978, 982 (1967), 4 YPA 187; Report to the President of the United States by the Director of the Bureau of the Budget (1927), Interdepartmental Board of Contracts & Adjustments, p. 93.
[74]177 Ct. Cl. 151 (1966).
[75]ASBCA No. 47733, 95-2 BCA ¶ 27,749.
[76]36 Fed. Cl. 347 (1996).
[77]*Id.* at 408.
[78]*See* FAR § 36.502.

if possible, a contractor should consider its intended approach or design solution when it evaluates the project site conditions contained in the solicitation and makes its site investigation.[79] However, in *Pitt-Des Moines, Inc.*,[80] where a design-builder was supplied with drawings and other information concerning the project site, the ASBCA concluded from evidence of trade practice that the design-build contractor was not obligated to retain a geotechnical engineer to examine the site during the proposal preparation phase. Documentation of a reasonable site investigation, which takes into account the contractor's intended approach, should provide a basis to rebut this possible argument by the government.

C. Notice Requirements

The purpose of the notice requirement for changed conditions is to alert the contracting officer of the existence of the condition and provide the government an opportunity to evaluate its potential impact on the project. Such an evaluation may cause the government to make changes in the design or alter the contractor's method of performance.

It is always advisable for a contractor to comply fully with the notice provisions of the contract. However, the lack of strict compliance may be excused. The underlying purposes of the contract requirements may be satisfied by *substantial compliance* with the terms of the notice requirement, or by *actual knowledge* of the condition by the contracting officer, or if the government has suffered *no prejudice* from the contractor's failure to give written notice.[81]

Three examples of cases where contractors were allowed to recover for a differing site condition despite the lack of strict compliance with the notice requirements are presented next.

(1) In *Parker Excavating, Inc.*,[82] the ASBCA sustained an appeal based on differing site conditions, finding that daily quality control reports placed the government on notice of the conditions encountered and that the contracting officer had written notice of the conditions. Additionally, the board found that the government was aware of the conditions from meetings and site visits. As a result, the burden was on the government to establish that it was prejudiced by absence of the required notice, and the government made no showing of prejudice from the passage of time or an inability to minimize extra costs resulting from any delay in receiving prompt written notice.

(2) In *Pat Wagner*,[83] the contract called for the installation of water meters and additional service lines to an existing water system. After the contractor began

---

[79]*See Conner Brothers Constr. Co., Inc. v. United States*, 65 Fed.Cl. 657 (2005).

[80]*Pitt-Des Moines, Inc.*, ASBCA No. 42838, 43514, 43666, 96-1 BCA ¶ 27,941.

[81]*Fru-Con Constr. Corp. v. United States*, 43 Fed. Cl. 306 (1999), where failure to notify of a differing site condition was found prejudicial to the government.

[82]ASBCA No. 54637, 06-1 BCA ¶ 33,217.

[83]IBCA No. 1612-A-82, 85-2 BCA ¶ 18,103.

work, it realized that the existing lines were copper rather than galvanized steel, as indicated by the contract documents. The contractor was forced to tie into the existing system with more expensive copper pipe. Although the contractor failed to provide written notice, the federal government's inspectors were fully aware of the discovery of copper pipe in the existing system. Therefore, the government was *not prejudiced* by lack of written notice, and recovery was allowed.

(3) In *Leiden Corp.*,[84] the board allowed recovery under the Differing Site Conditions clause where the contracting officer had constructive notice of the conditions at the site because *actual knowledge* of the changed conditions was imputed to the contracting officer's construction representative on site and thereby to the contracting officer.

When a contractor fails to strictly comply with the notice requirements of the Differing Site Conditions clause, its claim may not be barred, but the contractor will face a higher burden of persuasion in order to succeed on that claim. In *Harper Development & Associates*,[85] the contractor, while digging a trench on a housing development project, hit artesian water far in excess of what was expected. Although the government attempted to have the differing site conditions claim barred because of the lack of notice, the board allowed the claim to go forward but placed a heavier burden of persuasion on the contractor. The imposition of a heavier burden in these cases has continued to be the practice at the boards in the twenty-first century.[86]

D.   Record-keeping Requirements: Proving Damages

Even when a Differing Site Conditions clause is present in the contract, the contractor still must prove how much the unanticipated condition cost in order to be compensated. Even if a differing site condition can be demonstrated, the contractor must support its quantum request with reliable cost information. For example, a failure to maintain separate utilization records supporting a claim for equipment costs on an earthwork project may result in the rejection of the contractor's cost submission. Even if the lack of contemporaneous records documenting the extra work and costs can be overcome with the assistance of an expert, the effort to reconstruct the project costs can be a massive and therefore expensive undertaking.[87] The importance of good record keeping and the documentation required by a board or court are discussed in **Chapter 14.**[88]

---

[84]ASBCA No. 26136, 83-2 BCA ¶ 16,612.
[85]ASBCA No. 34719, 90-1 BCA ¶ 22,534 (citing *C.H. Leavell & Co*, ASBCA No. 16099, 72-2 BCA 9694).
[86]*See Monster Gov't Solutions, Inc., v. U.S. Department of Homeland Security,* DOTBCA No. 4532, 06-2 BCA ¶ 33,312.
[87]*Servidone Constr. Corp. v. United States*, 19 Ct. Cl. 346 (1990).
[88]*See also Baldi Brothers Constructors v. United States,* 50 Fed. Cl. 74 (2001), where the court determined that the adverse effect of the differing site conditions was so pervasive that the total cost method was the appropriate measure of damages.

➤ LESSONS LEARNED AND ISSUES TO CONSIDER

- Under *traditional contract law,* the contractor generally bears the risk (cost and time) of differing site conditions of which neither party is aware. A Differing Site Conditions or Changed Conditions clause operates to *shift or reallocate* that risk.

- *When the contractor assumes the risk* of differing site conditions, the contractor may include *a contingency* in its bid/proposal to cover the risk, which means that the contractor will receive a *windfall* (and the owner will pay too much) if no differing site condition is encountered, or the contractor could suffer a *loss* if the contingency is not sufficient to cover the cost when a significant differing site condition is encountered.

- In recognition of this fact, the federal government (as well as many other public and private owners) elects to use a *version of a Differing Site Conditions clause* in construction contracts, which places the risk of differing site conditions on the owner in exchange for the contractor not including a contingency for subsurface conditions in its bid.

- Public and private owners often use versions of Differing Site Conditions clauses in form contracts. These form contracts often have *great variation in the types of conditions* addressed in the Differing Site Conditions clause. Many of the form clauses do not provide the same protection as the standard FAR Differing Site Conditions clause.

- The FAR Differing Site Conditions clause recognizes two distinct types of differing site conditions: a *Type I changed condition* is a condition that differs materially from the conditions indicated in the contract documents; and a *Type II changed condition* is an unknown physical condition of an unusual nature that differs materially from the conditions ordinarily encountered in performing the type of work called for by the contract.

- Weather or conditions created by weather *usually* do not provide a basis for recovery under the Differing Site Conditions clause.

- Quantity variations may provide a basis for a recovery under the Differing Site Conditions clause if the variation was *unforeseen* and the *result* of a differing site condition.

- Some construction proposal or bid documents contain a *Site Investigation clause,* which requires the contractor to perform a thorough site investigation and examine the existing conditions prior to submitting its proposal or bid.

- Where the request for proposal or invitation for bid and the contract contain both a Site Investigation clause and a Differing Site Conditions clause, the contractor's ability to recover costs resulting from unanticipated conditions will depend on whether the condition was one that a reasonable, intelligent contractor would be expected to encounter based on a *reasonable site investigation* including referenced information that is reasonably made available to the contractor.

- Based on the circumstances related to the preparation of the proposal or bid and the nature of the work, a *sight* investigation *may satisfy* the requirements of a Site Investigation clause.
- A contractor needs to be able to demonstrate that it made a *reasonable site investigation* given the nature of the project and the available information and that it relied on that information in preparing its estimate and price for the work.
- A contractor needs to review its *expectations of the site conditions* with its on-site project management staff to ensure that the basis of the estimate is clearly understood.
- To *recover for a Type I changed condition,* a contractor generally must show that: (1) certain conditions are indicated by the contract documents; (2) the contractor relied on those indications of the physical conditions; (3) the actual conditions encountered differed materially from those indicated; (4) proper notice was given; and (5) the change in condition resulted in additional performance costs and/or time, as documented by appropriate documentation.
- To *recover for a Type II changed condition,* a contractor generally must show that: (1) the physical conditions encountered were unknown and unusual and differed materially from those ordinarily encountered and generally recognized as inherent in the work provided for in the contract; (2) proper notice was given; and (3) the change resulted in additional performance costs and/or time, as demonstrated by appropriate documentation.
- The Differing Site Conditions clause contains *notice requirements,* stating that the contractor should stop work and give the contracting officer prompt written notice upon encountering a differing site condition, before it is disturbed, so that the government's representatives will have an opportunity to inspect and evaluate the condition. Failure to give the required notice may *jeopardize a contractor's ability* to receive an equitable adjustment for the additional costs and/or added performance time incurred as a result of a differing site condition.
- A contractor should review the *requirements and the procedures* outlined in the Differing Site Conditions clause with its on-site project management staff in order to ensure that proper notice is provided and that contemporaneous records of the project costs are maintained. Both can be critical to recovering a proper adjustment to the contract price and time.
- *Good record keeping* is critical to a contractor's ability to receive a price adjustment for the costs and additional performance time resulting from a differing site condition. After-the-fact efforts to reconstruct the costs associated with a differing site condition can be very expensive and may have limited persuasive value.
- A contractor should *never assume* that it either will or will not receive an adjustment for a Differing Site Conditions clause without a careful review of the contract and the specific conditions involved.

APPENDIX 7A: SITE INVESTIGATION CHECKLIST

Solicitation (RFP or IFB) Documents

	Yes	No	??
Does the solicitation contain a Differing Site Conditions clause?			
Does the solicitation contain any provision that purports to limit the relief provided by the Differing Site Conditions clause?			
What reports, soil borings, as-built plans, or other information are provided in the solicitation that indicate existing conditions?			
Does the solicitation reference but not provide any other documents or sources of information on the possible site conditions?			
Does the solicitation indicate that environmental permits have been obtained and environmental restrictions observed?			
Do the solicitation documents purport to identify all existing utilities?			
Do the solicitation documents depict existing work to be modified, renovated, or demolished?			
Are there site access restrictions noted on the solicitation documents?			

Evaluation of Site and Solicitation Documents

	Yes	No	??
Has the company performed work similar to that in the solicitation in the vicinity of the project?			
Who are the individuals (estimators, consultants, engineers, etc.) who will be responsible for reviewing information provided in the solicitation documents or obtained during a site investigation? Has the company clearly identified the experience and qualifications of these individuals related to a review of this type of information?			
If the company has not performed similar work in the vicinity of the project, has the company contacted experienced potential employees, geotechnical consultants, or subcontractors regarding conditions (soil, rock, water table, weather, etc.) that might affect the work?			
Have company representatives made an on-site inspection?			
If permitted, has the company determined whether to obtain and test soil, rock samples, etc.?			
Have the plans and specifications been compared to the project site in an effort to identify unusual conditions, conflicts, etc.? (For example, did the site visit reveal the existence of manholes at locations where no utility lines are depicted on the drawings?)			
If conflicts were noted, were written preproposal questions submitted to the government?			
Was the site inspection documented with video, still photographs, memoranda?			
If portions of the work that could be affected by the site conditions will be subcontracted, do the potential subcontractors have experience with similar work in the vicinity of the project?			
If portions of the work that could be affected by the site conditions will be subcontracted, did the potential subcontractor(s) review with your firm all information obtained from the site investigation performed by the subcontractor(s)?			

If the site inspection observations were hindered by problems such as a lack of space, lack of adequate light (e.g., above-ceiling spaces), or other restrictions, has a written inquiry be made to the government seeking additional information on those conditions?			
Has a written inquiry been submitted to the government regarding the availability of any other information regarding the site, subsurface or soil data, as-built conditions, or environmental assessments?			
Has the information regarding the site investigation been reviewed by and with those estimating the project?			
Can the company clearly demonstrate that it utilized the information provided by the government and developed during its site investigation in preparing its estimate?			
Has the company considered how it might structure its project documentation system and cost reporting systems to track additional or unexpected costs due to a differing site condition? (For example does the company have a system to separately track equipment utilization by work activity?)			
Has the company reviewed its expectations of the site conditions with its on-site project management?			
Has the company reviewed the notice requirements of the Differing Site Conditions clause with its on-site project management?			

8

CONTRACT CHANGES

I. PURPOSE OF THE CHANGES CLAUSE

A. Historical Perspective

Both in the present as well as throughout history, changes are a common event on all construction projects. This can be seen in a very real sense in the so-called Bent Pyramid at Bahshur built in Egypt during the reign of King Snefru, founder of the Fourth Dynasty (2680–2560 B.C.). Construction was begun on a pyramid for King Snefru that was apparently planned to have smooth sides, rather than the steps or series of bench-shaped mounds (mastabas) used on earlier pyramids. The pyramid at Bahshur was started with an angle of incline of over 51 degrees. However, halfway up the side of the pyramid, the incline suddenly decreased to 43 degrees, resulting in the sides rising less steeply, hence the name "Bent Pyramid." The likely reason for the design change was to provide more stability. History does not record the change order to this government-directed construction project, but the result remains apparent to this day.

Many projects are eventually built differently than originally intended. A multitude of factors can result in the need for the modification to the project ranging from revised user requirements to evolution of technology to design problems. Since changes may be all but inevitable in government construction projects, the key question for all concerned is the effective and efficient management of that process in the context of the parties' contract. For the construction professional, this process requires an appreciation of the operation of the applicable Changes clause and the parties' rights and obligations.

B. Departure from Common Law Principles

A Changes clause in a construction contract departs from the established principle of contract law that a contract requires the mutual agreement of the contracting parties. Under the general common law applicable to contracts, a contract is created when two parties reach an agreement on an identified undertaking. Each party agrees to perform specific contract duties in exchange for some performance (consideration) by the other party. Once that agreement is reached, the terms of the contract define and limit the obligations of each party, and neither party can unilaterally change or modify the contract. Generally, in a construction contract lacking a Changes clause, if changes or modifications become necessary, both parties would have to reach a separate agreement incorporating any negotiated changes, which agreement is supported by an exchange of consideration between the parties.[1]

C. Evolution of the Changes Clause

The government has not always utilized a Changes clause that contains all of the elements of the current clause. For example, in an 1880 maintenance dredging contract for the Delaware River, the contract contained this provision on modifications to the contract:

> If, at any time during the progression of the work, it be found advantageous or necessary to make any change or modification in the project, and this change or modification should involve such change in the specifications as to character and quantity, whether of labor or material, as would either increase or diminish the cost of the work, then such change or modification *must be agreed upon in writing by the contracting parties,* the agreement setting forth fully the reasons for such change, and giving clearly the quantities and prices of both material and labor thus substituted for those named in the original contract, and before taking effect must be approved by the Secretary of War: *Provided,* That no payments shall be made unless such supplemental or modified agreement was signed and approved before the obligation arising from such modification was incurred. . . .[2] [Emphasis added.]

This provision required mutual agreement by the parties on the scope of the change, the reasons for the change, and the cost for the contract modification. In addition, before the change could take effect, the contract required that it be approved by the Secretary of War. This clause reflects a strict adherence to the common law principles regarding modifications to a contract.

Rigid application of these common law principles would create practical problems in government construction projects where there may not be time to allow for agreement and approvals prior to execution of the work. During times of emergency or war, requirements and needs often quickly change during the course of a long construction project. Small details left out of the work scope can affect the practical

---

[1]Changes clauses are so widely used in both government and private construction contracts that modern decisions regarding the absence of such a clause in a construction contract are rare. Generally, absent a clause providing for a change order process, a modification to the contract would require mutual agreement of the parties and consideration. *See* Samuel Williston and Richard Lord, Williston on Contracts § 8:9 (4th ed. Thomson/West, 1992); *U.S. v. Stump Home Specialties Mfg., Inc.,* 905 F.2d 1117, 1121 (7th Cir. 1990); *Werner v. Ashcraft Bloomquist, Inc.,* 10 S.W. 3d 575 (Mo. App. 2000) (general contractor could not delete work from a subcontractor's scope when the subcontract did not contemplate changes).
[2]*Ferris v. United States,* 28 Ct. Cl. 332 (1893).

use of the entire project and must be timely addressed. Problems could be caused by adherence to the common law rule due to the technological developments, the ever-changing needs of the government, or unanticipated site conditions. The widespread use of contract provisions similar in purpose to the modern Changes clause grew out of the practical need for a contract tool to address changes required by the government after contract formation. Examples of this type of clause have been used in government contracts for over 100 years and particularly in times of war.[3]

The Changes clause in a twenty-first-century government construction contract allows one party (the government) to implement changes in the work while performance is ongoing. A Changes clause may also come into play when issues arise involving defective specifications, differing site conditions, impossibility of performance, acceleration, or inspection, acceptance, and warranties. In short, the Changes clause is often an umbrella provision, involving numerous aspects of performance under the contract. No other clause more clearly illustrates the uniqueness and complexity of a government construction contract.

Changes clauses are found in construction contracts (government or private) more often than in other types of contracts. However, it is not correct to conclude that the clause and its use is inconsistent with principles of the common law of contracts. Through a Changes clause, the contracting parties agree in advance that one of the parties has the right to revise the work under defined terms. The requirement for legal consideration to support contract promises is satisfied by the provision requiring that the contract sum and/or time be adjusted if the change requires extra work or an extended period of performance. The Changes clause used in government contracts provides the framework for consideration of the issues and problems commonly encountered on construction projects procured by federal agencies. It and the Disputes clause **(see Chapter 15)** may well be the two most important provisions in twenty-first-century federal construction contracts.

II. FAR CHANGES CLAUSE

A. Basic Clause

The current version of the standard federal construction contract Changes clause is found at Federal Acquisition Regulation (FAR) § 52.243-4 and provides:

(a) The Contracting Officer may, at any time, without notice to the sureties, if any, by written order, designated or indicated to be a change order, make changes in the work within the general scope of the contract, including changes—

    (1) In the specifications (including drawings and designs);

---

[3]*McCord v. United States,* 9 Ct. Cl. 155 (1873); *aff'd sub. nom. Choteau v. United States,* 95 U.S. 61 (1877) (Civil War contract).

    (2) In the method or manner of performance of the work;

    (3) In the Government-furnished property or services; or

    (4) Directing acceleration in the performance of the work.

(b) Any other written or oral order (which as used in this paragraph (b), includes direction, instruction, interpretation, or determination) from the Contracting Officer that causes a change shall be treated as a change order under this clause; *provided,* that the Contractor gives the Contracting Officer written notice stating (1) the date, circumstances, and source of the order and (2) that the Contractor regards the order as a change order.

(c) Except as provided in this clause, no order, statement, or conduct of the Contracting Officer shall be treated as a change under this clause or entitle the Contractor to an equitable adjustment.

(d) If any change under this clause causes an increase or decrease in the Contractor's cost of, or the time required for, the performance of any part of the work under this contract, whether or not changed by any such order, the Contracting Officer shall make an equitable adjustment and modify the contract in writing. However, except for an adjustment based on defective specifications, no adjustment for any change under paragraph (b) of this clause shall be made for any costs incurred more than 20 days before the Contractor gives written notice as required. In the case of defective specifications for which the Government is responsible, the equitable adjustment shall include any increased cost reasonably incurred by the Contractor in attempting to comply with the defective specifications.

(e) The Contractor must assert its right to an adjustment under this clause within 30 days after (1) receipt of a written change order under paragraph (a) of this clause or (2) the furnishing of a written notice under paragraph (b) of this clause, by submitting to the Contracting Officer a written statement describing the general nature and amount of proposal, unless this period is extended by the Government. The statement of proposal for adjustment may be included in the notice under paragraph (b) above.

(f) No proposal by the Contractor for an equitable adjustment shall be allowed if asserted after final payment under this contract.

FAR § 43.205(d) mandates that all federal government agencies include this clause in their construction contracts if a fixed-price contract is contemplated and the contract amount is expected to exceed the simplified acquisition threshold.[4] The only variation in the text of the clause that is authorized in the FAR is the 30-day period found in paragraph (e).

---

[4]Generally $100,000 for contracts performed in the United States. *See* FAR § 2.101 Definitions. If the Changes clause were omitted from a government construction contract exceeding the applicable threshold, it is very likely that a court or board would read the clause into the contract under the *Christian* doctrine. *See G.L. Christian & Assocs. v. United States,* 160 Ct. Cl. 1 (1963).

B. Permissible Changes

The Changes clause specifies certain limits on the types of changes that the contracting officer is authorized to make. As a basic principle, the change must be "within the general scope" of the contract. Examples of such changes include:

- Specifications, drawings, and the design of the project
- Method or the manner of performance of the work
- Materials, equipment, facilities, and services furnished by the government
- Directed acceleration of the schedule

The combination of the phrase "within the general scope" and the listing of permissible types of changes under the word "include" implies that the government enjoys a fairly broad degree of latitude in directing a construction contractor to make changes in the work.[5] However, under certain circumstances, a contractor may question whether a particular directive or change order falls within the "general scope of the contract."[6] Alternatively, another contractor (potential competitor) may seek to challenge the change because it feels that the added or new work should be the subject of a new competition.[7] Depending on the facts of a particular project and the proposed contract modification, the challenged modification may be deemed an impermissible change.

C. Possible Impermissible Changes

Even though the "within the general scope" language of the Changes clause is relatively broad, there still exist questions regarding the types of changes that the contracting officer may direct. Traditionally, these *questionable* changes involve quantities, the schedule, and the contract terms and conditions. Each of these is discussed in the next sections.

---

[5]*Freund v. United States,* 260 U.S. 60 (1922). For example, the Changes clause used in government supply contracts omits the word "include." That omission arguably limits the government to making changes to only those categories specifically listed in that clause. *See* FAR § 52.243-1. The timing of the issuance of the change is generally not a factor in determining whether the change is within the scope of the contract. *J.D. Hedin Constr. Co. v. United States,* 347 F.2d 235 (Ct. Cl. 1965) (changes issued after contract completion held to be within the scope of the contract).

[6]Contractor challenges to the validity of a change order or series of changes are generally premised on the concept of a "cardinal change" meaning a change beyond the scope of the contract or a breach of contract. *See* **Section VI of this chapter**.

[7]The Competition in Contracting Act (CICA), 41 U.S.C. § 253(a)(1)(A), requires agencies to obtain "full and open competition" in contracting. Although this law does not preclude the use of change orders to add/revise work, *AT&T Commc'ns, Inc. v. Wiltel, Inc.,* 1 F.3d 1201, 1205 (Fed. Cir. 1993), it does permit a third party (potential competitor) to assert that the modifications changed the contract sufficiently that the government circumvented CICA. For example, in 2004, a federal government agency issued modifications shortly after award that added and deleted work to a service contract and resulted in a net increase to the contract price of 80 percent. A competitor challenged this action, and these modifications were found to be beyond the scope of the contract. *See Cardinal Maint. Serv., Inc. v. United States,* 63 Fed. Cl. 98, 106 (2004).

1. Quantity Changes

a. General Rule As reflected by the *Cardinal Maintenance* decision,[8] the government does not have unfettered discretion to make large additive changes to the quantity of work in the contract on even unit-priced construction contracts.[9] Similarly, the deletion of a building in a multifacility hospital construction contract has been held to be beyond the scope of the Changes clause.[10] However, although changes in quantities resulting from modifications within the scope of the Changes clause are permissible,[11] the government has an obligation to use good faith and reasonable care in computing the estimated quantity.[12]

b. Variation in Estimated Quantity Clause Given the potential question regarding the government's authority to directly change quantities of work items in a construction contract, contracts with significant quantities of excavation, grading, paving, or other work items that may vary significantly during construction often utilize unit-price pay items.[13] The use of a unit-priced pay item creates a potential of a substantial inequity for one of the contracting parties under certain circumstances. If the cost of the unit-price work item is based on an estimated quantity,[14] a substantial overrun or underrun could produce a windfall or severe loss for the contractor.

A substantial quantity variation may not necessarily qualify for relief under the Differing Site Conditions clause.[15] One contractual technique to address the possibility of a substantial overrun or underrun in a unit-priced item is the Variation in Estimated Quantity clause.[16] That clause provides:

[8]63 Fed. Cl. 98 (2004).

[9]*P.L. Saddler Constr. Co. v. United States,* 287 F.2d 411 (Ct. Cl. 1961); *Manis Drilling,* IBCA No. 2658, 93-3 BCA ¶ 25,931. The use of unit prices together with the Variation in Estimated Quantity clause, FAR § 52.211-18, in some contracts provides the government more flexibility in dealing with quantity variations.

[10]*General Contracting Co. v. United States,* 84 Ct. Cl. 570 (1937); *but see P.J. Dick, Inc. v. General Services Administration,* GSBCA No. 12215, 95-1 BCA ¶ 27,574 *modified on recons.* 96-1 BCA ¶ 26,017 (parties treated the deletion of the renovation work involving several floors in an older federal building as a deductive change). The GSBCA accepted the parties' joint treatment of the action as a change. The use of the standard Termination for Convenience clause, FAR § 52.249-2 (Alternate I), eliminates most issues of the government's authority to delete substantial quantities of work. However, the different approaches to pricing work deletions under the Changes clause and the Termination for Convenience clause can be important to the contractor. *See* **Chapter 11.**

[11]*See Morrison-Knudsen Co. v. United States,* 397 F.2d 826 (Ct. Cl. 1968).

[12]*See Engineered Demolition, Inc. v. United States,* 70 Fed. Cl. 580 (2006) (government confirmed erroneous quantity following a prebid inquiry, which identified a conflict between the bid schedule and drawings).

[13]FAR § 36.207 reflects a preference for lump-sum pricing unless quantities cannot be estimated with sufficient confidence without a substantial contingency or offerors (bidders) would have to expend an unusual effort to develop an adequate estimate for lump-sum price.

[14]Fixed costs often are spread over an estimated quantity as part of the process of developing a unit price.

[15]FAR § 52.236-2 (Apr. 1984). *Compare Perini Corp. v. United States,* 381 F.2d 403 (Ct. Cl. 1967) *and Gregg, Gibson & Gregg, Inc.,* ENGBCA No. 3041, 71-1 BCA ¶ 8677. *See* **Chapter 7** for a discussion of quantity variations as differing site conditions.

[16]FAR § 52.211-18.

If the quantity of a unit-priced item in this contract is an estimated quantity and the actual quantity of the unit-priced item varies more than 15 percent above or below the estimated quantity, an equitable adjustment in the contract price shall be made upon demand of either party. The equitable adjustment shall be based upon any increase or decrease in costs due solely to the variation above 115 percent or below 85 percent of the estimated quantity. If the quantity variation is such as to cause an increase in the time necessary for completion, the Contractor may request, in writing, an extension of time, to be received by the Contracting Officer within 10 days from the beginning of the delay, or within such further period as may be granted by the Contracting Officer before the date of final settlement of the contract. Upon the receipt of a written request for an extension, the Contracting Officer shall ascertain the facts and make an adjustment for extending the completion date as, in the judgment of the Contracting Officer, is justified.

In addition to the questions related to the relationship of this clause to the effect of a differing site condition, there are also questions as to its application when a change in requirements results in a variation from the estimated quantity. Generally, the Changes clause *trumps* the Variation in Estimated Quantity clause when the quantity variation results from a modified design requirement and the equitable adjustment is computed under the Changes clause.[17]

If the Variation in Estimated Quantity clause applies, two important elements of the clause warrant attention:

(1) *Quantity triggers.* The current standard clause operates only if the actual quantity varies by more than 15 percent from the estimated quantities. Contractors should check the clause in their contracts to determine if that percentage is varied by the agency.

(2) *Limitations on price adjustment.* The current standard clause provides that the equitable adjustment "shall be based upon any increase or decrease in costs *due solely* to the variation over 115 percent or under 85 percent of the estimated quantity." This language can present significant proof problems for the party seeking the price adjustment.[18]

2. Changes to Terms and Conditions

Generally, the boards have held that the Changes clause does not authorize the government unilaterally to modify a standard contract clause.[19] However, if the provision is found in a section of the contract documents that is listed as part of the "specifications," the contracting officer may have the right to change that provision.[20]

---

[17]*Morrison-Knudsen v. United States,* 397 F.2d 826 (Ct. Cl. 1968); *C.H. Leavell & Co.,* ENGBCA No. 3492, 75-2 BCA ¶ 11,596.

[18]*Victory Constr. Co. v. United States,* 510 F.2d 1379 (Ct. Cl. 1957); *Foley Co. v. United States,* 26 Cl. Ct. 936 (1992) *aff'd,* 11 F.3d 1032 (Fed. Cir. 1993).

[19]*B.F. Carvin Constr. Co.,* VABCA No. 3224, 92-1 BCA ¶ 24,481 (change in payment terms not permitted).

[20]*Melrose Waterproofing Co.,* ASBCA No. 9058, 1964 BCA ¶ 4119 (revision of a provision in special conditions authorized as it was included in the contract as part of the specifications).

3. Schedule Delays or Suspensions

Traditionally, schedule delays have been held to be beyond the scope of the Changes clause.[21] However, as acceleration of the schedule is specifically listed as a permissible change, that type of schedule change is authorized. Changes that alter the sequence of work on a construction project can accelerate one portion of the project while delaying another. Often the boards treat these changes as falling within the "method or manner of performance" language of the Changes clause.[22] In some, but not all, cases, it may be advantageous for the contractor to treat the government's action as a change under the Changes clause rather than a delay under the Suspension of Work clause because the former clause allows for the recovery of profit as a part of the equitable adjustment while the Suspension of Work clause expressly precludes the addition of profit.[23]

D. Determining if a Change Is Permissible

The *Cardinal Maintenance* decision[24] confirms that there is no simply stated bright-line test to determine if a proposed change to the contract falls within the scope of the work. Changes to quantities, schedule, or terms and conditions may or may not be authorized under the Changes clause, depending on a variety of factors, such as the magnitude of the cost, quantity, and the like. If the parties jointly treat the action as a valid exercise of the authority granted to the government under the Changes clause, a board or court may well accept that treatment if no third party imposes a legitimate objection.[25] If a contractor wishes to challenge the government's right to make a particular change and combine that challenge with a refusal to proceed with the work, it accepts a clear risk that it may be terminated for default for that refusal to perform the work. The combination of the lack of a clear standard for changes "beyond the scope of the contract" and the risks attendant with a refusal to proceed with the work is a clear deterrent to contractor challenges to the validity of a particular change.

E. Subcontracting Considerations and Checklist

Most subcontracts contain versions of a Changes clause that typically provide that the general contractor may make changes to the work, provide for adjustments in

---

[21]Schedule delays are generally covered by the Suspension of Work clause, FAR § 52.242-14. *See* **Chapter 9.**

[22]*Commercial Contractors, Inc.*, ASBCA No. 30675, 88–3 BCA 20,877; *Pan Arctic Corp.*, ASBCA No. 20133, 77-1 BCA ¶ 12,514.

[23]This traditional analysis can be altered if the contract contains a clause that limits the recovery of job site general conditions as part of the equitable adjustment under a Changes clause. *See* **Section VIII.D.2** of this chapter and **Chapter 9, Section IX.**

[24]63 Fed. Cl. 98 (2004).

[25]*P.J. Dick, Inc. v. General Services Administration*, GSBCA No. 12215, 95-1 BCA ¶ 27,574.

contract price and time, and address the subcontractor's obligation to proceed with the revised work. In addition, many general contractors utilize "flow-down" clauses with the intent of incorporating (flowing down) the terms of the government contract into the subcontract.

In that context, a general (prime) contractor can create an unintended conflict between the terms of the Changes clause and those found in the subcontract Changes clause. While it may not be practical to cross check every provision in a subcontract to the applicable clause in the government contract, cross checking the Changes clause and the subcontract Changes clause warrants consideration. Using a checklist similar to the next one can facilitate the effort of comparing clauses.

CHECKLIST: COMPARISON OF CHANGES CLAUSES

Scope:
- Is the scope of the permissible changes in the subcontract and prime contract identical? If not, identify the differences and determine if the subcontract provision needs revision.

Authority:
- Who is authorized to make changes to the subcontract—the government, the prime contractor, or both?

Timing:
- Are the time frames for submission of change proposals identical or complimentary? The prime contractor needs to receive its subcontractors' proposals sufficiently in advance of the government's due date to allow time for review and timely submission.

Constructive Changes:
- Do both Changes clauses contemplate constructive changes to the extent directed by the government, the contractor, or both?

Duty to Proceed:
- Are the two contracts consistent regarding the subcontractor's duty to proceed in the event of a dispute regarding a change?

Compensation for Changes:
- Does the subcontract limit the prime contractor's liability for a government ordered change to the equitable adjustment allowed by or paid by the government for the subcontractor's work?

Careful evaluation of these topics at the inception of a project may facilitate contract administration and help both prime contractors and subcontractors avoid or minimize later disputes.

III. AUTHORITY TO ISSUE CHANGES

A. Contracting Officer

The Changes clause for fixed-price federal construction contracts provides that changes are to be issued by the contracting officer. The precise definition of the term "contracting officer" may vary with the type of contract and the procuring activity.

The identity and scope of authority of the contracting officer may be unclear where federal agency designations are imprecise. For example, while the Department of Defense (DOD) has specific certificates of authority denoting the contracting officer, the FAR allows designation by position as well as by name.[26] In addition, many agencies require the use of Standard Form 1402, Certificate of Appointment, specified in FAR § 1.603-3.[27] As a basic rule, the Certificate of Appointment is required to be in writing, and it must set forth any limitations on authority, other than those contained in regulation or law. In accordance with FAR § 1.602-1, this government is obligated to make this information readily available to the public.

The basic principles and rules related to the issue of authority and government contracting are reviewed in detail in **Section I** of **Chapter 2.** Achieving an understanding of those critical principles is an essential step in the orientation of any contractor's project management team. Every contractor must appreciate the fundamental importance of the concept of *actual authority*[28] in government contracting and develop effective procedures to address authority questions that may develop in the administration of any contract. The next discussion provides suggested steps that can be adapted and used by any contractor as part of its premobilization orientation activities.

CHANGE ORDER AUTHORITY PROTOCOL

- Obtain and review all available documentation defining the limits of the *actual authority* of the *contracting officer* administering the contract.
- Obtain and review all available documentation related to the limits of the *actual authority* of other *government representatives* participating in the administration of the project. *See* **Appendix 2A to Chapter 2.**
- If the information is not freely offered by the agency, make a written request for the documentation. If there are changes to the government's representatives during performance of the work, this process must be repeated.[29]

---

[26] FAR § 1.601(a).

[27] *See* **Chapter** 2. For example, the Department of Veterans Affairs and the Department of State specify the use of Standard Form 1402, Certificate of Appointment. *See* 48 C.F.R. §§ 801.690-8; 48 C.F.R. § 601.603-3.

[28] *See* FAR § 1.602-1(a); *Federal Crop Ins. Corp. v. United States,* 332 U.S. 380 (1947).

[29] *See, e.g.,* 48 C.F.R. § 801.603-70 (Department of Veterans Affairs). Some authorized representatives of the contracting officer may have monetary limits on the dollar value of a single change order that can be issued by that individual as well as the total value of change orders in a given period of time: for example, a $100,000 limit on a single change order with a $100,000 cap on the total value of change orders per month. These limitations can create a potential for problems if a project, such as a renovation project, is affected by numerous design changes over a relatively short period of time.

- Carefully review the contract clauses to identify any provision which states that a *government representative* or *contracting officer's representative* does not have authority to order a change to the contract.[30]
- Establish a procedure to obtain written clarification from the government if there is any question regarding the authority of a government representative to order changes to the contract.
- Establish a procedure to provide appropriate written notification to the contracting officer if another representative of the government directs performance of a change to the work. *See* **Appendix 2B** to **Chapter 2.**

In addition, the contracting officer's authority to order a change is not absolute. It must be exercised within the context of the provisions of the Changes clause, any other applicable contract provision, as well as within any financial or funding limitations.[31]

A contracting officer's authority is also limited to administration of those contracts that have been assigned to him or her. Although the contracting officer can make appropriate changes in the specifications of any contract under that person's authority, there is no authority to bind the government to those changes in connection with future procurements.[32]

Finally, the contracting officer has no authority to order or withdraw changes after final payment to the contractor. Final payment, without a reservation or exception, is a positive bar to any further modifications, and the contracting officer may not waive this requirement.[33]

B. Authorized Representatives

Although the standard FAR clauses, such as the Changes clause, clearly designate the contracting officer as the key person in the administration of a government contract, it would be incorrect to anticipate that the contracting officer will have extensive day-to-day involvement in the administration of a construction project. The opposite is often the case as the contracting officer may have responsibility for scores of contracts and projects at any one time. Moreover, a particular contracting officer may have limited design or engineering expertise and will necessarily rely on others in the agency to provide that knowledge and experience.

To provide day-to-day administration of the project and contract, federal agencies typically will designate a *contracting officer's representative.* The actual title

---

[30]FAR § 52.246-12, Inspection of Construction; DFARS § 252.201-7000(b) (DOD contracts). (These clauses may be incorporated by reference.)

[31]*See* FAR § 1.602-2(a); FAR § 43.105(a) states that a contracting officer "shall not execute" a contract modification that causes an increase in funds without having first obtained a certification that funds are available.

[32]*Magna Indus., Electromotive Div.,* ASBCA No. 22381, 80-2 BCA ¶ 14,633.

[33]*Missouri Research Labs., Inc.,* ASBCA No. 12355, 69-1 BCA ¶ 7762. *See also Cecil Carr Constr. Co., Inc.,* DOTCAB No. 1859, 90-1 BCA ¶ 23,317; *Forel Films West,* ASBCA No. 23071, 79-2 BCA ¶ 13,913.

will vary from agency to agency.[34] That person *may* or *may not* have authority to issue change orders or otherwise bind the government as the actual authority may be far different from that implied by the person's title. Although these individuals have substantial authority to administer a construction contract, most federal agencies limit the authority of its representatives such as inspectors to change the contract in any manner.[35] In addition DFARS § 201.602-70 provides that all DOD contracts must contain the clause at Sec. 252.201-7000 whenever the agency anticipates the designation of a contracting officer's representative. That clause provides:

CONTRACTING OFFICER'S REPRESENTATIVE (DEC 1991)

(a) Definition. Contracting officer's representative means an individual designated in accordance with subsection 201.602-2 of the Defense Federal Acquisition Regulation Supplement and authorized in writing by the contracting officer to perform specific technical or administrative functions.

(b) If the Contracting Officer designates a contracting officer's representative (COR), the Contractor will receive a copy of the written designation. It will specify the extent of the COR's authority to act on behalf of the contracting officer. *The COR is not authorized to make any commitments or changes that will affect price, quality, quantity, delivery, or any other term or condition of the contract.* [Emphasis added]

Change order proposals or contractor claims for additional compensation or time may result from the directions issued by government representatives other than the contracting officer. These may also result from the day-to-day administration of the contract by the contracting officer's representative or as a result of a response provided to a contractor's request for clarification or request for information.

Most federal agencies' regulations require that the appointment of the contracting officer's representative be in writing, specify the scope and limits of that person's authority, and that a copy be made available to the contractor.[36] Part of the contractor's premobilization process should include obtaining copies of these written statements of authority, clarifying any questions regarding their meaning,[37] and review them with all of the contractor's project management staff. If the government changes personnel during the course of the project, this process should be repeated.

---

[34]For example, the Department of Veterans Affairs may designate a *Resident Engineer* (RE) or *Senior Resident Engineer* (SRE); the U.S. Army Corps of Engineers may designate *Area Engineers* or *Resident Engineers*; the Naval Facilities Engineering Command may designate *Resident Officer in Charge of Construction* (ROICC) or one or more *Assistant Resident Officer(s) in Charge of Construction* (AROICC).

[35]*See* FAR § 52.246-12, Inspection of Construction.

[36]*See, e.g.,* 48 C.F.R. § 801.603-70 (Department of Veterans Affairs).

[37]*See Winter v. Cath-DR/Balti,* 497 F.3d 1339 (Fed. Cir. 2007). ("Prepares and/or coordinates . . . modifications" held not to mean authority to issue change orders. DFARS § 252.201-7000 strictly enforced.)

The contractor should not assume that the limits of authority are identical even if the titles are identical.[38]

Finally, a contractor should anticipate the possibility that a contracting officer's representative or a government inspector will direct the performance of work that the contractor believes to be a change to the contract. This situation can present a significant payment risk for the contractor if the government's employee or representative does *not* have the authority to issue a change order.[39] The contractor's on-site management needs to be sensitive to this risk and have a protocol for responding to such directives. See **Section** I of **Chapter 2** for additional information on this topic and the management of this risk.

C. Exceptions to the Strict Authority Rule

These limits on authority and the obligation placed on the contractor to ascertain the actual authority of a representative of the government have the potential for harsh results. However, the Court of Federal Claims (and its predecessor courts) and the boards of contract appeals have developed legal theories that *may* form a potential basis for relief to the contractor despite the fact that the party ordering the "change" had no express authority. These theories include ratification by the contracting officer and implied authority of the ordering party.[40]

1. Ratification and Imputed Knowledge

Express or implied ratification by the contracting officer is sometimes the basis of decisions allowing equitable adjustments on account of changes originally ordered by unauthorized representatives.

In one case, the contracting officer was fully aware of a disagreement between the contractor and the government's inspection personnel regarding what were later found to be excessive and unreasonable inspection standards. The failure of the contracting officer to independently investigate the situation or take corrective action was held to amount to a ratification of the improper inspection practices, entitling the contractor to an equitable adjustment.[41]

Ratification may occur when the government accepts the benefits of the work as changed. When government technical supervisors requested a contractor to correct an engineering deficiency in a government specification and performance proceeded without written approval of the contracting officer, the subsequent acceptance of the work by the government entitled the contractor to an equitable adjustment.[42]

---

[38]Some authorized representatives of the contracting officer may have monetary limits on the dollar value of a single change that may be issued by that person as well as the total value of change orders that may be issued in a given time period. For example, a $100,000 limit on a single change order may be combined with a $100,000 cap on the total value of change orders per month. This can create the potential for problems if a project, such as a renovation project, is affected by numerous changes over a relatively short period of time.

[39]*See Winter v. Cath-DR/Balti,* 497 F.3d 1339 (Fed. Cir. 2007).

[40]See **Chapter 2** for a further discussion of these concepts. *See also Harbert/Lummus Agrifuels Projects v. United States*, 142 F.3d 1429 (Fed. Cir.1998).

[41]*G.W. Galloway Co.,* ASBCA No. 16656, 73–2 BCA ¶ 10,270.

[42]*The Hallicrafters Co.,* ASBCA No. 7097, 68–1 BCA ¶ 6950.

However, a contractor was denied compensation for the installation of wood paneling and glass partitions, as ordered by unauthorized government employees (in lieu of the metal partitions required by the contract), because the only government representative with authority to order the change continued to refuse to pay for the cost of the changed work. Thereafter, the building manager's decision to allow completion of the work did not amount to a ratification.[43]

A related line of cases provides relief under the Changes clause where the contracting officer should have known of the on-site representative's actions. In one situation, a contracting officer's reliance on an inspector to perform daily supervision of a contractor's work and the constant communication between the contracting officer and the inspector led the ASBCA to impute the inspector's knowledge to the contracting officer.[44]

In *U.S. Federal Engineering & Manufacturing, Inc.*,[45] the board acknowledged a contractor's entitlement to an equitable adjustment for certain changes to a sling-test facility despite the fact that the contractor acted with the knowledge and approval of an individual who was known to lack authority to order changes to the contract. The board first stated that the additional reinforcing steel, which had been ordered by the unauthorized government employee, was necessary to correct defects in the contract drawings and specifications and should have been the subject of a change order. Noting that the contracting officer's project manager and subordinates were "his eyes and ears (if not his voice) and their knowledge [of the contractor's corrective work] is treated as for all intents and purposes as his," the board concluded that the contracting officer had constructive knowledge of the additions and could have chosen a more suitable corrective method if one were desired by the government. However, in the same appeal, the board declined to reach a similar result regarding additions made for purposes other than to correct drawing defects, since the contracting officer was under no corresponding duty to issue a change order for that work.

2. Implied Authority

Problems frequently arise in cases where there has been a long acquiescence in a course of conduct in which changes have been ordered by other than formally authorized representatives. Such a course of conduct may result in the implication of authority to an individual as a representative of the contracting officer.

For example, a 1955 decision by the Court of Claims found that a government resident engineer on the site was an authorized representative of the contracting officer.[46] Similarly, it has been held that when an official of a contracting agency has been sent by the contracting officer for the express purpose of giving guidance in connection with a contract, the contractor is justified in relying on that person's representations.[47]

---

[43]*Paul A. Demusz, Inc.*, GSBCA No. 5148, 80-1 BCA ¶ 14,378.

[44]*Southwestern Sheet Metal Work, Inc.*, ASBCA No. 22748, 79-1 BCA ¶ 13,744.

[45]ASBCA No. 19909, 75-2 BCA ¶ 11,578. *See also B.V. Constr., Inc.*, ASBCA No. 47766, 04-1 BCA ¶ 32,604; *Walter Straga*, ASBCA No. 26134, 83-2 BCA ¶ 16,611.

[46]*General Cas. Co. v. United States*, 127 F. Supp. 805 (Ct. Cl. 1955).

[47]*Fox Valley Eng'r, Inc. v. United States*, 151 Ct. Cl. 228 (1960). *See also American Elecs. Labs. v. United States*, 774 F.2d 1110, 1116 (Fed. Cir. 1985).

In *UrbanPathfinders, Inc.*,[48] a moving contractor's equitable adjustment request was sustained on the basis of implied authority in the project manager to order additional work. Although the board found that the contracting officer did not expressly delegate such authority, authority was reasonably implied on the basis of the project manager's presence on the site, wide responsibilities in administration of the contract, and the need for "expeditious action" to avoid frustration of contract objectives. In certain circumstances, it has also been held that a resident engineer, who was an employee of the government's architect/engineer contractor, had authority to approve changes in the specifications.[49] However, none of these favorable implied authority decisions appears to have involved contracts containing a clause such as DFARS § 252.201-7000 expressly stating that the contracting officer's representative could not be delegated the authority to order changes to the contract.

In any event, a contractor must be cautious in performing under orders of individuals other than the contracting officer. While the boards and the Court of Federal Claims often attempt to achieve a fair result where constructive changes have occurred at the direction of unauthorized employees, negative decisions frequently are predicated on the failure of the contractor to make a timely protest of the order.[50] In order to avoid problems concerning lack of authority, the contractor should notify the contracting officer in writing of the requested "extra" work and that the contractor will proceed with the extra work unless otherwise directed. A factual and businesslike notice letter, faxed memo, or e-mail can save all parties expense and disputes at a later time.[51]

IV. EXPRESS CHANGES: BILATERAL AND UNILATERAL

A. Significance of Standard Form 30

Changes to government construction contracts are categorized as either bilateral or unilateral.[52] FAR § 43.103 provides these descriptions regarding each type of change.

---

[48] ASBCA No. 23134, 79-1 BCA ¶ 13,709.

[49] *T.F. Scholes of Ark., Inc.*, ASBCA No. 5629, 60-1 BCA ¶ 2534. *But see Charles G. Williams Constr., Inc.*, ASBCA No. 24967, 81-1 BCA ¶ 14,893.

[50] *WRB Corp. v. United States*, 183 Ct. Cl. 409 (1968).

[51] *See* **Chapter 14, Project Documentation Techniques**.

[52] If the construction contract is issued as a "commercial" item procurement under FAR Part 12, there is no requirement in the FAR mandating that either the standard Changes clause, FAR § 52.243-4, or the Differing Site Conditions clause, FAR § 52.236-2, be included in the resulting contract. Rather, FAR § 52.212-4, Contract Terms and Conditions—Commercial Items (Sept. 2005), states that changes to the contract may be made only by written agreement of the parties. This would seem to preclude the use of unilateral change orders. Notwithstanding a requirement for changes pursuant to mutual agreement, boards have considered claims by a contractor that actions by the government unilaterally changed a contract for a "commercial" item. *See SAWADI Corp.*, ASBCA No. 53073, 01-1 BCA ¶ 31,357 (ID/IQ Construction and Services contract). Construction contracts, which are treated as so-called commercial items under FAR Part 12, warrant careful preproposal review by contractors, given the absence of important standard FAR clauses as noted.

TYPES OF CONTRACT MODIFICATIONS

Contract modifications are of the following types:

(a) *Bilateral.* A bilateral modification (supplemental agreement) is a contract modification that is signed by the contractor and the contracting officer. Bilateral modifications are used to—

(1) Make negotiated equitable adjustments resulting from the issuance of a change order;

(2) Definitize letter contracts; and

(3) Reflect other agreements of the parties modifying the terms of contracts.

(b) *Unilateral.* A unilateral modification is a contract modification that is signed only by the contracting officer. Unilateral modifications are used, for example, to—

(1) Make administrative changes;

(2) Issue change orders;

(3) Make changes authorized by clauses other than a Changes clause (e.g., Property clause, Options clause, or Suspension of Work clause); and

(4) Issue termination notices.

FAR § 43.301 requires that Standard Form (SF) 30, Amendment of Solicitation/Modification of Contract, be used to issue change orders under the Changes clause.[53] Section 13E of SF 30 reads:

E. IMPORTANT: Contractor ❑ is ❑ is not required to sign this document and return_____ copies to the issuing office.

This section provides a clear indication if the government considers the change to be unilateral (contractor is not required to sign) or bilateral (contractor is required to sign). In many cases, a letter of transmittal may accompany the SF 30 and identify the change as unilateral or bilateral. However, the critical provision for the contractor to review is Section 13E of SF 30.

---

[53]SF 30 is authorized for use for a variety of purposes, and use of that form may not establish that a modification has been made. *See Southwest Marine, Inc.,* DOTCAB No. 1497, 93-3 BCA ¶ 26,170. While the absence of a SF 30 may not enable a party to avoid an otherwise valid written agreement, *Robinson Contracting Co. v. United States,* 16 Cl. Ct. 676, 688 (1989), the absence of an SF 30 can be important to the determination of whether the government is bound. *See Solar Turbines v. United States,* 23 Cl. Ct. 142 (1991) (written modification signed by a government employee who was not a contracting officer). However, at least one board decision has refused to justify a contractor's delay to the work when it would not proceed with a change because the contracting officer did not issue it on an SF 30. *P & M Indus., Inc.,* ASBCA No. 36625, 90-2 BCA ¶ 22,839.

B. Direction to the Contractor

1. Affirmative Direction by the Government

Affirmative direction in the form of an order that the government has the contractual right to issue is a condition precedent to recovery for a change.[54] For example, in *Schriock Construction, Inc.*,[55] the board denied compensation to a contractor that used more cover aggregate than the coverage specified in the contract, since no direction had been received from the government to increase the quantity.

The decisions also differentiate between an order by the government and the granting of a request by the contractor for a deviation from the contract requirements. In *Ventilation Cleaning Engineers, Inc.*,[56] a contractor was denied recovery under a contract for the supply of furnaces equipped with factory-installed components when it had difficulty in procuring furnaces meeting the specifications and asked for permission to install the components on site. The board found significance in the fact that the contractor's request was granted without any commitment to pay the contractor for the cost of this change in the method of performance.

2. Written Order Generally Required

The standard Changes clause requires that change orders be *in writing*. In a landmark case, *Plumley v. United States*,[57] the U.S. Supreme Court reversed a decision of the Court of Claims and held that the failure to reduce a change order to writing in accordance with the contract barred recovery by the contractor for work even though the work was "extra" and of value, and had been ordered by a representative of the government.

The ruling and result in *Plumley* represent a harsh and literal application of the requirement for a written change order. The current Changes clause still requires that the change be in writing. Even though the *Plumley* decision has never been reversed,[58] subsequent decisions by the Court of Claims recognized theories of recovery to mitigate its result. For example, in *Armstrong & Co. v. United States*,[59] the court held that a contractor's performance of work without a written change order was compensable upon the basis that the government's receipt of benefits from the performance raised an implied duty to pay. Similar cases have established a contractor's entitlement to compensation as a means of preventing unjust enrichment to the government.[60]

---

[54]*Century Eng'g Corp.*, ASBCA No. 2932, 57-2 BCA ¶ 1419.

[55]IBCA No. 961-3-72, 73-1 BCA ¶ 9850.

[56]ASBCA No. 18505, 74-1 BCA ¶ 10,417; *see also George C. Punton, Inc.*, ASBCA No. 9767, 65-2 BCA ¶ 5007. *But see Monaco Builders, Inc.*, PSBCA No. 323, 78-1 BCA ¶ 12,924.

[57]226 U.S. 545 (1913).

[58]*See General Bronze Corp. v. United States*, 338 F.2d 117, 123 (Ct. Cl. 1964); *Engineered Demolition, Inc. v. United States*, 70 Fed. Cl. 580, 590 (2006) (court stated that a written "memorandum that simply memorializes an oral discussion . . . falls well short of constituting a written order under the Changes Clause." *Id.* at 590).

[59]98 Ct. Cl. 519 (1943).

[60]*See, e.g., Williams v. United States*, 127 F. Supp. 617 (Ct. Cl. 1955).

The boards and the Court of Federal Claims and its predecessor courts also have specifically recognized the applicability of the "constructive change" doctrine as an exception to a requirement for a written change order.[61] This doctrine affords relief to a contractor that has performed changed work not pursuant to an express written change order but on the basis of informal representations or directions by the contracting officer or its representatives.[62] However, a contractor's ability to invoke this doctrine depends on the specific wording of the Changes clause in the construction contract.[63] In effect, the concept of a constructive change appears to be wed to the language of the specific Changes clause in the contract.[64] If a nonstandard Changes clause is included in the contract, any differences in the wording of that clause need to be evaluated carefully before submitting a proposal or bid.

Although the concept of a constructive change clearly contemplates oral orders as changes,[65] the court and board decisions reflect the problems associated with verbal directives that are not confirmed in writing. The lack of a written order may not prove fatal to a contractor's claim when there is no lack of clear evidence that a change to the contract's requirements was directed, but the absence of any writing may defeat the claim for lack of proof. Thus, a contractor that proceeds with a change "without express or implied approval of the Contracting Officer or his representative" may have no basis to recover under the Changes clause.[66]

If the contractor or the government seeks to establish that the other party has agreed to a contract modification, the lack of a written change order can be fatal to that assertion. For example, in *Conner Brothers Construction Co. v. United States*,[67] the court stated that negotiations between the parties regarding a contract modification did not trigger application of the Changes clause because the contracting officer never issued either a written or an oral order. Similarly, in *Mil-Spec Contractors v. United States*,[68] the fact that the government issued a bilateral modification following negotiations with the contractor did not make the change order binding on the contractor until it was signed by both parties. Reliance on a verbal agreement, particularly with representatives of the contracting officer, can be risky. The representative may be authorized to recommend acceptance of the agreement but not actually bind the government. Verbal understandings are also subject to vagaries of accurate recollection over time when there is no record of the change or agreement.[69]

---

[61] *See* **Section V** of this chapter for a discussion of constructive changes.

[62] *Randall H. Sharpe*, ASBCA No. 22800, 79-1 BCA ¶ 13,869; *Midwest Spray and Coating Co. v. United States*, 176 Ct. Cl. 1331 (1966).

[63] *See Len Co. v. United States*, 385 F.2d 438, 448 (Ct. Cl. 1967) (Changes clause used in Capehart Act housing project spoke of "proposed changes and did not give the contracting officer unilateral authority to alter the contract").

[64] *Id.* at 442.

[65] *Holt Hauling & Warehousing Sys., Inc.*, ASBCA No. 19136, 76-2 BCA ¶ 12,185.

[66] *Comspace Corp.*, GSBCA No. 3550, 72-2 BCA ¶ 9674.

[67] 65 Fed. Cl. 657 (2005). *See also Kato Corp.*, ASBCA No. 54162, 06-2 BCA 33,293 (oral argument to settle claims unenforceable); *Trawick Contractors, Inc.*, ASBCA No. 55097, 07-1 BCA ¶ 33,499 (oral settlement agreement on liquidated damages unenforceable).

[68] 835 F. 2d 865 (Fed. Cir. 1987); *See also TRW, Inc.*, ASBCA No. 51003, 00-2 BCA ¶ 30,992.

[69] *Sargent v. Department of Health & Human Services*, 229 F.3d 1088 (Fed. Cir. 2000); *Adams Constr. Co.*, VABCA No. 4669, 97-1 BCA ¶ 28,801 (oral agreement to extend the completion date of the contract enforced when board found sufficient written indications of the agreement).

Where possible, a contractor should obtain some confirmation from the contracting officer, in advance of doing additional work, that the contractor will be paid for it. However, a constructive change may occur when the contractor and the government disagree on the contract's requirements. Where the circumstances indicate that an oral change may have been issued by the government, the contractor should give written notice and make it clear that it intends to present a claim for an increase in price. Failure to give such notice may result in the contractor being considered to have done the work voluntarily.[70] Where a contractor has voluntarily performed work beyond the requirements of the contract documents, without any direction (written or oral), the contractor bears the risk that relief will be denied.[71]

V. CONSTRUCTIVE CHANGES

A. Origin of Concept

The concept of a "constructive change" has its roots in the disputes procedure applicable during the period from World War II into the late 1960s. During that period, the jurisdiction of the boards of contract appeals was technically limited to claims that could be addressed under a remedy-granting clause of the contract. If no clause applied to the contractor's claim, a breach of contract claim had to be brought in the Court of Claims.[72] This bifurcation of remedies could create a costly procedural morass for contractors and their counsel.

As representatives of the secretaries of their various departments, the boards overcame the lack of a written change order when it was proven that a change to the scope of work had been ordered and that a change order should have been issued. These changes in fact were treated as "constructive changes" by the boards, and relief was granted under the Changes clause.[73] The chief significance of the constructive change doctrine was its function as an alternative to an action for breach of contract. In the 1950s and 1960s, the Court of Claims and the boards frequently treated a government breach of contract as a constructive change entitling the contractor to an equitable adjustment under the contract.[74]

The definition of the term "constructive change order" is very broad, referring to any conduct by a contracting officer (or other government representative authorized to order changes) that is not a formal change order and has the effect of requiring the contractor to perform work different from that prescribed by the original terms

---

[70]*The Jordan Co.*, ASBCA No. 10874, 66-2 BCA ¶ 6030. *See also John H. Moon, Inc.*, IBCA No. 815-72-69, 72-2 BCA ¶ 9601.

[71]*Reeves Instrument Co.*, ASBCA No. 11534, 68-2 BCA ¶ 7078. *See also Smith*, DOTCAB No. 1381, 82-2 BCA ¶ 16,780.

[72]*United States v. Utah Constr. and Mining Co.*, 384 U.S. 394 (1966); *Phoenix Bridge Co. v. United States*, 85 Ct. Cl. 603 (1937); *Spencer B. Lane Co.*, WDBCA No. 449-451, 2 CCF ¶ 500 (1944).

[73]*Ryall Eng'g Co.*, IBCA-1, 6 CCF 61,639; *J.W. Hurst & Son Awnings, Inc.*, ASBCA No. 4167, 59-1 BCA ¶ 2095; *Lillard's*, ASBCA No. 6630, 61-1 BCA ¶ 3053. *See* Robert C. Gusman, "Constructive Change—A Theory Labeled Wrongly," 6 Pub. Cont. L.J. 229 (Jan. 1974).

[74]*See, e.g., Polan Indus., Inc.*, ASBCA No. 3996, *et al.*, 58-2 BCA ¶ 1982.

of the contract. A constructive change, like a formal change, may entitle the contractor to relief under the Changes clause. However, both the Court of Claims (and its successor courts) and the boards have pointed out that such relief depends on the performance of nonvoluntary additional work resulting from a direction by an authorized government officer.[75]

The concept of a constructive change can be rather broad, but it is likely that a board or the Court of Federal Claims will analyze a constructive changes claim using the two basic elements articulated by the board in *Industrial Research Associates, Inc.*:

> As we see it, the constructive change doctrine is made up of two elements—the "change" element and the "order" element. To find the change element we must examine the actual performance to see whether it went beyond the minimum standards demanded by the terms of the contract. But, this is not the end of the matter.
>
> The "order" element also is a necessary ingredient in the constructive change concept. To be compensable under the Changes clause, the change must be one that the Government ordered the contractor to make. The Government's representative, by his words or his deeds, must require the contractor to perform work which is not a necessary part of his contract. This is something which differs from advice, comments, suggestions, or opinions which Government engineering or technical personnel frequently offer to a contractor's employees.[76]

In 1968, the standard Changes clause used in government construction contracts was extensively modified. One of the significant revisions was the addition of language addressing the concept of a constructive change. Finally, with the adoption of the Contract Disputes Act of 1978, the boards were given jurisdiction to address breach of contract claims. However, the concept of a constructive change survives and still can affect the measure of a contractor's recovery if the claim is considered to be covered by the Changes clause rather than a damages claim for breach of contract.[77]

---

[75]*The Len Co. & Assocs. v. United States*, 385 F.2d 438 (Ct. Cl. 1967); *Industrial Research Assocs.*, DOT-CAB No. WB-5, 68-1 BCA ¶ 7069; *C.H. Leavell & Co.*, ASBCA No. 4899, 59-2 BCA ¶ 2291; *Blake Constr. Co., Inc.*, ASBCA No. 3046, 57-1 BCA ¶ 1281; *compare Watson, Rice & Co.*, HUDBCA No. 89-4468-C6, 90-1 BCA ¶ 22,499 *with Lott Constructors, Inc.*, ENGBCA No. 5852, 93-1 BCA ¶ 25,449 ("suggestions" by the government may or may not be constructive changes as the nature of the communication by the government and the lack of protest or notice from the contractor can be critical to the analysis).
[76]DCAB No. WB-5, 68-1 BCA ¶ 7069, pp 32,685–86; *see also Engineered Demolition, Inc., v. United States*, 70 Fed. Cl. 580 (2006).
[77]*See* **Section VIII.D.2** of this chapter for discussion of contractual limits on change order equitable adjustments.

B. Applications of the Doctrine

1. Oral Orders

The Court of Federal Claims and the boards generally hold that where the government authorizes and accepts the benefit of additional work, the matter may be treated as if a written change order had been issued despite the absence of an actual writing.[78]

Where there is a difference of opinion as to proper interpretation of contract specifications and the contracting officer's interpretation is incorrect, the result is often a constructive change in the contract requirements.[79] Such cases frequently involve interpretation of standards of quality.[80]

The specific facts and circumstances must be evaluated on a case-by-case basis. Proper notice is often critical to establishing that the contractor considered the action by the government's representative to be an order (change) rather than mere advice.[81] Similarly, supporting documentation of the extra or different work and additional expense is essential.

2. Impossibility of Performance

The boards also have applied the constructive change doctrine to provide relief for a contractor that finds it cannot perform in a manner contemplated by the parties at the time of contracting.[82] Such factual situations might be normally classified under the heading of breach of contract, but the boards have granted relief under the Changes clause by declaring that the contracting officer could and should have issued a change order. This is a classic illustration of the use of the constructive changes concept to allow a board to grant relief under the contract.

According to the Court of Claims, "the doctrine of legal impossibility does not demand a showing of actual or literal impossibility."[83] Rather, the standard is that of commercial impracticability (i.e., something can be done only at an excessive and unreasonable cost).[84] However, the court restricted the applicability of the commercial impracticability doctrine by holding that it "may be utilized only when the promissor has exhausted all its alternatives."[85] Affirmative acts by the government such as misrepresentation, negligent omission of information, or the withholding of information have been identified as contributing to situations of practical or actual impossibility of performance.[86]

---

[78]*Kingsport Utils., Inc.*, ASBCA No. 5666, 60-2 BCA ¶ 2710. *See also Armstrong & Co. v. United States*, 98 Ct. Cl. 519, 530 (1943).

[79]*Blake Constr. Co. Inc.* GSBCA No. 2477, 71-1 BCA ¶ 8870.

[80]*See, e.g., J.W. Bateson Co., Inc.*, ASBCA No. 19823, 76-2 BCA ¶ 12,032.

[81]*Industrial Research Assocs.*, DOTCAB No. WB-5, 68-1 BCA ¶ 7069. See **Chapter** 14 for a further discussion of notice.

[82]*J.W. Hurst & Son Awnings, Inc.*, ASBCA No. 4167, 59-1 BCA ¶ 2095.

[83]*Natus Corp. v. United States*, 371 F.2d 450, 456 (Ct. Cl. 1967).

[84]*Spindler Constr. Corp.*, ASBCA No. 55007, 06-2 BCA ¶ 33,376.

[85]*Jennie-O Foods, Inc. v. United States*, 580 F.2d 400, 409 (Ct. Cl. 1978).

[86]*Evans Constr. Co.*, ASBCA No. 22077, 78-1 BCA ¶ 13,018; *Helene Curtis Indus., Inc. v. United States*, 312 F.2d 774 (Ct. Cl. 1963).

3. Defective and Misleading Plans or Specifications

The constructive change doctrine often is invoked to provide relief where the contractor has been found to have been misled by government plans or specifications. In *Evans Construction Co.,*[87] a contractor's reliance on misleading government specifications led to a constructive change where the original contract documents significantly understated the amount of material required to complete a construction project.

A "constructive change" that arises in connection with defective plans and specifications has its basis in what is referred to as the *Spearin* doctrine.[88] This doctrine provides that, when the government supplies the plans and specifications for a construction project, the contractor cannot be held liable for an unsatisfactory final result attributable solely to defects or insufficiencies in those plans and specifications. This doctrine assumes the absence of any negligence on the contractor's part or assumption of the design risk or responsibility by the contractor, that the contractor followed the government's plans and specifications in executing the work, and that the contractor made no express warranty with regard to the suitability of those plans and specifications. Under this principle, an implied warranty exists for government-furnished plans and specifications that, if the contractor complies with them, a satisfactory product will result.[89] The delivery of defective plans and specifications is a breach of that warranty, absolving the contractor from responsibility for any resulting delays to project completion caused by the defective plans and specifications.[90] Similarly, the contractor is entitled to recover its reasonable costs incurred due to the defective plans and specifications and any necessary remedial work.

Some contracts contain disclaimers that have been drafted in an effort to negate the *Spearin* doctrine. For example, in *White v. Edsall Construction Co., Inc.,*[91] the contract drawings contained a note stating that the contractor was to verify certain aspects (details) of the design *before* bidding and notify the architect of all conditions that would require a change in plans. The court rejected the argument that this language shifted the risk of design errors which were not obvious to the contractor. In the court's view, verifying details was a different obligation from verifying the overall adequacy of the design. However, the court did not hold that it was not possible to draft sufficiently specific language to shift the risk of a design defect to a contractor.

4. Misinterpretation of Plans and Specifications

A fourth category under the constructive change concept relates to misinterpretation of the plans and specifications by the contracting officer or its representatives. This

---

[87] ASBCA No. 22077, 78-1 BCA ¶ 13,018.

[88] *United States v. Spearin,* 248 U.S. 132 (1918).

[89] *R. M. Hollingshead Corp. v. United States,* 111 F. Supp. 285 (Ct. Cl. 1953).

[90] *But see Centex Constr., Inc. v. United States,* 49 Fed. Cl. 790 (2001) (omitted specification applicable to a drawing detail did not constitute a basis for recovery given the provisions in the Specifications and Drawings for Construction clause). See **Chapter 6** for a discussion of that clause.

[91] 296 F.3d 1081 (Fed. Cir. 2002).

type of constructive change flows from the implied duty not to hinder or delay the contractor in the performance of its work, which is an implied obligation contained in every contract.[92]

This type of constructive change arises where, for example, the contract specifies a particular method of performance or allows the contractor to select the method, but the government's representative requires a different, more expensive method than that contemplated by the contractor at the time it prepared its bid. For example, in *H. I. Homa Co.*,[93] a constructive change was found where the contracting officer rejected a bar-type progress chart that satisfied the contract's progress of work clause and instead required the contractor to provide a critical path method schedule. Similarly, misinterpretation of the plans and specification can result in the imposition of unreasonable inspection standards or rejection of acceptable work by a government inspector.[94] Again, providing timely notice and objection to the requirement is essential to address questions of authority to direct the change and to avoid contentions that the lack of a contemporaneous objection is an indication that the contractor did not view the requirement as unreasonable or as causing it to perform extra work and incur extra cost.

This type of constructive change also can arise from the government's interpretation of a contract ambiguity in its favor. In *American Asphalt, Inc.*,[95] a contractor was entitled to recover the additional costs of removing excavated soil where the government-prepared plans were ambiguous as to whether the excavated materials were to remain on site. The government's plans appeared to indicate the material would remain on site as fill material, but the government intended to show the finish grade of paving, not the level for additional fill. The contractor established that it had made no allowance in the bid for the removal of the excavated soil.

Interpretations of contract specification requirements are a repetitive source of constructive change claims and disputes. Typically, the contractor must establish that its interpretation fell within the scope of reasonableness and that it relied on that interpretation during the bid/proposal phase. Failure to prove the latter element can defeat an otherwise valid claim for additional compensation.[96]

5. Constructive Acceleration

The constructive change doctrine also has been invoked where the government unjustifiably orders the contractor to speed up the performance of the work or implement a "recovery schedule"—that is, a constructive acceleration—thus entitling the contractor to an equitable adjustment under the Changes clause.[97] Constructive acceleration occurs in the absence of a government-directed acceleration, such as where the

[92]*See George A. Fuller Co. v. United States,* 69 F. Supp. 409 (Ct. Cl. 1947).
[93]ENGBCA Nos. PCC-41, PCC-42, 82-1 BCA ¶ 15,651.
[94]*See* **Chapter 10.**
[95]ASBCA No. 37349, 91-2 BCA 23,722.
[96]*Blueridge Gen., Inc.,* ASBCA No. 53663, 03-2 BCA ¶ 32,339.
[97]*Canon Constr. Corp.,* ASBCA No. 16142, 72-1 BCA ¶ 9404; *Robust Constr., LLC,* ASBCA No. 54056, 05-2 BCA ¶ 33, 019.

contracting officer has refused a valid request for time extensions or threatened other action that requires the contractor to accelerate its work to avoid liquidated damages or other loss or risk of loss. The classic case is when a valid request for a time extension for excusable delay is denied and the contract provides liquidated damages for late completion. This may be construed as an order to complete performance within the originally specified completion date, a shorter period of time and at higher cost than the contractor is entitled to be given under the terms of the contract. The constructive acceleration doctrine allows recovery for the additional expenses the contractor can establish where the government refuses to give the contractor a time extension to which the contractor is contractually entitled, thereby forcing the contractor to "accelerate" its work efforts in an attempt to maintain the original work schedule. Under those circumstances, the contractor is entitled to recover the costs related to the acceleration.[98]

Case law has identified five elements normally required to establish a claim for constructive acceleration.[99] Those elements are:

(1) An excusable delay must exist.
(2) Timely notice of the delay and a proper request for a time extension should have been given.
(3) The time extension must have been postponed or refused.
(4) The government must have ordered (by coercion, direction, or some other manner) that the project be completed within its original performance period.
(5) The contractor must actually accelerate its performance, thereby incurring additional costs.

The contractor should give the contracting officer the appropriate written notice of the delay and request an extension of time in order to establish part of the factual basis needed to recover its demonstrated acceleration costs.[100] This notice is important, especially when the contractor believes acceleration is occurring without a specific order to accelerate its performance, so that it is clear that the acceleration is not being undertaken voluntarily and, further, that the contractor expects the government to pay the additional costs incurred in that effort.

This notice of a "forced" acceleration will assist the contractor in recovering on an equitable adjustment claim under the Changes clause. This notice is not absolutely necessary if (a) the acceleration has been expressly directed; (b) the government has indicated no time extensions will be permitted; or (c) the government has waived the need for notice. Moreover, if the contracting officer has specific knowledge of excusable delays and unequivocally orders the contractor to complete on the contract completion date without regard to excusable delays, the notice requirement is satisfied. In these circumstances, it is important that the contractor give detailed information on the delay to the contracting officer so that it is clear that the government

[98]*Norair Eng'g Corp. v. United States,* 666 F.2d 546 (Ct. Cl. 1981).
[99]*Utley-James, Inc.,* GSBCA No. 5370, 85-1 BCA ¶ 17,816, *aff'd,* 14 Cl. Ct. 804 (1988).
[100]*Commercial Contractors Equip., Inc.,* ASBCA Nos. 59230, *et al.,* 03-2 BCA ¶ 32,381.

had the data to determine the reasonableness of the time extension request. It is imperative that the contractor carefully and fully document all delays and other factors that will aid in proving its entitlement to additional compensation. The Court of Federal Claims and the boards will recognize the potential validity of contractors' constructive acceleration claims, but if not carefully documented, the extra costs may not be recovered.[101]

The refusal to grant a time extension can be expressed either by a clear rejection of a time extension *or by the postponement of a decision concerning the request*. The government owes a duty to the contractor to timely respond and either grant or deny the request.[102] By failing to respond to the request, or even putting off the decision until the completion of the contract, the contracting officer places the contractor in a precarious position. If the contractor acts as if the time extension will be granted and continues at a pace that will complete the work after the contract's completion date, and the time extension is not granted, it is exposed to the imposition of liquidated or actual damages for completion delays. If the contractor acts as if the time extension will not be granted and accelerates in order to complete the work at the completion date, the contractor incurs additional expenses that might not be recovered. The contractor's updated schedule showing the contractor's reasonable expectations at the time of the request becomes especially important to support the decision to accelerate. If placed on appropriate notice, it is possible that the government eventually will pay the cost created by postponing its decision on time extensions.

An excellent example of constructive acceleration occurred in *Constuctors-Pamco*.[103] The board found that the government made it abundantly clear that work had to be completed without any extension of time by: (1) the use of language in the contract to the effect that the contract date would be strictly enforced; (2) the use of language in the contract to the effect that liquidated damages would be imposed for late-completed work; (3) the refusal to grant a time extension for a blizzard, which was an obvious excusable delay; (4) the refusal to respond to the contractor's request for instruction on how to proceed after it had complained about delays and extra costs; and (5) the daily communications with contractor personnel that no time extensions would be permitted. The board found that the contractor was required to accelerate performance of the work and to work under adverse conditions by direction of the government.

In *Norair Engineering Corp. v. United States*,[104] the Court of Claims described three key elements that must be proven to recover for the increased costs of acceleration under the Changes clause:

(1) Any delays giving rise to the order were excusable (thus entitling the contractor to a time extension).

---

[101]*See, e.g., Hemphell Contracting Co.*, ENGBCA No. 5698, 94-1 BCA ¶ 26,491, *and Allen L. Bender, Inc.*, PSBCA No. 2322, 91-2 BCA ¶ 23,828 (no time extension requests).

[102]*William Lagnion*, ENGBCA No. 3778, 78-2 BCA ¶ 13,260.

[103]ENGBCA No. 3468, 76-2 BCA ¶ 11,950.

[104]666 F.2d 546 (Ct. Cl. 1981).

(2) The contractor was ordered or required to accelerate.

(3) The contractor in fact accelerated performance and incurred extra costs.

The *Norair* court held that the contractor could recover its acceleration costs even though the project was over 500 days "late" because the contractor was entitled to approximately 700 days of time extensions and had to accelerate to finish "only" 500 days behind schedule.[105]

This type of constructive change also can arise when the government's incorrect interpretation of the contract requires the contractor to accelerate. For example, in *Rogers Excavating*,[106] an earthwork contract required the contractor to start work within 4 days after receipt of the Notice to Proceed and finish all work within 90 days. The contractor submitted a proposed schedule showing mobilization to start within 4 days but actual excavation not starting until 40 days after the date of the Notice to Proceed. The government refused to accept this schedule and required the contractor to start excavation work within the 4-day period. The board considered this to be an acceleration justifying additional compensation because mobilization was found to be "work," as it was defined in the contract. However, the contractor ultimately lost the case because it had failed to give the required notice that it considered the government's action as causing a constructive acceleration.

VI. CARDINAL CHANGES

A. Overview

The Changes clause in a government construction contract does not give the contracting officer unrestricted right to order extra work. Changed work or extra work must be "within the general scope of the original contract."[107]

If, for example, an agency contracts for the construction of one office building, it cannot require the contractor, by change order, to build a second office building. Such extra work is totally beyond the scope of the original contract. This would be an example of a "cardinal change." However, a change order requiring the addition of a room or the performance of finish work for space that was initially depicted as unfinished probably would be valid under the Changes clause. The difficult questions, of course, involve those cases that fall somewhere between the two extremes.

The term "cardinal change" refers to a change or changes ordered by the government that are beyond the scope of the contract and therefore constitute a material breach of contract.[108] If a change is a cardinal change, the government is in breach

---

[105]*Id. See also Continental Heller Corp.,* GSBCA No. 7140, 84-2 BCA ¶ 17,275; *but see Fraser Constr. Co. v. United States,* 384 F.3d 1354 (Fed. Cir. 2004) (constructive acceleration claimed denied where time extensions were issued within a reasonable time.).

[106]AGBCA No. 79-180-4, 83-2 BCA ¶ 16,701.

[107]FAR § 52.243-4(a).

[108]*Keco Indus., Inc. v. United States,* 364 F.2d 838 (Ct. Cl. 1966).

of the contract. In that situation, the contractor can, in theory, either refuse to perform or can perform and be paid the reasonable value for the work. However, if a contractor refuses to perform a proper change, incorrectly thinking it to be a cardinal change, it is subject to being terminated for default for refusing to perform disputed work and follow the Disputes procedure of the contract.[109] The risk of an incorrect decision is extremely high.

If the contractor is confronted with an undertaking *substantially different* from that originally contemplated due to the extensive changes ordered by the government or dictated by the government actions, then a cardinal change may exist. The contractor may have the right to avoid the terms of the contract agreement that may limit compensation in a cardinal change situation and seek compensation for the reasonable value of all services and materials provided.

A cardinal change also may result from physical conditions encountered by the contractor that were not expected and that fundamentally change the nature of the work. For example, in *Universal Contracting & Brick Painting Co. v. United States*,[110] a contractor entered into an agreement with the government calling for paint removal. After contract award, the contractor discovered that the paint contained asbestos. The contractor claimed that this constituted a cardinal change entitling the contractor to damages for breach of contract. The government moved for summary judgment, contending that the presence of asbestos in the paint was not a substantial enough condition to constitute a cardinal change. The Claims Court denied the government's motion, ruling that the contractor's claim presented a legitimate claim of cardinal change that could be decided only by a trial.

The question of whether a particular change (or group of changes) is sufficient to constitute a cardinal change is a matter of degree—and often is very subjective. The two basic tests for a cardinal change are:

(1) Whether the type of work was within the contemplation of the parties when they entered into the contract; and
(2) Whether the project, as modified, was still the same basic project.

No exact formula or criteria exist for determining whether the extra work constitutes a cardinal change. The general test is whether the change leaves the work "essentially the same" as what the parties bargained for at the time of contracting, or whether an essential deviation has occurred.[111]

It is the contractor's undertaking, that is, the total work done, rather than the product itself that is crucial in determining the existence of a cardinal change. For example, when a partially built hanger collapsed during construction due to faulty specifications and the contractor was directed to completely rebuild the hangar, a

---

[109]*See, American Dredging Co. v. United States*, 207 Ct. Cl. 1010 (1975); FAR § 52.233-1 (Disputes clause).
[110]19 Cl. Ct. 785 (1990).
[111]*See, e.g., Air-A-Plane Corp. v. United States*, 408 F. 2d 1030 (Ct. Cl. 1969).

cardinal change existed despite the fact that the finished product was basically the same structure.[112]

The test for cardinal change cannot simply be a comparison of the original contract price and the amount claimed as an equitable adjustment. Rather, it is important to consider the number of changes, the number of elements of the work changed, the number left unchanged, the character and timing of the changes, and the extent of the additional engineering, research, and development the contractor had to perform.[113]

B. Types of Possible Cardinal Changes

1. Quantity Variations

Variations in quantity of major items under the contract have been considered to be outside the scope of the contract and hence impermissible under the Changes clause. For example, the Court of Claims ruled that the deletion of 1 of the 17 buildings in a complex was not within the scope of the contract and hence not covered by the clause.[114] However, it should be noted that under a Termination for Convenience clause, a deletion of this nature would be authorized.

In addition to the Changes clause, federal agencies are authorized to use a Variation in Quantity clause to deal with situations in construction contracts where the actual quantity of unit-priced work varies from the contract estimate by a specified amount.[115]

2. Multiple Changes

The fact that many changes are made does not necessarily mean that a cardinal change has occurred.[116] It has been stated that the number of changes is less important than the impact, magnitude, or quality of the changes made. For example, a single change that more than doubles the quantity of work may be beyond the scope of the contract.[117] Conversely, the Court of Claims in *James F. Seger, et al. v. United States*[118] applied the "contemplation of the parties" test to deny a contractor's breach of contract claim based on numerous changes required by the contracting officer. The court held that because the end product was not "essentially different" from the structure called for in the original contract, no cardinal change had occurred.

---

[112]*Edward R. Marden Corp. v. United States*, 442 F.2d 364 (Ct. Cl. 1971).
[113]*Air-A-Plane Corp. v. United States*, 408 F.2d 1030 (Ct. Cl. 1969).
[114]*General Contracting and Constr. Co. v. United States*, 84 Ct. Cl. 570 (1937).
[115]*See* **Section II.C.1** of this chapter.
[116]*J.D. Hedin Constr. Co., Inc. v. United States*, 347 F.2d 235 (Ct. Cl. 1965).
[117]*Saddler v. United States*, 287 F.2d 411 (Ct. Cl. 1961).
[118]469 F.2d 292 (Ct. Cl. 1972).

VII.   VALUE ENGINEERING CHANGES

A.   Purpose of VE Changes

The government has long recognized that contractors may recognize ways to accomplish the work at less cost than the specific requirements set forth in the plans and specifications for a project. If a contractor proposes a cost-savings alternative to a requirement in the contract documents, one possibility is a deductive change under the Changes clause. If that process is followed, then the amount of the equitable adjustment would be based on the contractor's cost savings. That result provides little, if any, incentive for a contractor to invest the time and expense in a change order proposal, which may increase the contractor's risk under the contract. The concept of value engineering (VE) changes reflects an effort by the government to reward the contractor for its initiative by permitting it to share in the reduced cost of the work.

The concept of value engineering dates from its use in Department of Defense supply contracts in the 1960s. Periodically, the Congress has criticized the federal agencies for less than effective use of the process. Currently, the FAR contains specific guidance regarding the use of value engineering in various types of contracts (supply, architect/engineer design, and construction) and sets forth specific VE clauses to be used in these types of contracts. Since the VE clauses have significant variations and have evolved over the past 40 years, it is especially important that the specific VE clause in a particular contract be carefully examined if a contractor is contemplating the submission of a VE proposal. In addition, many of the important concepts affecting a VE proposal are defined terms in FAR §§ 48.001 and 52.248-3. Those definitions need to be understood when reviewing the VE clause in a specific contract.

B.   Construction VE Clause

FAR § 48.202 requires the insertion of the Value Engineering—Construction clause at FAR § 52.248.3 in all construction solicitations and contracts exceeding the simplified acquisition threshold.[119] A copy of that lengthy clause is included with this chapter at **Appendix 8A.**

C.   Key Issues for Contractors

1.   *Effect on Spearin Warranty*

As discussed in **Section V.B.3** of this chapter, the government warrants the adequacy of the detailed plans and specifications furnished to the contractor under the *Spearin* doctrine. However, if a contractor's VE proposal is incorporated into a contract and substantially affects the design or specifications, the contractor may be precluded

---

[119]Generally $100,000 for contracts performed in the United States.

from relying on the *Spearin* doctrine if the resulting defects relate to the contractor's proposed modifications.[120] However, in *Tranco, Industries, Inc.*,[121] the government elected to modify its specifications by adopting a contractor's proposed VE proposal, which the contractor had voluntarily withdrawn before it was accepted. In that case, the government remained responsible for the resulting problems in the detailed specification.

2. Scope of Shared Savings

The construction VE clause details the type of savings and the percentage of savings to be shared. FAR § 48.001 and the VE clause address and define "instant contract savings" and "collateral savings."

The phrase "instant contract savings" basically means the estimated reduction in the contractor's cost of performance resulting from the acceptance of the VE proposal less the cost of the development and implementation of the proposal. "Collateral savings" refers to savings in the agency's collateral costs (i.e., the agency's cost of operations, maintenance, logistical support, or government-furnished property). Collateral savings are capped at $100,000.[122]

FAR Part 48 also defines the phrases "concurrent contract savings"[123] and "future contract savings."[124] The VE clause authorized for use in construction contracts does not contemplate a sharing in those two savings categories, if any, with the contractor. Contractors need to be aware of these differences when drafting subcontract or purchase order conditions addressing VE proposals as firms that have supplied goods or materials for government supply contracts may not realize that these are differences in the treatment of savings under the various VE clauses. Similarly, any potential question regarding the scope of the shared savings should be clearly addressed in the VE proposal and resulting modification in order to avoid a very prolonged disputes process.[125]

Both the construction VE clause and FAR § 48.103(c) provide that the determination of the amount of the collateral savings is a unilateral decision of the contracting officer, who is described in the VE clause as the "sole determiner" of these savings. Although this language would appear to limit a contractor's rights to appeal the government's determination of the savings, there is authority for a review of the determination under an "arbitrary or capricious" standard.[126]

---

[120]*See Austin Co. v. United States*, 314 F.2d 518 (Ct. Cl. 1963), *cert. denied*, 375 U.S. 830.

[121]ASBCA No. 22379, 78-2 BCA ¶ 13,307, *recons. denied*, 78-2 BCA ¶ 13,522.

[122]FAR § 48.104-5.

[123]FAR § 48.001.

[124]*Id.*

[125]*Applied Cos. v. Harvey*, 456 F.3d 1380 (Fed. Cir. 2006) (2006 decision related to a 1990 value engineering change proposal (VECP).

[126]*ICSD Corp.*, ASBCA No. 28028, 90-3 BCA ¶ 23,027 *aff'd* 934 F.2d 313 (Fed. Cir. 1991).

3.   Discretion to Accept or Reject VE Proposal

The VE clause provides that the acceptance or rejection of a VE change proposal is a "unilateral decision made solely at the discretion of the Contracting Officer." It is very likely that the boards or the Court of Federal Claims will not review that decision[127] absent an allegation that the decision was "arbitrary and capricious." However, if the contracting officer rejects the VE proposal but later adopts the underlying idea, contractors have been able to obtain relief at the boards on a theory of constructive acceptance of the VE proposal by the government.[128]

4.   Delay Pending Consideration of a VE Proposal

The construction VE clause provides that the contracting officer notify the contractor of the status of the VE proposal within 45 calendar days of its receipt. If additional time is needed to consider the proposal, the contracting officer is obligated to notify the contractor within the 45-day period and provide the reason for the delay and the expected date for the decision. In an effort to preclude contractor claims for delay, the clause states that the government is not liable for any delay in acting on a VE proposal. In addition, the clause expressly obligates the contractor to perform in accordance with the existing contract until it receives either a notice to proceed with the VE proposal or a contract modification incorporating the VE proposal into the contract.

Potentially offsetting this allocation of the risk of delay to the contractor, the VE clause expressly states that the contractor may withdraw a VE proposal, in whole or in part, at any time before it is accepted.[129] Notwithstanding this effort to allocate the risk of delay to the contractor, a contractor should notify the government of any potential delay impact related to action on the VE proposal. In particular, if the government provides a preliminary indication of acceptance but delays final implementation or issuance of a notice to proceed, the contractor should document any potential delay as soon as it becomes likely.

5.   Subcontracting Considerations

FAR 52.248-3(h) requires a contractor to include an appropriate VE clause in any "subcontract" of $55,000 or more.[130] There is limited guidance in the FAR on the appropriate terms for that subcontract/purchase order clause, except that the subcontractor's VE payments shall not reduce the government's share of the savings. In tailoring a subcontract or purchase order VE clause, a general contractor should, at a minimum, address these points:

---

[127]*See NI Indus., Inc. v. United States,* 841 F.2d 1104 (Fed. Cir. 1988).

[128]*North Am. Rockwell Corp.,* ASBCA No. 14485, 71-1 BCA ¶ 8773.

[129]FAR § 52.248-3 (e)(2).

[130]The basic FAR definition of a "subcontract" found at FAR § 44.101 includes purchase orders. Contractors should include a suitable VE clause in purchase orders exceeding this dollar threshold.

- Sharing of savings paid by the government
- Responsibility for any engineering analysis/design risk and the cost of supporting the VE proposal
- Delay and timing of proposals
- Data rights as addressed in FAR § 52.248.3(i), as these could be important issues for a subcontractor's lower-tier vendor
- Confirmation of the discretion provided the government to accept/reject a proposal and determine the collateral savings

VIII. PREREQUISITES TO RECOVERY

A. Duty to Proceed

Older versions of the Changes clause in government construction contracts contained language stating that "nothing provided in this clause shall excuse the contractor from proceeding with the prosecution of the work as changed." Although this language is no longer included in the current clause, its omission does not alter a contractor's obligation to proceed when clearly directed to do so by the contracting officer. A failure to proceed as directed provides a basis for a termination of default.[131] If a contractor believes that the conditions stipulated in the change order are not acceptable, the prudent procedure, in most situations, is to proceed with an express reservation of rights and provide written notice of that reservation to the contracting officer.[132]

B. Notice Requirements

Notice requirements are common in government contracts.[133] The Changes clause contains two separate notice provisions applicable to contractor claims for additional compensation for changes. The clause provides the proposal for an equitable adjustment should be submitted within 30 days after receipt of a written change order or the furnishing of a written notice by the contractor that the government has made a constructive change to the contract. In each case, the clause requires that a written statement must be submitted setting forth "the general nature and amount of proposal."

There is a second notice requirement involving constructive change orders. The clause provides that no adjustment (time or money) for any such change is allowed

---

[131]*Discount Co. v. United States,* 554 F.2d 435 (Ct. Cl. 1977), *cert. denied,* 434 U.S. 938 (1977); *but see Wilner v. United States,* 26 Cl. Ct. 260, 267-8 (1992), *rev'd on other grounds,* 24 F.3d 1397 (Fed. Cir. 1994) (directive to proceed interpreted as a proposed bilateral change order).

[132]*Utley-James, Inc.,* GSBCA No. 5370, 85-1 BCA ¶ 17,816, *aff'd,* 14 Cl. Ct. 804 (1988).

[133]*See* **Chapter 14** for a detailed review of notice requirements in government construction contracts.

for any costs incurred more than 20 days before written notice of the constructive change order is given to the government. This latter provision explicitly excepts adjustments for increased costs resulting from defective specifications from the 20-day notice requirement.

Where a contractor fails to assert its proposed adjustment within the 30-day time limit or give the required 20-day notice, this failure may be raised as a defense by the government. However, a claim for an equitable adjustment will not be barred, prior to final payment, if formal adherence to the notice requirement could serve no useful purpose.[134] Generally, a contractor's claim or proposal is considered on its merits unless the government can demonstrate that it was prejudiced by the contractor's failure to give timely notice.[135] Even where the government is able to demonstrate some prejudice, the contractor may be entitled to an equitable adjustment where the government is not prevented from investigating and defending the claim.[136] In addition, late notice will not bar a contractor's claim if the contracting officer actually considered the claim on its merits without rendering an objection to lack of timely notice.[137]

Final payment traditionally has been considered a bar to assertion of claims under a Changes clause, but where a claim is filed by a contractor prior to final payment and is still pending at the time of such payment, a few older decisions held that the contract would "remain open" to allow the assertion of unrelated claims beyond that time.[138] However, the Armed Services Board of Contract Appeals has held that the timely assertion of a constructive change claim before a final payment reserves a contractor's right to later consideration of that claim only as final payment bars assertion of other claims.[139]

In general, the 30-day notice provision for changes in writing has not been literally enforced. In contrast to these rulings, the 20-day notice provision relating to constructive changes generally has been interpreted more strictly.[140] However, under certain circumstances, the government may be estopped from asserting the constructive change notice provision. For example, in *Ionics, Inc.*,[141] the contractor's failure to provide the required notice of a constructive change was based on a government assurance that a formal order would be issued. Therefore, the ASBCA held that the government was estopped from denying the claim on the basis of the notice provision.

---

[134]*Copco Steel and Eng'g Co. v. United States*, 341 F.2d 590 (Ct. Cl. 1965); *Powers Regulator Co.*, GS-BCA No. 4668, 80-2 BCA ¶ 14,463; *J.D. Abrams*, ENGBCA No. 4332, 89-1 BCA ¶ 21,379 (contractor told resident engineer that it did not agree with interpretation advanced by the government).

[135]*Parcoa, Inc.*, AGBCA No. 76-130, 77-2 BCA ¶ 12, 658; *Industrial Constructors, Corp.*, AGBCA No. 84-348-1, 90-2 BCA 22,767; *but see H.A. Anderson Co.*, ENGBCA No. 3724, 77-2 BCA ¶ 12,712.

[136]*See C.H. Leavell and Co.*, ASBCA No. 16099, 72-2 BCA ¶ 9694.

[137]*Dittmore-Freimuth Corp. v. United States*, 390 F.2d 664 (Ct. Cl. 1968).

[138]*See, e.g., Progressive Metal Equip., Inc.*, ASBCA No. 15954, 72-1 BCA ¶ 9301.

[139]*Gulf & Western Indus., Inc.*, ASBCA No. 22204, 79-1 BCA ¶ 13,706.

[140]*See, e.g., Preferred Contractors, Inc.*, ASBCA No. 15569, 15615, 72-1 BCA ¶ 9283.

[141]ASBCA No. 16094, 71-2 BCA ¶ 9030; *see also Davis Decorating Ser.*, ASBCA No. 17342, 73-2 BCA ¶ 10,107.

C. Release/Reservation of Rights

1. Change Orders

Change order releases and reservation of rights are often a sensitive topic on government contracts. Contracting officers seek closure. Contractors, uncertain about the impact of multiple changes and possible subcontractor claims, often seek to avoid closure. Sometimes it is difficult to avoid disagreements on these topics.

The Federal Acquisition Regulation provides specific guidance on change order releases in FAR § 43.204(c). That section provides this guidance to contracting officers:

Complete and final equitable adjustments. To avoid subsequent controversies that may result from a supplemental agreement containing an equitable adjustment as the result of a change order, the contracting officer should—

(1) Ensure that all elements of the equitable adjustment have been presented and resolved; and

(2) Include, in the supplemental agreement, a release similar to the following:

CONTRACTOR'S STATEMENT OF RELEASE

In consideration of the modification(s) agreed to herein as complete equitable adjustments for the Contractor's_____ (describe)_____ "proposal(s) for adjustment," the Contractor hereby releases the Government from any and all liability under this contract for further equitable adjustments attributable to such facts or circumstances giving rise to the "proposal(s) for adjustment" (except for _____).

In some situations, the U.S. Army Corps of Engineers uses a different form of a change order release. That version provides:

This Modification constitutes compensation in full on behalf of the contractor and its subcontractors and suppliers for all costs and markups directly or indirectly attributable to the changes ordered herein, for all delays related thereto, and for performance of the changes within the time stated.[142]

Sometimes contractors may consider this and similar provisions as routine or "boilerplate" and give them limited consideration. Contractors may conclude that delay or impact costs are reserved if they are not discussed. Others may assume that they are required to execute the release to obtain payment of the agreed-upon elements of the change order. Both conclusions are substantially in error.

---

[142]EP 415-1-3 (July 2, 1979).

As demonstrated by the decision in *Triple "A" South*,[143] these "boilerplate" releases are enforceable. In that appeal, the contractor performing a ship repair contract had received several hundred changes. Pricing for the revised work was confirmed in a series of bilateral supplemental agreements. Each of these agreements, signed by the contractor, provided that the price was "in full and final settlement of all claims" arising out of the change order, "including all claims for delays and disruption."

The contractor later submitted a substantial impact claim based on the same changes. One of the government's primary defenses was the "waiver and release" provision in the supplemental agreements. The contractor responded that because the government representatives had refused to discuss impact costs during the negotiations, the change orders were not full and final settlements. The ASBCA found the release language clear on its face and generally sufficient for the contractor to waive its impact claim. Lack of discussion did not necessarily reserve the claim from the release's coverage.

Seeking to avoid the effect of the release, the contractor argued that the government's representatives had misinformed it regarding the provision's effect. According to the contractor, the government's negotiators had stated that because the provision was merely "boilerplate," it would not affect the contractor's rights to later claim and recover impact costs. The ASBCA held that the contractor could overcome the clear language of the releases if it could show that *both parties* did not intend the change orders to be the final agreements. The board also held that the contractor could defeat the release if it could show that it accepted the release due to *intentional, factual misrepresentations* by the government's representatives.

These exceptions are not easily demonstrated. One of the potential exceptions requires the contractor to prove that the government actually intended something different from what was stated in the language the government drafted. The second requires the contractor to convince a board or the Court of Federal Claims that the government employees were engaged in conscious wrongdoing. Neither exception is easily or inexpensively established. If established, the exceptions allow only the right to have the impact claim considered on the merits.

Other government agencies may include a provision in the contract that is intended to provide closure on change order pricing. For example, some Navy construction contracts contain a clause entitled "Equitable Adjustments: Waiver and Release of Claims." Typically, this clause is found within a lengthy set of general provisions. Its title, not its placement, reflects its significance. The text of that clause reads:

(a) Whenever the Contractor submits a claim for equitable adjustment under any clause of this contract which provides for equitable adjustment of the contract, such claim shall include all types of adjustments in the total amounts to which the clause entitles the Contractor, including but not limited to adjustments arising out of delays or disruptions or both caused by such change. Except as the parties may otherwise expressly agree, the

---

[143] ASBCA No. 35824, 90-1 BCA ¶ 22,567.

Contractor shall be deemed to have waived (i) any adjustments to which it otherwise might be entitled under the clause where such claim fails to request such adjustments, and (ii) any increase in the amount of equitable adjustments additional to those requested in its claim.

(b) Further, the Contractor agrees that, if required by the Contracting Officer, he will execute a release, in form and substance satisfactory to the Contracting Officer, as part of the supplemental agreement setting forth the aforesaid equitable adjustment, and that such release shall discharge the Government, its officers, agents and employees, from any further claims, including but not limited to further claims arising out of delays or disruptions or both, caused by the aforesaid change.

The ASBCA has enforced this clause even though the resulting change orders contained no release language. The contractor's argument that it was impossible to evaluate the impact or delay costs at the time of submission of the initial change order proposal provided no relief from the effect of this clause.[144] However, the ASBCA also recognized the contractor's right to specifically reserve elements of cost even though the clause gives no indication that the contractor has that option.[145] If the government attempts to preclude a contractor from reserving its rights, and refuses to make payment on the agreed portions of a change unless the contractor accepts a release as part of the modification, the contractor should consider invoking the formal claim process under the Disputes clause.[146]

Absent an effective reservation of rights, a later assertion of a claim for delay or impact costs is likely to be rejected.[147] To avoid problems with waivers on government contracts, change order proposals should contain an appropriate reservation of rights.[148] However, if the reservation of rights is a vague or broad exception, there is a risk that a court or board may find that the release was not sufficiently specific to cover all the elements of the claim. For example, a reservation of rights to submit an impact claim because of the multiplicity of changes was not sufficient to reserve rights for cost elements such as crew relocation costs, idle time, and so forth.[149] However, if the government directs the contractor not to include *impact costs* in its change order proposals, change order releases waiving impact claims may be given no effect.[150]

It is not unusual for a contractor to include release language in change orders issued to its subcontractors during the performance of a government contract. On

---

[144]*Dyson & Co.*, ASBCA No. 21673, 78-2 BCA ¶ 13,482.

[145]*Id. See also R.C. Hedreen Co.*, ASBCA No. 20599, 77-1 BCA ¶ 12,328.

[146]*See* **Chapter 15.**

[147]*See Kato Corp.*, ASBCA No. 51462, 06-2 BCA ¶ 33,293.

[148]*Valenzuela Eng'g, Inc.*, ASBCA No. 54490, 06-2 BCA ¶ 33,399 (contractor's letters served as an exception to release language in a contract modification); *Corning Constr. Corp. v. Department of Treasury*, GS-BCA No. 16127-TD, 03-2 BCA ¶ 32,367 (subcontractor's price qualification on its change order proposal operated to nullify release language).

[149]*Bechtel Nat'l, Inc.*, NASA BCA No. 1186-7, 90-1 BCA ¶ 22,549.

[150]Compare *Serv. Eng'g Co.*, ASBCA No. 40274, 93-2 BCA ¶ 25,885 (release ineffective) with *Atlantic Dry Dock Corp. v. United States*, 773 F. Supp. 335, 338 (M.D. Fla. 1991) and *Southwest Marine, Inc.*, ASBCA Nos. 34058 *et al.*, 91-1 BCA ¶ 23,323 (releases barred impact claims).

occasion, the government may assert that the release or accord and satisfaction language in a subcontract change order operates to bar the contractor's submission of a claim to the government on behalf of the subcontractor under the *Severin* doctrine.[151] In that situation, a board will examine the communications between the contractor and its subcontractor to determine if the release language was limited by an effective reservation of rights.[152]

A recent decision by the Federal Circuit illustrates the necessity for contractors to obtain specific exceptions to releases in change orders. In *Bell BCI Co. v. United States*,[153] the contractor agreed to construct a five-story laboratory for the National Institutes of Health (NIH). During the course of construction, the NIH decided to add an additional floor to the building, and a change order was entered to reflect the addition. The change order included a release stating:

> The modification agreed to herein is a fair and equitable adjustment for the Contractor's direct and indirect costs. This modification provides full compensation for the changed work, including both Contract cost and Contract time. The Contractor hereby releases the Government from any and all liability under the Contract for further equitable adjustment *attributable* to the Modification.[154]

At the trial before the Court of Federal Claims, the contractor succeeded in obtaining a $2 million cumulative impact award because of the impact of that modification and other changes on other aspects of the work, particularly labor inefficiency.[155] The Federal Circuit partially vacated this award ruling that the language in that modification released the government from additional claims for impact costs attributable to the modification. The court remanded the case back to the Court of Federal Claims for a determination of which claims were attributable to that modification and consequently governed by the release language.[156] The ultimate outcome of this case at the Court of Federal Claims is uncertain but may be unfavorable to the contractor because of the broad "attributable" language in the modification.

Given the potentially broad interpretation placed on the release in *Bell BCI* by the Federal Circuit, a contractor should take advantage, whenever possible, of the provisions in the FAR to except claims from general releases of liability. The sample release to be included in an agreement containing an equitable adjustment allows for parties to agree to what claims are not covered by the release of liability.[157] Similarly, at the time of final payment, a contractor can except specified claims from a release

---

[151]*See Severin v. United States,* 99 Ct. Cl. 435 (1943), *cert. denied,* 322 U.S. 733 (1944).

[152]*M. A. Mortenson Co.,* ASBCA No. 53761, 06-1 BCA ¶ 33,180 (government had burden of establishing that the release was "iron clad").

[153]570 F.3d 1337 (Fed. Cir. 2009).

[154]*Id.* at 1339. (Emphasis added.)

[155]*Bell BCI Co. v. United States,* 81 Fed. Cl. 617 (2008), *aff'd in part, vacated in part and remanded,* 570 F.3d 1337 (Fed. Cir. 2009).

[156]570 F.3d 1337 at 1343.

[157]FAR § 43.204(c)(2).

of liability, if the contractor can give specified stated amounts for those claims.[158] These two provisions can and should be utilized to protect a contractor against losing otherwise valid claims if the Court of Federal Claims or a board interprets and enforces a release of liability similar to that done by the Federal Circuit's in *Bell BCI.*

2. Final Payment Releases

The Changes clause also addresses the effect of final payment on a contractor's right to submit an equitable adjustment proposal by providing no equitable adjustment shall be allowed if asserted after final payment.[159] In addition, the government typically requests a general release as part of the final payment process. An unconditional release provided at the time of final payment generally operates to bar all claims not expressly reserved.[160] Although the government does not have the right to impose a "no reservations or no exceptions" as a condition of final payment,[161] the contractor must provide exceptions or reservations that are reasonably specific, as illustrated by the decisions that follow.

In *Mingus Constructors, Inc. v. United States,*[162] the contractor provided the contracting officer two written notices of its intent to file a claim for "harassment," for "losses due to work that was misrepresented by the contract documents," "outside the scope of the original design," and/or "changed conditions." No dollar amount was stated. Subsequently the contractor executed the final payment release form and inserted in the space provided for exceptions this statement:

Pursuant to correspondence we do intend to file a claim(s)—the amount(s) of which is undetermined at this time.

Approximately eight months later, the contractor requested from the contracting officer a time extension for the submission of its formal written claim. The contracting officer denied the request and advised the contractor that no claim existed under the contract due to the passage of time and the lack of documents supporting the intended claim. When the certified claim was eventually submitted, the contracting officer returned it without action and stated that no "claim" as defined by the Disputes clause had been submitted at the time of final payment.

In reviewing the contracting officer's decision, the court held that the claim was barred by the general release. Moreover, the burden is on the contractor, not the government, to assure that a release is sufficiently specific. Noting that exceptions to releases are strictly construed against the contractor, the court termed the contractor's letters as "nothing more than a 'blunderbuss exception' which does nothing

---

[158]FAR § 52.232-5(h)(3).
[159]FAR § 52.243-4(f).
[160]*B.D. Click & Co. v. United States,* 614 F.2d 748 (Ct. Cl. 1980).
[161]*See Joseph Fresco Constr. Co.,* GSBCA No. 5717, 81-1 BCA ¶ 14,837.
[162]812 F.2d 1387 (Fed. Cir. 1987).

to inform the government as to the source, substance, or scope of the contractor's specification contentions." The court concluded that such "vague, broad exceptions" were "insufficient as a matter of law to constitute 'claims' sufficient to be excluded from the required release." Although the court stated that it was not necessary to submit their final, certified claims prior to executing the release required to obtain final payment, contractors were admonished to investigate the existing facts rather than merely list a vague intention to file a claim on the release. The lack of an explicit reservation of right and the submission of a reasonably specific claim to the contracting officer may not always bar a claim, but the absence of both will create serious problems and obstacles to the consideration of any claim on its merits.[163]

Following the *Mingus* decision, the United States Claims Court in *Miya Bros. Construction Co. v. United States*[164] construed the meaning of the release language that requires the contractor to state the value of the specific claims excepted from the release. Rejecting the government's contention that the contractor was precluded from increasing the stated value of the excepted claim, the court held that the claim's exact amount is irrelevant to the general principle that a monetary claim has been properly excepted from the release. The court concluded that an amendment of the amount claimed does not constitute the submission of a new claim any more than the adjustment of the amount of a previously certified claim under the Contract Disputes Act constituted a new claim.

3. Reservation of Rights Guidelines

In order to reserve rights, these items should be considered:

- Review the contract for change order waivers contained in the general or special provisions.
- When submitting the change order proposal, specifically state the cost elements that are reserved.
- Expressly exclude from the change order those elements of cost that have not been settled.
- Avoid vague or overly general reservations of rights.
- If there is an agreement that is, in fact, a settlement of certain issues related to the change order, that should be acknowledged.
- Act to promptly and reasonably reserve your rights if the government demands a release that is broader than the actual agreement.

---

[163]*See also P.I.O. GmbH Bau und Ingenieurplanung v. Int'l Broadcasting Bureau,* GSBCA No. 15934-IBB, 04-1 BCA ¶ 32,592 (contracting officer's knowledge or imputed knowledge that a contractor maintained a right to additional compensation *may* prevent a final release from operating as a bar even though a formal claim had not been filed).
[164]12 Cl. Ct. 142 (1987).

D. Documenting the Changed Work

1. Tracking the Cost

A basic element of recovery for a change is a demonstration that the change caused the contractor extra expense or time. The topic of pricing equitable adjustments is addressed in **Chapter 13**.

As stated in *Farnsworth & Chambers Co., Inc.*,[165] besides demonstrating that a change was ordered by an authorized government official, a contractor must show that the required work actually was performed. The contractor must also prove that additional effort or expense was required by demonstrating the increased cost resulting from performance under the change order.[166]

Assuming the contractor meets its burden of proof, the result will be an equitable adjustment in the contract price and/or the performance schedule. When responding to a change order or to a possible constructive change, a contractor should evaluate whether it is feasible to establish separate cost codes to track those costs that will be requested as part of the equitable adjustment.

Some government construction contracts require that the contractor separately track the cost of changed work. For example, FAR § 43.205(f) permits, but does not require, a contracting officer to include this clause found at FAR § 52.243-6 in a construction contract.

CHANGE ORDER ACCOUNTING (APR. 1984)

The Contracting Officer may require change order accounting whenever the estimated cost of a change or series of related changes exceeds $100,000. The Contractor, for each change or series of related changes, shall maintain separate accounts, by job order or other suitable accounting procedure, of all incurred segregable, direct costs (less allocable credits) of work, both changed and not changed, allocable to the change. The Contractor shall maintain such accounts until the parties agree to an equitable adjustment for the changes ordered by the Contracting Officer or the matter is conclusively disposed of in accordance with the Disputes clause.

Often this clause is incorporated by reference. Contractors need to review the list of incorporated clauses for this provision as well as other clauses that set forth change order cost documentation requirements. If that or other similar clauses are included in the contract, a contractor needs to evaluate its effect on its job cost accounting system and determine if the cost documentation obligation has been appropriately flowed down to its subcontractors and suppliers.

---

[165]ASBCA No. 5488, 60-1 BCA ¶ 2525.
[166]*McBride Elec. Co. v. United States*, 51 Ct. Cl. 448 (1916).

2. *Limitations on Recovery*

Some federal agencies utilize clauses that can operate to substantially limit the rights of contractors and subcontractors to recover field general conditions cost and home office expense associated with changes that extend the project's duration. An example of this provision is the Contract Changes—Supplement (July 2002) utilized by the Department of Veterans Affairs.[167] That lengthy clause provides, in part:

> The clauses entitled "Changes" in FAR 52.243-4 and "Differing Site Conditions" in FAR 52.236-2 are supplemented as follows:
>
> * * *
>
> (b) Paragraphs (b)(1) through b(11) apply to proposed contract changes costing $500,000 or less:
>
> * * *
>
> (4) Allowances not to exceed 10 percent each for overhead and profit for the party performing the work will be based on the value of labor, material, and use of construction equipment required to accomplish the change. As the value of the change increases, a declining scale will be used in negotiating the percentage of overhead and profit. Allowable percentages on changes will not exceed the following: 10 percent overhead and 10 percent profit on the first $20,000; 7~HF percent overhead and 7~HF profit on the next $30,000; 5 percent overhead and 5 percent profit on balance over $50,000. Profit shall be computed by multiplying the profit percentage by the sum of the direct costs and computer overhead costs.
>
> (5) The prime contractor's or upper-tier subcontractor's fee on work performed by lower-tier subcontractors will be based on the net increased cost to the prime contractor or upper-tier subcontractor, as applicable. Allowable fee on changes will not exceed the following: 10 percent fee on the first $20,000; 7~HF percent fee on the next $30,000; and 5 percent fee on the balance over $50,000.
>
> (6) Not more than four percentages, none of which exceed the percentages shown above, will be allowed regardless of the number of tiers of subcontractors.
>
> * * *
>
> (10) *Overhead and contractor's fee percentages shall be considered* to include insurance other than mentioned herein, field and office supervisors and assistants, security police, use of small tools, incidental job burdens and general home office expenses and *no separate allowance will be made therefor*. Assistants to office supervisors include all clerical, stenographic

---

[167]VAAR § 852.236-88.

and general office help. Incidental job burdens include, but are not necessarily limited to, office equipment and supplies, temporary toilets, telephone and conformance of OSHA requirements. Items such as, but not necessarily limited to, review and coordination, estimating and expediting relative to contract changes are associated with field and office supervision and are considered to be included in the contractor's overhead and/or fee percentage. [Emphasis added]

This type of clause has been consistently interpreted to preclude the recovery of field (job site) overhead in excess of the stated limit.[168] Under this clause, for example, a general contractor's maximum field (job site) overhead on a $30,000 change, which added 30 days to the project's duration, would be capped at $2,750.00 (10 percent of the first $20,000 plus 7.5 percent of the next $10,000).[169] This provision can substantially change the equity in an equitable adjustment.

IX. NOVATIONS

Under certain circumstances, a contractor may need to adopt a name change or have the government recognize a successor in interest following a merger or acquisition. Although it might appear that either of these could be accomplished by a contract modification, the FAR provides separate provisions addressing the procedure for recognizing novations[170] involving sales of assets, merger and the like, and name changes.[171]

The requirements and procedures set forth in these sections are quite specific, including the prescribed forms for Novation Agreement and Change-of-Name Agreement.

➤ LESSONS LEARNED AND ISSUES TO CONSIDER

- The standard Changes clause is critical to the *effective management* of a government construction project and to provide a means to adapt to current requirements and unforeseen conditions.
- A prudent contractor needs to anticipate the distinct probability that government will direct the performance of changes during the construction of the project and

---

[168]*Santa Fe Eng'rs, Inc. v. United States,* 801 F.2d 379 (Fed. Cir. 1986).
[169]*See* VAAR § 852.236-88(b)(4).
[170]FAR § 42.1204.
[171]FAR § 42.1205.

needs to determine whether its *project management and cost accounting systems are adequate* to respond to proposed modifications and appropriately identify and track the costs associated with a modification. Since those systems reflect a cost of contract performance, that *initial determination* should be made at the proposal or bid stage, not after receipt of a change.

- Since the contracting officer is the key government representative in the change order process, it is essential that a contractor's project management staff obtain and carefully review the documentation addressing the *authority limitations* of the contracting officer and its representatives.

- Although there are limitations on the types of changes contemplated by the Changes clause, a challenge that a particular change is not *within the scope* of the contract is problematic and risky.

- Although the concept of a *constructive* change is rooted in the historical disputes process for government contracts, it is now an integral part of the operation of the standard Changes clause. Contractors need to determine how the concept of a constructive change is flowed down to subcontractors and suppliers and managed effectively.

- A contractor must understand and manage *the notice requirements* of the Changes clause and the coordination of those requirements with subcontracts and purchase orders.

- The Changes clause provides for equitable adjustment in the contract price and time. However, some federal agencies *"supplement"* the Changes clause with clauses that may limit the amount of the equitable adjustment that a contractor can recover.

- Contractors need to anticipate that v*alue engineering* change proposals can create issues of delay, design responsibility, as well as the parties' rights related to the acceptance and rejection of a VE proposal.

- *Quantity variations* in unit-price work items may not be addressed by the standard Changes clause. A Variation in Estimated Quantity clause provides for limited repricing of unit-price work items under specific conditions. These conditions and related limitations need to be evaluated at the time of a bid/proposal.

- To minimize the extent of disputes related to equitable adjustments, contractors need to establish a process to *document the costs and time* effect of changes and constructive changes.

- *Change order releases* and *final payment releases* are intended to bring closure on issues related to modifications and potential claims. Contractors need to appreciate the effectiveness of such releases on their ability to reserve their rights to submit claims at a later date.

- It is very probable that reservations of right by a contractor in a change order or at the time of final payment will be *strictly (narrowly) construed* by a board or the courts.

APPENDIX 8A: VALUE ENGINEERING—CONSTRUCTION (SEPT. 2006)

(a) *General.* The Contractor is encouraged to develop, prepare, and submit value engineering change proposals (VECP's) voluntarily. The Contractor shall share in any instant contract savings realized from accepted VECP's, in accordance with paragraph (f) of this clause.

(b) *Definitions.* "Collateral costs," as used in this clause, means agency costs of operation, maintenance, logistic support, or Government-furnished property.

"Collateral savings," as used in this clause, means those measurable net reductions resulting from a VECP in the agency's overall projected collateral costs, exclusive of acquisition savings, whether or not the acquisition cost changes.

"Contractor's development and implementation costs," as used in this clause, means those costs the Contractor incurs on a VECP specifically in developing, testing, preparing, and submitting the VECP, as well as those costs the Contractor incurs to make the contractual changes required by Government acceptance of a VECP.

"Government costs," as used in this clause, means those agency costs that result directly from developing and implementing the VECP, such as any net increases in the cost of testing, operations, maintenance, and logistic support. The term does not include the normal administrative costs of processing the VECP.

"Instant contract savings," as used in this clause, means the estimated reduction in Contractor cost of performance resulting from acceptance of the VECP, minus allowable Contractor's development and implementation costs, including subcontractors' development and implementation costs (see paragraph (h) of this clause).

"Value engineering change proposal (VECP)" means a proposal that—

(1) Requires a change to this, the instant contract, to implement; and

(2) Results in reducing the contract price or estimated cost without impairing essential functions or characteristics; *provided,* that it does not involve a change—

(i) In deliverable end item quantities only; or

(ii) To the contract type only.

(c) *VECP preparation.* As a minimum, the Contractor shall include in each VECP the information described in paragraphs (c)(1) through (7) of this clause. If the proposed change is affected by contractually required configuration management or similar procedures, the instructions in those procedures relating to format, identification, and priority assignment shall govern VECP preparation. The VECP shall include the following:

(1) A description of the difference between the existing contract requirement and that proposed, the comparative advantages and disadvantages of each, a justification when an item's function or characteristics are being altered, and the effect of the change on the end item's performance.

(2) A list and analysis of the contract requirements that must be changed if the VECP is accepted, including any suggested specification revisions.

(3) A separate, detailed cost estimate for (i) the affected portions of the existing contract requirement and (ii) the VECP. The cost reduction associated with the VECP shall take into account the Contractor's allowable development and implementation costs, including any amount attributable to subcontracts under paragraph (h) of this clause.

(4) A description and estimate of costs the Government may incur in implementing the VECP, such as test and evaluation and operating and support costs.

(5) A prediction of any effects the proposed change would have on collateral costs to the agency.

(6) A statement of the time by which a contract modification accepting the VECP must be issued in order to achieve the maximum cost reduction, noting any effect on the contract completion time or delivery schedule.

(7) Identification of any previous submissions of the VECP, including the dates submitted, the agencies and contract numbers involved, and previous Government actions, if known.

(d) *Submission.* The Contractor shall submit VECP's to the Resident Engineer at the worksite, with a copy to the Contracting Officer.

(e) Government action.

(1) The Contracting Officer will notify the Contractor of the status of the VECP within 45 calendar days after the contracting office receives it. If additional time is required, the Contracting Officer will notify the Contractor within the 45-day period and provide the reason for the delay and the expected date of the decision. The Government will process VECP's expeditiously; however, it will not be liable for any delay in acting upon a VECP.

(2) If the VECP is not accepted, the Contracting Officer will notify the Contractor in writing, explaining the reasons for rejection. The Contractor may withdraw any VECP, in whole or in part, at any time before it is accepted by the Government. The Contracting Officer may require that the Contractor provide written notification before undertaking significant expenditures for VECP effort.

(3) Any VECP may be accepted, in whole or in part, by the Contracting Officer's award of a modification to this contract citing this clause. The Contracting Officer may accept the VECP, even though an agreement on price reduction

has not been reached, by issuing the Contractor a notice to proceed with the change. Until a notice to proceed is issued or a contract modification applies a VECP to this contract, the Contractor shall perform in accordance with the existing contract. The decision to accept or reject all or part of any VECP is a unilateral decision made solely at the discretion of the Contracting Officer.

(f) Sharing—

    (1) *Rates.* The Government's share of savings is determined by subtracting Government costs from instant contract savings and multiplying the result by—

        (i) 45 percent for fixed-price contracts; or

        (ii) 75 percent for cost-reimbursement contracts.

    (2) *Payment.* Payment of any share due the Contractor for use of a VECP on this contract shall be authorized by a modification to this contract to—

        (i) Accept the VECP;

        (ii) Reduce the contract price or estimated cost by the amount of instant contract savings; and

        (iii) Provide the Contractor's share of savings by adding the amount calculated to the contract price or fee.

(g) *Collateral savings.* If a VECP is accepted, the Contracting Officer will increase the instant contract amount by 20 percent of any projected collateral savings determined to be realized in a typical year of use after subtracting any Government costs not previously offset. However, the Contractor's share of collateral savings will not exceed the contract's firm-fixed-price or estimated cost, at the time the VECP is accepted, or $100,000, whichever is greater. The Contracting Officer is the sole determiner of the amount of collateral savings.

(h) *Subcontracts.* The Contractor shall include an appropriate value engineering clause in any subcontract of $55,000 or more and may include one in subcontracts of lesser value. In computing any adjustment in this contract's price under paragraph (f) of this clause, the Contractor's allowable development and implementation costs shall include any subcontractor's allowable development and implementation costs clearly resulting from a VECP accepted by the Government under this contract, but shall exclude any value engineering incentive payments to a subcontractor. The Contractor may choose any arrangement for subcontractor value engineering incentive payments; *provided,* that these payments shall not reduce the Government's share of the savings resulting from the VECP.

(i) *Data.* The Contractor may restrict the Government's right to use any part of a VECP or the supporting data by marking the following legend on the affected parts:

These data, furnished under the Value Engineering—Construction clause of contract_____, shall not be disclosed outside the Government or duplicated, used, or disclosed, in whole or in part, for any purpose other than to evaluate a value engineering change proposal submitted under the clause. This restriction does not limit the Government's right to use information contained in these data if it has been obtained or is otherwise available from the Contractor or from another source without limitations.

If a VECP is accepted, the Contractor hereby grants the Government unlimited rights in the VECP and supporting data, except that, with respect to data qualifying and submitted as limited rights technical data, the Government shall have the rights specified in the contract modification implementing the VECP and shall appropriately mark the data. (The terms "unlimited rights" and "limited rights" are defined in Part 27 of the Federal Acquisition Regulation.)

9

DELAYS, SUSPENSION, AND ACCELERATION

Delays and suspensions, as well as work accelerations required to overcome delays and suspensions, are primary contributors to cost overruns that affect many construction projects. Some delays are caused by occurrences beyond the control of any party (contractor, subcontractor, government). Many delays, however, result from a party's failure to fulfill its contractual obligations. Consequently, it is paramount for any entity involved in the construction process to understand its rights and responsibilities in each type of delay situation. This chapter discusses (1) the different types of delay claims, (2) the typical causes of delay, (3) the time extensions associated with delay, and (4) the acceleration of the contractor's work.

I. TYPES OF DELAY: BASIC PRINCIPLES

A. Excusable versus Nonexcusable

Typically, the parties' contract dictates whether a delay is excusable. As a general principle, most construction contracts provide that a contractor is excused from meeting the established contract completion date when the delay in performance is caused by unforeseeable events or conditions beyond the control and without the fault of the contractor and its subcontractors and suppliers. Examples of excusable delays to a contractor's work include: differing site conditions, design problems, changes to work, unusually severe weather, strikes, and acts of God.

Absent a provision defining excusable delays, a contractor performing a fixed-price contract generally is not excused from performance because it encountered unforeseen difficulties.[1] Under the basic law of contracts, performance is excused only when performance is deemed to be made impossible by an act of God, the law, or the other party.[2] Although these seemingly harsh rules of contract law have been

---

[1]*Day v. United States*, 245 U.S. 159 (1917); *see also United States v. Spearin*, 248 U.S. 132, 136 (1918); *American Gulf Companies*, ASBCA No. 49919, 99-2 BCA ¶ 30,452.
[2]*Carnegie Steel, Co. v. United States*, 240 U.S. 156, 165 (1916).

tempered by legal doctrines such as commercial impracticability, contractors need to appreciate this fundamental risk allocation concept and that the scope of *excusable delays* is basically a creature of the agreement between the parties.

Government construction contracts include a default termination clause which specifically recognizes that, under certain circumstances, the failure of a contractor to proceed with the project work in accordance with the approved schedule may be excused. The standard Default clause for government fixed-price contracts, FAR § 52.249-10, defines the term "excusable delays" in this way:

DEFAULT (FIXED-PRICE CONSTRUCTION) (APR. 1984)

* * *

(b) The Contractor's right to proceed shall not be terminated nor the Contractor charged with damages under this clause, if—

  (1) The delay in completing the work arises from unforeseeable causes beyond the control and without the fault or negligence of the Contractor. Examples of such causes include (i) acts of God or of the public enemy, (ii) acts of the Government in either its sovereign or contractual capacity, (iii) acts of another Contractor in the performance of a contract with the Government, (iv) fires, (v) floods, (vi) epidemics, (vii) quarantine restrictions, (viii) strikes, (ix) freight embargoes, (x) unusually severe weather, or (xi) *delays of subcontractors or suppliers at any tier* arising from unforeseeable causes beyond the control and without the fault or negligence of *both the Contractor and the subcontractors or suppliers*; and

  (2) The Contractor, within 10 days from the beginning of any delay (unless extended by the Contracting Officer), notifies the Contracting Officer in writing of the causes of delay. The Contracting Officer shall ascertain the facts and the extent of delay. If, in the judgment of the Contracting Officer, the findings of fact warrant such action, the time for completing the work shall be extended. The findings of the Contracting Officer shall be final and conclusive on the parties, but subject to appeal under the Disputes clause. [Emphasis added]

Although the Default clause and others like it set forth the typical examples of excusable delay, the clause is not intended to provide an exclusive list of excusable delay events. Similarly, a delay is not *per se* excusable because it is the result of one of the listed conditions. In the end, any delay analysis will focus on whether the delay is unforeseeable and beyond the control of the contractor as well as its subcontractors and suppliers at any tier.[3]

---

[3]Subsequent sections of this chapter generally refer to delays "beyond the contractor's control," a concept that includes subcontractors and suppliers at all tiers.

Nonexcusable delays are those that are within the contractor's control and for which the contractor is responsible. The causes for nonexcusable delays vary and include: lack of adequate manpower, slow progress due to unqualified workers, poor planning, defective work, failure to forward submittals in a timely manner, and the like. In such cases of delay, the contractor will not receive additional compensation or time to complete the work and may be liable for any damages caused by the delay.[4]

B. Compensable

Compensable excusable delays are delays for which the contractor is entitled to both a time extension and monetary compensation for costs incurred as a result of the delay. For example, delays caused by faulty or negligent acts or omissions of the government in contravention of the contractor's rights will not only result in an excusable delay but in many cases also will entitle the contractor to recover any increased costs related to the delay. Notably, timely completion of the project does not necessarily preclude recovery of delay damages. Where a contractor provides a reasonable as-planned schedule that would have produced an early completion of the project, the Court of Federal Claims and the boards *may* still hold the government liable for preventing the contractor from achieving an early finish.[5]

Under certain circumstances and depending on the nature of the government's conduct, the contractor may also be able to pursue a breach of contract action. As is discussed in **Section III** of this chapter, common examples of government acts or failures to act resulting in compensable excusable delays include: delayed notice to proceed, failure to furnish adequate plans and specifications, to provide access, to coordinate multiple prime contractors, to timely provide government furnished property, to timely approve shop drawings or submittals, to provide timely direction, to timely inspect or accept work, or to make timely payments, and suspensions.

C. Unforeseeable

As noted, a contractor is excused from meeting the established contract completion date when the delay in performance is caused by unforeseeable events. Generally, unforeseeable causes of delay refer to future events, not existing ones.[6] An event is foreseeable if the contractor knew or had reason to know of the event prior to submitting its proposal or bid on the contract.[7] Significantly, "conditions in an industry are presumed to be within the contractor's knowledge and contemplation in accepting

---

[4]*See Pete Vicari Gen. Contractors, Inc.*, ASBCA No. 54982, 06-1 BCA ¶ 33,136.

[5]*See, e.g., Jackson Constr., Co. v. United States*, 62 Fed. Cl. 84 (2004) (proof of intent to complete early not demonstrated); *Manuel Bros., Inc. v. United States*, 55 Fed. Cl. 8 (2002); *Emerald Maint., Inc.*, ASBCA No. 43929, 98-2 BCA ¶ 29,903; *Weaver-Bailey Contractors, Inc., v. United States*, 24. Cl. Ct. 576 (1991); *Metropolitan Paving Co. v. United States*, 325 F.2d 241 (Ct. Cl. 1963).

[6]*See Local Contractors, Inc.*, ASBCA No. 37108, 92-1 BCA ¶ 24,491; *Chas. I. Cunningham*, IBCA No. 242, 60-2 BCA ¶ 2816.

[7]*See Woodington Corp.*, ASBCA No. 37885, 91-1 BCA ¶ 23,579, *Arthur Venneri Co.*, GSBCA No. 851, 1964 BCA ¶ 4010.

the contract."[8] The mere possibility that an event might occur does not render it foreseeable.[9] Similarly, a contactor is not generally expected to have prophetic insight and to take extraordinary action to prevent the occurrence of a delay.[10] However, in asserting a claim for excusable delay, the contractor bears the burden of demonstrating that the delay causing event was unforeseeable.[11]

D. No Contractor Fault or Negligence

A delay in contract performance, even if caused by one of the conditions enumerated in the excusable delay section of the Default clause, is not excusable if the delay-causing event is within the control of the contractor or the result of the contractor's fault or negligence. For example, in *Pat-Ric Corp.,* the contractor asserted an excusable delay of 32 days resulting from a fire in a material supplier's plant.[12] In deciding whether the contractor was entitled to an extension for the delay, the board focused its analysis on whether the fire was caused by the contractor's negligence.[13] Ultimately, the board determined that the contractor took all normal and reasonable precautions and, thus, awarded it a contract time extension.[14]

E. Role of Subcontractors and Suppliers in Project Delay

In 1967, the Court of Claims held that a prime contractor could be held responsible only for the delays of its first-tier subcontractors or suppliers.[15] Thereafter, the then current Default clause was amended to include delays caused by subcontractors and suppliers at any level within the scope of the prime contractor's delay responsibility. In other words, the prime contractor's responsibility for delays was no longer limited to the first tier but effectively included all subcontractors and suppliers. Thus, under the Default clause as currently written, a contractor is not excused despite its lack of fault if the delay is attributable to a supplier's unexcused failure to provide necessary materials.[16] Board decisions have extended the prime contractor's responsibility for its subcontractors and suppliers to include delays caused by the fault or negligence of a government-designated sole source subcontractor or supplier.[17]

---

[8]*Demusz Mfg. Co., Inc.,* ASBCA No. 55311, 07-1 BCA ¶ 33,463.

[9]*See J.D. Hedin Constr. Co. v. United States,* 187 Ct. Cl. 45 (1969).

[10]*Id.*

[11]*See M.A.W. Co.,* AGBCA No. 95-226-1, 97-1 BCA ¶ 28,759; *Orbas & Assoc.,* ASBCA No. 35832, 89-3 BCA ¶ 22,023.

[12]*See* ASBCA No. 10581, 66-2 BCA ¶ 6026.

[13]*See id.*

[14]*See id.*

[15]*See Schweigert, Inc. v. United States,* 388 F.2d 697 (Ct. Cl. 1967).

[16]*See Thurmont Constr. Co., Inc.,* ASBCA No. 13473, 69-1 BCA ¶ 7604; *William Logan & Son, Inc.,* GSBCA No. 3597, 72-2 BCA ¶ 9759; *Fairchild Scientific Corp.,* ASBCA No. 21152, 78-1 BCA ¶ 12,869; *Wescor Forest Prods., Co.,* AGBCA No. 96-154-1, 97-2 BCA ¶ 29,242 (subcontractor death was not an excuse for delayed performance).

[17]*See Dennis Berlin, D/B/A Spectro Sort and as Spectro Sort Mfg. Co.,* ASBCA No. 51919, 02-1 BCA ¶ 31,675; *Joseph J. Bonavire Co.,* GSBCA No. 4819, 78-1 BCA ¶ 12,877; *Norit Constr. Corp.,* ASBCA No. 20584, 76-1 BCA ¶ 11,890; *Federal Television Corp.,* ASBCA No. 9836, 1964 BCA ¶ 4392.

One area of possible confusion under the current Default clause involves the scope of the term "subcontractor." The old Corps of Engineers Board of Contract Appeals (ENGBCA) restricted the term "subcontractor" to the Miller Act definition: one who "takes a significant part in the prime contractor's job on the site." Under this interpretation, common carriers were not included as a *subcontractor* for which the prime contractor was responsible.[18] However, the Armed Services Board of Contract Appeals (ASBCA) interpreted the term "subcontractor" to include common carriers.[19]

II. CAUSES OF EXCUSABLE DELAY

A. Weather

In order for weather to be considered an excusable cause of delay entitling the contractor to an extension of the contract completion date, the contractor must encounter adverse weather that is unusual for the time of year and the place it occurred.[20] Among the types of weather that have been recognized as excusable causes of delay are humidity, fog, and heavy rainfall. The number of days of contract extension to which a contractor will be entitled due to adverse weather is established by determining the impact of that weather on the progress of the job.

The mere fact that the weather is harsh or destructive is not sufficient to establish excusable delay if the contractor reasonably should have anticipated that type of weather at the time and place it occurred. Consequently, to recover a time extension for delays caused by adverse weather, the contractor has the burden of satisfying a two-step analysis: (1) that certain work which controlled the overall completion of the project was delayed by the weather; and (2) the adverse weather was unforeseeable and unusually severe. An essential factor to be considered in an adverse weather delay analysis is the nature of the contract work. As explained by the ASBCA:

> An exceptionally heavy one-day rain could have a serious adverse effect on a construction site highly subject to erosion. However, the same exceptionally heavy rain would cause less delay than a lighter rain continuously falling over a period of several days would cause on activities such as exterior painting. A light wind would normally have little if any effect on a construction project but if dust laden it could preclude activities such as painting or installation of sensitive electronic equipment. In construction work particularly sensitive to freezing, such as some paving and most masonry construction, a temperature of 10 below zero probably would have no greater adverse effect than one of 10 above.

---

[18]*See W.A. Roger,* ENGBCA No. PCC-25, 76-2 BCA ¶ 12,195.

[19]*See Hogan Mech., Inc.,* ASBCA No. 21612, 78-1 BCA ¶ 13,164. The ENGBCA was merged into the ASBCA, which effectively eliminated this conflict in decisions. With the creation of the new Civilian Board of Contract Appeals from the merger of multiple agency boards, the potential for conflicting decisions may become an issue in disputes. *See* **Chapter 15, Section II.D**.

[20]*Broome Constr., Inc. v. United States,* 492 F.2d 829, 835 (Ct. Cl. 1974); *Charles G. Williams Constr., Inc.,* ASBCA No. 42592, 92-1 BCA ¶ 24,635.

In order to apply the standard time extension provisions reasonably it is necessary that the parties consider not only the severity of the weather but the type of work being performed and the effect of the weather on the work. To do otherwise creates the same possibility of misapplication or misinterpretation as that which occurs when a sentence is taken out of context.[21]

The boards have shown a preference for determining whether weather is "unusually severe" on the basis of a historical standard. For example, the ASBCA reaffirmed its holding that the contractor must prove that the weather was unusually severe on the basis of average weather patterns for that season and locale.[22] Thus, under this type of analysis, it is not enough merely that the weather is severe enough to prevent work on the project. An excusable delay results only when the weather surpasses in severity the weather usually encountered or reasonably expected in the particular locality during the same period of the year involved in the contract.[23]

This test has been applied in a slightly different manner in *Bateson-Cheves Construction Co.*[24] Specifically, the board stated:

The weather data for the disputed period demonstrates that during each of the months involved it would not be unusual to have at least a few days of extremely cold weather and precipitation. *The contractor is therefore only entitled to an extension of time in the amount by which the number of days of severe weather exceeded the number of days that such weather was normal and should have been anticipated.*[25]

Regardless of the subtle distinctions in the application of the analysis, it is clear that the contractor is presumed to have based its proposal on the expectation that it will encounter the average weather conditions for that locale throughout the duration of the project.[26]

B. Acts of God

Under the Default clause, acts of God are always an excusable cause of delay entitling the contractor to an extension of the contract completion date. That is, by definition, acts of God are beyond the control and without the fault or negligence of the contractor. The Comptroller General has defined an "act of God" as "some inevitable accident

---

[21]*Essential Constr. Co., Inc. and Himount Constr., Ltd., A Joint Venture,* ASBCA No. 18491, 78-2 BCA ¶ 13,314 (emphasis added); *see also Fru-Con Constr. Corp v. United States,* 44 Fed. Cl. 298, 314 (1999)
[22]*See M. Zanis Contracting Corp.,* DOTBCA No. 2756, 96-2 BCA ¶ 28,439; *Diversified Marine Tech., Inc.,* DOTBCA No. 2455, 93-2 BCA ¶ 25,720; *Allied Contractors, Inc.,* IBCA No. 265, 1962 BCA ¶ 3501.
[23]*See Alpha Roofing and Sheet Metal Corp.,* GSBCA No. 1115, 1964 BCA ¶ 4461.
[24]IBCA No. 522-10-65, 67-2 BCA ¶ 6466.
[25]*Id.* (Emphasis added.)
[26]*Fairbanks Builders, Inc.,* ENGBCA No. 2634, 66-2 BCA ¶ 5865.

which cannot be prevented by human care, skill, or foresight, but results from natural causes such as lightning, tempest, floods and inundations."[27]

Consequently, unless the contract specifically provides otherwise, any natural disaster such as an earthquake, tornado, flood, or a similar event will provide the basic grounds for the contractor to request an extension of contract performance.[28] However, the contractor still must establish that the event adversely affected its progress or performance.[29]

C. Labor Problems/Strikes

Generally, delays caused by strikes are excusable. As with all causes of excusable delays, however, the controlling test as to whether labor problems or strikes give rise to an excusable delay is whether the delay was *unforeseeable* and *beyond the control* and *without the fault or negligence* of the contractor, its subcontractors or suppliers.[30] For example, in *Allied Contractors, Inc.*,[31] a contractor delay caused by a steel strike was not excusable, as the board deemed the delay foreseeable given that the strike had begun several months before submission of bids on the project.

In addition, an excusable delay may be granted for labor-related actions that impact the contractor in a similar manner as a strike. Examples include: organizational strikes;[32] jurisdictional strikes;[33] pickets protesting another contractor at or near the site;[34] a walkout in protest of contractor's accommodation, at the government's direction, of nonunion employees of another contractor;[35] and informational strikes.[36] When confronted with a strike or similar labor problem, a contractor must mitigate the effect of the labor-related event by attempting to obtain supplies or performance from other sources.[37]

[27]Comp. Gen. Dec. B-169473 (Apr. 24, 1970).

[28]*See Security Nat'l Bank of Kansas City v. United States,* 397 F.2d 984 (Ct. Cl. 1968) (highway contractor was granted a time extension for delay encountered when the government unreasonably refused to alleviate flooding on the work site caused by heavy rains); *see also Pat-Ric Corp.,* ASBCA No. 10581, 66-2 BCA ¶ 6026 (fire); *McKenzie Marine Constr. Co., Inc.,* ENGBCA No. 3245, 74-2 BCA ¶ 10,673 (floods); *Ardell Marine Corp.,* ASBCA No. 7682, 1963 BCA ¶ 3991 (high winds).

[29]*See Sach Sinha and Assoc., Inc.,* ASBCA No. 47594, 00-1 BCA 30,735 (contractor failed to prove that earthquake caused any significant damage).

[30]*See Gilbane Bldg. Co. v. United States,* 333 F.2d 867 (Ct. Cl. 1964).

[31]IBCA No. 265, 1962 BCA ¶ 3501.

[32]*See Fred A. Arnold, Inc.,* ASBCA No. 16506, 72-2 BCA ¶ 9608.

[33]*See Manufacturers Cas. Ins. Co. v. United States,* 63 F. Supp. 759 (Ct. Cl. 1946).

[34]*See Montgomery Ross Fisher, Inc.,* ASBCA No. 16843, 73-1 BCA ¶ 9799.

[35]*See Santa Fe Eng'rs, Inc.,* PSBCA No. 902, 84-2 BCA ¶ 17,377.

[36]*See Andrews Constr., Co.,* GSBCA No. 4364, 75-2 BCA ¶ 11,598.

[37]*See New England Tank Cleaning* Co., ASBCA No. 10208, 66-1 BCA 5654; *Alabama Bridge and Iron Co.,* ASBCA No. 6124, 61-1 BCA ¶ 2970.

D. Material Shortages

Generally, the contractor assumes the risk of obtaining materials necessary for performing the contract work.[38] As noted by the ASBCA:

> Market shortage is not an excusable cause for nonperformance. The fact that supplies cannot be obtained except at a cost in excess of the contract price is no excuse. Unwillingness to perform, although at some loss, does not relieve [the contractor] from [its] contractual obligation.[39]

Accordingly, although the contractor may have other viable claims under the contract, material shortages will not entitle it to a contract extension nor excuse untimely completion of the project.

III. CAUSES OF COMPENSABLE DELAY

A. Delayed Notice to Proceed

If the government fails to issue the notice to proceed within the time frame set forth in the contract or within a reasonable time if the contract does not specify a time, the government will generally be liable for delay. Where the contract is silent as to when the notice to proceed should be issued, it is implied that the notice to proceed will be issued within a reasonable period of time after the award of the contract. Moreover, an unreasonable delay in issuing the notice to proceed may amount to a breach of contract or a constructive suspension of work.[40] Similarly, if the contractor could not proceed with the work until after a preconstruction conference, a delay by the contracting officer in scheduling that conference might entitle the contractor to a time extension and expose the government to liability for that delay.[41]

In determining what is a reasonable period of time within which to issue the notice to proceed, the Court of Federal Claims and the boards will consider the time required to resolve any pending bid protest.[42] If a contractor wants the notice to proceed to be issued on an expedited basis, it should make a specific request in that regard.[43]

---

[38]*See Southwest Constr. Corp.,* ENGBCA No. 5286, 94-3 BCA ¶ 27,120; *Aargus Poly Bag,* GSBCA No. 4314, 76-2 BCA ¶ 11,927.

[39]*Betsy Ross Flag Co.,* ASBCA No. 12124, 67-2 BCA ¶ 6688; *see also SAI Indus. Corp.,* ASBCA No. 49149, 98-1 BCA ¶ 29,662 (refusal to supply materials in required quantity not an excuse).

[40]*See Nicon, Inc. v. United States,* 331 F.3d 878 (Fed. Cir. 2003) (recognizing government's implied obligation to issue notice to proceed within reasonable time); *Marine Constr. & Dredging,* ASBCA No. 38412, 95-1 BCA 27,286 (government breached implied duty to issue notice to proceed within a reasonable time); *Goudreau Corp.,* DOTBCA No. 1895, 88-1 BCA ¶ 20,479 (three-month delay between furnishing of performance bonds and payment bonds and the issuance of notice to proceed was unreasonable); *Freeman Electric Constr. Co.,* DOTBCA No. 74-23A, 77-1 BCA ¶ 12,258.

[41]*See, e.g., Tidewater Contractors, Inc. v. Department of Transportation,* CBCA 50, 07-1 BCA ¶ 33,525.

[42]*See DeMatteo Constr. Co. v. United States,* 600 F.2d 1384 (Ct. Cl. 1979); *see generally Louis Leustek & Sons, Inc., v. United States,* 41 Fed. Cl. 657 (1998) (recognizing that whether a particular delay is reasonable depends on the particular circumstances of a case).

[43]*See Freeman Elec. Constr. Co.,* DOTCAB No. 74-23A, 77-1 BCA ¶ 12,258.

B. Failure to Furnish Adequate Plans and Specifications

All delays resulting from defective and inadequate plans and specifications furnished by the government are *per se* excusable.[44] Federal courts and the boards follow the doctrine announced by the Supreme Court in *United States v. Spearin*.[45] The *Spearin* doctrine recognizes the duty of the government to the contractor to provide specifications, plans, and drawings free of defects and omissions.[46] If the government-furnished plans and specifications are defective or inadequate to allow the contractor to perform the work in accordance with the intended design, and the defects or inadequacies cause delay to the project work, the government will be liable for time extensions and delay damages.[47] However, a contractor will not be entitled to contract extension if the defects or inadequacies in the plans are not related to the delay.[48]

C. Failure to Provide Access

The government is generally required to provide the contractor and its subcontractors access to the work site in a timely and properly sequenced fashion.[49] In order to recover under this theory of liability, the contractor must show either government fault or an express government warranty that the site would be accessible at a time certain.[50] Government fault can be found under various factual scenarios but is often found in the breach of the implied duty to cooperate.[51] Generally, no government fault will be found where the government has been diligent and persistent in attempting to eliminate the causes of delayed site access.[52]

D. Failure to Coordinate Multiple Prime Contractors

When the government elects to employ multiple prime contractors to perform the project work, the government has a duty to coordinate the work of each prime contractor so that there is no conflict between the work of two contractors. In evaluating this type of delay claim, the test is whether the government "took reasonable steps to

---

[44]*See Chaney and James Constr. Co. v. United States*, 421 F.2d 728 (Ct.Cl. 1970); *Minmar Builders, Inc.*, GSBCA No. 3430, 72-2 BCA ¶ 9599.

[45]248 U.S. 132 (1918). *See, e.g., AAB Joint Venture v. United States*, 75 Fed. Cl. 414 (2007); *Travelers Cas. and Sur. v. United States*, 74 Fed. Cl. 75 (2006); *Weststar Revivor, Inc.*, ASBCA No. 52837, 06-1 BCA ¶ 33,288.

[46]*Id. See* **Chapter 8** for a further discussion of the *Spearin* doctrine.

[47]*Bechtel Envtl., Inc.*, ENGBCA No. 6137, 97-1 BCA ¶ 28,640 (delay due to design issues extended work into a season where the weather adversely affected performance and, as a result, contractor recovered the resulting extra costs).

[48]*Blinderman Constr. Co., Inc.*, ASBCA No. 21966, 79-1 BCA ¶ 13,875.

[49]*Blinderman Constr. Co., Inc. v. United States*, 695 F.2d 552 (Fed. Cir. 1983).

[50]*H.E. Crook Co. v. United States*, 270 U.S. 4 (1926); *United States v. Foley*, 329 U.S. 64 (1946).

[51]*See, e.g., Green Thumb Lawn Maint.*, ENGBCA No. 6249, 98-1 BCA ¶ 29,688; *Peter Kiewit Sons' Co. v. United States*, 138 Ct. Cl. 668 (1957) (government breached its implied duty of cooperation when issuing a notice to proceed with the knowledge that the prior contractor encountered difficulties performing its work).

[52]*See, e.g., Arvid E. Benson*, ASBCA No. 1116, 67-2 BCA ¶ 6659 (government made diligent efforts to achieve timely performance from prior contractor); *Asheville Contracting Co.*, DOTCAB No. 74-6, 76-6A, 76-2 BCA ¶ 12,027 (government made diligent effort to encourage prior contractor to perform in a timely manner).

coordinate the work and promote cooperation in order to insure that the work of one contractor does not interfere with the work of the other."[53] In most cases involving the failure of the government to coordinate multiple contractors, the contractor will not be entitled to a compensable delay if the government can show it acted diligently.[54]

E. Failure to Timely Provide Government Furnished Property

When the government has a contractual obligation to furnish material to the project, it has a corresponding duty to provide such government-furnished property in a timely manner.[55] Even where a specific date is not expressly set forth in the contract, the government impliedly warrants that government-supplied material will be furnished to the contractor in sufficient time to permit the contractor to finish on schedule.[56] However, the contractor must establish cause and effect by demonstrating that the government's failure delayed its work.[57]

F. Failure to Timely Approve Shop Drawings/Submittals

The submittal of shop drawings, product technical data brochures, or "cut sheets" for government review is an important part of the construction process. For contractors, the submittal and review process creates a potential source of delay because the contractor is not able to proceed with the work until the government's agent has approved the submittal. Consequently, when the contract provides that the contractor must obtain government approvals, a compensable delay of work will occur if the approvals are not given within a reasonable time period.[58] However, where approval of submittals are delayed or withheld because the contractor's submission is incomplete, it is unlikely that a time extension will be granted.[59] In asserting an approval delay claim, the contractor is generally required to establish the approval time that it could have reasonably anticipated.[60]

G. Failure to Provide Timely Direction

Similar to an unreasonable delay in issuing a notice to proceed, delays in responding to requests for information and any unreasonable failure to approve materials may

---

[53]*Asheville Contracting Co.,* DOTCAB Nos. 74-6, 76-6A, 76-2 BCA ¶ 12,027; *see also Robert B. Marquis, Inc.,* ASBCA No. 38438, 92-1 BCA ¶ 24,692 (government failed to resolve conflict between two contractors working on the same site).
[54]*Star Commc'ns, Inc.,* ASBCA No. 8049, 1962 BCA ¶ 3538; *see also Asheville Contracting Co.,* DOTCAB Nos. 74-6, 76-6A, 76-2 BCA ¶ 12,027.
[55]*Electro Plastic Fabrics, Inc.,* ASBCA No. 14761, 71-2 BCA ¶ 9118.
[56]*See Precisions Dynamics, Inc.,* ASBCA No. 50519, 05-2 BCA ¶ 33,071; *Peter Kiewit Sons' Co. v. United States,* 151 F.Supp. 726 (Ct. Cl. 1957).
[57]*Leonhard Weiss GmbH & Co.,* ASBCA No. 37574, 93-1 BCA ¶ 25,443.
[58]*See Sydney Constr. Co., Inc.,* ASBCA No. 21377, 77-2 BCA ¶ 12,719.
[59]*See, e.g., Constuction Servs., Inc.,* GSBCA No. 2423, 68-2 BCA ¶ 7154; *Sea Crest Constr. Corp. v. United States,* 59 Fed. Cl. 615 (2004); *Carousel Dev., Inc.,* ASBCA No. 50719, 01-1 BCA ¶ 31,262.
[60]*R.J. Crowley, Inc.,* ASBCA No. 35679, 88-3 BCA ¶ 21,151.

entitle a contractor to a compensable excusable delay. For example, a contractor was entitled to a contract extension where the government failed to provide adequate direction by waiting 15 days to advise the contractor whether an ordered suspension would be lifted or whether the contract would be terminated for convenience.[61]

H. Failure to Timely Inspect/Accept Work

Under a typical contract, the government has the right or, in some instances, a duty to inspect the contractor's work as it progresses. The government may be liable to the contractor for inspections that are unreasonably intensive or repetitious, or for failure to timely and promptly inspect the contractor's work.[62] For example, in *Piracci Construction Co., Inc.*,[63] a contractor was given an extension of time where the government, through its project architect, improperly rejected a preconstruction scale model and thereby delayed approval of shop drawings from the contractor.

I. Failure to Make Timely Payments

Failure to make required payments in a timely manner in accordance with the terms of the contract may entitle the contractor to a compensable excusable delay.[64] If excess liquidated damages are withheld by the government and the contractor's ability to perform is impaired, those facts may excuse the contractor's failure to make progress.[65] Moreover, where the government wrongfully withheld payment on a contract, the contractor was entitled to stop work on other ongoing contracts with the same agency.[66] Alternatively, the contractor may elect to continue with the contract work and seek compensation in accordance with the terms of the contract. A government failure to make payment, however, will not be considered to be an excusable cause of delay where there is a *bona fide* disagreement as to the contractor's right to receive payment or the amount due.[67]

J. Suspensions

A suspension is a form of delay that results from the government's purposeful interruption of the work. The Suspension of Work clause, FAR § 52.242-14, provides in part:

---

[61] *See River Equip. Co.*, ENGBCA No. 5856, 93-2 BCA ¶ 25,654.

[62] *See* Thomas E. Abernathy IV & Thomas J. Kelleher, Jr., Inspection Under Fixed-Price Construction Contracts, Briefing Papers (Federal Publications, Dec. 1976) at 6–8.

[63] GSBCA No. 3477, 74-1 BCA ¶ 10,647.

[64] *See Q.V.S., Inc.*, ASBCA No. 3722, 58-2 BCA ¶ 2007.

[65] *Abcon Assocs., Inc. v. United States*, 49 Fed. Cl. 678 (2001).

[66] *See William Green Constr. Co. v. United States*, 477 F.2d 930 (Ct. Cl. 1973).

[67] *See Precision Tool Co.*, ASBCA No. 5048, 60-2 BCA ¶ 2739.

(a) The Contracting Officer may order the Contractor in writing, to suspend, delay or interrupt all or any part of the work of this contract for the period of time that the Contracting Officer determines appropriate for the convenience of the Government.

The presence of this type of clause benefits both the government and the contractor. The government, on one hand, is provided the right to halt construction temporarily, if, for example, the government experiences funding difficulty or some other unforeseeable and unanticipated problem. The contractor, on the other hand, benefits from this clause as it ensures it will be compensated for those proven and allowable delay costs and extended performance time. Even where the government does not issue a formal affirmative suspension of work order, the clause expressly allows for a recovery for the delays created by the government's actions or inaction that effectively suspends the contract work.[68]

In addition, the Suspension of Work clause provides a mechanism for an administrative settlement of contractor claims for constructive delays and interruptions of the contract work. This clause is triggered when the contracting officer's act or omission results in an interference with the contractor's performance. Specifically, the clause provides, in relevant part:

(b) If the performance of all or any part of the work is, *for an unreasonable period of time,* suspended, delayed or interrupted (1) by an act of the Contracting Officer in the administration of this contract or, (2) by the Contracting Officer's failure to act within the time specified in this contract (or within a reasonable time if not specified), an adjustment shall be made for any increase in the cost of performance of this contract (excluding profit), necessarily caused by the unreasonable suspension, delay, or interruption, and the contract modified in writing accordingly. However, no adjustment shall be made under this clause for any suspension, delay or interruption to the extent that performance would have been so suspended, delayed, or interrupted by any other cause, including the fault or negligence of the Contractor, or for which an equitable adjustment is provided for or excluded under any term or condition of this contract.[69]

Pursuant to this provision, contractors are entitled to recover costs related to constructive suspensions including delay: (1) in issuing a notice to proceed;[70] (2) in making the site available;[71] (3) in issuing change orders;[72] and (4) caused by defective

---

[68] *See, e.g., Blinderman Constr. Co. v. United States,* 695 F.2d 552 (Fed. Cir. 1983).

[69] FAR § 52.242-14 (emphasis added).

[70] *Alvarez & Assocs. Constr. Co.,* ASBCA No. 50185, 97-2 BCA ¶ 29,320.

[71] *Blinderman Constr. Co., Inc. v. United States,* 695 F.2d. 552 (Fed. Cir. 1982); *P&A Constr. Co.,* ASBCA No. 29901, 86-3 BCA ¶ 19,101; *Wheatley Assocs.,* ASBCA No. 24629, 80-2 BCA ¶ 14,639.

[72] *Weldfab, Inc.,* IBCA No. 268, 61-2 BCA ¶ 3121; *but see Decker & Co. GmbH,* ASBCA No. 35051, 88-3 BCA ¶ 20,871 (board distinguished between reasonable and unreasonable periods of time to issue a change order).

plans and specifications.[73] However, a key element is the necessity to establish that the suspension, delay, or interruption was for an "unreasonable period of time."

IV. SCHEDULING AND DELAYS

A. Types of Project Schedules

Scheduling techniques have made tremendous advances in the past 30 years in large part due to advances in technology and widespread use of computers on construction projects. Although Gantt charts (i.e., bar charts identifying start and completion dates for critical activities of the project work) still are used on some projects or as short-term schedules, it is common to find computer-generated critical path method (CPM) schedules used to plan and monitor construction activities. In fact, many contracts require the use of electronic schedules on construction projects. These computer-generated schedules allow the construction team to plan and logically sequence activities and determine which of the activities are essential to the timely completion of the project.

The "critical path" in a schedule is the longest sequence of interrelated activities through a project that establishes the completion date for the project. If one of the critical activities is delayed, then—by definition—the project completion date will be delayed unless the contractor is able to overcome the delay by resequencing or accelerating its work. The most effective project schedules are those that are realistic, contain well-thought-out sequencing plans, have achievable resource requirements and interim milestones, and do not contain an unmanageable number of activities.

B. Use of Schedules to Establish Cause of Delay

The CPM schedules are useful not only in project planning but also serve as a good tool for monitoring progress and events occurring throughout the course of a project. For example, events that may affect critical path activities can be plugged into the schedule to determine if, and to what extent, they will alter the project's completion date.

Because construction projects are very detailed and complex, rarely are critical tasks—activities along the critical path—the only activities being performed at any given point in time. Typically, there are numerous activities being pursued on noncritical potions of the project. By definition, these noncritical activities include work activities for which the contractor's schedule allows more time for performance than is required. The difference between the scheduled time for a planned activity and the time it actually takes to perform the activity is referred to as *float*. To illustrate the float concept, when the construction schedule provides three days of float time for a given activity, a one-, two-, or three-day delay in the completion of that activity will

---

[73]*See Blinderman Constr. Co., Inc.,* ASBCA No. 21966, 79-1 BCA ¶ 13,875 (contractor must relate delay to defective specification or drawing).

not cause a delay in completion of the overall project. That is, the delay in completing the activity is absorbed by the float time included in the schedule for that activity. When the delay in completing the activity exceeds the amount of float time, the critical path may change and include the previous noncritical activity. As with all critical activities, once a noncritical activity becomes critical, any subsequent delay to that activity causes a delay to the completion of the project as a whole.[74]

Using CPM schedules, the boards, the Court of Federal Claims, and its predecessor courts have been able to segregate concurrent delays and reach a more equitable resolution to a delay dispute. For example, if a government-caused delay occurs along the critical path and a concurrent contractor delay only affects activities with float, the impact of the respective delays can be isolated and the contractor may be able to recover.[75]

Schedules are not perfect and may not accurately predict the actual completion date for the project. The agency or contractor that prepares the schedule may overestimate the amount of time required for a particular task, or it may deliberately incorporate extra time into the schedule to provide greater flexibility and to ensure avoidance of liquidated damages. Other factors could enhance a contractor's actual performance duration as compared to the schedule, for example, unusually good weather, an efficient workforce, or productive subcontractors and prompt suppliers. Whatever the cause, contractors do finish ahead of schedule on occasion. When it does, a contractor is in a position to save considerable sums of money in job site general conditions and variable expenses.

No contract, board, or court would expressly penalize the contractor for its efforts to finish the project early, unless, of course, it violates some implied duty of cooperation with other contractors still on the job. Nevertheless, a "constructive penalty" could be imposed on the contractor that is in a position to finish early but is prevented from doing so by the actions or inaction of the government. If a government-caused delay does not extend performance beyond the original completion date, the government will argue that because the contractor finished "on time," either the contractor was not delayed or it suffered no damage as a result of the delay. Critical path analyses make it possible to overcome the first argument, and case law suggests that a contractor is entitled to a recovery when it is precluded by government action from finishing the project work early.[76]

---

[74]*See Weaver-Bailey Contractors, Inc. v. United States*, 19 Cl. Ct. 474 (1990) (contractor concentrated manpower in areas that it thought needed earliest attention, leaving "unclassified excavation," a noncritical path activity, for later, properly using up float time; when the government's estimate as to the amount to be excavated was revealed to be 40 percent inaccurate, the resulting additional work caused a delay to the overall project because all float time had been used; contractor was entitled to an equitable adjustment for the delay).

[75]*See* **Section VI** of this chapter for a discussion of concurrent delay.

[76]*See, e.g., Manuel Bros., Inc. v. United States*, 55 Fed. Cl. 8 (2002); *Emerald Maint., Inc.*, ASBCA No. 43929, 98-2 BCA ¶ 29,903; *Weaver-Bailey Contractors, Inc., v. United States*, 24 Cl. Ct. 576 (1991); *Metropolitan Paving Co. v. United States*, 325 F.2d 241 (Ct. Cl. 1963). *See* **Section V** of this chapter for a discussion of delayed early completion claims.

Delay may not only impact the critical path, but it may independently consume float time as well. Loss of this valuable float time may reduce the contractor's flexibility in sequencing work activities and allocating resources. As a result, it may increase the costs of performance of certain activities even if there is no overall project delay. To the extent the contractor uses its own float time, it cannot complain, but government-caused delays may consume float as well and ultimately have an impact on the critical path. When all of the float time is used, any further contractor-caused delays are critical (i.e., they directly impact the critical path) and is likely to have an adverse impact on the project and the contractor. will cost the contractor money. Because of this reality, there can be a difference of opinion between the government and the contractor as to who owns the float.[77]

There are two schools of thought as to who owns the float. Some take the position that the contractor owns float. From that perspective, the government may use the float without cost unless the contractor has a need for it. To the extent the government has used float time needed by the contractor, the government must compensate the contractor for its use. Others take the position that the float belongs to the party that uses it first. Thus, once float is gone for one, it is gone for all. Some government agencies have begun to incorporate clauses specifically aimed at the "ownership of float" issue. For example, a recent U.S. Postal Service facility specification included this clause:

> Float or slack is not time for the exclusive use or benefit of either the Government or the contractor. Extensions of time for performance will be granted only to the extent that equitable time adjustments for the activity or activities affected exceed the total float or slack along the channels involved.

This type of clause is becoming more prevalent and is slowly becoming the rule rather than the exception. Contractors must be aware of the impact this and similar provisions can have on a project and their cost ofperformance.

V. RIGHT TO COMPLETE EARLY

It is well established that a contractor may recover for government-caused delay even though the contract was completed within the contract completion date.[78] As noted by the Court of Claims:

> While it is true that there is not an "obligation" or duty of [the government] to aid a contractor to complete prior to completion date, from this it does not follow that [the government] may hinder and prevent a contractor's early

---

[77]A comprehensive treatment of float ownership issues is the paper "Who Owns the Float?" Federal Publications, Construction Briefing No. 91-7 (May 1991) by John C. Person.

[78]*See, e.g., Manuel Bros., Inc. v. United States,* 55 Fed. Cl. 8 (2002); *Emerald Maint., Inc.,* ASBCA No. 43929, 98-2 BCA ¶ 29,903; *Weaver-Bailey Contractors, Inc., v. United States,* 24 Cl. Ct. 576 (1991); *Metropolitan Paving Co. v. United States,* 325 F.2d 241 (Ct. Cl. 1963).

completion without incurring liability. It would seem to make little difference whether or not the parties contemplated an early completion, or even whether or not the contractor contemplated an early completion. Where [the government] is guilty of "deliberate harassment and dilatory tactics" and a contractor suffers damages as a result of such action, we think that [the government] is liable.[79]

In order for a contractor to recover for delays that prevented early completion, the contractor must demonstrate that the contractor: (1) intended to complete the contract early; (2) had the capability to do so; and (3) would have completed early but for the government's actions.[80] In applying this test, the boards or courts look for direct and concrete evidence of the contractor's actual intent to complete early and its actual anticipated work schedule prior to the delay.

For example, in *Interstate General Government Contractors v. West,* the contractor finished 13 days early despite a 136-day delay in the issuance of the notice to proceed. The contractor argued that the actual early completion date proved that it had the intention and capability to finish at least six months (the length of the delay) ahead of schedule and would have done so but for the delay.[81] The court rejected this argument, finding that there was no evidence of an intention to finish early. To the contrary, there was testimony that the contractor accelerated the work after the delay by adding personnel not originally contemplated when the job was first estimated. In the court's view, the need for acceleration belied the contractor's argument that it intended or planned to complete early all along. Similarly, the contractor failed to present any work plan or schedule developed prior to the delay that would have feasibly resulted in early completion. Finally, the court held that the contractor's "post-facto, conclusory, self-serving assertion" that it would have completed earlier but for the delay was insufficient to satisfy the third prong of the analysis.[82] In applying the logic set forth in *Interstate General,* it appears that only those contractors that can present documents or other direct evidence of an actual intent to finish early that predate the occurrence of the delay may be able to recover delay costs despite an early completion date.[83] However, at least one board decision has held that the contractor need not provide the government early notification of its intent to complete ahead of schedule as a necessary prerequisite to recover for a delayed early completion.[84]

---

[79]*Metropolitan Paving Co. v. United States,* 325 F.2d 241 (Ct. Cl. 1963).

[80]*See Interstate Gen. Gov't Contractors v. West,* 12 F.3d 1053 (Fed. Cir. 1993) (citing *Elrich Contracting, Inc.,* GSBCA No. 10936, 93-1 BCA ¶ 25,316; *Frazier Fleming Co.,* ASBCA No. 34537, 91-1 BCA ¶ 23,378).

[81]*See Interstate Gen. Gov't Contractors v. West,* 12 F.3d 1053, 1059 (Fed. Cir. 1993).

[82]*Id.*

[83]*See RobGlo, Inc.,* VABCA No. 2879, 91-1 BCA ¶ 23,357 (holding that the contractor finishing early despite delay must prove that (1) the contractor intended to finish early, (2) the government was aware of the accelerated schedule and agreed to it or acquiesced in it, and (3) delays occurred during the time period in issue for which the government was solely responsible).

[84]*Rampart Waterblast, Inc.,* IBCA No. 3658-96 *et al.,* 98-2 BCA ¶ 29,894.

VI. CONCURRENT DELAY

"Concurrent delay" is a term of art that generally refers to a situation in which an excusable compensable delay and an unexcusable delay overlap for some period of time. Concurrent delay analysis is an analytical framework for identifying and evaluating construction delays. Concurrent delays are multiple delays to project work that occur, in part, at the same time and that impact the critical path of a planned sequence of events.[85] An example of a concurrent delay is where a contractor cannot commence work on the second phase of a project because the owner has failed to obtain a necessary right-of-way while at the same time the contractor is precluded from beginning work on the second phase due to its failure to timely complete phase one of the work.

Concurrent delay creates complex legal issues in determining which party ultimately is responsible for overall project delay. The analysis of concurrent delays may be further complicated if the delay periods: (1) are different lengths, (2) do not totally overlap, or (3) have different impacts on the number and types of work activities they affect and the severity of the impact on the affected work activities is different for each of the delays.[86]

Traditionally, a concurrent delay of the project performance precluded either party from obtaining an affirmative recovery from the other. The courts took the view that when a party proximately contributed to the delay, the law would not provide for the apportionment of damages between the parties.[87] The current trend in concurrent delay analysis, however, is moving toward the apportionment of the delay between the parties. This shift in philosophy is due, in part, to the use of the critical path method techniques for evaluating delay claims. These techniques make it possible to more accurately segregate and quantify the impact of concurrent delays.[88]

Logically, if the impact of one delay exceeds that of the other, the party responsible for the lesser impact should be allowed to recover damages for the excess impact. Apportionment analysis, at least on its face, would seem to allow for more equitable results than the traditional nonapportionment analysis. In apportioning delays, if the effects of concurrent delay cannot be accurately segregated and quantified, the court likely will revert to the traditional approach and bar either party from recovering delay damages. [89]

---

[85]*Singleton Contracting Corp. v. Harvey,* 395 F.3d 1353 (Fed. Cir. 2005).

[86]*See* Bidgood, Reed, & Taylor, *Cutting the Knot on Concurrent Delays,* Construction Briefings No. 08-2 (Thomson/West 2008) for a more detailed discussion of the concept of concurrent delay.

[87]*See Engineered Sys., Inc.,* DOTCAB No. 76-12, 78-1 BCA ¶ 13,074 (recognizing that recovery will generally be denied where the evidence shows the government caused delays are concurrent or intertwined with other delays for which the government is not responsible); *see also Commerce Int'l, Inc. v. United States,* 167 Ct. Cl. 529 (1964).

[88]*R.P. Wallace, Inc. v. United States,* 63 Fed. Cl. 402, 409 (2004); *Sunshine Constr. & Eng'g, Inc., v. United States,* 64 Fed. Cl. 346, 372 (2005).

[89]*See, e.g., SIPCO Servs. & Marine, Inc. v. United States,* 41 Fed. Cl. 196, 225–226 (1998); *Blinderman Constr. Co., Inc. v. United States,* 39 Fed. Cl. 529, 543–544 (1997).

In *Essex Electro Engineers, Inc. v. Danzig*,[90] the Federal Circuit reversed an ASBCA decision denying a contractor's claim for delay damages. The court of appeals determined that each party's delay was apportionable and, thus, should have been allocated to each responsible party. In so doing, the Federal Circuit noted:

> [T]he contractor generally cannot recover for concurrent delays for the simple reason that no causal link can be shown: a government act that delays part of the contract performance does not delay "the general progress of the work" when the "prosecution of the work as a whole" would have been delayed regardless of the government's act.[91]

However, the court recognized that "in recent cases, the principle has been characterized as requiring the government's act to have affected activities on the critical path."[92] Accordingly the court concluded "if 'there is in the proof a clear apportionment of the delay and the expense attributable to each party,' then the government will be liable for its delays."[93]

VII. SUSPENSION OF WORK

A. Effect of No Contract Clause

Under traditional principles of contract law, if a contract contains no clause giving the government the right to suspend the contract work, the issuance of such an order would be a breach of a duty owed by the government to the contractor.[94] However, in *Robert A. & Sandra B. Moura*,[95] the board found that the government had an inherent right to order a temporary suspension given its contractual right to terminate for default. According to the board, this contractual authority encompassed the lesser right to suspend performance of the project work.[96]

B. Suspension of Work Clause

In February 1968, the government developed a Suspension of Work clause for mandatory use in all government construction contracts. The current version of that clause,

---

[90]*Essex Electro Eng'rs, Inc. v. Danzig*, 224 F.3d 1283 (Fed. Cir. 2000).

[91]*Id.* (citing *Coath & Goss, Inc. v. United States*, 101 Ct. Cl. 702, 714–715 (1944)).

[92]*Id.* (citing *Mega Constr. Co. v. United States*, 29 Fed. Cl. 396, 424 (1993)).

[93]*Id.* (citing *Coath & Goss, Inc. v. United States*, 101 Ct. Cl. 701, 714–715 (1944)); *see also United States v. Killough*, 848 F.2d 1523, 1531 (11th Cir. 1988).

[94]*See generally Empire Gas Eng'g Co.*, ASBCA No. 7190, 1962 BCA ¶ 3323 (recognizing that in the absence of the contract's Suspension of Work clause, the government's act in suspending the project work would have been in violation of a duty owed by the government to the contractor, e.g., the duty not to interfere with the contractor's performance of the contract).

[95]PSBCA No. 3460, 96-1 BCA ¶ 27,956.

[96]*Id.*

which is set forth in FAR § 52.242-14, was developed to provide an administrative remedy for recovery of costs caused by government interruptions, delays, or suspensions of work. A copy of the text of the clause is found in **Appendix 9A**. Thus, pursuant to the Suspension of Work clause, a contractor will be compensated where a suspension of the contract work is for an unreasonable length of time and causes the contractor additional cost or expense not due to its own fault or negligence.

As will be discussed, the Suspension of Work clause provides a mechanism for recovery of costs resulting from both express orders suspending the contract work and constructive suspensions of the work.

C. Ordered Suspensions of Work

Subsection (a) of the Suspension of Work clause provides, in part:

(a) The Contracting Officer may order the Contractor in writing, to suspend, delay or interrupt all or any part of the work of this contract for the period of time that the Contracting Officer determines appropriate for the convenience of the Government.

Generally, only the contracting officer or a representative of the contracting officer is authorized to order a suspension of the contractor's work. However, in limited circumstances, the actions and documentation of another government representative may be properly deemed a suspension of work order triggering relief under the Suspension of Work clause. For example, in *Stamell Construction Co., Inc.*,[97] the contractor suspended its precast operations after receiving a field memorandum from the government's inspector.[98] The field report, which strongly suggested that the contractor suspend its precast work, was issued at the same time the contracting officer sent a letter threatening to reject the contractor's work and to withhold progress payments. Given these circumstances, the board determined that the contractor reasonably viewed the combination of the government actions as amounting to a stop work order on that part of the project.[99]

As is evident from its express terms, section (a) of the Suspension of Work clause is limited to situations where there is a written order from the contracting officer or a representative of the contracting officer suspending the contract work. That provision affords no relief where the contractor's performance is delayed through actions of the government that effectively suspends the contract work but is not accompanied by an express written directive from the contracting officer. These types of suspensions, commonly known as constructive suspensions, are addressed in section (b) of the clause and are reviewed in **Section VII.D**.

---

[97]DOTCAB No. 68-27J, 75-1 BCA ¶ 11,334.
[98]*Id.*
[99]*Id.*

D. Constructive Suspensions of Work

The Suspension of Work clause explicitly recognizes the constructive suspension of work doctrine. That is, section (b) provides, in relevant part:

(b) If the performance of all or any part of the work is, for an unreasonable period of time, suspended, delayed or interrupted (1) by an act of the Contracting Officer in the administration of this contract or, (2) by the Contracting Officer's failure to act within the time specified in this contract (or within a reasonable time if not specified), an adjustment shall be made for any increase in the cost of performance of this contract (excluding profit), necessarily caused by the unreasonable suspension, delay, or interruption, and the contract modified in writing accordingly. However, no adjustment shall be made under this clause for any suspension, delay or interruption to the extent that performance would have been so suspended, delayed, or interrupted by any other cause, including the fault or negligence of the Contractor, or for which an equitable adjustment is provided for or excluded under any term or condition of this contract.[100]

As illustrated by this text of the clause, this provision provides a mechanism for an administrative resolution of contractor claims based on unreasonable delays and interruptions of the contract work where there is no express government order to suspend the work. The clause is triggered and becomes operative whenever there is an act or omission of the government that results in an unreasonable interference with or delay of the contractor's performance. Examples of constructive suspensions include: unreasonable delay in issuing the notice to proceed,[101] delay in making the site available,[102] delay in issuing change orders,[103] and delay in providing information.[104]

---

[100] FAR § 52.242-14.

[101] See *Nicon, Inc. v. United States,* 331 F.3d 878 (Fed. Cir. 2003) (recognizing government's implied obligation to issue notice to proceed within reasonable time); *Marin Constr. & Dredging,* ASBCA No. 38412, 95-1 BCA ¶ 27,286 (government breached implied duty to issue notice to proceed within a reasonable time); *Goudreau Corp.,* DOTBCA No. 1895, 88-1 BCA ¶ 20,479 (three-month delay between furnishing of performance bonds and payment bonds and the issuance of notice to proceed was unreasonable); *Freeman Elec. Constr. Co.,* DOTCAB No. 74-23A, 77-1 BCA ¶ 12,258.

[102] See *Strand Hunt Constr. v. General Services Administration,* GSBCA No. 12859, 95-2 BCA ¶ 27,690 ("where government warrants that a job site will be available to the contractor by a particular date, and then denies access, it is liable for the cost of delay under the Suspension of Work clause"); *Singleton Contracting Corp.,* GSBCA No. 9614, 90-3 BCA ¶ 23,125 (recognizing that where work is delayed due to unavailability of job site a constructive suspension occurs); *Eickhof Constr. Co.,* ASBCA No. 20049, 77-1 ¶ 12,398; *Louis M. McMaster, Inc.,* AGBCA No. 76-156, 79-1 BCA ¶ 13,701.

[103] See *River Equip. Co., Inc.,* ENGBCA No. 5856, 93-2 BCA ¶ 25,654 (recognizing the "governing principle" that claims for delays in issuing change order falls under the Suspension of Work clause); *Decker & Co., GmbH,* ASBCA No. 35051, 88-3 BCA ¶ 20,871 (claims for delays in issuing change orders or modifications fall under the Suspension of Work clause); *Demauro Constr. Corp.,* ASBCA No. 12514, 73-1 BCA ¶ 9830.

[104] See *Amelco Elec.,* VABCA No. 3785, 96-2 BCA ¶ 28,381 (unreasonable delay that precedes determination of how to deal with a differing site condition is compensable under the Suspension of Work clause); *Berrios Constr. Co., Inc.,* VABCA No. 3152, 92-2 BCA ¶ 24,828 (where contractor is legitimately awaiting important information from government, and that information is needed in order to proceed with meaningful work, any delay damages related thereto are compensable under the Suspension of Work clause); *Royal Painting Co., Inc.,* ASBCA No. 20034, 75-1 BCA ¶ 11,311.

E. Reasonable versus Unreasonable Delays

Recovery of costs under the Suspension of Work clause is granted only where the resulting delays are unreasonable in duration. In this regard, there is no set formula for determining whether a period of delay is reasonable or unreasonable. The contractor seeking an equitable adjustment to its contract generally has the burden of proving the unreasonableness of the period of delay given the circumstances. However, where the government has exclusive knowledge of the facts surrounding the suspension, it is obligated to prove the reasonableness of the delay period.[105]

F. Preaward Delays

The Suspension of Work clause does not apply to preaward delays.[106] For example, in *K-W Construction, Inc. v. United States*,[107] the low bidder on a government contract appealed from a final decision of the ASBCA denying its damages allegedly incurred because of the government's delay in awarding contract. Specifically, the contractor's original bid was rejected as nonresponsive.[108] The contractor protested the rejection of its bid to the Comptroller General, who subsequently determined that the rejection was improper. However, "because the delay in this case occurred prior to the administration of the contract, [the contractor's] claim for delay damages is not cognizable under the suspension of work clause."[109] Accordingly, the contractor's claim for damages related to the delay in awarding the contract was denied.

G. Sovereign Acts

Generally, a sovereign act that interferes with the contractor's work will not be considered a constructive suspension. A recent case, *Conner Bros. Construction Co., Inc. v. United States*,[110] which developed in the context of the response to the terrorist attacks on the World Trade Center and the Pentagon, demonstrates the additional risk that government contractors, especially those working on defense contracts, face. In 2000, Conner Bros. entered a contract to build an Army Rangers headquarters facility at Fort Benning, Georgia.[111] Work extended until September 11, 2001, at which point Conner Bros. had completed approximately 70 percent of the work.[112] Based on the assertion of the Army Rangers' need for a secure post-9/11 training area on the base, Conner Bros. was excluded from the project site even after contractors and other civilians were allowed to return to other parts of the installation.[113]

---

[105]*See P & A Constr. Co., Inc.*, ASBCA 29901, 86-3 BCA ¶ 19,101; *M.A. Santander Constr., Inc.*, ASBCA No. 15822, 76-1 BCA ¶ 11,798.
[106]*M.A. Mortenson Co.*, ENGBCA No. 4780, 87-2 BCA ¶ 19,718.
[107]671 F.2d 481 (Ct. Cl. 1982).
[108]*Id.* at 482.
[109]*Id.* at 484.
[110]550 F.3d 1368 (Fed.Cir. 2008).
[111]*Id.* at 1370.
[112]*Id.*
[113]*Id.* at 1371.

The contracting officer granted Conner Bros. additional time due to the delay but denied its monetary claim.[114] After the ASBCA denied the claim, the Federal Circuit ruled against the contractor, invoking the principle of sovereign acts.[115] The court ruled that exclusion from the area of the installation used by the Army Rangers was a public and general act and was not directed at Conner Bros' contract rights.[116] In addition, the exclusion was "incidental to the achievement of a broader government objective relating to national security."[117] Contractors working in and around military installations should recognize that their ability to complete work on a project may be subject to factors similar to the ones that Conner Bros. faced.

Although it is well settled that the government can agree by contract to compensate the contractor for delays caused by sovereign acts,[118] the Suspension of Work provision by itself does not constitute such an agreement.[119] That said, if a sovereign act causes the contracting officer to suspend the project, the clause will be triggered and the contractor will be entitled to a recovery.[120]

VIII. TIME EXTENSIONS

A. Notice

Most construction contracts require the contractor to submit written notice to the government, within a definite period of time after the delay-causing event, prior to submitting any claim for additional compensation or for an extension of contract time. Such notice requirements are imposed to protect the interest of the government, which may be unaware of the causes of a particular delay and may thereby be precluded from taking immediate measures to rectify the situation and mitigate the cost of the delay. Failure to give prompt notice may result in a waiver of the contractor's rights or result in a time-consuming litigation effort that ultimately may prove unsuccessful.[121]

Formal notice may be unnecessary when the government has actual or constructive knowledge of the problem or when the lack of notice does not prejudice a legitimate

---

[114]*Id.*

[115]*Id.*

[116]*Id.* at 1375.

[117]*Id.* at 1379.

[118]*See D&L Constr. Co. & Assoc. v. United States,* 402 F.2d 990 (Ct. Cl. 1968).

[119]*See Amino Bros. Co. v. United States,* 178 Ct. Cl. 515 (1967); *but see Henderson, Inc.,* DOTBCA No. 2423, 94-2 BCA ¶ 26,728 (contract affirmatively represented that contractor would be allowed to work during specific time frame).

[120]*See Empire Gas Eng'g Co.,* ASBCA No. 7190, 1962 BCA ¶ 3323.

[121]*See Paul Hardeman, Inc.,* ENGBCA No. 2889, 69-2 BCA ¶ 7833 (ruling that the written notice requirement was clear and explicit and, therefore, the time limitation on recovery of costs will be strictly enforced).

government interest.[122] The contractor, however, should never knowingly forgo written notice on the assumption that one of those conditions is present. The contractor that gives the government prompt written notice of delays and disruptions that are the government's responsibility increases its opportunity to recover the costs generated by those problems.[123]

B. Documentation of Delay

A CPM analysis often is used to assess or present a delay claim. However, whether a CPM is used to establish a delay to the project or not, the basic project documentation is often critical to establishing the validity of the analysis.[124] This checklist itemizes many of the sources of information to be evaluated when preparing or attempting to rebut a delay claim.

CHECKLIST FOR DELAY DOCUMENTATION

- Estimates
- Original schedules
- Schedules used on the project, including look-ahead schedules, CPM logic diagrams, and tabular printouts
- As-built schedules
- Daily reports
- Diaries
- Manpower and manloading reports
- Cost-accounting records
- Scheduling meeting minutes
- Material and equipment delivery tickets
- Job photographs and videotapes
- As-built drawings
- Shop drawing logs
- Project correspondence
- Change orders
- Contract documents
- Pay applications
- Internal memoranda

---

[122]*See Hoel-Steffen Constr. Co. v. United States,* 456 F.2d 760 (Ct. Cl. 1972) (in discussing a contract notice provision, the Court of Claims cautioned "[t]o adopt [a] severe and narrow application of the notice requirements . . . would be out of tune with the language and purpose of the notice provisions, as well as with this court's wholesome concern that notice provisions in contract adjustment clauses not be applied too technically and illiberally where the Government is quite aware of the operative facts); *see also P.I.O. GmbH Bau und Ingenieurplanung v. Int'l Broadcasting Bureau,* GSBCA No. 16649, 04-1 BCA ¶ 32,592.
[123]*See* **Chapter 14** for a further discussion of notice.
[124]*See Whitesell-Green, Inc.,* ASBCA Nos. 53938, 53939, 54135, 06-2 BCA ¶ 33,323.

When evaluating a potential claim for delay, the next checklist is a useful tool for reviewing the pertinent factual information.

CHECKLIST FOR DELAY ANALYSIS

- *Accuracy.* Are the schedules used for the project accurate? Were they agreed on and used by the parties, or were they issued for "internal purposes" only? A board or Court of Federal Claims may give more weight to schedules to which the parties have previously agreed.
- *Abandonment of schedule.* Was the selected scheduling technique abandoned during performance? If so, why? A contractor that committed by contract to a particular scheduling technique might be precluded from proving its claim with that technique if it did not meet its scheduling commitments.
- *Schedule updates.* Was the schedule updated and kept current on a regular basis? Schedules often change dramatically over the course of a project.
- *Changes.* Was the schedule revised to reflect the effect of change orders? The schedule should show whether the change impacted work along the critical path or consumed float. A change order impact analysis is a handy tool for negotiating the price of a change.
- *Change order compensation.* Was additional overhead included in the change order? Even if the changed work affects only float, it may result in less effective resource utilization or involve unforeseen overhead or job staffing.
- *Cross-references.* Are the project records tied into the project schedule by work activity code or designation? Doing so provides data for subsequent updates and for the preparation of an accurate as-built analysis.
- *Float ownership.* Does the contract bar the contractor from seeking compensation, time extensions, or both for delays that consume only float time?
- *Coordination responsibility.* Which party is responsible for coordinating and scheduling?
- *Scheduling experts.* Obtain expert or in-house scheduling assistance early. This may aid in efficient record keeping and assist in minimizing the effect of a delay.

When pursuing a delay claim, it is essential that the claimant select a scheduling expert who is intimately familiar with the construction industry, how a construction project is run, and the scheduling methodology used by that federal agency on the specific contract. Specifically, an expert should be well versed in:

(1) The theory and output of commonly used scheduling techniques
(2) The theory and output of all scheduling techniques used
(3) The estimating process used and its relation to the contractor's resources
(4) The contractual relationships among the government, designers, contractors, subcontractors, and suppliers
(5) Good project record keeping and cost accounting
(6) The design and construction of the type of work involved

C.   Measuring Delay

To prove or refute delay claims, contractors and the government typically perform a detailed examination of project records and project schedules. Delay claims typically are based on contemporaneous CPM schedules that were generated during the course of the project or on after-the-fact as-built CPM schedules as reconstructed by a scheduling expert. The relative weight to be given to either approach has been vigorously contested over the last three decades. The cases that follow illustrate some of the key issues related to using CPM schedules to prove delay claims.

For example, the Department of Veterans Affairs Board of Contract Appeals (VABCA) addressed the question of using after-the-fact CPMs in two decisions in 1984 and 1987, both bearing the same name, *Santa Fe, Inc.*[125] The *Santa Fe* decisions indicate the board's clear preference for contemporaneous schedules when assessing delay claims.

In 1984, the VABCA denied a contractor's claim for time extensions because "the delayed activities had not been on the critical path for completion of the project."[126] The VABCA held that the "critical path calculations had been made for each task involved, and these calculations were continuously updated during construction."[127] The board further elaborated that "it is the very existence of the contractually agreed upon CPM procedure which, when properly utilized, allows the contracting officer and subsequent bodies to determine with greater exactitude whether, and to what extent, a particular change order affects critical path and hence delays ultimate performance."[128]

In the 1987 *Santa Fe, Inc.* decision, the VABCA determined that contract extensions "shall be based upon the computer-produced calendar dated schedule for the time period in question and all other relevant information."[129] In this decision, the VABCA held that "submission of proof based on revised activity logic durations and cost is obligatory to any approvals."[130] The board placed reliance on the most current CPM schedule developed when performing an analysis as to whether the contractor incurred delay on a VA contract or not. According to this decision, "there is a rebuttable presumption of correctness attached to CPM schedule upon which the parties have previously mutually agreed."[131] "To put it another way," the board stated, "in the absence of compelling evidence of actual errors in the CPMs, we will let the parties 'live or die' by the CPM applicable to the relevant time frames."[132] The *Santa Fe* decisions illustrate the rationale and preference for contemporaneous schedules when calculating delay claims.

---

[125]*See Santa Fe, Inc.*, VABCA No. 1943 *et al.*, 84-2 BCA ¶ 17,341; *Santa Fe, Inc.*, VABCA No. 2168, 87-3 BCA ¶ 20,104.
[126]*Santa Fe, Inc.*, VABCA No. 1943 *et al.*, 84-2 BCA ¶ 17,341.
[127]*Id.*
[128]*Id.*
[129]*Santa Fe, Inc.*, VABCA No. 2168, 87-3 BCA ¶ 20,104.
[130]*Id.*
[131]*Id.*
[132]*Id.*

In *P.J. Dick*,[133] the VABCA reiterated its preference for a contemporaneous schedule as its baseline for delay claims. In this case, the board was faced with a "rather unique situation." The CPMs used during the course of the project were, according to the board:

> properly constituted in its logic and assiduously and properly maintained throughout contract performance. This circumstance is in sharp contrast to the usual problems we encounter in dealing with CPMs where warring as-built schedules are constructed by the parties after the fact because the CPM was either never properly or timely prepared or was not updated in accordance with the contract scheduling requirements.[134]

The VABCA determined that "since there is no dispute concerning the validity of a CPM, these appeals present the circumstance where we have said in the past that we will let the parties 'live or die' by analysis of the CPM to determine the number of days of additional contract performance time."[135]

The *P.J. Dick* decision underscores the recent theme discussed in several boards of contract appeals decisions: the requirement of extensive documentary evidence that demonstrates the alleged claims before awarding the appropriate relief.[136] Boards are becoming more demanding and requiring more concrete evidence in their analysis of a contractor's delay claim.[137]

These increased demands should have a direct, practical impact on those contractors that rely on government contracts as a source of business. Project schedules should be updated routinely, they should reflect all changes, and they should be approved and shared with all interested parties through the life of the project.

Does the *P.J. Dick* decision end the use of after-the-fact as-built schedules in the analysis of delay claims? Possibly not. The strong preference expressed by some tribunals for contemporaneous analysis notwithstanding, the quantification of delays may also be developed retrospectively. If contemporaneous schedules are flawed or do not exist, an accurate and realistic after-the-fact schedule can be useful in a delay claim.

---

[133]*P. J. Dick, Inc.*, VABCA Nos. 5597, 5836, 01-2 BCA ¶ 31,647.
[134]*Id.*
[135]*Id.*
[136]*See, e.g., Thomas & Sons Bldg. Contractors, Inc.*, DOTBCA 3013, 01-1 BCA ¶ 31,386 (denying claim in part because contractor failed to offer any documentary evidence in support of its numerous delay claims); *Gavosto Assocs., Inc.*, PSBCA 4058, 01-1 BCA ¶ 31,389 (denying claim in part because "without contemporaneous records or other persuasive evidence, the Board [could not] determine the incidence of any delaying actions and the impact, if any on project completion").
[137]*Id.*

D. Burden of Proof

It is well settled that the contractor bears the burden of proof in demonstrating that a delay is excusable.[138] The contractor must satisfy this standard by a preponderance of the evidence (i.e., more than 50 percent).[139]

IX. RELATIONSHIP OF DELAY REMEDIES UNDER THE SUSPENSION OF WORK AND CHANGES CLAUSES

The Suspension of Work clause specifically provides that "no adjustment shall be made under this clause for any suspension, delay or interruption . . . for which an equitable adjustment is provided for or excluded under any term or condition of this contract."[140] Most clauses in construction contracts authorizing equitable adjustments include compensation for delays caused by the event triggering the clause. As a result, other clauses in the contract frequently will resolve a delay claim that would otherwise be governed by the Suspension of Work clause. In particular, claims that would otherwise be governed by the Suspension of Work clause frequently are resolved under the Changes clause.[141]

This rule notwithstanding, the inclusion of a Changes clause in a contract does not always preclude recovery under the Suspension of Work clause. As noted by the board in *Piracci Construction Co.*:[142]

> In our opinion, the Government has misinterpreted the clause. The [legislative] "history" makes it clear that, where a claim can be made under the Changes and the Suspension of Work clauses, use of the Changes clause is preferred to use of the Suspension clause. It does not, as the Government wants us to conclude, provide that availability of relief under the Changes clause precludes relief under the Suspension clause.

Thus, under the reasoning of the board in *Piracci Constuction Co.*, if a contractor fails to satisfy all the prerequisites for recovery under the Changes clause, it still may be entitled—in theory—to recover on that claim under the Suspension of Work clause. As a practical matter, most contractors choose to advance a claim under the Changes clause because it allows for profit as an element of recovery. However, the perceived advantage of the recovery of profit under the Changes clause needs to

[138]*See Williamsburg Drapery Co.*, ASBCA No. 5484, 61-2 BCA ¶ 3111.

[139]*See Mil-Craft Mfg., Inc.*, ASBCA No. 19305, 74-2 BCA ¶ 10,840.

[140]FAR § 52.242-14 (b).

[141]*Carpenter Constr. Co.*, NASABCA No. 18, 1964 BCA ¶ 4452; *Mech-Con Corp.*, GSBCA 1373, 65-1 BCA ¶ 4574.

[142]GSBCA No. 3477, 74-2 BCA ¶ 10,800.

be evaluated in the context of agency supplements to the Changes clause that can operate to substantively reduce the recovery for extended job site general conditions expense by limiting the amount of the general conditions expense to a percentage of the direct cost of the change.[143]

One topic that illustrates the relationship of the two clauses is the treatment of delays preceding the issuance of a change order. Basically, the Suspension of Work clause is applicable to delays preceding the issuance of a change order.[144] That is, while the government is entitled to a *reasonable* period of time to issue a change order,[145] an *unreasonable* delay in acting on a change will trigger the Suspension of Work clause.[146] Moreover, the contractor may be entitled to relief under that clause if the government issues multiple changes delaying the work.[147]

X. ACCELERATION

When the government requires the contractor to complete the work by a date earlier than the contract completion date, an "acceleration" occurs. For the purpose of determining whether the contractor's work has been accelerated, the contract completion date should reflect time extensions due the contractor for excusable delays to its work.

An acceleration of the contractor's work occurs under two different circumstances: (1) actual (or directed) acceleration or (2) constructive acceleration. Actual acceleration occurs when the government expressly directs its contractor to complete the project earlier than the contract completion date. Constructive acceleration occurs when the government fails to grant its contractor time extensions to which it is entitled and the contractor is required to achieve, or strive for, a completion date that is earlier than the properly extended contract completion date. Thus, acceleration may be a by-product of delay or other factors that justify a time extension that is not formally granted by the government.

Acceleration damages usually include premium time pay in the form of overtime or shift work, the cost of added crews or increased crew sizes, the cost of additional tools and equipment required for added crews, the cost of additional supervision and job site overhead, and the cost of labor inefficiency that may occur due to longer hours or increased crew sizes.

A. Directed Acceleration

In government contracts, the Changes clause used in fixed-priced construction contracts specifically allows the contracting officer to direct an acceleration in the performance of the work:

---

[143]See **Chapter 8, Section VIII.D.2**; VAAR § 852.236-88.
[144]*Weldfab, Inc.*, IBCA No. 268, 61-2 BCA ¶ 3121.
[145]See *Chaney & James Constr. Co. v. United States*, 421 F.2d 728 (Ct. Cl. 1970); *Pathman Constr. Co.*, ASBCA No. 22003, 82-1 BCA ¶ 15,790.
[146]*Triple "A" South*, ASBCA No. 43684, 94-2 BCA ¶ 26,609.
[147]*Id.*

The Contracting Officer may, at any time, without notice to the sureties, if any, by written order designated or indicated to be a change order, make changes in the work within the general scope of the contract, including changes—

* * *

(4) Directing acceleration in the performance of the work.[148]

Applying this standard, the government's order to a contractor to accelerate the modification of an air-conditioning system in a medical facility so that it could accelerate the completion of the operating rooms was a compensable change because the government was aware that excess costs for overtime would be incurred.[149]

B. Constructive Acceleration

In the seminal case *Fermont Division, Dynamic Corp. of America,*[150] the five basic elements of a constructive acceleration claim based on a government failure to grant a time extension for excusable delay were summarized:

(1) existence of a given period of excusable delay.
(2) contractor notice to the Government of the excusable delay, and request for extension of time together with supporting information sufficient to allow the Government to make a reasonable determination.

EXCEPTIONS:

    (a) such notice, request, and information are not necessary if the Government's order directs compliance with a given schedule expressly without regard to the existence of any excusable delay;
    (b) the supporting information is unnecessary if it is already reasonably available to the Government; and

(3) failure or refusal to grant the requested extension within a reasonable time.
(4) a Government order, either express or implied from the circumstances, to
    (a) take steps to overcome the excusable delay, or
    (b) complete the work at the earliest possible date, or
    (c) complete the work by a given date earlier than that to which the contractor is entitled by reason of the excusable delay. Circumstances from which such an order may be implied include expressions of urgency by the Government especially when coupled with

---

[148]FAR § 52.243-4.
[149]*E.C. Morris and Son, Inc.,* ASBCA No 20697, 77-2 BCA ¶ 12,622; *see also Gibbs Shipyard, Inc.,* ASBCA No. 9809, 67-2 BCA ¶ 6499.
[150]ASBCA No. 15006, 75-1 BCA ¶ 11,139.

(i) a threat of default or liquidated damages for not meeting a given accelerated schedule, or

(ii) actual assessment of liquidated damages for not meeting a given accelerated schedule.

(5) reasonable efforts by the contractor to accelerate the work, resulting in added costs, even if the efforts are not actually successful.[151]

If these elements are proven, the contractor is entitled to recover the costs incurred in accelerating its performance.[152]

A recent line of cases illustrates the difficulty of proving constructive acceleration and recovering acceleration costs in the context of alleged concurrent delays. In *Hemphill Contracting Co.*,[153] the contractor agreed to clear an island in the Mississippi in order for it to be submerged. The burning operation was delayed when the contractor and the government discovered that the specified method of burning violated a state law. The board did not allow recovery of the contractor's claimed acceleration costs because the board found that the burning delay was concurrent with other delays and the contractor was not able to isolate the delay period, which was solely the responsibility of the government. In *R.J. Lanthier Co., Inc.*, the board ruled against the a contractor presenting a pass-through claim and stated that the acceleration costs were not recoverable unless the contractor could clearly support apportionment and expenses attributable to each party.[154]

These cases could be read to signal a departure from the traditional rule of recovering acceleration costs due to any excusable, but not necessarily compensable, delay because a concurrent delay could never be apportioned as these two board cases require.[155] Although there might be situations where the delay is truly concurrent and apportionment is impossible, the boards in *Hemphill* and *R.J. Lanthier* ruled primarily based on the lack of evidence to support the claims of the contractors. However, in *Clark Construction Group, Inc.*, the board, citing both *Hemphill* and *R.J. Lanthier*, utilized CPM analysis to apportion the delay and allowed recovery as to the portion of the delay caused by the government.[156] Although this line of cases illustrates that recovery of acceleration costs can be difficult, a contractor that thoroughly documents a project will greatly improve its chances to recover acceleration costs in the event that a delay occurs.

---

[151]*Id.* at 52,999.

[152]*See Norair Eng'g. Corp. v. United States,* 666 F.2d 546 (Ct. Cl. 1981), *but see Fraser Constr. v. United States,* 384 F.3d 1354 (Fed. Cir. 2004) (claim for constructive acceleration denied where time extensions were issued within a reasonable time period). Since acceleration is a constructive change order in government contracts, this topic is addressed in greater detail in **Chapter 8.**

[153]ENGBCA Nos. 5698, 5776, 5840, 94-1 BCA ¶ 26,491.

[154]ASBCA, No. 51636, 04-1 BCA ¶ 32,481.

[155]*See A Government Windfall: ASBCA's Attack on Concurrent Delays as a Basis for Constructive Acceleration,* 44 The Procurement Lawyer 4 (2009).

[156]JCL BCA, No. 2003-1, 05-1 BCA ¶ 32,843.

➤ LESSONS LEARNED AND ISSUES TO CONSIDER

- Absent a clause *shifting the risk* for delays in completion, a fixed-price contractor is generally not excused from meeting the specified completion date due to unexpected difficulties unless performance is made *impossible* by an act of God, law, or the other party to the contract.
- The concept of *excusable delays* is essentially a creature of the contract between the parties.
- Most construction contracts contain an *express obligation* that the contractor will complete the work by a given date or within a specified time frame, which is accompanied by an *implied obligation* that neither party will do anything to delay, hinder, or interfere with the performance of the other.
- *Common causes* of delays include inclement weather, labor disputes, untimely equipment delivery, defective specifications, changes, and differing site conditions. Which party bears the risk for each of these types of delays depends on the language of the contract and the surrounding circumstances.
- An understanding of several key terms is important in evaluating who bears responsibility for construction delays:

    (1) An *excusable delay* is a delay that entitles the contractor to a time extension under the terms of the contract. Moreover, an excusable delay may be either compensable or noncompensable.

        (a) A *compensable* excusable delay is a delay that is not only excusable (entitling the contractor to a time extension) but also entitles the contractor to additional compensation for the resulting cost.

        (b) A *noncompensable* excusable delay is a delay for which the contractor is entitled to a time extension but no additional compensation.

    (2) A *nonexcusable* delay is a delay that does not entitle the contractor to a time extension and may subject the contractor to liability for delay damages arising out of the delay.

- To *determine* whether a given delay is nonexcusable or excusable and possibly a compensable delay, one must carefully study the contract, the nature of the delay, and all surrounding circumstances.
- Examples of delays that may constitute *excusable noncompensable delays* include: unusually severe weather, acts of God, and labor problems.
- Examples of some of the causes of delay that may constitute *excusable compensable delays* include: delayed notice to proceed; failure to: furnish adequate plans and specifications, provide access, coordinate multiple prime contractors, timely provide government furnished property, timely approve shop drawings/submittals, provide timely direction, timely inspect/accept work, make timely payments; and suspensions of the work.

- A contractor may recover for a compensable delay even though the contract was completed within the contract completion date. In order for a contractor to recover, it must *demonstrate* that the contractor: (1) intended to complete the contract early; (2) had the capability to do so; and (3) would have completed early but for the government's actions.
- "Concurrent delay" is a term of art that refers to the situation where two different delays, caused by different parties, occur *simultaneously or in overlapping time periods*, and one of the delays is a compensable delay while the other is a nonexcusable delay.
- The early view stated by the courts and administrative boards regarding concurrent delay was that *neither party* was allowed any affirmative recovery from the other in the case of concurrent delay.
- The modern trend is to *apportion responsibility* for project delays between the parties whenever it is possible, using modern, sophisticated scheduling techniques such as critical path method (CPM) scheduling, to segregate the impact of the concurrent delays.
- Pursuant to the Suspension of Work clause, a contractor will be compensated where a suspension of the contract work is for an *unreasonable length* of time and causes the contractor additional cost or expense not due to its own fault or negligence.
- The Suspension of Work clause provides a mechanism for recovery of costs resulting from both *express* orders suspending the contract work and *constructive* suspensions of the work.
- Most construction contracts require the contractor to provide the government prompt *written notice* of any excusable delay. A contractor's failure to provide such notice may jeopardize the contractor's right to a time extension and/or additional compensation.
- A persuasive and *credible scheduling analysis* and supporting documentation are critical to the successful presentation of any delay claim.
- Acceleration is another common source of construction claims and disputes. Acceleration may take two forms:
  (1) *Directed* acceleration, where the government explicitly directs the contractor to complete the work earlier than the contractually required completion date.
  (2) *Constructive* acceleration, where the government fails to grant a contractor a time extension to which it is entitled, thereby requiring the contractor to complete, or attempt to complete, the work by a date earlier than contractually required. Thus, constructive acceleration is, in effect, a possible by-product of an excusable delay.

- The six *essential elements* of a claim for constructive acceleration are:
  (1) An excusable delay
  (2) A timely request for a time extension
  (3) Failure or refusal by the government to grant the request for time extension
  (4) Conduct by the government that is reasonably construed as requiring the contractor to complete on a schedule that has not been properly extended
  (5) Effort by the contractor to accelerate performance
  (6) Additional costs incurred by the contractor as a result of the acceleration

APPENDIX 9A: SUSPENSION OF WORK (APR. 1984)

(a) The Contracting Officer may order the Contractor, in writing, to suspend, delay, or interrupt all or any part of the work of this contract for the period of time that the Contracting Officer determines appropriate for the convenience of the Government.

(b) If the performance of all or any part of the work is, for an unreasonable period of time, suspended, delayed, or interrupted (1) by an act of the Contracting Officer in the administration of this contract, or (2) by the Contracting Officer's failure to act within the time specified in this contract (or within a reasonable time if not specified), an adjustment shall be made for any increase in the cost of performance of this contract (excluding profit) necessarily caused by the unreasonable suspension, delay, or interruption, and the contract modified in writing accordingly. However, no adjustment shall be made under this clause for any suspension, delay, or interruption to the extent that performance would have been so suspended, delayed, or interrupted by any other cause, including the fault or negligence of the Contractor, or for which an equitable adjustment is provided for or excluded under any other term or condition of this contract.

(c) A claim under this clause shall not be allowed—

    (1) For any costs incurred more than 20 days before the Contractor shall have notified the Contracting Officer in writing of the act or failure to act involved (but this requirement shall not apply as to a claim resulting from a suspension order); and

    (2) Unless the claim, in an amount stated, is asserted in writing as soon as practicable after the termination of the suspension, delay, or interruption, but not later than the date of final payment under the contract.

10

INSPECTION, ACCEPT-
ANCE, COMMISSIONING,
AND WARRANTIES

Inspection, acceptance, commissioning, and warranties are closely related topics in construction. Inspection and acceptance are of great importance during the construction of the project, but they also affect the contractor's rights and obligations under warranties which may extend for years beyond project completion. Whereas warranty disputes which arise after project completion tend to be more legalistic in nature, disagreements regarding inspection standards and procedures as well as project completion and acceptance must be handled by project personnel on a practical basis each day. As with most topics, the provisions of a particular contract and the interpretation of these provisions are of great importance in resolving these issues. The purpose of this chapter is to provide a general overview of the parties' rights and obligations in regard to inspection, acceptance, commissioning, and warranties.

The government will in almost every instance utilize representatives to determine, by periodic inspections, if the contractor has met the required standards for material and workmanship. On a government construction project, the inspection function may be handled by government employees from the agency that awarded the contract, representatives of the agency that will ultimately use the project, or by employees of an outside architect/engineer.[1] It is almost inevitable that these inspection functions will involve questions of contract interpretation. A contractor may perceive that an inspector's interpretation of specification or drawing is, in reality, a change to the contract. In that context, the contractor needs to clearly understand who has *authority* to order changes and the importance of appropriate written notice to the contracting officer as discussed in **Chapters 2** and **8** of this book.

---

[1]As each of these entities may have a different perspective on the project, a contractor should seek to determine which entity will perform the inspections prior to submitting a proposal for the project. In addition, a contractor should seek to document the inspection process as the work proceeds to minimize the potential for misunderstandings in the event that the government's inspection team changes in the latter stages of the project.

While the government has the *right* to conduct inspections, that right cannot be exercised in such a manner as to delay or disrupt the contractor's work or to alter the requirements of the contract. Nonetheless, the government has no *duty* to inspect: inspection is generally for the protection of the government and not the contractor. Failure to effectively inspect, however, may affect the government's rights under warranties once the project is accepted. In addition to government inspection, some of the burden of assuring compliance with the contract requirements may be placed on the contractor through contractor quality control plan requirements and the like.

I. FAR CLAUSES AFFECTING INSPECTION AND ACCEPTANCE

A. Standard Inspection Clause

The rights and responsibilities of the government and contractor in a typical construction project are illustrated by the standard provisions regarding inspections. The standard Federal Acquisition Regulation (FAR) clause used in government construction contracting entitled Inspection of Construction[2] is a lengthy clause addressing many of the issues related to inspection and acceptance of the work. That clause provides:

INSPECTION OF CONSTRUCTION (AUG. 1996)

(a) *Definition.* "Work" includes, but is not limited to, materials, workmanship, and manufacture and fabrication of components.

(b) The Contractor shall maintain an adequate inspection system and perform such inspections as will ensure that the work performed under the contract conforms to contract requirements. The Contractor shall maintain complete inspection records and make them available to the Government. All work shall be conducted under the general direction of the Contracting Officer and is subject to Government inspection and test at all places and at all reasonable times before acceptance to ensure strict compliance with the terms of the contract.

(c) Government inspections and tests are for the sole benefit of the Government and do not—

(1) Relieve the Contractor of responsibility for providing adequate quality control measures;

(2) Relieve the Contractor of responsibility for damage to or loss of the material before acceptance;

(3) Constitute or imply acceptance; or

(4) Affect the continuing rights of the Government after acceptance of the completed work under paragraph (i) below.

[2]FAR § 52.246-12.

(d) The presence or absence of a Government inspector does not relieve the Contractor from any contract requirement, nor is the inspector authorized to change any term or condition of the specification without the Contracting Officer's written authorization.

(e) The Contractor shall promptly furnish, at no increase in contract price, all facilities, labor, and material reasonably needed for performing such safe and convenient inspections and tests as may be required by the Contracting Officer. The Government may charge to the Contractor any additional cost of inspection or test when work is not ready at the time specified by the Contractor for inspection or test, or when prior rejection makes reinspection or retest necessary. The Government shall perform all inspections and tests in a manner that will not unnecessarily delay the work. Special, full size, and performance tests shall be performed as described in the contract.

(f) The Contractor shall, without charge, replace or correct work found by the Government not to conform to contract requirements, unless in the public interest the Government consents to accept the work with an appropriate adjustment in contract price. The Contractor shall promptly segregate and remove rejected material from the premises.

(g) If the Contractor does not promptly replace or correct rejected work, the Government may—

(1) By contract or otherwise, replace or correct the work and charge the cost to the Contractor; or

(2) Terminate for default the Contractor's right to proceed.

(h) If, before acceptance of the entire work, the Government decides to examine already completed work by removing it or tearing it out, the Contractor, on request, shall promptly furnish all necessary facilities, labor, and material. If the work is found to be defective or nonconforming in any material respect due to the fault of the Contractor or its subcontractors, the Contractor shall defray the expenses of the examination and of satisfactory reconstruction. However, if the work is found to meet contract requirements, the Contracting Officer shall make an equitable adjustment for the additional services involved in the examination and reconstruction, including, if completion of the work was thereby delayed, an extension of time.

(i) Unless otherwise specified in the contract, the Government shall accept, as promptly as practicable after completion and inspection, all work required by the contract or that portion of the work the Contracting Officer determines can be accepted separately. Acceptance shall be final and conclusive except for latent defects, fraud, gross mistakes amounting to fraud, or the Government's rights under any warranty or guarantee.

The Inspection of Construction clause specifies that the government has the right of inspection "at all reasonable times and at all places prior to acceptance" of all contract work including "materials, workmanship, and manufacture and fabrication of components." This clause also makes clear that the inspection is solely for the government's benefit and does not constitute or imply acceptance of the contractor's work.

Therefore, the contractor is not relieved of the responsibility of ensuring compliance with contract requirements by virtue of the fact that the government has conducted inspections.[3] No contractor can ignore this basic allocation of rights and responsibilities related to inspection and acceptance of the work when subcontracting work or purchasing materials and equipment or during the performance of the work.[4]

B. Other FAR Clauses

Other standard clauses in government construction contracts relate to the contractor's performance and ultimately the inspection and acceptance of the work performed by the contractor. For example, the Material and Workmanship clause[5] provides that materials employed are to be new and of the most suitable grade unless the contract provides otherwise, that references to products by trade name are intended to set a standard of quality and not limit competition, that anything installed without required approval may be rejected, and that work must be performed in a skillful and workmanlike manner.

The Permits and Responsibilities clause[6] in government contracts requires the contractor to take proper safety and health precautions for the purpose of protecting the work, the workers, and the persons and property of others. It also states that the contractor is responsible for damage to persons or property caused by the contractor's fault or negligence, and places responsibility upon the contractor for all materials and work prior to acceptance by the government.

The Use and Possession Prior to Completion clause[7] provides that the government may take possession of or use a partially or totally completed portion of a project without being deemed to have accepted the work. Prior to such possession or use it is necessary for the contracting officer to provide the contractor with a list of work remaining to be done on the relevant portion of the project; however, the contractor still must comply with the contract terms even where the government fails to list a particular defect or item of work.

II. CONTRACTOR QUALITY CONTROL SYSTEMS

A. A Contractor's Basic Obligations

The contractor's inspection duties in the routine performance of a construction contract include not only the inspection of the work in place but an inspection of job conditions, including job cleanup, potential safety hazards, progress of the work,

---

[3]*S & M Mgmt. Inc. v. United States,* 82 Fed. Cl. 240, 249 (2008).
[4]The inspection, acceptance, and warranty provisions of purchase order forms and subcontract agreements need to be consistent with the rights and obligations set forth in the contract with the government.
[5]FAR § 52.236-5. (A copy of this clause is set forth in **Section VI** of **Chapter 6**.)
[6]*See, e.g.,* FAR §§ 36.507 and 52.236-7 (A copy of this clause is set forth in **Section VI** of **Chapter 6**.)
[7]*See, e.g.,* FAR § 52.236-11. (A copy of this clause is set forth in **Appendix 10A** to this chapter.)

and scheduling. The contractor must inspect subcontractors' and material suppliers' performance as well as the contractor's own performance. Most large construction contracts impose specific requirements on the contractor to perform such inspections. However, even if there is no specific contract requirement, prudence dictates that such inspection should be carried out routinely at all levels. It should include a regular process for reporting and exchanging information in order for the contractor to promptly, expeditiously, and economically complete the project.

The contractor may also be required to obtain test results of work in place or materials to be used. These usually are obtained through designated independent testing laboratories. For example, one usually must make test cylinders of the concrete mix for compressive strength testing. Sometimes such tests are required by the specification; in other cases they are imposed by industry standards incorporated expressly or impliedly in the contract documents. Such inspections not only satisfy the contractor's obligations to the government but also help it monitor the work. They also establish empirical data in the event disputes later develop or there is a failure.

In government construction contracts, the Inspection of Construction clause places the primary responsibility for contract compliance on the contractor. Section (b) of that clause obligates the contractor to *perform* such inspections to *ensure* that the work complies with the plans and specifications. That clause provides, in part:

(b)   The Contractor shall maintain an adequate inspection system and perform such inspections as will ensure that the work performed under the contract conforms to contract requirements The Contractor shall maintain complete inspection records and make them available to the Government.[8]

A contractor's failure to maintain an adequate inspection system may provide a basis for a government claim against the contractor even though the government is unable to establish the grounds to deny final acceptance of the work.[9] Similarly, the contract provisions and applicable regulations may also include various contractor record-keeping requirements. The contract specifications may also include provisions setting forth special inspection obligations, standards of performance, and required certifications of compliances with those standards.

B.   Contractor CQC Staffing Requirements

There is no standard FAR provision detailing the requirements for a contractor quality control (CQC) plan and staff. Consequently, an essential step in the preparation of the estimate for any government construction contract is the careful analysis of the special conditions and specifications to ascertain the requirements for the CQC staff. Due to variations in the technical nature of the work as well as varying approaches to CQC staffing followed by the different procuring activities even within the same agency, such as the Corps of Engineers, it is not safe to assume that the staffing

---

[8]FAR § 52.246-12.
[9]*See Kaminer Constr. Co. v. United States,* 488 F.2d 980, 987 (Ct. Cl. 1973).

requirements will be the same even on projects of similar size or duration. If, on a relatively small or noncomplex project, it appears that the CQC staff specification was simply copied from a more complicated or large project, an effort to obtain clarification or revision of those requirements should be made before the proposal or bid is submitted.

As part of the estimate preparation a checklist of CQC staff requirements can be useful in the evaluation of the requirements and related cost for each contract. A **Checklist for CQC Staff Requirements**, which can provide a framework for evaluating the scope of the staff requirements in each contract as part of the estimating and buyout process, is presented next.

CQC STAFF REQUIREMENTS CHECKLIST

- Is there a requirement for a separate CQC staff distinct from the project management team?
- How many individuals are required in the CQC staff organization?
- What are the required levels of education and/or experience for each member of the CQC staff?
- Can experience in a particular work discipline be substituted for formal education?
- To whom must each member of the CQC staff report? To whom does the senior member (manager) of the CQC staff report?
- May any member of the CQC staff be provided by/employed by a subcontractor or vendor?
- When must each member of the CQC staff be physically present at the job site? Does the contract require *full-time* presence or when certain work is *ongoing*?
- Are there special testing requirements that would necessitate the presence of additional manufacturers' technical representatives during the installation or startup of their equipment?
- Are there restrictions on the duties that can be assigned to the CQC staff members in addition to their quality control functions?
- Can CQC staff members perform essential project documentation related to extra work, delays, differing site conditions, and so on?

The CQC staff can present a major cost for the contractor or the various subcontractors. If it is not possible to ascertain clear answers to these questions from the solicitation, it may be appropriate to seek a written clarification of the CQC requirements from the government before submitting a proposal or bid in order to avoid a later, expensive surprise. At the same time, the project documentation routinely maintained by the CQC staff can prove invaluable in the event of a dispute or claim.[10]

---

[10]*See Whitesell-Green, Inc.,* ASBCA Nos. 53938, 06-2 BCA ¶ 33,323.

Related to the cost issue is the need to determine whether any member of the CQC staff may be an employee of a subcontractor or vendor. In *M.A. Mortenson Co.*,[11] the applicable specification expressly required that the CQC System Manager "shall be employed by the Contractor." There was no similar express requirement regarding the employment status for the electrical or mechanical members of the quality control staff. Based on its interpretation of the requirement that the contractor had to ensure that its subcontractor's work complied with the contract documents, the Armed Services Board of Contract Appeals (ASBCA) decided that the electrical and mechanical members of the CQC staff had to be employed by the contractor. In the board's view allowing the subcontractors to *police themselves* defeated the very purpose of requiring the contractor to have a CQC staff.

If there is any doubt about the agency's expectations regarding the composition, education, and employment status of the CQC staff, those questions should be addressed to the contracting officer prior to submission of the bid/proposal. If the questions are not clearly answered, then it would be appropriate to ask the questions at a preproposal conference or by formal written request for clarification in a manner so that every potential offeror/bidder has the same information.

III. STRICT COMPLIANCE VERSUS SUBSTANTIAL COMPLIANCE

It is well established that the government is entitled to strict compliance with its specifications.[12] This doctrine of *strict compliance* is somewhat softened by the concept of *substantial compliance*.[13] Substantial compliance is a judicially created concept; however, it does not require the government to accept defective work. Greater protection may be afforded a contractor by the doctrine of *economic waste* where the contractor has substantially complied with all contract requirements but there are minor defects in its work that will be very costly to repair.[14] This doctrine provides that nonconforming work will not be rejected if it is suitable for its intended purpose and the cost of correction would far exceed the gain that would be realized by correction. However, the government may be entitled to a reduction in price to reflect the diminished value.

The *economic waste* concept is well illustrated by the decision in *Granite Construction Co. v. United States*.[15] Granite entered into a contract with the Army Corps of Engineers (Corps) for the construction of a lock and dam. The dam and lock walls consisted of a series of concrete monoliths that were 60 feet high, 42 feet long,

[11]ASBCA No. 53349, 05-2 BCA ¶ 33,014.
[12]*See Blake Constr. Co., Inc. v. United States,* 28 Fed. Cl. 672 (1993), *aff'd,* 29 F.3d 645 (Fed. Cir. 1994).
[13]*See Technical Sys. Assoc., Inc.,* GSBCA Nos. 13277 and 14538, 00-1 BCA ¶ 30,684, *and Radiation Tech., Inc. v. United States,* 366 F.2d 1003 (Ct. Cl. 1966).
[14]*See Technical Sys. Assoc., Inc.,* GSBCA Nos. 13277 and 14538, 00-1 BCA ¶ 30,684.
[15]962 F.2d 998 (Fed. Cir. 1992). *See also Ball, Ball & Brosamer, Inc.,* IBCA No. 2103-N, 93-1 BCA ¶ 25,287; *Toombs & Co.,* ASBCA No. 34590, 91-1 BCA ¶ 23,403.

and 30 feet wide. The contract required Granite to embed a polyvinyl chloride (PVC) waterstop in the vertical joints between each monolith to prevent water leakage. After some of the waterstops were permanently embedded in the monoliths, the Corps tested the waterstops and determined that they did not meet the contract requirements. The Corps required Granite to remove and replace all the installed waterstops. Granite filed a $3.8 million claim with the Corps.

Granite's claim was denied, and it filed an appeal with the Corps of Engineers Board of Contract Appeals (ENGBCA).[16] Denying Granite's claim in its entirety, the board concluded that the contract required Granite to inspect and test the waterstops to *ensure* that they met contract specifications. The board further determined that the government had a right to insist on *strict compliance* with the contract specifications and to require the complete removal of the embedded, noncompliant waterstops.

Granite filed an action in the United States Claims Court challenging the board's decision. The court agreed with the board's ruling and granted the government's motion for summary judgment.

In reversing the Claims Court, the Federal Circuit addressed the right to require strict compliance as the only remedy for noncompliant work:

> We recognize that the government generally has the right to insist on performance in strict compliance with the contract specifications and may require a contractor to correct non-conforming work. However, there is ample authority for holding that the government should not be permitted to direct the replacement of work in situations where the cost of correction is economically wasteful and the work is otherwise adequate for its intended purpose. In such cases the government is only entitled to a downward adjustment in the contract price.[17] [Citations omitted.]

There are few decisions applying the doctrine of *economic waste* to inspection issues in government contracts. Therefore, it is difficult to predict whether that doctrine could be used successfully by contractors as a basis for an affirmative claim following a government directive to replace noncompliant work or as grounds to assert that the government should accept the in-place work with an appropriate credit. In any event, if a contractor believes that a government directive to replace work amounts to economic waste, that assertion with appropriate justification should be provided to the government *before* the work is replaced, not later. Even if the economic waste doctrine in the *Granite Construction* case is applicable, it is very likely that the government is still entitled to an appropriate reduction in the contract price.[18]

---

[16]*Granite Constr. Co.*, ENGBCA No. 4496, 89-1 BCA ¶ 21,447.
[17]962 F.2d at 1006–1007. *See also Ball, Ball & Brosamer, Inc.*, IBCA No. 2103-N, 93-1 BCA ¶ 25,287 *and Toombs & Co.*, ASBCA No. 34590, 91-1 BCA ¶ 23,403.
[18]*Granite Constr. Co.*, 962 F.2d 998.

IV. GOVERNMENT'S RIGHT TO INSPECT

The government's right and authority to inspect are set out in the clause at FAR § 52.246-12.[19] The same contract provision obligates the contractor to maintain an adequate system for inspecting the work and to inspect it for compliance with the plans and specifications.

A. No Duty to Inspect

Contemporaneous government inspection of the ongoing work for defects or omissions can operate to the contractor's benefit. If the contracting officer or the other government representatives reasonably object to the in-place work, it is often possible to modify the work to make it acceptable at a lower cost as compared to correcting the work after it is completed. If the government fails to object, waiver of the omission or deficiency *may* be inferred or, in the event of a doubtful or ambiguous requirement, the government's acquiescence may help establish that the government agreed with the contractor's interpretation at the time of performance.[20]

Aware of the risks of overlooking defects during inspection, the government seeks to minimize the contractor's ability to rely on government's inspection. The standard Inspection of Construction clause states that "Government inspections and tests are for the sole benefit of the Government and do not relieve the Contractor of responsibility for providing adequate quality control measures."[21] This provision makes clear that no inspection duty is imposed on the government but rather the government has the right to inspect should it elect to do so.

Notwithstanding this clause, the government may elect to impose on itself an affirmative duty to inspect the work. In those circumstances, contract specifications may specifically contemplate or require government tests during performance.[22] If the government then fails to inspect or test in accordance with the contract terms, that failure may cause the government to lose some of its specific rights and remedies, such as the right to reject the work or have defects corrected when a reasonable inspection, as set forth in the contract, would have disclosed such defects.

B. Effect of Government's Failure to Inspect

The government may inspect for itself, direct the contractor to maintain a quality control system that does the inspection, or both. The decision as to which system to employ often depends on the type of product or service being acquired. For example, when the government procures *commercial items*, these are generally purchased with

---

[19]See **Section I** of this chapter.

[20]*See J.R. Cheshier Janitorial*, ENGBCA No. 5487, 91-3 BCA ¶ 24,351; *Wilkinson & Jenkins Constr. Co.*, ENGBCA No. 5176, 87-2 BCA ¶ 19,950.

[21]FAR § 52.246-12(c)(1).

[22]*Cone Bros. Contracting Co.*, ASBCA No.16078, 72-1 BCA ¶ 9444. *But see Kelley Control Sys., Inc.*, VABCA No. 2337, 87-3 BCA ¶ 20,064.

no government inspections prior to delivery and with reliance only on the contractor's quality assurance program.[23] In contrast, contracts for government-unique, noncommercial items will almost always permit government in-process inspection and may mandate a quality control program meeting specific government requirements.[24] The degree and frequency of government inspection and quality assurance requirements varies with the nature of the item or work being acquired. The more critical the item, the more detailed the requirements. Ordinarily, government construction contracts require the contractor to implement a quality control (QC) inspection program while the government monitors with a quality assurance (QA) program.

V. LIMITS ON GOVERNMENT'S INSPECTION RIGHTS

The government has broad powers of inspection. For example, the scope of the government's right of inspection under the standard clause covers "all work," including "materials, workmanship, and manufacture and fabrication of components." In addition, inspection may be made at "all places and all reasonable times." Generally this means that the government may inspect at virtually any time within normal business hours, so long as the work is not unreasonably disrupted. The government may reinspect the same work, or may test or inspect at any place where work related to the project is being performed, whether at the construction site, a supplier's plant, or a contractor's or subcontractor's place of business. Nonetheless, a contractor has various rights and remedies resulting from constructive changes to the contract work, delays, and disruptions caused by the government's actions that are beyond the scope of its inspection rights.

A. Different Standards of Performance

The government has the discretion to use any reasonable inspection test. However, it may not obtain a higher standard of performance through the use of more stringent inspection procedures or tests than called for by the contract or which are not consistent with industry practice.[25] If a higher standard of performance is imposed as a consequence of the inspection procedure, the contractor may be entitled to additional compensation. Similarly, if the contractor is required by the inspector to use materials or construction methods that are not required by the contract and are more expensive than the contractor's chosen method, a compensable change may result.[26]

---

[23]FAR § 52.212-4(a). "Commercial item" is a defined term in FAR § 2.101. The definition can be somewhat subjective. In addition, some government agencies assert that an item of product that might qualify as a "commercial item" if purchased directly by the government loses that status as it becomes incorporated into a portion of the work being constructed by the contractor.

[24]FAR § 52.246-12(b).

[25]*Aerodyne Eng'g, Inc.*, VACAB No. 1420, 80-2 BCA ¶ 14,803; *Warren Painting Co.*, ASBCA No. 6511, 61-2 BCA ¶ 3199.

[26]*See Mann Constr. Co., Inc.* AGBCA No. 444, 76-1 BCA ¶ 11,710; *John Murphy Constr. Co.*, AGBCA No. 418, 79-1 BCA ¶ 13,836; *Jack Graham Co.*, ASBCA No. 4585, 58-2 BCA ¶ 1998.

Problems often occur in cases where the contract does not clearly define either the required standard of workmanship or the standard of inspection to be employed. In such cases, inspectors often will rely on *industry standards* and *trade customs,* or even on subjective standards such as *skillful and workmanlike.* Where such criteria actually require a level of performance in excess of that reasonably contemplated at the time the parties entered into the contract, the contractor may be entitled to additional compensation. The fact that construction projects usually draw trades and subcontractors from the local area means that it is very possible that the plans and specifications have been interpreted and priced in accordance with the trade usage and custom in that particular area.[27]

B. Rejection of Acceptable Work

Erroneous rejection of acceptable work by the government involves issues similar to the imposition of stricter standards of performance. If work that should have been accepted is *corrected* to a higher standard of quality and additional costs are incurred in the process, a compensable constructive change has occurred.[28] If acceptable work is rejected by the government, timely notice by the contractor is often critical to protection of the contractor's rights under the Changes clause.[29]

Where specifications are ambiguous, an inspector's silent acquiescence while the contractor performs in accordance with its own interpretation of the performance standards may establish that the contractor's approach was reasonable and the work was acceptable.[30] Also, if the government submits to the contractor what purports to be a complete list of defects in the work, the government may later be prevented from rejecting work that had been corrected pursuant to the list on the grounds that its list amounted to a binding interpretation of ambiguous specifications.[31]

The government has the right to reject defective work at any time prior to acceptance of the work, and a government inspector's observation of nonconforming work does not necessarily preclude later rejection.[32] However, if the government's delay in rejection of nonconforming work substantially prejudiced the contractor, the government may be estopped from later rejecting the work.[33] Estoppel will be more likely if the contractor clearly communicates its understanding of the required standards and methods of performance that later became the subject of the dispute. This communication may

---

[27]*John W. Johnson, Inc. v. J. A. Jones Constr. Co.,* 369 F. Supp. 484 (E.D. Va. 1973); *see* **Chapter 6** for a discussion of the use of trade usage and custom to assist in the interpretation of a contract.

[28]*Acme Missiles & Constr. Co.,* ASBCA No. 13671, 69-1 BCA ¶ 7698. *See* **Chapter 8** for a further discussion of constructive changes.

[29]*See* **Chapters 8** and **14** for a further discussion of notice and the recovery for a constructive change.

[30]*White Buffalo Constr.,* AGBCA 93-133-1 and 92-199-1, 94-3 BCA ¶ 27,176; *Dondlinger & Sons Constr. Co., Inc.,* ASBCA No. 13651, 70-2 BCA ¶ 8603.

[31]*H&H Enters., Inc.,* ASBCA No. 26864, 86-2 BCA ¶ 18,794; *Frederick P. Warrick Co.,* ASBCA No. 9644, *et al.,* 65-2 BCA ¶ 5169.

[32]*S & M Mgmt. Inc. v. United States,* 82 Fed. Cl. 240, 249 (2008); *Forsberg & Gregory, Inc.,* ASBCA No. 18457, 75-1 BCA ¶ 11,293.

[33]*Baltimore Contractors, Inc.,* ASBCA No. 15852, 73-2 BCA ¶ 10,281.

be set forth in meeting minutes, a prework checklist, or the like. Although a formal letter is not essential, written documentation is far better than recollections of oral discussions. A well-trained CQC staff can be invaluable in contemporaneously documenting these understandings.

C. Delay and Disruption

In each construction contract there is an implied obligation on the part of the government not to unduly delay or hinder the work. This duty extends to the government's exercise of its inspection rights. Consistent with that duty, the Inspection of Construction clause states:

> The Government shall perform all inspections and tests in a manner that will not unnecessarily delay the work.[34]

The determination of whether a particular delay related to a government inspection was unreasonable or not is a factual decision that is essentially controlled by the status of the work.[35] If the project utilizes a critical path method (CPM) of scheduling, the information provided by the CPM regarding critical activities and delays may control that determination. That factual determination will depend on the nature of the work, the time and type of inspection, follow-on activities as well as the effect of any inspection related delay on the contractor's progress. In that context, contemporaneous documentation is often critical.

The duration of the inspection related delay is but one important factor to be considered. The key is often the effect of the delay on the contractor's operations as illustrated by the next examples.

- A two-and-one-half-year delay between the discovery of latent defects in precast concrete panels and the rejection of those panels by the government was considered unreasonable.[36]
- In one case, a one-day delay in providing the results from the test of concrete cylinders was found to be an unreasonable delay because the contractor was pouring concrete on a daily basis.[37]
- In contrast, a 23-day delay by a testing service in advising a contractor that concrete cylinders being cured for 28 days had failed a strength test was not unreasonable.[38]
- In the context of a contractor's delayed on-site mobilization, a five-month delay by the government in notifying the contractor that the excavation work was unacceptable was not unreasonable.[39]

---

[34]FAR § 52.246-12(e).

[35]*Ball-Healy (JV)*, ENGBCA No. 5892, 96-2 BCA ¶ 28,580.

[36]*Utley-James, Inc.*, GSBCA No. 6831, 88-1 BCA ¶ 20,518.

[37]*Cone Bros. Contracting Co.*, ASBCA No. 16078, 72-1 BCA ¶ 9444.

[38]*William F. Klingensmith, Inc.*, GSBCA No. 5451, 83-1 BCA ¶ 16,201.

[39]*Washington Constr. Co.*, ENGBCA No. 5299, 89-3 BCA ¶ 22,077, *aff'd*, 907 F.2d 158 (Fed. Cir. 1990).

- A four-month delay in making a final inspection of fuel tanks was unreasonable because it prevented the contractor from making corrections within a reasonable time and exposed painted surfaces of the tanks to deterioration due to adverse weather conditions.[40]

VI. COST OF INSPECTION

A. Contractor Duty to Assist Government Inspection

Although the government usually bears its own costs of inspection, the contractor is required to provide, at no additional cost to the government,[41] the inspector with the facilities, labor, or material that are reasonably necessary to perform the test or inspection. However, there are circumstances in which the contractor may be reimbursed for expenses incurred for inspection or testing. For example, if the government increases the cost of conducting the inspection or test by changing the location or requiring special inspection devices, the contractor may recover those additional costs resulting from that type of action or directive.[42]

B. Reinspection Costs

The cost of reinspection is generally assigned to the party whose action or inaction resulted in the need for a reinspection. If, for example, the contractor's work was not sufficiently complete at the time of the original inspection, the contractor must pay the costs of reinspection.[43] Similarly, if the reinspection is the result of an earlier rejection, the contractor is responsible for the additional costs.[44] Before any reinspection, however, the government must provide reasonable notice and allow sufficient time for the contractor to correct or complete the work.[45]

C. Tearing Out Completed Work

The government may examine completed work and require the contractor to remove or tear out work that appears to be nonconforming. If the work is defective or does not conform to the requirements of the contract, the contractor must pay the costs of

---

[40]*Tranco Indus., Inc.,* ASBCA Nos. 26305, 26955, 83-1 BCA ¶ 16,414, *aff'd on recon.,* 83-2 BCA ¶ 16,679.
[41]FAR § 52.246-12(e); *J. L. Ewell Constr. Co.,* ASBCA No. 37746, 90-1 BCA ¶ 22,485; *Tecon-Green v. United States,* 411 F.2d 1262 (Ct. Cl. 1969).
[42]*See J.L. Ewell Constr. Co.,* ASBCA No. 37746, 90-1 BCA ¶ 22,485.
[43]*Toombs & Co. v. United States,* 4 C. Ct. 535 (1984); *Dawson Constr. Co., Inc.,* VABCA No. 3310, 3311, 91-2 BCA ¶ 23,889; *Okland Constr. Co., Inc.,* GSBCA No. 3557, 72-2 BCA ¶ 9675.
[44]*Coastal Structures, Inc.,* DOTBCA 1787, 88-3 BCA ¶ 21,016; *Minnesota Mining & Mfg. Co.,* GSBCA 4054, 75-1 BCA ¶ 11,065.
[45]*Moustafa Mohamed,* GSBCA No. 5760, 83-1 BCA ¶ 16,162, *recons. denied,* 83-2 BCA ¶ 16,805; *Baifield Indus. Div. of A-T-O, Inc.,* ASBCA No. 14582, 72-2 BCA ¶ 9676.

both the inspection and correction of the work. If the work is found to be satisfactory, however, the contractor is entitled to a price adjustment for the additional costs and a time extension if project completion is delayed.[46]

D.   Safety Inspections

The Accident Prevention clause, FAR § 52.236-13, which is a mandatory clause for fixed-price construction contracts, sets forth the contractor's basic duties in this way;

(a)   The Contractor shall provide and maintain work environments and procedures which will (1) safeguard the public and Government personnel, property, materials, supplies, and equipment exposed to Contractor operations and activities; (2) avoid interruptions of Government operations and delays in project completion dates; and (3) control costs in the performance of this contract.

* * *

(c)   If this contract is for construction or dismantling, demolition or removal of improvements with any Department of Defense agency or component, the Contractor shall comply with all pertinent provisions of the latest version of U. S. Army Corps of Engineers Safety and Health Requirements Manual, EM 385-1-1, in effect on the date of the solicitation.

* * *

A careful review of EM 385-1-1 is an essential step in estimating a project as well as managing the day-to-day safety operations at the project site. This publication can be accessed at *www.usace.army.mil/Pages/Default.aspx* and then clicking on the "library" link, which contains a link to Corps of Engineers' publications. Safety inspections by the government are for its benefit rather than for the contractor's. Similar to the inspection of the work, the government has the right to conduct safety inspections but no duty to do so. Inspection methods and the frequency of inspections are discretionary,[47] and the government generally is not liable for negligently performing a discretionary inspection, as provided in 28 U.S.C. § 2680(a).[48] Despite government safety inspections, the contractor remains responsible for providing a safe workplace.[49]

Disputes relating to the government's inspection of compliance with safety requirements are often concerned with (1) whether the government was obligated to conduct such inspections and (2) whether the government is liable for negligence in conducting the inspections.

---

[46]*See* FAR § 52.246-12(h).

[47]*See, e.g., Layton v. United States,* 984 F.2d. 1496 (8th Cir. 1993) (delegation of safety inspection to contractor with government spot checks was discretionary act).

[48]If the government is negligent in conducting its *mandatory* safety responsibilities, it may remain liable for that failure. *See Phillips v. United States,* 956 F.2d 1071 (11th Cir. 1992) (Corps of Engineers assumed duties beyond "spot checks").

[49]*Superior Abatement Servs., Inc.,* ASBCA 47118, *et al.,* 94-3 BCA ¶ 27,278.

VII. ACCEPTANCE

A. Types of Acceptance: Formal versus Constructive

1. Authority to Accept Work

The contracting officer has the authority to accept work, although authority to accept can be delegated by the contracting officer.[50] The FAR Inspection of Construction clause makes the acceptance of the contractor's work final and conclusive except for (1) latent defects, (2) fraud, (3) gross mistakes amounting to fraud, and (4) the government's rights under any warranty or guarantee.

Acceptance can be implied in certain circumstances even though the formalities of a final acceptance have not been observed. Where the government acts inconsistently with the contractor's ownership of the contract items, acceptance may be implied. The government's alteration of the completed product may also result in a determination that acceptance had occurred.[51] Generally, acceptance will not be implied from the government making partial payments for work in place.[52] Determining if and when the work has been accepted can be difficult on large construction projects, particularly if construction is performed in multiple phases and portions of the work are inspected by the government as construction proceeds.[53] To avoid acceptance issues with its subcontractors and suppliers, contractors need to determine that the acceptance terms in subcontracts and purchase orders mirror the government's rights.

2. Acceptance and Payment

Acceptance of a work on a government contract has great significance. As a general rule, acceptance of the work by the government limits its ability to complain of certain defects and reject the affected work. In addition, acceptance of the work or project usually commences the warranty period.[54] At acceptance, the contractor has the right to be paid the contract price. However, contractors and the government may dispute when the project is complete, sometimes resulting in the government's withholding formal acceptance. The theory of constructive acceptance has developed to help contractors avoid the harsh consequences of unreasonably withholding formal acceptance.

It is a well-recognized rule of contract law that strict performance may be waived by the party entitled to performance.[55] In the context of the construction contract,

---

[50]FAR § 46.502.

[51]*John C. Kohler Co. v. United States,* 498 F.2d 1360 (Ct. Cl. 1974).

[52]*Industrial Data Link, Corp.,* ASBCA No. 31570, 91-1 BCA ¶ 23,382, *recons. denied,* 91-1 BCA ¶ 23,570.

[53]*See M.A. Mortenson Co. v. United States,* 40 Fed. Cl. 389 (1998); *Southwest Welding & Mfg. Co. v. United States,* 413 F.2d 1167 (Ct. Cl. 1969).

[54]On projects with distinct phases, the work *may* be accepted in phases. In those cases, the contract should be carefully reviewed to determine if the warranty periods on equipment and so on commence with the acceptance of each phase or acceptance of the last phase (*final* acceptance). To avoid warranty gaps, subcontracts and purchase orders should be modified as necessary to conform to those acceptance and warranty provisions.

[55]*See generally* J. Calamari and J. Perillo, §11-37 (1977); *McQuagge v. United States,* 197 F. Supp. 460 (W.D. La. 1961); *see also Conrad Weihnacht Constr. Co., Inc.,* ASBCA No. 20767, 76-2 BCA ¶ 11,963; *Bromley Contracting Co., Inc.,* DOTCAB 78-1. 81-2 BCA ¶ 15,191.

this means that the government may acquiesce in the contractor's failure to perform strictly according to the terms of the agreement. Such waiver or acquiescence may result in acceptance.

3. Possession and Use

In government contracting, the concept of *constructive acceptance* is better developed than in state and local construction and is recognized in the FAR, which defines *acceptance* as:

> . . . the act of an authorized representative of the Government by which the Government, for itself or as agent of another, assumes ownership of existing identified supplies tendered or approves specific services rendered as partial or complete performance of the contract.[56]

Similarly, the Use and Possession Prior to Completion clause provides that "the Government shall have the right to take possession of or use any completed or partially completed part of the work . . . ," but that such "possession or use shall not be deemed an acceptance of any work under the contract."[57] However, in certain cases where that clause has not been included in the contract, government use, possession, or control, coupled with failure to notify the contractor that the work was not complete, amounted to an acceptance.[58] In government construction, *constructive acceptance,* like formal acceptance, usually commences the warranty period under the contract.[59]

The specific facts and circumstances supporting a claim must be viewed in light of the applicable contract language to determine if the government's actions are tantamount to acceptance. An additional factor that can affect the determination of whether there has been either a formal or a constructive acceptance is the authority of the individual on whose action or failure to act is relied. Acceptance is binding only if made by a person authorized to accept on behalf of the government.[60]

In cases where there has been no positive action, the delay or lack of any action that results in a failure to reject within a reasonable time may also imply an acceptance. That is, the government must notify the contractor of rejection within a reasonable period of time or it may be deemed to have accepted the nonconforming work and then be precluded from thereafter rejecting it.[61]

Any unreasonable delay may result in a waiver of the right to reject and constitute an acceptance of nonconforming materials. In *Tranco Industries, Inc.*,[62] a constructive

---

[56]FAR § 46.101.

[57]FAR § 52.236-11.

[58]*Walsky Constr. Co.,* ASBCA No. 36940, 90-2 BCA 22,934; *Bell & Flynn, Inc.,* ASBCA No. 11038, 66-2 BCA 5855.

[59]*Paul Tishman Co., Inc.,* GSBCA No. 1099, 1964 BCA ¶ 4256. However, on projects with phased construction, the start of the warranty period may not be strictly tied to the completion of a phase.

[60]*Wolverine Diesel Power Co.,* ASBCA No. 5079, 59-2 BCA ¶ 2327.

[61]*Tranco Indus., Inc.,* ASBCA Nos. 26305, 26955, 83-2 BCA ¶ 16,679.

[62]ASBCA Nos. 26305, 26955, 83-2 BCA ¶ 16,679.

acceptance was found where the government failed to inspect the painting of fuel tanks for three months. The board ruled that the proper standard for timely acceptance or rejection is "a reasonable time for prompt action under the circumstances." In this case, the board held that a three-month delay from paint sample approval to final inspection was unreasonable. Although a change in contracting officers justified a two-week delay, it did not justify the balance of the government's tardiness.

B. Limitations on Finality of Acceptance

In order for the government to waive strict performance or to acquiesce in a deviation from contract requirements through acceptance, its authorized representative[63] must be aware of the defect or the deviation. Whether particular conduct amounts to acceptance so as to waive the requirement of strict performance is a fact question that is controlled by the circumstances of each case. The government sometimes attempts to avoid the legal effect of acceptance with contract provisions that qualify its significance.

The standard FAR Inspection of Construction clause sets forth four categories of exceptions to the finality of acceptance. These are: latent defects, fraud, gross mistakes amounting to fraud, and the government's rights under any warranty or guarantee.[64] These warranty or guarantee rights may be set forth in the specifications as well as in the general or special provisions. (See **Section IX** of this chapter.)

Defects that are not apparent and that cannot be discovered until a later date (i.e., *latent* defects) are not accepted. This is especially important on government construction projects.[65] Where a contractor knowingly misrepresents the condition or quality of its work with the intention to deceive, the government is considered to have been induced to accept defective work as a result and may, in addition to other potential remedies, recover from the contractor the costs of repairing such defects.[66]

C. Substantial Completion

A doctrine closely related to constructive acceptance is *substantial completion*. This doctrine recognizes the point at which the government has basically received the benefit of its bargain, although every requirement has not been fulfilled. Perfection in compliance with the requirements of a construction contract is rarely achieved, so the concept of substantial completion has developed.

---

[63]*S & M Mgmt. Inc. v. United States*, 82 Fed. Cl. 240, 249 (2008) (alleged acceptance by an unauthorized representative of the VA did not bind the government).

[64]FAR § 52.246-12(i).

[65]*Kaminer Constr. Co. v. United States*, 488 F.2d 980 (1973); *H. B. Zachry Co.*, ASBCA 42266, 95-2 BCA ¶ 27,616; *ABM/Ansley Bus. Materials v. General Services Admin.*, GSBCA 9367, 93-1 BCA ¶ 25,246; *Herley Indus., Inc.*, ASBCA 13727, 71-1 BCA ¶ 8888.

[66]*See Chilstead Bldg. Co., Inc.*, ASBCA No. 49548, 00-2 BCA ¶ 31,097; *Bar Ray Prods., Inc. v. United States*, 340 F.2d 343 (1964).

Generally, when the government has the use and benefit of the contractor's work, substantial completion has occurred.[67] The work has been substantially completed when there are only minor punch-list items remaining to be completed and the project can be used for its intended purpose. The date of substantial completion is important in a construction project. First, the government's right to terminate the contractor's right to proceed with the work usually ends at time of substantial completion, although it may still retain funds to correct minor deficiencies if the contractor fails to do so. When the work has been substantially completed, the government has received essentially the benefit of its bargain, and the contractor has substantially performed its obligations; thus, the contractor usually is entitled to the balance of the contract price, less the cost of remedying minor defects. In addition, liquidated damages may not be assessed after the date of substantial completion.[68]

A typical step in the substantial completion process is the development of a *punch list* that identifies those items of incomplete or defective work to be performed by the contractor before final acceptance of the project and final payment. Generally, these items are relatively minor and are addressed by the contractor to the government's satisfaction. However, if some of the punch-list items become a matter of dispute, the government may elect to complete those alleged punch-list items. If the government elects to correct the defect or punch-list item and charge the cost to the contractor, the government has the burden to establish liability, causation, and the resulting costs. For example, in *Mitchell Enterprises, Inc.*,[69] the board essentially rejected a government claim for costs of correcting hundreds of punch-list items because the government failed to provide any allocation of costs for items for which the contractor was not responsible. The ASBCA concluded that it could not make an allocation and refused to make a *jury verdict*[70] decision on the government's claim.

D. Final Completion

Upon the contractor's final completion of the work (generally defined as when the punch-list work has been completed and the project is ready for final inspection and acceptance), the contractor's work should be finally accepted. There usually will be a final inspection to assure that correction of all punch-list items has been done and the contracting officer will officially accept the work as complete. At this point, the contractor may submit its final pay application for 100 percent of the contract price, including retainage, less any claims that are excepted or reserved from the release, which is a part of the final pay application.

---

[67]*See Dixon Contracting, Inc.*, AGBCA No. 98-191-1, 00-1 BCA ¶ 30,766; *Dimarco Corp.*, ASBCA No. 28529, 85-2 BCA ¶ 18,002.
[68]*See Kato Corp.*, ASBCA No. 51462, 06-2 BCA ¶ 33,293; *Continental Ill. Nat. Bank & Trust Co. v. United States*, 101 F. Supp. 755 (Ct. Cl. 1952).
[69]ASBCA Nos. 53202 *et al.*, 06-1 BCA ¶ 33,277.
[70]*See* **Chapter 13** for a discussion of *jury verdict* decisions on claims.

E. Setting Aside Final Acceptance

The Inspection of Construction clause states that an acceptance pursuant to the clause becomes final and binding on the government except in limited circumstances. Section (i) of that clause provides, in part:

Acceptance shall be final and conclusive except for latent defects, fraud, gross mistakes amounting to fraud, or the Government's rights under any warranty or guarantee.[71]

Latent defects are errors or omissions in the contractor's work existing at the time of government acceptance of the work, which were not and could not have been reasonably discovered by a government inspection that was made with ordinary care.[72] Latent defects are distinguished from *patent defects,* or defects that were apparent at the time of acceptance or could have been discovered through a reasonable and competent inspection. Although latent defects give the government the right to revoke acceptance, the government has only warranty rights against the contractor under the Inspection of Construction clause after acceptance of work containing patent defects. The question of whether a defect is latent or patent essentially turns on whether the defect should have been discovered through a reasonable inspection performed with ordinary care. What constitutes a *reasonable* inspection varies with the facts and circumstances of each case.[73]

Contractor quality control inspection requirements may play a part in the determination of whether a defect is latent. A defect, even if discoverable through a reasonable government inspection, may be latent if the contractor has the primary responsibility to inspect.[74] The government relies on the contractor's obligation to inspect and therefore reasonably may conduct less stringent inspection on its own. The government's quality assurance procedure function may be limited to checking the contractor's inspection (QC) procedure to see if it appears to be adequate while only spot-checking the work. Thus, the government's failing to detect a defect may be reasonable under the circumstances.

The government has two options upon discovery of a latent defect. It may revoke its previous acceptance and demand that the contractor correct the defect or omission. Alternatively, the government may correct the work itself and assert a claim for the costs against the contractor. In either case, the burden is on the government to prove that the defect was unknown at the time of acceptance. To have a viable post acceptance claim against the contractor, the government must show injury and a connection between the defect and the injury. Proving each of these elements can be difficult.

---

[71]FAR § 52.246-12(i).

[72]*See ABM/Ansley Bus. Materials v. General Servs. Admin.,* GSBCA No. 9367, 93-1 BCA ¶ 25,246; *Kaminer Constr. Corp. v. United States,* 488 F.2d 980 (Ct. Cl. 1973); *F.W. Lang Co.,* ASBCA No. 2677, 57-1 BCA ¶ 1334.

[73]*See Kaminer Constr. Co. v. United States,* 488 F.2d 980, 987 (Ct. Cl. 1973).

[74]*See Tricon-Triangle Contractors,* ENGBCA No. 5553, 92-1 BCA ¶ 24,667; *Kaminer Constr. Co. v. United States,* 488 F.2d 980, 987 (Ct. Cl. 1973).

The passage of time, the departure of witnesses, intervening events, subsequent construction work, improper maintenance or repairs, and natural wear and tear are only some of the problems that the government may have to sort out.

Once the government discovers a latent defect, it must act promptly to demand correction from the contractor. In one case, a two-year delay between the time the government discovered the defect and the time it demanded correction by the contractor was found to be a constructive acceptance of the defect.[75]

Fraud may also justify revocation of acceptance. Where a contractor knowingly misrepresents the condition or quality of its work, with the intent to deceive, and the government is induced to accept defective work as a result, the government may revoke its acceptance and recover the costs of repairing such defects.[76]

The government may also revoke acceptance of work where the contractor makes a "gross mistake amounting to fraud." This exception to final acceptance involves a contractor performance error or deficiency "so serious or uncalled for it was not to be reasonably expected, or justifiable, in the case of a responsible contractor" or a mistake that "cannot be reconciled in good faith."[77] The decision in *American Renovation and Construction Co.*,[78] is an example of acceptance being revoked due to gross mistakes amounting to fraud. This project, involving two contracts, was a design/build project for housing units at Malmstrom Air Force Base. The specifications called for placement of backfill in 8-inch lifts as well as defining the size and type of materials that could be used as backfill. The ASBCA found that during performance, the contractor placed fill in lifts of up to 3 feet in some areas of the project site. Additionally, the contractor disregarded the requirements of the type of backfill to be used as subsequent remedial work revealed fence posts, lumber, and steel stakes buried in the backfill. Although the government had accepted the housing units at the end of construction, many of the units suffered serious structural problems due to poor workmanship of the earthwork. The board found that the revocation of acceptance of part of the project was done within a reasonable time and upheld the default termination of one of the contracts.

VIII. PROJECT COMMISSIONING

A. Introduction

It has been a common practice on government construction projects to inspect the work when it is completed and to start up all systems and equipment to assure that

---

[75] *See Utley-James, Inc.,* GSBCA No. 6831, 88-1 BCA ¶ 20,518.

[76] *See Henry Angelo & Co.,* ASBCA No. 30502, 87-1 BCA ¶ 19,619 (in addition to recovering the costs of repair, the government may also pursue the contractor under both civil and criminal fraud statutes).

[77] *See Catalytic Eng'g & Mfg. Corp.,* ASBCA No. 15257, 72-1 BCA ¶ 9342. *Chilstead Building Co., Inc.,* ASBCA No. 49548, 00-2 BCA ¶ 31,097; *Bender GMBH,* ASBCA No. 52266, 04-1 BCA ¶ 32,474. *See also* the Certificate of Conformance required by the Payments Under Fixed-Price Construction Contracts, FAR § 52.232-5. This certification carries potential False Claims Act and False Statements Act exposures. *See* **Chapter 1** for a further discussion of certifications and related liabilities.

[78] ASBCA Nos. 53723, 54038, 09-2 BCA ¶ 34,199.

the project meets the design criteria and functions as expected. However, the emerging practice of *project commissioning* is more intensive and detailed than the conventional concepts inspection and start-up of systems. Commissioning is an integrated *process* that identifies design goals and ensures a complete and adequate design, proper construction in accordance with the plans and specifications, and maintenance and operation of the project during its life.

B. Commissioning Required on Federal Projects

Federal agencies are required by Executive Order No. 13423 to employ total building commissioning practices tailored to the size and complexity of the building and its system components.[79] In order to implement this executive order, a guidance document with sample specification language to be inserted in project specifications was prepared by the Environmental Protection Agency (EPA), in partnership with the Federal Environmental Executive and the Whole Building Design Guide of the National Institute of Building Sciences. Executive Order No. 13101 gave EPA authority to provide guidance to federal agencies in meeting pollution prevention and other green building requirements.[80]

This sample specification at Paragraph 1.2B states: "Commissioning is a comprehensive and systematic process to verify that the building systems perform as designed to meet the Owner's requirements." Agencies are to appoint a commissioning agent for each project to provide overall coordination and management of the commissioning program, as specified in the commissioning specification.[81]

The commissioning agent participates as part of a team that includes the contractor, subcontractors, owner's representative, green consultant (advises on environmental and energy efficiency issues), and architect/engineer.[82] The specification sets out at Paragraph 3.1A the commissioning tasks and the general order in which they occur:

- Design review and documentation
- Commissioning scoping meeting
- Commissioning plan
- Submittals review
- Start-up/prefunctional checklist
- Functional performance testing
- Short-term diagnostic testing
- Deficiency report and resolution record

---

[79]Executive Order No. 13423, "Strengthening Federal Environmental, Energy and Transportation Management" (Jan. 24, 2007).
[80]Federal Green Construction Guide for Specifiers, 01 91 00 (01810) Commissioning (*www.wbdg.org*, link to "Federal Green Construction Guide for Specifiers," open Section 01 "General Requirements," open specification 01 91 00 (01810). (A copy of this specification is set forth in **Appendix 10C** to this chapter.)
[81]*Id.* at 1.3B.
[82]*Id.* at 1.3C.

- Operations and maintenance training
- Record documents review
- Final commissioning report and LEED documentation
- Deferred testing
- If the contract contains a commissioning specification and references various industry standards, a careful review of those requirements during the estimating phase is an essential step. Not only must the cost of this effort be ascertained, but the anticipated schedule of performance must reflect these activities. Finally, subcontract and purchase order terms and conditions must be coordinated with the obligations set forth in the commissioning specification.

C. Industry Commissioning Standards

Other guides to project commissioning are similar to the sample specification discussed in **Section B**. One of the leading industry guides to assist in commissioning government projects is published by the American Society of Heating, Refrigeration, and Air Conditioning Engineers (ASHRAE). Its guide, ASHRAE Guideline 0, can be accessed at *www.ashrae.org* Guideline 0 advocates a *process* to use with any building system. It is a "quality-oriented process for achieving, verifying, and documenting that the performance of facilities, systems, and assemblies meet defined objectives and criteria."[83] Guideline 0 also has been adopted by the National Institute of Building Sciences, which publishes and sponsors the Whole Building Design Guide (WBDG). At the WBGD Web site (*www.wbdg.org/*), there are links to papers and documents that are useful in planning the commissioning process of a construction project.

The General Services Administration (GSA) is the "owner" of many government buildings that are used by other federal agencies. For example, the GSA is the "owner" of federal courthouses that are used by the Justice Department and non–Department of Defense office buildings. These buildings are constructed under contract with the GSA.[84]

The GSA has published *The Building Commissioning Guide,* a 100-page book available from GSA and from the Construction Criteria Base at the Whole Building Design Guide Web site.

> *The Building Commissioning Guide* provides the overall framework and process for building commissioning from project planning through tenant occupancy, keys to success within each step and the ways that each team member supports the process of commissioning. While recognizing that every project is unique and that the required activities will vary on every project, this Guide provides recommendations, minimum requirements, and best practices based upon industry guidance and GSA experience.[85]

---

[83] ASHRAE Guideline 0-2005.
[84] The Army Corps of Engineers and the Naval Facilities Engineering Command contract for the construction of buildings for Department of Defense agencies.
[85] *www.wbdg.org/ccb/browse_doc.php?d=5434* (accessed June 19, 2009).

This guide focuses much attention on selecting a commissioning agent and developing the commissioning plan at the outset before the project design is prepared. This emphasis helps to assure that the design development, construction, systems operation, operations and maintenance training, and functional testing are thought through at the outset.

D. Summary

Building commissioning is now required by Executive Order No. 13423 for all new construction and major renovations and for 15 percent of existing federal capital asset building inventory by 2015.[86] A White House Summit on Federal Sustainable Buildings was held in January 2006, resulting in the "Federal Leadership in High Performance and Sustainable Buildings Memorandum of Understanding" (MOU). This MOU sets out five guiding principles:

1.   Employing integrated design
2.   Optimizing energy performance
3.   Protecting and conserving water
4.   Enhancing indoor environmental quality
5.   Reducing the environmental impact of materials[87]

Executive Order No. 13423 makes mandatory the application of these guiding principles for all new construction and major renovations. Therefore, owners, contractors, subcontractors, and architect/engineer design professionals must become familiar with the commissioning process and must be prepared to commit to it in construction contracting.

These Internet sites contain useful information and resources on project commissioning that are available from the United States Department of Energy, other federal agencies, and various associations:

American Society of Heating, Refrigeration and Air Conditioning Engineers: *www.ashrae.org*

Building Commissioning Association: *www.bcxa.org*

Federal Energy Management Program (among other information, full text of GSA "Draft Building Commissioning Guide"): *www1.eere.energy.gov/femp*

Housing and Urban Development ("Building Commissioning, the Key to Quality Assurance"): *www.hud.gov/offices/pih/programs/ph/phecc/resources.cfm*

National Institute of Building Sciences (NIBS): *www.nibs.org*

President's Council on Sustainable Development (policies and strategies for sustainable communities): *www.usda.gov/oce/sustainable*

U.S. Green Building Council: *www.usgbc.org*

Whole Building Design Guide (a program of NIBS): *www.wbdg.org*

---

[86]Federal Green Construction Guide for Specifiers, Sample Specification para. 3.1, Specifier Note.
[87]*Id.*

IX. WARRANTIES

In government contract law, there are two kinds of warranties applicable to any contractual arrangement: (1) express warranties, which are specifically set forth in the contract documents, and (2) implied warranties, which are implied from the nature of the transaction between the parties unless the contract expressly provides that such warranties are inapplicable.

A. Contractual Warranties

Express warranties relating to construction contract performance can be complex and do not have to be labeled as a warranty or guaranty in order to have the effect of an express warranty. In that regard, it is essential that an offeror or bidder on a construction contract carefully review all the contract plans and documents to determine whether there are requirements and language that may have the effect of creating an express warranty and whether it has been factored into its bid or proposal price. Without this type of careful review and analysis, the offeror or bidder may find that it warranted a certain result or performance and that the risk attending such a warranty was not considered in the preparation of the bid or proposal for the work. The same review is also essential to ensure that subcontract and purchase order scopes of work and warranty clauses are tailored to include the appropriate flow-down of warranty obligations.

The contractor usually *expressly warrants* that the material and workmanship furnished by it are free from defects.[88] Liability under these express warranties expands the scope of the contractor's responsibility for defective work to extend beyond the date of final acceptance. This warranty is limited only by the six-year statute of limitation.[89] The express warranties are in addition to, and not in substitution for, other responsibilities of the contractor. The contractor usually also gives a repair warranty for a specified period of time, commonly one year.

Additional express warranties may be required in connection with equipment supplied by the contract. Such specific warranties usually are spelled out in the specifications to which they apply rather than in the general conditions. They often are in the nature of guarantees of performance and agreements to repair defects for a specified period of time.

The warranties required by the contract documents generally begin to run from the date of substantial completion. Other special warranties may commence at delivery or installation of the machinery or equipment or at commencement of operations. These special equipment or system warranties, other than the repair warranty, are limited by their terms. In some instances the warranty obligation may run from a

---

[88]*Henry Angelo & Co., Inc.,* ASBCA No. 30502, 87-1 BCA ¶ 19,619.
[89]41 U.S.C. § 605(a) (except for claims involving fraud). *See also* FAR § 52.233-1(d)(1) and FAR § 33.206. 41 U.S.C. § 605(a) (except for claims involving fraud). *See also* FAR § 52.233-1(d)(1) and FAR § 33.206.

supplier directly to the government even though there is no other contractual relationship between those two entities.[90]

In most government contracts, the warranty terms are set out in the request for proposal or invitation for bids. There is seldom any opportunity to negotiate scope or duration of the warranties. From a contractor's perspective, it needs to determine if there are dates or events that could trigger the beginning of a warranty period or that could create warranty issues with subcontractors or suppliers. For example, in a phased construction project, the government may state that *all* warranties run from the date of acceptance of the last phase. In that situation, the work or equipment installed in earlier phases may have been placed in operation well before overall *final acceptance* of the entire project. This could create an expensive *warranty* gap for the contractor. However, if this is identified when the work was initially bought out, it may be possible to obtain the added warranty coverage at little or no cost.

FAR Subpart 46.7, Warranties, does not mandate the use of a particular warranty clause in all construction contracts. Rather these clauses may be used when authorized by agency procedures. Moreover, the agencies are authorized to modify the terms of the warranty clause to fit the requirements of the particular contract.[91] An example of a possible provision for use in construction contracts is the Warranty of Construction[92] clause.

In determining whether a warranty is appropriate for a specific acquisition, FAR § 46.703 requires the contracting officer to consider several factors, including (1) the nature and use of the project, (2) the warranty costs, (3) the government's ability to enforce and administer the warranty, (4) trade practice, and (5) any reduced requirement for the government's contract quality assurance.[93]

B. Implied Warranties

Implied warranties are commonly considered in the context of the sale of goods and supplies in the commercial market place. The implied warranty of merchantability[94] and the implied warranty of fitness for a particular purpose[95] are the two most invoked implied warranties in private commercial contracts. Commercial contracts and purchase orders may contain extensive provisions excluding those warranties or placing limits on the available remedies for breach of either or both.

Government contracts address implied warranties in various ways. The warranty clauses authorized for government supply contracts expressly exclude both the implied warranty of merchantability and the implied warranty of fitness for a particular purpose.[96] In contrast, the standard Warranty of Construction clause provides:

---

[90]*Lee Lewis Constr., Inc. v. United States,* 54 Fed. Cl. 88, 92 (2002) (contractor not obligated under roofing warranty, which the contract specifications stated would be provided by the "roofing contractor").
[91]FAR §§ 46.710; 46.704; 46.710.
[92]FAR § 52.246-21. (A copy of this clause is set forth in **Appendix 10B** to this Chapter).
[93]FAR § 46.703.
[94]*See* Uniform Commercial Code (UCC) § 2-314.
[95]*See* UCC § 2-315.
[96]See FAR § 52.246-17(B)(4), Warranty of Supplies of a Noncomplex Nature *and* FAR § 52.246-18(B)(6), Warranty of Supplies of a Complex Nature.

(g) With respect to all warranties, express or implied, from subcontractors, manufacturers, or suppliers for work performed and materials furnished under this contract, the Contractor shall—

(1) Obtain all warranties that would be given in normal commercial practice;

(2) Require all warranties to be executed, in writing, for the benefit of the Government, if directed by the Contracting Officer; and

(3) Enforce all warranties for the benefit of the Government, if directed by the Contracting Officer.[97]

Consequently, if there are implied warranties applicable to the materials or supplies incorporated into a construction project, this language would appear to obligate the contractor to assist the government in the enforcement of them. In the context of government construction projects, the government has asserted claims based on implied warranties for turbines that were purchased for incorporation in a dam construction project[98] and materials used in the construction of rail lines for a subway.[99]

Another example of an implied warranty is in design-build contracts, where the design-builder has the implied duty to furnish an adequate and sufficient design. The design-builder has the implied duty, instead of the government, because it is responsible for furnishing the design.[100]

C. Government Remedies

Not only are the government's remedies under Warranty of Construction clause[101] cumulative, these remedies do not operate to limit the government's rights under the Inspection of Construction clause relating to latent defects, fraud, and gross mistakes amounting to fraud. The standard Warranty of Construction clause provides that the government's warranty rights do not limit the application of the rights reserved the Inspection of Construction clause. See FAR § 52.246-21(j), which states:

> This warranty shall not limit the Government's rights under the Inspection and Acceptance clause of this contract with respect to latent defects, gross mistakes, or fraud.

Thus, certain limited government rights related to defective work continue after the expiration of an applicable warranty provision since such rights are not limited to a specific period.[102]

---

[97]FAR § 52.246-21(g).

[98]*Newport News Shipbuilding and Dry Dock Co.*, ENGBCA No. 3117, 72-1 BCA ¶ 9210.

[99]*Transit Prods. Co., Inc.*, ENGBCA No. 4796, 88-2 BCA ¶ 20,673.

[100]*Aleutian Contractors v. U.S.*, 24 Cl. Ct. 372 (1991).

[101]FAR § 52.246-21.

[102]*See Charles G. Williams Constr., Inc.*, ASBCA No. 24967, 81-1 BCA ¶ 14,893; *Keco Indus., Inc.*, ASBCA No. 13271, 71-1 BCA ¶ 8727.

➤ LESSONS LEARNED AND ISSUES TO CONSIDER

- *Inspection* and *acceptance* are important during the construction of the project, and they also affect the contractor's rights and obligations under warranties that may extend for years beyond project completion.

- With federal government construction, the standard inspection clause places *primary responsibility* for contract compliance with the contractor.

- A contractor must review the terms and conditions of both its *subcontracts* and *purchase orders* to ensure that there is a consistent flow down of inspection, acceptance, and warranty obligations to these lower-tier firms.

- *Contractor Quality Control* (CQC) staffing requirements can be detailed and costly. The scope of these obligations must be evaluated during the preparation of the estimate, any questions clarified, and thereafter coordinated with the terms of subcontracts and purchase orders. Use of a *CQC Checklist* can be of substantial assistance in this process.

- Although the CQC staffing requirements may be detailed and costly, a well-trained CQC staff can provide invaluable assistance in developing and maintaining *critical project documentation.*

- The government has the right to conduct inspections, but it may not exercise that right in a way that *delays or disrupts* the contractor's work or alters the requirements of the contract.

- Although the government has no duty to inspect, inspection is generally for the *protection* of the government and not the contractor.

- In addition to government inspection, some of the burden of assuring compliance with the contract requirements may be assigned to the contractor through *quality control* requirements of the contract.

- Although the government usually bears its own *costs* of inspection, the contractor is required to bear the expense of providing the government with whatever facilities, labor, or materials are reasonably necessary to perform the test or inspection.

- *Acceptance* by the government is very significant and limits its ability to complain of defects and reject work and also commences the running of warranties. However, the finality of acceptance can be set aside for *latent defects, fraud,* or *gross mistakes amounting to fraud.*

- *Building commissioning* is a comprehensive and systematic process to verify that the building systems perform as designed to meet the owner's requirements. As this can be a costly and time-consuming process, the contractor must anticipate this cost in its estimate, in its anticipated project schedule, and in the terms and conditions of its subcontracts and purchase orders.

(Continued)

- The FAR contains a variety of *warranty* clauses that may be included in construction contracts in the general or special provisions. In addition, specific warranties, usually for equipment or machinery, may be found in the detailed specifications.
- On projects with phased completion of facilities, the contractor's warranty obligations may commence only on the acceptance of the last phase of the project. This type of requirement has the potential for placing the contractor in a *warranty gap* if the terms of the subcontracts and purchase orders are not carefully reviewed and coordinated with the government's warranty terms.

APPENDIX 10A: USE AND POSSESSION PRIOR TO COMPLETION (APR. 1984)

(a) The Government shall have the right to take possession of or use any completed or partially completed part of the work. Before taking possession of or using any work, the Contracting Officer shall furnish the Contractor a list of items of work remaining to be performed or corrected on those portions of the work that the Government intends to take possession of or use. However, failure of the Contracting Officer to list any item of work shall not relieve the Contractor of responsibility for complying with the terms of the contract. The Government's possession or use shall not be deemed an acceptance of any work under the contract.

(b) While the Government has such possession or use, the Contractor shall be relieved of the responsibility for the loss of or damage to the work resulting from the Government's possession or use, notwithstanding the terms of the clause in this contract entitled "Permits and Responsibilities." If prior possession or use by the Government delays the progress of the work or causes additional expense to the Contractor, an equitable adjustment shall be made in the contract price or the time of completion, and the contract shall be modified in writing accordingly.

APPENDIX 10B: WARRANTY OF CONSTRUCTION (MAR. 1994)

(a) In addition to any other warranties in this contract, the Contractor warrants, except as provided in paragraph (i) of this clause, that work performed under this contract conforms to the contract requirements and is free of any defect in equipment, material, or design furnished, or workmanship performed by the Contractor or any subcontractor or supplier at any tier.

(b) This warranty shall continue for a period of 1 year from the date of final acceptance of the work. If the Government takes possession of any part of the work before final acceptance, this warranty shall continue for a period of 1 year from the date the Government takes possession.

(c) The Contractor shall remedy at the Contractor's expense any failure to conform, or any defect. In addition, the Contractor shall remedy at the Contractor's expense any damage to Government-owned or controlled real or personal property, when that damage is the result of—

   (1) The Contractor's failure to conform to contract requirements; or

   (2) Any defect of equipment, material, workmanship, or design furnished.

(d) The Contractor shall restore any work damaged in fulfilling the terms and conditions of this clause. The Contractor's warranty with respect to work repaired or replaced will run for 1 year from the date of repair or replacement.

(e) The Contracting Officer shall notify the Contractor, in writing, within a reasonable time after the discovery of any failure, defect, or damage.

(f) If the Contractor fails to remedy any failure, defect, or damage within a reasonable time after receipt of notice, the Government shall have the right to replace, repair, or otherwise remedy the failure, defect, or damage at the Contractor's expense.

(g) With respect to all warranties, express or implied, from subcontractors, manufacturers, or suppliers for work performed and materials furnished under this contract, the Contractor shall—

   (1) Obtain all warranties that would be given in normal commercial practice;

   (2) Require all warranties to be executed, in writing, for the benefit of the Government, if directed by the Contracting Officer; and

   (3) Enforce all warranties for the benefit of the Government, if directed by the Contracting Officer.

(h) In the event the Contractor's warranty under paragraph (b) of this clause has expired, the Government may bring suit at its expense to enforce a subcontractor's, a manufacturer's, or a supplier's warranty.

(i) Unless a defect is caused by the negligence of the Contractor or subcontractor or supplier at any tier, the Contractor shall not be liable for the repair of any defects of material or design furnished by the Government nor for the repair of any damage that results from any defect in Government-furnished material or design.

(j) This warranty shall not limit the Government's rights under the Inspection and Acceptance clause of this contract with respect to latent defects, gross mistakes, or fraud.

(End of clause)

Alternate I (Apr. 1984). If the Government specifies in the contract the use of any equipment by "brand name and model," the contracting officer may add a paragraph substantially the same as the following paragraph (k) to the basic clause:

(k) Defects in design or manufacture of equipment specified by the Government on a "brand name and model" basis, shall not be included in this warranty. In this event, the Contractor shall require any subcontractors, manufacturers, or suppliers thereof to execute their warranties, in writing, directly to the Government.

APPENDIX 10C:
WHOLE BUILDING DESIGN GUIDE

FEDERAL GREEN CONSTRUCTION GUIDE FOR SPECIFIERS

This is a guidance document with sample specification language intended to be inserted into project specifications on this subject as appropriate to the agency's environmental goals. Certain provisions, where indicated, are required for U.S. federal agency projects. Sample specification language is numbered to clearly distinguish it from advisory or discussion material. Each sample is preceded by identification of the typical location in a specification section where it would appear using the SectionFormat™ of the Construction Specifications Institute; the six-digit section number cited is per CSI Masterformat™ 2004 and the five-digit section number cited parenthetically is per CSI Masterformat™ 1995.

SECTION 01 91 00 (SECTION 01810)—COMMISSIONING

Part 1. General

1.1 Summary

(A) Section includes:

  (1) Building commissioning of the following systems:

    (a) HVAC components and equipment.

    (b) HVAC system: interaction of cooling, heating, and comfort delivery systems.

    (c) Building Automation System (BAS): control hardware and software, sequence of operations, and integration of factory controls with BAS.

    (d) Lighting Control System and interface with day lighting.

  (2) Building commissioning activities and documentation in support of the U.S. Green Building Council (USGBC) LEED™ rating program.

    (a) Commissioning activities and documentation for the LEED™ section on "Energy and Atmosphere" prerequisite of "Fundamental Building Systems Commissioning."

    (b) Commissioning activities and documentation for the LEED™ section on "Additional Commissioning."

  (3) Building commissioning activities and documentation in support of the Building Research Establishment (BRE) Green Globes—US rating system.

    (a) Commissioning activities and documentation for the Green Globes Commissioning Plan - Documentation.

(B) The Owner, Green Consultant, Architect/Engineer, and Commissioning Agent are not responsible for construction means, methods, job safety, or management function related to commissioning on the job site.

(C) Related Sections:

(1) 01 30 00 (01300) - Administrative Requirements

(2) 01 40 00 (01400) - Quality Requirements

(3) 01 57 19.11 (01352) - Indoor Air Quality (IAQ) Management

(4) 01 57 19.13 (01354) - Environmental Management

(5) 01 78 23 (01830) - Operation & Maintenance Data

(6) 01 78 53 (01780) - Sustainable Design Close-Out Documentation

(7) 22 05 00 (15050) - Common Work Results for Plumbing

(8) 23 05 00 (15050) - Common Work Results for HVAC

(9) 26 05 00 (16050) - Common Work Results for Electrical

1.2 Definitions

(A) Basis of Design—The basis of design is the documentation of the primary thought processes and assumptions behind design decisions that were made to meet the design intent. The basis of design describes the systems, components, conditions and methods chosen to meet the intent. Some reiterating of the design intent may be included.

(B) Commissioning—Commissioning is a comprehensive and systematic process to verify that the building systems perform as designed to meet the Owner's requirements. Commissioning during the construction, acceptance, and warranty phases is intended to achieve the following specific objectives:

• Verify and document that equipment is installed and started per manufacturer's recommendations, industry accepted minimum standards, and the Contract Documents.

• Verify and document that equipment and systems receive complete operational checkout by installing contractors.

• Verify and document equipment and system performance.

• Verify the completeness of operations and maintenance materials.

• Ensure that the Owner's operating personnel are adequately trained on the operation and maintenance of building equipment.

The commissioning process does not take away from or reduce the responsibility of the system designers or installing contractors to provide a finished and fully functioning product.

(C) Commissioning Plan—An overall plan that provides the structure, schedule and coordination planning for the commissioning process.

(D) Deficiency—A condition in the installation or function of a component, piece of equipment or system that is not in compliance with the Contract Documents, does not perform properly or is not complying with the design intent.

(E) Design Intent—A dynamic document that provides the explanation of the ideas, concepts and criteria that are considered to be very important to the Owner. It is initially the outcome of the programming and conceptual design phases.

(F) Functional Performance Test—Test of the dynamic function and operation of equipment and systems using manual (direct observation) or monitoring methods. Functional testing is the dynamic testing of systems (rather than just components) under full operation (e.g., the chiller pump is tested interactively with the chiller functions to see if the pump ramps up and down to maintain the differential pressure set point). Systems are tested under various modes, such as during low cooling or heating loads, high loads, component failures, unoccupied, varying outside air temperatures, fire alarm, power failure, etc. The systems are run through all the control system's sequences of operation and components are verified to be responding as the sequences state. Traditional air or water test and balancing (TAB) is not functional testing, in the commissioning sense of the word. TAB's primary work is setting up the system flows and pressures as specified, while functional testing is verifying that which has already been set up. The Commissioning Agent develops the functional test procedures in a sequential written form, coordinates, oversees and documents the actual testing, which is usually performed by the installing contractor or vendor. Functional Performance Tests are performed after prefunctional checklists and startup activities are complete.

(G) Manual Test—Using hand-held instruments, immediate control system readouts or direct observation to verify performance (contrasted to analyzing monitored data taken over time to make the "observation").

(H) Monitoring—The recording of parameters (flow, current, status, pressure, etc.) of equipment operation using dataloggers or the trending capabilities of control systems.

(I) Non-Compliance—See Deficiency.

(J) Non-Conformance—See Deficiency.

(K) Prefunctional Checklist—A list of items to inspect and elementary component tests to conduct to verify proper installation of equipment, provided by the Commissioning Agent to the contractor. Prefunctional checklists are primarily static inspections and procedures to prepare the equipment or system for initial operation (e.g., belt tension, oil levels OK, labels affixed, gages in place, sensors calibrated, etc.). However, some prefunctional checklist items entail simple testing of the function of a component, a piece of equipment or system (such as measuring the voltage imbalance on a three-phase pump motor of a chiller system). The word "prefunctional" refers to before functional testing. Prefunctional checklists augment and are combined with the manufacturer's start-up checklist.

(L) Seasonal Performance Tests—Functional Performance Test that are deferred until the system(s) will experience conditions closer to their design conditions.

(M) Warranty Period—warranty period for entire project, including equipment components. Warranty begins at Substantial Completion and extends for at least one year, unless specifically noted otherwise in the Contract Documents and accepted submittals.

1.3 Coordination

(A) Perform commissioning services to expedite the testing process and minimize unnecessary delays, while not compromising the integrity of the procedures.

(B) Commissioning Agent shall provide overall coordination and management of the commissioning program as specified herein.

(C) Commissioning Team: The commissioning process will require cooperation of the Contractor, subcontractors, vendors, Architect/Engineer, Commissioning Agent, Green Consultant, and Owner. The commissioning team shall be comprised of the following.

(1) Contractor

(a) Project Manager

(b) Test Engineer

(2) Subcontractors: As appropriate to product or system being commissioned.

(3) Commissioning Agent

(a) Project Manager

(b) Project Engineers

(4) Owner Representative(s)

(5) Green Consultant

(6) Architect/Engineer

(a) Architect

(b) MEP engineers

(c) Specialty Consultant(s)

(D) Progress Meetings: Attend construction job-site meetings, as necessary, to monitor construction and commissioning progress. Coordinate with contractor to address coordination, deficiency resolution and planning issues.

(1) Plan and coordinate additional meetings as required to progress the work.

(E) Site Observations: Perform site visits, as necessary, to observe component and system installations.

(F) Functional Testing Coordination:

(1) Equipment shall not be "temporarily" started for commissioning.

(2) Functional performance testing shall not begin until pre-functional, start-up and TAB is completed for a given system.

(3) The controls system and equipment it controls shall not be functionally tested until all points have been calibrated and pre-functional checklists are completed.

(G) Indoor Air Quality (IAQ) baseline evaluation: Coordinate with IAQ baseline evaluation as specified in Section 01 57 19.11 (01352)—Indoor Air Quality (IAQ) Management.

1.4 Quality Control

(A) Qualifications for Commissioning Agents: Engage commissioning service personnel that specialize in the types of inspections and tests to be performed.

(1) Inspection and testing service agencies shall be members of the Building Commissioning Association (BCA).

1.5 Submittals

(A) Commissioning Agent shall submit the following:

(1) Basis of Design and Design Intent.

(a) Update as necessary during the work to reflect the progress on the components and systems. Forward updates to the Green Consultant in a timely manner.

(2) Scoping Meeting Minutes.

(3) Commissioning Plan: Submit within 30 calendar days of authorization to proceed.

(a) Update as necessary during the work to reflect the progress on the components and systems. Forward updates to the Green Consultant in a timely manner.

(4) Commissioning Schedule: Submit with Commissioning Plan.

(a) Update as necessary during the work to reflect the progress on the components and systems. Forward updates to the Green Consultant in a timely manner.

(5) Functional performance test forms: Submit minimum 30 calendar days prior to testing.

(6) Deficiency Report and Resolution Record: Document items of non-compliance in materials, installation or operation. Document the results from start-up/ pre-functional checklists, functional performance testing, and short-term diagnostic monitoring. Include details of the components or systems found to be non-compliant with the drawings and specifications. Identify adjustments

and alterations required to correct the system operation, and identify who is responsible for making the corrective changes.

    (a) Update as necessary during the work to reflect the progress on the components and systems. Forward updates to the Green Consultant in a timely manner.

(7) Final Commissioning Report: Compile a final Commissioning Report. Summarize all of the tasks, findings, conclusions, and recommendations of the commissioning process. Indicate the actual performance of the building systems in reference to the design intent and contract documents. Include completed pre-functional inspection checklists, functional performance testing records, diagnostic monitoring results, identified deficiencies, recommendations, and a summary of commissioning activities.

(8) O&M Submittals:

    (a) Training plan: Training plan shall include for each training session:

        (1) Dates, start and finish times, and locations;

        (2) Outline of the information to be presented;

        (3) Names and qualifications of the presenters;

        (4) List of texts and other materials required to support training.

    (b) O&M Database.

(9) **[LEED™][Green Globes—US] [xxxxx]** Documentation related to commissioning. Format as required by **[USGBC] [GBI] [xxxx]** for submittal under the referenced green building rating system.

Part 2. Products

2.1 Test Equipment

(A) Instrumentation shall meet the following standards:

    (1) Be of sufficient quality and accuracy to test and measure system performance within the tolerances required to determine adequate performance.

    (2) Be calibrated on the manufacturer's recommended intervals with calibration tags permanently affixed to the instrument being used.

    (3) Be maintained in good repair and operation condition throughout the duration of use on this project.

(B) All standard testing equipment required to perform startup and initial checkout and required functional performance testing shall be provided by the contractor for the equipment being tested.

(C) Datalogging equipment or software required to test equipment will be provided by the Commissioning Agent, but shall not become the property of the Owner.

Part 3. Execution

3.1 Commissioning Process

(A) The following activities outline the commissioning tasks and the general order in which they occur. The Commissioning Agent shall coordinate all activities.

    (1) Design Review and Documentation.

        (a) Documentation of Basis of Design and Design Intent.

        (b) Design Development Review.

        (c) Construction Document Review.

    (2) Commissioning Scoping Meeting.

    (3) Commissioning Plan.

    (4) Submittals Review.

    (5) Start-Up/Pre-Functional Checklists.

    (6) Functional Performance Testing.

    (7) Short-Term Diagnostic Testing.

    (8) Deficiency Report and Resolution Record.

    (9) Operations and Maintenance Training.

        (a) O&M Manual.

        (b) Training.

        (c) O&M Database.

    (10) Record Documents Review.

    (11) Final Commissioning Report and **[LEED™] [Green Globes—US][xxxx]** Documentation.

    (12) Deferred Testing.

        (a) Unforeseen Deferred Tests.

        (b) Seasonal Testing.

        (c) End-of-Warranty Review.

3.2 Design Review and Documentation

(A) Documentation of Basis of Design and Design Intent: Document basis of design and design intent as they relate to environmentally responsive characteristics, including: functionality, energy performance, water efficiency, maintainability, system cost, indoor environmental quality and local environmental impacts.

(B) Design Development Review: Review design documents to verify that each commissioned system meets the design intent.

(C) Construction Document Review: Review construction documents to verify that commissioning is adequately specified, that each commissioned system can be commissioned and is likely to meet the design intent.

3.3 Commissioning Scoping Meeting

(A) Commissioning Scoping Meeting:

(1) Schedule, coordinate, and facilitate a scoping meeting.

(2) Review each building system to be commissioned, including its intended operation, commissioning requirements, and completion and start-up schedules.

(3) Establish the scope of work, tasks, schedules, deliverables, and responsibilities for implementation of the Commissioning Plan.

(B) Attendance: Commissioning Team members.

3.4 Commissioning Plan

(A) Commissioning Plan: Develop a commissioning plan to identify how commissioning activities will be integrated into general construction and trade activities. The commissioning plan shall identify how commissioning responsibilities are distributed. The intent of this plan is to evoke questions, expose issues, and resolve them with input from the entire commissioning team early in construction.

(1) Identify who will be responsible for producing the various procedures, reports, Owner notifications and forms.

(2) Include the commissioning schedule.

(3) Describe the test/acceptance procedure.

3.5 Submittals Review

(A) Submittal Review: Review the contractor submittals to verify that the equipment and systems provided meet the requirements of the Contract Documents and Design Intent.

3.6 Start-Up/Pre-Functional Checklists

(A) Start-Up/Pre-Functional Checklists: Coordinate start-up plans and documentation formats, including providing contractor with pre-functional checklists to be completed during the startup process.

(1) Manufacturer's start-up checklists and other technical documentation guidelines may be used as the basis for pre-functional checklists.

(B) Start-Up/Pre-Functional Checklist shall help verify that the systems are complete and operational, so that the functional performance testing can be scheduled.

3.7 Functional Performance Testing

(A) Functional Performance Testing: Test procedures shall fully describe system configuration and steps required for each test; appropriately documented so that another party can repeat the tests with virtually identical results.

(1) Test Methods: Functional performance testing and verification may be achieved by direct manipulation of system inputs (i.e. heating or cooling sensors), manipulation of system inputs with the building automation system (i.e. software override of sensor inputs), trend logs of system inputs and outputs using the building automation system, or short-term monitoring of system inputs and outputs using stand alone data loggers. A combination of methods may be required to completely test the complete sequence of operations. The Commissioning Agent shall determine which method, or combination, is most appropriate.

(2) Setup: Each test procedure shall be performed under conditions that simulate normal operating conditions as closely as possible. Where equipment requires integral safety devices to stop/prevent equipment operation unless minimum safety standards or conditions are met, functional performance test procedures shall demonstrate the actual performance of safety shutoffs in real or closely-simulated conditions of failure.

(3) Sampling: Multiple identical pieces of non-life-safety or non-critical equipment may be functionally tested using a sampling strategy. The sampling strategy shall be developed by the Commissioning Agent. If, after three attempts at testing the specified sample percentage, failures are still present, then all remaining units shall be tested at the contractors' expense.

(B) Develop functional performance test procedures for equipment and systems. Identify specific test procedures and forms to verify and document proper operation of each piece of equipment and system. Coordinate test procedures with the contractor for feasibility, safety, equipment and warranty protection. Functional performance test forms shall include the following information:

(1) System and equipment or component name(s).

(2) Equipment location and ID number.

(3) Date.

(4) Project name.

(5) Participating parties.

(6) Instructions for setting up the test, including special cautions, alarm limits, etc.

(7) Specific step-by-step procedures to execute the test.

(8) Acceptance criteria of proper performance with a Yes / No check box.

(9) A section for comments.

(C) Coordinate, observe and record the results of contractor's functional performance testing.

(1) Coordinate retesting as necessary until satisfactory performance is verified.

(2) Verify the intended operation of individual components and system interactions under various conditions and modes of operation.

3.8 Short-Term Diagnostic Testing

(A) Short-Term Diagnostic Testing: After initial occupancy, perform short-term diagnostic testing, using data acquisition equipment or the building automation system to record system operation over a two to three week period.

(1) Investigate the dynamic interactions between components in the building system.

(2) Evaluate the scheduling, the interaction between heating and cooling, and the effectiveness of the HVAC system in meeting the comfort requirements.

3.9 Deficiency Report and Resolution Record

(A) Deficiency Report and Resolution Record: Document items of non-compliance in materials, installation or operation.

(B) Non-Conformance. Non-conformance and deficiencies observed shall be addressed immediately, in terms of notification to responsible parties, and providing recommended actions to correct deficiencies.

(1) Corrections of minor deficiencies identified may be made during the tests at the discretion of the Commissioning Agent. In such cases the deficiency and resolution shall be documented on the procedure form.

(2) For identified deficiencies:

(a) If there is no dispute on the deficiency and the responsibility to correct it:

(1) The Commissioning Agent documents the deficiency and the adjustments or alterations required to correct it. The contractor corrects the deficiency and notifies the Commissioning Agent that the equipment is ready to be retested.

(2) The Commissioning Agent reschedules the test and the test is repeated.

(b) If there is a dispute about a deficiency or who is responsible:

(1) The deficiency is documented on the non-compliance form and a copy given to the Green Consultant.

(2) Resolutions are made at the lowest management level possible. Additional parties are brought into the discussions as needed. Contractor shall have

responsibility for resolving construction deficiencies. If a design revision is deemed necessary and approved by Owner, Architect/Engineer shall have responsibility for providing design revision.

(3) The Commissioning Agent documents the resolution process.

(4) Once the interpretation and resolution have been decided, the appropriate party corrects the deficiency and notifies the Commissioning Agent that the equipment is ready to be retested. The Commissioning Agent reschedules the test and the test is repeated until satisfactory performance is achieved.

(3) Cost of Retesting: Costs for retesting shall be charged to the Contractor.

3.10 Operations and Maintenance Training

(A) O&M Manual: Review the operation and maintenance manuals compiled by the contractor for completeness and for adherence to the requirements of the specifications.

(1) Obtain additional materials from contractor as necessary to stress and enhance the importance of system interactions, troubleshooting, and long-term preventative maintenance and operation.

(B) Training: Develop a Training Plan. Coordinate and review the training programs for Owner's personnel.

(1) Obtain additional materials from contractor as necessary to stress and enhance the importance of system interactions, troubleshooting, and long-term preventative maintenance and operation.

(C) O&M Database: Develop a database from the O&M manual that contains the information required to start a preventative maintenance program.

3.11 Record Documents Review

(A) Record Documents: Review record documents to verify accuracy.

3.12 Final Commissioning Report and Leed™ Documentation

(A) Final Commissioning Report: Compile final commissioning report. Summarize all of the tasks, findings, conclusions, and recommendations of the commissioning process.

(B) Documentation. Compile **[LEED™ Documentation] [Green Globes - US Documentation] [xxxx Documentation].** Format as required by [USGBC] **[GBI] [xxxx]** for submittal under the referenced green building rating system.

3.13 Deferred Testing

(A) Unforeseen Deferred Tests: If a test cannot be completed due to the building structure, required occupancy condition, or other deficiency, the functional testing may be delayed upon recommendation of the Commissioning Agent and the approval of the Owner. These tests are conducted in the same manner as the seasonal tests as soon as possible.

(B) Seasonal Testing:

(1) Schedule, coordinate, observe, and document additional testing for seasonal variation in operations and control strategies during the opposite season to verify performance of the HVAC system and controls. Complete testing during the warranty period to fully test all sequences of operation.

(2) Update O&M manuals and Record Documents as necessary due to the testing.

(C) End-of-Warranty Review: Conduct end of warranty review prior to the end of the warranty period. Review the current building operation with the facility maintenance staff. The review shall include outstanding issues from original or seasonal testing. Interview facility staff to identify concerns with building operation. Provide suggestions for improvements and assist owner in developing reports or documentation to remedy problems.

(1) Update O&M manuals and Record Documents as necessary due to the testing.

3.14 Equipment & System Schedule

(A) The following equipment shall be commissioned in this project.

SPECIFIER NOTE:

Edit below to suit project

System	Equipment	Check
HVAC System	Chillers	
	Pumps	
	Cooling tower	
	Variable frequency drives	
	Air handlers	
	Packaged AC units	
	Terminal units	
	Unit heaters	
	Heat exchangers	
	Fume hoods	

(Continued)

System	Equipment	Check
	Lab room pressures	
	Exhaust fans	
	Supply fans	
Electrical System	Sweep or scheduled lighting controls	
	Daylight dimming controls	
	Lighting occupancy sensors	
BAS System		

11

CONTRACT TERMINATIONS

I. INTRODUCTION

This chapter provides an overview of general principles and procedures involved in the termination of a government contract for default or for the convenience of the government. At common law, an unjustified default termination is a material breach of the contract, which entitles the contractor to recover damages, including anticipatory profits. The damages awarded are intended to place the contractor in as good a position as it would have been had the termination not occurred. A termination for the convenience of the government, however, precludes the contractor from recovering anticipated profits.

The termination for convenience concept was developed after the Civil War to avoid unnecessary spending of public funds on goods and services that were no longer needed by the government. The same policy is reflected in the Termination for Convenience (T4C) clause in the Federal Acquisition Regulation (FAR) at § 52.249-2.

This chapter addresses the rights and remedies of the government, the contractor, and its surety. Other topics related to terminations are also discussed including liquidated damages, constructive terminations, termination settlements after a T4C, and the effect of these procedures on subcontractors.

II. TERMINATIONS FOR DEFAULT

The right to terminate a construction contract for default arises only when a material provision of the contract has been breached or a party has failed to perform a material obligation required by the contract. Common law standards and the contract terms define the contracting parties' rights and obligations and set forth the parameters for a default termination. At common law, a default termination was not proper if based on a minor deviation in performance or failure to satisfy a contract requirement that was not material. Although a minor deviation or performance failure may give the nonbreaching party a right to damages, it does not rise to the level of significance justifying a complete termination of the contract. The determination of whether the breach is material depends on the significance of the event or act as compared to the overall purpose of the contract.

The terms of the contract can expand, limit, or redefine the common law grounds for termination for default. They can also provide for the payment of damages and other obligations after the contract is terminated for default. The termination for default provision in a government contract provides an extensive statement of the parties' respective rights and obligations.

A. Standard FAR Clause

In government construction contracts, the contract terms and conditions provide the basic statement of the parties' rights and duties in the event of a default. A study of those provisions and the cases which interpret them demonstrates the types of performance issues involved in contracting with the government. The standard FAR clause addressing the termination for default (T4D) of a construction contract is found at FAR § 52.249-10 and provides:

DEFAULT (FIXED-PRICE CONSTRUCTION) (APR. 1984)

(a) If the Contractor refuses or fails to prosecute the work or any separable part, with the diligence that will insure its completion within the time specified in this contract including any extension, or fails to complete the work within this time, the Government may, by written notice to the Contractor, terminate the right to proceed with the work (or the separable part of the work) that has been delayed. In this event, the Government may take over the work and complete it by contract or otherwise, and may take possession of and use any materials, appliances, and plant on the work site necessary for completing the work. The Contractor and its sureties shall be liable for any damage to the Government resulting from the Contractor's refusal or failure to complete the work within the specified time, whether or not the Contractor's right to proceed with the work is terminated. This liability includes any increased costs incurred by the Government in completing the work.

(b) The Contractor's right to proceed shall not be terminated nor the Contractor charged with damages under this clause, if—

   (1) The delay in completing the work arises from unforeseeable causes beyond the control and without the fault or negligence of the Contractor. Examples of such causes include—

      (i) Acts of God or of the public enemy,

      (ii) Acts of the Government in either its sovereign or contractual capacity,

      (iii) Acts of another Contractor in the performance of a contract with the Government,

     (iv)   Fires,

     (v)   Floods,

     (vi)   Epidemics,

     (vii) Quarantine restrictions,

     (viii) Strikes,

     (ix)   Freight embargoes,

     (x)   Unusually severe weather, or

     (xi)   Delays of subcontractors or suppliers at any tier arising from unforeseeable causes beyond the control and without the fault or negligence of both the Contractor and the subcontractors or suppliers; and

  (2) The Contractor, within 10 days from the beginning of any delay unless extended by the Contracting Officer), notifies the Contracting Officer in writing of the causes of delay. The Contracting Officer shall ascertain the facts and the extent of delay. If, in the judgment of the Contracting Officer, the findings of fact warrant such action, the time for completing the work shall be extended. The findings of the Contracting Officer shall be final and conclusive on the parties, but subject to appeal under the Disputes clause.

(c) If, after termination of the Contractor's right to proceed, it is determined that the Contractor was not in default, or that the delay was excusable, the rights and obligations of the parties will be the same as if the termination had been issued for the convenience of the Government.

(d) The rights and remedies of the Government in this clause are in addition to any other rights and remedies provided by law or under this contract.

The contract affords the government several remedies for default. For example, if a contractor fails to perform by the date specified, produces defective or nonconforming work, or refuses or fails to prosecute the work in such a way as to ensure its timely completion, the government may terminate the contract and complete the work at the contractor's expense. The contractor is also liable for any liquidated or actual damages caused by unexcused delays in completing the work.

The Default clause also gives the contractor the right to contest the termination for default. Under the common law of contracts, a wrongful termination would be a breach of contract, entitling the contractor to recover damages, including anticipated profit. However, the Default clause precludes a breach of contract claim by providing that a wrongful default termination shall be treated as a termination for the convenience of the government. Consequently, a wrongfully terminated contractor's relief under a government contract is compensation under the Termination for Convenience clause,[1] which provides a far more limited recovery than common law breach of contract damages.

---

[1]FAR § 52.249-2, Alternate I.

A brief analysis of the major elements of the Default clause follows.

Paragraph (a) provides the *basic grounds* for a default termination, that is, where the contractor refuses or fails to prosecute any or all of the work with sufficient diligence to insure timely completion or fails to complete the work within the specified time. In such circumstances, the government may terminate the contract in accordance with the provisions of the clause, take over the contractor's work and complete it using a variety of means. Paragraph (a) also establishes the government's right to recover *breach of contract damages* from the contractor and its surety.

Paragraph (b) sets forth the criteria for determining *excusable delay* and lists examples of delays that would not justify a termination for default. This paragraph excuses the failure of the contractor to prosecute or complete the work when the delay results from "unforeseeable causes beyond the control and without the fault or negligence of the Contractor . . ." if timely notice of the delay is given to the contracting officer. Delays arising from subcontractors or suppliers at any tier are excusable where unforeseeable and without the fault or negligence of *both* the contractor and the subcontractor or supplier.[2]

Paragraph (c) provides that, if the contractor is found not to be in default because the delay was excusable, the termination shall be treated as if it had been issued as "for the convenience of the Government." This *precludes the contractor's recovery* for breach of contract damages in the event the government's termination for default action is improper for any reason.

Paragraph (d) states that the government's remedies set out in the Default clause are not exclusive but are *in addition* to any other rights and remedies otherwise available, whether under the contract or under statutory or common law.

B. Grounds for Default Termination

The standard Default clause provides that the government may terminate the contract for default, in whole or in part, if the contractor (1) refuses or fails to prosecute any or all of the work with sufficient diligence to insure timely completion, or (2) fails to complete the work within the specified time. Contractors have also been terminated for repudiation of the contract, for failure to comply with other provisions of the contract, and for failure to maintain acceptable standards of skill and workmanship.

1. *Failure to Proceed*

Pursuant to the Disputes clause, a contractor has a duty to proceed with the work while a request for relief or a claim is pending resolution, including any appeal of a contracting officer's final decision.[3] Failure to proceed when directed by the contracting

---

[2]This provision effectively reversed the decision of the United States Court of Claims in *Schweigert, Inc. v. United States*, 388 F.2d 697 (Ct. Cl. 1967*). See* **Chapter 9** for a further discussion of excusable delays.

[3]*All State Constr., Inc.*, ASBCA No. 50586, 06-2 BCA ¶ 33,344 (contractor failed to perform any work for five weeks on the basis that the government was withholding disputed liquidated damages). *See* **Chapter 15** for a discussion of the disputes process.

agency can justify a default termination. A classic example of this principle is *American Dredging Co. v. United States*,[4] in which the court decided that the contractor was rightfully terminated for failing to proceed with its dredging operations even though it was entitled to a contract adjustment for the differing site condition that had been denied by the contracting officer. Similarly, a contractor cannot "abandon performance because the contracting officer failed to negotiate requests for equitable adjustments (REAs), issue a final decision on its claims, or modify the contract to include the cost of changed work. Nor [can] it condition resuming contract performance on the Board rendering a requested decision."[5]

2. Failure to Make Progress

FAR § 52.249-10(a) provides "if the Contractor refuses or fails to prosecute the work or any separable part, with the diligence that will insure its completion within the time specified in [the] contract," the contract may be terminated for default. However, the government must show not only a significant lag in performance, but also that it was reasonable for the government to conclude that, at the time of termination, there was no reasonable likelihood that the contractor could have completed performance on time.[6]

The mere fact that the contractor makes little or no progress during one stage of the work may not, alone, support termination. For example, in *Strickland Co.*,[7] a termination for default was overturned as premature because sufficient time remained in the contract schedule for the contractor to complete performance. However, in *MMI Capital, LLC v. General Services Administration*,[8] a defaulted lease-build contractor did not have control of the property on which the building was to be constructed. On that basis, the board upheld a default termination as there was no reasonable likelihood that the building would be completed by the lease start date. It is important to recognize that in reviewing a default termination, the board will consider the entire project record. This is significant given the "well settled principle that a termination for default may be sustained on grounds other than those cited by the CO [contracting officer] in the termination notice even if they were not known to the CO at the time of termination."[9] For example, in *M.E.S., Inc.*,[10] the board upheld a default even though it rejected the contracting officer's analysis regarding the contractor's ability to complete the work within the contract time. The board made its own delay and time analysis, which resulted in a decision to sustain the termination for default.

---

[4]207 Ct. Cl. 1010 (1975).

[5]*C. H. Hyperbarics Inc.*, ASBCA No. 49401, 04-1 BCA ¶ 32,568. *See also F&D Constr. Co., Inc. & D&D Mgmt., Consulting, & Constr. Co., Inc.*, ASBCA No. 41444, 91-2 BCA ¶ 23,983; *Howell Tool and Fabricating, Inc.*, ASBCA No. 47939, 96-1 BCA ¶ 28,225 at 140,941; *Brenner Metal Products Corp.*, ASBCA No. 25294, 82-1 BCA ¶ 15,462.

[6]*Lisbon Contractors, Inc. v. United States*, 828 F.2d 759 (Fed. Cir. 1987); *Morganti Nat'l, Inc. v. United States*, 49 Fed. Cl. 110 (2001).

[7]ASBCA No. 9840, 67-1 BCA ¶ 6193.

[8]GSBCA No. 16739, 2006 WL 2170507(Aug. 2, 2006).

[9]*In re Trinity Installers, Inc.*, AGBCA No. 2004-139-1, 05-1 BCA ¶ 32,868.

[10]PSBCA No. 4462, 06-1 BCA ¶ 33,184.

In deciding whether to pull the trigger on a termination for default decision, the contracting officer is often placed in a difficult position. If the contract cannot be terminated for failure to make progress until such time as the contractor will no longer be able to timely complete the work, there is little chance that the work can be completed by anyone by the original completion date. However, if the contracting officer terminates the contract any earlier, the termination may be overturned as premature. In addition to evaluating the amount of work to be completed and the available time for performance of that work, there may be excusable causes for the delay that must be considered in determining the adequacy of progress.

Although the factual determination varies from project to project, many cases have interpreted the Default clause to require a showing that timely completion was clearly in jeopardy. For example, in *Discount Co., Inc. v. United States*,[11] the Court of Claims held that the Default clause did not require a finding that completion within the original time frame was impossible. Rather, the termination for default was upheld when the government established that it was "justifiably insecure" that the contractor could complete on time. Similarly, the contracting officer's assessment of the contractor's ability to complete the work need not be "correct" in an objective sense if the determination was reasonable.[12] However, where the decision to terminate is based on an inaccurate analysis of the percent of contract completion and a flawed assessment of the contractor's ability to complete the work, the termination will not be upheld.[13]

To overturn a default termination for failure to make progress, a contractor has to demonstrate that the government's analysis was inaccurate, that it could have completed the work on time,[14] or that it had an excuse for the delay.[15] This approach reflects a shifting of the burden of proof as each party introduces evidence regarding the progress of the work and the possibility of completing the work within the original or extended schedule. The contractor must take an active role in explaining and justifying construction progress. This proof should include a showing of excusable delays as well as a realistic plan and schedule for completion of the remaining work.

3. Failure to Complete by Required Date

The contract may be properly terminated if the contractor fails to complete performance on time absent an excusable delay. As a general rule, time is of the essence, or fundamental, to the terms of a contract that contain fixed or specific performance dates. If timely performance does not occur, the government can immediately terminate without notice or providing an opportunity to cure.[16] However, where time is not

---

[11]554 F.2d 435 (Ct. Cl. 1977). *See also Lisbon Contracting Co., Inc. v. United States,* 828 F.2d 759 (Fed. Cir. 1987); *Cal. Dredging Co.,* ENGBCA No. 5532, 92-1 BCA ¶ 24,475.

[12]*FFR-Bauelemente+Bausanierung GMBH,* ASBCA No. 52152, 07-2 BCA ¶ 33,627.

[13]*Kostmayer Constr., LLC,* ASBCA No. 55053, 08-2 BCA ¶ 33,869.

[14]*Ventilation Cleaning Engineers, Inc.,* ASBCA No. 16678, *et al.,* 72-2 BCA ¶ 9537.

[15]*Interstate Gen. Gov't Contractors, Inc. v. United States,* 40 Fed. Cl. 585 (1998).

[16]*National Farm Equip. Co.,* GSBCA No. 4921, 78-1 BCA ¶ 13,195.

of the essence, late performance is only one of a number of factors to be considered in determining the adequacy of performance and the justification for a default termination.[17] In *Franklin E. Penny Co.*,[18] the Court of Claims rejected the traditional rule that time is of the essence in any government contract establishing fixed performance dates and observed that there may be situations where substantial delay in final completion will not constitute a default if the contractor's work is *sufficiently complete* that it can be put to use by the government.

Substantial completion does not preclude the possibility of a default termination. Even if a contractor has substantially completed the project, the government nevertheless may terminate the uncompleted portion of the contract for default in the event that the contractor refuses to complete punch-list items or fails to complete them within a reasonable period of time.[19]

4. Failure to Comply with Other Contract Requirements

A contractor's breach of other contract provisions may authorize the government to invoke the Default clause and terminate the contract for default. To invoke the Default clause on this basis, the government must establish that the contract provision at issue is of such significance that the failure to perform is a material breach of the contract. Some contract clauses—for example, the Inspection of Construction clause[20]—expressly state that default termination is an available remedy if the contractor fails to comply.

Other contract provisions, however, may not expressly warn the contractor that default termination is a potential sanction for noncompliance. A contractor must not view the absence of an express statement referencing default termination as an indication that the government does not have the right to default a contractor for its failure to comply with that provision. For example, the Davis-Bacon Act[21] and Withholding of Funds[22] clauses do not expressly refer to a default termination as a potential consequence for noncompliance. However, the Contract Termination-Debarment clause[23] and FAR § 22.406.11 Contract Terminations clearly and expressly contemplate that action by the government if the contractor breaches one of the labor standard clauses.

A default termination has been affirmed when the quality of a contractor's work is not consistent with required standards of workmanship and skill.[24] In such cases, the government must present evidence to support a finding that the contractor failed

---

[17]*Franklin E. Penny Co. v. United States,* 524 F.2d 668 (Ct. Cl. 1975).
[18]*Id.*
[19]*Southland Constr. Co.,* VABCA Nos. 2217, 2543, 89-1 BCA ¶ 21,548. *See also U.S. Royal Indus.,* PSBCA Nos. 1026, 1027, 83-2 BCA ¶ 16,673.
[20]FAR § 52.246.12. *See Keith Crawford & Assoc.,* ASBCA No. 46893, 95-1 BCA ¶ 27,388.
[21]FAR § 52.222-6.
[22]FAR § 52.222-7.
[23]FAR § 52.222-12.
[24]*Mega Constr. Co. v. United States,* 29 Fed. Cl. 396 (1993); *Ulibarri Constr. Co., Inc.,* VABCA Nos. 1780 and 1784, 87-3 BCA ¶ 20,169; *Edward E. Davis Contracting, Inc.,* VABCA No. 1108, 76-1 BCA ¶ 11,651.

to meet certain specific contract requirements.[25] Standard industry practice is not an acceptable measure of performance (i.e., is not a valid defense to a termination) where the contract specifications clearly differ from such practice.[26] A contractor's undisputed failure to provide a required submittal, which the court found was neither optional not incidental, justified a default termination even that performance failure was not the original basis for the contracting officer's default action.[27]

In addition, a contractor's failure to furnish Miller Act bonds pursuant to 40 U.S.C. §§ 3131–3134 has been accepted as a valid ground justifying default termination.[28] For example, in *Antonio Santisteban & Co, Inc.*,[29] a contractor's failure to furnish a performance bond constituted a material breach.

5. Anticipatory Repudiation

A termination based on a contractor's repudiation or anticipatory breach of the contract is closely related to the contractor's refusal or failure to prosecute the work. However, when a contractor repudiates the contract, it is subject to default termination regardless of its previous rate of progress on the project.

In this situation, the government is not required to provide notice to the contractor that the contract has been terminated since the contractor's breach by repudiation ends the contract. It has been stated that, under these circumstances, notifying the contractor of a termination would be a "vain and futile act."[30] However, a cautious contracting officer may send a *cure notice* and a termination notice to ensure that the government's rights under the Default clause are not diminished or lost.

To terminate a contract on the basis of a contractor's repudiation, the government must have a clear indication from the contractor that it cannot or will not perform the contract. This standard requires a "positive, definite, unconditional and unequivocal manifestation of intent, by words or conduct, on the part of a contractor" not to perform.[31] A failure to make progress alone does not constitute anticipatory repudiation absent some objective manifestation of intent not to perform. For example, a contractor may be considered to have repudiated the contract where it refuses to perform

---

[25]*Santee Dock Builders,* AGBCA No. 96-161-1, 99-1 BCA ¶ 30,190; *Richard W. Goff,* AGBCA No. 77-111, 78-1 BCA ¶ 12,891.

[26]*Metric Constructors, Inc.,* ASBCA No. 48423, 96-2 BCA ¶ 28,459.

[27]*Takota Corp. v. United States,* __Fed. Cl. ___, 2009 WL 3574132 (October 28, 2009).

[28]*Airport Indus. Park, Inc. d/b/a P.E.C. Contracting Eng'rs v. United States,* 59 Fed. Cl. 332 (2004); *see also Cole's Constr. Co., Inc.,* ENGBCA No. 6074, 94-3 BCA ¶ 26,995; *F&D Constr. Co., Inc. & D&D Mgmt., Consulting & Constr. Co., Inc.,* ASBCA No. 41441, 91-2 BCA ¶ 23,983; *Ruffin's A-1 Constr., Inc.,* ASBCA No, 38343, 90-3 BCA ¶ 23,243; *H. L. & S. Contractors, Inc.,* IBCA No. 1085-11-75, 76-1 BCA ¶ 11,878.

[29]ASBCA No. 5586 *et al.,* 60-1 BCA ¶ 2497.

[30]*Imperial Van & Storage,* ASBCA No. 11462, 67-2 BCA ¶ 6621. *See also Fairfield Scientific Corp.,* ASBCA No. 21151, 78-1 BCA ¶ 13,082 *aff'd on recon.* 78-2 BCA ¶ 13,429; *Scott Aviation,* ASBCA No. 40776, 91-3 BCA ¶ 24,123; *Reddy-Buffaloes, Pump, Inc.,* ENGBCA No. 6049, 96-1 BCA ¶ 28,111.

[31]*United States v. DeKonty Corp.,* 922 F.2d 826 (Fed. Cir. 1991); *Twigg Corp.,* NASABCA No. 62-0192, 93-1 BCA ¶ 25,318 (refusal to replace concrete); *Cox & Palmer Constr. Corp.,* ASBCA No. 38739, 92-1 BCA ¶ 24,756; *Mountain State Constr. Co., Inc.,* ENGBCA No. 3549, 76-2 BCA ¶ 12,197.

during a dispute about contract interpretation.[32] Similarly, a contractor is not entitled to stop work while waiting on a contract modification or resolution of a request for price escalation for inflation.[33]

A contractor's expressed concern to the contracting officer about the negative effect of a proposed modification on the contractor is not, standing alone, justification for a default termination since a contractor's repudiation must be unequivocal. There must also be a corresponding abandonment of performance.[34]

In a few circumstances, a contractor's apparent refusal to perform may *not* be regarded as an anticipatory breach or repudiation of the contract.[35] For example, in *McKenzie Marine Construction Co.*,[36] the contractor's letter to the government stated that it had "no moral or contractual obligation" to perform "such an insane project." After considering all of the circumstances related to the contractor's efforts to perform, the board rejected the government's contention that the letter constituted an unequivocal refusal to perform. The burden of proof on the government is relatively high when asserting that the contractor repudiated performance under the contract. However, if the contractor abandons performance, the facts justifying the contractor's refusal to continue performance must be clearly proven.[37]

6. Subcontractor Performance Failures

A contractor's performance obligations include the entire scope of work. Regarding its subcontractors' work, a performance delay is excusable only if it is without the fault or negligence of the subcontractors (or suppliers) at all tiers.[38] If the contract provides for government approval of a proposed subcontractor, the contracting officer's unreasonable delay in acting on an approval request or refusal to approve a proposed subcontractor may excuse a termination for default.[39] However, the contractor must seek approval in a timely manner.[40]

---

[32]*Twigg Corp.*, NASABCA No. 62-0192, 93-1 BCA ¶ 25,318; *F&D Constr. Co., Inc. & D&D Mgmt., Consulting & Constr. Co., Inc.*, ASBCA No. 41441, 91-2 BCA ¶ 23,983; *The Tester Corporation*, ASBCA No. 21312, 78-2 BCA ¶ 13,373.

[33]*See F&D Constr. Co., Inc. & D&D Mgmt., Consulting & Constr. Co., Inc.*, ASBCA No. 41441, 91-2 BCA ¶ 23,983; *Wear Ever Shower Curtain Corp.*, GSBCA 4360, 76-1 BCA ¶ 11,636; *Fraenkische Parkettverlegung R.*, ASBCA No. 18453, 75-2 BCA ¶ 11,388.

[34]*Brenner Metal Products Corp.*, ASBCA No. 25294, 82-1 BCA ¶ 15,462; *Sharjon, Inc.*, ASBCA No. 22954, 79-1 BCA ¶ 13,585; *Denison Research Foundation*, ASBCA No. 7653, 1963 BCA ¶ 3651.

[35]*Marine Constr. & Dredging, Inc.*, ASBCA No. 39246, 95-1 BCA ¶ 27,286 (government failure to provide clear direction); *Todd-Grace, Inc.*, ASBCA No. 34469, 92-1 BCA ¶ 24,742 (government breached implied obligation not to hinder performance); *Kahaluu Constr. Co., Inc.*, ASBCA No. 31187, 89-1 BCA ¶ 21,308 (government failed to clarify specifications following a valid request by the contractor).

[36]ENGBCA No. 3245, 74-2 BCA ¶ 10,673.

[37]*See D&D Mgmt., Consulting & Constr. Co., Inc.*, ASBCA No. 41441, 91-2 BCA ¶ 23,983; *Kahaluu Constr. Co., Inc.*, ASBCA No. 31187, 89-1 BCA ¶ 21,308.

[38]*General Railway Signal Co.*, ENGBCA No. 6309, 97-2 BCA ¶ 29,170 (subcontractor bankruptcy not an excusable delay); *Decker & Co. Gmbh*, ASBCA No. 41089, 94-2 BCA ¶ 26,759; *M&T Constr. Co.*, ASBCA No 42750, 93-1 BCA ¶ 25,223. *See also* 48 C.F.R. § 352.249-14 (contract provision with Health and Human Services excluding "failures of subcontractors" from definition of excusable delays).

[39]*A. Dubois and Sons*, ASBCA No. 3265, 57-1 BCA ¶ 1174.

[40]*Vlier Contracting Co.*, ASBCA No. 9427, 1963 BCA ¶ 3994.

As a general rule, delays by government designated sole-source subcontractors are not excusable if the subcontractor is at fault.[41] The government's designation of a sole-source subcontractor or supplier can introduce a variety of performance risks for the contractor and the government. An illustration of such risks is presented in the next list.

- Specified components are not compatible or fail to achieve expected performance requirements.[42]
- Sole-source subcontractor's materials are defective.[43]
- Specified or sole-source subcontractor/supplier refuses or is not able to supply materials.[44]
- Sole-source subcontractor delays overall project.[45]

Assertions that the government had superior knowledge of the sole-source firm's performance capabilities may or may not shift the risk of the subcontractor's failure to perform to the government. A key fact involves the nature of the information available to the contractor at proposal time and the contractor's efforts to investigate the sole-source subcontractor's capabilities prior to submitting its bid or proposal.[46]

A subcontractor's default may not result in a default of the contractor as the government may agree to permit the contractor to replace that subcontractor with another. For example, in *McLain Plumbing & Electrical Services v. United States,*[47] a contractor was declared in default because its principal subcontractor failed to perform. The government offered to accept substitute performance rather than terminate the prime contractor. The government proposed that the contractor terminate and replace the subcontractor, remove work installed by the subcontractor, and have the new subcontractor perform the original scope of work. The contractor agreed to this proposal and terminated the subcontractor. Later the terminated subcontractor prevailed in its arbitration claim against the contractor for wrongful termination of the subcontract. However, the contractor's effort to recover from the government the amount of damages awarded in the arbitration was rejected. The court held that the agreement with the government was an accord and satisfaction barring any claim against the government related to the termination of that subcontractor.[48]

---

[41]*Joseph J. Bonavire Co.,* GSBCA No. 4819, 78-1 BCA ¶ 12,877; *see also Systems & Electronics, Inc.,* ASBCA No. 41113, 97-1 BCA ¶ 28,671 (government designation of subcontractor was not a guarantee of timely performance); *Ainslie Corp.,* ASBCA No. 29303, 89-2 BCA ¶ 21,811; *Cascade Elec. Co.,* ASBCA No. 28674, 84-1 BCA ¶ 17,210.

[42]*Turner Constr. Co., et al.,* ASBCA No. 25447, 90-2 BCA ¶ 22,649 (government risk).

[43]*Environmental Tectonics Corp.,* ASBCA No. 21657, 79-1 BCA ¶ 13,796 at 67,578 (contractor's risk).

[44]*James Walford Constr. Co.,* GSBCA No. 6498, 83-1 BCA ¶ 16,277; *Frontier Contracting Co.,* ENGBCA No. 4333, 80-2 BCA ¶ 14,528 (generally contractor's risk).

[45]*Joseph J. Bonavire Co.,* GSBCA No. 4819, 78-1 BCA ¶ 12,877; *see also Systems & Electronics, Inc.,* ASBCA No. 41113, 97-1 BCA ¶ 28,671; *N. S. Meyer,* ASBCA No. 27144, 83-1 BCA ¶ 16,214 (generally contractor's risk).

[46]*Compare Alabama Dry Dock & Shipbuilding Corp.,* ASBCA No. 39215 90-2 BCA ¶ 22,855 *with Franklin E. Penny Co.,* 524 F.2d 668 (Ct Cl. 1975).

[47]30 Fed. Cl. 70 (1993).

[48]*Id.* at 82.

C. Termination Process

This section examines the principles and procedures when the government default terminates a contractor and the effect of the government's failure to comply with these procedures. Compliance with these procedures can be critical to the resolution of the parties' respective rights and obligations. The boards and the Court of Federal Claims view a default termination as a drastic sanction[49] in the nature of a forfeiture.[50] Thus the government has the initial burden of providing "good grounds and solid evidence"[51] to support a termination for default. However, once the government satisfies that initial burden, the contractor bears the burden of establishing that the performance failure was excusable.[52]

1. Exercise of Discretion

Prior to issuing a termination for default notice, the contracting officer must carefully consider the various factors listed in the FAR. In 1967, the Court of Claims considered the motives of a contracting officer and the discretion exercised when issuing a default termination. The court held that the Default clause did not require the contracting officer's decision to be a personal one, but it was necessary that some discretion be exercised by the procuring activity, and that factors relevant to and weighing on the default and the termination should be evaluated in making the termination decision.[53]

In *J. D. Hedin Construction Co. v. United States,*[54] the court applied the rationale of *Schlesinger*[55] in the default termination of a construction contract. However, a challenge to a default termination on the basis of the contracting officer's subjective motivation is extremely difficult. For example, in *Hydraulic Systems Co.,*[56] the Armed Services Board of Contract Appeals (ASBCA) sustained a termination for default, stating that the "motives of the procuring activity are immaterial and not subject to review by this Board . . . , if the termination is legally permissible."[57]

---

[49]*Lisbon Contractors, Inc. v. United States,* 828 F.2d 759, 765 (Fed. Cir. 1987); *Necco, Inc. v. General Services Administration,* GSBCA No. 16354, 05-1 BCA ¶ 32,902; *FFR-Bauelemente+Bausanierung GMBH,* ASBCA No. 52152, 07-2 BCA ¶ 33,627.

[50]*DeVito v. United States,* 413 F.2d 1147, 1153 (Ct. Cl. 1969).

[51]*J.D. Hedin Constr. Co. v. United States,* 408 F.2d 424, 431 (Ct. Cl. 1969); *Necco, Inc. v. General Services Administration,* GSBCA No. 16354, 05-1 BCA ¶ 32,902.

[52]*DCX, Inc. v. Perry,* 79 F.3d 132, 134 (Fed. Cir. 1996); *Necco, Inc. v. General Services Administration,* GSBCA No. 16354, 05-1 BCA ¶ 32,902.

[53]*Schlesinger v. United States,* 390 F.2d 702 (Ct. Cl. 1967).

[54]408 F.2d 424 (Ct. Cl. 1969).

[55]*Schlesinger v. United States,* 390 F.2d 702 (Ct. Cl. 1967); *see also Mountain State Constr. Co., Inc.,* ENGBCA No. 3549, 76-2 BCA ¶ 12,197.

[56]ASBCA No. 16856, 72-2 BCA ¶ 9742.

[57]*Id.* at p. 45,534. *See also FFR-Bauelemente+Bausanierung GMBH,* ASBCA No. 52152, 07-2 BCA ¶ 33,627 (clear and convincing evidence of specific intent to injury the contractor required).

FAR § 49.402-3(f) sets forth a detailed list of factors that a contracting officer is required to consider in determining whether to terminate a construction contract for default. The provisions included in the FAR provide, in part:

(f) The contracting officer shall consider the following factors in determining whether to terminate a contract for default:

 (1) The terms of the contract and applicable laws and regulations.

 (2) The specific failure of the contractor and the excuses for the failure.

 (3) The availability of the supplies or services from other sources.

 (4) The urgency of the need for the supplies or services and the period of time required to obtain them from other sources, as compared with the time delivery could be obtained from the delinquent contractor.

 (5) The degree of essentiality of the contractor in the Government acquisition program and the effect of a termination for default upon the contractor's capability as a supplier under other contracts.

2. Cure Notices—Contractor Responses

The FAR contains several different default clauses. Many of these clauses obligate the government to issue a cure notice when contemplating a default termination.[58] However, no cure notice or cure period is required before the termination of a government construction contract.[59] Notwithstanding the absence of a mandated cure notice in the FAR, a contractor usually has relatively early information or notice of the government's dissatisfaction with its performance because it is rare that an agency remains passive in the face of schedule slippage or other perceived failures to comply with the contract. It is more likely that the contracting officer will utilize the authority under the contract to hold retainage for unsatisfactory performance to gain the contractor's attention. Another tool that may be employed by the government is to issue an unsatisfactory Interim Past Performance Evaluation.[60] It is not unusual to see these tools utilized in combination before there is consideration of sending a cure notice.

As a practical matter, many agencies issue cure notices on construction contracts and copy the performance bond surety before formally defaulting the contractor. A cure notice provides the contractor and its surety an opportunity to address excusable delays and performance concerns by providing a new or revised schedule evidencing an ability to complete the work within the remaining performance period. In *Danzig v. AEC Corp.*,[61] the Federal Circuit articulated the basic purposes of the cure notice and the contractor's response in a construction project in this manner:

---

[58]*See* FAR § 52.249-8 Default (Fixed-Price Supply and Service); FAR § 52.249-9 Default (Fixed-Price Research & Development).

[59]*Professional Services Supplier, Inc., v. United States*, 45 Fed. Cl. 808 (2000) (contractor's argument that the common law required a cure notice in a government construction contract rejected by the court).

[60]*See* **Chapter 3, Section IV** for a discussion of the past performance evaluation process.

[61]224 F.3d 1333 (Fed. Cir. 2000).

When the government has reasonable grounds to believe that the contractor may not be able to perform the contract on a timely basis, the government *may* issue a cure notice as a precursor to a possible termination of the contract for default. When the government justifiably issues a cure notice, the contractor has an obligation to take steps to demonstrate or give assurances that progress is being made toward a timely completion of the project, or to explain that the reasons for any prospective delay in completion of the contract are not the responsibility of the contractor.[62]

Having issued a cure notice, at least one board has held that the government needs to consider the contractor's response prior to terminating the contractor for default.[63] With respect to the contractor's response, the Federal Circuit has adopted the doctrine that the contractor must give reasonable assurances of performance when addressing a valid cure notice.[64] The court has further stated that a contractor's failure to provide the requested assurances may be treated as a breach of contract justifying termination for default.[65] In assessing the contractor's response to a cure notice, the ASBCA has stated the contracting officer's conclusion regarding the contractor's ability to timely complete need not be objectively correct. Rather, for the default termination to be sustained the contracting officer need only be "justifiably insecure" about the contractor's timely completion.[66]

Receipt of a cure notice is a clear signal that a default termination is under consideration by the government. The serious consequences related to a termination cannot be ignored. However, at the same time, the contractor's staff must initiate certain affirmative actions to permit the contractor either to address the government's concerns or to mitigate the consequences if a termination for default is issued. Attached to this chapter at **Appendix 11B** is a checklist addressing topics and actions that should be considered or implemented by a contractor if faced with a default termination threat.

3. Termination Notice

The contracting officer's termination for default notice should comply with the guidelines set forth in FAR § 49.402-3(g):

If, after compliance with the procedures in Paragraphs (a) through (f) of this 49.402-3, the contracting officer determines that a termination for default is proper, the contracting officer shall issue a notice of termination stating—

---

[62]*Id.* at 1337 (emphasis added).
[63]*See Omni Dev. Corp.,* AGBCA No. 97-203-1, 01-2 BCA ¶ 31,487.
[64]*Danzig v. AEC Corp.,* 224 F.3d 1333, 1338 (Fed. Cir. 2000).
[65]*Id.,* at 1340.
[66]*FFR-Bauelemente+Bausanierung GMBH,* ASBCA No. 52152, 07-2 BCA ¶ 33,627 citing *McDonnell Douglas Corp. v. United States,* 323 F.3d 1006,1017 (Fed. Cir. 2003).

(1) The contract number and date;

(2) The acts or omissions constituting the default;

(3) That the contractor's right to proceed further under the contract (or a specified portion of the contract) is terminated;

(4) That the supplies or services terminated may be purchased against the contractor's account, and that the contractor will be held liable for any excess costs;

(5) If the contracting officer has determined that the failure to perform is not excusable, that the notice of termination constitutes such decision, and that the contractor has the right to appeal such decision under the Disputes clause;

(6) That the Government reserves all rights and remedies provided by law or under the contract, in addition to charging excess costs; and

(7) That the notice constitutes the decision that the contractor is in default as specified and that the contractor has the right to appeal under the Disputes clause.

Even though FAR § 49.402-3(g) states that the contracting officer "shall" issue a termination notice addressing the seven elements just set forth, the government's failure to precisely follow this regulation may not provide the basis to overturn a termination for default unless the contractor can prove that it was prejudiced by that failure.[67] The notice of the default termination is usually effective upon delivery to the contractor's place of business.[68]

4. Alternative Grounds for Termination

A default termination usually will be upheld on any valid basis even if that ground was not known to the contracting officer at the time the contract was terminated. The government's right to "reach back" exists as long as the basis for the default existed when the termination decision was made. For example, in *Joseph Morton Co., Inc. v. United States*,[69] a contract was terminated for default following the issuance of several cure notices related to performance issues. Thereafter, the termination was sustained by the court on the basis of the contractor's conviction for fraud involving a single change order to the contract. At the time of the termination, the contracting officer was unaware of the contractor's fraudulent act. Similarly, in *Glazer Construction Co. v. United States*,[70] a default was upheld for violations of the Davis-Bacon Act by the contractor even though the violations were discovered following the contracting officer's issuance of the termination for default decision. In *Takota*

---

[67]*See Philadelphia Regent Builders, Inc. v. United States*, 634 F.2d 569 (Ct. Cl. 1980).

[68]*Fred Schwartz*, ASBCA No. 20724, 76-1 BCA ¶ 11,916; *recon. denied*, 76-2 BCA ¶ 11,976.

[69]3 Cl. Ct. 120 (1983). *See also Keith Crawford & Assoc.*, ASBCA No., 46893, 95-1 BCA ¶ 27,388; *Shostak Constr. Co.*, VABCA No. 3671, 94-2 BCA ¶ 26,791; *F&D Constr. Co, Inc. & D&D Mgmt., Consulting & Constr., Inc.*, ASBCA No. 41441, 91-2 BCA ¶ 23,983.

[70]52 Fed. Cl. 513 (2002).

Corp. v. United States,[71] the Court of Federal Claims sustained a default termination on the grounds that the contractor had clearly failed to provide a required submittal. In its termination for default action the government had justified the default on the grounds that the contractor failed to make progress, The court concluded that it did not have to reach that contested issue.

5. Contract Reinstatement

Even if the termination for default is valid, the government retains the authority to reverse that action and reinstate the contract. FAR § 49.102(d) authorizes a contracting officer to reinstate a terminated contract by mutual agreement of the parties if that action is advantageous to the government. The Government Accountability Office (GAO) has stated that a government agency has the "inherent authority" to reinstate a contract that was terminated for default if there is a good faith determination "that the basis for the default no longer exists and that it is reasonable to expect satisfactory completion of contract performance upon reinstatement."[72]

6. "No-Cost" Terminations

In lieu of a termination for default with the resulting litigation, the parties may consider a "no-cost" termination—a rescission of the contract that simply allows both parties to walk away from the contract without further loss, risk, or consequences.[73] The no-cost settlement is an effective way by which the costs, litigation, and risks of the litigation arising from a default termination can be substantially mitigated by both parties. In a closely contested termination for default case in which neither party is clearly correct, a no-cost settlement may be a fair resolution of the dispute. Samples of a no-cost termination settlement agreement can be found in FAR §§ 49.603-6 and 49.603-7. In considering the potential advantages of a no-cost termination, a contractor should carefully evaluate the effect of the termination on its potential contract liabilities to its subcontractors and suppliers and should carefully coordinate the details of this action with its surety. Similarly, if a past performance evaluation will be prepared by the agency, the contractor should seek to negotiate an acceptable evaluation in the context of a negotiated no-cost settlement.

D. Limitations on Termination Rights

1. Substantial Completion

The concept of *substantial completion* may operate to limit the government's default termination rights when the contractor has substantially performed the project's scope of work. As a general rule, *substantial completion* is achieved when the

---

[71]__Fed. Cl. ___, 2009 WL 3574132 (October 28, 2009).
[72]*Stancil-Hoffman Corp.,* Comp. Gen. Dec. B-193001.2, 80-2 CPD ¶ 226.
[73]*See* FAR § 49.402-4(c).

contract work is "satisfactorily completed to the extent that facilities might be occupied or used by the Government for the purposes for which they were intended."[74] In the construction industry, *substantial completion* generally means that a building or facility has been completed except for minor punch-list items.[75]

Generally, the government may not terminate the contractor for failure to complete the project or impose liquidated damages once substantial completion is achieved. However, the contractor remains obligated to complete the remaining work ("punch-list" items) in compliance with the contract documents. If the contractor fails to perform the punch-list work, the government may reduce the contract price or under certain circumstances terminate the contractor for default as to the remaining, incomplete contract work.[76] If the contract work is *substantially complete* when the contractor is terminated for default, a termination for failure to complete the overall project is improper. When determining whether the contractor achieved substantial completion, a board or the Court of Federal Claims considers "(1) the quantity of work remaining to be done; and (2) the extent to which the project was capable of adequately serving its intended purpose."[77]

Contractors bear the burden of proving that the project is essentially complete and capable of being used for its intended purpose.[78] Payment applications may or may not be compelling evidence of the status of the project's overall completion. If the contractor's assertions at the hearing regarding the project's stage of completion conflict with its contemporaneous documents (e.g., schedule updates, internal reports, daily reports, etc.), it may be difficult, if not impossible, for the contractor to sustain its burden of proof.[79] Although no fixed percentage of completion is a prerequisite to establishing substantial completion, it has been observed that a building less than 85 percent complete will rarely be held to be "substantially complete."[80]

2. Obligations Following Substantial Completion

Following substantial completion, the right to terminate for default for performance failures is limited to the contract's incomplete scope of work. In that situation the right to default a contractor has certain limitations. For example, in *Donat Gerg Haustechnick*,[81] the contractor inexcusably failed to provide records for an audit.

---

[74]*See Central Ohio Bldg. Co., Inc.*, PSBCA No. 2742, 92-1 BCA ¶ 24,399; *Matthew Andrew Kalosinakis*, ASBCA No. 41337, 91-2 BCA ¶ 23,744 at 118,909; *Robert E. McKee, Inc.*, ASBCA No. 33643, 90-1 BCA ¶ 22,391 at 112,511. *See also Thoen v. United States*, 765 F.2d 1110, 1115 (Fed. Cir. 1985).

[75]*See Central Ohio Bldg. Co., Inc.*, PSBCA No. 2742, 92-1 BCA ¶ 24,399.

[76]*See Keith Crawford & Assoc.*, ASBCA No. 46893, 95-1 BCA ¶ 27,388; *D & D Mgmt., Consulting & Constr., Co., Inc.*, ASBCA No. 41444, 91-2 BCA ¶ 23,983 at 120,030.

[77]*Blinderman Constr. Co., Inc. v. United States*, 39 Fed. Cl. 529, 573 (1997) (*quoting Electrical Entrs., Inc*, IBCA No. 972-9-72, 74-1 BCA ¶ 10,400).

[78]*Fidelity Constr. Co.*, DOTCAB No. 75-19, 75-19A, 77-2 BCA ¶ 12,831; *Capitol City Constr.*, DOTCAB No. 72-49, 75-1 BCA ¶ 11,012.

[79]*See Dixson Contracting, Inc.*, AGBCA No. 98-191-1, 00-1 BCA ¶ 30,766.

[80]*K & M Constr.*, ENGBCA No. 3115, 73-2 BCA ¶ 10,034.

[81]ASBCA No. 41197, 97-2 BCA ¶ 29,272.

The board ruled that the contract could not be terminated for default on that basis. However, even though contractor achieves substantial completion, the government still has the right to require correction of defective work and default the contractor for a refusal or failure to correct the defective work.[82]

E. Waiver of Right to Terminate

1. Demonstrating Waiver by the Government

When responding to a threatened or actual default termination, a contractor may contend that the government is estopped from terminating the contract or has waived its right to default terminate the contractor. The legal principle that the government might, under certain circumstances, be estopped from terminating a contract for default originated in government supply contracts. In *DeVito v. United States,*[83] the Court of Claims overturned a default termination when the government failed to terminate the contractor performing a supply contract within a reasonable time after a required delivery date had passed. In *DeVito,* the government was aware that contractor was continuing to perform in an effort to overcome the delay with the government's actual or implied consent. Consequently, the government's subsequent termination action was improper.

Since the government's action may waive the completion date, the government may be estopped from terminating the contractor. In a strict sense, *estoppel* and *waiver* are different legal theories or principles. In *Olson Plumbing & Heating Co.,* the ASBCA described the difference in the two legal concepts in this manner:

> A waiver requires the intent to waive. It [referring to the government's inaction] is rather an estoppel. The inaction of the contracting officer after the passing of the due date induces the contractor to change its position by continuing performance at its expense; thus, the Government is estopped to terminate until it has again asserted that time is of the essence, and gives the contractor a reasonable time to complete performance.[84]

The concept of a *waiver* of the right to terminate can be difficult to define with precision and it often is hard to distinguish it from the concept of *estoppel.* In addition, absent conduct clearly indicating the government's intent to waive the right to terminate a contract for default, certain government conduct (actions or inactions) may be labeled by a board or court as a constructive waiver. Since the label "waiver" often is applied to either type of situation and since the facts are far more important than labels, the terms often are used synonymously.

---

[82]*See* FAR § 52.246-12, Inspection of Construction (that clause permits the government to default a contractor if it "does not promptly" replace or correct rejected work and provides the grounds upon which the government may set-aside a prior acceptance). *See also Keith Crawford & Assoc.,* ASBCA No. 46893, 95-1 BCA ¶ 27,388; *Sentell Bros., Inc.,* DOTBCA No. 1824, 89-3 BCA ¶ 21,904, and **Chapter 10.**
[83]413 F.2d 1147 (Ct. Cl.1969).
[84]ASBCA No. 17965 *et al.,* 75-1 BCA ¶ 11,203, *aff'd,* 601 F.2d 950 (Ct. Cl. 1979).

In general, to establish that the government has waived or lost its default termination right, the contractor must prove:

Unreasonable Forbearance. The government fails to terminate the contract within a reasonable period of time after the grounds for a default termination were apparent. and

Continued Performance. With the government's knowledge or consent, the contractor continued working in reliance on the government's failure to terminate it.[85]

In an effort to avoid having either concept (waiver or estoppel) invoked against it, the government typically will communicate and document in a written cure notice that it is reasonably *forbearing* from exercising its right to issue an immediate termination for default in order to give the contractor an opportunity to submit further evidence of excusable delays or to provide assurances of satisfactory performance going forward.[86] Similarly, a cure notice may expressly reserve the government's rights to default terminate the contractor and to assess liquidated damages notwithstanding discussions regarding a revised completion date and the like.

In construction contracts, the courts and the ASBCA generally have not applied *DeVito* and its concept of constructive waiver (unreasonable forbearance or delays in exercising the default remedy).[87] A basic factor for the varying treatment of the two types of contracts relates to differences in the payment processes. In most supply contracts, the contractor is paid only after delivering acceptable end products. In construction contracting, the contractor is paid for partially completed work on a percentage of completion basis. As part of the proof that the government's conduct constituted waiver of the delivery date or estoppel, the contractor must demonstrate that it relied to its detriment on the government's intent that the contractor continue performance.[88] Focusing on the significance of the progress payments, the ASBCA explained:

[O]ne reason why the *DeVito* . . . principle normally does not apply to construction contracts [is that] the construction contract provides for payment or credit for partially completed work and materials delivered to the site and the transfer of title thereto to the Government. The work performed after the due date is not normally wasted . . . [citation omitted]. In other words, the contractor has not changed his position to his detriment.[89]

---

[85]*Id. See also A.R. Sales Co., Inc. v. United States.* 51 Fed. Cl. 370 (2002) (no evidence of unreasonable forbearance or contractor reliance).

[86]*Raytheon Service Co.,* ASBCA No. 14746, 70-2 BCA ¶ 8390 (period of reasonable forbearance is generally no longer than the time needed for the contracting officer to investigate the facts and to determine a course of action that is in the best interest of the government).

[87]*See Indemnity Ins Co. of North America v. United States,* 14 Cl. Ct. 219, 224-5 (1988); *Nisei Constr. Co.,* ASBCA No. 51464 *et al.,* 99-2 BCA ¶ 30,448; *Arens Corp.,* ASBCA No. 50289, 02-1 BCA ¶ 31,671.

[88]*See also MPT Entrs.,* ASBCA No. 25835 *et al.,* 83-2 BCA ¶ 16,769, *aff'd on recon.,* 84-2 BCA ¶ 17,252.

[89]*Olson Plumbing and Heating Co.,* ASBCA No. 17965, 75-1 BCA ¶ 11,203 at 53,336, *aff'd,* 602 F.2d 950 (Ct. Cl. 1979). *See also Technocratica,* ASBCA Nos. 47992, 47993, 48054, 48060, 48061, 06-2 BCA ¶ 33,316.

The availability of routine progress payments in construction makes proving waiver (estoppel) more difficult. The *DeVito* principle may still apply in the context of construction projects when the government acts in a manner that is clearly inconsistent with the enforcement of the contract completion date.[90] The ASBCA has recognized the potential application of this principle in "unusual circumstances."[91] For example, in *Corway, Inc.*,[92] the ASBCA applied *DeVito* in a case where the original completion date passed and, *without* establishing a new reasonable completion date, mentioning or assessing liquidated damages, or manifesting any concern with the obvious late completion of the project, the government allowed the contractor to remain working.[93] Similarly, in *Martin J. Simko Construction, Inc. v. United States*,[94] the Court of Claims held that, by directing the contractor to continue performance long after the contract completion date had passed, the government lost its right to default the contractor.[95]

In summary, even in the context of a construction project, the government can waive its termination right when it allows a delinquent contractor to work past the contract completion date without issuing a timely termination notice or advising the contractor that the government intends to impose liquidated damages and is reserving its right to terminate the contract.[96] Such actions or inactions by the government may demonstrate that the contract no longer has an enforceable completion date.[97] In that context, the right to default the contractor may be waived due to the government's failure to establish a new realistic completion date or reserve its right to default, on one hand, and the contractor's reliance on the government's failure to declare a default, on the other.[98]

---

[90]*Brent L. Sellick*, ASBCA No. 21869, 78-2 BCA ¶ 13,510. *Cf. B. V. Constr. Inc.*, ASBCA No. 47766, 04-1 BCA ¶ 32,604; *Jess Howard Elec. Co.*, ASBCA No. 44437, 96-2 BCA ¶ 28,345; *La Grow Corp.*, ASBCA No. 42386, 91-2 BCA ¶ 23,945.

[91]*Technocratica*, ASBCA Nos. 47992, 06-2 BCA ¶ 33,316 (contracting officer decided to allow the contractor to "muddle through a bit longer").

[92]ASBCA No. 20683, 77-1 BCA ¶ 12,357.

[93]*See also Milo Werner Co.*, IBCA No. 1202-7-78, 82-1 BCA ¶ 15,698; *Sentinel Standard*, ASBCA No. 26199, 83-1 BCA 16,517. *But see Indemnity Ins. Co. of North America v. United States*, 14 Cl. Ct. 219 (1988) (right to terminate for default reserved and liquidated damages assessed); *Abcon Assoc., Inc. v. United States*, 44 Fed. Cl. 625 (1999) (the government's 10-month delay in terminating the contractor was *not* a waiver since the government advised the contractor that it was reserving its rights, that assistance or forbearance was *not* waiver, and levied liquidated damages).

[94]11 Cl. Ct. 257 (1986).

[95]*See also Sun Cal, Inc. v. United States*, 21 Cl. Ct. 31 (1990) (government encouraged the contractor to continue performance and agreed to a contract modification waiving liquidated damages to perform); *Overhead Elec. Co.*, ASBCA No. 25656, 85-2 BCA ¶ 18,026 at 90,472-73 (government permitted the completion date to pass without action, did not set a new completion date (unilaterally or bilaterally), and made no effort to assess liquidated damages).

[96]*Arens Corporation, Inc.*, ASBCA No. 50289, 02-1 BCA ¶ 31,671.

[97]*Technocratica*, ASBCA Nos. 47992, 47993, 48054, 48060, 48061, 06-2 BCA ¶ 33,316.

[98]*B.V. Constr., Inc.*, ASBCA Nos. 47776, 49337, 50553, 04-1 BCA ¶ 32,604; *Robert E. Moore Constr.*, AGBCA No. 85-262-1, 90-2 BCA ¶ 22,803.

2. Reviving the Right to Terminate

Even if the government waived its right to terminate a contractor for failure to complete the work on schedule, the right is not completely lost. The government, however, must take some affirmative action demonstrating its intent to reestablish its termination rights. For example, the right to terminate the contract may be regained by the government if it establishes a new completion date by unilateral notice to the contractor or by mutual agreement.[99] If the new completion date is set unilaterally by the government, the date must be specific[100] and reasonable.[101] In *McDonnell Douglas*, the court held that the new completion date could be reasonable even if it was based on the contracting officer's subjective evaluation of the facts and circumstances.[102]

F. Assignment of Subcontracts to Surety

When the general contractor is terminated for default, it is often more efficient and cost effective to have its subcontractors and suppliers complete the work. Completing the work with these entities is advantageous because they are familiar with the work, are already mobilized to the project, and typically have the necessary equipment and material on-site or available. The subcontracts (and purchase orders) of the general contractor, however, are typically terminated when general contractor is terminated and, thus, the subcontractors or suppliers are no longer contractually obligated to perform. Subcontractors and suppliers, therefore, can have significant leverage in negotiating with the replacement contractor or surety. This leverage is used to negotiate more favorable subcontract terms as a condition to their continued performance. The subcontractors or suppliers also will frequently demand resolution of all disputes prior to recommencing work. As a result, the terminated contractor faces significant exposure in this regard as all "reasonable" costs to complete the work are its responsibility.

The contractor can minimize the leverage of its subcontractors and suppliers by including in its subcontracts or purchase orders a provision authorizing their assignment to the performance bond surety. The assignment provision should be carefully drafted and explicitly provide that the contractor's right to assign survives its termination for default. The assignment provision gives the surety the option to retain the "benefit of the bargain" and to proceed to complete the work with little, or no, interruption. That is, the subcontractors or suppliers will be required to complete their scope of work in accordance with the original terms of their subcontracts or purchase orders. The failure to comply with the assigned agreements could subject the subcontractors or suppliers to back charges or a claim for reimbursement for the cost to complete their respective scope of work. It is important to note, however, that

---

[99]*Martin J. Simko Constr., Inc. v. United States,* 11 Cl. Ct. 257 (1986); *Lumen, Inc.,* ASBCA No. 6431, 61-2 BCA ¶ 3210.

[100]*International Tel. & Tel. Corp. v. United States,* 206 Ct. Cl. 37 (1975).

[101]*McDonnell Douglas Corp. v. United States,* 50 Fed. Cl. 311, 316 (2001), *aff'd in part and rev'd in part,* 323 F.3d 1006 (Fed. Cir. 2003).

[102]*Id.*

the government has discretion in accepting the surety's completion plan and may object to one or more subcontractors or vendors. As discussed more fully in **Section III.C**, FAR § 49.404 provides that the contracting officer should consider—but is not required to accept—the surety's proposal for completing the work.

Notwithstanding the benefits, the surety may decline the assignment of these subcontracts as it may expose it to liability beyond its bond obligations. Indeed, once a subcontract is assigned to the surety, the surety steps into the shoes of the terminated contractor and assumes all its rights *and obligations* under the assigned subcontract.[103] The surety's liability, therefore, will no longer be limited to its bond obligations as the subcontractors and suppliers will now have a contractual basis to assert a claim against the surety.

While providing no guarantee of subcontractor/vendor cooperation in the event of a default termination, a contractor is well served by including an assignment provision in its subcontracts and purchase orders. This provision provides the surety with a mechanism to complete the project work in an efficient and cost-effective manner.

III. GOVERNMENT DAMAGES

A. Overview

In the event of a default termination, the government's right to recover damages includes the excess costs of completion and any damages (liquidated or actual) due to delays in completion. Absent a liquidated damages provision, the government is entitled to actual damages caused by the contractor's delay in completing the project. Actual damages include, for example, the cost of keeping the government inspector on the job after the specified completion date.[104] When a contractor is terminated for default, the government may (1) complete the contract work itself; (2) let a contract to another contractor to complete the work; or (3) allow the defaulted contractor's surety to complete construction, pay the cost of completion, or pay the penal sum of the performance bond.

In government construction contracts, the liquidated damages provision functions as the substitute measure of damages for the government's actual damages caused by the delayed completion.[105] In *Johnson Controls World Services, Inc.,* the board rejected the government's argument that a liquidated damages clause was not the exclusive remedy available to the government for the contractor's performance failure and held that liquidated damages clauses are exclusive in that a party "may not recover 'actual' damages when those damages have been 'liquidated' in advance by

---

[103] *See generally Employers Ins of Wausau v. Bright Metal Specialties, Inc.,* 251 F.3d 1316, 1323 (11th Cir. 2001) (by executing ratification agreement guaranteeing performance of subcontract work "according to the terms and conditions of the Subcontract," Surety gained the benefits and obligations of the agreement.)

[104] *B & E Constructors, Inc.,* IBCA No. 526-11-65, *et al.,* 67-1 BCA ¶ 6239.

[105] *J. P., Inc.,* ASBCA No. 32327, 87-1 BCA ¶ 19,453.

agreement."[106] The parties may agree in advance, however, that some damages will be liquidated and that others will not.[107] Furthermore, the government does, however, have the right to assess actual damages for any breach or harm not covered by the liquidated damages clause.[108]

B. Completion/Reprocurement Costs

As set forth in **Section III.A** above, the standard Default clause[109] specifies the government's remedies. In cases where performance is substantially complete at the time of the contractor's default, government agencies sometimes may finish the job with their own personnel and charge the cost of completion to the contractor.[110] Normally the government will not attempt to complete the work on its own without the acquiescence or specific approval of the contractor's surety.

The government often will utilize the materials and supplies brought to the site by the contractor, and it may even use the contractor's equipment, if it completes the work. At the same time, the government is under a duty to mitigate the additional costs incurred.[110] Therefore, the government may not be allowed to recover from the contractor or surety the cost of transporting government personnel to the work site from far away when local labor could have been obtained more cheaply. Also, where the work could have been completed more expeditiously and at less cost by a commercial contractor (even the defaulting contractor), recoverable costs may be limited to the difference between the original contract price and what it would have cost a commercial contractor to complete.[112] The government bears the burden of proving that the claimed excess costs were necessary and reasonable.[113]

When a construction contract is terminated for default, the contracting agency usually enters into a new contract to complete the work. In such cases, reasonable steps must be taken to minimize additional costs. This duty to mitigate damages limits the recovery the government can obtain for breach of contract by requiring that reasonable efforts must be undertaken to have the contract work completed in an economical and efficient manner.[114] Reflecting that duty, FAR § 49.405 provides:

---

[106] ASBCA No. 46691, ASBCA No. 46838, 96-2 BCA ¶ 28,590. *See also Sorensen v. United States,* 51 Ct. Cl. 69 (1916); *Fidelity and Deposit Co. of Maryland, Inc.,* ASBCA No. 32710, 87-1 BCA ¶ 19,356; *Tracor Data Sys.,* GSBCA Nos. 4662, 4791, 78-1 BCA ¶ 13,143.

[107] *Id.*

[108] *See United States v. American Sur. Co.,* 322 U.S. 96 (1944). *See* **Section III.D** of this chapter for a further discussion of liquidated damages.

[109] FAR § 52.249-10(a).

[110] *Datronics Engineers, Inc.,* ASBCA No. 10355, 66-2 BCA ¶ 6069.

[111] FAR § 49.402-6(a) (reprocure "same or similar" services at "as reasonable a price as practicable"); *Barrett Refining Corp.,* ASBCA No. 36590, 91-1 BCA ¶ 23,566.

[112] *R. C. Allen Business Machines, Inc.,* ASBCA No. 12932, 72-1 BCA ¶ 9325.

[113] *Brent L. Sellick,* ASBCA No. 21869, 78-2 BCA 13,510. *See also Surf Cleaners,* ASBCA No. 20197, 77-2 BCA ¶ 12,687.

[114] *E&J Trucking,* PSBCA No. 5092, 09-1 BCA ¶ 34,073 (government did not meet its burden of proof in seeking recovery of reprocurement costs when it did not adequately justify the reason for the termination of the original, lower-priced, reprocurement contract).

If the surety does not arrange for completion of the contract, the contracting officer normally will arrange for completion of the work by awarding a new contract based on the same plans and specifications. The new contract may be the result of sealed bidding or any other appropriate contracting method or procedure. *The contracting officer shall exercise reasonable diligence to obtain the lowest price available for completion.* [Emphasis added.]

In *Marley v. United States,* the Court of Claims held that, to protect the defaulting contractor, the government will be awarded excess costs under the Default clause only upon proving that its reprocurement action was "reasonably diligent to minimize [such] costs."[115] A *reasonably diligent* reprocurement may be a resolicitation of the offerors on the original procurement. It is not necessary for the government to solicit the original second low offeror if the contracting officer acts in other respects like a prudent businessperson in managing the reprocurement.[116]

The reasonableness of the government's actions will be determined in light of the particular facts and circumstances of each case. Therefore, while the government may negotiate a reprocurement on a sole-source basis when circumstances require, in *Manhattan Lighting Equipment Co., Inc.,*[117] the board held that the government had failed to take reasonable steps to mitigate excess costs where it gave no explanation for handling reprocurement on a sole-source basis. Relevant considerations in determining the reasonableness of a reprocurement action include (1) the necessity for expeditious reprocurement, (2) the relationship of the price of the reprocurement contractor to the original bid, and (3) a comparison with other competitive prices. The boards generally will reduce an award of excess reprocurement costs where the government has failed to mitigate damages.[118] Where the newly procured work exceeds the scope of the terminated contract, the government may recover from the defaulted contractor only those additional costs that are attributable to the work in the original contract.[119] However, if the reprocurement contract contains significant deviations from the original contract, no excess costs may be assessed against the defaulting contractor.[120]

When the government terminates the contractor for default, it generally issues a final decision that is subject to appeal under the Disputes clause.[121] If the contractor fails to file a timely appeal or institute an action in the Court of Federal Claims, that decision becomes final. However, if the government issues a subsequent final decision assessing its excess reprocurement cost claim, the contractor may be able

---

[115]423 F.2d 324, 333 (Ct. Cl. 1970). *See also Cascade Pacific Int'l v. United States,* 773 F.2d 287, 293-4 (Fed. Cir. 1985).

[116]*Zoda v. United States,* 180 F. Supp. 419 (Ct. Cl. 1960).

[117]ASBCA No. 7419, 61-2 BCA ¶ 3223.

[118]*Rhocon Constructors,* AGBCA No. 86-125-1, 91-1 BCA 23,308; *A & W Gen. Cleaning Contractors,* ASBCA Nos. 15010, 14809, 71-2 BCA ¶ 8994.

[119]*M.S.I. Corporation,* VACAB No. 599, 67-2 BCA ¶ 6643.

[120]*Blake Constr. Co., Inc.,* GSBCA No. 4013 *et al.,* 75-2 BCA ¶ 11,487; *see also Louis Martinez,* AGBCA No. 86-148-1, 87-3 BCA ¶ 20,219.

[121]*See* FAR § 49.402-3(g)(5). *See* **Chapter 15** for a discussion of the disputes process.

to challenge both of the final decisions under the Fulford doctrine if it takes a timely appeal of the excess cost final decision.[122]

C. Surety's Obligations and Rights

Under the standard performance bond, the defaulting contractor's surety is obliged to indemnify the government up to the amount of the bond. At the time of termination, the surety may discharge its bond obligations by forfeiting the penal sum of the bond or undertaking one of these completion strategies: (1) enter into a "takeover agreement" with the government, wherein the surety effectively steps into the shoes of the contractor and agrees to complete the project work for the original contract price; (2) select a new contractor, subject to the approval of the government, to complete the project work with any costs incurred above the original contract price to be paid by the surety; (3) reinstate the defaulted contractor, subject to the approval of the government, with any costs incurred above the original contract price to be paid by the surety; and (4) play no role in the reprocurement process with liability under the bond to be determined at completion of project. As is evident in these enumerated items, the surety has no formal obligation to the government beyond its guarantee, but it may propose a plan to complete the project. Because the surety is ultimately liable for the damages resulting from the default, the "contracting officer must consider carefully the surety's proposals for completing the contract."[123] Nevertheless, the contracting officer's overriding consideration is to "take action on the basis of the Government's interest."[124]

Although not obligated to complete the contract work, most sureties will cooperate with the government in arranging for completion of the contract, except in unusual circumstances. In most cases, the surety will be allowed to take over the work unless there is reason to believe that the contractor proposed by the surety is incompetent or unqualified so that the interests of the government would be substantially prejudiced by its efforts.[125]

In recognition of the fact that sums earned by a defaulting contractor but held by the government may be subject to claims by the surety as subrogee, the FAR provides for an agreement between the government and the surety regarding payment of such sums before the surety takes over completion of the contract. By entering into a takeover agreement, the surety risks incurring costs greater than the amount of the performance bond as it effectively commits to complete the project work within the contract amount.[126] The surety is also "bound by contract terms governing liquidated damages for delays in completion work".[127] FAR § 49.404 sets forth the essentials elements of such takeover agreements.

---

[122]*Fulford Manufacturing Corp.,* ASBCA Nos. 2143, 2144, Cont. Case Fed. (CCH) ¶ 61,815 (May 20, 1955) (digest only); *American Telecom Corp. v. United States,* 59 Fed. Cl. 467 (2004); *D. Moody & Co. v. United States,* 5 Cl. Ct. 70, 72 (1984); *Deep Joint Venture v. General Services Administration,* GSBCA No. 14,511, 02-2 BCA 31,914. *See also* **Chapter 15, Section VI.B**. Notwithstanding the potential relief provided by the Fulford doctrine, the more prudent course is to appeal the initial final decision.
[123]FAR § 49.404(b).
[124]*Id.*
[125]FAR § 49.404(c).
[126]FAR § 49.404(e); *see also Golden Gate Bldg. Maint. Co.,* ASBCA 12202, 68-1 BCA ¶ 6739.
[127]FAR § 49.404(e)(2).

The takeover agreement should provide that the surety undertakes to complete the work, while the government agrees to pay the surety the balance of the contract price that remained unpaid at the time of termination. However, such payment cannot exceed the surety's costs and expense of completion, excluding profit. The unpaid earnings of the defaulted contractor are subject to setoff against government claims, except to the extent of costs and expenses incurred by the surety in completion of the work (exclusive of payments and obligations under the payment bond). Such unpaid earnings will not be paid over to the surety where there is an outstanding assignment to any third party, unless such assignee agrees in writing. Also, the government retains the right to liquidated damages for unexcused delays in completion of the work which may be paid out of the contract proceeds before disbursement to the surety.

If a surety fails to enter into a takeover agreement, it is typically precluded from pursuing claims under the Disputes clause. For example, in *Royal Indemnity*, the board determined that a surety did not have standing to assert claims for price adjustments and excusable delays (both governed by the Disputes clause) because it lacked privity of contract with the government.[128] In the absence of a takeover agreement, the surety "is said to be on the same footing as a subcontractor."[129] Likewise, in *Sentry Insurance*, the surety—despite recommending the reprocurement contractor and financing the completion of the work—was denied the opportunity to challenge the amount of completion costs and delay damages assessed against the terminated contractor.[130] In reaching its decision the board recognized that "[p]arties, however interested, other than 'the Contractor' have no standing to pursue [claims under the Disputes clause].[131] Consequently, the "only manner by which a surety may appeal under the Disputes clause of its principal's defaulted contract is in a representative capacity, with the consent of its principal."[132]

A surety is entitled to expenses incurred in discharging liabilities under the payment bond when dealing with the federal government. However, such payments can be made only on the basis of an agreement among the government, the contractor, and the surety, a Comptroller General's decision as to the payee and amount, or a court order.[133]

In general, the rights of the surety are based on its subrogation to the contractor's rights and hence are subject to a setoff by the government's claims. Any agreement that purports to grant the surety rights in excess of those provided by the FAR has been held to be unauthorized.[134]

Where the surety tenders another contractor to complete the work, the bonds of the surety under the original contract are still effective. In addition, the government has the benefit of the surety bonds provided by the completion contractor, if any are required.

---

[128]ENGBCA No. PCC-8, 70-2 BCA ¶ 8546.
[129]*Royal Indemnity, Co.*, ENGBCA No. PCC-8, 70-2 BCA ¶ 8546.
[130]ASBCA 21918, 77-2 BCA ¶ 12,721.
[131]*Sentry Ins*, ASBCA 21918, 77-2 BCA ¶ 12,721.
[132]*Id.*
[133]FAR § 49.404
[134]*Security Ins Co. of Hartford v. United States*, 428 F.2d 838 (Ct. Cl. 1970).

D. Liquidated Damages

1. Basic Concept

When completion is delayed, the Default clause attempts to secure for the government a common law measure of recovery, putting it in as good a position as it would have enjoyed had the delay not occurred. If the contract includes an enforceable liquidated damages clause, the government may recover delay damages in an amount calculated by multiplying the liquidated damages per diem rate by the number of days of delay. Depending on the language of the contract, liquidated damages may be assessed for delays in completing phases of the contract and for delays in substantial completion of the entire project.[135] Liquidated damages may not be assessed after the date on which the work is substantially complete.

The typical government construction contract provides that the government is entitled to recover liquidated damages until completion of the work. The number of days used to calculate the total liquidated damages amount runs from the specified date for completion of the project plus any added days for excusable delays through the date of when the work is substantially complete.[136] The contractor may not be liable for the entire time utilized to complete the project. The reasonableness of the government's reprocurement actions may be challenged and the total assessed liquidated damages may be reduced if the government failed to act in a reasonably prompt manner. In addition, the government may not recover liquidated damages unless the contract work is actually completed.[137] As noted by the board in *Standard Coating Service, Inc.*, the "Government may not collect liquidated damages for work that it has decided that it does not want."[138]

2. FAR Policy

FAR § 11.501 details for the contracting officer the basic factors and guidelines to be considered in determining whether to use a liquidated damages clause and in establishing the rate for the liquidated damages. That section provides:

(a) The contracting officer must consider the potential impact on pricing, competition, and contract administration before using a liquidated damages clause. Use liquidated damages clauses only when—

    (1) The time of delivery or timely performance is so important that the Government may reasonably expect to suffer damage if the delivery or performance is delinquent; and

    (2) The extent or amount of such damage would be difficult or impossible to estimate accurately or prove.

(b) Liquidated damages are not punitive and are not negative performance incentives. (See 16.402-2.) Liquidated damages are used to compensate

---

[135]*See* **Section II.D.1** of this chapter for a discussion of substantial completion.
[136]*Glenn*, ASBCA Nos. 31260, 31628, 37901, 91-3 BCA ¶ 24,054.
[137]*Standard Coating Service, Inc.*, ASBCA No. 48611, ASBCA No. 49201, 00-1 BCA 30,725.
[138]*Id.*

the Government for probable damages. Therefore, the liquidated damages rate must be a reasonable forecast of just compensation for the harm that is caused by late delivery or untimely performance of the particular contract. Use a maximum amount or a maximum period for assessing liquidated damages if these limits reflect the maximum probable damage to the Government. Also, the contracting officer may use more than one liquidated damages rate when the contracting officer expects the probable damage to the Government to change over the contract period of performance.

(c) The contracting officer must take all reasonable steps to mitigate liquidated damages. If the contract contains a liquidated damages clause and the contracting officer is considering terminating the contract for default, the contracting officer should seek expeditiously to obtain performance by the contractor or terminate the contract and repurchase. (See Subpart 49.4.) Prompt contracting officer action will prevent excessive loss to defaulting contractors and protect the interests of the Government.

(d) The head of the agency may reduce or waive the amount of liquidated damages assessed under a contract, if the Commissioner, Financial Management Service, or designee approves (see Treasury Order 145-10).

In construction contracts, the clause set forth at FAR § 52.211-12 is used:

LIQUIDATED DAMAGES—CONSTRUCTION (SEPT. 2000)

(a) If the Contractor fails to complete the work within the time specified in the contract, the Contractor shall pay liquidated damages to the Government in the amount of _____ [*Contracting Officer insert amount*] for each calendar day of delay until the work is completed or accepted.

(b) If the Government terminates the Contractor's right to proceed, liquidated damages will continue to accrue until the work is completed. These liquidated damages are in addition to excess costs of repurchase under the Termination clause.

Under the standard liquidated damages clause, liquidated damages are assessed for each calendar day of delay. Both contract performance and liquidated damages are measured in calendar days—including weekends and holidays—not just weekdays.[139]

3. Agency Manuals

Some government agencies have their own internal guidelines for establishing liquidated damages rates. Liquidated damages rates consistent with such guidelines are presumed to be reasonable measures of the foreseeable actual damages that the government will sustain due to late completion of the project.[140] It is important to note, however, that the methodology for calculating the liquidated damages amount varies

---

[139] *J. H. Strain and Sons, Inc.,* ASBCA No. 34432, 90-2 BCA ¶ 22,770.
[140] *Marshall,* ENGBCA No. 6066, 00-1 BCA ¶ 30,730.

from agency to agency. Some agencies, for example, follow the criteria set forth in FAR Subpart 11.5, which authorizes the use of liquidated damages in limited circumstances and calculates the rate to include "the estimated daily cost of Government inspection and superintendence" as well as "renting substitute property," "paying additional allowance for living quarters," and "other expected expenses associated with delayed completion."[141]

The Naval Facilities Engineering Command (NAVFAC) Contracting Manual, however, has a much different approach to liquidated damages. The Navy manual requires the inclusion of a liquidated damages provision and provides a formula for establishing the per diem rate. In fact, the manual provides a series of tables for determining the liquidated damages rate for a given project. Table 1, which applies to "General Construction Projects," sets liquidated damages based on "Project Cost," as follows:

Project Cost	Estimated [LDs] Per Calendar Day
$2,000–25,000	$80
$25,000–50,000	$110
$50,000–100,000	$140
$100,000–500,000	$200
Each additional $100,000	—add $50[142]

Table 2, which applies to "Family Housing Units," sets liquidated damages per housing unit based on the average daily basic allowance for quarters (BAQ) to which service members with dependents are entitled under Navy regulations, plus average applicable housing allowances or temporary living allowances, if appropriate.[143] Table 3 sets liquidated damages for Bachelor Officers Quarters (BOQ) and Bachelor Enlisted Quarters (BEQ) based upon the applicable BAQ rate for service members assigned to live in the BOQ or BEQ, multiplied by the number of rooms available.[144] Tables 4 and 5 set liquidated damages for Storage Space and Office Space based on the amount of square feet per calendar day that is subject to construction delay.[145] The NAVFAC Contracting Manual also authorizes the contracting officer to adjust the liquidated damages amount by up to 50 percent upon a "written determination that the Government's anticipated loss from delayed completion is less or greater than [the above] amounts."[146]

Since there is no standard liquidated damages amount and method of calculation for the per diem rate, it is essential that the contractor review the liquidated damages clause and ascertain the liquidated damages amount prior to submitting its proposal or bid. The per diem rate, although typically not negotiable, may cause a contractor to alter the amount of its estimate or the manner in which it intends to perform the

[141]FAR § 11.502.
[142]Naval Facilities Engineering Command Construction Manual § 11.502, Table 1.
[143]Naval Facilities Engineering Command Construction Manual § 11.502, Table 2.
[144]Naval Facilities Engineering Command Construction Manual § 11.502, Table 3.
[145]Naval Facilities Engineering Command Construction Manual § 11.502, Tables 4–5.
[146]Naval Facilities Engineering Command Construction Manual § 11.502(c).

contract work, or reconsider its decision to propose or bid on the project. A contractor that simply assumes that a very high stated rate will be later adjusted does so at its own risk as the amount derived in accordance with the agency internal guidelines are presumed *reasonable*.

4. Challenging the Liquidated Damages Rate

For a liquidated damages rate to be valid and enforceable, the amount fixed must represent a fair and reasonable attempt to establish equitable compensation for anticipated losses.[147] Typically, in many construction contracts, liquidated damages rates are based on the estimated daily costs of administration in the event of delayed completion, including increased inspection, supervision, engineering, and associated overhead expenses.[148]

There is a general presumption that liquidated damages clauses are reasonable.[149] If a contractor elects to challenge the daily rate set forth in the contract, the contractor must show that the liquidated damages rate "bear no reasonable relationship" to the losses that the government would incur due to a delayed completion.[150] Moreover, unless a contractor challenges the liquidated damages rates before bidding, it may have a hard time disputing the assessment of liquidated damages afterward. In a 2006 case where the contractor had not "protested the specified liquidated delay damages rates as unreasonable when it bid the contract," the board held that the liquidated damages clause was an agreed term of the contract.[151] According to the board, the time for a contractor to challenge a liquidated damages per diem rate is prior to submitting a bid or proposal on the contract.

Even after-the-fact inquiries into the propriety of a liquidated damages rate must focus on whether the rate was a fair and approximation at the *time of contracting*, not at the time the liquidated damages were assessed.[152] Thus, liquidated damages provisions may be held to be valid although they do not prove to be accurate approximations of the damages actually incurred by the government, or even if the government does not incur any actual damages.[153] The government need not actually incur any costs as a condition precedent to recovery of liquidated damages.[154] It is only necessary that at the time the contract was awarded, the liquidated damages were reasonable forecasts of damages likely to be caused by delayed completion.[155]

---

[147]*P & D Contractors, Inc. v. United States*, 25 Cl. Ct. 237 (1992) (citing *Priebe & Sons, Inc. v. United States*, 332 U.S. 407 (1947)); *Southwest Eng'g Co. v. United States*, 341 F.2d 998 (8th Cir. 1965), *cert. denied*, 382 U.S. 819 (1965). *See also Central Ohio Bldg. Co., Inc.*, PSBCA No. 2742, 92-1 BCA ¶ 24,399.

[148]*Marshall*, ENGBCA No. 6066, 00-1 BCA ¶ 30,730.

[149]*Idela Constr. Co.*, ASBCA No. 45070, 01-2 BCA ¶ 31,437.

[150]*Spiess Constr. Co., Inc.*, ASBCA No. 48295, 95-2 BCA ¶ 27,767.

[151]*Pete Vicari Gen. Contractors, Inc.*, ASBCA No. 54982, 06-1 BCA ¶ 33,136.

[152]*Southwest Eng'g Co. v. United States*, 341 F.2d 998 (8th Cir. 1965), *cert. denied*, 382 U.S. 819 (1965).

[153]*Id.*

[154]*William F. Klingensmith, Inc.*, ASBCA No. 52028, 03-1 BCA ¶ 32,072, citing *U.S. Floors, Inc.*, ASBCA No. 45915, 94-2 BCA ¶ 26,636.

[155]*Id.*

A provision for liquidated damages may be considered an unenforceable penalty when: (1) the damages to be anticipated from a delay are neither uncertain in amount nor difficult to prove, (2) the parties did not intend to liquidate them in advance, or (3) the amount stipulated is greatly disproportionate to the presumable loss or injury.[156] A liquidated damages clause is most likely to be considered an unenforceable penalty if the rate is so greatly disproportionate as to have been an unreasonable forecast of the probable expected loss in the event of delayed completion.[157] Such a finding is rare, however, unless the government admits the rate was unreasonable or the rate departs greatly from an agency's standards. For example, the board found a liquidated damages clause to be an unenforceable penalty where the contracting agency erroneously applied its own internal guidelines and used a grossly excessive and unreasonable liquidated damages rate.[158] The board characterized the government's mistake as a "failure to correct an obvious and material error at the time of contracting."[159] In such cases, the liquidated damages clause may be characterized as an unenforceable penalty and stricken altogether.[160]

5. Multiple Liquidated Damages Rates

If the probable damage to the government is expected to change over the course of the contract, the contracting officer may specify different liquidated damages rates for different phases of the contract.[161] When the government uses phased liquidated damages rates, it cannot simply apply a rate specified for one phase of the contract to any delays in substantial completion.[162] To recover for delays in substantial completion, the government must specify a liquidated damages rate for such in the contract.[163] The ASBCA has rejected the government's attempt to assess liquidated damages for activities not within one of the phases for which liquidated damages were specified in the contract, "even if non-completion of those activities leaves the project less than substantially complete."[164]

---

[156]*George Ledford Constr., Inc.,* ENGBCA No. 6218, 97-2 BCA ¶ 29,172.

[157]*George F. Marshall,* ENGBCA No. 6066, 00-1 BCA ¶ 30,730.

[158]*Id.*

[159]*Id.*

[160]*Id.* (citing *Proserv. Inc.,* ASBCA No. 20768, 78-1 BCA ¶ 13,066;); *D.E.W. Inc,* ASBCA No. 38392, 92-2 BCA ¶ 24,840 (rate unenforceable where erroneously based on projected damages for delayed completion of a different type of building; government admitted that rate was improperly computed); *Dave's Excavation,* ASBCA No. 35956, 88-3 BCA ¶ 20,911 (rate exceeded amount properly computable under agency's internal guidelines by 150 percent); *Orbas & Assoc.,* ASBCA No. 33569, *et al.,* 87-3 BCA ¶ 20,051 (where rate was more than twice as great as specified in Navy manual, contractor established prima facie case that liquidated damages were not reasonably related to probable loss); *Coliseum Constr. Inc.,* ASBCA No. 36642, 89-1 BCA ¶ 21,428 (contracting officer reduced "unreasonable" rate of $1,820 to $220; board held that contracting officer's reduction was an admission that original rate was an unenforceable penalty).

[161]See FAR § 11.501(b).

[162]*William F. Klingensmith, Inc.,* ASBCA No. 52028, 03-1 BCA ¶ 32,072.

[163]*Id.*

[164]*Id.*

Following this same principle, when a contract includes phased liquidated damages rates, a contractor may not rely on its on-time substantial completion of an entire contract to avoid the assessment of liquidated damages for delays in completing particular phases of the contract.[165] In the *Vicari* appeal, the contract specified differing rates for the three phases of a contract and did not set a single rate for the entire contract. The contractor completed the first phase 62 days late, the second phase 33 days late, and the third phase—and the entire contract—41 days early.[166] The board found the timing of substantial completion irrelevant and upheld the assessment of liquidated damages for delays in completion of phases one and two.[167]

6. Relief from Enforcement

The standard Default clause relieves the contractor of liability for liquidated damages for excusable delays by granting extensions to the completion date. Disputes commonly arise when the government declines to grant an extension of the completion date, assesses liquidated damages, and the contractor seeks an adjustment to the contract's completion date. The board must determine which party is responsible for the delays. When the government has caused part of the delay to project completion, liquidated damages are either waived or apportioned between the government and the contractor.[168]

If the delays are concurrent, the liquidated damages will likely be waived. Waiving liquidated damages is consistent with the "rule against apportionment," which holds that "where delays are caused by both parties to the contract the court will not attempt to apportion them, but will simply hold that the provisions of the contract with reference to liquidated damages will be annulled."[169] In a case where the excusable delays and the delays for which the contractor "could have been responsible" could not be separated, the board denied both the government's request for liquidated damages and the contractor's claim for delay damages.[170] Likewise, where critical path delays were intertwined and could not be apportioned with any certainty between the contractor and the government agency, the assessment of liquidated damages was not valid, and the board ordered the government to return the withheld liquidated damages amount to the contractor.[171]

However, if the excusable delay is separate in time from the delay caused by the contractor, "there could hardly be a contention that the provision for liquidated damages

---

[165] *Pete Vicari Gen. Contractors, Inc.*, ASBCA No. 54982, 06-1 BCA ¶ 33,136.
[166] *Id.*
[167] *Id.*
[168] *George Sollitt Constr. Co. v. United States*, 64 Fed. Cl. 229, 243 (2005).
[169] *Sunshine Constr. & Eng'g, Inc. v. United States*, 64 Fed. Cl. 346, 372 (2005) (citing *Acme Process Equip. Co. v. United States*, 347 F.2d 509, 535 (Ct. Cl. 1965)).
[170] *Karcher Environmental, Inc.*, PSBCA Nos. 4282, 4093, 4085, 00-1 BCA ¶ 30,843 (citing *Blinderman Constr. Co. v. United States*, 695 F.2d 552 (Fed. Cir. 1982); *Commerce Int'l, Co. v. United States*, 338 F.2d 81 (Ct. Cl. 1964); *John McShain, Inc. v. United States*, 412 F.2d 1281 (Ct. Cl. 1969); *Port-A-Built*, PSBCA No. 3134, 94-2 BCA ¶ 26,694; *ADCO Constr., Inc.*, PSBCA Nos. 2355, 2465, 2480, 90-3 BCA ¶ 22,944).
[171] *George Sollitt Constr. Co. v. United States*, 64 Fed. Cl. 229, 274 (2005).

should not apply."[172] Thus, if the delays clearly are sequential or responsibility can be determined using a critical path method scheduling analysis, the board is more likely to apportion the liquidated damages. Apportioning liquidated damages is consistent with the judicially recognized rule requiring "a party asserting that liquidated damages were improperly assessed to bear the burden of showing the extent of the excusable delay to which it is entitled."[173] As a practical matter, when the delays are sequential rather than concurrent, the contractor may more easily show which portion of the delay is excusable.

In government contracts, the Secretary of the Treasury has the statutory authority to remit liquidated damages. This authority previously rested with the Comptroller General) in accordance with 41 U.S.C. § 256a. That section provides:

> Whenever any contract made on behalf of the Government by the head of any Federal Agency or by officers authorized by him to do so, includes a provision for liquidated damages, the Secretary of the Treasury upon recommendation of such head is authorized and empowered to remit the whole or any part of such damages as in his discretion may be just and equitable.

The Secretary of the Treasury basically defers to the agencies on the issue of remitting of liquidated damages, just as the Comptroller General did previously, and only exercises its authority upon receipt of a recommendation from the head of the contracting agency.[174] As a consequence, obtaining relief directly from the agency is usually more expeditious and less costly than going to the Secretary of the Treasury.

IV. TERMINATIONS FOR CONVENIENCE

During the Civil War, the Department of War and the Department of Navy entered into contracts to acquire weapons, ships, and other necessary items for prosecution of the war. At the conclusion of the war, these items were no longer needed, and the government sought to terminate or suspend performance of the contracts. One of the contracts awarded by the Department of the Navy was with the Corliss Steam-Engine Company for machinery connected with the construction, armament, and equipment of warships. In 1869, the Navy suspended performance of the contract while the work was incomplete.

Corliss offered to keep the machinery that had been partially completed and be paid $150,000 or to deliver it in its incomplete condition to the Navy Yard in Charleston for payment of $259,068. The Navy elected to take the machinery and agreed to pay the amount requested. A question arose concerning the authority of the

---

[172]*Sunshine Constr. & Eng'g, Inc. v. United States,* 64 Fed. Cl. 346, 372 (2005) (citing *Robinson v. United States,* 261 U.S. 486 (1923)).

[173]*Sauer Inc. v. Danzig,* 224 F.3d 1340 (Fed. Cir. 2000).

[174]*See J. Murray,* Comp. Gen. Dec. B-236673, 90-1 CPD ¶ 55.

government to make such a settlement. The Supreme Court held that the Secretary of the Navy had the duty to procure items needed in connection with the naval establishment of the United States, requiring him to enter into contracts. This was authorized by the act creating the Navy Department.[175] Therefore, the Court reasoned that this contracting authority must necessarily include the power to "suspend work contracted for . . . when from any cause the public interest requires such suspension."[176] It followed that the Secretary of the Navy also had the authority to settle claims related to such contracts suspended in the public interest.

This reasoning has evolved into the present-day authority for termination for the convenience of the government. The contract provision exercising this authority operates to avoid the unnecessary expenditure of public funds when requirements change or work should not be completed for other reasons.

A. Broad Termination Rights

As a general rule, convenience termination is permitted whenever the contracting officer determines that such action is in the government's interest.[177] A termination for convenience clause is a mandatory clause in nearly all government contracts.[178] Given its fundamental purpose, that clause and the related termination rights are deemed to be incorporated into a government contract by operation of law. [179]

Although a termination for convenience clause grants very broad termination rights to the government, this seemingly unlimited contract right has been judicially circumscribed by the implied duty of good faith and fair dealing, which limits its use only for good faith reasons.[180] The exercise of the right to terminate for convenience, either in bad faith or in abuse of discretion, constitutes a material breach of contract that allows, in theory, for the recovery of common law damages and anticipated profits beyond the recovery limitations in the clause.[181] Following *Torncello*, the Federal Circuit reviewed this termination for convenience standard on two occasions and reaffirmed the bad faith/abuse of discretion test.[182] *Changed circumstances* remains a primary factor to be considered in reviewing a termination for convenience decision. The ultimate issue is not whether circumstances have changed but whether the decision to terminate for convenience was motivated by bad faith or constituted an abuse of discretion.[183]

---

[175]1 Stat. 553.

[176]*United States v. Corliss Steam-Engine Co.*, 91 U.S. 321 (1875).

[177]FAR § 52.249-2(a).

[178]*See* FAR § 49.502.

[179]*See Advance Window Sys., Inc.*, VACAB No. 1276, 78-1 BCA ¶ 13,126. Even if that clause is omitted from a contract, it will be read into the contract pursuant to the Christian doctrine set forth in *G.L. Christian & Assoc. v. United States*, 312 F.2d 418, *rehearing denied*, 320 F.2d 345 (Ct. Cl. 1965). *See* **Chapter 1, Section II.C,** for discussion of the policy underlying the Christian doctrine.

[180]*See Torncello v. United States*, 681 F.2d 756 (Ct. Cl. 1982).

[181]*See Northrop Grumman Corp. v. United States*, 46 Fed. Cl. 622 (2000).

[182]*See Salsbury Indus. v. United States*, 905 F.2d 1518 (Fed. Cir. 1990), *rehearing denied*, (July 6, 1990); and *Caldwell & Santmyer, Inc. v. Glickman*, 55 F.3d 1578 (Fed. Cir. 1995).

[183]*Krygoski Constr. Co. v. United States*, 94 F.3d 1537 (Fed. Cir. 1996).

The breadth of discretion afforded the government to terminate a contract (and the high hurdle facing the contractor alleging wrongful termination) is colorfully illustrated by the board's decision in *Oregon Woods, Inc. v. Department of the Interior*:

> The Fish and Wildlife Service (FWS), a bureau of the Department of the Interior, bungled virtually every step of the way in conducting a procurement for the construction of a boardwalk. Like the old saw about Christopher Columbus, it didn't know where it was going, didn't know where it was when it arrived, and once it had left, didn't understand where it had been. Along the way, however, the FWS somehow found a rational reason for terminating for the convenience of the Government the contract it had awarded to [the contractor]. We therefore grant the agency motion for summary relief and consequently deny the contractor's appeal.[184]

Even if bad faith is demonstrated, however, a contractor cannot have the subject contract reinstated. In *Legislative Resources, Inc.*,[185] the board held that it did not have the authority to reinstate the contract in such a situation.

B. Limits on Termination Rights

Absent bad faith or malice, the government may terminate a contract for convenience to prevent or eliminate unnecessary future loss or expense to the government related to the performance of a contract.[186] In *Noland Brothers, Inc. v. United States*,[187] the Court of Claims stated:

> The [Government] could properly wish to cut unnecessary losses as early as possible and to consider what drastic changes might be needed in its plans.

Subsequently, in *Colonial Metals Co. v. United States*,[188] the Court of Claims ruled that a convenience termination was justified to obtain lower prices for supplies.[189] In *Colonial Metals*, the court stated it had repeatedly approved terminations for convenience issued to enable the government to take advantage of a better price even though the contracting officer was aware of a better price elsewhere when the original contract was awarded so long as the termination was not based on malice or conspiracy against the terminated contractor.

The government's ability to invoke the Termination for Convenience clause as a shield regarding its liability to a contractor may have some limits. For example, in *Ardco, Inc.*,[190] the contractor performing an indefinite-delivery indefinite-quantity

---

[184]*Oregon Woods, Inc. v. Department of the Interior*, CBCA No. 1072, 09-1 BCA ¶ 34,014.

[185]DCAB No. OMBE-16-74, 76-2 BCA ¶ 11,951.

[186]*Askenazy Constr. Co.*, HUD BCA 7802, 78-2 ¶ 13,402.

[187]405 F.2d 1250, 1253 (Ct. Cl. 1969).

[188]*Colonial Metals Co. v. United States*, 494 F.2d 1355 (Ct. Cl. 1974).

[189]*Id.*

[190]AGBCA No. 2003-183-1, 06-2 BCA ¶ 33,352.

(ID/IQ) contract for fire-fighting aircraft services filed a claim for lost profits resulting from the loss of use of an aircraft that was damaged by an employee of the
United States Forest Service. The government sought to have the claim dismissed on
the basis that it could have invoked the Termination for Convenience clause, which
precluded the recovery of lost anticipated profits. The board rejected that contention and held that the Termination for Convenience clause could not be used to limit
recovery for a breach of contract that was not related to a termination or for breaches
alleged after completion of contract performance.

C. Standard FAR Clause

Most government construction contracts contain the clause at FAR § 52.249-2, which
provides, in part:[191]

> **TERMINATION FOR CONVENIENCE OF THE GOVERNMENT
> (FIXED-PRICE) (MAY 2004)**
>
> (a) The Government may terminate performance of work under this contract
> in whole or, from time to time, in part if the Contracting Officer determines
> that a termination is in the Government's interest. The Contracting Officer
> shall terminate by delivering to the Contractor a Notice of Termination
> specifying the extent of termination and the effective date.

* * *

If the termination is total, the contractor is required to stop all work on the terminated contract as of the effective date of the termination.[192] Costs incurred after that
date may be disallowed, except that recovery of reasonable continuing costs incurred
because of the termination, may be allowed under FAR § 31.205-42(b). At the same
time, the contractor is obligated to continue work on any portions of the contract that
are not terminated.[193]

The balance of the clause (paragraphs (b)–(n)) provides detailed guidance regarding the parties' rights and responsibilities following a convenience termination.[194]

> Paragraph (b) details the contractor's duties following receipt of a notice of ter-
> mination for convenience including cessation of the work specified in the
> termination notice and the termination of all affected purchase orders and
> subcontracts.
>
> Paragraphs (c) and (d) address termination inventories and disposal of property to
> the extent that those related to the project.

[191]FAR § 52.249-2, Alternate I (attached to this chapter at **Appendix 11A**).
[192]FAR § 52.249-2(b)(1).
[193]FAR § 52.249-2(b)(7).
[194]*Id.*

Paragraph (e) requires the contractor to submit its final termination settlement proposal to the contracting officer within one year from the effective date of the termination. If the contractor cannot comply with this provision, a written extension must be obtained from the contracting officer. Paragraph (e) also provides for a contracting officer's determination of the settlement amount if the contractor fails to submit a proposal within the time allowed.

Paragraph (f) provides that the contractor and the contracting officer may agree on the whole or any part of the amount to be paid because of the termination including a reasonable allowance for profit. The total agreed amount may not exceed the contract price, less payments and the value of the terminated work.

Paragraph (g) provides a "formula" for the termination settlement in the event that a settlement cannot be negotiated by the parties. It allows for payment of legal, clerical, and other expenses incurred in preparing the settlement proposal, subcontract termination costs, and other reasonable costs related to the termination inventory.

Paragraph (h) requires that the fair value of destroyed, lost, or stolen property to be deducted from the amount due the contractor.

Paragraph (i) requires the application of the Cost Principles and Procedures set forth in FAR Part 31.

Paragraph (j) affords the contractor the right to appeal the contracting officer's determination of the settlement amount under the Disputes clause, unless the termination claim was not submitted within the required period.

Paragraph (k) provides for specified deductions from the termination settlement, including unliquidated progress payments or other offsetting claims.

Paragraph (l) addresses partial terminations and provides that the contractor is entitled to an equitable adjustment for the continued portion of the contract as well as a settlement of claims relating to the terminated part.

Paragraph (m) provides for partial payments to be made at the discretion of the contracting officer and recovery by the government in the event of an overpayment.

Paragraph (n) requires the contractor to maintain all records and documents related to the terminated portion of the contract for a period of three years from the date of final settlement.

D. Partial Termination or Deductive Change Order

When the government decides to delete work from a contract, an initial question is whether the deduction should be treated as a deductive change order or as a partial termination. As previously discussed, the standard Termination for Convenience clause allows the government to terminate a portion of the work for its convenience. When this clause is included in the contract along with the standard Changes clause, the government must decide whether to use a deductive change order under the Changes clause or a partial termination for convenience. Two broad considerations may affect this decision.

The first consideration is whether the deletion is major or minor. If the deletion is major, it should be accomplished by a partial termination for convenience.[195] If the deletion is minor, it probably should be accomplished by a deductive change order.[196]

The second consideration is whether revised work will be substituted for the deleted work. If revised work will be substituted for the original work, a change order accomplishing both the deletion of the original work and the addition of the substitute work would be the more appropriate mechanism. If work is to be deleted only and no other work is to be added as a substitute, a partial termination for convenience of the government may be the more appropriate method.[197]

The government has the burden of proving the value of a downward equitable adjustment in price due to the deletion of the original contract work.[198] Generally, the proper measure of a deductive change order is the reasonable cost to the contractor if it had performed the deleted work,[199] while a termination settlement amount is derived as the Termination for Convenience clause and FAR Subpart 49.2 prescribe.[200] If both the government and the contractor price the deletion of the work in a particular manner, a board or the Court of Federal Claims may decide not to change the parties' approach absent a compelling reason to do so.[201]

Depending on the actual cost of the deleted work, pricing the amount of the credit owed the government can vary substantially depending on whether the calculation is made under the Changes clause or the Termination for Convenience clause. A price adjustment under the Changes clause focuses on how much the deleted work *would have cost* the contractor while the price adjustment under the Termination for Convenience clause focuses on the contract value of the work that was not terminated (deleted).[202] If the actual cost for the deleted work varies significantly from the amount originally estimated by the contractor, the amount of credit to the government will depend on the clause selected for determining the price adjustment.

---

[195] *J.W. Bateson Co. v. United States*, 308 F.2d 510 (5th Cir. 1962).

[196] *See John N. Brophy Co.*, GSBCA 5122, 78-2 BCA ¶ 13,506; *Bromley Contracting Co.*, HUDBCA 75-8, 77-1 BCA ¶ 12,232.

[197] *Frederick Constr. Co.*, ASBCA Nos. 12108, 12241, 68-1 BCA ¶ 6832; *Manis Drilling*, IBCA No. 2658, 93-3 BCA ¶ 25,931.

[198] *Singleton Contracting Corp.*, GSBCA 8546, 90-2 BCA ¶ 22,879.

[199] *See generally Dawson Constr. Co.*, VABCA 3558, 94-1 BCA ¶ 26,362.

[200] *See* **Section VI** on termination settlements.

[201] *See P.J. Dick, Inc. v. General Services Administration*, GSBCA No. 12215, 95-1 BCA ¶ 27,574 *modified on recons.*, 96-1 BCA ¶ 28,017 (both parties priced a partial termination under the Termination for Convenience clause as a deductive change but included the *settlement* costs normally permitted only under FAR 49 to the calculation).

[202] *Id.* (Contractors should consider the alternative pricing approaches before submitting a credit proposal when a partial termination for convenience might be appropriate.)

E. Subcontract Termination Issues

Following the receipt of a total termination notice, the contractor may not enter into any further subcontracts. In addition, termination notices should be sent immediately to all subcontractors and vendors on the terminated portion of the project. Thereafter, the contractor should attempt to settle all claims resulting from the termination. If the contractor intends to include the amount of the subcontractor/vendor settlement in its termination proposal, the government reserves the right to approve the settlement.[203]

FAR § 49.108 provides guidance for settlement of subcontractor settlement proposals. It provides that a subcontractor has no contractual rights against the government, although it may have rights against the prime contractor or an upper-tier subcontractor with whom it has contracted. It is the prime contractor's responsibility to settle the termination claims of its subcontractors and vendors. If required by the contracting officer, the contractor must assign terminated subcontracts and purchase orders to the government. Generally, however, the government prefers to have the contractor settle with its subcontractors and vendors. In that case, the terminated subcontractors and suppliers have no right to payment from the government, and their only recourse is against the contractor and its surety. However, the Comptroller General has held that the subcontractor may seek recovery from the government where there has been an express guarantee of payment to the claimant.[204]

In order for terminated subcontractors to avoid problems in receiving payment from prime contractors, the ASBCA has held that a prime contractor is not entitled to payment by the government for subcontractor's settlement costs unless evidence is provided that the subcontractors and vendors actually have been paid.[205]

The FAR recommends that a prime contractor seek to include a termination for convenience clause in its subcontracts and purchase orders for its own protection. Failure to make a *reasonable effort* to include such a clause in subcontracts or purchase orders may prevent a contractor from including the full amount of a final judgment in favor of a subcontractor or supplier in the contractor's termination for convenience settlement proposal.[206] Suggestions regarding use of such clauses are set forth in FAR Subpart 49.5. Similarly, CONSENSUSDOCS 752, *Standard Subcontract Agreement for Use on Federal Government Construction Projects*, includes this provision:

> 10.4 TERMINATION BY OWNER Should the Owner terminate its contract the Contractor or any part which includes the Subcontract Work, the Contractor shall notify the Subcontractor in writing within three (3) business

---

[203]FAR § 52.249-2(b)(5).
[204]52 Comp. Gen. 377, 381 (1972).
[205]*Atlantic, Gulf & Pacific Co. of Manila, Inc.* ASBCA No. 13533, 72-1 BCA ¶ 9415.
[206]*See* FAR § 49.108-5(a).

Days of the termination and upon written notification, this Agreement shall be terminated and the Subcontractor shall immediately stop the Subcontract Work, follow all of Contractor's instructions, and mitigate all costs. In the event of Owner termination, the Contractor's liability to the Subcontractor shall be limited to the extent of the Contractor's recovery on the Subcontractor's behalf under the Subcontract Documents., The Contractor agrees to cooperate with the Subcontractor, at the Subcontractor's expense, in the prosecution of any Subcontractor claim arising out of the Owner termination and to permit the Subcontractor to prosecute the claim, in the name of the Contractor, for the use and benefit of the Subcontractor, or assign the claim to the Subcontractor. . . .

This and similar subcontract clauses protect the contactor by limiting its liability to its subcontractors to the amount it recovers from the government on the subcontractors' behalf.

Generally, a contractor must settle with subcontractors in the same way the government settles with the contractor. All settlements with subcontractors must be supported by information and data and are subject to review by the government. Under certain conditions, however, the termination contracting officer may authorize the contractor to settle subcontracts without approval or ratification when the amount of the settlement is $100,000.00 or less, provided the conditions set out in FAR § 49.108-4 are satisfied.

If a subcontractor or supplier obtains a final judgment against the contractor on a claim arising from a termination for convenience, the government shall treat the amount of the judgment properly allocable to the terminated portion of the prime contract as a cost of settling with the contractor, provided the conditions of FAR § 49.108-5 are satisfied. When the contractor is unable to settle with a subcontractor and that delays the settlement of the prime contract, the contracting officer may except the subcontractor settlement proposal from the settlement with the contractor and reserve the rights of the government and the contractor.

V. CONSTRUCTIVE TERMINATIONS

A contracting officer makes an express termination by issuing a modification pursuant to the authority provided by a Termination clause. Any other act that prevents the contractor from completing performance of the contract work is treated as a constructive termination if the contracting officer could have reached the same result by an express termination under the clause. Otherwise, preventing completion of contract performance would be considered to be a breach of contract. In government contracting, constructive terminations arise in a variety of contexts as described in the next three sections.

A. Wrongful Termination for Default

A wrongful termination for default is converted into a termination for convenience in order to avoid government liability for breach of contract and to limit the contractor's damages. Paragraph (c) of the standard Default clause at FAR § 52.249-10 states:

> If, after termination of the Contractor's right to proceed, it is determined that the Contractor was not in default, or that the delay was excusable, the rights and obligations of the parties will be the same as if the termination had been issued for the convenience of the Government.

Although the Default clause provides the contractor an administrative remedy for a wrongful default termination, it precludes recovery of anticipated profit (which might otherwise be an element of common law damages for breach of contract) due to the Termination for Convenience clause. A board or the Court of Federal Claims will convert a termination for default to a termination for convenience when the government fails to carry its burden of establishing the contractor's default, when the contractor's default was excusable or was waived by the government, when the contracting officer fails to follow the prescribed procedures in the FAR, or when there is a termination for almost any other wrongful reason.[207]

B. Failure to Order under Requirements, Multiyear, or ID/IQ Contracts

Indefinite delivery/indefinite quantity contracts are used when government agencies cannot predetermine the precise quantity of supplies or services that they will require during the contract period. The ID/IQ contract must provide both a "stated minimum quantity" and "reasonable maximum quantity" of the supplies or services to be furnished under the contract.[208] The minimum quantity, which must be more than a nominal amount, serves as the government's base obligation to order supplies or services under the contract.[209]

The government's failure to order the stated minimum quantity during the contract period entitles the contractor to recover damages commensurate with the loss resulting from the unordered items. The measure of the loss to the contractor will vary depending on the unique circumstances of the dispute and the forum. For example, the Federal Circuit has ruled that the contractor is entitled to the contract price for the unordered quantity.[210] The ASBCA, however, takes a more narrow approach,

---

[207]*See Specialty Constr. Co.*, ASBCA No. 21132, 78-2 BCA ¶ 13,348; *Electromagnetic Finishers, Inc.*, GSBCA No. 5035, 79-1 BCA ¶ 13,697; *Carter Indus.*, DOTBCA 4108, 02-1 BCA ¶ 31,738.

[208]FAR § 16.504 (a)(1).

[209]FAR § 16.504 (a)(2).

[210]*Maxima Corp. v. United States*, 847 F.2d 1549 (Fed. Cir. 1988).

limiting recovery to the actual cost incurred plus anticipated profits on the minimum quantity *less* payment received.[211]

It is important to note that if the government directs a termination for convenience during the contract period without satisfying the minimum quantity obligation, the contractor's damages will be limited by the Termination for Convenience clause. That is, the contractor's recovery will be subject to FAR Part 31 Cost Principles and Procedures. "The contractor, instead of receiving compensation for governmental breach of contract based on classical measures of damages, is limited to recovery of 'costs incurred, profit on work done and the costs of preparing the termination settlement proposal'."[212] The Termination for Convenience clause precludes recovery for anticipated profits.[213] As noted, a termination for convenience is proper when there are *changed circumstances* and the decision to terminate is not motivated by bad faith. This is a fairly low threshold, given that the government's inability to satisfy the minimum quantity requirement would arise only from a changed circumstance altering the government's need for the contracted supply or service.

For example, in *Hermes Consolidated, Inc., d/b/a/ Wyoming Refining Co.,*[214] the contractor received two contracts to furnish fuel to the government. The contracts were ID/IQ contracts with a stated ("guaranteed") minimum order quantity. As the contract performance period neared its end, the government had not ordered the minimum quantity of fuel under either contract. A few days before expiration of the contract performance period, the contracting officer issued a partial termination for convenience for both contracts. This termination reduced the total estimated quantity to be ordered in each contract. By operation of the contract terms, the stated minimum quantity in each contract was also reduced. The contractor filed a breach of contract claim due to the government's failure to order the original minimum quantities. The board held that the government did not breach the contracts because it had ordered more than the reduced minimum quantity. Consequently, the terminations were not contract breaches entitling the contractor to recover lost anticipated profits.

In *Varo, Inc.,*[215] the government failed to fund the third year of a multiyear contract for bomb racks. Thereafter, the government awarded a contract for a larger quantity of the identical items to a different firm. The ASBCA held that the government's *cancellation* of the third year's requirements, when funds were available and the requirement for the bomb racks still existed, was a constructive termination for convenience by the government. Although the contractor could not recover its lost anticipated profits on that portion of the contract, it was entitled to recover the amounts permitted under the Termination for Convenience clause.

---

[211]*PHP Healthcare Corp.*, ASBCA 39207, 91-1 BCA ¶ 23.647 (calculating damages so that the contractor will be placed in as good a position as it would have been by performance of the contract); *see also Merrrimac Mgmt. Inst., Inc.,* ASBCA 45291, 94-3 BCA ¶ 27, 251 (finding that the "proper measure of damages is not the full amount of the minimum quantity, but the amount the [contractor] lost as a result of the Government's failure to order that quantity . . . [t]he cost that [contractor] would have incurred had the full amount been ordered must be taken into account in determining its actual loss]."

[212]*Maxima Corp. v. United States*, 847 F.2d 1549, 1552 (Fed. Cir. 1988).

[213]*Id.*

[214]ASBCA Nos. 52308, 52309, 02-1 BCA ¶ 31,767.

[215]ASBCA No. 13739, 70-1 BCA ¶ 8099.

C. Wrongful Cancellation of Award

Typically, when an action by the government otherwise would be considered to be a breach of contract, it will be treated as a termination for convenience if the contracting officer could have achieved the same result under the Termination for Convenience clause. For example in *John Reiner Co. v. United States*,[216] a disappointed bidder prevailed on a bid protest to the GAO. The GAO recommended cancellation of the contract that had been awarded to Reiner. The contracting officer followed that recommendation. Thereafter, Reiner asserted a breach of contract claim because the contracting officer failed to follow the termination for convenience procedures. However, the court held that the contracting officer's action would have been entirely valid if exercised under the Termination for Convenience clause. Therefore, the action was not a breach but a *constructive termination for convenience*, and the contractor was not entitled to recover lost anticipated profits. In a similar situation when faced with a potential bid protest to the GAO after a notice of award had been issued, but before receipt by the contactor, cancellation of the award was not a breach but a constructive termination for convenience.[217]

VI. TERMINATION SETTLEMENTS

The preferred method of determining the amount due a contractor following a termination for convenience is an agreement between the government and the contractor. If the parties cannot reach an agreement on the settlement amount, the contracting officer is authorized to make a unilateral determination following the guidance set forth in FAR Part 49.[218]

A. Timing Requirements

The contractor is obligated to stop work as prescribed in the notice of termination, cancel affected subcontracts and purchase orders, and submit its termination settlement proposal. Preparing an effective termination settlement proposal often requires the collection and organization of the contractor's cost documentation. The contractor must submit a settlement proposal no later than one year from the effective date of the termination.[219] The effective date of termination is either: (1) the date on which the notice of termination requires the contractor to stop performance; or (2) the date the contractor receives the notice of termination, whichever is later.[220] Contractors must carefully note this deadline as it is strictly enforced.[221] In some cases, however, the contracting officer will allow an extension of time beyond that

---

[216]325 F.2d 438 (Ct. Cl. 1963) *cert. denied,* 377 U.S. 931 (1964).

[217]*G. C. Casebolt Co. v. United States,* 421 F.2d 710 (Ct. Cl. 1970). *See also Albano Cleaners, Inc. v. United States,* 455 F.2d 556 (Ct. Cl. 1972); *and Dairy Sales Corp. v. United States,* 593 F.2d 1002 (Ct. Cl. 1979).

[218]*See* FAR § 49.103.

[219]FAR § 52.249-2(e).

[220]FAR § 2.101. *See also, Ryste & Ricas, Inc. v. Harvey,* 477 F.3d 1337, 1341 (Fed. Cir. 2007).

[221]*Ryste & Ricas, Inc. v. Harvey,* 477 F.3d 1337 (Fed. Cir. 2007).

date. Any request for an extension must also be submitted within one year of the notice of termination.[222]

FAR § 49.206-2 sets forth guidelines for the preparation of a settlement proposal. The methods differ depending on whether the termination is partial or total. The inventory basis is the preferred method where a construction contract has been partially terminated. That method involves the pricing of the cost of completing the remaining work. In that regard, it differs from a credit for a deductive change order, which focuses on the cost of the deleted work. The total cost basis is the required method where a construction contract is completely terminated.

B. Calculating the Settlement Amount

Subject to the applicability of the Contract Price Cap Rule[223] and the Loss Rule,[224] the contractor whose contract is wholly or partially terminated for convenience should be made financially whole for the direct consequences of the government's termination. Negotiated termination settlements are limited only by the requirements of reasonableness of cost and the original contract price. Individual cost items need not be negotiated.

If the parties are not able to agree on a termination settlement, the contracting officer is authorized by the regulations to make a unilateral determination of the amount to be awarded the contractor. The Termination for Convenience clause sets forth guidelines for the contracting officer's determination. The basic elements of the settlement amount are detailed in the Termination for Convenience clause found at FAR § 52.249-2 including these categories of reasonable, allowable, and allocable costs:

- Cost of the work performed.
- Subcontractor/supplier settlements.
- Reasonable profit on the incurred costs when the contractor would have earned a profit on the contract if it had been completely performed.
- Costs (legal, accounting clerical, etc.) associated with the preparation of the settlement proposal and reaching settlements with subcontractors/suppliers.
- Costs associated with the disposition of the termination inventory.

The current Termination for Convenience clause incorporates the Cost Principles of FAR Part 31 by reference.[225] The purpose of the Cost Principles is to achieve uniformity in pricing similar work.[226] To that end, FAR Part 31 provides detailed guidelines to evaluate costs and determine whether they are allowable. These guidelines, however, are not strictly applied and are subject to other principles aimed at "ascertaining fair compensation."[227]

---

[222]*The Swanson Group, Inc.*, ASBCA No. 54863, 07-2 BCA ¶ 33,672.
[223]FAR§ 49.207 (contractors should ensure that the contract price is adjusted for all contact modifications to ensure that the correct adjusted price is available for consideration when addressing the Contract Price Cap rule.)
[224]FAR § 49.203.
[225]FAR § 49.113.
[226]FAR § 31.101.
[227]FAR §§ 49-113; 49.201.

In addition FAR Subpart 31-205-42 enumerates "cost principles peculiar to termination situations."[228] This subpart addresses costs unique to terminations including: common items, costs continuing after termination, initial costs, loss of useful value of equipment, rental costs under unexpired contracts, alterations to leased property, settlement expenses, and subcontractor claims.[229]

In general, costs arising from the contract that have not been reimbursed are allowable to the extent they were reasonable when incurred.[230] However, only costs related to work authorized by the contract are recoverable in a termination settlement. For example, a contractor was precluded from recovering costs related to its unsuccessful value engineering change proposal as there was no "value engineering clause" authorizing such efforts.[231] In other words, these efforts were viewed as voluntary and performed at the contractor's risk. Similarly, a contractor could not recover costs related to site work performed prior to the issuance of the notice to proceed because such work was not authorized at the time it was performed.[232] The contract specifically provided: "Contractor shall perform no work at the contract site except pursuant to a notice to proceed given by the Contracting Officer."[233]

The next categories of costs are recognized as recoverable under the FAR Part 31 cost principles:

- Costs related to the termination inventory.
- Post-termination costs that were unavoidable with reasonable diligence.
- Bid/proposal preparation costs and start-up costs.
- Employee severance and relocation expense.
- Costs of idle special tools and equipment.
- Rental costs under unexpired leases.
- Unabsorbed overhead allocable to the continuing part of the work, if any.
- Subcontractor settlement costs and expenses.
- Legal, accounting, clerical, and other costs associated with the preparation of the settlement proposal.

With regard to the last item, the cost principles allow only costs involved in preparation and presentation of the termination claim to the contracting officer.[234] Costs associated with appeals from that decision are not recoverable.[235] Contingency fees are also likely unallowable, but fees paid on retainer for preparing the termination claim are allowable.[236]

---

[228]FAR § 31.205-42.

[229]FAR § 31.205-42(a)–(h).

[230]*Systems & Computer Information, Inc.*, ASBCA No. 18458, 78-1 BCA ¶ 12,946.

[231]*Derrick Elec. Co.*, ASBCA 21246, 77-2 BCA ¶ 12,643.

[232]*Sherkade Constr. Corp.*, DOTCAB 68-29, 68-2 BCA ¶ 7365.

[233]*Id.*

[234]*See Richerson Constr., Inc. v. General Services Administration*, GSBCA No. 11045, *et al.*, 93-1 BCA ¶ 25,239.

[235]*Q.V.S., Inc.*, ASBCA No. 7513, 63 BCA ¶ 3699.

[236]*R-D Mounts, Inc.*, ASBCA No. 17422, et al, 75-1 BCA ¶ 11,077.

Additionally, the contractor is entitled to a *fair and reasonable* profit on work performed, unless it can be demonstrated that the contractor would have lost money on the contract.[237] Anticipatory profits are *not* allowed. The profit is computed only on the work performed. Although any reasonable method may be used to arrive at a fair profit, the regulations set forth factors that should be considered by the contracting officer in negotiation or unilateral determination of profit. FAR § 49.202 requires that nine factors be considered in determining profit. These factors include consideration of the extent and difficulty of the work completed by the contractor, the contractor's efficiency, the extent and difficulty associated with the award and management of subcontracts, the rate of profit that the contractor would have earned if the contract had been completed, and the rate of profit contemplated by the parties at the time of award.

C. Loss Contracts

The Termination for Convenience clause has an "adjustment for loss" provision. It provides that if the contractor would have suffered a loss on the contract if it had been completed, the termination recovery will be reduced by a percentage of the loss that would have been realized on full completion. The procedure is set out in FAR § 49.203.

The government has the burden of demonstrating that it is entitled to a loss adjustment.[238] The government's burden is twofold. First, it must demonstrate that the contractor operated at a loss. Second, it must prove the amount of the loss. A contract is a "loss contract" when the total cost of the project (actual and projected) exceed the contract price. Because contract price is a significant variable in the analysis, a contractor can potentially avoid a loss adjustment by ensuring that all open requests for equitable adjustments are resolved and included in the calculation. That is, any equitable adjustments to which the contractor is entitled will increase the contract price.

In an effort to establish the second part of its burden, the government may direct the contractor to provide an estimate of the remaining portion of the project.[239] The DCAA Contract Audit Manual, for example, provides:

Request the contractor, through the contracting officer, to furnish an estimate of the cost required to complete the terminated portion of the contract. Review the estimate with necessary help from technical representatives. The contractor's estimate to complete may be conservative and show that no loss would have occurred. Make a concerted effort to evaluate the contractor's projected profit.[240]

---

[237]FAR § 49.203; *C.W. McGrath, Inc.*, GSBCA No. 4586, 77-1 BCA ¶ 12,379.
[238]*McDonnel Douglas Corp. v. United States*, 37 Fed. Cl. 270, 273 (Fed. Cl. 1996).
[239]DCAM ¶ 12-307(a)(2) (June 29, 2007).
[240]DCAM ¶ 12-307(a)(2) (June 29, 2007).

As illustrated by this quoted provision, any cost to complete information provided by the contractor will be carefully scrutinized by the government. A contractor should use caution in deciding whether to comply with the request, and consider the consequences in providing this information—especially since the information provided will be used to satisfy the government's burden. More important, the contractor must recognize that typically "[t]here is no contractual requirement for the contractor to furnish an estimate to complete."[241]

Finally, the government is not entitled to a loss adjustment where it is responsible for or contributed to the circumstances resulting in a loss.[242] For example, in *R.H.J. Corp.*, the government's failure to make progress payments contributed to the contractor's enhanced costs. Accordingly, the board determined that the government was not entitled to a loss adjustment.[243]

➤ LESSONS LEARNED AND ISSUES TO CONSIDER

- The right to terminate a construction contract for default generally arises only when a *material or substantial provision* of the contract has been breached or a party has failed to perform a material obligation.
- By the terms of the contract, the parties can *expand, limit, or redefine the grounds* for termination for default that exist in common law.
- Under the standard Default clause for fixed-price construction contracts, the federal government may terminate the contract for default if the contractor (1) refuses or fails to prosecute any or all of the work with *sufficient diligence* to ensure timely completion, or (2) fails to complete the work within the original or extended completion date.
- The contract may be terminated *in whole or in part*.
- The doctrine of *substantial completion* prevents the government from assessing liquidated damages or terminating the contract without giving the contractor reasonable time to correct deficiencies.
- Contractors bear the *burden* of proving substantial completion.
- The government may also default terminate if the contractor fails to comply with other *material contract clauses* even if those clauses do not expressly provide for termination.
- The FAR requires the contracting officer to consider various factors including reasons for delay in determining whether to default terminate.
- While the standard Default clause does not require that the government issue a *cure notice*, many contracting officers will issue such a notice prior to default

---

[241]DCAM ¶ 12-307(a)(3) (June 29, 2007).
[242]*R.H.J. Corp.*, ASBCA 12404, 69-1 BCA 7587.
[243]*Id.*

terminating the contract. In addition, the government may start holding retention based on its conclusion that the contractor's rate of progress is not satisfactory and may issue an Interim Unsatisfactory Performance Evaluation.

- If threatened with default, the contractor should document and explain to the contracting officer the *status of its completion* and any excusable reasons for delay in order to avoid termination.

- If threatened with default, the contractor should carefully *review the possible grounds* for a default, develop a well-considered cure plan, document the status of its work and that of its subcontractors including equipment and materials stored on site, assess its staff needs, and take steps to protect its project records.

- If *threatened* with a default, the contractor should take steps to advise the performance bond surety and obtain its concurrence in the planned response.

- If the contract is terminated and the contractor cannot demonstrate excusable delays, the government may recover its damages such as the *excess cost* of completing the terminated work and any *actual or liquidated damages* due to delays in completion.

- If the default termination is challenged and found not justified, then the Default clause states that the parties' rights and obligations will be the same as if the termination had been issued for the *convenience of the government.*

- The contracting officer also has the right to terminate a contract in whole or in part for the convenience of the government if that action is determined to be in the *government's interest.*

- The Termination for Convenience clause provides detailed guidance for the preparation, date for submission, and resolution of the contractor's *settlement proposal,* which will also include its subcontractors and suppliers.

- The convenience settlement allows profit on completed work but *prohibits profit* on the terminated work.

- If the contract would have been completed at a *loss*, then the settlement should also reflect the contractor's loss position.

APPENDIX 11A: TERMINATION FOR CONVENIENCE OF THE GOVERNMENT (FIXED-PRICE) (ALTERNATE I) (SEP. 1996)

(a) The Government may terminate performance of work under this contract in whole or, from time to time, in part if the Contracting Officer determines that a termination is in the Government's interest. The Contracting Officer shall terminate by delivering to the Contractor a Notice of Termination specifying the extent of termination and the effective date.

(b) After receipt of a Notice of Termination, and except as directed by the Contracting Officer, the Contractor shall immediately proceed with the following obligations, regardless of any delay in determining or adjusting any amounts due under this clause:

   (1) Stop work as specified in the notice.

   (2) Place no further subcontracts or orders (referred to as subcontracts in this clause) for materials, services, or facilities, except as necessary to complete the continued portion of the contract.

   (3) Terminate all subcontracts to the extent they relate to the work terminated.

   (4) Assign to the Government, as directed by the Contracting Officer, all right, title, and interest of the Contractor under the subcontracts terminated, in which case the Government shall have the right to settle or to pay any termination settlement proposal arising out of those terminations.

   (5) With approval or ratification to the extent required by the Contracting Officer, settle all outstanding liabilities and termination settlement proposals arising from the termination of subcontracts; the approval or ratification will be final for purposes of this clause.

   (6) As directed by the Contracting Officer, transfer title and deliver to the Government—

      (i) The fabricated or unfabricated parts, work in process, completed work, supplies, and other material produced or acquired for the work terminated; and

      (ii) The completed or partially completed plans, drawings, information, and other property that, if the contract had been completed, would be required to be furnished to the Government.

   (7) Complete performance of the work not terminated.

   (8) Take any action that may be necessary, or that the Contracting Officer may direct, for the protection and preservation of the property related to this contract that is in the possession of the Contractor and in which the Government has or may acquire an interest.

   (9) Use its best efforts to sell, as directed or authorized by the Contracting Officer, any property of the types referred to in paragraph (b)(6) of this

clause; *provided,* however, that the Contractor (i) is not required to extend credit to any purchaser and (ii) may acquire the property under the conditions prescribed by, and at prices approved by, the Contracting Officer. The proceeds of any transfer or disposition will be applied to reduce any payments to be made by the Government under this contract, credited to the price or cost of the work, or paid in any other manner directed by the Contracting Officer.

(c) The Contractor shall submit complete termination inventory schedules no later than 120 days from the effective date of termination, unless extended in writing by the Contracting Officer upon written request of the Contractor within this 120-day period.

(d) After expiration of the plant clearance period as defined in Subpart 49.001 of the Federal Acquisition Regulation, the Contractor may submit to the Contracting Officer a list, certified as to quantity and quality, of termination inventory not previously disposed of, excluding items authorized for disposition by the Contracting Officer. The Contractor may request the Government to remove those items or enter into an agreement for their storage. Within 15 days, the Government will accept title to those items and remove them or enter into a storage agreement. The Contracting Officer may verify the list upon removal of the items, or if stored, within 45 days from submission of the list, and shall correct the list, as necessary, before final settlement.

(e) After termination, the Contractor shall submit a final termination settlement proposal to the Contracting Officer in the form and with the certification prescribed by the Contracting Officer. The Contractor shall submit the proposal promptly, but no later than 1 year from the effective date of termination, unless extended in writing by the Contracting Officer upon written request of the Contractor within this 1-year period. However, if the Contracting Officer determines that the facts justify it, a termination settlement proposal may be received and acted on after 1 year or any extension. If the Contractor fails to submit the proposal within the time allowed, the Contracting Officer may determine, on the basis of information available, the amount, if any, due the Contractor because of the termination and shall pay the amount determined.

(f) Subject to paragraph (e) of this clause, the Contractor and the Contracting Officer may agree upon the whole or any part of the amount to be paid or remaining to be paid because of the termination. The amount may include a reasonable allowance for profit on work done. However, the agreed amount, whether under this paragraph (f) or paragraph (g) of this clause, exclusive of costs shown in paragraph (g)(3) of this clause, may not exceed the total contract price as reduced by (1) the amount of payments previously made and (2) the contract price of work not terminated. The contract shall be modified, and the Contractor paid the agreed amount. Paragraph (g) of this clause shall not limit, restrict, or affect the amount that may be agreed upon to be paid under this paragraph.

(g) If the Contractor and Contracting Officer fail to agree on the whole amount to be paid the Contractor because of the termination of work, the Contracting Officer shall pay the Contractor the amounts determined as follows, but without duplication of any amounts agreed upon under paragraph (f) of this clause:

  (1) For contract work performed before the effective date of termination, the total (without duplication of any items) of—

   (i) The cost of this work;

   (ii) The cost of settling and paying termination settlement proposals under terminated subcontracts that are properly chargeable to the terminated portion of the contract if not included in subdivision (g)(1)(i) of this clause; and

   (iii) A sum, as profit on subdivision (g)(1)(i) of this clause, determined by the Contracting Officer under 49.202 of the Federal Acquisition Regulation, in effect on the date of this contract, to be fair and reasonable; however, if it appears that the Contractor would have sustained a loss on the entire contract had it been completed, the Contracting Officer shall allow no profit under this subdivision (g)(1)(iii) and shall reduce the settlement to reflect the indicated rate of loss.

  (2) The reasonable costs of settlement of the work terminated, including—

   (i) Accounting, legal, clerical, and other expenses reasonably necessary for the preparation of termination settlement proposals and supporting data;

   (ii) The termination and settlement of subcontracts (excluding the amounts of such settlements); and

   (iii) Storage, transportation, and other costs incurred, reasonably necessary for the preservation, protection, or disposition of the termination inventory.

(h) Except for normal spoilage, and except to the extent that the Government expressly assumed the risk of loss, the Contracting Officer shall exclude from the amounts payable to the Contractor under paragraph (g) of this clause, the fair value, as determined by the Contracting Officer, of property that is destroyed, lost, stolen, or damaged so as to become undeliverable to the Government or to a buyer.

(i) The cost principles and procedures of Part 31 of the Federal Acquisition Regulation, in effect on the date of this contract, shall govern all costs claimed, agreed to, or determined under this clause.

(j) The Contractor shall have the right of appeal, under the Disputes clause, from any determination made by the Contracting Officer under paragraph (e), (g), or (l) of this clause, except that if the Contractor failed to submit the termination settlement proposal or request equitable adjustment within the time provided in paragraph (e) or (l), respectively, and failed to request a time extension, there is no right of appeal.

(k) In arriving at the amount due the Contractor under this clause, there shall be deducted—

   (1) All unliquidated advance or other payments to the Contractor under the terminated portion of this contract;

   (2) Any claim which the Government has against the Contractor under this contract;

   (3) The agreed price for, or the proceeds of sale of, materials, supplies, or other things acquired by the Contractor or sold under the provisions of this clause and not recovered by or credited to the Government.

(l) If the termination is partial, the Contractor may file a proposal with the Contracting Officer for an equitable adjustment of the price(s) of the continued portion of the contract. The Contracting Officer shall make any equitable adjustment agreed upon. Any proposal by the Contractor for an equitable adjustment under this clause shall be requested within 90 days from the effective date of termination unless extended in writing by the Contracting Officer.

(m) The Government may, under the terms and conditions it prescribes, make partial payments and payments against costs incurred by the Contractor for the terminated portion of the contract, if the Contracting Officer believes the total of these payments will not exceed the amount to which the Contractor will be entitled.

   (1) If the total payments exceed the amount finally determined to be due, the Contractor shall repay the excess to the Government upon demand, together with interest computed at the rate established by the Secretary of the Treasury under 50 USC App. 1215(b)(2). Interest shall be computed for the period from the date the excess payment is received by the contractor to the date the excess is repaid. Interest shall not be charged on any excess payment due to a reduction in the Contractor's termination settlement proposal because of retention or other disposition of termination inventory until 10 days after the date of the retention or disposition, or a later date determined by the Contracting Officer because of the circumstances.

(n) Unless otherwise provided in this contract or by statute, the Contractor shall maintain all records and documents relating to the terminated portion of this contract for 3 years after final settlement. This includes all books and other evidence bearing on the Contractor's costs and expenses under this contract. The Contractor shall make these records and documents available to the Government, at the Contractor's office, at all reasonable times, without any direct charge. If approved by the Contracting Officer, photographs, microphotographs, or other authentic reproduction may be maintained instead of original records and documents.

APPENDIX 11B: CHECKLIST: POTENTIAL DEFAULT TERMINATION

If a contractor believes that it may be terminated for default, it should consider at least these points in assessing its legal position.

IDENTIFY THE PROCEDURAL STEPS TO TERMINATION
- The standard Default clause at FAR § 52.249-10 does not require a cure notice.
- If a cure notice is received, respond to all issues identified in it. Document facts, not feelings.
- If directed or required to take certain actions, consider doing so with an express reservation of rights.
- Establish cost codes and documentation procedures to track actions taken and expenses incurred in responding to default threat.

IDENTIFY THE POTENTIAL GROUNDS FOR A DEFAULT TERMINATION
- On a government contract, a default may be justified on grounds that existed at termination but were not initially identified by the contracting officer.
- Review the project documentation to identify all complaints or possible grounds for complaints regarding its performance as well as its subcontractors/vendors at any tier.
- Consider whether the default grounds been waived by the government by a prior course of performance or course of dealing between the parties.
- Consider whether the default may be excusable under the contract.
- If not previously requested, ask for all legitimate extensions of time.

CURE PLANS
- Observe any time limits and stated requirements for taking action to cure an alleged default.
- Provide a written cure plan and seek the opportunity to meet with the contracting officer and representatives to review it.
- Once submitted, take affirmative steps to implement the cure plan and communicate with the contracting officer regarding those efforts.
- If subsequent events adversely affect the implementation of a cure plan, communicate with the government regarding the steps to mitigate those problems.
- Reserve claim rights when offering a plan to cure the alleged default.

DOCUMENT THE SCOPE OF WORK PERFORMED AND REMAINING
- Prepare as-built plans for work completed and work in progress.
- Make a photographic/video record of work performed and work in progress.
- Make a photographic record and inventory of stored materials.
- Make a photographic record and inventory of equipment and tools on site.
- Make a record of the status of the work of any other contractors on site.
- Make a record of any circumstances or conditions affecting the efficient performance of the work.
- Make a photographic record of site conditions (the need for cleanup, the absence of damage to other work, etc.).
- Store copies of this as-built information and photographic records in a secure, off-site location.

EVALUATE THE CONSEQUENCES OF THE LIKELY LOSS OF PROJECT STAFF
- Take steps to secure the continued employment of key project staff members.
- Consider engaging your attorney to take statements from project staff members who may leave the company after termination.
- Secure diaries, logs, and other project records from project staff responsible for their safekeeping.
- Ensure that the project staff has organized and properly labeled photographic and video records.
- Secure assistance from project staff in documenting and supporting claim positions while memories are fresh.

ASSESS SUBCONTRACTOR AND SUPPLIER TERMINATION RIGHTS
- Review subcontracts/purchase orders to identify termination rights.
- Determine if the subcontracts and purchase orders include termination for convenience provisions.
- Determine if the subcontracts and purchase orders provide for their assignment to another entity (surety or contractor).
- Take steps to minimize the potential impact of a termination on subcontractors and suppliers.
- Determine if the subcontracts and purchase orders contain pass-through provisions protecting your firm in the event of a termination.
- Consider the possible use of claim cooperation or liquidation agreements.

DOCUMENT THE STATUS OF SUBCONTRACTOR AND SUPPLIER PERFORMANCE

- Document the as-built record of the extent of each subcontractor or supplier's contract performance.
- Make a photographic record of subcontractor and supplier materials, tools, and equipment on site.
- Make a photographic record of the completeness, cleanliness, and adequacy of each subcontractor's work.

TAKE STEPS TO PROTECT PROJECT RECORDS

- Are copies of critical project records maintained off site?
- Are confidential or sensitive records secured in an off-site location?
- Have you advised employees of the need for care in documenting project events as a result of the likely disclosure of project records in litigation?
- Have you made adequate written responses to all allegations of contract performance deficiencies?
- Are your cost records adequate to segregate and calculate the consequences of contract breaches or an improper termination?
- Are your cost records adequate to segregate and capture costs associated with your steps to cure the alleged default?

CONSIDER THE POSSIBLE TAKEOVER OF MATERIALS, EQUIPMENT, AND TOOLS

- The standard Default clause allows the government to use the contractor's materials, equipment, and tools in completing the project.
- Ensure that there is an adequate record of the nature and condition of on-site materials, equipment, and tools.
- Have you considered the advisability of removing materials, equipment, or tools from the site?
- Is the equipment on site company owned, or is it either rented or the property of subcontractors?
- Can you afford to be without the use of the site equipment during the project completion?

DEMONSTRATE THE WEAKNESSES IN THE GOVERNMENT'S POSITION

- Have you developed the factual history that would support your defense to a wrongful termination?
- Do you have in place cost records to demonstrate the significant consequences of a wrongful termination?
- Have you developed a persuasive legal/contractual defense to the threatened termination?
- Have you attempted to discourage the threatened termination by actions designed to convince the opposition of the strength of your position?

REASSURE YOUR PERFORMANCE BOND SURETY

- Have you advised your surety of the termination threat?
- Have you taken steps to assure your surety that your position is factually and legally justified?
- Have you discussed the impact of a wrongful termination on your bonding capacity?
- Have you advised your staff that correspondence with your surety may not be protected from disclosure during litigation?
- Have you gained your surety's agreement to support your legal position?
- If you are terminated, and if your surety plans to participate in the project completion, have you considered having the surety retain you as the completion contractor?

12

PAYMENT AND PERFORMANCE BONDS

I. INTRODUCTION

In government construction contracts, contractors provide three types of bonds: (1) bid bonds, which guarantee the contractor's bid or proposal; (2) payment bonds, which guarantee payment of lower-tier subcontractors and suppliers; and (3) performance bonds, which guarantee performance on the project. These bonds represent a promise to the bond obligee (or beneficiary) that the surety will perform some duty on the part of the bond principal (in this case, usually the general contractor). If the principal fails to perform its duty (whether it is payment or contract performance), the obligee can demand that this duty be fulfilled by the surety. Although the obligee's claim is guaranteed by the bond, the principal—the general contractor—is jointly liable with the surety. Since the sureties typically require the contractor, as well as its major stockholders or owners, to execute company and personal indemnity agreements as a condition of providing the bond, the indemnitors are liable to the surety for any monies or performance the surety provides under a bond. In this respect, surety bonds are fundamentally different from insurance policies where the insured typically is not liable to repay an insurer's payment or loss.

This chapter addresses payment and performance bonds, which generally involve issues that develop during the performance of a government construction contract. *Bid bonds* are closely related to the contract formation process. Consequently, that topic is addressed in **Chapter 3**, **Section IV** of this book.

Bond coverage is litigated frequently in government construction contracts. These bonds involve a third party, the surety, that is outside of the contractor/government relationship. In the dispute process, the surety may take positions that are adverse to both the contractor's and government's positions. Similarly, a surety's positions may reflect the fact that its relationship with the contractor is derived from agreements that are based on state law rather than government contract law. In addition, payment bond claims involve contractors, certain subcontractors/suppliers, as well as the surety and are resolved outside of the forums (boards or Court of Federal Claims) that address disputes between the contractor and the government. Understanding the complexity of these relationships, the resulting three-party disputes, and the bonds themselves is crucial for the contractor dealing with a government contract or claims that arise out of it.

II. MILLER ACT PAYMENT BONDS

A. Introduction

In a private construction contract, if a subcontractor was not paid for its work on the project, it likely would be able to take advantage of state law and file a mechanic's lien on the property. However, liens cannot be placed against property of the federal government so this form of remedy for nonpayment is unavailable to any party performing work on or furnishing materials for a government construction project.[1] However, the Miller Act payment bond guarantees payment to certain parties supplying labor and materials to contractors or subcontractors engaged in the construction, alteration, or repair of any public building or public work of the United States. The Miller Act protects certain subcontractors and suppliers that otherwise would be without a remedy because there are no lien rights against the property of the federal government.

The payment bond makes the surety the guarantor of payment to the general contractor's lower-tier subcontractors and suppliers according to the terms of the bond and the Miller Act. The surety's obligation to payment bond claimants is separate from the surety's performance bond obligation to the government, and the surety must respond to payment bond claims even though it may have a valid defense to performance bond obligations.[2] In order to evaluate a potential payment bond claim and obtain basic information regarding the surety, an unpaid supplier or subcontractor first should obtain a copy of the actual bond. To facilitate this process, Federal Acquisition Regulation (FAR) § 28.106-6(d) provides specific guidance to contracting officers regarding requests for information related to Miller Act payment bonds. Upon either written or oral request by a subcontractor/supplier or prospective subcontractor/supplier, the contracting officer "shall promptly provide" information on the name and address of the surety, the penal sum of the bond, and a copy of the actual payment bond. A copy of Standard Form 25A - Payment Bond is included at the end of this chapter in **Appendix 12A** and on the support Web site at www.wiley.com/go/federalconstructionlaw.

B. Parties Covered under the Miller Act

40 U.S.C. § 3131, Bonds of Contractors of Public Buildings or Works, provides this direction to contracting officers regarding the coverage and penal sum of a Miller Act payment bond:

---

[1] *United States ex rel. Westinghouse Elec. Supply Co. v. Sisson*, 927 F.2d 310, 312 (7th Cir. 1991).
[2] *See generally Morrison Assur. Co., Inc. v. United States*, 3 Cl. Ct. 626, 632 (1983).

(b) Type of Bonds Required—Before any contract of more than $100,000 is awarded for the construction, alteration, or repair of any public building or public work of the Federal Government, a person must furnish to the Government the following bonds, which become binding when the contract is awarded:

* * *

(2) Payment bond—A payment bond with a surety satisfactory to the officer *for the protection of all persons supplying labor and material* in carrying out the work provided for in the contract for the use of each person. The amount of the payment bond shall equal the total amount payable by the terms of the contract unless the officer awarding the contract determines, in a writing supported by specific findings, that a payment bond in that amount is impractical, in which case the contracting officer shall set the amount of the payment bond. The amount of the payment bond shall not be less than the amount of the performance bond. [Emphasis added]

Since many parties work on a construction project at various tiers, the broad language of the Miller Act—"all persons supplying labor and materials in carrying out the work"—suggests that any party performing labor or materials in connection to the project could recover under that payment bond. However, the court decisions define the extent of the parties protected under that statute, which may make it difficult to determine the precise limits of the parties entitled to Miller Act payment bond coverage.

In *Clifford F. MacEvoy Co. v. United States ex rel. Tomkins Co.*,[3] the United States Supreme Court identified two general classes of claimants entitled to protection under the Miller Act. The first class is composed of all laborers, materialmen, subcontractors, and suppliers that deal *directly* with the prime contractor—the "first tier." The Court also defined a second class of Miller Act claimants composed of all materialmen, subcontractors, and laborers that have a *direct* contractual relationship with a first-tier subcontractor—the "second tier." As such, the *MacEvoy* case limited Miller Act protection to only those parties that have direct relationships with the prime or a first-tier subcontractor.

The Supreme Court's *MacEvoy* decision did not precisely define the term "subcontractor" for purposes of Miller Act payment bond coverage. Thirty years later, in *F. D. Rich Co. v. United States ex rel. Industrial Lumber Co.*,[4] the Supreme Court reiterated *MacEvoy*'s functional definition of "subcontractor" and emphasized that whether one is a subcontractor relates to "the substantiality and importance of his relationship with the prime contractor."[5]

---

[3] 322 U.S. 102 (1944).
[4] 417 U.S. 116 (1974).
[5] *Id.* at 123.

Traditionally, a party that only supplied materials would fail to qualify as a *subcontractor*, while a party that *installed* materials was more likely more to meet the *subcontractor* test. However, in recent years, the broad criteria established in *MacEvoy* and *F.D. Rich* have resulted in courts liberally construing the definition of *subcontractor* for Miller Act purposes. For example, in *United States ex rel. E & H Steel Corp. v. C. Pyramid Enterprises, Inc.,* the court held that a steel fabricator that supplied materials but did no installation work was a subcontractor under the Miller Act.[6] In reaching its determination that the steel fabricator was a subcontractor and not a supplier for the purposes of Miller Act, the court focused on the fact that the steel fabricator arranged for fabrication and delivery of materials, prepared shop drawings, designed connectors, and performed some design-assist engineering.[7]

In determining whether a claimant qualifies as a subcontractor, courts may consider such factors as: (1) the nature of the material or service supplied by the alleged subcontractor to the prime contractor; (2) the financial magnitude of the goods or services provided in relation to the total federal contract; (3) the payment terms and exchange of information between the prime contractor and alleged subcontractor; and (4) the overall relationship between the prime contractor and the alleged subcontractor.[8]

The distinction between a supplier and subcontractor at the second tier is critical because it determines whether the Miller Act protects the claimant. The Act protects suppliers to prime contractors and subcontractors but not suppliers to suppliers. For example, in *United States ex rel. Gulf States Enterprises, Inc. v. R. R. Tway, Inc.,*[9] the court found that the provider of a bulldozer and operator was a supplier, not a subcontractor, because the provider was not responsible for performing a definable part of the contract work. The agreement with the entity providing the bulldozer and operator was not to do specific work but was to provide those items at an agreed-on hourly rate "as needed." As a supplier and not a subcontractor, the company that actually owned and leased the dozer to the provider could not recover unpaid lease payments from the prime contractor's surety under the Miller Act.

Other factors supporting that a party is a subcontractor, and not a supplier, under the Miller Act include whether: (1) the product supplied is custom fabricated; (2) the product supplied is a complex integrated system; (3) a close financial interrelationship exists between the companies; (4) a continuing relationship exists with the prime contractor; (5) the supplier is required to perform on site; (6) there is a contract for labor in addition to materials; (7) the term "subcontractor" is used in the agreement; (8) the materials supplied do not come from existing inventory; (9) the supplier's contract constitutes a substantial portion of the prime contract; (10) the supplier is required to furnish all the material of a particular type; (11) the supplier is required to post a performance bond; (12) there are back charges for the cost of correcting

[6]509 F.3d 184 (3rd Cir. 2007).
[7]*Id.* at 190.
[8]*Id.* at 188.
[9]938 F.2d 583 (5th Cir. 1991).

the supplier's mistakes; and (13) there is a system of progressive or proportionate payment.[10]

The question of *who is a subcontractor* sometimes involves the issue of *telescoping* contractual relationships. A prime contractor will not be allowed to insert a dummy subcontractor between itself and the subcontractors that actually perform the work in order to avoid Miller Act liability. If a dummy subcontractor is inserted into the contractual arrangements, lower-tier subcontractors or suppliers, which are not paid by the dummy subcontractor, still may sue on the payment bond.[11] However, absent evidence of bad faith—for example, use of a dummy subcontractor—or fraud, the courts generally will not look beyond formal contractual relationships.

C. Claims Covered

After determining *who* is protected by a payment bond, the next step is to determine *what items* qualify for protection. The Miller Act provides payment protection only for labor and materials furnished *in the prosecution of the work.* Courts generally have interpreted this language liberally, holding that it covers not only work incorporated in the project but also other work done for the benefit of the project.[12] However, performing administrative duties such on-site project management probably will not qualify as labor under the Miller Act because there must be some physical toil or manual work performed.[13]

In *United States ex rel. National U.S. Radiator Corp. v. D. C. Loveys Co.*,[14] a claimant obtained recovery from a Miller Act surety for material that was damaged in transit and then delivered to the project. Although the material was not incorporated into the project itself, the court held that the material was nevertheless furnished *in the prosecution of the work,* because the subcontractor had assumed the risk of loss or damage during shipment.[15]

Similarly, in another case, a supplier recovered for delivery of materials to a subcontractor in spite of the subcontractor's subsequent removal of the material from the job site and use on another project.[16] Following this general line of federal cases, state courts also have ruled that a supplier does not need to prove that the materials

---

[10] *United States ex rel. Maryland Minerals v. United States Fidelity and Guaranty Co.*, 2007 WL 1687572, *7 (N.D. W.Va. 2007) citing *United States ex rel. Conveyor Rental & Sales Co. v. Aetna Cas. & Sur. Co.*, 981 F.2d 448, 451 (9th Cir. 1992).

[11] *United States ex. rel. Ferguson Enterprises, Inc. v. St. Paul Fire & Marine Ins. Co.*, 2000 WL 1880313 (D. Kan. 2000).

[12] *See United States ex rel. Sunbelt Pipe Corp. v. United States Fidelity & Guar. Co.*, 785 F.2d 468 (4th Cir. 1986); *United States ex rel. Westinghouse Elec. Supply Co. v. Endebrock-White Co.*, 275 F.2d 57 (4th Cir. 1960); *United States ex rel. Skip Kirchdorfer, Inc. v. Aegis/Bublin Joint Venture*, 869 F. Supp. 387 (E.D. Va. 1994).

[13] *See United States ex rel. Constructors, Inc. v. Gulf Ins. Co.*, 313 F. Supp. 2d 593, 597 (E.D. Va. 2004).

[14] 174 F. Supp. 44 (D. Mass. 1959).

[15] *Id.*

[16] *See Glassell-Taylor Co. v. Magnolia Petroleum Co.*, 153 F.2d 527 (5th Cir. 1946); *see also United States ex rel. Carlson v. Continental Cas. Co.*, 414 F.2d 413 (5th Cir. 1969).

delivered to the site were actually incorporated into the project.[17] As a general rule, the Miller Act remedy is available if the material is substantially consumed or rendered useless in the prosecution of the work.[18] This includes parts and equipment necessary to and wholly consumed by the project and material used in construction but not incorporated into the project, such as concrete formwork.[19]

Guarantee work or repairs that become necessary and are performed after substantial completion of the project also may be the subject of a Miller Act payment bond claim. The cost of incidental repairs necessary to maintain equipment during its use on the project also may be recovered under a bond.[20] However, substantial *replacement* repairs are not covered, on the theory that they add value to the construction equipment by extending its useful life beyond the project in question.

The fair rental value of equipment leased for use in the prosecution of the contract work may be covered by a Miller Act bond.[21] Miller Act payment bonds also have been construed to cover transportation and delivery costs.[22] Food and lodging have been found to be covered when they are a necessary and integral part of performance.[23] Also, coverage typically will include union contributions and contributions to employee funds, such as health and welfare.[24] Work done by a qualifying claimant under a change order is generally within the payment bond's protection, and recovery for extra work under the Miller Act does not depend on the prime contractor's ability to recover its additional costs from the government.[25]

---

[17]*See, e.g., American Safety Cas. Ins. Co. v. C.G. Mitchell Constr. Inc.,* 601 S.E.2d 633 (Va. 2004); *see also Key Constructors, Inc. v. H & M Gas Co.,* 537 So. 2d 1318 (Miss. 1989); *Mid Continent Cas. Co. v. P&H Supply, Inc.,* 490 P.2d 1358 (Okla. 1971) (evidence that materials were delivered to the project site creates a rebuttable presumption that the materials actually were consumed in the construction).

[18]*United States ex rel. Tom P. McDermott, Inc. v. Woods Constr. Co.,* 224 F. Supp. 406 (N.D. Okla. 1963).

[19]*United States ex rel. Sunbelt Pipe Corp. v. United States Fidelity & Guar. Co.,* 785 F.2d 468 (4th Cir. 1986); *United States ex rel. Skip Kirchdorfer, Inc. v. Aegis/Bublin Joint Venture,* 869 F. Supp. 387 (E.D. Va. 1994); *United States ex rel. Chemetron Corp. v. George A. Fuller Co.,* 250 F. Supp. 649 (Mont. 1966).

[20]*See Finch Equip. Corp. v. Frieden,* 901 F.2d 665 (8th Cir. 1990); *Massachusetts Bonding & Ins. Co. v. United States ex rel. Clarksdale Mach. Co.,* 88 F.2d 388 (5th Cir. 1937); *Maryland Cas. Co. v. Ohio River Gravel Co.,* 20 F.2d 514 (4th Cir. 1927), *cert. denied,* 275 U.S. 570 (1927); *but see Transamerica Premier Ins. Co. v. Ober,* 894 F. Supp. 471 (D. Me. 1995).

[21]*United States ex rel. Miss. Rd. Supply Co. v. H.R. Morgan, Inc.,* 542 F.2d 262 (5th Cir. 1976), *cert. denied,* 434 U.S. 828 (1977); *Friebel & Hartman, Inc. v. United States ex rel. Codell Constr. Co.,* 238 F.2d 394 (6th Cir. 1956); *United States ex rel. D&P Corp. v. Transamerica Ins. Co.,* 881 F. Supp. 1505 (D. Kan. 1995) (may recover rental value of owned equipment).

[22]*See United States ex rel. Benkurt Co. v. John A. Johnson & Sons, Inc.,* 236 F.2d 864 (3d Cir. 1956); *United States ex rel. Carlisle Constr. Co., Inc. v. Coastal Structures, Inc.,* 689 F. Supp. 1092 (M.D. Fla. 1988).

[23]*Brogran v. National Sur. Co.,* 246 U.S. 257 (1918); *United States ex rel. T.M.S. Mech. Contractors, Inc. v. Millers Mut. Fire Ins. Co. of Texas,* 942 F.2d 946 (5th Cir. 1991).

[24]*United States ex rel. Sherman v. Carter,* 353 U.S. 210, 218 (1957).

[25]*See Mai Steel Serv. Inc. v. Blake Constr. Co.,* 981 F.2d 414 (9th Cir. 1992); *United States ex rel. Warren Painting v. J. C. Boespflug Constr. Co.,* 325 F.2d 54 (9th Cir. 1963); *United States ex rel. Kilsby v. George,* 243 F.2d 83 (5th Cir. 1957).

Traditionally, delay damages have been viewed as outside the scope of Miller Act coverage.[26] However, a restriction on a payment bond claim for delay damages does not affect the claimant's general right to collect delay damages from the contractor or subcontractor causing the delay.[27]

This prohibition of delay damage claims against Miller Act payment bonds has been eroded, and the federal courts are split on whether such damages are covered. For example, in *United States ex rel. Mandel Bros. Contracting Corp. v. P. J. Carlin Construction Co.*,[28] a federal district court held that the payment bond claimant could recover its costs of delay on a *quantum meruit* ("value added") theory. In *Mandel Bros.*, the claimant alleged that the general contractor's failure to provide access to the work site restrained the claimant's performance and was such a substantial interference with the claimant's progress that it amounted to an abandonment of the contract. The court rejected the traditional doctrine that breaches of contract predicated on delays were not compensable under the Miller Act, reasoned that a *quantum meruit* claim is one for labor and materials actually furnished in the prosecution of the work, and therefore concluded that delay related expenses were within the scope of the payment bond.

In *Metric Electric Inc.*,[29] the court held that a subcontractor could recover increased costs attributable to delay, as long as the subcontractor did not create the delay. However, the subcontractor could not recover on the payment bond for overhead and profit on work that was not performed.[30]

In another case, *United States ex rel. Pertun Construction Co. v. Harvester's Group, Inc.*,[31] the subcontract contained a *no-damages-for-delay clause,* but the court read the clause as conditioned on the subcontractor being granted reasonable time extensions for delays. The court found that the prime contractor wrongfully and prematurely terminated the subcontractor, and as a result, neither the contractor nor its surety could claim protection under the no-damages-for-delay clause.[32]

D. Surety Defenses

In any claim against a payment bond, the surety is entitled to assert any defenses that its contractor/principal has, including the defense of offset or recoupment. The surety also may have additional defenses of its own. Generally, such independent defenses are found in the applicable bond statute and/or the terms of the bond itself. The most common surety defenses are the claimant's failure to comply with notice requirements and time limitations. When asserting technical defenses relating to

---

[26]*See McDaniel v. Ashton-Median Co.*, 357 F.2d 511 (9th Cir. 1966); *United States ex rel. Pittsburgh Des Moines Steel Co. v. MacDonald Constr. Co.*, 281 F. Supp. 1010 (E.D. Mo. 1968).

[27]*See United States Fidelity & Guar. Co. v. Ernest Constr. Co.*, 854 F. Supp. 1545 (M.D. Fla. 1994).

[28]254 F. Supp. 637 (E.D.N.Y. 1966).

[29]*United States ex rel. Metric Elec. Inc. v. Enviroserve, Inc.*, 301 F. Supp. 2d 56, 68 (D. Mass. 2003).

[30]*Id.* at 70.

[31]918 F.2d 915 (11th Cir. 1990).

[32]*Id.*

the timing and sufficiency of the required notice, the surety should deal with bond claimants in good faith.[33]

In recent years, courts have held that a surety cannot rely on an otherwise enforceable contingent payment clause as a defense to a Miller Act claim.[34] In *Walton Technology*, the court reasoned that if a surety was permitted to use a "pay when paid clause" to delay payment to a subcontractor for more than one year, the subcontractor's Miller Act rights would be forfeited.[35]

It is imperative that potential claimants review the terms of the payment bond, any applicable statutes, and case law construing them, to determine the exact timing, nature, recipient of the notice, and any other requirements necessary to secure their rights under the bond. Any deviation from these requirements may defeat an otherwise valid claim. This careful review is equally important when responding to a payment bond claim.

E. Notice

Depending on a firm's contractual relationship with the prime contractor, Miller Act bond claimants must satisfy the Act's notice requirements. Claimants (subcontractors or suppliers) in direct privity of contract with the prime contractor have no notice requirement under the Miller Act. Those entities are considered as *first-tier* bond claimants. *Second-tier* claimants refer to those entities that are not in a direct contractual relationship with the prime contractor but have a contractual relationship with a first-tier subcontractor. Under the Miller Act, notice by a *second-tier* bond claimant must be received by the prime contractor within 90 days of the last date on which that bond claimant performed work or supplied materials.[36] However, the Miller Act's notice requirements are fixed by statute, and the bond may not be amended to add additional notice requirements, such as a preliminary notice within a specified number of days of the potential bond claimant's starting work on the project.[37]

Although the Miller Act notice must be sent to the prime contractor, the surety does not have to receive notice.[38] The Miller Act requires notice to be sent "by any

---

[33]*See Datastaff Tech. Group, Inc. v. Centex Constr. Co., Inc.,* 528 F. Supp. 2d 587 (E.D. Va. 2007) (recognizing that a surety may be barred under theory of equitable estoppel from relying on a technical defense such as a statute of limitations when it acts in bad faith); *see also United States ex rel. Ehmcke Sheet Metal Works v. Wausau Ins. Co.,* 755 F. Supp. 906, 909 (E.D. Cal. 1991) (state law may provide a cause of action against Miller Act surety for breach of covenant of good faith and fair dealing); *see also K-W Indus. v. National Sur. Corp.,* 855 F.2d 640 (9th Cir. 1988).

[34]*See United States ex. rel. McKenney's Inc. v. Government Technical Services, LLC,* 531 F. Supp. 2d 1375 (N.D. Ga. 2008); *United States ex. rel. Walton Tech., Inc. v. Weststar Eng'g, Inc.,* 290 F.3d 1199, 1209 (9th Cir. 2002).

[35]*Walton Tech., Inc.* 290 F.3d at 1208.

[36]40 U.S.C. § 3133(b)(2)(A).

[37]*See Nagel Constr., Inc v. Crest Constr. & Excavating, LLC,* 2006 WL 1806487 (W. D. Mich. 2006) citing *United States ex rel. S & G Excavating v. Seaboard Sur. Co.,* 236 F.3d 883, 884 (7th Cir. 2001).

[38]*See United States ex. rel. N.E.W. Interstate Concrete, Inc. v. EUI Corp.,* 93 F. Supp. 2d 974, 981 (S.D. Ind. 2000) (holding that a supplier's state court complaint which did not name surety was sufficient notice under Miller Act).

means which provides written third party verification of delivery."[39] Oral notice by itself, however, generally will be insufficient.[40] Since notice is intended to protect the contractor that provided the payment bond, the written notice must expressly or impliedly inform the contractor that the claimant is looking to it or the surety for payment.[41] Pursuant to 40 U.S.C. § 3133(b)(2), the notice must also state with substantial accuracy the amount claimed and the name of the party to whom the materials or services were provided.

F. Time for Enforcement

A lawsuit to enforce the provisions of a Miller Act payment bond generally must be brought within one year of "the day on which the last of the labor was performed or material was supplied by [claimant]."[42] A surety may be barred from relying on the statute of limitations defenses in certain instances such as when a surety admits liability and promises to pay a claimant on the bond and then terminates negotiations after the Miller Act one-year limitation period has run.[43]

A substantial body of law has developed defining "the day on which the last of the labor was performed or material was supplied by [claimant]." For example, in *General Insurance Co. of America v. United States ex rel. Audley Moore & Son*,[44] the court refused to include the act of inspecting within the definition of labor as used in the Miller Act. However, the correction of prior work has been held to constitute labor where the government has refused to accept the project until such work has been completed. The correction of defects or warranty work done after completion of the original subcontract work most likely will not constitute the furnishing of labor or materials for purposes of the Miller Act's time limitation.[45]

These Miller Act cases distinguish *guarantee work* from *punch-list work.* In other words, work that the government demands be finished in accordance with the contract plans and specifications by a punch list or other similar device is considered to be contract work. Performance of this work normally will stop the relevant time limits of the Miller Act governing when notice must be given and suit must be filed. Work performed under a warranty or to repair latent defects, however, is regarded

---

[39]40 U.S.C. § 3133(b)(2)(A).

[40]*See United States ex. rel. Martinez v. Encon Int'l, Inc.*, 571 F. Supp. 2d 754, 758 (W.D. Tex. 2008) (sufficient notice found where sub-subcontractor presented a written invoice during meeting with prime contractor to discuss outstanding balance owed).

[41]*See United States ex. rel. Viking Disposal Corp. v. Western Sur. Co.*, 2007 WL 5287926, *5 (W.D. Wis. 2007).

[42]40 U.S.C. § 3133(b)(4).

[43]*See Datastaff Tech. Group, Inc. v. Centex Constr. Co., Inc,*, 528 F. Supp. 2d 587 (E.D. Va. 2007) (recognizing that a surety may be barred under theory of equitable estoppel from relying on a technical defense such as a statute of limitations when it acts in bad faith).

[44]406 F.2d 442 (5th Cir. 1969), *cert. denied*, 396 U.S. 902 (1969).

[45]*See United States ex. rel. Interstate Mech. Contractors, Inc. v. Int'l Fidelity Ins. Co.*, 200 F.3d 456, 462 (6th Cir. 2000); *United States ex. rel. PRN Assocs., Inc. v. K&S Enterprises, Inc.*, 2007 WL 925267, *3 (S.D. Ind. 2007).

by the courts as being noncontract work and, as such, outside of the term "labor" as used in the Miller Act.[46]

G. Waiver of Bond Rights

On government projects, a waiver of Miller Act rights by a party otherwise protected must be specific. For there to be an effective and clear waiver of Miller Act payment bond rights, the Miller Act must be mentioned. Courts do not favor a finding that a subcontractor has contractually waived its rights under the Miller Act.[47] This is evidenced by the cases holding that a contingent payment clause, which would be otherwise enforceable under state law, will not bar Miller Act claims.[48]

The Miller Act also addresses the timing of any waiver of the right to sue on the payment bond. 40 U.S.C. § 3133(c) provides:

> (c) A waiver of the right to bring a civil action on a payment bond required under this subchapter is void unless the waiver is—
>   (1) in writing;
>   (2) signed by the person whose right is waived; and
>   (3) executed after the person whose right is waived has furnished labor or material for use in the performance of the contract.

This text reflects the substance of a 2002 amendment to the Miller Act.

Prime contractors should consider the possible effect of this provision on the terms of their subcontracts and purchase orders, especially terms addressing recovery for changes and extra work, delays, and those subcontract provisions addressing subcontractor participation in the disputes process. In addition, prime contractors may require waiver forms to be submitted with each pay application. Waiver forms vary considerably in content. Often the form will include a waiver of lien and/or payment bond rights through a certain date or for work performed up to a certain date. In defending a payment bond suit, the surety will examine the underlying contract, pay applications, and monthly waiver forms to determine if a waiver has occurred.

III. MILLER ACT PERFORMANCE BONDS

A. Declaration of Default

The requirement for performance bonds in government contracts is also found in the Miller Act, 40 U.S.C. §§ 3131–3134. On any contract in excess of $100,000,

---

[46]*See, e.g., United States ex rel. State Elec. Supply Co. v. Hesselden Constr. Co.,* 404 F.2d 774 (10th Cir. 1968); *United States ex rel. Hussman Corp. v. Fidelity & Deposit Co. of Md.,* 999 F. Supp. 734 (D.N.J. 1998).

[47]*See H.W. Caldwell & Son, Inc. v. United States for Use and Benefit of John H. Moon & Sons, Inc.,* 407 F.2d 21, 23 (5th Cir. 1969); *United States for the Use and Benefit of DDC Interiors, Inc. v. Dawson Constr. Co., Inc.,* 895 F. Supp. 270 (1995), *aff'd,* 82 F.3d 427 (1996).

[48]*See United States ex. rel. McKenney's Inc. v. Government Technical Services, LLC,* 531 F. Supp. 2d 1375 (N.D. Ga. 2008); *United States ex. Rel. Walton Tech., Inc. v. Weststar Eng'g, Inc.,* 290 F.3d 1199, 1209 (9th Cir. 2002).

the Miller Act requires the general contractor to furnish a performance bond in an amount equal to the contract price, unless the contracting officer determines that it is impractical to obtain a bond in that amount and specifies an alternative amount for the bond.[49]

In the private market as well as on many state/local projects, many of the performance bond forms condition the surety's obligations on the receipt of a notice of default or a *declaration of default*.[50] In contrast, the Miller Act performance bond form does *not* contain a requirement for any advance or preliminary notice from the government to the surety prior to issuing a default termination notice. Similarly, although there is no requirement in the FAR for the government to issue a *cure notice* before terminating a contractor for default,[51] the government typically provides a cure notice to both the contractor and its surety prior to a default action. This may be different from bond forms used in private or state/local construction contracts. Once a contract is terminated for default, the performance bond surety often becomes a key participant in completing the project.[52] A performance bond claim and the surety's response to that claim generally are triggered by the contractor's default or an alleged default. In that context, an understanding of the principles related to default terminations as discussed in **Chapter 11** is essential, and those materials should be reviewed in the context of any consideration of potential claims against the performance bond.

From the surety's perspective, an analysis of the propriety of the default is necessary so that the surety may determine its future course of action. The surety must determine: (1) whether the principal's actions or failure to act constituted a breach of contract, (2) whether any breach of contract by its principal was sufficiently material to warrant a termination for default, and (3) whether the government has satisfied its obligations under the contract.

The central point in assessing the surety's liability under a performance bond is whether the contractor/principal was in default under the contract which gave rise to the termination.[53] A performance bond surety cannot be held liable for the default of its principal where the government materially breaches the contract.[54] For example, the courts will not affirm the default termination of a construction contract where the government materially breaches its payment obligation to the principal. In this regard, the Armed Services Board of Contract Appeals (ASBCA) has held that the government's failure to make adequate payments to a contractor constituted a material breach of the contract that excused further performance of work by the contractor.[55]

---

[49]On very-large-dollar, multiyear projects, the government may consider employing *stacked* performance bonds or other forms of successive performance bonds to mitigate the effect of tying up a contractor's bonding capacity for inordinate periods of time.

[50]*See L & A Contracting Co. v. Southern Concrete Services, Inc.*, 17 F.3d 106 (5th Cir. 1994).

[51]*See* **Chapter 11, Section II.C.**

[52]*See* FAR § 49.404.

[53]See **Chapter 11** for a more detailed discussion of the grounds for a default termination and those conditions or circumstances that excuse an apparent default.

[54]*See Nexus Constr. Co.*, ASBCA No. 31070, 91-3 BCA 24,303.

[55]*Id.; see also Wolfe Constr. Co.*, ENGBCA Nos. 3607, 3608, 3609, 84-3 BCA 17,701.

B. Sureties' Options upon Default

The performance bond surety has certain obligations and options if its principal defaults. For example, in private construction contracts using the AIA 312 Performance Bond, there are several actions that the surety may take, such as:

- The surety may agree to take over and complete the bonded contract;
- The surety may *buy back* or repurchase the bond through a cash settlement with the government;
- The surety may tender a replacement contractor to the government to complete the remaining contract work and pay the excess cost of completing the work; or
- The surety may offer to have the defaulted contractor complete the bonded contract work.

In lieu of exercising any of these options, the surety also can decide to take no action and effectively deny the bond claim.

In contrast to performance bonds often utilized in private construction, the Standard Form 25 Performance Bond used on government construction contracts is a type of statutory bond that only provides for *payment* in the event of the contractor's default. (A copy of the Standard Form 25 Performance Bond is included as **Appendix 12B** of this chapter and the support Web site at www.wiley.com/go/federalconstructionlaw.)[56] However, despite this limitation, government contracting officers are authorized to consider options allowing the surety to arrange for completion.[57] As such, although the bond as written only provides the surety with the option of payment upon default, the practical result is that the surety's options are only limited to what the government will accept.[58]

C. Takeover Agreements

The surety's option to take over and complete the contract work following contractor default is a remedy available in many termination scenarios.

The so-called *takeover agreement* is a vehicle for the surety and obligee to define the surety's obligations to complete the remaining work. Although this agreement often is just between the government and the surety, the contracting officer is instructed to include the terminated contractor in the discussion so the contractor can minimize any subsequent exposure.[59]

The terms of the takeover agreement typically provide for the surety to complete the work according to the terms and conditions of the original contract. The government

---

[56]*See* FAR §§ 53.228(b), 53.301-25.
[57]*See* FAR § 49.404.
[58]*See Preferred National Ins. Co. v. United States,* 54 Fed. Cl. 600 (Fed. Cl. 2002) (government's insistence that surety either enter into takeover agreement pursuant to its performance bond or suffer removal from the list of government-approved sureties did not constitute duress).
[59]*See* FAR § 49.404(d).

will pay the balance of the contract price not to exceed to the surety's costs.[60] Although the completing surety knows the government's stated bases for the default termination of the contractor, the surety in its completion role often requires a more specific statement of the specific items to be addressed in completing the contract. A specifically defined scope for completing the work is key factor for the surety to identify its risk and allow it to secure replacement contractors to complete the work.

An often-litigated issue involves whether the takeover agreement gives the surety the right to bring claims against the government relating to the project work. Generally, when there is a takeover agreement with the government, the completing surety may maintain an action against the government under the Contract Disputes Act.[61] However, if there is no formal takeover agreement, the surety's claims against the government may be limited to the balance owed to the original contractor under the terminated contract.

D.   Damages Available

The measure of damages allowed against a performance bond surety are the costs in excess of the available contract proceeds to complete the construction or remedy any defective work.[62] However, the government's allowable damages under a perform-ance bond depend on the terms of the bond and the contract. As a result, the surety generally is liable for any damages that its principal would be liable for in completing the construction contract.

In addition, under the Default clause, the federal government is entitled to recover its *increased costs* from the surety—that is, those additional costs in excess of the original contract price that are necessarily incurred by the government in completion of the work. Although administrative costs have not always been recoverable by the government, these costs are now recoverable if supported by reasonable estimates.[63]

When the scope of the newly procured work exceeds the coverage of the termi-nated contract, the government may recover from the surety only those additional costs that are attributable to the work in the original contract.[64] Also, the Court of Federal Claims or a board of contract appeals generally will reduce an award of excess reprocurement costs where the government has failed to mitigate damages.[65] Finally, if the reprocurement contract contains significant deviations from the original contract, no excess costs may be assessed against the surety.[66]

---

[60]*See United States Sur. Co. v. United States,* 83 Fed. Cl. 306, 310 (Fed. Cl. 2008).

[61]*See Insurance Co. of the West v. United States,* 55 Fed. Cl. 529, 538 (Fed. Cl. 2003).

[62]*See United States Sur. Co. v. United States,* 83 Fed. Cl. 306 (Fed. Cl. 2008) (surety on Miller Act per-formance bond guarantees performance of contract and completion of project if principal defaults).

[63]*Evans,* ASBCA No. 10951, 66-1 BCA ¶ 5316 (denying recovery of administrative costs); *Arctic Corner,* ASBCA No. 38075, 94-1 BCA ¶ 26,317 (allowing administrative costs if supported by estimates); *see also ARCO Eng.* ASBCA No. 52450, 01-1 BCA ¶ 31,218.

[64]*M.S.I. Corp.,* VACAB No. 599, 67-2 BCA ¶ 6643.

[65]*See, e.g., A & W Gen. Cleaning Contractors,* ASBCA No. 14809, 71-2 BCA ¶ 8994.

[66]*Blake Constr. Co.,* GSBCA No. 4013, *et al.,* 75-2 BCA ¶ 11,487.

In addition to reprocurement costs, the government default clause also provides that the contractor and its sureties will be liable for damages resulting from the contractor's refusal or failure to complete the work within the specified time. If there is no liquidated damages provision, the government is entitled to recover its actual damages caused by the contractor's delay. Actual damages include, for example, the cost of keeping a government inspector on the job after the specified completion date.[67]

When the contract contains a liquidated damages provision, the government is entitled to damages for the period between the contract completion date and the actual date of completion, regardless of whether the contract is completed by the contractor, the surety, or a reprocurement contractor. The Default clause[68] provisions attempt to secure for the government a common law measure of recovery by putting it in as good a position as it would have had if the contractor's breach had not occurred. Liquidated damages may be recovered in addition to the excess costs of reprocurement.

E. Surety Defenses

A surety's liability under the performance bond is considered to be coextensive with that of its principal. That is, the surety's liability for the principal cannot be greater than the principal's liability.[69] Similarly, if a principal would be barred from asserting a defense to a suit, the surety will be barred from asserting that defense as well.[70]

Additionally, the surety may have independent defenses, arising from the language of the performance bond itself (also known as technical defenses) or from the circumstances that give rise to the claim. For example, a surety generally is not liable for the acts of the principal that occurred prior to the posting of the bond.[71]

Under the general common law related to a surety's performance bond liability, condition precedent clauses or limitations clauses in a performance bond are generally enforceable.[72] Therefore, the obligee also has an obligation to perform faithfully and comply with any conditions precedent in order to recover under the performance bond.[73] A failure to do so can result in the discharge of any liability the surety may have otherwise had under the bond. In determining the extent of any waiver, discharge, or conditions precedent to liability under the bond, the terms of the bond as well as the terms of the principal's construction contract must be reviewed.[74]

---

[67]*B & E Constructors, Inc.*, IBCA No. 526-11-65, 67-1 BCA ¶ 6239.

[68]FAR § 52.249-10.

[69]*United States ex. Rel. Walton Tech., Inc. v. Weststar Eng'g, Inc.*, 290 F.3d 1199, 1209 (9th Cir. 2002); *see also* Cal. Civ. Code § 2809; *National Fire Ins. Co. v. Fortune Constr. Co.*, 320 F.3d 1260 (11th Cir. 2003).

[70]*Indemnity Ins. Co. v. United States*, 74 F.2d 22 (5th Cir. 1934). *See also Rhode Island Hosp. Trust Nat'l Bank v. Ohio Cas. Ins. Co.*, 789 F.2d 74 (1st Cir. 1986).

[71]*Morton Regent Enters., Inc. v. Leadtec California, Inc.*, 141 Cal. Rptr. 706, 708 (Cal. Ct. App. 1977).

[72]*Decca Design Build, Inc. v. American Automobile Ins. Co.*, 77 P.3d 1251, 1252 (Ariz. Ct. App. 2003).

[73]*U.S. Fidelity and Guar. Co. v. Braspetro Oil Services Co.*, 369 F.3d 34, 51 (2nd Cir. 2004).

[74]*Pacific Employers Ins. Co. v. City of Berkeley*, 204 Cal. Rptr. 387 (Cal. Ct. App. 1984).

However, a review of the Miller Act Performance Bond, Standard Form 25, reveals that the bond form does not express any limitations or conditions on the surety's liability as might be found in other industry standard bond forms. Consequently, the surety's defenses and liability limitations are controlled by FAR § 52.249-10 Default (Fixed-Price Construction) clause, as discussed in **Chapter 11.** Basically, the surety's liability is limited to the penal sum of the bond unless it undertakes to complete the project.

A *material alteration* in the principal's performance obligation resulting in an *increase in risk* for the surety may also provide the surety with an independent defense.[75] Where the surety does not consent to a material change to the contract that it has financially guaranteed, the surety may be discharged, either in whole or to the extent of injury caused by a material alteration. This discharge is based on the theory that the bond binds the surety only to certain risks, and consent of the surety is necessary in order to expand its liability beyond the terms of the bond.[76] In recognition of this doctrine, most bond forms permit the owner and the principal to alter the terms of the underlying construction contract by change orders and provide that the surety consents to such modifications in advance.[77] To assert an independent defense, the surety must show that some harm was caused by the material alteration.

If the principal engages in fraud to induce the surety to issue a bond, the surety cannot assert that defense to a claim by the government obligee.[78] However, if the obligee has perpetrated a fraud on the surety or even participated in it, the surety will be discharged from its obligation.[79]

➤ LESSONS LEARNED AND ISSUES TO CONSIDER

- There are *three main types* of bonds in government construction contracts: bid bonds, payment bonds, and performance bonds. Payment and performance bonds are governed by the federal Miller Act. Bid bonds are addressed in **Chapter 3.**
- *Miller Act payment bonds* are required for government construction contracts over $100,000, and they provide protection because contractors may not file liens on government property. Their payment protection extends only to firms in

(Continued)

---

[75]*See, e.g., Ramada Dev. Co. v. United States Fidelity & Guar. Co.,* 626 F.2d 517 (6th Cir. 1980); *Continental Bank & Trust Co. v. American Bonding Co.,* 605 F.2d 1049 (8th Cir. 1979), *aff'd in part, rev'd in part on other grounds,* 630 F.2d 606 (8th Cir. 1980).

[76]*See United States ex rel. Army Athletic Ass'n v. Reliance Ins. Co.,* 799 F.2d 1382 (9th Cir. 1986); *Maryland Cas. Co. v. City of South Norfolk,* 54 F.2d 1032 (4th Cir.), *modified,* 56 F.2d 822, *cert. denied,* 286 U.S. 562 (1932).

[77]*See, e.g., Trinity Universal Ins. Co. v. Gould,* 258 F.2d 883 (10th Cir. 1958); *Massachusetts Bonding & Ins. Co. v. John R. Thompson Co.,* 88 F.2d 825 (8th Cir.), *cert. denied,* 301 U.S. 707 (1937).

[78]*Kvaerner Constr. Inc. v. American Safety Cas. Ins. Co.,* 847 So. 2d 534, 539 (Fla. 5th DCA 2003).

[79]*Filippi v. McMartin,* 188 Cal. App. 2d 135 (1961); *St. Paul Fire & Marine Ins. Co. v. Commodity Credit Corp.,* 646 F.2d 1064 (5th Cir. 1981).

a direct contractual relationship with the prime contractor (*first-tier claimants*) and second-tier subcontractors and suppliers (*second-tier claimants*).

- *Suppliers to suppliers* and parties below the *sub-subcontractor* level are generally not covered by a Miller Act payment bond.
- Whether an entity is considered a subcontractor or a supplier for Miller Act purposes depends on number of factors including the *relation of the entity's work* with the specific nature of the project.
- *Second-tier claimants* must satisfy certain notice requirements in order to assert a valid claim on the payment bond. However, neither a surety nor a contractor can add *additional notice requirements* to those set forth in the Miller Act.
- Actions seeking payment under the Miller Act must be brought within *one year* of the claimant's last work on the project.
- Although waiver or limitations on a Miller Act payment bond claimant's rights are possible, the waiver must comply with the *conditions* set forth in the Miller Act.
- *Performance bonds* are required on government construction contracts over $100,000. They protect the government from the prime contractor's failure to perform.
- After a *valid termination* of the prime contractor, the surety must analyze the default, its liability under the performance bond, and the options that it may take.
- As to payment and performance bonds, the surety has all of the *defenses* available to its contractor/principal in addition to some surety-specific defenses.
- If the default termination is valid, then the performance bond surety is liable for *completion* of the contract and for *liquidated or actual damages* resulting from the delayed completion of the work up to the *penal sum* of the bond.
- If the surety *elects to complete* the contract, its financial obligation for the cost of completion may exceed the penal sum of the bond. For that reason, a completing surety typically engages a completion contractor that provides the surety with separate payment and performance bonds.

APPENDIX 12A

STANDARD FORM 25A—PAYMENT BOND

53.301-25-A

53.301-25-A Payment Bond

48 CFR Ch. 1 (10–1–03 Edition)

PERFORMANCE BOND *(See instructions on reverse)*	DATE BOND EXECUTED *(Must be same or later than date of contract)*	OMB No.: 9000-0045

Public reporting burden for this collection of information is estimated to average 25 minutes per response, including the time for reviewing instructions, searching existing data sources, gathering and maintaining the data needed, and completing and reviewing the collection of information. Send comments regarding this burden estimate or any other aspect of this collection of information, including suggestions for reducing this burden, to the FAR Secretariat (MVR), Federal Acquisition Policy Division, GSA, Washington, DC 20405.

PRINCIPAL *(Legal name and business address)*	TYPE OF ORGANIZATION *("X" one)*
	☐ INDIVIDUAL ☐ PARTNERSHIP
	☐ JOINT VENTURE ☐ CORPORATION
	STATE OF INCORPORATION

SURETY(IES) *(Name(s) and business address(es)*	PENAL SUM OF BOND

	MILLION(S)	THOUSAND(S)	HUNDRED(S)	CENTS

CONTRACT DATE	CONTRACT NO.

OBLIGATION:

We, the Principal and Surety(ies), are firmly bound to the United States of America (hereinafter called the Government) in the above penal sum. For payment of the penal sum, we bind ourselves, our heirs, executors, administrators, and successors, jointly and severally. However, where the Sureties are corporations acting as co-sureties, we, the Sureties, bind ourselves in such sum "jointly and severally" as well as "severally" only for the purpose of allowing a joint action or actions against any or all of us. For all other purposes, each Surety binds itself, jointly and severally with the Principal, for the payment of the sum shown opposite the name of the Surety. If no limit of liability is indicated, the limit of liability is the full amount of the penal sum.

CONDITIONS:

The Principal has entered into the contract identified above.

THEREFORE:

The above obligation is void if the Principal -

(a)(1) Performs and fulfills all the undertakings, covenants, terms, conditions, and agreements of the contract during the original term of the contract and any extensions thereof that are granted by the Government, with or without notice to the Surety(ies), and during the life of any guaranty required under the contract, and (2) performs and fulfills all the undertakings, covenants, terms conditions, and agreements of any and all duly authorized modifications of the contract that hereafter are made. Notice of those modifications to the Surety(ies) are waived.

(b) Pays to the Government the full amount of the taxes imposed by the Government, if the said contract is subject to the Miller Act, (40 U.S.C. 270a-270e), which are collected, deducted, or withheld from wages paid by the Principal in carrying out the construction contract with respect to which this bond is furnished.

WITNESS:

The Principal and Surety(ies) executed this performance bond and affixed their seals on the above date.

		PRINCIPAL		
SIGNATURE(S)	1.	2.	3.	Corporate Seal
	(Seal)	(Seal)	(Seal)	
NAME(S) & TITLE(S) *(Typed)*	1.	2.	3.	

		INDIVIDUAL SURETY(IES)	
SIGNATURE(S)	1.	2. (Seal)	(Seal)
NAME(S) *(Typed)*	1.	2.	

			CORPORATE SURETY(IES)		
SURETY A	NAME & ADDRESS		STATE OF INC.	LIABILITY LIMIT $	Corporate Seal
	SIGNATURE(S)	1.	2.		
	NAME(S) & TITLE(S) *(Typed)*	1.	2.		

AUTHORIZED FOR LOCAL REPRODUCTION
Previous edition not usable

STANDARD FORM 25 (REV. 5-96)
Prescribed by GSA-FAR (48 CFR) 53.228(b)

CORPORATE SURETY(IES) *(Continued)*

			STATE OF INC.	LIABILITY LIMIT		
SURETY B	NAME & ADDRESS			$		Corporate Seal
	SIGNATURE(S)	1.		2.		
	NAME(S) & TITLE(S) *(Typed)*	1.		2.		
SURETY C	NAME & ADDRESS		STATE OF INC.	LIABILITY LIMIT $		Corporate Seal
	SIGNATURE(S)	1.		2.		
	NAME(S) & TITLE(S) *(Typed)*	1.		2.		
SURETY D	NAME & ADDRESS		STATE OF INC.	LIABILITY LIMIT $		Corporate Seal
	SIGNATURE(S)	1.		2.		
	NAME(S) & TITLE(S) *(Typed)*	1.		2.		
SURETY E	NAME & ADDRESS		STATE OF INC.	LIABILITY LIMIT $		Corporate Seal
	SIGNATURE(S)	1.		2.		
	NAME(S) & TITLE(S) *(Typed)*	1.		2.		
SURETY F	NAME & ADDRESS		STATE OF INC.	LIABILITY LIMIT $		Corporate Seal
	SIGNATURE(S)	1.		2.		
	NAME(S) & TITLE(S) *(Typed)*	1.		2.		
SURETY G	NAME & ADDRESS		STATE OF INC.	LIABILITY LIMIT $		Corporate Seal
	SIGNATURE(S)	1.		2.		
	NAME(S) & TITLE(S) *(Typed)*	1.		2.		

BOND PREMIUM ▶	RATE PER THOUSAND ($)	TOTAL ($)

INSTRUCTIONS

1. This form is authorized for use in connection with Government contracts. Any deviation from this form will require the written approval of the Administrator of General Services.

2. Insert the full legal name and business address of the Principal in the space designated "Principal" on the face of the form. An authorized person shall sign the bond. Any person signing in a representative capacity (e.g., an attorney-in-fact) must furnish evidence of authority if that representative is not a member of the firm, partnership, or joint venture, or an officer of the corporation involved.

3. (a) Corporations executing the bond as sureties must appear on the Department of the Treasury's list of approved sureties and must act within the limitation listed therein. Where more than one corporate surety is involved, their names and addresses shall appear in the spaces (Surety A, Surety B, etc.) headed "CORPORATE SURETY(IES)." In the space designated

"SURETY(IES)" on the face of the form, insert only the letter identification of the sureties.

(b) Where individual sureties are involved, a completed Affidavit of Individual Surety (Standard Form 28) for each individual surety, shall accompany the bond. The Government may require the surety to furnish additional substantiating information concerning their financial capability.

4. Corporations executing the bond shall affix their corporate seals. Individuals shall execute the bond opposite the word "Corporate Seal", and shall affix an adhesive seal if executed in Maine, New Hampshire, or any other jurisdiction requiring adhesive seals.

5. Type the name and title of each person signing this bond in the space provided.

STANDARD FORM 25—PERFORMANCE BOND

53.301-2548   CFR Ch. 1 (10–1–03 Edition)

53.301-25 Performance Bond4.

PAYMENT BOND *(See instructions on reverse)*	DATE BOND EXECUTED *(Must be same or later than date of contract)*	OMB No.: 9000-0045

Public reporting burden for this collection of information is estimate to average 25 minutes per response, including the time for reviewing instructions, searching existing data sources, gathering and maintaining the data needed, and completing and reviewing the collection of information. Send comments regarding this burden estimate or any other aspect of this collection of information, including suggestions for reducing this burden, to the FAR Secretariat (MVR), Federal Acquisition Policy Division, GSA, Washington, DC 20405

PRINCIPAL *(Legal name and business address)*

TYPE OF ORGANIZATION *("X" one)*

☐ INDIVIDUAL   ☐ PARTNERSHIP

☐ JOINT VENTURE   ☐ CORPORATION

STATE OF INCORPORATION

SURETY(IES) *(Name(s) and business address(es)*

PENAL SUM OF BOND

MILLION(S)	THOUSAND(S)	HUNDRED(S)	CENTS

CONTRACT DATE	CONTRACT NO.

OBLIGATION:

We, the Principal and Surety(ies), are firmly bound to the United States of America (hereinafter called the Government) in the above penal sum. For payment of the penal sum, we bind ourselves, our heirs, executors, administrators, and successors, jointly and severally. However, where the Sureties are corporations acting as co-sureties, we, the Sureties, bind ourselves in such sum "jointly and severally" as well as "severally" only for the purpose of allowing a joint action or actions against any or all of us. For all other purposes, each Surety binds itself, jointly and severally with the Principal, for the payment of the sum shown opposite the name of the Surety. If no limit of liability is indicated, the limit of liability is the full amount of the penal sum.

CONDITIONS:

The above obligation is void if the Principal promptly makes payment to all persons having a direct relationship with the Principal or a subcontractor of the Principal for furnishing labor, material or both in the prosecution of the work provided for in the contract identified above, and any authorized modifications of the contract that subsequently are made. Notice of those modifications to the Surety(ies) are waived.

WITNESS:

The Principal and Surety(ies) executed this payment bond and affixed their seals on the above date.

PRINCIPAL					
SIGNATURE(S)	1.	2.	3.		Corporate Seal
	(Seal)	(Seal)	(Seal)		
NAME(S) & TITLE(S) *(Typed)*	1.	2.	3.		
INDIVIDUAL SURETY(IES)					
SIGNATURE(S)	1.	2.			(Seal)
		(Seal)			
NAME(S) *(Typed)*	1.	2.			

CORPORATE SURETY(IES)				
SURETY A	NAME & ADDRESS		STATE OF INC.	LIABILITY LIMIT $	Corporate Seal
	SIGNATURE(S)	1.	2.	
	NAME(S) & TITLE(S) *(Typed)*	1.	2.	

AUTHORIZED FOR LOCAL REPRODUCTION
Previous edition is usable

STANDARD FORM 25A (REV. 10-98)
Prescribed by GSA-FAR (48 CFR) 53.2228(c)

CORPORATE SURETY(IES) *(Continued)*

			STATE OF INC.	LIABILITY LIMIT	
SURETY B	NAME & ADDRESS		STATE OF INC.	LIABILITY LIMIT $	Corporate Seal
	SIGNATURE(S)	1.	2.		
	NAME(S) & TITLE(S) *(Typed)*	1.	2.		
SURETY C	NAME & ADDRESS		STATE OF INC.	LIABILITY LIMIT $	Corporate Seal
	SIGNATURE(S)	1.	2.		
	NAME(S) & TITLE(S) *(Typed)*	1.	2.		
SURETY D	NAME & ADDRESS		STATE OF INC.	LIABILITY LIMIT $	Corporate Seal
	SIGNATURE(S)	1.	2.		
	NAME(S) & TITLE(S) *(Typed)*	1.	2.		
SURETY E	NAME & ADDRESS		STATE OF INC.	LIABILITY LIMIT $	Corporate Seal
	SIGNATURE(S)	1.	2.		
	NAME(S) & TITLE(S) *(Typed)*	1.	2.		
SURETY F	NAME & ADDRESS		STATE OF INC.	LIABILITY LIMIT $	Corporate Seal
	SIGNATURE(S)	1.	2.		
	NAME(S) & TITLE(S) *(Typed)*	1.	2.		
SURETY G	NAME & ADDRESS		STATE OF INC.	LIABILITY LIMIT $	Corporate Seal
	SIGNATURE(S)	1.	2.		
	NAME(S) & TITLE(S) *(Typed)*	1.	2.		

INSTRUCTIONS

1. This form, for the protection of persons supplying labor and material, is used when a payment bond is required under the Act of August 24, 1935, 49 Stat. 793 (40 U.S.C. 270a-270e). Any deviation from this form will require the written approval of the Administrator of General Services.

2. Insert the full legal name and business address of the Principal in the space designated "Principal" on the face of the form. An authorized person shall sign the bond. Any person signing in a representative capacity (e.g., an attorney-in-fact) must furnish evidence of authority if that representative is not a member of the firm, partnership, or joint venture, or an officer of the corporation involved.

3. (a) Corporations executing the bond as sureties must appear on the Department of the Treasury's list of approved sureties and must act within the limitation listed therein. Where more than one corporate surety is involved, their names and addresses shall appear in the spaces (Surety A, Surety B, etc.) headed "CORPORATE SURETY(IES)." In the space designated

"SURETY(IES)" on the face of the form, insert only the letter identification of the sureties.

(b) Where individual sureties are involved, a completed Affidavit of Individual Surety (Standard Form 28) for each individual surety, shall accompany the bond. The Government may require the surety to furnish additional substantiating information concerning their financial capability.

4. Corporations executing the bond shall affix their corporate seals. Individuals shall execute the bond opposite the word "Corporate Seal", and shall affix an adhesive seal if executed in Maine, New Hampshire, or any other jurisdiction requiring adhesive seals.

5. Type the name and title of each person signing this bond in the space provided.

STANDARD FORM 25A (REV. 10-98) **BACK**

13

EQUITABLE ADJUSTMENTS AND COSTS

I. OVERVIEW

The concept of equitable adjustments or price adjustments and costs are clearly addressed in the standard Federal Acquisition Regulation (FAR) clauses used in government construction contracts as illustrated by Table 13.1.

The principal remedy-granting clauses in government contracts routinely employ the concepts of "equitable adjustments" or "adjustments" in the contract price rather than the term "damages." In addition, the contract price adjustment concept is almost always related to an increase or decrease in *costs* due or caused by a change, delay, differing site condition, or quantity variation.

In seeking an "equitable adjustment" for a change, differing site condition, or delay, the contractor must establish three essential elements:[1] (1) The contractor must demonstrate that the work or site condition was different from what was reasonably anticipated; (2) the contractor must prove that the change, differing site condition, or delay adversely impacted its contract work; (3) the contractor must show a resulting injury (i.e., amount of additional cost). An otherwise viable claim may have diminished or no value if the contractor is unable to reasonably demonstrate the cost impact of the change, differing site condition, or delay.[2] Consequently, even on relatively noncomplex fixed-price construction contracts, it is essential for contractors or subcontractors to develop and implement an effective project cost documentation system from the inception of a project. If properly implemented, such a system will allow the contractor to quantify the financial impact of the change in the work and provide the appropriate documentation in support thereof.

---

[1]*See Wunderlich Contracting Co. v. United States,* 351 F.2d 956, 968 (Ct. Cl. 1965); *Servidone Constr. Corp. v. United States,* 931 F.2d 860, 861 (Fed. Cir. 1991) (to recover on an equitable adjustment, the contractor must show liability, causation, and resultant injury).
[2]*See, e.g., Silver Enters. v. Department of Transportation,* CBCA No. 63-C, 07-1 BCA ¶ 33,496 (board allowed the claimant a limited recovery on a jury verdict basis and denied recovery of legal fees under the Equal Access to Justice Act (EAJA) because the contractor had not maintained adequate records of the additional costs caused by the government's action).

Table 13.1   Federal Acquisition Regulation Price Adjustment Concepts

FAR Clause	Price Adjustment Concept
FAR § 52.243-4(d) Changes	Equitable adjustment if change causes increase or decrease in *costs*
FAR § 52.236-2(b) Differing Site Conditions	Equitable adjustment if conditions cause increase or decrease in *costs*
FAR § 52.242-14(b) Suspension of Work	Adjustment shall be made for any increase in *cost* (excluding profit) necessarily caused by the suspension
FAR § 52.211-18 Variation in Estimated Quantity	Equitable adjustment based on increase or decrease of *costs* due solely to variation in quantity

Contractors must also appreciate that there is often a significant difference between the government's policy on pricing adjustments and the reality facing both the government and the contractor during the performance of the work. For example, FAR § 43.103(b) sets forth the *policy* on pricing contract modifications:

Contract modifications, including changes that could be issued unilaterally, shall be priced before their execution if this can be done without adversely affecting the interest of the Government. If a significant cost increase could result from a contract modification and time does not permit negotiation of a price, at least a maximum price shall be negotiated unless impractical.

This policy reflects a philosophy that definitive pricing should be accomplished before the work is performed. In that context, the change order would, in many cases, be priced on the basis of estimates. Similarly, the Changes clause[3] states that the contractor "must assert its right to an adjustment" within 30 days of receipt of a written change order.

Literal compliance with these policies can be very difficult. Project constraints may dictate that changes to the work can be priced and negotiated only after the modified work is performed. If a contactor receives multiple change orders within a relatively short period of time, it may not be practical to price all of these proposals within a 30-day period or before the work must proceed. If the potential modification relates to a constructive change or disputed differing site condition, it is likely that the contractor will be directed to proceed and submit a claim in accordance with the disputes procedure. As a result of these circumstances, many contract modifications and claims are priced and resolved after the work has been completed or the delay has been experienced. Consequently, it is essential that a contractor appreciate and consider all of the impacts of a given change or delay and develop a practical approach to managing and documenting its requested price adjustment or adjustments. Similarly, the contractor needs to understand the effect of any change order release that it executes on its rights to obtain subsequent equitable adjustments on

---

[3]FAR § 52.243-4(e).

behalf of itself or its subcontractors.[4] If there is any doubt regarding the effect of the proposed release, the contractor should insert an express reservation of rights.

II. EQUITABLE ADJUSTMENT THEORY

A. Basic Concept

The *equitable adjustment* concept has a relatively long history in government contracts. In a 1942 decision involving a contract for the construction of a levee on the Mississippi River, the United States Supreme Court provided this description of an "equitable adjustment":

> An "equitable adjustment" of the respondent's additional payment for extra work involved *merely the ascertainment of the cost* of digging, moving, and placing earth, and the *addition to that cost of a reasonable and customary allowance for profit.*[5] [Emphasis added]

This basic standard for valuing an equitable adjustment (cost plus reasonable allowance for profit) continues to be applied in the twenty-first century.[6]

During the 1950s and 1960s, there were two competing theories or approaches to determining the cost component of an equitable adjustment. One approach advanced the so-called *objective* test, which based the equitable adjustment on the *reasonable cost* of the changed work.[7] Under this approach, the contractor's actual cost experience was not conclusive. This approach stressed the *value* of the change from the perspective of a prudent and reasonable contractor.[8] The alternate theory stressed a *subjective* approach, which placed primary reliance on the contractor's *actual cost* for performance of the change. Since many equitable adjustments are determined after the fact, the availability of actual costs tended to support the use of the *subjective* approach.

In its 1963 decision in *Bruce Construction Corp., et al. v. United States,*[9] the Court of Claims sought to articulate a rule that would resolve the debate between the two schools of thought. The modification at issue in *Bruce* involved a change in the specifications from a concrete block to a sand block. Although the contractor was able to effect the change in block types at no additional cost, it asserted that it was entitled to a substantial equitable adjustment because the current and fair market value for the sand block was greater than for a concrete block.

---

[4]*See Bell BCI Co. v. United States,* 570 F.3d 1337 (Fed. Cir. 2009). *Compare Service Eng'g Co.,* ASBCA No. 40274, 93-2 BCA ¶ 25,885, *and JT Constr. Co., Inc.,* ASBCA No. 54352, 06-1 BCA ¶ 33,182 (releases did not waive additional claims) *with Kato Corp.,* ASBCA No. 51462, 06-2 BCA ¶ 33,293; *Southwest Marine, Inc.,* ASBCA No. 34058 *et al.,* 91-1 BCA ¶ 23,323 *and Atlantic Dry Dock Corp. v. United States,* 773 F. Supp. 335 (M.D. Fla. 1991) (impact claims waived in a release).

[5]*United States v. Callahan Walker Constr. Co.,* 317 U.S. 56, 61 (1942).

[6]*Hi-Shear Tech. Corp. v. United States,* 356 F.3d 1372 (Fed. Cir. 2004).

[7]*United States v. Rice,* 317 U.S. 61, 64 (1942).

[8]*Montgomery Ross Fisher, Inc.,* ASBCA No. 9983, 65-1 BCA ¶ 4633.

[9]324 F.2d 516 (Ct. Cl. 1963).

The Court of Claims denied the contractor's claim for compensation on a fair market value basis. In so doing, the court articulated the basic principles governing the pricing of equitable adjustments and resolved the conflict between the competing approaches for valuing changed work. The court stated:

Though the price which plaintiffs actually paid for the "sand block" was the same as they would have paid for the original block selected, they contend that the fair market value of the sand block was greater than the purchase price. Essentially then, plaintiffs argue that defendant should not benefit from the bargain price plaintiffs secured from their supplier, but should pay for the actual value of the sand block received by defendant, not merely its actual cost.

Though there is substantial controversy as to the market value of the sand block as of the time of the transaction between plaintiffs and their supplier, for purposes of defendant's motion for partial summary judgment, we are called upon only to decide the narrow question whether "cost" or "fair market value" controls in the award of an equitable adjustment.

Equitable adjustments in this context are simply corrective measures utilized to keep a contractor whole when the Government modifies a contract. Since the purpose underlying such adjustments is to safeguard the contractor against increased costs engendered by the modification, it appears patent that the measure of damages cannot be the value received by the Government, but must be more closely related to and contingent upon the altered position in which the contractor finds himself by reason of the modification.... But fair market value is not the measure of damages in this case. This is not to say that in all cases, historical cost is to be the gauge. The more proper measure would seem to be a "reasonable cost."

* * *

But the standard of reasonable cost "must be viewed in the light of a *particular* contractor's cost ... " [emphasis added], and not the universal, objective determination of what the cost would have been to other contractors at large.

To say that "reasonable cost" rather than "historical cost" should be the measure does not depart from the test applied in the past, for the two terms are often synonymous. And where there is an alleged disparity between "historical" and "reasonable" costs, the historical costs are presumed reasonable.[10]

Although the concept of *reasonable cost* still applies, the presumption of reasonableness that attached to a contractor's actual incurred cost was negated by the 1987

---

[10]324 F.2d at 517–519.

revision to the FAR Cost Principles, which stated that "no presumption of reasona-bleness shall be attached" to the contractor's incurred costs.[11] This language remains in the FAR Contract Cost Principles and Procedures. Therefore, to the extent that the contract incorporates the FAR Part 31, Cost Principles, for the purpose of pricing equitable adjustments, the presumption of reasonableness attached to a contractor's actual cost, as articulated in *Bruce Construction,* is inapplicable.[12]

B. Equitable Adjustments versus Damages for Breach of Contract

The remedy-granting clauses utilized in most government contracts typically restrict the contractor's recovery for changed work to only those extra costs incurred in the performance of the contract under which the changes were directed or the work delayed. Extra costs incurred in the performance of work unrelated to the contract usually are not recoverable on a claim *arising under* a contract.[13] However, these types of consequential damages are recoverable in breach of contract actions, if the contractor can show that the adverse effect was "reasonably foreseeable."[14]

Several basic premises underlie the theory of damages that can produce a differ-ent recovery from that provided by an equitable adjustment. For example, when a claimant seeks to recover additional costs and/or damages it has incurred as a result of another party's breach of contract, the court will attempt to put the contractor in the same position that it would have been had the contract been performed accord-ing to its terms.[15] This principle for determining damages applies to all breach of contract actions, not just those arising from construction contracts.[16] In other words, the law of contract damages is purely compensatory in nature.

There are two basic types of damages resulting from a breach of a construction contract. *General* or *direct* damages are those resulting from the direct, natural, and immediate impact of the breach and are recoverable in all cases where proven.[17] In the contractor's case, such damages include items such as idle labor and machin-ery, material and labor escalations, and extended job site and home office overhead. Some of these direct damages are computed according to standardized formulas, which are discussed later in this chapter.

The second category of breach of contract damages is termed *special* or *conse-quential damages.* Special damages do not flow directly from the alleged breach but are an indirect or consequential source of injury. Such losses, which are indirectly related to the breach, may include lost profits or lost bonding capacity. These dam-ages are more difficult to prove because the causal link between cost incurred and the breach is likely to be tenuous and uncertain.

---

[11]FAR § 31.201-3(a).
[12]*See* FAR § 31.105(c)(5); DFARS § 252.243-7001. See **Section I.C** of this chapter.
[13]*General Dynamics Corp. v. United States,* 585 F.2d 457 (Ct. Cl. 1978).
[14]*See Specialty Assembly and Packing Co. v. United States,* 355 F. 2d 554 (Ct. Cl. 1966).
[15]*Bennett v. Associated Food Stores, Inc.,* 165 S.E. 2d 581 (Ga. App. 1968).
[16]*Meares v. Nixon Constr. Co.,* 173 S.E. 2d 593 (N.C. App. 1970).
[17]*Spang Indus. v. Aetna Cas. and Sur. Co.,* 512 F.2d 365 (2d Cir. 1975).

Recovery of consequential damages by a claimant is subject to three general conditions. First, the party seeking recovery must show that this particular type of injury was reasonably foreseeable to the other party at the time of contracting. The *reasonably foreseeable* test, which was first articulated in the venerable English case *Hadley v. Baxendale*,[18] has since been widely adopted by American courts.[19]

Second, the party must prove that the damages flowed *naturally* or *proximately* from the breach. In laymen's terms, this means that the injury must be the result of the breach rather than some other cause.[20] This requirement, like the first, has its origin in *Hadley v. Baxendale.*

The third limitation on the recovery of consequential damages is that the damages sought must not be too remote or speculative.[21] This general requirement frequently is codified under the state law. For example, Georgia Code § 13-6-8 provides:

> Remote or consequential damages are not recoverable unless they can be traced solely to the breach of the contract or unless they are capable of exact computation, such as the profits which are the immediate fruit of the contract, and are independent of any collateral enterprise entered into in contemplation of the contract.

Arguments focused on the *remote and speculative* nature of the damages frequently arise where a claimant seeks the recovery of lost profits sustained due to the breach—for example, lost profit as a result of tied-up capital or reduced bonding capacity. Although statutes, such as Section 13-6-8 of the Georgia Code, require "exact computation" of consequential damage, some courts take a less stringent approach.[22]

Compared to the rather settled rules governing the measure of an *equitable adjustment,* the recovery of lost profits and other consequential damages in government contracts is relatively rare. For example, in *Padbloc Co. v. United States,*[23] the contractor recovered lost profits when the government misused the contractor's confidential data. Also, the Court of Claims has stated that a work stoppage due to a material breach by the government (failure to make progress payment) is justified and does not give the government the right to terminate the contract for default. Under the traditional theory of recovery for breach of contract, the contractor would be able to recover anticipated profits on the balance of the contract work.[24]

Recovery of lost profits that are reasonably foreseeable and caused by the breach is consistent with the concept of expectation damages.[25] However, with respect to breach of contract claims against the government, the recovery of lost profits is very

[18]156 Eng. Rep. 145 (1854).

[19]*See Bumann v. Maurer,* 203 N.W. 2d 434 (N.D. 1972).

[20]*Kline Iron & Steel Co. v. Superior Trucking Co.,* 201 S.E. 2d 388 (S.C. 1973).

[21]*Baker v. Riverside Church of God,* 453 S.W. 2d 801 (Tenn. 1970).

[22]*Bitler v. Terri Lee, Inc.,* 81 N.W. 2d 318 (Neb. 1957).

[23]161 Ct. Cl. 369 (1963).

[24]*See, e.g., Northern Helex Co. v. United States,* 524 F.2d 707 (Ct. Cl. 1975).

[25]*Bluebonnet Savings Bank, F.S.B. v. United States,* 266 F.3d 1348, 1355 (Fed. Cir. 2001).

rare due to the terms and conditions in most government contracts. For example, if a contractor demonstrates that a default termination by the government is wrongful, that termination is converted to a termination for the convenience of the government.[26] Under the Termination for Convenience clause, the contractor's damages is limited to: (1) its reasonable cost incurred through the date of termination; and (2) a reasonable profit on that work. The contractor *is not* permitted to recover lost anticipated profits on the unperformed work.[27]

In *breach of contract* actions not based on a wrongful termination, the Federal Circuit and its predecessor courts have allowed profit only in limited circumstances. Summarized next are examples illustrating the limited circumstances in which the contractor was allowed to recover profit in conjunction with breach of contract damages:

- Damages due to delays related to a government breach of contract: no profit allowed.[28]
- Damages awarded on a *jury verdict* basis[29] due to the government's breach when it refused to permit a substitution of a subcontractor under a Subcontractor Listing clause: no profit allowed.[30]
- Damages awarded for a claim that could have been addressed under a contract clause that allows for profit on the increased cost of performance: profit allowed on increased cost.[31]
- Damages related to a faulty estimate of quantities in a requirements type contract: no lost anticipated profit allowed.[32]
- Damages related to diversion of orders under a requirements contract to other sources: profit allowed on diverted work or quantities.[33]

In summary, although not completely rejecting the concept that a contractor might recover lost profits, the Federal Circuit has limited the basis for any recovery of lost profits and has ruled that the claimant must also establish that, without the government's breach, it would have earned a profit under the contract.[34]

C. Cost Recovery Criteria—The Cost Principles

The FAR establishes three basic criteria for recovery of costs under a government contract. The costs must be reasonable, allocable to the contract or pricing action, and otherwise allowable under the FAR Cost Principles.[35]

---

[26]FAR § 52.249-10(c).

[27]*See* FAR § 49.202(c) and **Chapter 11.**

[28]*J.D. Hedin Constr. Corp. v. United States,* 347 F.2d 235 (Ct. Cl. 1965).

[29]*See* **Section III.D** *of this chapter.*

[30]*Meva Corp. v. United States,* 511 F.2d 548 (Ct. Cl. 1975).

[31]*Bennett v. United States,* 371 F.2d 859 (Ct. Cl. 1967).

[32]*Rumsfeld v. Applied Companies, Inc.,* 325 F.3d 1328 (Fed. Cir. 2003).

[33]*Ace-Federal Reporters, Inc. v. Barram,* 226 F.3d 1329 (Fed. Cir. 2000).

[34]*Rumsfeld v. Applied Companies, Inc.,* 325 F.3d 1328, 1340 (Fed. Cir. 2003).

[35]FAR § 31.201-2.

1. What Are "Reasonable" Costs?

Although the Court of Claims in *Bruce Construction Corp.*[36] established that the basic measure of an equitable adjustment is the "reasonable cost" related to the change, differing site condition, or delay, the court did not define the term. Instead, it established a presumption that the contractor's actual cost was the appropriate measure of reasonable cost. That presumption was negated in a 1987 revision to the FAR.

FAR § 31.201-3 provides guidance in determining whether a particular cost is *reasonable*. That section provides:

(a) A cost is reasonable if, in its nature and amount, it does not exceed that which would be incurred by a prudent person in the conduct of competitive business. Reasonableness of specific costs must be examined with particular care in connection with firms or their separate divisions that may not be subject to effective competitive restraints. No presumption of reasonableness shall be attached to the incurrence of costs by a contractor. *If an initial review of the facts results in a challenge of a specific cost by the contractor officer* or the contracting officer's representative, the burden of proof shall be upon the contractor to establish that such cost is reasonable.

(b) What is reasonable depends upon a variety of considerations and circumstances, including—

(1) Whether it is the type of cost generally recognized as ordinary and necessary for the conduct of the contractor's business or the contract performance;

(2) Generally accepted sound business practices, arm's-length bargaining, and Federal and State laws and regulations;

(3) The contractor's responsibilities to the Government, other customers, the owners of the business, employees, and the public at large; and—

(4) Any significant deviations from the contractor's established practices. [Emphasis added]

The basic standard for *reasonableness* is that of a prudent person operating in a competitive business environment. Although actual costs are no longer presumed reasonable, the contractor's burden of demonstrating reasonableness is triggered after the government's fact-based challenge to the actual costs (when an initial review of the facts results in a challenge by the contracting officer or contracting officer's representative). This suggests that the government's challenge to the cost reasonableness must have some factual basis.

2. How Are Costs Allocated?

A cost is *allocable* to a government contract when it can be assigned or charged to a contract or cost objective (e.g., a change order) on the basis that it benefits that contract or cost objective, directly or indirectly.[37]

---

[36]324 F.2d 516 (Ct. Cl. 1963).
[37]FAR § 31.201-4.

3. Which Costs Are Allowable?

The FAR Cost Principles contain detailed guidance on the allowability of approximately 46 categories of selected costs under government contracts.[38] The cost categories that are addressed are varied and include, but are not limited to:

- Advertising and public relations (allowable in part)[39]
- Bad debts (unallowable)[40]
- Bonding costs (allowable)[41]
- Compensation for personal services (generally allowable unless it is a distribution of profits)[42]
- Depreciation (generally allowable)[43]
- Entertainment costs (unallowable)[44]
- Insurance (generally allowable with special provisions on captive insurers)[45]
- Interest (generally unallowable)[46]
- Lobbying and political activities (generally unallowable)[47]
- Material costs (generally allowable)[48]
- Consultant costs (limited allowability)[49]
- Rental costs (generally allowable)[50]
- Federal income and excess profit taxes (unallowable)[51]
- Training costs (generally allowable)[52]
- Travel costs (generally allowable)[53]
- Claim prosecution costs against the federal government (unallowable)[54]
- Alcoholic beverage costs (unallowable)[55]

---

[38]*See* FAR §§ 31.205-1-31.205-52. The number of categories of cost addressed does not equal the number of sections in FAR § 31.205 because certain numbered subsections are "reserved."
[39]FAR § 31.205-1.
[40]FAR § 31.205-3.
[41]FAR § 31.205-4.
[42]FAR § 31.205-6.
[43]FAR § 31.205-11.
[44]FAR § 31.205-14.
[45]FAR § 31.205-19.
[46]FAR § 31.205-20; see **Section II.D.5** of this chapter for a further discussion of the recovery of interest as an expense.
[47]FAR § 31.205-22.
[48]FAR § 31.205-26.
[49]FAR § 31.205-33.
[50]FAR § 31.205-36.
[51]FAR § 31.205-41.
[52]FAR § 31.205-44.
[53]FAR § 31.205-46
[54]FAR § 31.205-47.
[55]FAR § 31.205-51.

With few exceptions (e.g., alcoholic beverage costs), the treatment of these specific costs in the FAR is quite detailed. Even if a cost such as personal services expenses (executive compensation and salaries) or insurance is generally allowable, the treatment of that cost category in the FAR can be very extensive. A cost that appears to be allowable may have specific requirements or conditions that must be satisfied. Therefore, it is important to review the version of the Cost Principles applicable to the particular contract carefully to determine if a cost is allowable and any conditions affecting allowability.

Historically, government construction contracts were awarded on a fixed-price basis and the price was based upon adequate price competition, whether the solicitation was a sealed bid or a negotiated procurement. In that context, the cost elements of the initial contract price were not subject to evaluation under the FAR Cost Principles.[56] Rather the Cost Principles applied to the pricing of equitable adjustments or claims. However, with the use of project delivery systems such as Early Contractor Involvement (ECI) as discussed in **Chapter 4,** contractors need to appreciate that the resulting firm fixed-price contract may be subject to a detailed audit and scrutiny under FAR Part 31. In that context, costs such as executive compensation and insurance may face more intense review and challenge by the government's auditors.

Many of these conditions on allowability provide a basis for a government challenge to the amount or composition of a particular cost. In addition, *reasonableness* is a general criterion applicable to all costs. However, certain of the Cost Principles can assist a contractor in establishing entitlement to recovery of a cost. For example, many general contractors bond major subcontractors and add the incremental cost of the subcontractors' bonds to the cost of an equitable adjustment. Some government agencies seek to avoid this added bond expense on the grounds that neither the contract nor the government agency specifically required the general contractor to bond its subcontractors. The cost principle applicable to bonding costs[57] directly addresses the issue:

> (c) Costs of bonding required by the contractor in the general conduct of its business are allowable to the extent that such bonding is in accordance with sound business practice and the rates and premiums are reasonable under the circumstances.

This provision, FAR § 31.205-4(c), can be cited by the contractor in responding to an objection to a bonding expense related to subcontracting. As illustrated by this example, basic knowledge of the Cost Principles is essential for any government construction contractor.

In addition to identifying *unallowable* costs, FAR Part 31 also addresses the accounting for unallowable costs. FAR § 31.201-6 requires a contractor to identify and exclude such costs from "any billing, claim, or proposal applicable to a Government contract" and provides guidance on the accounting for such costs.

---

[56]*See* **Section IV** of this chapter.
[57]FAR § 31.205-4.

4. Special Cost Principles for Construction Contracts

In addition to the Cost Principles that are generally applicable to every commercial contractor, FAR § 31.105 contains specific principles applicable to construction and architect-engineer contracts. These principles apply to all contracts and *contract modifications* negotiated on the "basis of cost." This section requires the contracting officer to incorporate the principles and procedures in FAR Subpart 31.2 in construction contracts for the purpose of pricing changes and other contract modifications. While stating that the principles in FAR Subpart 31.2 have general application, FAR § 31.105(d) contains specific guidance on the treatment of costs associated with construction contracts on these topics:

- Advance agreements[58] on items such as home office overhead, partners' compensation, consultants' costs, and equipment usage are specifically encouraged. Such agreements are suggested for consideration, as a means to simplify negotiation of modifications and avoid disputes.
- Determination of equipment ownership and equipment costs based on actual costs when available, or a schedule of predetermined costs.[59]
- Equipment rental costs.[60]
- Job site expenses, such as superintendents, clerical staff, engineering, material handling, and cleanup, as direct or indirect project expenses.[61]

As discussed, *costs* are the key component of equitable adjustments. Every contractor performing a government construction contract needs to appreciate that the rules and procedures governing the concept of cost are detailed and complicated. Having the resources available to understand these principles and procedures at the inception of a project to document those costs can assist in avoiding expensive and time-consuming disputes during performance of the work.

D. Specific Elements of Recovery

1. Direct Costs—Additive/Deductive Changes

The first step in making an equitable adjustment is the determination of the increase or decrease in costs incurred by the contractor as a result of a change, constructive change, differing site condition, or delay. Ordinarily this involves determining the cost of performing changed work and deducting from that amount the reasonable cost of

---

[58]*See* FAR § 31.109. Advance agreements can be used to address issues of allowability, allocability, and reasonableness. However, an advance agreement cannot provide for a treatment of a cost inconsistent with FAR Part 31. For example, an advance agreement making interest an allowable cost, notwithstanding FAR § 31.205-20, is not authorized.

[59]FAR §§ 31.105(d)(2), 31.205-11(f). *See Union Boiler Works, Inc. v. Caldera,* 156 F. 3d 1374 (Fed. Cir. 1998) (no "rental cost" allowed on property fully depreciated by the contractor).

[60]FAR § 31.105(d)(2)(ii).

[61]FAR § 31.105(d)(3).

the work without the change.[62] Generally, both *direct* and *indirect* costs are recoverable. However, as discussed, the "cost" concept in a government contract must consider the principles and procedures set forth in the FAR Cost Principles.

Direct costs are best established by means of a detailed, original entry record system that clearly segregates or identifies actual costs pertaining to the changed work, differing site condition, or delay.[63] Secondary records such as ledgers and computer printouts are sufficient if supported by the underlying basic records, which can be made available for inspection and audit.

Construction firms generally have daily job reports containing data such as the number of personnel (including subcontractor personnel) on the project, the type of work performed and the location of work, and unusual events or difficulties encountered in performing the work; however, these customary records may require adaptation in order to effectively document extra costs. For example, if changed work is being performed simultaneously with other contract work not affected by the change, the contractor should make a special effort to segregate or identify those labor, equipment, and material costs relating to the changed work. Even then this task may be inherently difficult as craft workers shift from one task to another and from base contract work to changed work.[64] However, if feasible, there is no better substitute than having available the underlying primary records kept on a daily basis to establish the validity and accuracy of any summary of the reasonable direct cost of a modification.

The basic purpose of the equitable adjustment (additive or deductive) is not to alter the contractor's profit or loss position.[65] Documentation of the direct cost effect of an additive change can be based on the amount paid by the contractor to its vendors or subcontractors.[66] When the work involves a deductive change, the amount for that work in the actual subcontract often is determinative of the reasonable cost of the change.[67]

If there are no records to support the reasonable or actual costs, estimates (original or revised) may be accepted as the best available evidence.[68] This proof may require resorting to estimating manuals.

In those cases, proof of the applicability of the particular estimating guide to the work in question may be necessary to establish that an estimate derived from

---

[62]*Great Lakes Dredge & Dock Co.*, ENGBCA No. 3657, 77-2 BCA ¶ 12,711. *See generally Nager Elec. Co. v. United States*, 442 F.2d 936, 951 (Ct. Cl. 1971).

[63]*Gary Constr. Co., Inc.*, ASBCA No. 19306, 77-1 BCA ¶ 12,461.

[64]*Service Eng'g Corp.*, ASBCA No. 40274, 93-2 BCA ¶ 25,885.

[65]*Pacific Architects & Engineers, Inc. v. United States*, 491 F.2d 734 (Ct. Cl. 1974).

[66]*Ensign-Bickford Co.*, ASBCA No. 6214, 60-2 BCA ¶ 2817.

[67]*Nager Elec. Co. v. United States*, 442 F.2d 936, 948 (Ct. Cl. 1971); *see also A.A. Beiro Constr. Co.*, GSBCA No. 3915, 74-2 BCA ¶ 10,860 (net cost savings to contractors); *Atlantic Elec. Co.*, GSBCA No. 6016, 83-1 BCA ¶ 16,484 (invoice for identical part used to price change).

[68]*Select Contractors, Inc.*, ENGBCA No. 3919, 82-2 BCA ¶ 15,869; *State Mech. Corp.*, VABCA No. 2797, 91-2 BCA ¶ 23,830.

the manual is reasonable.[69] Ultimately, the actual effect of the change on the performing contractor's cost will be of critical, if not controlling, weight.[70]

2. General Conditions and Overhead Costs

a. Job Site General Conditions Some categories of costs may be referred to as "overhead" costs, general conditions, or perhaps as general and administrative (G&A) expenses. In government contracts, these expenses often are placed in two categories. Costs or expenses incurred at the job site often are labeled as job site overhead or general conditions costs. The expense of the home office and regional company office is labeled as G&A or home office overhead. Each of these categories of cost is treated differently in pricing adjustments under federal government contracts.

Prior to 1968, the basic principle governing the recovery of indirect costs associated with changes was enunciated by the United States Supreme Court in *United States v. Rice*.[71] This decision interpreted the language of the then current Changes clause in the context of extensive project delays related to design changes and differing site conditions.

In *Rice*, the government delayed the commencement of a project for a number of months while it revised the project's structural design to overcome a differing site condition. The Court granted an equitable adjustment for the direct costs of performing the changes but denied compensation for the costs resulting from the delay. The Court ruled that such costs were *consequential damages* to be taken care of by a time extension.

This decision resulted in a 25-year effort by the construction industry to revise the standard Changes clause to effectively reverse this decision. In 1968, the government adopted a revised standard Changes clause for construction contracts. (*See* **Chapter 8.**) Simultaneously with the publication of the revised Changes clause, this explanation (legislative history) for the modification of the Changes clause was provided:

(i) A significant revision in the clause is the adoption of additional text designed to eliminate the application of the "Rice" Doctrine (which reflected interpretive rulings relating to the meaning of the clause previously prescribed). The elimination of the "Rice" Doctrine has been accomplished primarily by adding the phrases "any part of the work" and "whether or not changed." These phrases now appear in the Changes clause of Standard Form 32, the general provisions for standard supply contracts. An equitable adjustment clearly encompasses the effect of a change order upon any part of the work,

---

[69]*Globe Constr. Co.*, ASBCA No. 21069, 78-2 BCA ¶ 13,337.
[70]*Harrison/Franki-Denys, Joint Venture*, ENGBCA No. 5506, 93-1 BCA ¶ 25,406; *Dawson Constr. Co.* GSBCA No. 5672, 81-2 BCA ¶ 15,387 *recon. denied*, 82-2 BCA ¶ 15,914 (contractor proved that it had included savings in its bid).
[71]317 U.S. 61 (1942).

including delay expense; provided, of course, that such effect was the necessary, reasonable, and foreseeable result of the change.

(ii) Except for defective specifications, the Changes clause will continue to have no application to any delay prior to the issuance of a change order. An adjustment for such type of delay, if appropriate, will be for consideration under the provisions of the Suspension of Work clause.[72]

This appeared to put the *Rice* doctrine to rest. However, in a series of decisions, primarily by the United States Court of Appeals for the Federal Circuit and the Armed Services Board of Contract Appeals (ASBCA), this doctrine has been, to a degree, revived in substance if not in name.

In practical terms, it now may be more difficult for a government contractor to obtain full and fair compensation for extended job site general conditions and home office overhead than in 1975. The most unfortunate aspect of that trend is that there have *not* been any significant revisions to the critical contract clauses or procurement regulations on these topics in the last three decades.

Many contractors track job site general conditions costs as a percentage of direct expense even when these costs are directly charged to a particular contract. When pricing a change order proposal on a federal government contract, job site general conditions often are expressed as a percentage of direct costs (labor, materials, and equipment). Using that approach, payment for job site general conditions is simply a factor of the sum of the direct labor, materials, or equipment. At other times, it may appear to be reasonable to express conditions on a per diem or daily rate basis. For example, if the value of the change order is relatively small, recovery for field general conditions on a percentage of direct costs basis may result in a significant shortfall if a small-dollar-value modification adds significant time to the duration of the project. An example of this might involve a change that has a long lead time for delivery of a key component. The converse is also true if general conditions are expressed on a daily rate basis and a modification is large dollar wise, but no time is added to the project.

To avoid a shortfall on changes that added time, one approach to the recovery or pricing of job site general conditions might involve blending the two approaches. For example, if a change added significant time, job site overhead might be priced by applying a daily rate. Government concerns about excess recovery could be addressed by crediting any amount for job site general conditions calculated[73] in the same change on the percentage basis against the daily rate total. An example of this calculation follows.

---

[72]32 Fed. Reg. 16269 (Nov. 29, 1967).

[73]If the added time and added direct cost of the modification were resolved at the same time, there would be no need to credit out any job site general conditions expense. Rather, no percentage markup for that expense would be included in the proposal.

Assume a fixed-price contract of $18,500,000 with duration of 540 calendar days. The contractor estimated that its job site/general conditions would be $825,000, or 4.46 percent, of the direct costs. Excluding the one-time expenses, the rate for job site general conditions is $1,400 per calendar day, or 4.09 percent. During performance of the work, the design undergoes multiple revisions. Many involve equipment changes with varying lead times for delivery. One design change is the result of a differing site condition.

Each of these changes delayed specific work activities until the change was finalized. However, at no time did the actual billings for work in place for a given month drop below 50 percent of the scheduled monthly value for in-place work. Compounding the effect of the changes on field work, delivery of the revised materials or equipment also extended the job. As the project neared completion, the project's overall duration was increased by 80 calendar days, reflecting the time needed to perform the revised work, the impact of equipment and materials delivery times, and the differing site condition.

The value of the added or revised work totaled $700,000. There were credits of $125,000 for equipment that was deleted. The net direct cost value of the modifications was $575,000. Assume further that the contractor reserved its rights to be compensated for extended job site general conditions.

A *blended approach* to the pricing of job site general conditions would seek to obtain payment for all of the extended job site general conditions while addressing concerns about double recovery. The calculation would be basically:

a.   Job site general conditions due to added duration

$$\$1400 \times 80 \text{ days} = \$112,000.00$$

b.   Less job site general conditions credit on the net increase to the contract price for modifications:

$$4.09\% \times \$575,000 = (\$23,517.50)$$
$$\text{NET RECOVERY } \$88,482.50$$

Unfortunately, recent ASBCA decisions appear to reject this methodology as prohibited by FAR § 31.105(d)(3), which provides:

(3) Costs incurred at the job site incident to performing the work, such as the cost of superintendence, timekeeping and clerical work, engineering, utility costs, supplies, material handling, restoration and cleanup, etc. are allowable as direct or indirect costs, provided the accounting practice used is in accordance with the contractor's established and consistently followed cost accounting practices for all work.

In effect, this section provides that job site general conditions may be charged by a contractor, at its election, as a direct cost (daily rate) or an indirect cost (percentage basis) provided the costs are consistently charged in accordance with the contractor's established accounting system.

This language has been in the FAR since at least 1985. Even though this cost principle appears to address how a contractor *charges* its cost to a particular project or contract, the ASBCA has interpreted that language as applying to the *pricing* of changes and the treatment of job site general conditions in those modification proposals.

In *M.A. Mortenson Co.*,[74] the ASBCA's Senior Deciding Group held that once a contractor elected to use one of the two permissible approaches to develop its job site general conditions cost in pricing a modification, the use of the alternative methodology was prohibited. Since *Mortenson* had priced its job site general conditions on a percentage basis for the initial contract modifications, that approach or election applied to all modifications.

To the extent that the various government agencies elect to follow the *Mortenson* job site (field) general conditions theory strictly, a contractor's recovery of field general conditions can be substantially diminished depending on the factual circumstances of each change, as described earlier. The Defense Contract Audit Agency (DCAA) should be expected to apply this rule rather strictly since it is clearly summarized in the DCAA *Contract Audit Manual*. DCAA's guidance or direction to its auditors provides:

> . . . Evaluate the proposed or claimed jobsite/field overhead costs to ensure that costs associated with the overall operation of the business (home office overhead) are not included. Jobsite/field overhead costs are allowable as direct or indirect costs provided the costs are charged in accordance with the contractor's established accounting system and consistently applied for all contracts (FAR 31.105(d)(3)).[75]

When comparing the FAR language with the DCAA's directive to its auditors, DCAA has made the assumption that the words "all work" in the FAR means "all contracts." That conclusion is not necessarily correct; "all work" could refer to the work in the contract being audited or *all work* done for the federal government. In addition, the process that a contractor uses to consistently charge or account for job site/field overhead costs could be very different from a markup process imposed or preferred by a particular federal government agency.

Given the probable DCAA approach to job site overhead, a contractor might consider the benefit of reaching an advance agreement with the agency of the treatment of job site general conditions in modifications. This would appear to be consistent with FAR § 31.109, which authorizes the use of advance agreements.

b. Home Office Overhead   Home office overhead or G&A expenses include those costs of the business that are associated with the overall company operation. In *C.B.C. Enterprises v. United States*,[76] the Federal Circuit identified payroll preparation,

---

[74]ASBCA Nos. 40750, *et al.*, 98-1 BCA ¶ 29,658.
[75]*See* DCAA *Contract Audit Manual*, p. 1248 (2007).
[76]978 F.2d 669 (Fed. Cir. 1992).

cost records, and the like as examples of such expenses. The court summarized the nature of that expense pool:

> Conventional wisdom has it that such costs cannot, by their nature, be specified and traced to any particular contract.[77]

In general, a contractor looks to the billings on each contract to bear (absorb) a portion of these G&A expenses. DCAA's *Contract Audit Manual* describes the effect of a project delay on the absorption of the home office expenses:

> The term "unabsorbed overhead" is actually a misnomer because all overhead costs are allocated to, and absorbed by, contracts in process. The term refers to the reallocation of fixed overhead costs among contracts because of the delay/ suspension. The delay/suspension results in a contract being allocated less fixed overhead costs than it would have been allocated absent the interruption (the contract under-absorbs). At the same time, other contract(s) are allocated a greater amount of fixed overhead costs than they would have been allocated absent the interruption (these contracts over-absorb). When unabsorbed overhead costs are allocated to other contracts, the cost of performing the remaining work on these contracts (work that was not delayed/suspended) increases. Without compensating upward contract price adjustments, the company's profitability is decreased.[78]

Home office overhead is the most common example of an indirect cost that usually must be allocated to more than one contract in government construction contracting. In the case of delays, a contractor may seek compensation for extended or unabsorbed overhead expenses.

The basic formula for allocating overhead among a contractor's various projects was described in *Eichleay Corp*, a 1960 ASBCA decision.[79] The *Eichleay* decision set forth a three-step procedure for allocation of overhead in delay situations.

STEP (1)

$$\frac{\text{Contract Billings}}{\text{Total Billings for Contract Period}} \times \frac{\text{Total Overhead for Contract Period}}{} = \text{Overhead Allocable to Cont}$$

STEP (2)

$$\frac{\text{Allocable Overhead}}{\text{Days of Contract Performance}} = \text{Daily Contract Overhead}$$

---

[77]978 F.2d at 672.
[78]*See* DCAA *Contract Audit Manual,* p. 1250 (2007).
[79]*Eichleay Corp.,* ASBCA No. 5183, 60-2 BCA ¶ 2688.

STEP (3)

Daily Overhead × Number of Days Delay = Allocable Overhead

In 1994, the Federal Circuit in *Wickham v. United States* ruled that the Eichleay formula was the "exclusive" means to calculate extended (unabsorbed) overhead on construction contracts.[80] Two years before the *Wickham*, the Federal Circuit in *CBC Enterprises*[81] held that a contractor was *not* entitled to recover extended G&A for time added to a project due to the performance of changed work.

In *CBC Enterprises,* the contractor performed a $927,300 contract for the Navy. The contract price was increased by a total of $10,846 by a series of modifications, and 24 days were added to contract duration due to those modifications. If the Eichleay formula had been applied, CBC would have been entitled to recover $15,317.54 for the added project duration caused by the changed work. The Navy allowed $1,512 based on a G&A rate of 13.94 percent of the overall direct costs. The Federal Circuit, after reviewing the cases that first set forth the Eichleay formula,[82] rejected the argument that the Eichleay formula should be used whenever the project duration is extended by compensable delays. In the court's opinion, the Eichleay formula could *not* be used to calculate extended home office overhead when additional work, not suspension of work, extended the duration of the contract. In *CBC,* the Federal Circuit summarized the basic rule in this way:

> CBC and *amicus curiae* argue that use of Eichleay should be permitted in any instance in which a contract modification results in an erosion of direct costs because a percentage mark-up of the decreased additional direct costs will not allocate a fair proportion of home office overhead to the contract.

***

> We decline the invitation to stand availability of the Eichleay formula on its head. The *raison d'etre* of Eichleay requires at least some element of uncertainty arising from suspension, disruption or delay of contract performance. Such delays are sudden, sporadic and of uncertain duration. As a result, it is impractical for the contractor to take on other work during these delays. *See George Hyman Constr. Co. v. Washington Metro. Area Transit Auth.,* 816 F.2d 753, 756-58 (D.C. Cir. 1987). By contrast, CBC negotiated a change order with the government which extended contract performance for a brief known period of time. CBC experienced no suspension of work, no idle time and no uncertain periods of delay during the agreed upon extended contract performance period. Where no element of uncertainty is imposed on the contractor,

---

[80]*Wickham Contracting Co., Inc. v. Fischer,* 12 F.3d 1574 (Fed. Cir.1994).

[81]978 F.2d 669 (1992).

[82]Many of these cases were decided *before* the Changes clause was amended in 1968 to allow a contractor to be fully compensated for the delay impact of changes and to effectively reverse the so-called Rice doctrine. *United States v. Rice,* 317 U.S. 61 (1942).

use of the Eichleay formula to calculate extended home office overhead is not permissible. Such a limitation on the use of the Eichleay formula is reasonable because, after all, the Eichleay formula only roughly approximates extended home office overhead. [Citation omitted.] Thus, computation of extended home office overhead using an estimated daily rate is an extraordinary remedy which is specifically limited to contracts affected by government-caused suspensions, disruptions and delays of work. Absent these circumstances, the Claims Court properly recognized that it is inappropriate to use the Eichleay formula to calculate home office overhead for contract extensions because adequate compensation for overhead expenses may usually be calculated more precisely using a fixed percentage formula.[83]

Whether this decision appears equitable or not, it reflects the current rule in government contracts.[84] To recover home office overhead using the Eichleay formula, the price adjustment may need to be asserted under the application of the Suspension of Work clause rather than the Changes clause. (See **Chapter 9.**) In effect, the contractor would need to show that the added project duration was not the result of performing extra work. One negative related to that approach is that recovery for profit is not permitted under the Suspension of Work clause.

3. Profit

With limited exceptions, profit is an allowable element of equitable adjustments.[85] The rate of profit allowed in a particular situation will vary depending on such factors as the rate of profit utilized by the contractor in making its proposal or bid, the profit allowed on other changes under the same contract, the degree of risk taken by the contractor, the efficiency of the contractor, and so on.[86] Finally, as discussed next, the profit rate also may be expressly limited[87] by particular contract clauses. Profit is not allowable where a specific clause, such as the Suspension of Work[88] clause, expressly excludes the recovery of profit.

4. Contractual Limitations on Recovery of Overhead and Profit

Limitations contained in some prime contracts are of primary importance not only in relation to changed work but, more significantly, in connection with work that has been delayed in its performance. These provisions can have a significant impact on the recovery of delay-related costs by both the prime contractor and its subcontractors and should be considered in the formation of subcontracts. For example, in *Irvin*

---

[83]*CBC Enters. v. United States*, 978 F.2d at 674.
[84]*Applied Companies, Inc.*, ASBCA No. 54506, 06-1 BCA ¶ 33,269 (government-caused delay deemed a "strict prerequisite" to an application of the Eichleay formula).
[85]*Keco Indus., Inc.*, ASBCA Nos. 15184, 15547, 72-2 BCA ¶ 9576.
[86]FAR § 15.404-4.
[87]*J. Harvey Crow*, ASBCA No. 4146, 75-2 BCA ¶ 11,423.
[88]FAR § 52.242-14.

Prickett & Sons, Inc.,[89] a Department of Labor contract provided in the "extra work" provision that reimbursement for such work would be on the basis of actual costs (including certain specified elements) but excluding overhead, plus 15 percent of that cost to cover profit and all indirect charges. When extra work was ordered and performed by a subcontractor, a dispute arose as to the amount to be added for overhead and profit, and the contractor claimed that the price increase should be computed by the subcontractor's actual costs, plus 10 percent for overhead and 10 percent for profit, and the total resulting figure would then be marked up 15 percent for the prime's profit and overhead. However, the board held that because the contract provision did not provide for allowance of the subcontractor's overhead or profit, such items should come out of or be considered a part of the 15 percent allowed to the contractor.[90]

5. Recovery of Interest Expenses

In *S&E Contractors, Inc. v. United States*,[91] Supreme Court Justice Blackmun stated:

> By accepting the disputes clause in his contract, the contractor bears the interim financial burden and gives up the right of rescission and the right to sue for damages. What he receives in return is the Government's assurances of speedy settlement and of prompt payment, not payment delayed for months, or, as here, for years.

This statement provides an excellent observation on the allocation of the burden of financing extra work and the burden of any delays in resolving the claims.

Recovery of interest expense as part of the cost of financing extra work has had a roller-coaster history in government contracting. The decision by the Court of Claims in *Bell v. United States*[92] clearly established the contractor's right to recover interest costs in an equitable adjustment if the interest was incurred in borrowing money to finance the change. Following that decision, the ASBCA held that if it was necessary for the contractor to borrow money to finance the performance of increased work, it could recover the amount of interest on the borrowing up to the date of the contracting officer's final decision.[93]

Recovery of interest on borrowings to finance a change or the cost of a disputed claim would not benefit a contractor that utilized its own equity capital. In a series

---

[89]IBCA No. 203, 60-2 BCA ¶ 2747.

[90]This is not an isolated decision or unique policy. For example, the Department of Veteran Affairs routinely uses a supplement to the Changes clause that contains contractual limits on the recovery of field general conditions, home office overhead, and profit. *See* 48 C.F.R. § 852.236-88, Contract Changes-Supplement (2002) as set forth in **Chapter 8.**

[91]406 U.S. 1 (1972).

[92]404 F.2d 975 (Ct. Cl. 1968).

[93]*Keco Indus., Inc.* ASBCA No. 15131, 72-1 BCA ¶ 9262; *Ingalls Shipbuilding Division, Litton Sys., Inc.*, ASBCA No. 17059, 78-1 BCA ¶ 13,038.

of decisions the ASBCA held that where the contractor financed the cost of changed work with its own money, interest could be recovered as a cost of financing the changed work as additional "imputed profit."[94] The additional profit approach provided a relatively low interest rate (6 percent) and allowed recovery from the date the costs were incurred through the date of the board's decision.

However, in 1977, the Court of Claims refused to accept this theory of recovery.[95] Since those decisions, the boards have rejected contractor claims for additional profit for use of equity capital to finance extra or delayed work.[96] Moreover, interest as an element of cost is not an allowable cost under the FAR Cost Principles.[97]

Under the Contract Disputes Act of 1978(CDA), Congress has provided that contractors are entitled to interest on "amounts found due" from the date the contracting officer *receives the claim until payment is made*. Interest usually is calculated from the date of receipt of a certified claim rather than the date on which the cost was incurred by the contractor.[98] However, interest should not start on a claim that is patently and grossly inaccurate until the claimant corrects the claim in writing.[99]

Even if a contractor prevails on a claim that results in the release of funds being held by the government, the CDA interest may not accrue unless there was an express, separate claim for those monies. For example, in *Tidewater Contractors, Inc. v. Department of Transportation,*[100] the contractor submitted a claim for an extension of time and later an appeal from the contracting officer's final decision denying the requested time extension. Thereafter, the government assessed liquidated damages. Even though the contractor substantially prevailed on its claim for a time extension, no interest was due on the retained liquidated damages because it failed to submit a distinct CDA claim seeking release of the liquidated damages or filed a motion to amend its pending appeal before the board.

6. Claim Presentation Costs/Legal Fees

When pursuing any claim other than a termination for convenience claim, contractors generally are not entitled to collect legal or administrative expenses incurred in presenting or preparing claims (see FAR § 31.205-47). An exception to this prohibition on the recovery of legal fees and expenses is found in the Equal Access to Justice Act (EAJA), 5 U.S.C. § 504(a), *et seq*. This act permits small business concerns, as defined by that Act, to recover legal fees incurred in the prosecution of

---

[94]*New York Shipbuilding Co., a Division of Merritt-Chapman & Scott Corp.*, ASBCA No. 16164, 76-2 BCA ¶ 11,979; *Aerojet Inc.*, ASBCA No. 17171, 74-2 BCA ¶ 10,863; *see also Ingalls Shipbuilding Div., etc.*, ASBCA No. 17059, 78-1 BCA ¶ 13,038.

[95]*See Framlau Corp. v. United States*, 568 F.2d 687 (Ct. Cl. 1977), *and The Singer Co. v. United States*, 568 F.2d 695 (Ct. Cl. 1977).

[96]*Technology, Inc.*, DCAB NBS-1-78, 79-1 BCA ¶ 13,752; *Owen L. Schwam Constr. Co., Inc.*, ASBCA No. 22407, 79-2 BCA ¶ 13,919.

[97]FAR § 31.205-20.

[98]*Servidone Constr. Corp. v. United States*, 931 F.2d 860, 862-3 (Fed. Cir. 1991).

[99]*Fidelity Constr. Co., Inc.*, DOT Nos. 1113 & 1219, 82-1 BCA ¶ 15,633.

[100]CBCA No. 982-C(50)-R, 08-2 BCA ¶ 33,974.

claims against the government. The contractor would be entitled to recover its legal fees and expenses if it is the "prevailing" party and the government's position was not "substantially justified."

In termination for convenience claims, settlement costs, such as accountants' and attorneys' fees, expended in *developing* and *presenting* claims to the contracting officer are allowable. However, no recovery is allowed for legal expenses attendant to an appeal to a board or the Court of Federal Claims except under the EAJA legal fee recovery process.[101]

7. Delay and Disruption

In the "typical" delay-disruption case, one of the contractor's tasks is to establish and isolate the period of delay attributable to the government. Once this is done, claim cost documentation involves itemizing those fixed (ongoing) costs that were incurred during that period of delay. If, however, the period of delay itself cannot be isolated in this manner, the ensuing problems can be difficult. For example, generally courts will not make any effort to apportion damages in a situation where both parties are found to have contributed to the delays in completion of the contract.[102] Thus, where each party proximately contributes to the delay, "the law does not provide for the recovery or apportionment of damages occasioned thereby to either party."[103] All of the decisions on government contracts, however, do not strictly adhere to this general rule.[104]

In addition to costs directly associated with the extended duration of a project, a project of extended duration also generates other costs that present difficult proof problems. Among these costs are labor inefficiency that results from demobilization and subsequent remobilization, loss of learning-curve efficiency, and loss of efficiency when workers work overtime during an acceleration period.[105] Although proving labor inefficiency is a difficult task, several studies have been undertaken, and the results published on many aspects of this problem. Often these reports can be used as support for a contractor's claim.

One of the more obvious indirect costs attributable to a delay are the extended job site general conditions and home office overhead cost incurred by the contractor. Given the importance of these topics on government construction projects and the rather complicated rules, these two topics are separately addressed in this chapter.

[101]*See* **Chapter 15.**

[102]*See United States v. United Eng'g Contracting Co.*, 234 U.S. 236 (1914).

[103]*Malta Constr. v. Henningson, Durham & Richardson*, 694 F. Supp. 902 (N.D. Ga. 1988); *J.A. Jones Constr. Co. v. Greenbrier Shopping Ctr.*, 332 F. Supp. 1336 (N.D. Ga. 1971), *aff'd*, 461 F.2d 1269 (5th Cir. 1972).

[104]*See United States ex rel. Heller Elec. Co. v. William F. Klingensmith, Inc.*, 670 F.2d 1227 (1982); *see also Wilner v. United States*, 23 Cl. Ct. 241 (1991); *Inversiones Aransu S.A.*, ENGBCA No. PCC-77, 92-1 BCA ¶ 24,584; *JEM Dev. Corp.*, VABCA No. 3272, 91-2 BCA ¶ 24,010.

[105]*Luria Bros. & Co. v. United States*, 369 F.2d 701 (Ct. Cl. 1966); *Youngdale & Sons Constr. Co. v. United States*, 27 Fed. Cl. 516 (1993).

Other delay and disruption types of damages include:

- Increased or protracted equipment rentals[106]
- Increased labor costs, including wage or benefit increments, such as when an owner-caused delay forces contract performance into a new labor contract period[107]
- Increased material costs[108]
- The costs of an idle workforce and equipment[109]

8. Acceleration

Acceleration arises from the requirement that the contractor complete performance of the contract, or a portion thereof, on a date earlier than that originally contemplated and specified by the contract or required by a properly adjusted schedule. Two types of acceleration may entitle the contractor to damages:

(1) "Directed acceleration" occurs when the owner requires the contractor to complete the contract before the scheduled completion date.

(2) "Constructive acceleration" occurs when the owner refuses to grant the contractor a time extension in the event of owner-caused or other excusable delays.

When the contractor must accelerate the pace of performance, the contractor's increased costs are generally compensable.[110] These costs may include, among other things, overtime and shift premiums, supervision costs, extra equipment costs, loss of efficiency, overhead, and profit.[111] If prolonged overtime is required, the effect may be an overall decrease in worker productivity for both overtime and regular (straight) time activities.[112]

The contractor must be able to demonstrate that it has incurred extra costs due to the accelerated efforts. For example, where the government clearly ordered the contractor to accelerate its work, but there was no evidence that the contractor did anything different or suffered any damage as a result of the acceleration order, the contractor's claim for acceleration was denied.[113]

---

[106]*See Weaver-Bailey Contractors, Inc. v. United States,* 19 Cl. Ct. 474 (1990); *Folk Constr. Co. v. United States,* 2 Cl. Ct. 681 (1983).

[107]*See Weaver Constr. Co.,* ASBCA No. 12577, 69-1 BCA ¶ 7455; *Excavation-Constr., Inc.,* ENGBCA No. 3858, 82-1 BCA ¶ 15,770; *Garcia Concrete, Inc.,* AGBCA No. 78-105-4, 82-2 BCA ¶ 16,046.

[108]*See Samuel N. Zarpas, Inc.,* ASBCA No. 4722, 59-1 BCA ¶ 2170.

[109]*See Cornell Wrecking Co. v. United States,* 184 Ct. Cl. 289 (1968); *Caddell Constr. Co.,* VABCA No. 5608, 03-2 BCA 32,257.

[110]*J. W. Bateson Co.,* ASBCA No. 6069, 1962 BCA ¶ 3529; *Tyee Constr. Co.,* IBCA No. 692-1-68, 69-1 BCA ¶ 7748.

[111]*See* Nash, *Government Contract Changes* § 18-14.2 (Federal Publications, 1975).

[112]*See* Business Roundtable, Constr. Indus. Cost Effectiveness Task Force, Scheduled Overtime Effect on Construction Projects (Nov. 1980).

[113]*Utley-James, Inc. v. United States,* 14 Cl. Ct. 804 (1988).

9. Defective Drawings or Specifications

Where the government supplies the plans and specifications to be used in construction, it usually is held to impliedly warrant that the plans and specifications will be adequate to achieve the purposes contemplated.[114] If the plans and specifications are defective or contain omissions, the contractor may incur substantially increased costs of performance. These costs, including the costs of identifying and correcting defects in the drawings or specifications, along with any delay costs related thereto, may be recovered if the contractor properly relied on such drawings in attempting to perform its contractual obligations.[115] In addition, where defective specifications create "wasted effort" and hinder the contractor's performance, the resulting delay is excusable.[116]

10. Inefficiency Claims

The loss of efficiency can be defined as the increased level of effort required to complete an activity above that reasonably estimated and anticipated by the contractor. If an activity that was reasonably estimated to take 8 hours requires 16 hours due to factors beyond the contractor's control, the overrun in labor (8 hours) is viewed as the inefficiency loss. This often is expressed in the form of a percentage, which, in this case, would be a 100 percent productivity loss. Although loss of efficiency claims most frequently involve significant labor overruns, dramatic overruns in equipment usage and costs can arise from inefficient operations.

Construction projects are perceived as having some inherent degree of confusion. An inefficiency claim therefore must demonstrate that the disruption experience went beyond that reasonably anticipated. Recovery of additional compensation for loss of efficiency requires that the claimant establish the other party's liability for the events giving rise to the claim and the amount of damages associated with those events. Some of the factors that may adversely affect a contractor's efficiency are:

- Excessive overtime
- Out-of-sequence work
- Restricted access to the working area
- Trade stacking
- Overmanning
- Excess changes that disrupt the planned sequence of the unchanged work
- Differing site conditions
- Adverse weather
- Acceleration

---

[114]*United States v. Spearin*, 248 U.S. 132 (1918).
[115]*La Crosse Garment Mfg. Co. v. United States*, 432 F.2d 1377 (Ct. Cl. 1970); *Hol-Gar Mfg. Corp. v. United States*, 360 F.2d 634 (Ct. Cl. 1966); *Celesco Indus., Inc.*, ASBCA No. 18370, 76-1 BCA ¶ 11,766.
[116]*J. W. Hurst & Son Awnings, Inc.*, ASBCA No. 4167, 59-1 BCA ¶ 2095.

Studies conducted by the Business Roundtable's Construction Industry Cost Effectiveness Task Force indicate that overtime impairs productivity because of physical fatigue, increased absenteeism, increased likelihood of accidents, and overall reduced quality of work installed.[117] As a general rule, courts and boards have accepted the proposition that overtime results in loss of efficiency.[118]

Weather has a significant impact on construction[119] if the work is exposed to the elements. The types of weather impacting productivity include rain, abnormal humidity, frozen ground, subfreezing temperatures, extreme heat, and excessive wind. Relief for weather usually is limited to an extension of contract time for completion without any monetary compensation.[120] If an earlier compensable delay forces the contractor to perform work during a period of inclement weather, then the contractor may be entitled to additional time for the delay as well as compensation for loss of efficiency suffered by its labor force.[121]

One of the best ways to demonstrate productivity losses is through the use of data specific to the project. However, the majority of construction contractors do not generally maintain such records. There are many ways to collect productivity data for a particular project at the job site. One example is illustrated by the inefficiency claim presentation in *Barrett Co.*[122] In that case, the contractor noted on a daily basis the events that adversely affected efficient performance and the consequences of those events. This contemporaneous record was described by the board as "well described and documented,"[123] and the contractor recovered the amounts claimed.

One effective way to establish the loss of efficiency is a comparison of actual production before and after the problem was encountered, referred to as a "measured mile" analysis.[124] This technique is especially effective for a contractor that routinely performs a particular division of the work with its own forces and has historical as well as job-specific records. The availability of this approach depends on the nature and validity of the project documentation system.

In a loss of efficiency claim, the cause-and-effect relationship may be less capable of exact scientific measurement. This does not mean that a loss of efficiency claim is fiction. The effect of a disruption on a construction project can be devastating in terms of labor dollars or equipment hours expended. In addition, due to the intricate and subtle dependencies among the various trades and activities (the ripple effect), there may be a need for an expert to offer an opinion as to the cause of a disruption

---

[117]Business Roundtable Constr. Indus. Cost Effectiveness Task Force, Scheduled Overtime Effect on Construction Projects (Nov. 1980).

[118]*See Maryland Sanitary Mfg. Corp. v. United States,* 119 Ct. Cl. 100 (1951); *Casson Constr. Co.,* GSBCA No. 4884, 83-1 BCA ¶ 16,523; *Metro Eng'g,* AGBCA No. 77-121-4, 83-1 BCA ¶ 16,143.

[119]National Electrical Contractors Ass'n, The Effect of Temperature on Productivity (Feb. 1987).

[120]*Turnkey Enter., Inc. v. United States,* 597 F.2d 750, 754 (Ct. Cl. 1979).

[121]See *Baldi Bros. Constructors v. United States,* 50 Fed. Cl. 74 (2001) (severe weather interacted with undisclosed site condition); *see also Abbett Elec. Corp. v. United States,* 162 F. Supp. 772 (Ct. Cl. 1958); *F. H. McGraw & Co. v. United States,* 82 F. Supp. 338 (Ct. Cl. 1949).

[122]ENGBCA No. 3877, 78-1 BCA ¶ 13,075 at 63,853.

[123]*Id.* at 63,854.

[124]*Goodwin Contractors, Inc.,* AGBCA No. 89-148-1, 92-2 BCA ¶ 24,931.

and its effect on the project. In *Luria Bros. & Co. v. United States*,[125] the court stated: "It is a rare case where loss of productivity can be proven by books and records; almost always it has to be proven by the opinion of expert witnesses." Thus, the opinion of an expert witness may be required to quantify the nature, the cause, and the amount of inefficiency experienced on the project.

Good documentation combined with appropriate historical data or measured mile studies provide an objective basis for the opinions of experts. Expert opinions and testimony may be extremely valuable in laying the foundation for an inefficiency claim and proving damages. However, without corroboration, the cases indicate that it is quite possible that the opinion testimony will be discounted if there is no other independent basis to support the opinions.[126]

III. METHODS OF PRICING EQUITABLE ADJUSTMENTS AND CLAIMS

There are several basic methods for pricing equitable adjustments or claims on government construction contracts. The most demanding in terms of project documentation is the discrete or *actual cost method*. This method requires the submission of contemporaneous documentation such as records showing expenditures for additional labor, materials, equipment, general conditions, and other costs actually expended by the contractor due to changed work, differing site condition, delay, and so on. The simplest method is the *total cost method*. The simplicity of the total cost method causes it to be frowned on and accepted only in extreme cases. The *modified total cost method* attempts to address those weaknesses. Finally, there is a *jury verdict* approach, which a board or the Court of Federal Claims may apply when it is not completely persuaded by the evidence or cost support provided by either the contractor or the government.

A. Actual Cost Method

The actual cost method of pricing claims is based on records of actual costs that are contemporaneously documented as the work is performed. Under this approach, the additional costs associated with the events or occurrences that gave rise to the claim are tracked separately from those incurred in the normal course of performance of the contract. For example, on an extra work claim, the pricing would reflect an identification of the specific additional labor, materials, and equipment used in performing the extra work as recorded on a contemporaneous basis. If a project has been delayed or the extra work generated a specific requirement for additional job site or

---

[125]369 F.2d 701 (Ct. Cl. 1966).
[126]*Compare Bechtel National, Inc.*, NASABCA No. 1186-7, 90-1 BCA ¶ 22,549 (recovery denied) *with Clark Constr. Group Inc.*, VABCA No. 5674, 00-1 BCA ¶ 30,870 *recon. denied* 00-2 BCA ¶ 30,997 (recovery allowed for impact).

home office administration, the resulting increased costs for job site general conditions or home office overhead also would be tracked and summarized.[127]

The use of this *cause-and-effect* methodology often yields an accurate, well-defined, and defensible presentation of cost. The data can be used to support a stand-alone presentation of a request for an equitable adjustment or may be used to support a *measured mile* calculation. It may, however, be extremely difficult to provide discrete cost documentation in the absence of detailed job cost record keeping and possibly sophisticated cost-control systems that segregate changed or impacted work. This method tends to have added credibility when the person presenting the cost or claim summary shows that the sum of all specifically identified costs does not equal the total difference between the bid proposal cost and total cost (incurred in performing the contract). The difference remaining represents the costs related to contractor-caused events that have been excluded from the equitable adjustment proposal or claim.

If it appears that a change, differing site condition, or delay may adversely affect the performance of the unchanged work, a contractor should consider adopting a means of contemporaneously tracking units of work installed as well as labor, job site general conditions, equipment costs, and the factors or events affecting productivity. Although it may be possible to determine progress from monthly pay applications, that process can be very difficult if the work was disrupted or affected by the need to rework (modify) partially completed work. Similarly, it may be necessary to establish new cost codes to track and record costs associated with changes. The level of sophistication can vary depending on the circumstances. For example, if the work involves the placement of reinforced concrete floors or slabs that is affected by the addition or revision of slab penetrations, it may be sufficient to color code a set of prints with notes on the areas formed, reinforcement placed, and the like, on a daily basis.

If the contracting officer anticipates that numerous changes are anticipated during performance of a contract, the FAR authorizes, but does not require, the use of a clause substantially the same as FAR § 52.243-6. That clause provides:

CHANGE ORDER ACCOUNTING (APR 1984)

The Contracting Officer may require change order accounting whenever the estimated cost of a change or series of related changes exceeds $100,000. The Contractor, for each change or series of related changes, shall maintain separate accounts, by job order or other suitable accounting procedure, of all incurred segregable, direct costs (less allocable credits) of work, both changed and not changed, allocable to the change. The Contractor shall maintain such accounts until the parties agree to an equitable adjustment for the changes ordered by the Contracting Officer or the matter is conclusively disposed of in accordance with the Disputes clause.

---

[127] *See Amec Constr. Mgmt., Inc. v. General Services Administration*, GSBCA No. 16233, 06-2 BCA ¶ 33,410.

Although the actual cost is preferred, the ASBCA has recognized that a change order accounting-type provision is to be used only for large-value changes (greater than $100,000) as the record keeping is burdensome and expensive.[128] This analysis is consistent with the holding of the Court of Claims in *Joseph Pickard's Sons Co. v. United States*[129] that it was *not* an "invariable prerequisite" to recovery that the extent of the damage must be established with "absolute certainty or precise mathematical accuracy."[130] However, while precise mathematical accuracy may not be required, the actual cost method is clearly preferred by the boards and courts.[131]

B. Total Cost Method

A total cost claim is simply what the name implies. It essentially seeks to convert a standard fixed-price construction contract into a cost-reimbursement arrangement. The contractor's total out-of-pocket costs of performance are tallied and marked up for overhead and profit. Payments already made to that contractor are deducted from that amount, and the difference is the contractor's claim. Of course, this approach can be refined or adjusted to meet particular needs and circumstances, but the basic components and approach remain: Costs associated with the basis for the claim are not segregated. The total cost method often is used for impact disruption claims when the segregation of costs may be more difficult.

The total cost approach, although preferred by some claimants because of the ease of computation, generally is discouraged by the boards and Court of Federal Claims because it assumes that the contractor was virtually fault-free and because it is fraught with uncertainties. For these reasons, numerous decisions have established four rigorous requirements for the presentation of total cost claims:

(1) Other methods of calculating damages are impossible or impractical.

(2) Recorded actual costs must be reasonable.

(3) The contractor's bid or estimate must have been realistic (i.e., contained no underbidding).

(4) The contractor is not responsible for any of the cost overruns.[132]

The second requirement, the reasonableness of recorded costs, is typically not a difficult assumption to prove. The claimant must demonstrate the appropriateness

---

[128]*Service Eng'g Co.*, ASBCA No. 40274, 93-1 BCA ¶ 25,520 *mod. on recon.* 93-2 BCA ¶ 25,885 *aff'd, United States v. Service Eng'g Co.*, 1994 WL 519501 (N.D. Cal.). *See* **Chapter** 17 for a discussion of the contractor's obligations to separately track and report expenditures related to work funded by the American Recovery and Reinvestment Act (ARRA).

[129]532 F.2d 739 (Ct. Cl. 1976).

[130]*Id.* at 742.

[131]See *Dawco Constr., Inc. v. United States*, 930 F.2d 872, 882 (Fed. Cir. 1991).

[132]*WRB Corp. v. United States*, 183 Ct. Cl. 409 (1968); *Baldi Bros. Constructors v. United States*, 50 Fed. Cl. 74 (2001).

of costs, the reliability of the contractor's accounting methods and systems, and a relationship to industry practices and standards. Ironically, the more detailed and well documented the claimant's costs are, the more vulnerable the claimant is to an argument that it can use another method for calculating damages, and therefore the claimant fails to satisfy the first requirement for use of the total cost method.

The remaining two requirements can be more difficult to meet. Proving that the contractor's estimate was realistic can be challenging. That proof might require a comparison of other bids/proposals and supplier and subcontractor quotes to bid/proposal amounts as well as the comparison of material quantity estimates to contract drawings. Presenting such an analysis is expensive, often difficult to follow, and easily refutable because most estimates rely, to a varying degree, on assumptions. Due to the nature of the estimating process, many of these assumptions are developed and applied in the absence of precise factual data.

Establishing that the claimant was blameless for any overruns is perhaps the most difficult aspect of a total cost claim and why the method so rarely succeeds. The claimant essentially attempts to prove causation by showing that the damages were not its own fault and therefore must be due to the acts of the other party. The premise is easily attacked by demonstrating only a single area of potential blame attributable to the contractor, which could erode the credibility of the entire total cost claim. That is why this method often has been defeated in practice[133] and why contractors may instead pursue a "modified" total cost method, as discussed next. The difficulties of establishing the prerequisites for use of a total cost calculation in a board or Court of Federal Claims proceeding, combined with the skepticism it can generate, counsel against use of the total cost method whenever possible.

C. Modified Total Cost Method

Another general approach to calculating and presenting damages borrows from the concepts of both the actual cost identification and total cost methods. The modified total cost method employs the inherent simplicity of the total cost approach but modifies the calculation to demonstrate more direct cause-and-effect relationships that exist between the costs and acts giving rise to liability. The success of the approach often depends on the extent of the modifications that demonstrate the cause-and-effect dynamics.

The initial step in calculating damages using the modified total cost method involves adjusting the contractor's estimate for any weaknesses uncovered during job performance, whether they were judgment or simple calculation errors. A reasonable estimate (an as-adjusted estimate) is thus established. The recorded project costs are then examined for reasonableness, and reductions are made for costs that cannot be attributed to the owner, such as unanticipated labor material cost escalations that are not tied to the alleged basis for liability.

---

[133]*Grumman Aerospace Corp.*, ASBCA No. 48006, 06-1 BCA ¶ 33,216.

Focusing on specific areas of work can further refine the modified total cost calculation or cost categories related to the claim issues. For example, if the claim relates to a differing site condition that affected only site work and foundations and not the balance of the project, the claimant should focus only on those areas in the calculation and eliminate extraneous costs and issues that complicate and dilute the credibility of the pricing.

Although the modified total cost approach sometimes is viewed as a method of avoiding the unfavorable scrutiny generally given to a total cost analysis, recent cases imply that courts and boards of contract appeals may apply the same standards of admissibility to modified total cost claims as they have in the past to traditional total cost claims.[134] If the claimant has modified the cost calculation properly and not relied on a simplistic approach, the modified total cost method is far more likely to withstand scrutiny and offer a credible means of quantifying a claim.

D. Jury Verdict Recovery

Normally, a contractor must prove with reasonable certainty the excess costs incurred from government acts or omissions by one of the methods previously discussed. Occasionally, however, a contractor will fail to offer sufficient specific evidence to establish the quantum of costs incurred as a result of a change and the total cost approach is not applicable. Although the contractor runs the risk of recovering no money at all, the board or Court of Federal Claims may invoke a jury verdict method of determining dollar recovery by exercising its judgment and arriving at a figure based on the overall record.[135]

Of course, the contractor must make at least a minimum showing in support of its claim. This will probably involve a demonstration that: (1) clear proof of injury by the government exists; (2) there is no more reliable way to establish the damages; and (3) the evidence is sufficient to enable a board or court to make a fair and reasonable approximation of the damages (costs).[136] In addition, the contractor must demonstrate a justifiable inability to substantiate the resulting costs or damages by direct and specific proof.[137]

However, in some cases where no means exist for establishing an adequate basis on which to measure recovery, a board may not render a jury verdict opinion but either will (1) remand the issue to the contracting officer for negotiation and possible settlement or (2) deny recovery.[138]

---

[134]*Servidone Constr. Corp. v. United States,* 19 Cl. Ct. 346 (1990) *aff'd* 931 F.2d 860 (Fed. Cir. 1991); *Grumman Aerospace Corp.,* ASBCA No. 48006, 06-1 BCA ¶ 33,216.

[135]*See, e.g., Grumman Aerospace Corp.,* ASBCA No. 48006, 06-1 BCA ¶ 33,216; *Silver Enters. v. U.S. Dep't of Transportation,* DOTBCA No. 4459, 4464, 06-2 BCA ¶ 33,370.

[136]*Dawco Constr. v. United States,* 930 F.2d 872, 881 (Fed. Cir. 1991).

[137]*Joseph Pickard's Sons Co. v. United States,* 532 F.2d 739, 742 (Ct. Cl. 1976); *Service Eng'g Co.,* ASBCA No. 40274, 93-1 BCA ¶ 25,520 (board found that it was inherently difficult to require craft workers who went from task to task and from one specification to another to distinguish base contract work from changed work).

[138]*Compare T. & B. Builders, Inc.,* ENGBCA No. 3664, 77-2 BCA ¶ 12,663 *with Soledad Enters., Inc.,* ASBCA No. 20376, 77-2 BCA ¶ 12,757.

IV. COST AND PRICING DATA

As is evident from the concepts discussed in this chapter, establishing the right to a price adjustment for changed work, delay, differing site conditions, or other conditions impacting a government contractor's work is only half the battle. A contractor also has the obligation of establishing the amount of the price adjustment in accordance with established rules and procedures. Although the price adjustment typically is settled through negotiation, contractors must have a clear understanding of the pricing techniques and procedures so that they can avoid the pitfalls that could jeopardize their claim or potentially expose them to liability under the False Claims Act. In particular, contractors must know when certified cost and pricing data is required, and the consequences of submitting certain information to the government.

Under the Federal Acquisition Regulation, "[t]he contracting officer should use every means available to ascertain whether a fair and reasonable price can be determined before requesting cost or price data."[139] This stated policy recognizes that the submission of such information is both time consuming and costly. Subject to limited exceptions, however, the contracting officer *must* obtain cost or pricing data for any price adjustment in excess of $650,000.[140] In calculating the amount of the price adjustment, both increases and decreases are considered (*e.g.*, "a $200,000 modification resulting from a reduction of $500,000 and an increase of $300,000 is a pricing adjustment exceeding $650,000").[141] In addition, the contracting officer may be authorized to obtain cost and pricing data for price adjustments or bids that are below the $650,000 threshold.[142] In such circumstances, the head of the contracting activity must provide a written finding justifying that cost and pricing data are necessary.[143] However, cost and pricing data shall not be required for any price adjustment or bid that is at or below the simplified acquisition threshold (i.e., $100,000).[144]

A. Basic Concepts of Cost and Pricing Data

In an effort to address concerns that offers submitted to the government agencies were not always based on the best information available at the time of negotiations—which ultimately led to inflated contract price awards—Congress passed the Truth in Negotiations Act.[145] This legislation, which is designed to put the government in a position equal to the contractor with respect to making judgments on pricing, requires government contractors to submit "cost or pricing data" for contract price adjustments or proposals exceeding certain dollar thresholds.[146] The contractor must

---

[139]FAR § 15.402(a)(3).
[140]The dollar threshold for cost or pricing data is subject to adjustment every five years.
[141]FAR § 15.403-4(a)(1)(iii).
[142]FAR § 15.403-4(a)(2).
[143]FAR § 15.403-4(a)(2).
[144]FAR § 15.403-1(a); FAR § 2.101.
[145]10 U.S.C. § 2306a.
[146]*Lockheed Aircraft Corp. v. United States*, 193 Ct. Cl. 86, 432 F.2d 801 (1971).

certify that the cost or pricing data submitted is "accurate, complete, and current as of the date of agreement on price. . . ."[147] Section 2.101 of the FAR defines Cost and Pricing Data in this way:

All facts that, as of the date of price agreement, or if applicable, an earlier date agreed upon between the parties that is as close as practicable to the date of agreement on price, prudent buyers and sellers would reasonably expect to affect price negotiations significantly. Cost or pricing data are data requiring certification in accordance with 15.406-2. Cost or pricing data are factual, not judgmental, and are verifiable. While they do not indicate the accuracy of the prospective contractor's judgment about estimated future costs or projections, they do include the data forming the basis for that judgment. Cost or pricing data are more than historical accounting data; they are all the facts that can be reasonably expected to contribute to the soundness of estimates of future costs and to the validity of determinations of costs already incurred. They also include such factors as—

(1) Vendor quotations;

(2) Nonrecurring costs;

(3) Information on changes in production methods and in production of or purchasing volume;

(4) Data supporting projections of business prospects and objectives and related operations costs;

(5) Unit-cost trends such as those associated with labor efficiency;

(6) Make-or-buy decisions;

(7) Estimated resources to attain business goals; and

(8) Information on management decisions that could have a significant bearing on costs.

Although this broad definition does not provide a precise framework for determining what constitutes cost or pricing data, it is clear the information must be factual, verifiable, and not judgmental. For example, purchase orders, labor rates, vendor quotations, subcontractor bids, and similar information used in preparing contract estimates or change proposals must be disclosed and certified. Reports and estimates containing information that is both factual and judgmental typically are considered cost and pricing data and must be disclosed. In *Texas Instruments, Inc.*, for example, computer-generated reports using actual cost data to project the cost of future work were held to be cost and pricing data and, thus, needed to be disclosed during negotiations.[148] Data relating to judgments, business strategies, or plans for the future are not cost and pricing data, but "any information relating to execution or implementation of any such strategies or plans" must be disclosed.[149]

[147]FAR § 15.403-4(b)(2).
[148]*Texas Instruments, Inc.*, ASBCA No. 23678, 87-3 BCA ¶ 20,195.
[149]*Lockheed Corp.*, ASBCA No. 39195, 95-2 BCA ¶ 27,722.

The disclosure requirement is aimed at precluding a contractor from withholding relevant pricing information from the government during negotiations.[150] Consequently, the disclosure requirement can be onerous to the contractor. That is, "in order that there be effective disclosure . . . either the government must be clearly advised of the relevant cost or pricing data or it must have actual, rather than imputed, knowledge thereof."[151] This places a significant burden on the contractor as it is not enough to simply make available or hand over information to the government that, if examined, would disclose differences between proposed costs and lower historical costs. Instead, in some instances, it is also necessary to advise the government of the contents of the cost and pricing data and their significance to the proposed price adjustment or bids.[152] Moreover, the contractor is required to disclose all cost and pricing data that are "reasonably available." The fact that the person(s) negotiating the contract or price adjustment or certifying the cost and pricing data are not aware of certain cost analyses is of no moment as long as others in the organization have access to, or are familiar with, the cost information.[153] In other words, access by others in the organization renders the information "reasonably available."

The submission or disclosure of cost and pricing data must be presented in accordance with the format prescribed in FAR Part 15, Table 15-2. Namely, in submitting its proposal, the contractor *must*: (1) include an index of all cost or pricing data referenced in the proposal; (2) clearly identify the cost or pricing data included in the proposal and explain its estimating process; and (3) show the relationship between contract line items and the cost or pricing data disclosed.[154]

Cost and pricing data submitted to the government that are "inaccurate, incomplete, or noncurrent" are considered defective and entitle the government to a contract "price adjustment, including profit or fee, of any significant amount by which the price was increased because of the defective data."[155] If there is a dispute as to whether the contactor satisfied its obligation, the government has the burden to demonstrate that: (1) the contractor failed to disclose accurate, complete, and current cost or pricing data; (2) the data not disclosed involved significant sums; and (3) the government relied on the inaccurate, incomplete, or noncurrent data in agreeing to the price adjustment or bid.[156] The last element establishes "the causal relationship between the incorrect data and the final negotiated price."[157] Significantly, the government, in asserting its claim, is aided by a rebuttable presumption that the "natural and probable consequence of nondisclosure is an increase in the contract price."[158] Once the contractor rebuts the presumption, however, the government is required to

---

[150] *Texas Instruments, Inc.*, ASBCA No. 23678, 87-3 BCA ¶ 20,195.
[151] *Id.*
[152] *Id.*
[153] *Aerojet Gen. Corp.*, ASBCA No. 12264, 69-1 BCA ¶ 7664 (nonresponsive subcontractor quotation).
[154] FAR Part 15, Table 15-2 (B)-(D).
[155] FAR § 15.407-1(b)(1).
[156] *Texas Instruments, Inc.*, ASBCA No. 23678, 87-3 BCA ¶ 20,195.
[157] *Id.*
[158] *Aerojet Solid Propulsion Co.*, ASBCA No. 44568, 00-1 BCA ¶ 30,855.

present affirmative evidence that it relied on the defective information to its detriment. The failure to do so precludes the government's claim for a contract price adjustment.[159]

In addition, contractors can avoid liability for submitting defective cost and pricing data (i.e., the information presented is inaccurate, incomplete, or noncurrent) by demonstrating that the information omitted from the submission would not have significantly affected the final negotiated price.[160]

B. Absence of Adequate Price Competition

The Truth in Negotiations Act was designed to level the playing field for the government in making pricing decisions on certain negotiated procurements and modifications to contracts. The contractor's disclosure requirement provides the government access to "current, accurate and complete" information relevant to a given price proposal. In theory, this places the government in a position equal to its commercial counterpart when negotiating contract prices.

It is important to recognize that disclosure of cost and pricing data is required only for *negotiated* contracts above the price threshold (or as directed by the contracting officer) when the contract price is not based on *adequate price competition*. That is, FAR § 15.403-1(b) states that the "contracting officer shall not require submission of cost or price data . . . when the contracting officer determines that prices agreed upon are based on adequate price competition."[161] Adequate price competition exists when: (1) two or more responsible offerors, competing independently, submit priced offers that satisfy the government's requirements; and (2) there is no finding that the price of the would-be successful bidder is unreasonable. Sealed bid contracts are exempt from the disclosure requirement because the government can evaluate the reasonableness of a given bid by comparing it to the competing bids submitted in response to the solicitation. In addition, sealed bid procurement ensures that proposals are competitively priced and artificially inflated bids are not likely to be selected.

A *modification* to either a sealed bid (FAR Part 14) contract or a negotiated (FAR Part 15) contract, however, is subject to the disclosure requirements if the price adjustment(s) effected by that modification exceed the threshold amount ($650,000).[162] This includes negotiated final pricing actions such as termination settlements and total final price agreements for fixed-price incentive and redeterminable contracts.[163] The disclosure requirement applies only to the cost and pricing data relevant to the contract modification.

---

[159] *Wynne v. United Technologies Corp.*, 463 F.3d 1261, 1267 (Fed. Cir. 2006).

[160] *Plessy Indus., Inc.*, ASBCA No. 16720, 74-1 BCA ¶ 10,603 (rejected vendor quotes were not cost or pricing data as "they would not reasonably have been expected by a prudent buyer or seller to have a significant effect on the price negotiations").

[161] FAR § 15.403-1(b).

[162] FAR § 15.403-1(b). This threshold is not the net amount of a change order. Rather, it is the *sum of the additive and deductive* amounts. For example, a modification that adds work priced at $400,000 and deducts work priced at $255,000 is subject to the cost or pricing data submission requirements, even though the net value for the final modification is $145,000.

[163] FAR § 15.403-4(a)(iii).

C. Role of the Defense Contract Audit Agency and Other Audit Agencies

The Defense Contract Audit Agency, under the direction of the Under Secretary of Defense, is responsible for performing all contract audits for the Department of Defense and providing accounting and financial advisory services in connection with negotiation, administration, and settlement of contracts and subcontracts.[164] The DCAA also provides contract audit services to other government agencies.[165] Although some government agencies have their own internal audit department, the DCAA is by far the largest. In fact, many civilian agencies retain the DCAA for contract proposal and price adjustment audits. However, some agencies may utilize the agency's Office of Inspector General to conduct audits

The DCAA's primary function is contract audit services. In this capacity, the DCAA provides preaward contract audits, postaward contract audits, and contractor internal control systems audits.[166] The goal of these audits is to ensure that the government pays a fair and reasonable price for goods and services and that contractors are in compliance with applicable laws and regulations. The DCAA *Contract Audit Manual* provides guidance, standards, policies, and procedures to be followed by DCAA personnel in performing contract audits.[167]

A DCAA audit can be triggered by various events including submission of contract proposals or requests for price adjustments. Contractors must be aware of the wide discretion afforded the auditors in reviewing the business affairs of a given entity and the subjective judgments they are called on to make in deciding whether to refer a situation for further investigation of potential fraudulent activity. Contractors also must know how certain information and documentation, or lack thereof, will be interpreted and what conditions are perceived as indicators of fraud. Armed with this knowledge, a contractor can adjust its documentation and business practices to avoid incurring the cost and expense of defending against fraud allegations.

D. Fraud Indicators

The DCAA's contract audit services include reviews of contract proposals or price adjustments to ensure compliance with the Truth in Negotiations Act. The goal of these audits is to ensure that the cost estimates contained in a contractor's proposal or claim are reasonable, allocable to the work at issue, and "in accordance with applicable cost limitations or exclusions as stated in the contract or in FAR."[168] That agency's Web site asserts:

> In FY 2008, DCAA audited $138 billion of costs incurred on contracts and reviewed 8,113 forward pricing proposals amounting to $313 billion.

---

[164]Defense Contract Audit Agency Web site: *www.dcaa.mil* (accessed August 4, 2009).
[165]*Id.*
[166]Defense Contract Audit Agency Web site, DCAA Products and Services: (accessed August 4, 2009).
[167]DCAM ¶ 0-002(a) (June 29, 2007).
[168]DCAM ¶ 1-104.2(b) (June 29, 2007).

Approximately $3.3 billion in net savings were reported as a result of audit findings. When compared to the $470 million expended for the Agency's operations, the return on taxpayers' investment in DCAA was approximately $7.00 for each dollar invested.[169]

Although the DCAA *Contract Audit Manual* instructs its auditors that "the detection of fraud or similar unlawful activity is not the primary function of contract audit," the manual affords the auditor broad authority "to obtain reasonable assurance about whether the contractor submissions and supporting data are free of material misstatement, whether caused by error or by fraud."[170] To that end, the auditors are required to be familiar with specific indicators of fraud as described in the *Handbook on Fraud Indicators for Contract Auditors* published by the Inspector General for the Department of Defense.[171] The handbook describes in detail the government's view of symptoms or characteristics of possible fraud and encourages the auditor to consider documentation and information beyond the government contract in question. For example, the auditor is reminded that "[k]nowledge of contractor estimating and charging practices, policies and procedures is essential to recognizing fraud indicators."[172] In addition, there must be a "thorough understanding of the company's internal controls" and consideration of the "total picture."[173] The handbook provides to the auditor substantial discretion in determining whether there is "a reasonable suspicion of fraud, corruption, or unlawful activity relating to a Government contract."[174]

The handbook is organized by three types of audits—Incurred Cost Audits, Forward Pricing Proposal Audits, and Defective Pricing Audits—and enumerates indicators of fraud encountered in each. For example, when performing incurred costs audits, the auditor is reminded to take note of:

- Sudden, significant shifts in charging costs.
- Decrease in labor charges to project or contracts experiencing cost overruns or approaching the contract amount.
- Weak internal controls over labor charges.
- Use of adjusting journal entries to shift costs between contracts.
- Transfer of material costs from ongoing jobs to open work orders for items previously delivered or items scheduled in the distant future.
- Transfers of costs from fixed price government contracts to commercial jobs.
- Transfers of material costs at costs substantially different than actual.
- Poor enforcement of existing contractor policies on conflicts of interests or acceptance of gratuities.

---

[169]Defense Contract Audit Agency Web site, DCAA Products and Services: (accessed August 4, 2009).
[170]DCAM ¶ 1-102(c) (June 29, 2007).
[171]DCAM ¶ 4-702.3(a) (June 29, 2007).
[172]*Handbook on Fraud Indicators for Contract Auditors*, p. II-6 (March 31, 1993).
[173]*Id.*
[174]DCAM ¶ 4-702.4(a) (June 29, 2007).

- Purchasing employees maintaining a standard of living obviously exceeding their income.
- Poor or nonexistent procedures for obtaining competitive bids from subcontractors or lack of competitive awards.
- Poor or nonexistent documentation supporting the award of subcontracts
- Failing to award subcontract to lowest bidder.
- Vaguely worded or nonexistent agreements for professional or consulting services employed on a project.
- Missing or nonexistent documents supporting claimed costs.

With respect to forward pricing proposal audits, these circumstances are noted as fraud indicators:

- Significant differences between proposed unit costs or quantities and actual unit costs and quantities with no corresponding changes in the scope of work.
- Task-by-task billings consistently at the upper level established in the contract.
- Specific employees identified in proposal as "key employees" not working on the contract.
- Employees typically charged indirectly by the company being charged directly to the contract.
- Proposed labor not based on existing workforce (i.e., massive new hires needed).
- Employees' skills do not match the skill requirements as specified for their labor category or the contract requirements.
- Supporting documentation is consistently poor, illegible, or nonexistent.
- Changes to original documentation that do not appear authentic (*e.g.*, different print or incorrect spacing).
- Inconsistent supporting documentation for same item of costs.

Likewise, with respect to defective pricing audits, these items are deemed indicators of fraudulent activity:

- High incidence of defective pricing.
- Repeated defective pricing involving similar patterns or conditions.
- Continued failure or refusal to correct known system deficiencies.
- Withholding relevant historical records or other data.
- Altered or false documents.

Although there is no way for a contractor to avoid an audit, contractors must take necessary precautions to avert fraud investigations triggered by poor document retention or lax enforcement of company policies and procedures. A government contractor must become familiar with the DCAA's audit manual and have knowledge of circumstances that are deemed indicators of fraud. Doing this will educate the contractor as which records to retain and for how long. In addition, it will allow the contractor to design a documentation system that will effectively document key issues during the project and explain any discrepancies that may arise. Implementing and maintaining an effective cost documentation system is essential to avoid triggering a fraud investigation or effectively defending against such an investigation. This is significant as costs for successfully defending against fraud allegations may be recoverable.[175]

A contractor also must take careful note that its actions in structuring a proposal or request for an equitable adjustment ultimately may trigger a fraud investigation and subject it to liability under the False Claims Act or the fraud provisions of the Contract Disputes Act. For example, in *Daewoo Engineering and Construction. Co. v. United States*, the contractor submitted a certified claim in the amount of $63,798,648.95.[176] Of that amount, $50,629,855.88 was categorized as "costs to be incurred."[177] The certifying official testified that the purpose of the inflated claim for "costs to be incurred" was intended to "get the Government's attention" and to facilitate the approval of a compaction method preferred by the contractor.[178] Under the fraud provision of the Contract Disputes Act: "[i]f a contractor is unable to support any part of his claim . . . [because of] misrepresentation of fact or fraud on the part of the contractor, he shall be liable to the Government for an amount equal to such unsupported part of the claim."[179] Based on the evidence, the contractor was assessed damages in the amount of $50,629,855.88 under the Contracts Dispute Act and $10,000 under the False Claims Act.

A secondary issue in the *Daewoo* case involved purported fraudulent representations in the contractor's proposal.[180] The government alleged that Daewoo "obtained the job other than on merit" by effecting a "bait and switch."[181] That is, the contractor's final proposal identified 14 key personnel as its management for the project, only two of whom came to the project.[182] The two who arrived acted in capacities different from those disclosed in the bid.[183] Although the "bait and switch" did not trigger any additional damages, the court determined:

[The contractor] indeed induced the Government fraudulently to award this contract. If [the contractor] were entitled to damages, which is not a possibility given the facts of this case, arguably such damages would be forfeited because plaintiff obtained its contract by fraud.[184]

The *Daewoo* case provides a good illustration of where an overly zealous contractor can expose itself to fraud liability and related damages.

[175]*The Boeing Co., Successor-In-Interest of Rockwell Int'l Corp.*, CBCA Nos. 337, 338, 339, 978, 09-1 BCA ¶ 34,026.
[176]73 Fed. Cl. 547, 595 (2006).
[177]*Daewoo Eng'g and Constr. Co. v. United States*, 73 Fed. Cl. 547, 595 (2006).
[178]*Id.*
[179]*Id.* (quoting 41 U.S.C. § 604).
[180]*Daewoo Eng'g and Constr. Co. v. United States*, 73 Fed.Cl. 547, 586–89 (2006).
[181]*Id.*
[182]*Daewoo Eng'g and Constr. Co. v. United States*, 73 Fed.Cl. 547, 587 (2006).
[183]*Id.*
[184]*Daewoo Eng'g and Constr. Co. v. United States*, 73 Fed. Cl. 547, 588 (2006).

➤ LESSONS LEARNED AND ISSUES TO CONSIDER

- An *equitable adjustment* is a term of art in government contracts entailing a recovery of cost and an allowance for profit.
- To demonstrate entitlement to an equitable adjustment, a contractor must establish government *liability, causation,* and cost *effect.*
- When negotiating a contract modification, carefully consider the effect of a *change order release* proposed by the agency before agreeing to that modification.
- *Cost* is a key element of any government contract price adjustment.
- The treatment of costs in government contracts is largely controlled by the *Contract Cost Principles and Procedures* in FAR Part 31, which address cost *allocation, reasonableness,* and *allowability.*
- If the contractor's actual costs are challenged by the government, there is no *presumption* that these cost are reasonable.
- The FAR contains *special cost principles* addressing construction contracts and costs such as job site general conditions, home office overhead, and equipment expense.
- Contractors should carefully explore using *advance agreements* to avoid disputes on the recovery of job site general conditions.
- Recovery of extended job site general conditions and home office overhead can be limited by *special contract clauses* even if the contractor can satisfy the legal prerequisites to recovery.
- The contractor's ability to provide *credible contemporaneous documentation* of the cost, as well as the time, associated with a change, differing site condition, or delay is often the primary difference between a reasonable recovery and a denial of the proposal or claim.
- The preferred method of pricing an equitable adjustment is the *actual cost method.* If that method cannot be utilized, a contractor still may recover by alternative methods, such as the *modified total cost* or *total cost methods.*
- The modified total cost or total cost methods may be used only if *specific conditions* are satisfied.
- Under some circumstances, a board or court may use a *jury verdict* approach if it established that the government is responsible for some portion of the claimed cost and the contractor was prevented from proving damages for reasons beyond its control.
- The government and its auditors focus on the identification of possible contractor fraud and the submission of overstated claims. Contractors need to be aware of the so-called *fraud indicators* related to the documentation of proposals and claims.

14

PROJECT DOCUMENTATION TECHNIQUES

I. DOCUMENTATION GENERALLY

One of the principal responsibilities of job personnel is to document the project. Done properly, this task requires skill, judgment, accuracy, thoroughness, and commitment. Although construction personnel rarely enjoy paperwork and far prefer building to writing, the quality of a party's documentation can be the difference between a successful, profitable project and a financial disaster.

Proper record keeping does not mean documenting every event, discussion, or decision. It means documenting those events where a historical record of the project is needed and creating a written record of activities that have a reasonable possibility of involving problems, claims, or legal issues in the future. Skill and experience are needed not only to record events accurately but also to identify what needs to be documented in the first place. Unnecessary or poor documentation may cause more problems than a lack of documentation.

Experience is often the best teacher of what should and should not be documented. A construction manager, superintendent, or owner's representative will learn through experience that documentation is not a chore but a critical skill needed on every project.

Experience also will teach what should not be documented. Wrong and poorly thought out documentation is not only a waste of critical project management resources; it also can be counterproductive to the interests of the contractor or subcontractor. Project personnel should avoid documenting complaints they have with the home office, whether it is poor decision making, lack of support, or personal disagreements. Similarly, persons should resist the temptation to record events to make themselves look good at the expense of other project or corporate interests.

This chapter addresses some of the more critical documentation aspects of government construction contracts and projects. These principles can (and should) be applied to subcontracts and to purchase orders on those projects as well.

II. NOTICE OBLIGATIONS IN GOVERNMENT CONTRACTS

Perhaps the most important issue involving documentation on a government construction project is notice. Notice provisions may be triggered by changes, differing site conditions, delays, terminations (actual or threatened), or other events affecting the contractor's work and the project. Many notice provisions impose obligations on the contractor. The converse is also true. The government's contracting officer also may be required to give the contractor notice before many of the government's contract rights are triggered.

The Federal Acquisition Regulation (FAR) contains several critical notice provisions that contractors should adhere to at the risk of waiving claims or defenses to government back charges. These notice provisions are summarized next.

A. Changes Clause

The Changes clause[1] for fixed-price federal construction contracts at FAR § 52.243-4 sets forth the elements of a valid change, including specific time frames for the contractor to provide notice to the government after encountering a directive or constructive change. That clause provides in relevant part:

(a) The Contracting Officer may, at any time, without notice to the sureties, if any, by written order designated or indicated to be a change order, make changes in the work within the general scope of the contract, including changes—

(b) Any other written or an oral order (which, as used in this paragraph (b), includes direction, instruction, interpretation, or determination) from the Contracting Officer, that causes a change shall be treated as a change order under this clause; *provided, that the Contractor gives the Contracting Officer written notice stating (1) the date, circumstances, and source of the order and (2) that the Contractor regards the order as a change order. . . .*

(d) . . . However, except for an adjustment based on defective specifications, no adjustment for any change under paragraph (b) of this clause shall be made for any costs incurred more than 20 days before the Contractor gives written notice as required.

(e) The Contractor must assert its right to an adjustment under this clause within 30 days after (1) receipt of a written change order under paragraph (a) of this clause or (2) the furnishing of a written notice under paragraph (b) of this clause, by submitting to the Contracting Officer a written statement describing the general nature and amount of such proposal, unless this period is extended by the Government. The statement of proposal for adjustment may be included in the notice under paragraph (b) above. [Emphasis added]

---

[1]Changes issues are discussed more fully in **Chapter 8.**

FAR § 43.205(d) mandates that all government agencies include the Changes clause in their construction contracts if a fixed-price contract is contemplated and the contract amount is expected to exceed the simplified acquisition threshold, which is generally $100,000.[2] The only authorized variation to the Changes clause is when the government chooses to extend the 30-day notice provision, as allowed by paragraph (e).

As is discussed more fully in **Section V** of this chapter, notice is important to preserve the contractor's position that a change has occurred, particularly when there is a constructive change and the government has not issued a change order for the work. Providing notice allows the government the opportunity to evaluate whether a situation is a change and gives the government the opportunity to correct any misinterpretation by the contractor, to take action that may avoid the change, or to try to minimize the expenses associated with the change. The contractor that fails to give notice to the government may be deemed to have volunteered its work and to have waived any claim for the change.[3]

When the government issues a written change order to the contractor under FAR § 52.243-4, the contractor has 30 days from receipt of that change order to submit a written notice that: (1) the contractor believes the change will affect the price or schedule for the project; (2) describes generally how the change will affect the price or time; and (3) describes the amount of monetary and time adjustment that the contractor feels is necessary as a result of the change. Similarly, where the contractor receives an oral or written direction from the government in a form other than a change order, the contractor must give the contracting officer written notice stating: (1) the date, circumstances, and source of the order; and (2) that the contractor regards the direction as a change order. The contractor then has 30 days after furnishing that notice to notify the contracting officer of the nature and effect of the change on the contractor's work and the amount of time and money adjustment that the contractor believes is appropriate.

The boards and courts have given contractors some leeway regarding the information that is needed to satisfy the Changes clause notice requirements. For example, the Armed Services Board of Contract Appeals (ASBCA) has stated that the notice needs to contain some indication that the contractor regards the event as a change order and should contain this information: (1) the date of the order; (2) the circumstances of the order; (3) the source of the order; and (4) a statement that the contractor considers the order to be a change to the contract.[4]

Thus, a contractor can comply with notice requirements in a variety of ways so long as the communication provides general notice of the date, source and context of the purported change, and communicates that the contractor believes a change, formal or constructive, has been directed by the government.

---

[2]FAR § 2.101 Definitions.

[3]*Corbett Tech. Co.*, ASBCA No. 49478, 00-1 BCA ¶ 30,801; *JGB Enters., Inc.*, ASBCA No. 49493, 96-2 BCA ¶ 28,498.

[4]*J.M. Covington Corp.*, ASBCA No. 15633, 73-2 BCA ¶ 10,235.

The Changes clause also provides that the contractor's claims will be limited to those costs incurred no earlier than 20 days before the contractor gives written notice of an alleged constructive change, except for adjustments due to defective specifications.[5] This time limitation has been relaxed by judicial and board decisions,[6] but the contractor should seek to submit its notice no later than 20 days after it begins to incur costs on the constructive change. Regarding defective specifications for which the government is responsible, the contractor is entitled to recover all reasonable costs that it incurs, whether those costs were incurred more than 20 days before the notice is submitted or not. But the contractor is well advised to submit its notice as promptly as possible even where a defective specification is encountered to minimize arguments regarding timeliness of the notice and possible prejudice to the government.

The contractor also needs to give notice of a change before final payment. Notice first given after final payment will bar the claim.[7] However, a claim submitted before final payment still may be considered after final payment.[8]

B.   Differing Site Conditions Clause

The Differing Site Conditions clause[9] at FAR § 52.236-2 contains these notice terms:

(a)   The Contractor shall promptly, and before conditions are disturbed, give a written notice to the Contracting Officer of

    (1)   subsurface or latent physical conditions at the site which differ materially from those indicated in this contract, or

    (2)   unknown physical conditions at the site, of an unusual nature, which differ materially from those ordinarily encountered and generally recognized as inhering in work of the character provided for in the contract.

(b)   The Contracting Officer shall investigate the site conditions promptly *after receiving the notice.*

(c)   No request by the Contractor for an equitable adjustment to the contract under this clause shall be allowed, unless the Contractor has given the written notice required; provided, that the time prescribed in (a) above for giving written notice may be extended by the Contracting Officer. [Emphasis added.]

---

[5]FAR § 52.243-4(d).

[6]*Hoel-Steffen Constr. Co. v. United States,* 197 Ct. Cl. 561, 456 F.2d 760 (1972); *Smith & Pittman Constr. Co.,* AGBCA No. 76-131, 77-1 BCA ¶ 12,381; *R. R. Tyler,* AGBCA No. 381, 77-1 BCA ¶ 12,227; *J. L. Pitts Constr. Co.,* AGBCA No. 311, 75-2 BCA ¶ 11,535; *Russell Constr. Co.,* AGBCA No. 379, 74-2 BCA ¶ 10,911; *Davis Decorating Service,* ASBCA No. 17342, 73-2 BCA ¶ 10,107.

[7]*Gulf & Western Iindus., Inc.,* ASBCA No. 22204, 79-1 BCA ¶ 13,706; *JT Constr. Co.,* ASBCA No. 54352, 06-1 BCA ¶ 33,182.

[8]*Gulf & Western Iindus., Inc.,* ASBCA No. 22204, 79-1 BCA ¶ 13,706; *Jo-Bar Mfg. Corp. v. United States,* 535 F.2d 62 (Ct. Cl. 1976).

[9]Differing site conditions are discussed more fully in **Chapter 7.**

Under the Differing Site Conditions clause, the contractor is required to give prompt written notice to the contracting officer of a differing site condition before it is disturbed.[10] The contracting officer is then responsible to undertake an investigation and, if he or she finds that the conditions are materially different than those indicated in the contract or than could have reasonably been anticipated by the contractor, an equitable adjustment should be issued. If the parties fail to agree on whether a differing site condition has been encountered or whether an adjustment is necessary in the contract price or time, then the matter is to be resolved pursuant to the Disputes clause as discussed in **Chapter 15.**

Notice procedures under the Differing Site Conditions clause are very similar to the procedures under the Changes clause. However, the Differing Site Conditions clause also requires the contracting officer to conduct an investigation of the allegedly different condition before it is disturbed by the contractor. Failure to allow the contracting officer an opportunity to investigate may result in prejudice to the government that could preclude an adjustment for the differing site condition.[11]

C. Suspension of Work Clause

The Suspension of Work clause[12] at FAR § 52.242-14 contains this notice provision:

(c) A claim under this clause shall not be allowed—

    (1) For any costs incurred more than 20 days before the Contractor shall have notified the Contracting Officer in writing of the act or failure to act involved (but this requirement shall not apply as to a claim resulting from a suspension order); and

    (2) Unless the claim, in an amount stated, is asserted in writing as soon as practicable after the termination of the suspension, delay, or interruption, but not later than the date of final payment under the contract. [Emphasis added]

The Suspension of Work clause indicates that a contractor is not entitled to compensation for costs incurred more than 20 days before the contractor notifies the contracting officer in writing of the act or failure to act, other than a suspension that results from a contracting officer's order. In addition, the contractor should submit its claim promptly after the end of the suspension, delay, or interruption, and must do so before final payment is made under the contract. As discussed in more detail in **Section IV** of this chapter, courts and boards might not strictly enforce the notice requirements under Subsection (c)(1) and have allowed oral notice, particularly where the government has actual or constructive notice of the suspension of the contractor's work.[13] If the government does not have knowledge of the act or failure

---

[10]FAR § 52.236-2; *Ace Constructors, Inc. v. United States,* 70 Fed. Cl. 253 (2006).

[11]*A & M Gregos, Inc.,* PSBCA No. 632, 81-1 BCA ¶ 15,083; *see also Hemphill Contracting Co, Inc.,* ENGBCA No. 5776, 94-1 BCA ¶ 32,595.

[12]Suspensions of work are discussed more fully in **Chapter 9.**

[13]*Hoel-Steffen Constr. Co. v. United States,* 456 F.2d 760 (Ct. Cl. 1972). *See also Leiden Corp.,* ASBCA No. 26136, 84-1 BCA ¶ 16,947; *Strate,* ASBCA No. 19914, 78-1 BCA ¶ 13,128; *J. J. Welcome Constr. Co.,* ASBCA No. 19653, 75-1 BCA ¶ 10,997; *M.M. Sundt Constr. Co.,* ASBCA No. 17475, 74-1 BCA ¶ 10,627.

to act in some form and it suffers prejudice due to the lack of notice, then the claim will be considered waived, unless the suspension results from a direct order by the contracting officer.[14]

Although the initial notice requirements under a government issued suspension order may not be strictly enforced, the notice provisions under Subsection (c)(2) are strictly enforced and a claim under the Suspension of Work clause first asserted after final payment is barred.[15]

D. Terminations

1. Termination for Convenience Clause

The Termination for Convenience clause[16] at FAR § 52.249-1 provides:

TERMINATION FOR CONVENIENCE OF THE GOVERNMENT (FIXED-PRICE) (SHORT FORM) (APR. 1984)

The Contracting Officer, *by written notice* may terminate this contract in whole or in part, when it is in the Government's interest. If this contract is terminated, the rights, duties, and obligations of the parties, including compensation to the Contractor, shall be in accordance with part 49 of the Federal Acquisition Regulation in effect on the date of this contract. [Emphasis added]

Under the Termination for Convenience clause, the government is responsible for providing a written termination notice to the contractor. However, even if the government fails to provide written notice and provides only verbal notice to the contractor at the time of termination, the contractor's recovery is limited by the Termination for Convenience clause.[17] Pursuant to FAR § 49.102(a), the government's termination notice should include:

(a) whether the contract is being terminated for convenience of the government or for default;

(b) the effective date of termination;

(c) the extent of termination (partial or total);

(d) any special instructions; and

(e) the steps the contractor should take to minimize the impact on personnel if the termination will lead to a significant reduction in the contractor's workforce.

After the contractor has been properly notified of the termination for convenience, the contractor should follow the regulations with regard to submitting its termination

---

[14]*Electrical Enters., Inc.,* IBCA No. 971, 74-1 BCA ¶ 10,528.

[15]*Southwest Eng'g Co.,* DOTCAB No. 70-26, 71-1 BCA ¶ 8818. *See* **Chapter 15** for a further discussion of contractor claim releases.

[16]Terminations are more fully discussed in **Chapter 11.**

[17]*Neathery,* DOTCAB No. 4177, 04-2 BCA ¶ 32,649; *Albano Cleaners v. United States,* 455 F.2d 556 (Ct. Cl. 1972).

cost proposal. In particular, the contractor must file a termination proposal within one year of the effective date of the termination.[18] Failure to provide a settlement proposal can result in a waiver of the contractor's claims.[19] However, boards also have recognized that the one-year deadline for the termination settlement proposal does not apply where there is a constructive termination for convenience following the conversion of a wrongful default termination.[20]

2. Termination for Default Clause

The Termination for Default clause at FAR § 52.249-10 provides:

DEFAULT (FIXED-PRICE CONSTRUCTION) (APR. 1984)

(a) If the Contractor refuses or fails to prosecute the work or any separable part, with the diligence that will insure its completion within the time specified in this contract including any extension, or fails to complete the work within this time, the Government may, by written notice to the Contractor, terminate the right to proceed with the work (or the separable part of the work) that has been delayed.

* * *

(b) The Contractor's right to proceed shall not be terminated nor the Contractor charged with damages under this clause, if—

(1) The delay in completing the work arises from unforeseeable causes beyond the control and without the fault or negligence of the Contractor.

* * *

(2) *The Contractor, within 10 days from the beginning of any delay (unless extended by the Contracting Officer), notifies the Contracting Officer in writing of the causes of delay.* [Emphasis added]

Under the Termination for Default clause, the contractor may be able to avoid a default termination if the contractor timely notifies the contracting officer in writing of the causes of delay and those causes arise from unforeseeable conditions beyond the control of, and without the fault or negligence of, the contractor and its subcontractors and suppliers at all tiers. Thus, a contractor seeking to prevent a termination for default must promptly notify the contracting officer of the reasons why the delay

---

[18]FAR § 49.206-1(a). That FAR provision permits, but does not require, the contracting officer to extend that one-year period.

[19]*Rivera Technical Products, Inc.*, ASBCA Nos. 48171, 49564, 96-2 BCA ¶ 28,564. *See also M-Pax, Inc.*, HUD BCA No. 81-570, 81-2 BCA ¶ 15,409; *R-D Mounts, Inc.*, ASBCA No. 17,667, 74-2 BCA ¶ 10,740; *Prestex, Inc.*, ASBCA No. 8663 and 9726, 1964 BCA ¶ 4348.

[20]*Earth Property Services, Inc.*, ASBCA No. 36764, 91-2 BCA ¶ 23,753.

is not the fault of the contractor and how the delay falls within the scope of the excusable delays recognized under the contract.

E. Notice of Claims

The Disputes clause at FAR § 52.233-1[21] sets out the requirement for notice and demand for payment of a claim.[22] It provides in part:

> (c) "Claim," as used in this clause, means a *written demand or written assertion* by one of the contracting parties seeking, as a matter of right, the payment of money in a sum certain, the adjustment or interpretation of contract terms, or other relief arising under or relating to this contract . . . [A] written demand or written assertion by the Contractor seeking the payment of money exceeding $100,000 is not a claim under the Act until certified. . . .

> (d)
> > (1) A claim by the Contractor shall be made in writing and, unless otherwise stated in this contract, submitted . . . to the Contracting Officer for a written decision. [Emphasis added]

Although no particular wording or format is required,[23] the contractor must provide sufficient information that the contracting officer has notice of the basis for (entitlement) and the amount (quantum) of the claim.[24] The elements of a claim under the Disputes clause can be complex; one element is consistent with the other clauses discussed in this chapter. Specifically, the party seeking an adjustment to the contract price or terms must assert its position by written notice.

F. Notice Checklists

Regardless of the project type or delivery system, preparation of notice checklists should be a mandatory requirement for the project staff prior to mobilization to the site. The project staff needs to be familiar with the contractor's notice obligations to the government and to its subcontractors and suppliers. The staff should screen contract documents for any special notice requirements and evaluate their effect on the project's planned documentation process. Although it may seem more efficient on government contracts to provide the staff with a basic list of notice requirements, a benefit in having this done for every project is that the preparation of a notice checklist essentially requires project personnel to review the contract carefully.

---

[21]The Disputes clause at FAR § 52.233-1 currently dates from July 2002.
[22]The claims and disputes procedures are addressed more fully in **Chapter 15.**
[23]*Contract Cleaning Maint., Inc. v. United States,* 811 F.2d 586, 592 (Fed. Cir. 1986); *Scott Timber Co. v. United States,* 333 F.3d 1358, 1365 (Fed. Cir. 2003).
[24]*Gauntt Constr. Co.,* ASBCA No. 33323, 87-3 BCA ¶ 20,221; *Mitchco, Inc.,* ASBCA No. 41847, 91-2 BCA ¶ 23,860; *Holk Dev., Inc.,* ASBCA No. 40579, 90-3 BCA ¶ 23,086.

Preparation of a meaningful notice checklist entails a diligent review of the contract, even if the final product fits on a single piece of paper. **Appendix 14A** to this chapter is a sample format for a Notice Checklist. An electronic version of that document is also found on the support Web site at www.wiley.com/go/federalconstructionlaw. **Appendix 14B** is a partially completed Notice Checklist for a typical government construction contract.[25]

III. NOTICE TO REPRESENTATIVES/AGENTS OF CONTRACTING OFFICER

The Changes, Differing Site Conditions, Suspension of Work, and Termination for Default clauses all require the contractor to provide certain notices to the *contracting officer* as the government's authorized representative. A contractor's failure to provide required notice to the contracting officer may result in a waiver of claims, even if the notice is given to other government personnel. For example, notice of a change proposal submitted to a government research laboratory was invalid because it had not been submitted to the contracting officer.[26]

In some circumstances, however, the notice provisions may be satisfied without providing notice specifically to the contracting officer.[27] In the absence of the contracting officer, the notice requirements may be satisfied if notice is given to the contracting officer's authorized representative prior to final payment.[28] Discussions between the contractor and the contracting officer's job site representative, and the knowledge of other personnel such as a contract specialist, base engineer, or other contracting officer technical representative, also may be sufficient to show that the contracting officer had sufficient imputed knowledge of the change to overcome the requirement for specific notice to the contracting officer.[29]

Notice to, or knowledge of, government personnel other than the contracting officer or an authorized representative also may be sufficient to satisfy notice requirements. The issue usually turns on whether the government has been prejudiced by a lack of written notice to the contracting officer. The boards generally have found that notice specifically to the contracting officer is not required if notice is provided to a government representative (other than the contracting officer) so that the government is not prejudiced, such as where the contracting officer has actual or imputed knowledge of the condition or the contracting officer considers the claim on its merits without asserting a lack of proper notice.[30]

[25]Consistent with the thought that the project staff should prepare a Notice Checklist for each project, **Appendix 14B** is *not* included on the support Web site.

[26]*Vogt Bros. Mfg. Co. v. United States,* 160 Ct. Cl. 687 (1963).

[27]Authority to bind the government is discussed more fully in **Chapter 2.**

[28]*Aerodex, Inc.,* ASBCA No. 7121, 1962 BCA ¶ 3492.

[29]*B. V. Constr., Inc.,* ASBCA No. 47766, 49337, 50553, 04-1 BCA ¶ 32,604; *Davis Decorating Service,* ASBCA No. 17342, 73-2 BCA ¶ 10,107.

[30]*Korshoj Constr. Co.,* IBCA No. 321, 1963 BCA ¶ 3848; *Harper Dev. Assocs.,* ASBCA No. 34719, 90-1 BCA ¶ 22,534; *SUFI Network Services, Inc.,* ASBCA No. 55306, 06-2 BCA ¶ 33,444.

The burden of demonstrating prejudice generally falls on the government.[31] The finding of prejudice is less likely where the government's employees directly involved with the project have actual or imputed knowledge of the condition encountered by the contractor and knowledge that the condition is adversely affecting the contractor's work. For example, where the government's resident engineer directed the contractor to place additional rock in a particular location, the contracting officer was deemed to have imputed knowledge of the work through the resident engineer.[32] Similarly, a field inspector's reports showed sufficient government notice of a contractor's claim.[33] Where the government's area engineer approved changes to the locations for steel beams due to errors in the government's drawings, the area engineer's knowledge of the condition overcame the need for written notice.[34] Finally, the requirement that notice be provided directly to the contracting officer may be waived where the contracting officer reviews the contractor's claim without asserting a lack of proper notice.[35]

The contractor bears the burden, however, of proving that the government's representative had knowledge of the actual condition to overcome the contractor's failure to provide the required notice. For example, a contractor's assertion that it had provided notice to a responsible government official and, therefore, had substantially complied with the notice requirement was rejected when the evidence indicated that the contractor had provided written notice to the contracting officer in all other instances involving changes.[36]

These fairly liberal notice decisions seem to imply that, under certain circumstances, providing notice to government representatives other than the contracting officer may be an acceptable way to satisfy a contractor's notice obligation. However, there is a substantive risk associated with a failure to provide written notice to the contracting officer. Contractors should not be misled by the title of contracting officer's representative (COR). Many CORs do not have any authority to change the contract price, time, or terms and conditions. Written notice to a COR may be completely futile if the government is able to successfully assert a defense to a claim for an equitable adjustment based on the COR's lack of authority to change the contract.[37] Although it may be appropriate to provide written notice of a change, delay, or differing site condition to a COR in order to maintain effective communications on a project, sending a copy of that notice to the contracting officer is an essential project documentation practice for every contractor. That practice minimizes the chances that resolution of the issue will turn on the lack of authority defense.

---

[31]*Gibbs Shipyard, Inc.*, ASBCA No. 9809, 67-2 BCA ¶ 6499; *Flathead Contracting, Inc.*, AGBCA No. 2005-130-1, 06-1 BCA ¶ 33,174; *SUFI Network Services, Inc.*, ASBCA No. 55306, 06-2 BCA ¶ 33,444.

[32]*J. D. Abrams*, ENGBCA No. 4332, 89-1 BCA ¶ 21,379.

[33]*Hartford Accident & Indemn. Co.*, IBCA No. 1139, 77-2 BCA ¶ 12,604.

[34]*A.M. Turner Co.*, ASBCA No. 13384, 71-1 BCA ¶ 8648.

[35]*Eggers & Higgins and Edwin A. Keeble Assocs., Inc. (a Joint Venture) v. United States*, 403 F.2d 225 (Ct. Cl. 1968); *Webb Elec. Co. of Fla., Inc.*, ASBCA No. 54293, 07-2 BCA ¶ 33,717.

[36]*Superior Asphalt & Concrete Co.*, AGBCA No. 78-162, 81-1 BCA ¶ 15,102.

[37]DFARS § 252.201-7000; *see also Winter v. Cath-dr/Balti Joint Venture*, 497 F.3d 1339 (Fed. Cir. 2007).

IV. WRITTEN VERSUS ALTERNATIVE FORMS OF NOTICE

A. General Considerations

It is always advisable for the contractor to provide notice of a change, differing site condition, or suspension of work in writing to the contracting officer.

However, the boards and courts have carved out a number of exceptions to the written notice requirements.[38] Generally, a lack of written notice will not preclude a claim where: (1) the government has not been prejudiced or injured by the failure of the contractor to issue a written protest or notice; (2) the government's records or testimony show that the contracting officer or an authorized representative of the contracting officer knew of the circumstances that formed the basis of the alleged extra, change or changed condition;[39] and (3) the contracting officer considered the contractor's claim on the merits without invoking the protest or notice requirement.[40] Conversely, where the contractor's failure to provide proper written notice prevents the government from exploring the condition or proposed change so that the government is prejudiced from pursuing an alternative method of performing the work, then a lack of proper notice will bar a claim.[41]

B. Alternative Notice under the Changes Clause

1. Written Notice Exceptions: No Prejudice to the Government

The government bears the burden of showing that it was prejudiced by a lack of proper notice: The failure to meet that burden can obviate the need for written notice. For example, in *Telecommunications Services, Inc.,*[42] the Veterans Affairs Board of Contract Appeals (VABCA) found that the government had changed a telecommunications equipment contract when it issued new drawings relocating a control room and reconfiguring the equipment in that room. The board rejected the government's argument that the contractor waived its claim for extra costs because it had not given timely notice. The board found that the government had not demonstrated any prejudice as a result of the lack of notice.

The government also must establish prejudice in order to defeat a claim for increased costs. Where the government is aware of unsuitable materials or other conditions that will adversely affect the contractor, and the government is aware that the contractor will incur additional costs as a result of that condition, the government is not entitled to rely on a defense of prejudice.[43] In addition, the government

[38]*See, e.g., Northrup Carolina, Inc.,* ASBCA No. 13958, 71-2 BCA ¶ 8970; *Jackson & Church Co.,* ASBCA No. 12229, 68-1 BCA ¶ 6815; *Aerodex, Inc.,* ASBCA No. 7121, 1962 BCA ¶ 3492.

[39]*Korshoj Constr. Co.,* IBCA No. 321, 1963 BCA ¶ 3848; *Kiewit Sons,* ASBCA No. 5600, 60-1 BCA ¶ 2580; *Flathead Contracting, Inc.,* AGBCA No. 2005-130-1, 06-1 BCA ¶ 33,174.

[40]*Monarch Lumber Co.,* IBCA No. 217, 60-2 BCA ¶ 2674; *Pittsburgh-Des Moines Corp.,* EBCA No. 314-3-84, 89-2 BCA ¶ 21,739; *Precision Standard, Inc.,* ASBCA No. 54027, 03-2 BCA ¶ 32,265.

[41]*Todd v. United States,* 292 F.2d 841 (Ct. Cl. 1961); *Calfon Constr., Inc. v. United States,* 18 Cl. Ct. 426 (1989); *Ace Constructors, Inc. v. United States,* 70 Fed. Cl. 253 (2006).

[42]VABCA Nos. 1218, 1219, 1185, 77-2 BCA ¶ 12,847.

[43]*Lockheed Aircraft Corp.,* ASBCA No. 9396, 65-1 BCA ¶ 4689.

cannot rely on mere allegations but must provide specific proof that the lack of notice caused it prejudice.[44] Prejudice generally is shown by (1) the government's inability to investigate or defend against a claim and (2) preventing the government from considering alternatives to actions taken by the contractor.[45]

2. Written Notice Exceptions: Actual Knowledge

Where the government is aware of the cause of extra work, the failure to give timely notice will not invalidate a claim. For example, where the government is aware of errors in its own drawings,[46] or additional costs result from a change that is a matter of common knowledge,[47] then the government will be deemed to have sufficient knowledge to obviate the need for written notice. Similarly, knowledge of a government inspector and resident engineer of a condition encountered by the contractor, and that the contractor would incur additional costs as a result of that condition, obviated the need for written notice of the claim to the contracting officer.[48]

3. Written Notice Exceptions: Consideration of the Claim

Where the contracting officer considers a claim under the Changes clause without asserting inadequate notice, the government may be deemed to have waived the notice defense. In *Fox Valley Engineering, Inc. v. United States*,[49] the court recognized that lack of proper, written notice is a good defense but held that the defense was not available to the government when the contracting officer and the board considered the claim on its merits and did not address a lack of proper notice. However, if the contracting officer does not consider the contractor's claims on the merits, the government does not waive the improper notice defense.[50]

C. Alternative Notice under the Differing Site Conditions Clause

1. Generally

Although the Differing Site Conditions clause calls for the contractor to give written notice promptly and before conditions are disturbed,[51] courts and boards have

---

[44]*Power Line Erectors, Inc.,* IBCA No. 637, 69-1 BCA ¶ 7417; *Valley Asphalt, Inc.,* AGBCA No. 97-118-1, 97-2 BCA ¶ 28,997; *SUFI Network Services, Inc.,* ASBCA No. 55306, 06-2 BCA ¶ 33,444.

[45]*Mel Williamson Constr. Co.,* VABCA No. 1199, 76-2 BCA ¶ 12,168; *Ace Constructors, Inc. v. United States,* 70 Fed. Cl. 253 (2006).

[46]*Lormack Corp.,* IBCA No. 652, 69-2 BCA ¶ 7989; *ACS Constr. Co.,* GSBCA No. 5665, 81-1 BCA ¶ 14,933.

[47]*Langevin v. United States,* 100 Ct. Cl. 15 (1943).

[48]*Valley Constr. Co.,* ASBCA No. 9819, 65-1 BCA ¶ 4753; *see also Dawco Constr., Inc. v. United States,* 18 Cl. Ct. 682 (1989).

[49]151 Ct. Cl. 228 (1960); *see also Dittmore-Freimuth Corp. v. United States,* 182 Ct. Cl. 507, 390 F.2d 664 (Ct. Cl. 1968).

[50]*Eggers & Higgins and Edwin A. Keeble Assocs., Inc. (a Joint Venture) v. United States,* 403 F.2d 225 (Ct. Cl. 1968); *LaDuke Constr. and Krumdieck, Inc., J.V.,* AGBCA No. 83-177-1, 90-1 BCA ¶ 22,302.

[51]FAR § 52.236-2(a).

allowed a broad array of oral and written notices to satisfy the notice provision. For example, a contractor's oral representations to the government's resident engineer have been found to satisfy notice requirements.[52] Similarly, oral notice was sufficient where it was made to the government representative immediately after discovery of the condition and the government had the opportunity to investigate the condition before it was disturbed.[53]

Written notice to the government of a differing site condition does not have to be extremely detailed. The notice need only provide a summary of the facts underlying the claim without any detailed substantiation.[54]

2. Written Notice Exceptions: Lack of Prejudice

Where the government is not prejudiced by the lack of written notice, failure to provide written notice will not preclude compensation to the contractor for a differing site condition. In *DeMauro Construction Corp.*,[55] the board stated that "[w]e have held for many years that this notice [obligation] should be waived if the lack of it did not result in prejudice to the government."[56] The government bears the burden of proving that prejudice.[57]

Although the requirements for notice under the Differing Site Conditions clause have been relaxed by the courts and boards interpreting those provisions, the contractor still has a duty to be sure that the government is aware of the differing site condition. The purpose of the written notice requirement is to allow the government sufficient time to investigate the condition and exercise some control over the time and cost expended to resolve the problem.[58] If the government does not have actual knowledge of the condition and notice is not provided in a timely manner, then prejudice may arise where the contractor delays notice so that the government cannot fully investigate the condition.[59]

3. Written Notice Exceptions: Actual Knowledge

As with notice under the Changes clause, the notice requirements under the Differing Site Conditions clause may be waived where the contracting officer or an authorized representative has actual knowledge or constructive notice of the differing site

---

[52]*General Cas. Co. of America v. United States,* 127 F. Supp. 805 (Ct. Cl. 1955); *see also Ace Constructors, Inc. v. United States,* 70 Fed. Cl. 253 (2006).

[53]*Edgar M. Williams, Gen. Contractor,* ASBCA No. 16058, 72-2 BCA ¶ 9734.

[54]*Heppner Eng'g Co., Inc.,* GSBCA No. 871, 65-1 BCA ¶ 4723; *Ace Constructors, Inc. v. United States,* 70 Fed. Cl. 253 (2006).

[55]ASBCA No. 17029, 77-1 BCA ¶ 12,511 at 60,650.

[56]*See also Tutor-Saliba-Perini,* PSBCA No. 1201, 87-2 BCA ¶ 19,775; *Shepherd v. United States,* 113 F. Supp. 648 (1953); *Rockwell Int'l Corp.,* EBCA No. C-9509187, 02-2 BCA ¶ 32,018.

[57]*Ace Constructors, Inc. v. United States,* 70 Fed. Cl. 253 (2006); *SUFI Network Services, Inc.,* ASBCA No. 55306, 06-2 BCA ¶ 33,444.

[58]*Parker Excavating, Inc.,* ASBCA No. 54637, 06-1 BCA ¶ 33,217.

[59]*A & M Gregos, Inc.,* PSBCA No. 632, 81-1 BCA ¶ 15,083; *Barnet Brezner,* ASBCA No. 9967, 65-2 BCA ¶ 4902; *Ace Constructors, Inc. v. United States,* 70 Fed. Cl. 253 (2006).

condition or the government's interest is not prejudiced or injured by the lack of notice. The Department of Transportation Contract Appeals Board (DOTCAB) found in *Building Maintenance Corp.*[60] that the government had actual knowledge of deficiencies in power generators because it had tested the on-site generators five months before contract award and knew that there would be problems generating the power the contractor would need. This knowledge demonstrated that notice from the contractor was unnecessary because the government already knew of the problem, in this case, even before contract award.[61] Moreover, if the government is familiar with the difficulties being experienced by the contractor and is working with the contractor on a solution to that problem, the government cannot maintain that it does not have proper notice, even if written notice is not provided.[62]

However, the simple fact that the government may be aware that the contractor deviated from the contract requirements and does not object to that deviation will not be sufficient to establish the government's actual knowledge of a differing site condition. Instead, the contractor either needs to demonstrate that the government had actual knowledge of the specific differing site condition or must make clear that it has encountered a differing site condition that has adversely affected its work and that it expects to be compensated for additional costs. The failure to give notice in this latter circumstance can result in the contractor's claim being lost.[63]

In addition, the contractor may be unable to rely entirely on the government's awareness of a condition as constituting constructive notice where the contractor fails to give actual notice—orally or in writing—to the contracting officer or the contracting officer's authorized representative. In *Leavell & Co.*,[64] the ASBCA found that the government had been prejudiced because lack of notice made it more difficult, but not impossible, for the government to investigate the condition. Despite finding some level of prejudice, the board found that lack of notice was not enough to bar the claim outright. However, the contractor faced a greater burden of persuasion to prove the cause and effect of the differing site condition than if it had given proper notice.[65]

4. Written Notice Exceptions: Consideration of Claims

If the contracting officer considers a claim based on an alleged differing site condition without adequate notice, it will effectively waive the notice requirement.[66]

---

[60]DOTCAB Nos. 76-42, 76-42A, 76-42B, 79-1 BCA ¶ 13,560.

[61]*See also Delphcon/J.A.M.E., J.V.,* PSBCA No. 1455, 88-3 BCA ¶ 21,107; *J&J Paving, Inc.,* DOTCAB No. 1570, 85-1 BCA ¶ 17,840; *R. R. Tyler,* AGBCA No. 381, 77-1 BCA ¶ 12,227; *Russell Constr. Co.,* AGBCA No. 379, 74-2 BCA ¶ 10,911.

[62]*George A. Fuller Co.,* ASBCA No. 8524, 1962 BCA ¶ 3619; *Peter Kiewit Sons' Co.,* ASBCA No. 5600, 60-1 BCA ¶ 2580.

[63]*M.S.I. Corp.,* VACAB No. 730, 68-2 BCA ¶ 7177; *Calfon Constr., Inc. v. United States,* 18 Cl. Ct. 426 (1989); *Eichberger Enters., Inc.,* VABCA No. 3923, 95-2 ¶ 27,693.

[64]ASBCA No. 16099, 72-2 BCA ¶ 9694, *reconsid. denied,* 73-1 BCA ¶ 9781.

[65]*See also Acme Missiles & Constr. Corp.,* ASBCA No. 13531, 71-1 BCA ¶ 8641; *Resource Conservation Corp. v. General Services Admin.,* GSBCA No. 13399, 97-1 BCA ¶ 28,776.

[66]*Morgan Constr. Co.,* IBCA No. 299, 1963 BCA 3855; *Dayton Constr. Co.,* HUDBCA No. 82-746-C34, 83-2 BCA ¶ 16,809.

However, if the contractor appeals the contracting officer's final decision, the waiver of notice by consideration of the claim is not binding on a court or board. In *Schnip Bldg. Co. v. United States,*[67] the court found that the government still could raise untimely waiver as a defense even though the contracting officer had not asserted that defense in its final decision because the contractor's appeal vacated the contracting officer's decision. Proceedings before the board or Court of Federal Claims are *de novo,* which allows the government to raise a lack of notice or untimely notice as a renewed defense.[68] However, where the contracting officer *and* the board consider the contractor's claims on their merits without raising lack of notice as a defense, then a lack of notice will not serve as a defense to the claim on appeal.[69]

D. Alternative Notice under the Suspension of Work Clause

Courts and boards do not always strictly enforce the written notice requirements under Subsection (c)(1) of the Suspension of Work clause and have allowed oral notice. This is particularly true in situations where the government has actual or constructive notice of the suspension of the contractor's work. In *Hoel-Steffen Construction Co. v. United States,*[70] the Court of Claims reversed an Interior Board of Contract Appeals decision that had limited the contractor's recovery of costs to only those incurred within 20 days of a written notice. The court held that the government had effectively been placed on notice of the suspension by written and oral communications from the contractor and that the Suspension of Work clause did not require a specific monetary claim to be made within the 20-day period. Similarly, in *Davis Decorating Service,*[71] the ASBCA found that a contractor's repeated oral protests to the contracting officer provided sufficient notice.

Contractors also have satisfied the notice requirements where government employees have constructive knowledge of a suspension situation. In *Raby Hillside Drilling, Inc.,*[72] the Agriculture Board of Contract Appeals (AGBCA) held that the notice provisions were satisfied by constructive notice because government inspectors who worked on the project on a daily basis were aware of the suspension.[73] Similarly, where the contractor made oral protests to the government's resident engineer who recorded the complaints in a diary, the notice provisions were considered to have been satisfied.[74]

---

[67]645 F.2d 950 (Ct. Cl. 1981).

[68]*See* **Chapter 15** for a discussion of the status of a final decision once an appeal is filed.

[69]*Blount Bros. Corp. v. United States,* 424 F.2d 1074 (Ct. Cl. 1970), *aff'g Blount Bros. Corp.,* GSBCA No. 1385, 66-1 BCA ¶ 5652; *Whittaker Corp. v. United States,* 443 F.2d 1373 (Ct. Cl. 1971) (denying the government's argument that the plaintiff did not exhaust administrative remedies when requesting a time extension because both the contracting officer and the board considered the plaintiff's request on its merits without raising an improper request for time extension defense).

[70]456 F.2d 760 (Ct. Cl. 1972).

[71]ASBCA No. 17342, 73-2 BCA ¶ 10,107.

[72]AGBCA No. 75-101, 78-1 BCA ¶ 13,026.

[73]*See also Pittsburgh-Des Moines Corp.,* EBCA No. 314-3-84, 89-2 BCA ¶ 21,739; *Singleton Contracting Corp.,* IBCA No. 1413-12-80, 81-2 BCA ¶ 15,269; *Gresham & Co. v. United States,* 470 F.2d 542 (Ct. Cl. 1972).

[74]*Burn Constr. Co.,* IBCA No. 1042, 78-2 BCA ¶ 13,405; *see also Kumin Assocs.., Inc.,* LBCA No. 94-BCA-3, 98-2 BCA ¶ 30,007; *A.R. Mack Constr. Co.,* ASBCA No. 50035, 01-2 BCA ¶ 31,593.

Although oral notice may be acceptable in some circumstances, the contractor must give notice in some form or else it may waive its claim, unless the suspension results from a direct order by the contracting officer.[75]

V. FAILURE TO GIVE NOTICE—CONSEQUENCES

Lack of timely notice is often a costly omission for contractors and can be fatal to a claim. A contractor that fails to provide required notice may find that it has waived its claim.

A. Lack of Notice: Potential Adverse Consequences

1. *Changes*

Where the government orally orders a change to the work and the contractor fails to give notice of its belief that the direction is a change, then the contractor may be deemed to have agreed to perform the work without compensation.[76] The contractor may be considered a volunteer that has waived its claims for additional costs.[77] A contractor that remains silent and fails to protest or notify the contracting officer that the contractor will claim an extra for changed work will be found to have waived any claim for extra costs.[78]

Timely notice is intended to give the government an opportunity to investigate the condition that the contractor maintains will result in extra costs or additional time beyond that contemplated by the contract. With regard to changes directed by the government, constructive changes resulting from defective specifications, or conditions encountered that the contractor could not have reasonably anticipated, the requirement that the contractor provide notice is designed to ensure that the government has an opportunity to investigate the merits of the claim. The government should then make a timely investigation. However, the contractor that waits for a long period of time before asserting a claim often will be deemed to have waived the claim.

Claims have been held barred where there is a showing of substantial prejudice to the government.[79] Boards have found that the failure to make a timely objection to the contracting officer's order and a long delay in asserting a claim raises a presumption against its validity.[80] Even if the lack of timely notice was deemed to have

---

[75]*Electrical Enters., Inc.*, IBCA No. 971, 74-1 BCA ¶ 10,528.

[76]*Science Mgmt. Corp.*, EBCA No. 289-5-83; 84-2 BCA ¶ 17,319; *see also Northrop Grumman Corp. v. United States*, 47 Fed. Cl. 20 (2000).

[77]*Science Mgmt. Corp.*, EBCA No. 289-5-83; 84-2 BCA ¶ 17,319; *see also Corbett Tech. Co.*, ASBCA No. 49478, 00-1 BCA ¶ 30,801; *Ford Constr. Co.*, AGBCA No. 252, 71-2 BCA ¶ 8966.

[78]*Corbett Tech. Co.*, ASBCA No. 49478, 00-1 BCA ¶ 30,801; *Ford Constr. Co.*, AGBCA No. 252, 71-2 BCA 8966.

[79]*DeMauro Constr. Corp.*, ASBCA No. 17029, 77-1 BCA ¶ 12,511; *Eggers & Higgins and Edwin A. Keeble Assocs., Inc. (a Joint Venture) v. United States*, 403 F.2d 225 (1968); *SUFI Network Services, Inc.*, ASBCA No. 55306, 06-2 BCA ¶ 33,444.

[80]*Thorn Constr. Co.*, IBCA No. 1254-3-79, 83-1 BCA ¶ 16,230; *Futuronics, Inc.*, DOTCAB No. 67-15, 68-2 BCA ¶ 7079.

adversely affected the government but not enough to bar the claim, some decisions have placed an increased burden on the claimant to the extent that untimeliness has made the government's defense more difficult.[81]

The notice requirements of the Changes clause may be waived by the government where there is no showing of prejudice to the government, where the government knows of the condition and its potential for extra costs, or where the contracting officer evaluates a claim without raising the inadequate notice defense. If one of these three exceptions does not exist, however, a contractor's improper notice may invalidate a claim. Failure to notify the government within the required 20 days that the building plans did not meet local zoning requirements resulted in the rejection of a contractor's claim under the Changes clause.[82] Some board decisions have allowed leeway in the timing of the claim by not requiring the contractor to comply with the strict time deadlines other than that the claim must be submitted prior to final payment.[83]

2. Differing Site Conditions

Similarly, a contractor that encounters a differing site condition and fails to give notice to the contracting officer or its authorized representative before taking action runs the risk of waiving its claim.[84]

3. Suspension of Work

With regard to the Suspension of Work clause, a contractor that fails to give notice of any kind will waive its claim unless the suspension results from a direct suspension order by the contracting officer.[85] Also, any suspension claim first asserted after final payment is barred.[86]

Although the contractor may be able to avail itself of one of the exceptions to show that the government was not injured by inadequate notice, it is far simpler and less costly for the contractor to provide notice to the contracting officer whenever it believes it has encountered a change, differing site condition, or delay, than to defend against its failure to provide written notice.

---

[81]*Chimera Corp., Inc.*, ASBCA No. 18690, 76-1 BCA ¶ 11,901; *C. H. Leavell and Co.*, ASBCA No. 16099, 72-2 BCA ¶ 9694, *aff'd on reconsid.*, 73-1 BCA ¶ 9781; *SUFI Network Services, Inc.*, ASBCA No. 55306, 06-2 BCA ¶ 33,444.

[82]*Beacon-Hicksville, Inc.*, POD BCA No. 325, 71-1 BCA ¶ 8624.

[83]*Compare Wilcox Elec., Inc.*, DOTCAB No. 73-14, 74-2 BCA ¶ 10,725 *with Russell Constr. Co.*, AGBCA No. 380, 75-1 BCA ¶ 10,991. (With the consolidation of the various boards, these potentially inconsistent notice decisions may not provide a clear indication of the board's approach on notice issues.)

[84]*David Boland, Inc.*, ASBCA Nos. 48715, 48716, 97-2 BCA ¶ 29,166; *McElroy Machine & Mfg. Co.*, ASBCA No. 46477, 99-1 BCA ¶ 30,185.

[85]*Electrical Enters., Inc.*, IBCA No. 971-8-72, 74-1 BCA ¶ 10,528.

[86]*Southwest Eng'g Co.*, DOTCAB No. 70-26, 71-1 BCA ¶ 8818.

VI. COORDINATING NOTICE REQUIREMENTS IN SUBCONTRACTS AND PURCHASE ORDERS

A. General Considerations

For the general contractor, notice provisions in subcontracts and purchase orders may have a direct bearing on the contractor's notice obligations to the government. Changes, differing site conditions claims, or other claims often originate with a subcontractor that encounters an unanticipated condition or is directed to take some action by the government through the contractor. Contractors need to be particularly careful to coordinate the notice obligations in their subcontracts and purchase orders with the notice requirements in the contract with the government. Contractors also should be careful to hold their subcontractors to at least the same documentation requirements that the contractors have to the government.

Subcontractors and suppliers need to know and follow the notice and other documentation requirements in their subcontracts and purchase orders. The cautions and suggestions discussed in this chapter regarding the contractor's notice and documentation requirements to the government also apply to subcontractors and suppliers dealing with contractors on government projects. Subcontractors need to closely follow the subcontract requirements regarding notice of changes, differing site conditions, delays or suspensions of work, claims, and terminations. Failure to follow those requirements can result in a waiver of claims for money and time extensions. In addition, subcontractors also should use the documentation tools discussed in **Section VII** of this chapter to memorialize job conditions, unanticipated events, costs, and scheduling matters.

B. Flow-down Clause

Perhaps the most important provision a contractor can include in its subcontracts and purchase orders is a flow-down clause that binds the subcontractors and suppliers to the same documentation requirements that the contractor has to the government. The prime contractor should bind the subcontractor's performance to the contractor in the same manner as the contractor is bound to the government. To do otherwise would leave the contractor exposed to liability and unable to require the same performance from the subcontractor that the government can demand of the contractor. Similarly, the contractor should be willing to give the subcontractor the same rights of making claims, filing protests, and providing notices to the contractor as the contractor has to the government. In other words, the rights and duties should flow equally down from the government through the contractor to the subcontractor as well as flowing upward from the subcontractor through the contractor to the government. This will enable all of the parties to be governed by the same general requirements, even though there is no privity of contract between the subcontractor or a supplier and the government.

An example of a flow-down clause is found at ConsensusDOCS 752, Standard Subcontract Agreement for use on Federal Government Construction Projects ¶ 3.1:

3.1 Obligations The Contractor and Subcontractor are hereby mutually bound by the terms of this Agreement. To the extent the terms of the Prime Contract apply to the Subcontract Work, then the Contractor hereby assumes toward the Subcontractor all the obligations, rights, duties, and redress that the Owner under the Prime Contract assumes toward the Contractor. In an identical way, the Subcontractor hereby assumes toward the Contractor all the same obligations, rights, duties, and redress that the Contractor assumes toward the Owner under the Prime Contract. In the event of an inconsistency among the documents, the specific terms of this Agreement shall govern.

This type of clause can automatically incorporate all of the provisions of the prime contact into a subcontract or purchase order with minimal effort. Courts have recognized the validity of such clauses and their effectiveness to incorporate prime contract terms into the subcontract or purchase order.[87]

The Federal Acquisition Regulation Clauses Exhibit to ConsensusDOCS 752 also reinforces the application of the FAR to the subcontract through specific incorporation:[88]

The Subcontractor, by signing this Agreement, agrees to abide by the provisions of the Federal Acquisition Regulation, which are applicable to this Agreement in accordance with the Prime Contract.

The contractor also may wish to include a subcontract provision specifically addressing notice requirements. For example, ConsensusDOCS 752 ¶ 5.3.2 requires:

5.3.2 CLAIMS RELATING TO OWNER The Subcontractor agrees to initiate all claims for which the Owner is or may be liable in the manner, process, and within the time limits provided in the Subcontract Documents for like claims by the Contractor upon the Owner and in sufficient time for the Contractor to initiate such claims against the Owner in accordance with the Subcontract Documents. At the Subcontractor's request and expense to the extent agreed upon in writing, the Contractor agrees to permit the Subcontractor to prosecute a claim in the name of the Contractor for the use and benefit of the Subcontractor in the manner provided in the Subcontract Documents for like claims by the Contractor upon the Owner.

C. Clauses Relating to Changes, Extras, Differing Site Conditions, Work Suspensions, and Claim Presentation

One of the most important items of the subcontract is the procedure providing for changes and requiring that the subcontractor provide notice of its claims for extras.

---

[87]*See, e.g., MCC Powers v. Ford Motor Co.,* 184 Ga. App. 487, 361 S.E.2d 716 (1987).

[88]In addition, the FAR contains numerous clauses which may be used in construction contracts that impose flow-down obligations on the contractor. *See* **Section VII** of **Chapter 6** and **Appendix 6A**.

The prime contractor needs subcontract provisions that obligate the subcontractor (supplier) to provide the same types of notice that the general contractor is required to provide to the government but, perhaps, shorter time frames so that the general contractor has enough time after receiving notice from the subcontractor to provide notice to the government. Thus, the subcontract may contain specific changes, differing site conditions, suspension of work, termination, and claims clauses mirroring those discussed in **Section II** of this chapter but providing a shorter notice period than required by the prime contract.

The subcontract also should clearly identify to whom the subcontractor is to provide notice so there is no dispute regarding who is to receive the notice on behalf of the contractor. For those same reasons, the subcontract also should specifically identify the subcontractor's representative to whom a general contractor should provide change orders, notices of terminations, notices of claims against the subcontractor, and the other notices required by the subcontract.

In addition to requiring that a subcontractor provide the same documentation to the contractor that the contractor is required to provide to the government, a general contractor also should make certain that the subcontract is in harmony with the prime contract when it comes to proceeding with disputed work. In a government contract, the general contractor usually is required to proceed with changed work during a pending dispute over compensation for that change. For that reason, the contractor will want a disputes clause that similarly requires the subcontractor to proceed with its work while a dispute is being resolved, and when it is directed to do so by the contractor. Conversely, the subcontractor may not want to proceed with changed or extra work until compensation in time and money is agreed on, and may attempt to negotiate a subcontract provision that allows it to proceed only after an agreed change order has been negotiated.

D. Effect of a Subcontractor's Failure to Give Proper Notice/ Waiver of Notice by the Contractor

The subcontractor that fails to comply with written notice requirements risks loss of its remedy against the contractor. For example, in *Associated Mechanical Contractors, Inc. v. Martin K. Eby Construction Co.*,[89] the court held that a subcontractor that provided 10 letters of notice detailing delays still had waived its claims because the notices were not submitted within 10 days of commencement of the alleged delay, as required by the subcontract.

In certain circumstances, however, a subcontractor may be able to pursue its claims despite failing to comply strictly with the notice provisions. This may occur if the contractor waives the notice provision.[90] If the contractor had actual or constructive knowledge or notice of the condition, then that knowledge may overcome the subcontractor's failure to comply strictly with the subcontract notice requirements.[91]

[89]983 F. Supp. 1121 (M.D. Ga. 1997).
[90]*See Ballenger Corp. v. Dresco Mech. Contractors, Inc.*, 274 S.E.2d 786 (Ga. App. 1980).
[91]*See id.*; *United States ex rel. Yonker Constr. Co. v. Western Contracting Corp.*, 935 F.2d 936 (8th Cir. 1991).

Also, a general contractor that acts in bad faith in the performance of the subcontract may be deemed to have waived the notice provisions or other exculpatory clauses of the contract.[92]

E. Notice of Claims

Subcontractors must timely submit their claims in accordance with the terms of the subcontract or risk waiver of the claim.[93] In a situation where the subcontract calls for the subcontractor's relief to be limited to the relief that the contractor may recover from the government on behalf of the subcontractor, the subcontractor should comply with the notice provisions of the subcontract. The general contractor, however, also must comply with the terms of the prime contract with the government for submission of claims. For example, if the general contractor fails to submit a timely appeal on the subcontractor's behalf, the subcontractor would be barred from taking an indirect appeal even though the general contractor, not the subcontractor, failed to give timely notice of the appeal.[94] Thus, a subcontractor whose relief is limited in such a manner should make sure that the general contractor complies with the notice and appeal deadlines applicable to the prime contract and the Contract Disputes Act, or risk the possibility of having its claims waived.

VII. DOCUMENTATION RECOMMENDATIONS

A. Proper Documentation Starts with Implementing Appropriate Project Controls and Procedures

Successful documentation starts with the development of project controls and procedures that should be followed during the course of the job. Published procedures are advisable in order to make the project team aware of expectations. A formal procedure has a far better chance of being utilized than an informal one. The military has long recognized the value of having detailed and published procedures for its personnel to follow. The same holds true in construction.

Even small companies need project procedures. Although job controls for smaller companies typically are less formal and detailed, they nevertheless should exist and be known by everyone with responsibility on the job.

Job procedures should not only be published; they also need to be followed. It does little good if procedures are published but never implemented or enforced. The entire project staff must understand their obligation to follow the procedures established by the company.

From a legal standpoint, established procedures often create favorable presumptions for the contractor. Daily reports that are prepared in an accurate and consistent

---

[92]*Cf. A.H.A. Gen. Constr., Inc. v. New York City Housing Auth.,* 92 N.Y.2d 20 (1999).

[93]*Associated Mech. Contractors, Inc. v. Martin K. Eby Constr. Co.,* 983 F. Supp. 1121 (M.D. Ga. 1997).

[94]*Mishara Constr. Co.,* IBCA No. 869-8-70, 72-1 BCA ¶ 9353; *Baltimore Constructors, Inc.,* GSBCA Nos. 3489, 3490, 73-1 BCA ¶ 9928.

manner not only set out what happened on a job but can help establish that something did not happen. If a board or court gains confidence that a contractor used established procedures to record important events, then that judge is more likely to believe that an event did or did not happen simply because it was or was not recorded.

Implementing a successful set of procedures should never be taken for granted. Every key person on a construction project needs to understand why systems and procedures are in place and why they are important. Superintendents and foremen also should be trained to follow established procedures. Thus, making sure that procedures are being followed should be a constant theme in the field and at internal job meetings.

Procedures can be company-wide or created to address specific job requirements. Company-wide procedures provide familiarity and continuity. However, jobs often have unique requirements that make use of previously established company procedures inappropriate. Once assigned to a project, one of the first things that a project manager must do is evaluate the specific job requirements to determine whether standard project controls should be altered.

Job procedures should ensure the efficient flow of information and documents to facilitate proper project management and coordination. Job documentation needs to be copied, routed, and delivered to all appropriate parties. Whether project documentation is done electronically or with hard copies, procedures are needed for receiving, logging in, and routing documentation. Problems can occur when people who need to see an incoming document do not because of a routing or filing breakdown.

Next is a filing or retrieval system that must be capable of organizing and filing job records so they are readily accessible. Project managers who keep all critical records in locations known only to them (i.e., in piles on the floor behind their desks) may have ready access to their files, but no one else does. A system should be in place so that anyone familiar with the company should have a good idea how the files are organized. A proper retrieval system generally includes some type of numbering system for common documents such as requests for information, change order requests, and transmittals.

Project controls also should establish the division of responsibilities between job personnel. These procedures should establish who will generate and maintain the necessary job documentation. The project staff should be familiar with the identity of persons who have authority to act on behalf of the other contracting parties on the project (the owner, general contractor, or subcontractor) and the extent of their authority. Obviously, project documentation starts with knowing not only what to communicate but with whom to communicate.

Other variables can affect documentation procedures. The contractual relationships between the parties will affect the need and type of documentation. A design-build project may not have the same set of formal submittal requirements that are utilized on a design-bid-build project. A cost-plus contract will require a different type of cost documentation from a fixed-price contract. Documenting the "cost" in a cost-plus contract is essential to the contractor that wants to be paid correctly. As discussed in **Chapter 13,** documenting cost is also a key element in the equitable adjustment process for a government contract. A reality of government contracting is that the change order process is often a matter of documentation of the *actual cost* associated with that modification to the contract.

Specific contract requirements likely also will affect the types and extent of job documentation. A careful review of the general conditions and special conditions in every contract is necessary to identify job specific documentation to be maintained. For example, many public entities at the federal, state, and local level now require that their public projects meet LEED™ or other green building standards.[95] A green project usually requires the contractor to provide proof of compliance by maintaining and submitting additional documentation at various stages of the project.[96] The importance of collecting and organizing this green documentation as a project progresses cannot be overstated because it may be difficult, if not impossible, to compile the necessary documents at project closeout. A green rating or certification could be jeopardized. Contracts with the federal government also require particular attention to the documentation of Davis-Bacon payrolls, Equal Employment Opportunity Commission compliance, small business subcontracting and minority employment goals, and so on.

Contracts often require that the contractor maintain certain documentation, with copies or access available to the contracting officer's representative or other agencies, such as the Government Accountability Office. Project documentation creates an accessible history of the project that serves two roles: (1) planning and managing the project and (2) aiding in resolving claims and disputes. It must be organized and maintained in such a manner that it is a help, not a hindrance, to effective project management and the prosecution or defense of claims. Making the documentation both routine and uniform is essential to an effective system of project documentation. The procedures should be standardized not only for the project but for the company as a whole.

Additionally, standard form letters should be prepared or modified, as needed, for a particular project. The appendices to this chapter contain sample notice letters that can be modified for use (**Appendix 14C**). Various form documents and logs also are provided in the text and on the support Web site in an electronic format and can be easily customized for any given project (**Appendices 14D-14L**). On any project, both the type and the content of standardized letters and forms should be carefully reviewed and modified, as needed, based on the contract's risk allocation and documentation provisions.

B. Types of Project Documentation

Standard procedures should be established to address the creation, maintenance, and organization of certain specific types of documentation including:

(1) Bid/estimate documents

(2) Contract documents and change orders

(3) Correspondence

---

[95]The General Services Administration currently requires all new construction projects and substantial renovations to be certified under the Leadership in Energy and Environmental Design (LEED™) rating system of the U.S. Green Building Council.

[96]Thomas E. Glavinich, *Contractor's Guide to Green Building Constr.* 180 (John Wiley & Sons 2008).

   (4) Meeting minutes

   (5) Daily reports and quality control logs

   (6) Standard forms and status logs

   (7) Photographs and videotapes

   (8) Cost accounting records

   (9) Scheduling documentation

  (10) E-mail and electronic documents

1. *Bid/Estimate Documents*

Documenting a job typically starts with the bid/proposal estimate. Estimates do more than simply compute quantities and tally subcontractor quotations. A bid or proposal estimate often will record or reflect the countless assumptions made when the contractor determined its price. The estimate file should contain information gathered during the site investigation as well as any evaluation of the geotechnical data referenced or included in the solicitation. Recovery on a differing site condition claim often depends on demonstrating that the contractor relied on the site or soils data in the solicitation. Successful contractors desire to know how their actual performance compares to the assumptions made at the time its proposal or bid was submitted to the government. Poor assumptions concerning productivity, equipment usage, overhead costs, and the like should be identified during performance so that the same mistakes are not made on the next job.

There is no general answer as to how detailed an estimate needs to be. The best estimate is the most accurate estimate, which is not necessarily the most detailed estimate. Optimistic estimates are not necessarily wrong but should be recognized for their limitations before they are relied on when submitting a price. Conservative estimates may price the bidder or offeror right out of a job. Accuracy in estimating is without a doubt one of the most important skills to being successful at either design-bid-build or design-build construction.

The estimate should set the baseline for the project budget. Properly prepared, a budget will assist the contractor in tracking performance during the course of the project. This budget should be used in conjunction with an accurate cost accounting system. An effective cost report on a project should compare actual performance against the job budget derived from the assumptions made in the original estimate. It makes little sense to keep job cost reports if they cannot effectively track this vital information.

Estimates are often critical when pursuing claims for additional compensation. Estimates can be used to establish reliance on the anticipated site conditions, production baselines, scheduling assumptions, and work sequences. Estimates also may help establish the contractor's original interpretation of the contract documents, if an ambiguity arises during performance, and may be invaluable for pricing many claims.

2. *Contract Documents and Change Orders*

The most basic documentation that any party to a contract must maintain is the contract, subcontract, and purchase order files. These documents—along with the project drawings, specifications, change directives, and change orders—form the basis of the obligations between the parties. To fully understand its contract obligations, the project team needs to read and have ready access to these documents. A contract file folder should be kept separately for each project subcontractor and vendor. For easy reference, an executed copy of the contract should be maintained in the file folder, followed by any executed change orders in sequential, numerical order. A bid or original set of project drawings and specifications also should be preserved to address any ambiguities that may arise in the future.

3. Correspondence

Clerical support staff should implement standard procedures for date-stamping, copying, routing, filing, and indexing both incoming and outgoing correspondence. The party responsible for responding to or acting on incoming correspondence should be identified. Even if there is a decision to utilize a browser-based or network-based project management system, a disciplined approach still must be used. In addition, each project participant should carefully consider the extent to which *hard copies* need to be maintained as backup.

As a matter of routine, project management personnel should be trained on the importance of complying with technical notice requirements in the contract. Likewise, discussions should be confirmed in writing and sent to the other party, with copies to the file. Such confirmation will help immediately resolve any misunderstandings and also will preserve the substance of the discussion if there is a dispute at some later date.

4. Meeting Minutes

On a cumulative basis, regular job-coordination meetings between the various project parties cover multiple issues and can contribute more to the exchange of information necessary to complete the work than all other project correspondence. Thus, what occurs at such meetings is of great importance. Someone should be designated to maintain the minutes or notes for each meeting, preferably the same person at each meeting. That person should record the subjects covered, the nature of the discussion, the future actions to be taken, who has responsibility for the future action, and the applicable deadline for action. The name, title, and affiliation of each participant should be listed. The minutes should be concise but accurate and informative. The items discussed should be indexed or designated in a manner so they can be located in the future. The minutes should be distributed to all participants and any other affected parties on a regular basis.

A computer can be invaluable when updating regular meeting minutes, as certain items likely will remain open for discussion through several meetings. At the opening of each regular meeting, the minutes from the previous meeting should be reviewed to confirm their accuracy and the mutual understanding of the participants. By identifying those items that remain outstanding, the previous meeting's minutes can also serve as an agenda for the current meeting.

5. *Daily Reports and Quality Control Logs*

Daily reports and quality control (QC) logs generally are maintained by the quality control staff or the project superintendent and can provide the best record of what happens in the field. They help keep management and office personnel informed of field progress and problems. In the event of a claim, they are often among the most helpful documents in re-creating job progress and as-built schedules. They can be critically important when evaluating lost-productivity and disruption claims.

The daily report must be made part of the project staff's daily routine. Entries should be precise and objective. Daily reports need to include more than the date and the weather conditions. At a minimum, the information covered should identify and include:

- Subcontractors and self-performed work crews on site, areas where they are working, and the major work activities being performed
- Manpower quantities, preferably broken down by trade and subcontractor
- Progress tracking that is consistent with activity descriptions on the project's critical path method (CPM)
- Work activities that started or were completed
- Specific progress made on project schedule's critical path and near–critical path activities
- Any delays or problems encountered, and the work activities affected
- Areas of work not available
- Extra or out-of-scope work performed
- Major equipment, both used and idle
- Noteworthy materials being used or stored on site
- Safety-related concerns, accidents, and injuries
- Job site visitors
- Oral instructions and informal meetings
- A weather summary
- Key inspections and their results
- Author and date of the report

By using a standard form, the clerical burden on the superintendent, project engineer, or QC manager may be eased and the information maintained in a more organized and uniform manner. The process can be expedited further simply by allowing the superintendent to dictate entries and then having the office staff type up the report. An e-mail format also may ease transmission of the information to the home office.

Computers can be invaluable in the preparation of daily reports. Computers provide a convenient format and can prompt the writer essentially to fill in the blanks. However, parties should guard against becoming lazy when utilizing a computer during the preparation of daily reports. The same work descriptions pasted in a daily report day after day may be easy for the superintendent or project engineer, but the information may not be accurate or complete. The report's credibility may be compromised. Thus, the efficiency provided by a computer should be taken advantage of but should not control what is documented.

All key project personnel, such as foremen, project engineers, quality control managers, and project managers, also should be encouraged to maintain personal daily logs and follow the established procedures to facilitate this effort. The information they record should be similar to the job log or daily report but may not need to be as extensive or detailed.

These types of routine, contemporaneous descriptions of work progress, site conditions, labor and equipment usage, and the contractor's ability (or inability) to perform its work can provide valuable information necessary to accurately reconstruct project events in the event of a claim or dispute. When maintaining these reports or logs, project personnel must be consistent in recording the events and activities on the job, particularly those relating to claims or potential claims. Once the responsibility of a daily report or log is undertaken, failure to record an event carries with it the implication that the event did not occur or was insignificant and also threatens the credibility of the entire log.[97]

6. Standard Forms and Status Logs

Using a variety of media, there is a constant flow of information between project participants. Drawings are revised; shop drawings are submitted, reviewed, and returned; field orders and change orders are issued; questions are asked; and clarifications are provided. Cumulatively and individually, these bits and pieces of information are essential for building the job and for reconstructing the progress of events on paper in the event of a claim. The standard procedures must include the means for providing, eliciting, recording, and tracking this mass of data so that it can be used during the course of the job and efficiently retrieved in an after-the-fact claim setting.

Routine transmittal forms should be *customized* not only to address specific, routine types of communications in order to expedite the process but also to ensure that required information is provided. For example, separate specialized forms can be prepared for transmittal of shop drawings and submittals, requests for information, drawing revisions, and, of course, field orders and change orders. When possible, the forms should provide space for responses, including certain standard responses that simply can be checked off or filled in. At a minimum, the forms should identify the individual sender, the date issued, and specific and self-descriptive references to the affected or enclosed drawings, submittals, or specifications. If a response is requested by a certain date, that date should be identified on the form. Again, while a network-based or browser-based system can expedite the prompt exchange of information significantly, the key to good project management is the development of a systematic and consistently followed routine. If a computer network system is employed, the need to make backups or hard copies cannot be ignored.

---

[97]*See Cape Tool & Die, Inc.*, ASBCA No. 46433, 95-1 BCA ¶ 27,465; *Fuel Economy Eng'g Co.*, ASBCA No. 6580, 61-1 BCA ¶ 3042; *see also* Federal Rule of Evidence 803(7).

Ideally, each discrete type of communication or specialized form should be *numbered* or somehow *identified* in a chronologically sequential manner based on the date it is initiated. Shop drawings and submittals, however, are best identified by specification section, with a suffix added to indicate resubmittals. This provides a basis for easy reference and orientation. Copies of the completed forms should be maintained in binders in reverse chronological or numerical order. Although various project staff members may require working copies, a master file should be maintained as a complete reference source and historical document.

In order to maintain the status of and track these numerous and varied communications, which can number many thousands, *logs* should be maintained. These logs need only address key information, such as number assigned, date, and a self-descriptive reference. Proposed change orders and change order logs also should identify any increase or decrease in contract amount as well as time extensions. Such logs can be kept on a personal computer using inexpensive, commercially available software. Logs should be maintained for internal record keeping and also for distribution to other parties on the project. The logs serve as a reminder of outstanding items and can highlight action required to keep the work progressing.

The contractor should use *standard forms* and *procedures* for communications with subcontractors as well as the owner and architect. Ideally, subcontractors also should be encouraged to standardize their communications so there will be a more integrated approach for the entire project. In that regard, the adoption of a network-based project management system can facilitate a standardized approach to project communications.

7. Photographs and Videotapes

Contractors should keep a photographic record of every project. Photographs have numerous uses, both on the project and for future reference. The old adage that "a picture is worth a thousand words" is true in construction. Photographs are inexpensive to create and maintain. Digital cameras make film obsolete and photo development instantaneous. Cameras capable of producing quality images are essential. Digital cameras offer an excellent method for taking, storing, and transmitting project images. A digital camera also allows for an immediate check on the photograph's content before conditions change.

Photos provide a historical record, are excellent for marketing purposes, and usually are compelling evidence when disputes or claims need to be resolved after the work is covered or otherwise inaccessible. Photos are invaluable in showing access problems, traffic conditions, unexpected obstructions, congestion, and the like. Photos taken at regular intervals show progress and are effective in supporting or defending against delay and lost-productivity claims. A photographic record at the time of any termination is absolutely essential to accurately record the status of the work for purposes of settling a convenience termination or simply to justify the termination.

One approach, incorporated in many contracts, is to accumulate a periodic pictorial diary of the job through a series of weekly or monthly photographs of significant milestones. This approach encourages personnel to take photographs of site conditions on a routine basis, perhaps concentrating on problem areas and those areas associated with crucial construction procedures and scheduling. Photographs are also the best evidence of defective work or problem conditions that are cured or covered up and cannot be viewed later.

Photographs always should be marked with notations as to time, date, location, conditions depicted, personnel present, and the photographer. Some of this information can be imprinted on the negative or stored digital image. A log should be kept as the photographs are taken, and the log should be checked immediately when the photographs are printed or stored. If digital images are printed that do not contain information on the date, location, photographer and so on, that information should be recorded on the back of print copy. . Without this information correlated to specific photographs, the utility of the entire effort can be substantially undermined. Digital files stored on a computer should be backed up to mitigate the risk of corrupted computer files. If utilized, negatives also should be retained in an organized, retrievable manner.

In some situations, videotape can be considerably more informative than a still photograph, such as when attempting to depict an activity or the overall status of the project. A monthly videotape is an excellent way of preserving and presenting evidence. Static conditions, however, are best photographed.

The availability of a contemporaneous narrative as part of the video can give an after-the-fact viewer a much better idea of what is being depicted and why. When narrating the video, however, the video operator needs to be careful not to make statements that can be used against the entity making the video. The video may well be subject to disclosure in a lawsuit or may otherwise find its way into an opposing party's hands. A narration describing how that entity failed to give necessary instructions, failed to cooperate with the other party, or otherwise failed to properly perform can become Exhibit A against the party that hoped to rely on the video. To avoid this problem, the video should be prepared by properly trained job-site personnel without narration. Later, witnesses can testify in conjunction while the videotape is played.

8. Cost Accounting Records

Construction firms need to track job costs for a variety of reasons. First, costs should be tracked for historical purposes. If a mistake was made when the job was estimated, tracking actual performance costs will identify where and how extensive the mistakes were. Successful contractors record actual costs for identified work activities for historical reasons and to improve bids on future work. Contractors need to know when their proposals are too high or overly conservative. As more and more actual information becomes available, the chances that the estimator will arrive at an accurate price improve.

Job cost records are important when pricing changes that initiate during job performance. An accurate set of job cost records not only will help compute the fair price of a change order but also will provide documentary support during negotiations to justify the price for the change.

Cost records are essential when pricing claims for extra work, differing site conditions, delays, and so on. The use of effective cost-accounting methods and the maintenance of appropriate cost records can minimize many of the proof problems inherently associated with construction claims.[98] Even though a claimant may be able to prove that an event has occurred which entitles it to additional compensation, it will be able to recover only that amount of damages it can prove with reasonable certainty. Proving the actual dollars expended is crucial to a claim.

Accurate cost records make computation of extended job-site general conditions claims both accurate and convincing when delays occur on a project. With such records, the opposing party will find it difficult to dispute a delay claim calculation. Similarly, accurate cost records can assist in tracking labor and/or equipment productivity during the project. Detailed unit labor costs can help accurately establish productivity losses when problems arise during the project.

Cost records need to be detailed enough to assist the project manager on the job as well as those in the home office estimating department. How detailed a cost accounting system should be depends on the nature of the project and, to a lesser degree, on the sophistication of the contractor. Generally, the cost accounting system should be patterned after the estimate, the bid schedule, the schedule of values, or a project-specific budget. Most cost systems are computerized. Thus, it is critical to input accurate data. Cost codes should be established for specific contract tasks and for specific changes, differing site conditions, and other unanticipated conditions. Whoever identifies the cost codes for individual charges must be consistent and accurate, and must update the information faithfully.

A number of federal agencies have embraced the movement toward computerized cost accounting and payment systems. The Department of Defense, for example, requires that all vendors use an electronic method for submitting invoices and supporting documentation.[99] To that end, it employs Wide Area Work Flow (WAWF), a web-based application designed to eliminate paper from the receipts and payment request process. Using WAWF, a vendor may create, correct, and submit payment-related documents. WAWF benefits include an expedited payment process, reduction in the number of duplicative or lost documents, and easy online access to vendor payment records. To self-register for the program, vendors should begin by visiting the Defense Finance and Accounting System Web site, selecting the Contractor/Vendor Pay tab, choosing Electronic Commerce, and then clicking Wide Area Work Flow.[100]

9. Scheduling Documentation

The general contractor or construction manager should closely monitor the work of all trade contractors to determine whether each is meeting its deadlines so that the work of other trades can proceed as originally scheduled. The owner must perform the same task when multiple prime contractors are involved. Even when the contractor has primary scheduling responsibility, which is most often the case, the owner should nonetheless monitor the progress of the work and the scheduling effort.

---

[98] *See Marshall Associated Contractors, Inc. & Columbia Excavating, Inc. (J.V.)*, IBCA No. 2088F, 2005 WL 1231813.

[99] DFARS § 252.232-7003(b).

[100] *See www.dfas.mil/contractorpay/electroniccommerce/wideareaworkflow.html* (accessed June 30, 2009).

Some prime contracts allow the preparation of a bar (Gantt) chart or progress schedule, which provides the easiest means of monitoring the work. Many contracts require a CPM schedule, which can be even more valuable as a scheduling tool if properly developed, updated, and utilized. Regardless of the form, a file should be maintained that includes the original ("baseline") schedule, all updates to the original schedule, and any interim look-ahead schedules.

Input from subcontractors and other affected project participants during the development and updating of any project schedule is critical to its usefulness. As a practical matter, a schedule that is developed without the input of the parties actually performing the work may result in an unworkable product, and the schedule as an instrument of coordination will be wasted. By obtaining the participation of all parties in the preparation of the schedule, it becomes a much more meaningful and productive project management device. Through their involvement in the process, each party has effectively admitted or acknowledged what was reasonable and expected of it. If a party later fails to perform or follow the schedule, its ability to dispute the relevance of the project schedule and what was required of that party can be reduced substantially.

A project schedule can be a double-edged sword for the prime contractor, particularly if it is a CPM that shows the interrelationship of all activities and trades. On one hand, a properly developed schedule can be used to demonstrate that a subcontractor is behind schedule and how its delayed performance is impacting the entire project.[101] Conversely, a subcontractor may use a project schedule against the prime contractor to show how the subcontractor reasonably expected and planned to proceed with the work, and how that plan was disrupted by the prime contractor, another subcontractor, or the owner, for which the affected subcontractor may claim additional compensation.[102]

There is nothing more compelling than the use of a properly updated project schedule to prove delays or to justify a time extension. In fact, many contracts authorize time extensions only when a delay event is shown by the schedule to have affected the critical path of the job. When it comes to justifying additional time, contractors must make certain that they are updating a schedule that accurately shows the critical path.

If the schedule is not properly maintained, updated, and enforced—and thus bears little relationship to actual work progress or the parties' contractual obligations—it may be dismissed as merely representing "theoretical aspirations rather than practical contract requirements."[103] The heavy use of scheduling information and analysis when resolving claims underscores the importance of preparing, and maintaining through updates, a realistic schedule that secures subcontractor involvement and agreement.

---

[101]*See, e.g., Illinois Structural Steel Corp. v. Pathman Constr. Co.,* 318 N.E.2d 232 (Ill. App. Ct. 1974); *Santa Fe, Inc.,* VABCA No. 2168, 87-3 BCA ¶ 20,104; *Santa Fe Eng'rs, Inc.,* ASBCA No. 24578, 94-2 BCA ¶ 26,872; *Kaco Contracting Co.,* ASBCA No. 44937, 01-2 BCA ¶ 31,584.

[102]*See United States ex. rel. R.W. Vaught Co. v. F.D. Rich Co.,* 439 F.2d 895 (8th Cir. 1971); *see also George Sollitt Constr. Co. v. United States,* 64 Fed. Cl. 229 (2005).

[103]*See Vaught,* 439 F.2d at 900.

10. E-mail and Electronic Documents

Computers continue to revolutionize the construction industry by improving communication, increasing productivity, and streamlining contract administration.[104] The proliferation of e-mail and electronic documents evidences a lasting trend away from paper documents. Under the E-Sign Act,[105] no contract, signature, or record can be denied legal effect solely because it is in electronic form. The purpose of this Act is to protect transactions from legal challenges that are based solely on the electronic form of the agreement. Although application of the Act is not mandatory in private-sector contracts, it does require persons to use or accept electronic records or electronic signatures when dealing with the federal government.[106] Thus, e-mail and electronic documents should be used with care.

E-mail should be used with heightened awareness because of its casual nature. E-mail is a convenient and quick way to send not only formal reports and correspondence but also informal messages. Many people have become so comfortable with e-mail that they use it as much or more than the telephone to communicate. There is a very big difference, however, between using the phone and using e-mail. E-mail leaves a written record of exactly what was said. E-mail abuse has resulted in litigation over sexual harassment, racial harassment, disclosure of protected company information and trade secrets, defamation, and cyberstalking.[107]

E-mail and other electronic documents are discoverable in litigation just like letters, notes, and memos. In fact, e-mail can be an especially valuable source of litigation evidence because people are often less careful during these exchanges than they would be when sending traditional hard-copy correspondence. People tend to be at least as frank and open in e-mails as they are on the phone. Furthermore, e-mail and other digital data are difficult to destroy completely. Many employees mistakenly believe that an e-mail message is gone forever once deleted, when reality has shown that deleted messages can be resurrected with various software and by computer forensic experts. For better or worse, e-mail has significantly expanded the record of events.

Construction companies must develop and *enforce* e-mail policies. An appropriate policy may allow reasonable personal use of e-mail and Internet so long as that use does not interfere with the employee's job performance while concurrently prohibiting use in any manner that may harm the business interests of the employer, subject the employer to liability, or be offensive to other employees. Although e-mail technology is valuable and should be encouraged, it also needs to be controlled. E-mail is just like any other type of project documentation and should be treated as such.

Construction personnel need to understand that e-mail is not equivalent to using the phone for proof purposes in litigation. Even deleted e-mail may become part of

---

[104]See Dana K. Smith & Michael Tardif, *Building Information Modeling, A Strategic Implementation Guide* 3-19 (John Wiley & Sons 2009).
[105]Electronic Signatures in Global and National Commerce Act, 15 U.S.C. §§ 7001-7006.
[106]*The Prudential Ins. Co. v. Prusky*, 413 F. Supp. 2d 489 (E.D. Pa. 2005).
[107]*Black's Law Dictionary* (7th ed. 2000) defines cyberstalking as "threatening, harassing, or annoying someone through multiple e-mail messages."

the project record. Before hitting the "Send" key, always consider whether the message would be appropriate if it were being sent via letter.

C. Documenting Disputes

Some of the most important reasons to document a project properly include avoiding disputes and resolving claims. Disputes and claims can arise on any job. The party that is prepared for these issues is more likely to be able to achieve a better and quicker resolution than a party that is not. Documentation is essential to support a position in a dispute or claim. Failure to maintain records of the work performed may result in denial of a claim for lack of documentation.[108] Seven steps are essential to be better prepared if a problem arises during construction of the project:

(1) *Know* the contract.

(2) Understand the *notice* requirements.

(3) Give *timely* and *proper* notice.

(4) Document *facts*, not *feelings*.

(5) Document a problem with the goal of *solving* it.

(6) Do not be belligerent; *do not write* like a lawyer.

(7) Promptly address the *cost and time* impact.

D. Evaluating the Project Documentation System and Retaining Records

Whether performing a government or private construction contract, the development of an organized approach to project documentation is critical to success. Systems that work well for one type of owner or a particular federal agency may need to be modified to meet the requirements of a particular contract. The contract documents should be reviewed carefully to evaluate their effect on the contractor's documentation practices and systems. In particular, the contractor should identify and consider documents that are incorporated by reference into the contract. These documents can impose or modify requirements that are not obvious from an initial review of the contract. As part of its prebid or initial proposal activities, a contractor should *prequalify* its project documentation system for a particular project. Use of a checklist approach can assist in that prequalification review, which should consider these topics.

QUALIFYING THE PROJECT DOCUMENTATION SYSTEM CHECKLIST

- Identify and review *every* document that will form the basis of the contract.
- Review the contract to determine if modifications to the firm's *standard* project documentation system are required.

---

[108] *Swanson Printing Co.*, GPO BCA No. 27-94, 27A-94, 1996 WL 812958.

- Determine if the contract imposes specific *cost accounting* requirements.
- Determine if *all members* of the project staff, including the quality control and safety personnel, understand their documentation responsibilities.
- Determine if the staff understands both its *upstream* and *downstream* notice obligation related to the government as well as to the subcontractors and suppliers.
- Determine if the notice provisions in subcontracts and purchase orders are *coordinated* with the notice requirements in the contract with the government.
- Determine if the contract contains *special notice requirements* beyond the standard FAR clauses.
- Evaluate any special *cost tracking* or cost accounting procedures imposed on change order work.
- Determine if the contract contains *unusual risk-shifting or design verification clauses* or agency supplements that may affect (diminish) rights under standard FAR clauses, such as the Differing Site Conditions or Changes clauses.
- Determine if *advice or assistance* from legal or accounting professionals is needed in light of contract documentation provisions.
- Determine if *in-house training* on documentation practices, the claims process, use of e-mail, cost tracking, and so on, is warranted.

In addition to evaluating the project-specific documentation program, a good records retention program can help avoid litigation and save costs. When litigation or a government investigation is "reasonably anticipated,"[109] a duty to preserve potentially relevant records—including electronic records—is triggered. A litigation hold must be implemented that advises employees in writing about the nature and the extent of the investigation and directs them to preserve required documents.

A business is on notice and must cease the destruction of relevant records anytime litigation is foreseeable.[110] A safe harbor provides some protection against sanctions for parties that have disposed of potentially discoverable data in the normal course of good-faith business operations.[111] Otherwise, a party that destroys potentially relevant evidence may face sanctions that include reimbursing the opposing party for discovery-related costs, construing evidence in favor of the opposing party, and payment of attorney fees.[112] The safe harbor applies only if a business has implemented a record retention policy prior to the anticipation of litigation arising.

Besides implementing procedures for early identification and monitoring of potential disputes, a document preservation and retention policy must be formalized and implemented in a consistent, good-faith manner. The program should proactively manage all forms of electronic data and physical documentation. A records retention schedule should be included that defines those records that need to be kept and for how long. Records that should not be kept should be identified as

---

[109]*Zubulake v. UBS Warburg LLC*, 220 F.R.D. 212 (S.D.N.Y. 2003).
[110]*Id.*; *Doe v. Norwalk Cmty. Coll.*, 248 F.R.D. 372 (D. Conn. 2007).
[111]*See* Fed. R. Civ. P. 37(e).
[112]*See United Med. Supply Co. v. United States*, 77 Fed. Cl. 257 (2007).

well, along with those eligible for deletion to reduce storage and management costs. A good records management program is communicated effectively to all employees, includes proper training, and is monitored regularly for compliance.

VIII. DOCUMENTING WITH PROJECT MANAGEMENT SOFTWARE

The construction industry has long been criticized for systemic inefficiencies that increase risk and inflict unnecessary costs onto project participants. Lack of communication and poor documentation often are cited as major inefficiencies that fuel claims and disputes. Project management software is a frequently used solution that helps facilitate communication and collaboration amongst team members.

Project management software, such as Meridian Systems' Prolog® Manager, Autodesk's Constructware®, and the U.S. Army Corps of Engineers' Quality Control System, have all contributed to improved documentation and communication. Such programs have facilitated the sharing of information and helped project participants achieve significant productivity gains as well.

A. General Considerations

Project management software can provide many benefits to project participants. Project staff may be efficiently reallocated and productivity increased because software automates processes and time-consuming tasks. Absent software, the project team is tasked with manually creating project-specific documents, forms, and reports. Software can allow project data, processes, and documents to be standardized across an entire organization. Confusion between the field and the main office can be reduced because documents are stored online and easily accessed. Duplicate information is nearly eliminated. Further, by managing multiple projects in the same database, data may be collected and reported across the organization. Individual project budgets may be tracked and collectively summarized, giving company officers the ability to gauge the organization's health.

Project management software also may decrease the number of claims and disputes by streamlining communication. Geographically dispersed team members can work on the most current information by accessing a central project database. Disputes that arise from poorly communicating a change to all of the affected parties can be kept to a minimum because of automated communication processes. Project data also may be shared across various software systems, such as with project scheduling programs or accounting software. This capability ensures accurate reporting and decision making based on the most current data.

Much like its hard-copy predecessor, project management software is not without its own risks and costs. The initial capital outlay may be substantial and may involve computer hardware, personnel training, and the software itself. Periodic training also should be viewed as a potential cost. The initial learning curve may be steep, especially for industry veterans who may be used to traditional forms and processes.

Further, even computer-savvy employees will face a learning curve when navigating new software and achieving uniformity. For example, a project team might use its own unique numbering system for such documents as requests for information, submittals, transmittals, and change orders. When various personnel either begin a new project or join an existing project, they will need to learn a project-specific numbering system for these documents, unless the organization has standardized this for all projects.

Computer data involve a certain level of inherent risk as well. If data input is not current, any project-specific or summary reporting will be inaccurate. Further, mistakes made during data entry are quickly dispersed across the project spectrum because of the centralized nature of the database. Loss of data should be a concern that can be mitigated through the use of backup systems. Further, personnel productivity may be affected if a system is overloaded with users or if network connectivity fluctuates. In sum, an organization or project team would be well advised to weigh the advantages, disadvantages, and capabilities of the various project management software systems before committing substantial resources.

B.   Software Documentation Capabilities

Typical project management software may improve documentation and communication in four specific areas: procurement, contract administration, field management, and cost control. An organization's procurement process can realize greater flexibility and experience a reduction in errors because program management software utilizes a central database to track all scopes of work and materials to procure. Confusion over which contract attachments were distributed to selected bidders can be reduced by logging these items in the central database. Communications can be sent directly from the database to bidders via e-mail or fax, significantly reducing hard-copy paperwork, which is more prone to error. Furthermore, a centralized database of contacts can be maintained within the software, used by the current project team, and carried forward for use on subsequent projects.

Contract administration documentation may be generated automatically and tracked easily using software. A list of the most current drawings and specifications can be reviewed easily in order to verify that the field has the most up-to-date documents. Meeting minutes can be generated and action item responsibility monitored. Requests for information, submittal package cover sheets, and transmittal documents can all be generated and tracked through the software. Reports can be generated that identify overdue items and party responsibility. These reports can be particularly valuable when addressing action items at project meetings. Further, many software programs are capable of tracking "hot lists," generating auto alerts, and notifying the project team when it needs to follow up with an issue.

Project management software does not just benefit those sitting in the field office. It also may help field forces be more productive. Nearly all software is capable of logging daily work activities, tests and inspections, safety issues, material deliveries, and productivity problems. Most software also can generate field work directives and notices to comply. Last, project engineers and superintendents may benefit from the ability of many programs to log, track, and sort punch-list items.

Cost control capabilities generally focus on contracts, purchase orders, applications for payment, and the change order process. A typical software system allows contracts to be logged against the budget and the actual contract documents to be linked through the software for easy viewing. Purchase orders can be generated and applied against the project budget. Invoices can be generated using current data. Change order requests can be logged, change orders generated, and the entire government change order approval process can be tracked. Similarly, the corresponding budgetary impact also can be monitored throughout the process.

Typical project management software may save a significant amount of time by generating form documents in each of these areas and allowing these documents to be populated easily. Project documentation can be dispatched, tracked, and sorted directly from the central database. A significant audit trail of documentation is produced automatically, which reduces the risk of litigation and increases productivity. Some software makers have even developed products that provide compliance assistance with the stimulus funding oversight requirements of the American Reinvestment & Recovery Act of 2009.[113]

The Internet has brought the collaborative capabilities of project management software into the mainstream. Web-based project management is now quite common on many construction projects because it allows project team members to send, share, update, track, review, and store project documents through the Internet. Easy collaboration also has helped fuel virtual design and construction, or building information modeling (BIM). BIM uses computer software to simulate the construction and operation of a facility.[114] This, in turn, allows the owner, designer, and contractor to determine whether all components of a building properly fit together before construction begins. The final BIM model also establishes a comprehensive project record, which can be invaluable when performing facility maintenance or undergoing future expansion. The interactive capabilities of project management software continue to expand.

Some government agencies require that project management software be used. Some agency solicitations mandate that a contractor have available and utilize a specific program while some agencies provide the software for a given project. The U.S. Army Corps of Engineers, for example, requires that contractors use the Quality Control System (QCS)—a component of the Corps' Resident Management System—to record, maintain, and submit various information throughout the contract period. QCS provides the means for the contractor to input, track, and electronically share information with the government in these areas: administration, finances,

---

[113]*See* Prolog® Oversight Pack by Meridian Systems®.

[114]*See* Chuck Eastman et al., *BIM Handbook: A Guide to Building Information Modeling* (John Wiley & Sons 2008). C. Government-Mandated Software Use

quality control, submittal monitoring, and scheduling. The purpose of this government-furnished software is to assist the contractor in performing quality control activities "more consistently and within the requirements specified by the Corps of Engineers."[115]

Agency software requirements may be diverse. For example, recent solicitations appearing on the FedBizOpps.gov Web site have requested that these criteria be met:

- A Department of Health and Human Services requirement that construction management firms submit a statement of qualifications highlighting the firm's capabilities with Prolog® Manager and Primavera scheduling software.
- A General Services Administration requirement that the construction manager "develop and maintain, in Microsoft® Project, an integrated master project schedule throughout the course of the project."
- A U.S. Army Corps of Engineers requirement that the contractor "use the Government-furnished Construction Contractor Module of the Resident Management System (RMS), referred to as QCS, to record, maintain, and submit various information throughout the contract period."
- A Department of the Navy requirement that only contractors registered in the central contractor register and Wide Area Work Flow program will be considered for award and that the only authorized method to electronically process vendor requests for payment is through WAWF.

A prudent offeror or bidder will verify any mandatory or preferred software requirements and plan accordingly, whether this means investing in new software or lining up additional training.

➢ LESSONS LEARNED AND ISSUES TO CONSIDER

- The project team needs to *review the contract* for documentation requirements before the project starts and monitor compliance with those requirements during construction.
- Require the project staff to prepare *notice checklists* and demonstrate an understanding of the *documentation procedures* needed to control and monitor the project.

---

[115]*Quality Control System (QCS) User Manual & Training Guide*, Sept. 18, 2008, *www.rmssupport.com/ qcs/guides.aspx*.

- Ensure that the notice and documentation obligations contained in subcontracts and purchase orders *complements*, not contradicts, the purpose and specific provisions in the contract with the government.
- Pay particular attention to *notice requirements* and obligations for documenting changes, differing site conditions, delays, suspensions of work, and claims.
- Document *facts*, not feeling or judgments.
- Develop a *systematic procedure* to ensure that written notice is sent to governmental representatives with actual authority to bind the government. Routinely *copy* the *contracting officer* with written notice of constructive changes, requests for the performance of extra work to the contract, differing site conditions, delays, and so on.
- Confirm that *project personnel* are familiar with how notices and other documentation are to be submitted, to whom they are to be provided, and what information needs to be included.
- Be prepared to show that the government *did not suffer prejudice* from the type of notice given; that the contracting officer or other authorized representative had *actual notice* of a condition and knowledge of its *adverse effect*; or that the government *considered the claim* without asserting defective notice.
- Proper project documentation involves procedures to *systematically* maintain bid or proposal and estimating documents, subcontractor and vendor files, correspondence, meeting minutes, daily reports and job site logs, schedules, cost accounting records, standard forms and status logs, photographs, and videotapes.
- Company operating procedures should be *standardized and consistent* from project to project.
- Project participants should establish and maintain *open lines of communication* by engaging in regular job meetings.
- Proper documentation should provide the information necessary to accurately *reconstruct* project events in the event of a claim.
- Because problems of proof are inherently associated with construction claims, maintenance of a *good cost accounting system* is crucial.
- A *realistic project schedule* that secures the involvement and agreement of all project participants should be prepared and routinely updated.
- Project personnel should remember that *e-mail and electronic documents are discoverable* during litigation just like paper documents.
- Prior to sending, consider whether the contents of an electronic document or e-mail would be *appropriate* if sent via letter.
- Project management software can facilitate project collaboration and improve productivity; any mandatory or preferred *software requirements* should be verified at during the proposal preparation phase of the project.
- A company-wide *records retention policy* should be implemented.
- When litigation or a government investigation is reasonably foreseeable, any relevant documents—including both electronic data and physical documents—need to be preserved and *a litigation hold* should be communicated.

APPENDIX 14A: NOTICE CHECKLIST

Clause Reference	Subject Matter of Notice	Time Requirement for Notice	Form of Notice	Stated Consequences of Lack of Notice
Changes Paragraph # _____	Proposal for adjustment	(Sent)(Rec'd) in ___ days Triggering Event: _____ _____ _____ Other Action Required:_____ _____	___ Written ___ Certified ___ Registered Sent to: _____	
Constructive Changes Paragraph # _____	Date, circumstances, and source of the order and that the contractor regards the order as a contract change	(Sent)(Rec'd) in ___ days Triggering Event: _____ _____ _____ Other Action Required:_____ _____	___ Written ___ Certified ___ Registered Sent to: _____	
Differing Site Conditions Paragraph # _____	Existence of unknown or materially different conditions affecting the contractor's cost	(Sent)(Rec'd) in ___ days Triggering Event: _____ _____ _____ Other Action Required:_____ _____	___ Written ___ Certified ___ Registered Sent to: _____	
Suspension of Work Paragraph # _____	The act or failure to act involved and the amount claimed	(Sent)(Rec'd) in ___ days Triggering Event: _____ _____ _____ Other Action Required:_____ _____	___ Written ___ Certified ___ Registered Sent to: _____	
Time Extensions Paragraph # _____	Causes of delay beyond contractor's control	(Sent)(Rec'd) in ___ days Triggering Event: _____ _____ _____ Other Action Required:_____ _____	___ Written ___ Certified ___ Registered Sent to: _____	

APPENDIX A (CONTINUED)

Claims Paragraph # _____	Notice of event or condition giving rise to a claim	(Sent)(Rec'd) in ___ days Triggering Event: _____ _____ _____ Other Action Required:_____ _____ _____	___ Written ___ Certified ___ Registered Sent to: _____	
Termination for Default Paragraph # _____	Notice of intent to terminate for default	(Sent)(Rec'd) in ___ days Triggering Event: _____ _____ _____ Other Action Required:_____ _____ _____	___ Written ___ Certified ___ Registered Sent to: _____	
Termination for Convenience Paragraph # _____	Notice of intent to invoke right to termi-nate for convenience	(Sent)(Rec'd) in ___ days Triggering Event: _____ _____ _____ Other Action Required:_____ _____ _____	___ Written ___ Certified ___ Registered Sent to: _____	
Injury or Damage to Person or Property Paragraph # _____	Claim of injury or damage to property caused by act or omis-sion of other party or agent	(Sent)(Rec'd) in ___ days Triggering Event: _____ _____ _____ Other Action Required:_ _____ _____	___ Written ___ Certified ___ Registered Sent to: _____	
Disputes Paragraph # _____	Appeal of A/E or C.O. Final Decision	(Sent)(Rec'd) in___ days Triggering Event: _____ _____ _____ Other Action Required:_____ _____	___ Written ___ Certified ___ Registered Sent to: _____	

APPENDIX 14B: SAMPLE (PARTIAL) NOTICE CHECKLIST: FEDERAL GOVERNMENT CONSTRUCTION CONTRACTS

Clause Reference	Subject Matter of Notice	Time Requirements for Notice	Writing Required	Stated Consequences of a Lack of Notice
Changes FAR § 52.243-4	Proposal for adjustment.	*30 days* from receipt of a written change order from the government or written notification of a constructive change by the contractor.	Yes	Claim may not be allowed. Notice requirement may be waived until final payment.
Constructive Changes FAR § 52.243-4	Date, circumstances, and source of the order and that contractor regards the government's order as a contract change.	No starting point stated, but notice within *20 days* of incurring any additional costs due to the constructive change fully protects the contractor's rights.	Yes	Costs incurred more than *20 days* prior to giving notice cannot be recovered, except in the case of defective specifications.
Differing Site Conditions FAR § 52.236-2	Existence of unknown or materially different conditions affecting the contractor's cost.	From the time such conditions are identified, notice must be furnished "promptly" and before such conditions are disturbed.	Yes	Claim not allowed. Lack of notice may be waived until final payment.
Suspension of Work FAR § 52.242-14	(1) Of "the act or failure to act involved,"	(1) Within *20 days* from the act or failure to act by contracting officer (not including a suspension order). (2) "As soon as practicable" after termination of the suspension, delay, or interruption.	(1) Yes (2) Yes	(1) Costs incurred more than *20 days* prior to notification cannot be recovered. (2) Claim not allowed but claim may be considered until final payment.
Termination for Default Damages for Delay—Time Extensions FAR § 52.249-10	Causes of delay beyond contractor's control.	*10 days* from the beginning of any delay.	Yes	Contractor's right to proceed may be terminated and the government may sue for damages.

APPENDIX B (CONTINUED)

Disputes **FAR § 52.233-1**	Appeal of any final decision by the contracting officer.	(1) Boards of Contract Appeals— *90 days* from receipt of contracting offic- er's final decision.	(1) Yes— Notice of Appeal	Contracting officer's decision becomes final and conclusive.
		(2) U.S. Court of Federal Claims—*1 year* from receipt of contracting officer's final decision.	(2) Yes— Filing of Complaint	Contracting officer's decision becomes final and conclusive.

General Notes on Preparation of a Checklist:
The two tables above are sample formats for Notice Checklists. Regardless of your familiarity with the contract, each contract should be reviewed carefully, as special notice requirements are often in "stand- ard" contracts. The checklist should identify the clause, time requirements for notice, the subject of the notice, whether notice must be in writing, and the stated consequences for failing to give notice. The checklist should not be provided to the project staff. Rather, those responsible for giving timely notice should prepare the checklist for every contract. The checklist can be contained on a single sheet of paper, three-hole punched, and retained in the project manual.

APPENDIX 14C: FORMS

SAMPLE NOTICE LETTER—EXTENSION OF TIME FOR DELAYS (AND EXTRA COSTS IF APPROPRIATE)

XYZ # _____

Addressee:

(To Contracting Officer)

Re: Contract No.

Dear:

We are continuing to pursue the completion of our work as rapidly as is reasonably possible under the current circumstances. We have, however, recently encountered certain delays to our performance through no fault of our own and which are beyond our control. We have continued to keep your job representatives informed of these delays and of their effect on overall job completion. You may be assured that we will diligently seek to reasonably minimize the effects of these delays on our work.

Specifically, we have been delayed in the following particulars:

Accordingly, we hereby request an extension of [_____ days]* to our contract completion to take into consideration the above delays in accordance with the contract provisions.

**[Such delays have also had a serious effect on costs of performance in that they have created additional time for performance with resultant additional costs for supervision, overhead, rentals, etc., and loss of efficiency for direct labor. Accordingly, this is to place you on notice that we are entitled to additional compensation for all costs flowing from these delays and interference that have been imposed on us through no fault of our own. We will provide you with the specific amount of additional compensation covered by this notice as soon as we research this matter and have computed it.]

Sincerely yours,

XYZ Construction Company, Inc.

By _____
(Title)

* To be inserted where specific time of delay is known.
** To be used where extra money is claimed for delay.

566

APPENDIX 14C (CONTINUED)

SAMPLE NOTICE LETTER—CLAIM FOR EXTRAS

XYZ # _____

Addressee:

(To Contracting Officer)

Re: Contract No. (Describe Extra Work)

Dear:

This is to notify you that (on _____ we will begin) (we are about to begin) this extra work and are expecting to be compensated for it. If you do not want us to perform this work as an extra to the contract, please immediately notify us before we incur additional costs in the preparation for performance of this extra work. If we do not hear from you right away, we will proceed on the basis that you agree with our plan to perform this work.

OR

This work was performed pursuant to your representative's requirement and entitles us to additional compensation. We have proceeded to complete this work so as to minimize the cost of the work and any delay to (our work) (the job). We will be pleased to review this matter with you at your convenience.

We will provide you with a detailed cost breakdown for this added work as soon as we are able to compute it.

Sincerely yours,

XYZ Construction Company, Inc.

By _____
   (Title)

APPENDIX 14C (CONTINUED)

SAMPLE NOTICE LETTER CONFIRMING CONSTRUCTIVE CHANGE DIRECTIVE

XYZ # _____

Addressee:

(To Contracting Officer)

Re: Contract No. (Describe Constructive Change)

Dear:

We were given instructions by (insert name) on (date) (put in time also if pertinent) to (describe work added or changed).

This change is for work not within the scope of our present contract, and we therefore request a written modification to cover the added (material, labor, equipment, etc.) required to perform the work as ordered. (Give notice of other factors involved such as delay, acceleration, diversion of men or equipment from contract work, material shortages, etc.)

Our proposal for the added cost resulting from this change order is being prepared and will be submitted for your approval as soon as possible. We cannot determine at this time the effect on contract completion date or other work under the contract, and will advise when a full analysis has been made.

As ordered, we (are proceeding) (have proceeded) at once to (procure materials) (perform the work) in order to complete this change order at the earliest possible time. In the event you do not approve of such action, please advise immediately in order that we may stop this effort and minimize the cost involved.

Very truly yours,

XYZ Construction Company, Inc.

By _____
(Title)

APPENDIX 14C (CONTINUED)

SAMPLE NOTICE LETTER—ORAL DIRECTIONS OF EXTRA WORK

XYZ # _____

Addressee:

(To Contracting Officer)

Re: Contract No. (Describe Extra Work)

Dear:

On the _____ day of _____, 20_____, we received certain oral instruction (or orders, approvals, changes, as the case may be) from (insert name). These instructions were confirmed by our _____, 20_____, letter and should have been given to us in writing under the terms of our agreement. Your (insert name) has refused to confirm the oral instructions (or orders, approvals, changes, as the case may be) that we have recited in our referenced letter. Accordingly, we must advise that we will not (proceed with) (continue to follow) these verbal instructions unless we receive your immediate written confirmation. In any event, we will expect reimbursement for all costs reasonably incurred in reliance upon your direction.

We understand that it may take time to go though all the steps necessary to bring about a written authorization for extra work, and that sometimes it is more practical to do the work before that written authorization can be obtained. It has been our past practice to try to recognize your need to follow this method of operation. However, in this case, and in order to avoid any misunderstanding, we think it appropriate that you first provide us with a formal written authorization for changed work.

Very truly yours,

XYZ Construction Company, Inc.

By _____

(Title)

NOTE: Where the work already has been performed, it may be important to establish a prior history of reliance by the parties on oral directives. If the work has been fully performed, then the second paragraph should be deleted and the last sentence of the first paragraph replaced with the following:

As you know, we proceeded immediately as directed to perform this additional work. We did so in order to minimize your extra cost, and in the same manner in which we have handled other verbal directives in the past. Consistent with that past practice, we will provide you with our costs as soon as they are fully known and expect your prompt reimbursement.

APPENDIX 14C (CONTINUED)

SAMPLE NOTICE LETTER—CONFIRMING CONSTRUCTIVE ACCELERATION DIRECTIVE

XYZ # _____

Addressee:

(To Contracting Officer)

Re: Contract No. (Describe Acceleration Directive)

Dear:

We were given direction by (name or letter dated _____) on (date) (put in time also if pertinent) to (describe work and specific location). This directive stipulates and orders that we are to complete this work by (date and time).

This directive necessarily (accelerates, delays, diverts men and equipment from contract work, involves inefficiencies, interrupts contract work, involves excessive working hours, shortages, causes manpower shortage for contract work, inefficient working conditions, involves work under hazardous conditions, etc.) and thereby will result in increased cost to _____ on this contract.

Our proposal for the added costs resulting from this directive is being prepared and will be submitted for your approval as soon as possible. We cannot determine at this time the effect on contract completion date and will advise after a full analysis has been made.

Very truly yours,

XYZ Construction Company, Inc.

By _____
(Title)

File No. _____

15

CONTRACT CLAIMS
AND DISPUTES

I. INTRODUCTION

Bid protests, contract claims, and disputes often are treated as distinct topics in materials addressing federal government contract law. That treatment reflects the fact that bid protests generally involve issues related to the competition for the contract award while disputes and claims typically arise in the context of the performance of the contract. However, the Federal Acquisition Regulation (FAR) addresses both bid protests and disputes in Part 33. In addition, both topics involve different types of legal proceedings, which have their own particular rules of procedure. The United States Court of Federal Claims is an available forum to resolve both bid protests and contract claims. Although both topics are included in Part 33 of the FAR and are legal proceedings, bid protests are addressed in **Chapter 3** of this book in the discussion of the award of government contracts.

The current treatment of contract claims and disputes in the FAR may obscure the fact that the twenty-first-century principles, forums, and procedures for the resolution of claims and contract disputes have evolved over the last century, as discussed in **Appendix 1A** of **Chapter 1** of this book. During those years, both the federal government agencies and the private sector have attempted to balance competing interests and provide procedures that are seen as reasonable, credible, and effective by all concerned. A dispute resolution process that is perceived to lack objective credibility, reasonable efficiency, or effectiveness will deter contractors from entering into the government marketplace. Reduced competition ultimately increases the cost for project delivery.

II. THE CONTRACT DISPUTES ACT

As noted in the discussion of the historical evolution of the disputes process in **Chapter 1**, the Contract Disputes Act (CDA or Act) addresses the processing of claims from their initiation to final disposition and payment. It describes the manner in which claims are asserted and provides time frames for decisions on contractor claims by the contacting officer. The Act also provides for a contractor's right to appeal the contracting officer's final decision or the lack of a final decision as well as the time frames for appeals to each of two alternative forums (a board of contract appeals or the Court of Federal Claims) and any appellate review of those decisions.[1] Since the primary focus of the CDA is the disposition of a claim, this chapter sequentially addresses the assertion of a claim and follows the processing of that claim through a decision and any appeal. It also reviews provisions of the CDA and related laws dealing with fraudulent or inflated claims,[2] small claims[3] interest on amounts found due the contractor,[4] payment of claims,[5] and recovery of attorney's fees by certain contractors pursuant to the Equal Access to Justice Act (EAJA).[6]

Under the contract provisions used prior to the CDA, the typical disputes clause extended only to disputes "arising under" the contract. Generally, this meant that breach-of-contract claims were not subject to the disputes clause and were outside the jurisdiction of the boards.[7] The CDA encompasses all disputes arising under or related to a contract. This broader formulation of the scope of the disputes process includes claims for breach of contract. However, by its terms and as its name implies, the CDA is applicable only to contract disputes, and tort claims or claims seeking specific performance are not subject to the CDA.[8]

Although the Act is specifically applicable to any "express or implied contract," the U.S. Court of Appeals for the Federal Circuit (Federal Circuit) held that the CDA does not give jurisdiction to the boards to hear claims based on an implied contract by the federal government to treat bids fairly and honestly.[9] This was historically the basis for bid protest and bid preparation cost actions before the adoption of the Competition in Contracting Act (CICA). The Federal Circuit held that an implied contract to treat bids fairly and honestly is not a contract for the procurement of "goods or services." Consequently, it did not fall within the definition of a contract contained in § 3(a) of the CDA.

---

[1]These forums dispose of the vast majority of all contract-related issues.
[2]41 U.S.C. § 604.
[3]41 U.S.C. § 608.
[4]41 U.S.C. § 611.
[5]41 U.S.C. § 612.
[6]5 U.S.C. § 504.
[7]*United States v. Utah Constr. & Mining Co.*, 384 U.S. 394 (1966).
[8]41 U.S.C. § 602(a); *Malnak Assoc. v. United States*, 223 Ct. Cl. 783 (1980); *Siska Constr. Co.*, VABCA No. 3524, 92-2 ¶ 24,825; *Maritime Equipment & Sales, Inc. v. General Services Administration*, GSBCA No. 15266, 00-2 BCA ¶ 30,987; *Midwest Properties, LLC v. General Services Administration*, GSBCA Nos. 15822, 15844, 03-2 BCA ¶ 32,344; *Aim Constr. and Contracting Corp.*, ASBCA No. 52540, 07-1 BCA ¶ 33,466; *but see Qatar Inter, Trading Co.*, ASBCA No. 55518, 08-2 BCA ¶ 33,948 (government defense introducing allegations of tortious conduct did not require dismissal of appeal).
[9]*Coastal Corp. v. United States*, 713 F.2d 728 (Fed. Cir. 1983); *but see LaBarge Prods., Inc. v. West*, 46 F.3d 1547 (Fed. Cir. 1995) (contractor entitled to assert a claim for reformation of contract based on government's allegedly improper preaward actions).

Although § 605(a) of the CDA provides that all claims related to a contract shall be submitted to the contracting officer for a decision, the Act also states that the authority of that subsection does not extend to "a claim or dispute for penalties or forfeitures prescribed by statute or regulation which another federal agency is specifically authorized to administer, settle, or determine." Therefore, disputes arising under the Walsh-Healey Act,[10] the Davis-Bacon Act[11] and the Service Contract Act of 1965[12] are not subject to the CDA even though these matters and issues arise in the context of contract performance. These acts involve labor laws that are administered by the Secretary of Labor, and disputes related to the enforcement of these statutes are generally beyond the jurisdiction of the boards or Court of Federal Claims.[13] Similarly, contracting officers are expressly precluded from settling, compromising, or paying any claim involving fraud.[14]

During the three decades since 1978, there has been a gradual consolidation of the boards of contract appeals, as the federal agencies found that the number of contract appeals in a particular agency did not justify the administrative expense of a full-time board in that agency. When the Act was passed in 1978, many federal agencies had their own board of contract appeals. As of January 6, 2007, two boards: the Armed Services Board of Contract Appeals (ASBCA) and the Civilian Board of Contract Appeals (CBCA or Civilian Board)[15] decide the appeals for nearly all federal agencies.[16]

---

[10]41 U.S.C. § 35 *et seq.*

[11]40 U.S.C. § 276(a).

[12]41 U.S.C. § 351 *et seq.*

[13]*Emerald Maintenance, Inc. v. United States,* 925 F.2d 1425 (Fed. Cir. 1991); *Tele-Sentry Sec., Inc.,* GSBCA No. 10945, 91-2 BCA ¶ 23,880.

[14]*See* FAR § 33.210(b).

[15]In Section 847 of the National Defense Authorization Act for Fiscal Year 2006, Pub. L. No. 109-163, Congress established the Civilian Board of Contract Appeals (CBCA) within GSA to hear and decide contract disputes involving executive agencies (other than the Department of Defense, the Department of the Army, the Department of the Navy, the Department of the Air Force, the National Aeronautics and Space Administration, the United States Postal Service, the Postal Rate Commission, and the Tennessee Valley Authority) under the provisions of the Contract Disputes Act of 1978 and the applicable regulations. The legislation establishing the Civilian Board provided that the CBCA has jurisdiction to decide contract appeals from any executive agency (other than the Department of Defense, the Department of the Army, the Department of the Navy, the Department of the Air Force, the National Aeronautics and Space Administration, the United States Postal Service, the Postal Rate Commission, the Federal Aviation Administration, and the Tennessee Valley Authority). Cases pending before a board of contract appeals affected by the legislation were transferred to the CBCA on January 6, 2007, and reassigned CBCA docket numbers. Prior to January 6, 2007, boards of contract appeals currently existed at the General Services Administration and the departments of Agriculture, Energy, Housing and Urban Development, Interior, Labor, Transportation, and Veterans Affairs. All of those boards in existence on that date (January 6, 2007) terminated and their cases, board judges, and other personnel were transferred to the new Civilian Board. *See* 71 Fed. Reg. 65825, Nov. 8, 2006. In *Business Mgmt. Research Assocs., Inc. v. General Services Administration,* CBCA 464, 07-1 BCA ¶ 33,486, the full board issued its first decision stating "that the holdings of our predecessor boards shall be binding as precedent in this Board." That holding presupposes uniformity of holdings on all issues among the predecessor boards.

[16]The U.S. Postal Service and the Federal Aviation Administration have separate boards of contract appeals.

III. CONTRACTOR CLAIMS UNDER THE CDA

A. Types of CDA Claims: Monetary and Nonmonetary

The CDA states that "[a]ll claims by a contractor against the government relating to a contract shall be in writing and shall be submitted to the contracting officer for a decision."[17] Claims may be either monetary or nonmonetary. Consequently, it is possible to assert a claim over an issue such as an interpretation of a specification.

The boards and the Court of Federal Claims do not always agree on the limits of their respective jurisdictions as granted by the CDA, especially with regard to non-monetary claims. Issues regarding past performance evaluations can have major significance for federal government contractors. As discussed in **Chapter 3**, the use of a centralized database containing evaluations of a contractor's past performance as a tool in the selection process for new projects is common in the era of best value procurements. Contractors are justifiably sensitive to these evaluations and may wish to challenge one that seems unfair. The Court of Federal Claims and the boards have split as to whether a contractor's challenge to a past performance evaluation can be addressed as a CDA claim. To date, the boards have not accepted jurisdiction over a direct challenge to a past performance evaluation;[18] however, the Court of Federal Claims has taken jurisdiction over challenges to past performance evaluations as a CDA claim.[19] A contractor should carefully determine the type of claim it wishes to raise and make certain that it has selected the proper forum to raise that type of claim.

The existence of a "claim in dispute" and the submission of that claim to the contracting officer for a decision are prerequisites to the contractor's ability to invoke the dispute resolution procedures of the CDA. Until a claim is submitted, the contracting officer has no obligation to issue a final decision, and the contractor has no right of access to either a board of contract appeals or to the Court of Federal Claims. The CDA contains an additional requirement on monetary claims. As discussed in **Sections III.D** of this chapter, the CDA provides that monetary claims in excess of $100,000 must be certified[20] by the contractor in order to be processed as a proper claim.

If a contractor initiates either a board or Court of Federal Claims proceeding prior to the submission of a proper claim, the proceeding will be either dismissed or stayed, depending on the deficiency in the contractor's claim submission. Even though a dismissal would be without prejudice to the right of the contractor to reinitiate the process,[21] the contractor must begin the process by submitting a proper claim

---

[17]41 U.S.C. § 605(a). Even if the alleged wrong that occurred in the context of a contract is a claim for tort damages or defamation, a board does not have jurisdiction over those types of claims. *See EDL Constr., Inc.,* ASBCA No. 34623, 88-1 BCA ¶ 20,313; *Aim Constr. and Contracting Corp.,* ASBCA No. 52540, 07-1 BCA ¶ 33,466.

[18]*Compare Sundt Constr. Inc.,* ASBCA No. 56293, 09-1 BCA ¶ 34,084 *with TLT Constr. Corp.,* ASBCA No. 53769, 02-2 BCA ¶ 31,969.

[19]*Todd Constr. L.P. v. United States* 85 Fed. Cl. 34 (2008); *BLR Group of America v. United States,* 84 Fed. Cl. 634 (2008); *Record Steel and Constr., Inc. v. United States,* 62 Fed. Cl. 508 (2004).

[20]41 U.S.C. § 605(c). When the CDA was enacted, the threshold for certification was any amount greater than $50,000. This threshold was increased to $100,000 in 1994.

[21]*Thoen v. United States,* 765 F.2d 1110, 1116 (Fed. Cir. 1985).

to the contracting officer for a final decision.[22] That costs time and money. CDA interest does not begin to accrue on a proposal or request for an equitable adjustment until the submission qualifies as a claim under the Contract Disputes Act. Failure to understand and to follow the requirements of the CDA can be very costly.

B. CDA Claims Distinguished from Requests for an Equitable Adjustment/Change Order or Settlement Proposals

The CDA states that all claims "shall be in writing and shall be submitted to the contracting officer for a decision."[23] However, the CDA does not set forth a definition of a claim. The term "claim" also has multiple meanings in federal government contracting. A contractor's submission to the contracting officer may be a *claim* under one law or regulation[24] even though it is *not* a claim for the purposes of the CDA. Change order proposals or requests for equitable adjustments (REAs) are fairly routine on federal government contracts. If the contract is terminated for the convenience of the government, the contractor has the right to submit a termination proposal.[25] Some of these submissions have distinct certification requirements[26] and requirements for detailed supporting data and pricing information. Notwithstanding these requirements, a priced proposal or REA is not necessarily a *CDA claim*. The latter is a term of art in federal government contracting.

The current FAR Disputes clause[27] does set forth a definition of a *claim* for the purposes of that clause. It provides in relevant part:

(c) *Claim,* as used in this clause, means a written demand or written assertion by one of the contracting parties seeking, as a matter of right, the payment of money in a sum certain, the adjustment or interpretation of contract terms, or other relief arising under or relating to this contract. . . . [A] written demand or written assertion by the Contractor seeking the payment of money exceeding $100,000 is not a claim under the Act until certified. . . .

(d) (1) A claim by the Contractor shall be made in writing and, unless otherwise stated in the contract, submitted within 6 years after accrual of the claim to the Contracting Officer for a written decision.

Although no particular wording or claim format is required,[28] the contractor must provide sufficient information so that the contracting officer has notice of the basis

---

[22]*Skelly and Loy. v. United States,* 685 F.2d 414, 419 (1982); *Technassociates, Inc. v. United States,* 14 Cl. Ct. 200, 212 (1988); *T.J.D. Servs., Inc. v. United States,* 6 Cl. Ct. 257, 260 (1984).

[23]41 U.S.C. § 605(a).

[24]False Claims Act, 18 U.S.C. § 287; *United States v. Neifert-White Co.,* 390 U.S. 228 (1968) (submission of an application for a loan treated as a *claim* under the False Claims Act); disallowance of costs of prosecution of a "claim" against the federal government, *Bill Strong Enters., Inc.,* ASBCA Nos. 42946, 43896, 93-3 BCA ¶ 25,961.

[25]*See* **Chapter 11.**

[26]*See* **Section III.D** of this chapter.

[27]FAR § 52.233-1.

[28]*Contract Cleaning Maintenance, Inc. v. United States,* 811 F.2d 586, 592 (Fed. Cir. 1986). *See* **Section III.C** of this chapter.

for (entitlement) and amount (quantum) of the claim.[29] Not every submission seeking the payment of money is a claim under the CDA. In 1995, the Federal Circuit expressly overruled an earlier "in dispute" standard[30] as an essential element of CDA claim and substituted a distinction based on whether the submission could be characterized as a *routine* claim for payment or a *nonroutine* claim for payment.[31] However, in *Reflectone*, the Federal Circuit refused to adopt a bright-line test between *routine* and *nonroutine* submissions to determine whether the contractor's submission was a CDA claim. Subsequently, the Court of Federal Claims[32] provided these examples of the distinction between *routine* and *nonroutine* requests for payment or other relief:

Routine. Invoices for completed work; requests for scheduled progress payments; and vouchers for disbursement under a cost-reimbursement contract.

Nonroutine. Request for equitable adjustment; a termination for convenience proposal; a submission disputing the government's planned setoff; a submission seeking the return of property, and a submission asserting a breach of contract by the federal government.

Although case law does not make a specific request for a final decision an absolute condition precedent for a CDA claim,[33] the failure to make a specific request for a final decision creates a risk that a board or the Court of Federal Claims will hold that there was no dispute between the parties sufficient to constitute a claim under the Act.[34] For this reason, correspondence from a contractor that contains information detailing costs and merely expresses a willingness to reach an agreement, rather than demanding or requesting that the contracting officer issue a final decision, may not constitute a claim.[35]

---

[29]*Gauntt Constr. Co.*, ASBCA No. 33323, 87-3 BCA ¶ 20,221; *Mitchco, Inc.*, ASBCA No. 41847, 91-2 BCA 23,860; *Holk Dev., Inc.* ASBCA No. 40579, 90-3 BCA ¶ 23,086.

[30]*Dawco Constr., Inc. v. United States*, 930 F.2d 872 (Fed. Cir. 1991) (preexistence of a dispute regarding a sum certain was critical to existence of CDA claim).

[31]*Reflectone, Inc. v. Dalton*, 60 F.3d 1572 (Fed. Cir. 1995). The court also stated that not every nonroutine submission was a CDA claim.

[32]*Scan-Tech Security, L.P. v. United States*, 46 Fed. Cl. 326 (2000).

[33]*Magnum, Inc.*, ASBCA No. 53890, 04-1 BCA ¶ 32,489 (contractor "implicitly" requested a final decision); *Lemay v. General Services Administration*, GSBCA No. 16093, 03-2 BCA ¶ 32,345.

[34]*M. Maropakis Carpentry, Inc. v. United States* 84 Fed. Cl. 182 (2008).

[35]*Hoffman Constr. Co. v. United States*, 7 Cl. Ct. 518, 525 (1985). *See Technassociates, Inc. v. United States*, 14 Cl. Ct. 200, 209-10 (1988) (letters the contractor sent in order to get the contracting officer to negotiate with it "on the future direction of the contract" did not constitute claims). *See also GPA-I, LP v. United States*, 46 Fed. Cl. 762 (2000) (a request for a final decision can be implied from the context of the submission; "'[M]agic words' need not be used"; the request need not be explicit, but only show "an 'expression of interest,' which may be made implicitly;" thus, where the plaintiff sent letters stating plaintiff's need for timely payment under a lease or the applicable late interest penalty, and demanded that the contracting officer address the issues "immediately," the letters were "expression[s] of interest" in the contracting officer's final decision and were thus sufficient to constitute a request for a final decision).

Whether a particular submission is a claim often depends on all of the circumstances. For example, a certified request for an equitable adjustment accompanied by a letter requesting a final decision did constitute a valid claim even though the letter also suggested the possibility or hope of a negotiated resolution of the dispute.[36] A prime contractor's cover letter that transmitted a subcontractor's detailed statement of its position and entitlement to relief was deemed to be a claim.[37] In contrast, a letter sent to the contracting officer during negotiations on a proposal requesting that the matter be referred to an auditor did not qualify as a claim because a contracting officer had not been asked to issue a final decision.[38] Finally, the contractor may not submit a claim in an unspecified amount or in an amount that is open-ended.[39] A proper monetary claim and request for a final decision exists only when the amount sought is either set forth in a "sum certain"[40] or is determinable by a simple mathematical calculation or from the information provided by the contractor.[41] For example, a delay claim in "an amount of no less than $1,072,957.05" was not a demand for a sum certain as required by the Act.[42] In contrast, a request for an interpretation of a specification without an accompanying statement of a sum certain may qualify as a CDA claim.[43]

In *Ellett Construction Co. v. United States*,[44] the Federal Circuit held that a termination for convenience (T4C) settlement proposal *may* at some point constitute a CDA claim. Although the court held that T4C proposals are nonroutine requests for payment, the T4C proposal must be a written demand seeking, as a matter of right, the

---

[36]*Isles Eng'g & Constr., Inc. v. United States*, 26 Cl. Ct. 240, 243 (1992). Failure by the contracting officer to respond for six months to a reasonably simple engineering change proposal permitted the contractor to convert the proposal into a claim and request a final decision. *S-Tron*, ASBCA No. 45890, 94-3 BCA ¶ 26,957.

[37]*Clearwater Constructors, Inc. v. United States*, 56 Fed. Cl. 303 (2003).

[38]*G.S. & L. Mech. & Constr., Inc.*, DOTBCA No. 1856, 87-2 BCA ¶ 19,882.

[39]*Metric Constr. Co. v. United States*, 14 Cl. Ct. 177, 179-80 (1988) (contractor's submissions made it clear that contractor was seeking to recover extended home office overhead and third-party indemnification fees, but the submissions did not constitute a claim because the amount was not specified).

[40]Disputes clause ¶ (c), FAR § 52.233-1. A claim for a "sum certain" that reserved the right to include additional line items to "modify the presentation" was, in fact, deemed to be a predicate for negotiations rather than a CDA claim. *McElroy Mach. & Mfg. Co., Inc.*, ASBCA No. 39416, 92-3 BCA ¶ 25,107; *see also Total Procurement Service, Inc.*, ASBCA No. 54163, 08-1 BCA ¶ 33,843 (claim asserting damages in excess of $66,000.00 was not a sum certain). The sum certain requirement is determined when the claim is filed, not when the complaint is filed. *Morgan & Son Earthmoving, Inc.*, ASBCA No. 53524, 02-2 BCA ¶ 31,874; *MDP Constr. Inc.*, ASBCA No. 52769 et al., 01-1 BCA ¶ 31,359. If the monetary claim does not explicitly state a sum certain, it may be acceptable if the amount could be determined from the submission by simple arithmetic. *Mulunesh Berhe*, ASBCA No. 49681, 96-2 BCA ¶ 28,339; *J&J Maintenance, Inc.*, ASBCA No. 50984, 00-1 BCA ¶ 30,784.

[41]*Metric Constr. Co. v. United States*, 1 Cl. Ct. 383, 392 (1983).

[42]*Sandoval Plumbing Repair, Inc.*, ASBCA No. 54640, 05-2 BCA ¶ 33,072. (The phrase "no less than" made the amount other than a sum certain.)

[43]*William K. Euille & Assocs., Inc. v. General Services Administration*, GSBCA No. 15261, 00-1 BCA ¶ 30,910. *See also Thomas Creek Lumber & Log Co.*, IBCA No. 4201-2000, 00-2 BCA ¶ 31,118.

[44]93 F.3d 1537 (Fed. Cir. 1996).

payment of a sum certain. In addition, the contractor must also request a final decision. In *Ellett*, the court held that, because there was no express request for a final decision, the settlement proposal ripened into a claim only after negotiations failed and the contractor impliedly requested a final decision.[45]

The ASBCA appears to have adopted the Federal Circuit's "impasse" analysis set forth in *Ellett Construction Co. v. United States*.[46] The board found that an "impasse" in negotiations had occurred four months after the contractor provided a CDA certification on its termination settlement proposal. As a result, the proposal had ripened into a CDA claim. Consequently, a typical change order proposal, termination settlement proposal, or REA may not be considered or treated as a CDA claim when it is first submitted because there has been no impasse.[47] If there is no action on it within a reasonable period of time, that delay may, by itself, operate to provide a basis to change the status of the submission to a CDA claim.[48] If the federal government disputes either the entitlement or the quantum (monetary element and its support), that can operate to convert the proposal or REA into a potential CDA claim, particularly if the contractor expressly or by implication requests a final decision.[49]

C. CDA Claim Submissions

A CDA claim can be a concise letter of a few pages or a multiple volume submission with extensive data and exhibits. However, there are certain common issues or questions that relate to any CDA claim. These are:

- Who may submit a CDA claim?
- What time deadlines apply?
- Claim submission preparation and elements

Each of these topics is addressed in the next subsections.

1. Who May Submit a CDA Claim?

Generally, under the CDA, only a prime contractor may assert claims against the government. The CDA refers to "contractor claims"[50] and states that the term "contractor means a party to a government contract other than the government. . . ."[51] Similarly, FAR § 33.201 defines a claim as a demand or assertion "by one of the contracting parties." This language reflects the requirement for "privity of contract" denoting a contractual relationship between the parties. This requirement is strictly

---

[45]*See also Med-America Eng'g & Mfg.*, ASBCA No. 48831, 96-2 BCA ¶ 28,558.

[46]93 F.3d 1537 (Fed. Cir. 1996); *Central Envt'l, Inc.*, ASBCA No. 51086, 98-2 BCA ¶ 29,912.

[47]However, that initial submission would most certainly be a "claim for the purposes of other statutes such as the Civil False Claims Act, 31 U.S.C. §§ 3729-3733.

[48]*Mid-America*, 96-2 BCA ¶ 28,558.

[49]*See Magnum, Inc.*, 04-1 BCA ¶ 32,489 *and* **Section III.D** of this chapter.

[50]41 U.S.C. § 605(a).

[51]41 U.S.C. § 601(4).

enforced in government contracts. Generally, subcontractors are not considered to be one of the contracting parties and are not in privity of contract with the government. Accordingly, subcontractors may not assert claims directly against the government under the Act.[52] Only in rare situations does privity of contract exist between a subcontractor and the government. Examples of these fairly unique factual situations include when a subcontractor establishes that it is entitled to submit an unsponsored claim as a third-party beneficiary of the contract with the government;[53] when the government utilizes an agent to enter into a contract "by and for" the government; or there is an assignment of a subcontractor's contract to the government pursuant to a clause, such as the termination for convenience clause.[54] Similarly, a noncompleting performance bond surety is not a contractor for the purposes of the CDA.[55] However, a surety that expressly or implicitly assumes contract performance is in privity with the government and may assert claims under the CDA.[56] Finally, unless the government is a party to the assignment transaction, the assignee of a contractor's rights under a government contract does not attain the status of a contractor and is not in the position to assert a claim under the CDA.[57]

Given the nature of most construction projects, subcontractor performance is often a key issue and subcontractor claims are common. Although subcontractors do not have the right to directly assert a claim against the government, subcontractor claims are routinely considered in the context of the disputes process. With the prime contractor's consent and cooperation, a subcontractor claim can be submitted to the contracting officer for decision and appealed through the disputes process. In that situation, the prime contractor acts as a "sponsor" for the subcontractor's claim.[58] However, a prime contractor must recognize that only it can certify a subcontractor's claim when it is submitted under the CDA. See **Section IV** of this chapter.

---

[52]*Lockheed Martin Corp. v. United States,* 50 Fed. Cl. 550 (2001); *Erickson Air Crane Co. of Wash., Inc. v. United States,* 731 F.2d 810, 813 (Fed. Cir. 1984); *United States v. Johnson Controls, Inc.,* 713 F.2d 1541 (Fed. Cir. 1983); *G. Schneider,* ASBCA No. 333021, 87-2 BCA ¶ 19,865; *Alpine Computers, Inc.,* ASBCA No. 54659, 05-2 BCA ¶ 32,997.

[53]*Floorpro, Inc.,* ASBCA No. 54143, 08-1 BCA ¶ 33,793 *rev'd Winter v. Floorpro, Inc.,* 570 F.3d 1367 (Fed. Cir. 2009) (Federal Circuit held that the ASBCA could not take jurisdiction over a third-party beneficiary claim but that the Court of Federal Claims could take jurisdiction.)

[54]*Kern-Limerick v. Scurlock,* 347 U.S. 110 (1954); *Deltec Corp. v. United States,* 326 F.2d 1004 (Ct. Cl. 1964); *Lockheed Martin Corp. v. United States,* 50 Fed. Cl. 550 (2001); *Globex-Corp. v. United States,* 54 Fed. Cl. 343 (2002); *Marine Contractors, Inc.,* ASBCA No. 54017, 03-1 BCA ¶ 32,240; *Coastal Drilling, Inc.,* ASBCA No. 54023, 03-1 BCA ¶ 32,241.

[55]*See Universal Sur. Co. v. United States,* 10 Cl. Ct. 794, 799-800 (1986).

[56]*See Fireman's Fund Ins. Co. v. United States,* 909 F.2d 495, 499 (Fed. Cir. 1990); *Balboa Ins. Co. v. United States,* 775 F.2d 1158, 1161 (Fed. Cir. 1985); *but see Atherton Constr., Inc.,* ASBCA No.56040, 08-2 BCA ¶ 34,011 (takeover contractor entitled to assert claim).

[57]*See Fireman's Fund/Underwater Constr., Inc.,* ASBCA No. 33018, 87-3 BCA ¶ 20,007.

[58]*Clearwater Constructors, Inc. v. United States,* 56 Fed. Cl. 303 (2003); *Erickson Air Crane Co. of Wash., Inc. v. United States,* 731 F.2d 810, 813-14 (Fed. Cir. 1985); *Planning Research Corp. v. Dept. of Commerce,* GSBCA No. 11286-COM, 96-1 BCA ¶ 27,954. However, if a prime contractor refuses to authorize prosecution of a claim in its name, the claim will be dismissed. *See Divide Constructors, Inc.,* IBCA No. 1134-12-76, 77-1 BCA ¶ 12,430.

2. What Time Deadlines Apply?

Timely submission of a claim is essential for both practical as well as legal considerations. If a matter is identified in its early stages, all parties may have more flexibility in reaching a mutually satisfactory resolution. Even if the claim is submitted within the time limits to be outlined, the passage of time may make it more difficult to obtain prompt resolution as personnel are transferred and memories are less clear. Consequently, when considering time frames or time deadlines associated with CDA claims, a contractor needs to consider:

- Notice requirements
- Timely submission of a claim
- Contractor claim releases, especially final payment releases
- CDA statute of limitations

a. Notice Requirements   A lack of timely notice can be a costly omission for contractors. Factual and objective written notice is good business because it allows the contracting officer and the federal agency's representatives an opportunity to address a problem before it evolves into a costly dispute. **Chapter 14** of this book provides more specific information on notice requirements in federal government construction contracts. To avoid the potential loss of rights under the contract, the project staff should develop a *notice checklist* for each project. That checklist should reflect a careful review of the standard FAR clauses in the contract as well as any specific additional notice or claim-related documentation requirements that are set forth in the general or special conditions of the contract. Even though the "front-end" documents in a government construction contract may appear to be similar, there can be significant variations due to the effect of agency supplements to the FAR or due to the use of special conditions. Thus, a careful review of every contract is as essential as a prebid/preproposal site visit. (See **Appendix 14A** for a sample notice checklist.)

b. Contractor Claim Releases   Previously executed claim releases can have a significant effect on a contractor's rights. As discussed in **Chapter 8, Section VIII,** FAR § 43.204(c) provides that contracting officer's "should" include a release in every supplemental agreement or change order. In evaluating a potential claim, a contractor should carefully review the language in any reservation of rights as well as the scope of any releases provided to the contracting officer with executed change orders.[59] A failure to appreciate the possible interpretation of a change order release on a contractor's rights may result in the expensive and unpleasant surprise of having subsequent requests for equitable adjustments or claims denied because of the wording of the release.[60]

---

[59]*Valenzuela Eng'g, Inc.*, ASBCA No. 54490, 06-2 BCA ¶ 33,399 (specific reservation of right in contractor's letters served as an exception to the general release language in a subsequently issued change order).
[60]*Bell/BCI Co. v. United States*, 570 F.3d 1337 (Fed. Cir. 2009).

As the project nears completion, potential issues related to cost overruns or schedule delays may become apparent. Similarly, the contractor may receive notice from a subcontractor that it intends to submit a claim, which the contractor might subsequently submit to the government. In both situations, the basis for and the amount of potential claims may not be capable of precise definition. At the same time, the contractor may seek to obtain final payment. Under these circumstances, final payment on a federal government contract may bar the later presentation of any claim that is not specifically reserved from the release provided to the government in conjunction with the request for final payment.

The Payments under Fixed-Price Construction Contracts (September 2002) (FAR § 52.232-5) clause contains these provisions regarding claims, releases, and final payment:

(h) *Final Payment.* The Government shall pay the amount due the Contractor under this contract after—

(1) Completion and acceptance of all work;

(2) Presentation of a properly executed voucher; and

(3) Presentation of release of all claims against the Government arising by virtue of this contract, other than claims, in stated amounts, that the Contractor has specifically excepted from the operation of the release. A release may also be required of the assignee if the Contractor's claim to amounts payable under this contract has been assigned under the Assignment of Claims Act of 1940 (31 U.S.C. § 3727 and 41 U.S.C. § 15).

The nature and wording of the claims reserved by contractor can be critical to the operation of this provision. A general or broad-brush reservation may not be effective.

In *Mingus Constructors, Inc. v. United States,*[61] the contractor provided the contracting officer two written notices of its intent to file a claim for "harassment," for "losses due to work that was misrepresented by the contract documents," "outside the scope of the original design," and/or "changed conditions." No dollar amount was stated. Subsequently, the contractor executed the final payment release form and inserted in the space provided for exceptions this statement:

Pursuant to correspondence we do intend to file a claim(s)—the amount(s) of which is undetermined at this time.

Following the submission of a certified claim many months later, the contracting officer returned it without action and stated that no "claim" as defined by the Disputes clause had been submitted at the time of final payment. In reviewing the contracting officer's treatment of the claim, the court held that the contractor's claim was barred by the general release, and the burden was on the contractor to assure that any reservation of rights set forth in the release was sufficiently specific. Noting that

---

[61]812 F.2d 1387 (Fed. Cir. 1987).

exceptions to releases are strictly construed against the contractor, the court characterized the contractor's letters as "nothing more than a 'blunderbuss exception' which does nothing to inform the government as to the source, substance, or scope of the contractor's specific contentions."[62] The court concluded that such "vague, broad exceptions" were "insufficient as a matter of law to constitute 'claims' sufficient to be excluded from the required release."[63]

Following the *Mingus* decision, the United States Claims Court in *Miya Brothers Construction Co. v. United States*[64] construed the meaning of the release language that requires the contractor to state the value of the specific claims excepted from the release. Rejecting an argument that the contractor did not have a right to increase the dollar amount of a reserved claim, the court held that the claim's exact amount is irrelevant to the general principle that a monetary claim has been properly excepted from the release. The court concluded that an amendment of the amount claimed did not constitute the submission of a new claim any more than the upward adjustment of the amount of a previously certified claim under the CDA constituted a new claim.

c. CDA Statute of Limitations   When originally enacted, the CDA did not contain a statute of limitations on claims by the contractor or the government. This issue was addressed in a 1994 amendment, and the CDA now states that "each claim" relating to a contract shall be submitted "within six years after the accrual of the claim.[65] The 1994 amendment to the CDA failed to define the term "accrual"; however, the final implementing regulations addressed the need for a definition of that term and also addressed the application of the six-year limitation to contracts in effect at the time of the passage of the 1994 amendment to the CDA.

The final implementing regulations[66] stated that the six-year period did not apply to contracts awarded prior to October 1, 1995. For purposes of the CDA, "accrual of a claim" was defined as:

> [T]he date when all events, which fix the alleged liability of either the Government or the contractor and permit assertion of the claim, were known or should have been known. For liability to be fixed, some injury must have occurred. However, monetary damages need not have been incurred.[67]

The implementing regulations made the term "accrual" applicable to claims by either the contractor or the federal government, except for federal government claims based on a contractor claim involving fraud. This definition still may stimulate questions of what constitutes "injury" when no monetary damages have been incurred. Contractors also need to bear in mind that there is a distinction between a CDA claim

---

[62]812 F.2d at 1394.
[63]*Id.*
[64]12 Cl. Ct. 142 (1987).
[65]41 U.S.C. § 605(a); *see Robinson Quality Constructors,* ASBCA No. 55784, 09-1 BCA ¶ 34,048.
[66]60 Fed. Reg. 48,225 (1995).
[67]*Id.*

and a proposal or a request for an equitable adjustment. Submission of the latter may not satisfy the six-year submission requirement if it is later determined that the equitable adjustment proposal or REA was not a "CDA claim."[68] Even if a contractor has a pending appeal on one dispute, that does not toll (stop the running) of the statute of limitations on other claims related to that contract.[69] Moreover, the government does not have the authority to waive the six-year statute of limitations.[70]

Even if the six-year statute of limitations is not applicable, a prolonged delay in the submission of a claim by a contractor may provide the basis for the government to assert a defense on the basis of "laches." In that context, the claim may be denied if the prolonged passage of time substantially prejudiced the government's ability to defend against the claim.[71] Similarly, failure to submit a claim for relief due to a mistake in bid until several years after award may be the basis for rejection of the claim on the grounds of waiver by the contractor, even though the federal government does not demonstrate any prejudice and no statute of limitations has expired.[72]

3. Claim Submission Preparation and Elements

a. Written Submission Requirement As indicated by the language of the Disputes clause, the claim for the purposes of the CDA must be in writing and must be submitted to the contracting officer for a written decision. Accordingly, oral demands or assertions seeking compensation are not claims for the purposes of the CDA. Similarly, a written demand or submission that is submitted to a person who is not the contracting officer is not a claim for the purposes of the Act. The CDA defines a contracting officer as "any person who, by appointment in accordance with applicable regulations, has the authority to enter into and administer contracts and make determinations and findings with respect thereto."[73] This definition also includes an authorized representative of the contracting officer, acting within the limits of that person's authority.[74]

Under certain circumstances, however, the submission of a written claim to a subordinate of the contracting officer has been held to be ineffective for the purposes of the Act.[75] A contractor should be able to avoid the problem of misdirected claims by

[68] *See Reflectone, Inc. v. Dalton,* 60 F.3d 1572 (Fed. Cir. 1995).

[69] *Environmental Safety Consultants, Inc.,* ASBCA No. 54615, 07-1 BCA ¶ 33,483.

[70] *John R. Sand & Gravel Co., v. United States,* 552 U.S. 130 (2008).

[71] *Systems Integrated,* ASBCA No. 55439, 07-1 BCA ¶ 33,575 (contractor unduly delayed submitting claim, but government fails to carry burden of proof to establish laches); *Anlagen und Sanierungstechnik GmbH,* ASBCA No. 49869, 97-2 BCA ¶ 29,168.

[72] *Turner-MAK (JV),* ASBCA No. 37711, 96-1 BCA ¶ 28,208.

[73] 41 U.S.C. § 601(3).

[74] *Magnum, Inc.,* ASBCA No. 53890, 04-1 BCA ¶ 32,489; *see Cath-dr/Balti Joint Venture v. Winter,* 497 F.3d 1339 (2007) (notifications to the government of an alleged or constructive change or delay needs to consider the requirement that a person authorized to issue that change to the contract or order a work suspension must be aware that an alleged change or delay order has been issued to the contractor); *States Roofing Corp.,* ASBCA No. 55500, 09-1 BCA ¶ 34,036.

[75] *Lakeview Constr. Co. v. United States,* 21 Cl. Ct. 269 (1990); *but see West Coast Gen. Corp. v. Dalton,* 39 F.3d 312 (Fed. Cir. 1994); *Gardner Zemke Co.,* ASBCA No. 51499, 98-2 BCA ¶ 29,997.

ascertaining at the outset of performance the specific individuals, in addition to the contracting officer, who are authorized to receive claims. Moreover, if there is any doubt regarding whether the submission to the contracting officer's subordinate will be deemed appropriate under the Act, the contracting officer should be copied on the submission. This removes any doubt regarding the date when the contracting officer received the submission for the purposes of the Act.

In submitting a claim, a contractor is not required to use any particular wording or format.[76] However, the contractor is required to give the contracting officer sufficient information so that the contracting officer has adequate notice of the basis for, and the amount of, the claim.[77] The two elements of any claim are commonly referred to as entitlement and quantum.

The *entitlement* portion establishes why, under the terms of the contract and the facts, the contractor is entitled to recover from the government. For example, in a claim involving the alleged misinterpretation of a specification by the government, the entitlement portion would describe how the government misinterpreted the specification as contrasted to the contractor's reasonable understanding of the contract's requirements. In addition, that section would also describe the extra work and delay, if any, caused by the misinterpretation.

If the claim seeks payment of money rather than an interpretation of a contract term, the second part of the claim is *quantum*. At a minimum, the quantum portion of a monetary claim requires that the contractor state a *sum certain* under the Act and the Disputes clause.[78] The contractor must describe the amount of money or time to which it is entitled and attempt to relate the cause (act by the government) to the effect (expenditure of money). The relation of cause and effect can be quite difficult, as it often encompasses issues of scheduling, cost accounting, and the support for estimates. As with any contract, project documentation is often critical to establishing cause and effect. The evaluation of a contractor's claims for time or money can be affected significantly by the nature and quality of the contemporaneous records maintained by the contractor. For example, some government contracts expressly require that the contractor document on a daily basis the specific critical path activities affected by delays such as adverse weather.[79] Even if there is no specific contract requirement for contemporaneous documentation of the extra time or expense incurred by the contractor, the absence of such documentation can create a substantial hurdle to recovery.[80]

There is no simple test to determine the degree of detail necessary to constitute a claim. In general, the boards of contract appeals and Court of Federal Claims have adopted a standard similar to notice pleadings. That is, the contractor must provide the contracting officer with adequate notice of the basis for and the amount of the claim and a request for the contracting officer to render a final decision.[81] In

[76]*Contract Cleaning Maintenance, Inc. v. United States,* 811 F.2d 586, 592 (Fed. Cir. 1986); *Clearwater Constructors, Inc. v. United States,* 56 Fed. Cl. 303 (2003).

[77]*Holk Dev., Inc.,* ASBCA No. 40579, 90-3 BCA ¶ 23,086.

[78]FAR § 52.233-1(c); *MDP Constr., Inc.,* ASBCA No. 52769 *et al.,* 01-1 BCA ¶ 31,359.

[79]*Consolidated Constr., Inc.,* ASBCA No. 46498, 99-1 BCA ¶ 30,148.

[80]*Bechtel Nat'l, Inc.,* NASA BCA No. 1186-7, 90-1 BCA ¶ 22,549; *Centex Bateson,* VABCA No. 4613 *et al.,* 99-1 BCA ¶ 30,153 (number of requests for information and changes are not enough by themselves to demonstrate loss of productivity); *Hensel Phelps,* ASBCA No. 49270, 99-2 BCA ¶ 30,531.

[81]*Metric Constr. Co. v. United States,* 1 Cl. Ct. 383, 392 (1983); *I.B.A. Co.,* ASBCA No. 37182, 89-1 BCA ¶ 21,576.

other words, the contractor should assert specific rights and request specific relief. Ultimately, whether a contractor's submission constitutes a claim depends on the totality of the circumstances and communications between the parties.[82] For example, these submissions have been found to be claims:

- A letter in which a contractor specified various items that a government audit had disallowed but to which the contractor claimed entitlement. The letter was viewed together with a prior letter from the contractor giving a detailed breakdown of the additional amounts to which the contractor believed it was entitled and referring to the contractor's previous request for "funding of [a] back-wage demand."[83]
- A letter sent by a company having a contract for transportation services at an Air Force base stated that the company viewed certain newly demanded bus service as beyond the contract's requirements and specifically sought "compensation of $11,000.04 per year, to be billed at $916.67 per month."[84]
- A letter from the contractor's attorney to the contracting officer that "expressed interest" in a final decision with respect to the contractor's request for contract reformation and stated that the contractor was seeking a decision so that it could pursue its appeal routes under the CDA, if necessary.[85]
- Letters that, when taken together, showed the contractor protesting a government requirement that the contractor pay additional sums under a contract to purchase crude oil from the government and demanding that certain identified wire transfer payments comprising those sums be returned to the contractor.[86]
- A contractor's cover letter, attaching a detailed argument by a subcontractor, that set forth the basis of the subcontractor's disagreement with the government regarding the effect of a contract modification.[87]

However, the basic claim must be made "by the contractor." Accordingly, a letter from the contractor's lawyer to the contracting officer has been held to be insufficient to establish a claim.[88] The burden is on the claimant to identify, specify, and perfect its claims.

Some decisions of the Court of Federal Claims and the Federal Circuit have emphasized the need for an explicit request for a final decision as part of the submission of a claim. However, in *Transamerica Insurance Corp. v. United States*,[89] the Federal Circuit expressly rejected a rule that a "claim" must include a specific request for a

---

[82]*Transamerica Ins. Corp. v. United States*, 973 F.2d 1572 (Fed. Cir. 1992); *Penn Envtl. Control, Inc.*, VA-BCA No. 3599, 93-3 BCA ¶ 26,021; *Atlas Elevator Co. v. General Services Administration*, GSBCA No. 11655, 93-1 BCA ¶ 25,216; *Winding Specialists Co.*, ASBCA No. 37765, 89-2 BCA ¶ 21,737.

[83]*Contract Cleaning Maintenance, Inc. v. United States*, 811 F.2d 586, 592 (Fed. Cir. 1986).

[84]*Tecom, Inc. v. United States*, 732 F.2d 935, 937 (Fed. Cir. 1984).

[85]*Paragon Energy Corp. v. United States*, 645 F.2d 966, 976 (Ct. Cl. 1981).

[86]*Alliance Oil & Refining Co. v. United States*, 13 Cl. Ct. 496, 499-500 (1987).

[87]*Clearwater Constructor's, Inc.*, 56 Fed. Cl. 303 (2003).

[88]*Construction Equip. Lease v. United States*, 26 Cl. Ct. 341 (1992). In contrast, the bankruptcy trustee for a contractor in bankruptcy has been held to be the proper party to assert a prebankruptcy claim against the government. *See Jerry Dodds d/b/a Dodds & Assocs.*, ASBCA No. 51682, 02-1 BCA ¶ 31,844.

[89]973 F.2d 1572, 1574 (Fed. Cir. 1992).

final decision. In particular, a formal demand for final decision has been held to be unnecessary if the claim gave a clear indication that a decision was desired.[90]

Even though there is case law that there is no absolute requirement for a specific request for a final decision, there is a risk that the failure to make a specific request for a final decision will be viewed as evidence that there was no dispute between the parties sufficient to constitute a "claim" under the Act. For this reason, correspondence from a contractor that contains information detailing costs and, rather than demanding or requesting that the contracting officer issue a final decision, merely expresses a willingness to reach an agreement may not constitute a claim.[91] Similarly, the passage of time may not convert a proposal into a CDA claim. In *Santa Fe Engineers, Inc. v. Garrett,*[92] the contractor's certified proposal had been pending for more than two years, during which time the parties met on several occasions to discuss the proposal and the contractor submitted additional information for the government's consideration. During that period of time, the contractor never asked for a contracting officer's final decision. The Federal Circuit affirmed the board decision,[93] which concluded that the failure to ask for a contracting officer's final decision was an indication that the parties had not reached an impasse in their negotiations and that the matter was not sufficiently in dispute to constitute a CDA claim.

Often the decision whether a particular submission is a claim depends on the totality of the circumstances. For example, in one case a certified request for an equitable adjustment accompanied by a letter requesting a final decision did constitute a valid claim even though the letter also suggested the possibility, or hope, of a negotiated resolution to the dispute.[94] In contrast, a letter sent to the contracting officer during negotiations on a proposal requesting that the matter be referred to an auditor did not qualify as a claim, because a contracting officer had not been asked to issue a decision.[95] Finally, the contractor may not submit a claim in an unspecified amount or in an amount that is open-ended.[96] A proper claim and request for a final

---

[90]*Cable Antenna Sys.,* ASBCA No. 36184, 90-3 BCA ¶ 23,203; *Ellett Constr. Co. v. United States,* 93 F.3d 1537 (Fed. Cir. 1996).

[91]*Hoffman Constr. Co. v. United States,* 7 Cl. Ct. 518 (1985). *See also Technassociates, Inc.,v. United States,* 14 Cl. Ct. 200, 209-10 (1988) (letters contractor sent in order to get contracting officer to negotiate "on the future direction of the contract" did not constitute claims).

[92]991 F.2d 1579, 1583-84 (Fed. Cir. 1993). However, in *D.H. Blattner & Sons, Inc.,* IBCA Nos. 2589, 2643, 89-3 BCA ¶ 22,230, the board held that a properly certified letter using the term "proposal" was not a claim. The Federal Circuit reversed this decision in an unpublished decision. *See Blattner & Sons, Inc. v. United States,* 909 F.2d 1495 (Fed. Cir. 1990).

[93]*Santa Fe Eng'rs, Inc.,* ASBCA No. 36292, 92-2 BCA ¶ 24,795.

[94]*Isles Eng'g & Constr., Inc. v. United States,* 26 Cl. Ct. 240, 243 (1992).

[95]*G.S. & L. Mech. & Constr., Inc.,* DOTBCA No. 1856, 87-2 BCA ¶ 19,882; *see also Huntington Builders,* ASBCA No. 33945, 87-2 BCA ¶ 19,898, at 100,654-655 (letters to contracting officer that, when taken together, alleged defective specifications, requested a 30-day time extension to contract and the release of monies withheld for liquidated damages did not constitute a *claim* because no specific monetary relief was requested for costs incurred as a result of contractor's having to comply with allegedly defective specifications).

[96]*Metric Constr. Co. v. United States,* 14 Cl. Ct. 177, 179-80 (1988) (contractor's submissions made it clear that contractor was seeking to recover extended home office overhead and third-party indemnification fees, but the submissions did not constitute a claim because the amount was not specified).

decision exist only when the amount sought is either set forth in a "sum-certain"[97] or is determinable by a simple mathematical calculation or from the information provided by the contractor.[98]

b. Claim Preparation Considerations   In any dispute, settlement at the earliest time and at the *first level*—that is, with the first person who has the authority to consider and settle a contractor's claim—is preferable because it saves the expense and avoids the inherent problems of late payment. For federal government contracts, this "first-level" individual is the contracting officer. The contracting officer and the project staff are the key people to convince if a contractor is to settle its claim without expensive litigation. The claim submission should be prepared in a manner to persuade the recipient of the merit of the claimant's position.

The contracting officer is probably *not* intimately familiar with the events leading to the dispute and thus will rely on the staff and representatives to explain the matter. Since these representatives often are directly involved in those events (or may be the *cause* of the problem), it is generally not realistic to expect that these individuals will be able to view the problem from a contractor's perspective and to best advance its arguments. Accordingly, the claim should be prepared in a professional and comprehensive manner for submission to the person making the decision followed by an in-person presentation and explanation of the claim.

One means of gathering, understanding, and persuasively marshaling and presenting the facts is the claim document. Such a document can be used as both a starting point and a reference source for settlement negotiations. A brief discussion of the components of a well-prepared claim (or responsive) document will detail the points that must be fully prepared in order to start knowledgeable and effective negotiations.

The claim document should be developed to promote a prompt solution of the financial difficulties caused by the unanticipated construction problems and should cover all relevant aspects of the dispute. Further, the written claim should have a format and organization that facilitates an easy review. Although the size of the document depends on the nature and the complexity of the claim, it must be sufficiently detailed to establish the contractor's entitlement to the requested additional money, contractor's interpretation of the contract documents, and the extension of time, if applicable. Consequently, the claim document should contain, at a minimum, these four elements:

---

[97]Disputes Clause (c), FAR § 52.233-1. A claim for a "sum certain" that reserved the right to include additional line items to "modify the presentation" was, in fact, deemed to be a predicate for negotiations rather than a CDA claim. *McElroy Mach. & Mfg. Co., Inc.*, ASBCA No. 39416, 92-3 BCA ¶ 25,107; *see also Total Procurement Service, Inc.*, ASBCA No. 54163, 08-1 BCA ¶ 33,843 (claim asserting damages in excess of $66,000.00 was not a sum certain). The sum certain requirement is determined when the claim is filed, not when the complaint is filed. *Morgan & Son Earthmoving, Inc.*, ASBCA No. 53524, 02-2 BCA ¶ 31,874; *MDP Constr., Inc.*, ASBCA No. 52769 *et al.*, 01-1 BCA ¶ 31,359. If the monetary claim does not explicitly state a sum certain, it may be acceptable if the amount could be determined from the submission by simple arithmetic. *Mulunesh Berhe*, ASBCA No. 49681, 96-2 BCA ¶ 28,339; *J&J Maintenance, Inc.*, ASBCA No. 50984, 00-1 BCA ¶ 30,784.

[98]*Metric Constr. Co.,v. United States*, 1 Cl. Ct. 383, 392 (1983).

(1) *Facts.* The claim should contain a well organized *statement of the facts* that references any significant pieces of correspondence and all relevant provisions of the contract. To establish credibility, this section of the claim document needs to be as comprehensive and accurate as is practicable under the circumstances, as the facts are often the most important aspect of any claim. Negative facts *must* be anticipated and covered.

(2) *Cost data.* If money is sought, the CDA claim should include an accurate and credible *cost analysis,* detailing the specific areas of damage and items of cost for which the contractor seeks recovery, documenting the entire amount being claimed, and referencing appropriate supporting documents. Although it may seem advisable or easier to defer the submission of all of the detailed costs, few people are willing to discuss liability in the abstract (i.e., with no idea of the monetary value of the claim). Similarly, a CDA claim seeking monetary relief requires a definite cost statement.

(3) *Legal authorities.* The submission should also include a written discussion of the applicable *legal principles* that support and illustrate the legal theories on which the claim is based. This will increase the possibility that the agency's legal counsel will review the claim, which may introduce more objectivity to the claim review process.

(4) *Exhibits.* Exhibits—such as charts, graphs, drawings, photographs, and various other items of *visual evidence*—can be the key to effectively communicating a complete understanding of the contractor's position. Similarly, all relevant and important expert reports should be attached to the claim document and perhaps referenced and discussed within the body of the factual narrative. Demonstrative evidence need not and should not be reserved for trial. A commercial artist, guided by the knowledge of key personnel, often can greatly facilitate communicating a complex construction claim issue in a more understandable form. Remember, a picture is often worth a thousand words.

In summary, these four elements are used together to demonstrate *entitlement* and *quantum.* The level of detail may vary depending on the circumstances. For example, the legal section may be no more than a page or two. The goal is to persuade in a professional manner while being, at all times, credible.

c. Use of Experts, Consultants, and Attorneys   Some construction problems require the involvement of experts and consultants to help solve performance problems and assemble and analyze the facts in preparation of the claim. An expert also may be called on during a board hearing or Court of Federal Claims trial to render opinions based on either actual knowledge of the facts of the case or assumed hypothetical facts.

As a general rule, it is wise to involve an expert during the actual construction process when a claim or dispute appears probable because that person may be able to (1) suggest ways of mitigating damages or reducing the impact of an injurious condition and (2) recommend methods of preserving evidence and of creating demonstrative exhibits for use during the claim presentation. Finally, involving the expert during construction allows that person to acquire firsthand knowledge of the problem that will make the views and conclusions more persuasive than if they are based solely on facts provided by others.

The services of a variety of experts may well be required to resolve technical and complex construction problems. For example, subsurface problems frequently require a soils engineer, hydrologist, or geologist, while structural engineers may be needed to determine structural failure problems, and scheduling experts may help identify or isolate the causes of delay, disruption, and acceleration. Many government construction contracts require the contractor to develop and update a critical path method (CPM) or similar schedule during construction of the project. Since these contract provisions often require the contractor to *prove* the delay by means of a CPM analysis, the involvement of an expert as soon as significant delays begin to occur can be crucial.

A construction attorney who is well versed in government contracting is another "expert" whose involvement may be desirable when problems begin to develop. An attorney can assist in organizing the facts and structuring them in a legally persuasive manner in order to comply with the various certification issues and conditions to the submission of a CDA claim. In addition, experienced government contract and construction lawyers often are able to suggest competent technical consultants in specialized areas such as accounting and scheduling.

Other contractors are valuable to establish custom and practice within the industry when it is necessary to explain the meaning of technical words in the specifications or to determine a standard of reasonable workmanship. Industry standards often can be established by reference to standard texts in the field, and other bidders may be able to establish the reasonableness of a particular contract interpretation.

If the contracting officer is relying on a technical expert, the contractor or its representative should review everything that person has written on the subject matter in dispute. That review may disclose written information that supports the contractor's position. The converse is also true. Whenever feasible, the prior writings or even testimony of the contractor's potential technical consultants or experts need to be reviewed or vetted during the engagement stage to reduce the potential for an unpleasant surprise.

d. Use of the Freedom of Information Act Often agency files contain documents that may provide support for a contractor's request for a contract adjustment, for the contractor's understanding of its contract obligations, or for the agency's responsibilities under the contract. Typically, these materials are available during discovery. However, this is after a claim has been submitted, a final decision has been issued, and an appeal or suit has been initiated. The Freedom of Information Act (FOIA)[99] may provide a means for a contractor to obtain access to relevant records and documents maintained by the government prior to the submission of a claim. Contractors and their counsel should consider using FOIA requests as a means of obtaining additional information pertaining to the contract. Agencies subject to the FOIA include: (1) any department or agency of the executive branch; (2) government corporations; (3) government-controlled corporations; or (4) any independent regulatory branch.[100] The FOIA does not apply to the federal courts or to the Congress.

---

[99]5 U.S.C. § 552.
[100]5 U.S.C. § 552(f).

Basically, information is made available to the public in three ways: (1) publication in the *Federal Register*;[101] (2) by sale to the public or availability for examination in public reading rooms;[102] or (3) upon request for documents that are reasonably described.[103] This third category of records is generally the best source of information pertaining to contract performance issues.

To obtain these documents in the third category, it is necessary to submit a written FOIA request. When making a FOIA request, each agency's procedures governing the submission should be carefully reviewed and followed.[104] The regulations generally are contained in the Code of Federal Regulations (CFR) and usually identify the FOIA official to whom a request should be sent and provide the appropriate address. In addition, these procedures set forth time limits for agency responses and appeal procedures if the request is denied. Following these procedures usually will save time in processing the request. In addition, it will avoid having a court decline jurisdiction over a suit to compel disclosure due to a failure to follow these rules.[105]

Generally, a request for agency documents usually will be processed more quickly if it clearly states that it is an FOIA request and acknowledges that the federal government may be entitled to be paid certain fees and costs for responding to the request.[106] In addition, it is necessary to provide a *reasonable description* of the desired records.[107] A *reasonable description* is one that enables a professional employee of the agency who is familiar with the subject area of the request to locate the record with a reasonable amount of effort.[108] However, broad categorical requests that make it impossible for the agency to reasonably determine what is sought are not permissible.[109] If the agency denies the initial request, the person seeking disclosure may file an action in a United States district court to compel disclosure after exhausting the applicable administrative procedures (including any appeal process) set forth in the agency's FOIA regulations.[110]

Appendix 15A of this chapter contains a sample FOIA letter. The processing of that letter by the federal government agency should be expedited if the applicable CFR provisions of the agency regulations are researched to determine the correct entity to whom the request is to be sent. Similarly, to limit issues about the date of receipt, send the request Certified Mail—Return Receipt Requested or by a commercial carrier that will obtain written confirmation of receipt.

---

[101] 5 U.S.C. § 552(a)(1); 1 C.F.R. Part 5.

[102] 5 U.S.C. § 552(a)(2).

[103] 5 U.S.C. § 552(a)(3)(A).

[104] 5 U.S.C. § 552(a)(3)(B).

[105] *Hamilton Securities Group, Inc. v. Department of Housing and Urban Development,* 106 F. Supp. 2d 23 (D.D.C. 2000); *Television Wis., Inc. v. NLRB,* 410 F. Supp. 999 (W.D. Wis. 1976).

[106] If the requesting party is not sure of the cost (scope) associated with the FOIA request, it is possible to advise the agency to contact the party making the request prior to conducting a search that is expected to cost more than a stated amount, e.g., $250.00.

[107] H.R. Rep. No. 93-876, 93d Congress, 2d Sess. (1974).

[108] 5 U.S.C. § 552(a)(3)(A). *See Jimenez v. F.B.I.,* 910 F. Supp. 5 (D.D.C. 1996).

[109] S. Rep. No. 93-854, 93rd Congress, 2d Sess. (1974). *See Jimenez v. F.B.I,* 910 F. Supp. 5 (D.D.C. 1996); *Fonda v. Central Intelligence Agency,* 434 F. Supp. 498 (D.D.C. 1977).

[110] 5 U.S.C. § 552(a)(6)(C); *Oglesby v. United States Dept. of the Army,* 920 F.2d 57 (D.C. Cir. 1990).

FOIA requests can be a means of expeditious, informal discovery to assist in the preparation of a claim. Even after an appeal or suit is pending, contractors and their counsel should consider the potential for appropriate contemporaneous FOIA requests and formal discovery requests. For example, many projects involve federal agencies other than the contracting agency. An appropriate FOIA request to the non-contracting agency may provide quicker access to records than the use of subpoenas on nonparty federal agencies. Similarly, in a case involving issues related to the performance by, or positions asserted by, an architect-engineer firm, a FOIA request to view the records on other projects designed or administered by the same firm may reveal contradictory views and positions on topics such as the "coordination" of the design, timely turnaround of submittals, and the like.

D. Certification Requirements

1. Contract Disputes Act Claim Certification

The CDA currently requires that a claim in excess of $100,000 be certified.[111] The accepted purpose of the certification requirement is to discourage the submission of unwarranted or inflated contractor claims, to decrease litigation, and to encourage settlements.[112] Prior to an amendment to the CDA in 1992, proper compliance with the certification requirement was a prerequisite to invoking the dispute resolution procedures of the Act.[113] If the requirements for a properly certified claim were not satisfied, any final decision on that claim was a nullity, and any appeal or suit was subject to dismissal.[114]

Section 907(a) of the Federal Courts Administration Act significantly modified many of the more rigid formalities of the Act pertaining to the certification of claims while leaving intact the basic policy safeguards underlying the requirement for the certification.[115] This amendment made the following changes to the law regarding the requirements for a proper CDA claim certification and the consequences if the certification was *defective* or improper in some manner:

- Broadened the class of individuals who could properly certify a claim to include anyone who was authorized to bind the contractor with respect to the claim.
- Expressly stated that a "defective" certification would not deprive a board or court of jurisdiction over that claim.

---

[111]41 U.S.C. §§ 605(c)(1); 605(c)(2).

[112]*Paul E. Lehman, Inc. v. United States,* 673 F.2d 352, 354 (Ct. Cl. 1982).

[113]*Id.* at 352.

[114]Prior to 1992, if a claim in excess of the monetary threshold was not properly certified, interest did not accrue, *Fidelity Constr. Co. v. United States,* 700 F.2d 1379, 1382-85 (Fed. Cir. 1983); no valid final decision could be issued, *Paul E. Lehman, Inc. v. United States,* 673 F.2d 352, 354 (Ct. Cl. 1982); *Conoc Constr. Corp. v. United States,* 3 Cl. Ct. 146, 147–48 (1983); and the contractor had no right to access to either a board of contract appeals or a court, *Skelly and Loy v. United States,* 685 F.2d 414 (Ct. Cl. 1982); *W. M. Schlosser Co. v. United States,* 705 F.2d 1336, 1338–39 (Fed. Cir. 1983).

[115]Pub. L. No. 102-572.

- Required that a "defective" certification be cured before a board or court could render a decision on that claim.
- Excused a contracting officer from issuing a final decision on a defectively certified claim if the contracting officer advised the contractor in writing of the basis for the conclusion that the certification was inadequate within 60 days of the date of the contracting officer's receipt of the claim.
- Allowed CDA interest to accrue on a claim even though the certification was defective.

These changes clearly liberalized the rules related to CDA claim certification. However, the fundamental requirement for the certification as an element of a claim in excess of $100,000 remains in effect. Consequently, the total absence of a required certification will not be treated as the equivalent of a "defective" certification.[116] If a claim is submitted without the required certification and a final decision is issued, that omission cannot be cured by the submission of a retroactive submission. The claim submission and the certification must occur at the same time.[117] The certification must be signed separately, as at least one board has ruled that the failure to sign the certification is "more akin" to a failure to provide any certification.[118] Similarly, submission of the certification by e-mail may result in the dismissal of a subsequent appeal for failure to provide a *signed* certification.[119] Moreover, the failure to properly certify a claim most likely will delay its resolution and increase the cost of resolving the dispute.

Therefore, it remains important to understand the monetary threshold at which a claim must be certified, how it must be certified, who can certify it, and the relationship of that certification to other government contract certifications.

a. Monetary Threshold   The Act requires a certification when the contractor asserts a claim exceeding $100,000. Therefore, it is not possible to bypass a certification requirement by breaking a claim into a series of separate claims each of which is $100,000 or less.[120] The test is whether there exists a "single, unitary claim based upon a common and related set of operative facts" that the contractor, unintentionally

---

[116]*Golub-Wegco Kansas City I, LLC v. General Services Administration,* GSBCA No. 15387, 01-2 BCA ¶ 31,553; *Weststar Eng'g, Inc.,* ASBCA No. 52484, 02-1 BCA ¶ 31,759. In *Weststar,* the board dismissed an uncertified claim labeled as a request for an interpretation upon concluding that it was a veiled quantum claim in excess of $100,000; *Schnider's of OKC Inc.,* ASBCA No. 53947, 03-1 BCA ¶ 32,160.

[117]*Golub-Wegco Kansas City I, LLC v. General Services Administration,* GSBCA No. 15387, 01-2 BCA ¶ 31,553; if a claim and its certification are separately submitted, it is treated as a new claim at the time of the certification. *See also Kenan Constr. Co. v, Department of State,* CBCA No. 807, 08-1 BCA ¶ 33,797; *J&J Maintenance, Inc.,* ASBCA No. 50984, 00-1 BCA ¶ 30,784.

[118]*Hawaii Cyberspace,* ASBCA No. 54065, 04-1 BCA ¶ 32,455.

[119]*Teknocraft, Inc.,* ASBCA No 55438, 08-1 BCA ¶ 33,846 (signature denoted by "//signed//" followed by a typed name was not sufficient to constitute a signed certification).

[120]*Fidelity & Deposit Co. of Md. v. United States,* 2 Cl. Ct. 137, 143 (1983). Initially the certification threshold was $50,000, and this amount is referenced in the earlier CDA decisions. Although the monetary threshold has been increased, that statutory amendment did not affect the basic analysis in those decisions regarding efforts to avoid the certification requirement by dividing the claim into separate parts, each less than the threshold amount for a certification. *See also Columbia Constr. Co., Inc.,* ASBCA No. 28536, 96-1 BCA ¶ 27,970; *D&K Painting Co., Inc.,* DOTBCA No. 4014, 98-2 BCA ¶ 30,064.

or otherwise, divided into separate and distinct claims.[121] However, if the claims are distinct and independent, with one claim having no relationship to the operative facts of the other claim, each independent claim of $100,000 or less need not be certified.[122] Even if the contractor submits a single letter to the contracting officer that reflects claims totaling more than $100,000, the certification is not required unless the claims arose from a common or related set of operative facts and are therefore truly a unitary claim.[123] In other words, the operative test is whether the claim or claims submitted by the contractor arose from the same or different events or causes of action.[124]

The next examples illustrate the application of this test. In one case, the contractor alleged that one differing site condition gave rise to three separate claims: one for additional paving costs; one for additional insurance, supervision, and maintenance costs; and one for loss of interest on funds spent to perform additional work. The court concluded that, in fact, there was just one claim.[125] Similarly, when a contract was terminated for the convenience of the government, the court concluded that a contractor's demand for "pre-termination and post-termination items" constituted one claim because both items were directly related to the government's termination of the contract and the resolution of both items depended on what, if any, liability the government incurred as a result of its action.[126] Similarly, in a case where the contract involved security guard services at five different locations in Boston, the court held that the contractor could not fragment its total dollar claim into separate claims based on each of the different locations. The rationale was that the amounts claimed from the various locations were based on the same operative facts (a total number of hours of services performed for which a total number of dollars allegedly was due).[127] In another case, however, the ASBCA determined that when 18 different claims arose from different causative events and were brought under different legal theories, such as differing site conditions and defective specifications, it was proper to separate the claims.[128]

b. Modification of Claim Amount   Sometimes a claim that initially does not exceed $100,000 (and therefore is not certified) increases in amount after a contracting officer's decision is issued. In these circumstances, the question arises whether the contractor still can proceed on the basis of the increased claim before the court or board of contract appeals, or whether it is necessary for the contractor to certify the claim in the increased amount and resubmit it to the contracting officer for a decision.

[121]*Warchol Constr. Co. v. United States,* 2 Cl. Ct. 384, 389 (1983). *See also LDG Timber Enters., Inc. v. United States,* 8 Cl. Ct. 445, 452 (1985).

[122]*Little River Lumber Co. v. United States,* 21 Cl. Ct. 527 (1990); *Walsky Constr. Co. v. United States,* 3 Cl. Ct. 615, 619 (1983); *Reliance Ins. Co. v. United States,* 27 Fed. Cl. 815, 822 (1993); *C.B.C. Enters., Inc.,* ASBCA No. 43496, 92-2 BCA ¶ 24,803.

[123]*Placeway Constr. Corp. v. United States,* 920 F.2d 903 (Fed. Cir. 1990); *Spirit Leveling Contractors v. United States,* 19 Cl. Ct. 84 (1989).

[124]*Zinger Constr. Co.,* ASBCA No. 28788, 86-2 BCA ¶ 18,920; *J.S. Alberici Constr. Co., Inc., and Martin K. Eby Constr. Co., Inc., a Joint Venture,* ENGBCA No. 6178, 98-2 BCA ¶ 29,875; *GLR Corp.,* VABCA No. 7018, 04-1 BCA ¶ 32,438.

[125]*Warchol Constr. Co. v. United States,* 2 Cl. Ct. 384 (1983).

[126]*Palmer & Sicard, Inc. v. United States,* 4 Cl. Ct. 420, 422–23 (1984).

[127]*Black Star Sec., Inc. v. United States,* 5 Cl. Ct. 110 (1984).

[128]*Zinger Constr. Co.,* ASBCA No. 28788, 86-2 BCA ¶ 18,920.

This question was addressed in *Tecom, Inc. v. United States.*[129] In *Tecom*, the contractor's claim was less than the threshold value for certification when it was submitted to the contracting officer. However, by the time the company filed its complaint before the ASBCA, the amount of the claim exceeded the monetary threshold for a certification. This increase was the result of two events that occurred after the contracting officer's decision: (1) a reevaluation of the claim by the contractor, and (2) the government's exercise of an option to extend the contract for an additional year. Under these circumstances, the court held that it was not necessary for the contractor to certify and resubmit its claim.

Tecom stands for the proposition that a monetary claim properly considered by a contracting officer "need not be certified or recertified if that very same claim (but in an increased amount reasonably based on further information) comes before a board of contract appeals or a court."[130] The Federal Circuit stated that it would be disruptive of normal litigation procedures "if any increase in the amount of a claim based on matters developed in litigation before the court [or board] had to be submitted to the contracting officer before the court [or board] could continue to final resolution on the claim."[131] In a footnote, however, the *Tecom* court pointed out that its decision should not be taken as an invitation to seek to evade the certification requirement.[132] Thus, a contractor that deliberately understates the amount of its original claim (with the intention of raising the amount on appeal on the basis of information that was readily available at the time the claim first was submitted) may well find its subsequent suit in the Court of Federal Claims or its board appeal dismissed for lack of

---

[129]732 F.2d 935 (Fed. Cir. 1984); *see also Todd Pacific Shipyards Corp.,* ASBCA No. 552126, 06-2 BCA ¶ 33,421.

[130]*Id.* at 938.

[131]*Id.* at 937–38 (quoting *J.F. Shea Co. v. United States,* 4 Cl. Ct. 46, 54 (parentheticals in *Tecom*)). *See Kunz Constr. Co. v. United States,* 12 Cl. Ct. 74, 79 (1987) (stating that contractor can enlarge dollar amount of its claim in court over what was presented to contracting officer under two conditions: (1) if increase is based on same set of operative facts previously submitted to contacting officer, and (2) if court finds that contractor neither knew, nor reasonably should have known, at time when claim was presented to contracting officer of factors justifying increase). *See also E.C. Schleyer Pump Co.,* ASBCA No. 33900, 87-3 BCA ¶ 19,986 (costs that were merely an additional area of damages from same facts alleged in claim could be brought before board though not presented to contracting officer). Also, in *Glenn v. United States,* 858 F.2d 1577, 1580 (Fed. Cir. 1988), the contractor submitted a claim to the contracting officer in the amount of $31,500. Because the claim was less than $50,000 (the applicable threshold for certification at that time), the contractor did not certify it. The contracting officer issued a final decision denying the claim, which the contractor appealed to the Armed Services Board of Contract Appeals. Thereafter, the contracting officer issued a second final decision. In that decision, the contracting officer stated that he was withholding $66,570.32 from the contractor (consisting of the $31,500 that the contractor previously had sought to recover and an additional $35,070.32). Relying on its prior decision in *Tecom,* the Federal Circuit held that it was not necessary for the contractor to certify its $66,570.32 claim before bringing suit in the Claims Court. "Because Glenn was not required to certify his $31,500 claim before the C.O. [contracting officer], he need not have certified the $66,570.32 resulting from the denial of his initial claim . . . and [the] additional setoffs."

[132]732 F.2d at 938, n.2. *See D.E.W., Inc.,* ASBCA No. 35173, 89-3 BCA ¶ 22,008.

any certification.[133] Dismissal occurs because the total absence of a certification is different from a defective (or inadequate) certification.

c. Certification Language The CDA sets forth the language to be used in the certification. The contractor must certify:

> [T]hat the claim is made in good faith, that the supporting data are accurate and complete to the best of [the contractor's] knowledge and belief, [and] that the amount requested accurately reflects the contract adjustment for which the contractor believes the Government is liable.[134]

The applicable FAR provision[135] and the Disputes clause utilized in contracts covered by the CDA[136] contain identical language. The 1992 amendment to the Act added a fourth element to the Disputes clause certification. This element requires that the person signing the certification state that "the certifier is duly authorized to certify the claim on behalf of the contractor."[137]

Under the pre-1992 statutory language, the boards, the United States Claims Court, and the Federal Circuit developed strict rules defining a defective or inadequate certification. These rules, in conjunction with the holding that the submission of a proper certification for any claim in excess of the monetary threshold was a jurisdictional requirement,[138] created extensive problems for claimants and their counsel. Although the 1992 amendment eliminated the rule that the submission of a valid certification was a jurisdictional requirement that could not be waived, potential problems for a contractor regarding the form of the certification remain. To the extent that the cases interpreting the preamendment Act provide guidance regarding the proper wording of a certification, it is possible that these decisions still will be relied on by the boards and the courts in determining whether a certification is "defective."

Prior to the 1992 amendment, there was a split in authority regarding the contractor's obligation to strictly track the statutory certification language in order to submit a valid certification. One line of cases took a very formalistic view and held that any deviation from the statutory language would be subject to strict scrutiny. In those cases, a contractor's attempt to deviate from the statutory language by substituting alternate language usually was held to invalidate the certification.[139] A second line of cases held

---

[133]*Id.* Even the reduction of a claim below the applicable threshold at the board will not eliminate the need for a certification if the claim, as submitted to the contracting officer, exceeded the certification threshold. *Building Sys. Contractors,* VABCA Nos. 2749 *et al.,* 89-2 BCA ¶ 21,678.

[134]41 U.S.C. § 605(c)(1).

[135]FAR § 33.207(a).

[136]FAR § 52.233-1.

[137]41 U.S.C. § 605(c)(7).

[138]*B.E.S. Environmental Specialists, Inc. v. United States,* 23 Cl. Ct. 751 (1991); *Skelly & Loy v. United States,* 685 F.2d 414, 419 (Ct. Cl. 1982); *Paul E. Lehman v. United States,* 673 F.2d 352 (Ct. Cl. 1982); *Kaco Contracting Co.,* ASBCA No. 43066, 92-1 BCA ¶ 24,603; *Giuliani Contracting Co.,* ASBCA No. 41435, 91-2 BCA ¶ 23,774.

[139]*Centex Constr. Co.,* ASBCA No. 35338, 89-1 BCA ¶ 21,259; *Liberty Envtl. Specialties, Inc.,* VABCA No. 2948, 89-3 BCA ¶ 21,982.

that substantial compliance was sufficient and the inadvertent omission of a few words in the certification and the omission of the claimed amount, which was stated elsewhere in the claim, were not fatal defects.[140] In general, any certification must simultaneously state all elements of the statutory requirements, and an effort to satisfy the certification requirement by reference to multiple letters or by piecemeal submissions has not been deemed to be sufficient to satisfy the statutory requirement.[141]

In endorsing the substantial compliance approach, the General Services Administration Board of Contract Appeals (GSBCA) accepted a certification that omitted any reference to "knowledge or belief." The board held that this unqualified certification "more fully exposed [the contractor] to potential liability" for false statements than "if it had mimicked the words of the statute."[142] In contrast, the ASBCA has held that a failure to state that the contractor "believes" the government is liable invalidates the certification.[143] In addition, these certifications have been held to be defective:

- A certification that varied from the language of the statute and the Disputes clause by referring to "all data used" instead of the "supporting data" for the claim (thereby restricting the certification to "unidentified data [that the contractor] chose to use while the statute requires certification of all data that support the claim").[144]

- A certification that omitted the assertion that the supporting data was accurate and complete.[145]

- A certification in which the contractor stated that it would not assume any legal obligations that it would not have without the certification, that the data submitted was "as accurate and complete as practicable," and that the contractor was not demanding a "particular amount."[146]

By contrast, in another case, a contractor whose certification did not contain the amount of the claim involved and did not have the words "the amount requested accurately reflects the contract adjustment for which the contractor believes the government is liable" still was found to be in substantial compliance with the certification requirement. The statement in which the certification was contained did have

---

[140]*United States v. General Elec. Corp.*, 727 F.2d 1567, 1569 (Fed. Cir. 1984); *Young Enters., Inc. v. United States*, 26 Cl. Ct. 858, 862 (1992); *P.J. Dick Contracting, Inc.*, GSBCA No. 11847 et al., 93-1 BCA ¶ 25,263. In *Metric Constructors, Inc.*, ASBCA No. 50843, 98-2 BCA ¶ 30,088, the ASBCA held that a signed termination for convenience settlement proposal on Standard Form 1438 contained certification language sufficiently similar to the CDA to constitute a correctable certification. However, in *Keydata Sys., Inc. v. Dept. of Treasury*, GSBCA No. 14281-TD, 97-2 BCA ¶ 29,330, the GSBCA held that the 1992 Amendment did not authorize contractors to cure defective certifications resulting from fraud, bad faith, or "negligent disregard" of the certification requirements.

[141]*W. H. Moseley Co. v. United States*, 677 F.2d 850, 852 (Ct. Cl. 1982); *Black Star Sec., Inc. v. United States*, 5 Cl. Ct. 110, 117 (1984); *Parrino Enters. v. United States*, 230 Ct. Cl. 1052 (1982).

[142]*P.J. Dick, Inc.*, GSBCA No. 11847, et al., 93-1 BCA ¶ 25,263 at 126,605.

[143]*C.F. Elecs.*, ASBCA No. 4077, 91-2 BCA ¶ 23,746.

[144]*Gauntt Constr. Co.*, ASBCA No. 33323, 87-3 BCA ¶ 20,221 at 102,412.

[145]*Raymond Kaiser Eng'rs, Inc./Kaiser Steel Corp., a Joint Venture*, ASBCA No. 34133, 87-3 BCA ¶ 20,140, at 101,940-41.

[146]*Cochran Constr. Co.*, ASBCA No. 34378, 87-3 BCA ¶ 19,993 at 101,280–81, *aff'd on reconsideration*, 87-3 BCA ¶ 20,114.

the remainder of the elements required by the Act; and when the statement was read in its entirety and together with documents that accompanied it, all of the information and statements required by the statute were found to be present.[147] Notwithstanding the degree of flexibility that may be allowed by the liberalized amendment to the Act, the prudent course is to track the language of the Act when certifying a claim.

d. Supporting Data   A contractor that certifies its claim by tracking the language of the statute still may find itself confronted with the argument that the data supporting its claim is inadequate for purposes of the certification requirement. For the most part, though, neither the courts nor the boards have taken an overly stringent attitude with respect to the extent of the supporting data.

In *Metric Construction Co. v. United States*,[148] the government argued that the contractor's certification was defective because the contractor had failed to attach copies of the pertinent change order modifications to its claim. In rejecting the government's argument, the court observed that the certification requirement "was not intended, nor should it be so construed, to require a full evidentiary presentation before the contracting officer."[149] The court noted that the contracting officer had not denied the contractor's claim for lack of supporting data and that the data that had been presented had assisted the contracting officer "in making a meaningful determination on the dispute before him."[150]

The Department of Energy Board of Contract Appeals (EBCA) took a similar position and cited *Metric* with approval in *Newhall Refining Co.*[151] In response to the government's argument that the contractors involved had not submitted accurate and complete supporting data when they certified their claims, the board noted that, on their face, the certifications met the requirements of the Act, that the claims were "articulated in a clear and concise fashion," that the contractors had notified the contracting officer of the basis for their claims prior to submitting them, and that the contracting officer already was in possession of information relating to the claims.[152] The board also noted the fact that the claims before it involved a legal issue of contract interpretation, and it found "highly persuasive" the fact that the contracting officer had not requested additional information from the contractors.[153] Under these circumstances, the board determined that the data submitted with the claims was "adequate."[154] These cases suggest that, when the language of the contractor's certification meets the requirements of the Act and the contracting officer is provided with the needed information to render a final decision and has issued a final decision, the contractor probably should not have its claim derailed by a government assertion at the board or court that it failed to submit adequate supporting data.

---

[147]*United States v. General Elec. Corp.*, 727 F.2d 1567, 1569 (Fed. Cir. 1984).
[148]1 Cl. Ct. 383 (1983).
[149]*Id.* at 391.
[150]*Id.*
[151]EBCA Nos. 363-7-86, *et al.*, 87-1 BCA ¶ 19,340.
[152]*Id.* at 97,583.
[153]*Id.*
[154]*Id.*

e. Who May Certify the Claim? The 1992 Amendment to the CDA effectively eliminated the prior questions regarding the authority of the person signing the certification and the extent of the personal knowledge that the person certifying the claim needed to have regarding the underlying facts set forth in the claim.

With the enactment of the 1992 amendment to the CDA,[155] questions regarding a person's authority to certify a claim should be minimal, so long as the certifier follows the language of the amended Act, which requires an express representation that the person signing the certification is authorized to do so. In that context, it is not necessary for the person certifying the claim to have personal knowledge of the facts and data supporting the claim. Rather, it is sufficient for the person certifying the claim to rely on data and facts developed by others within the contractor's organization.[156]

2. Other Certifications

The complexity of the CDA certification is only compounded by the fact that there are at least two other claim-related certifications that a contractor may be required to submit under other statutes.

Contracts with agencies of the Department of Defense (DOD) are subject to an additional statutory certification requirement.[157] This law requires that any request for equitable adjustment to the contract terms or request for relief under Public Law 85-804 that exceeds the "simplified acquisition threshold"[158] may not be paid unless it is certified by a person authorized to bind the contractor at the time of submission. This certification must state that the request is made in good faith and that the supporting data are accurate and complete to the best of that person's knowledge and belief. The Department of Defense FAR Supplement Regulation implementing this law expands the scope of the certification.[159] The certification must state that the request for an equitable adjustment includes only the cost for performing the change, does not include any costs that have already been reimbursed or separately claimed, and all claimed indirect costs are properly allocable to the change.

A third certification is required by the Truth in Negotiations Act.[160] There are significant differences between the Truth in Negotiations certificate and the two certificates previously discussed. Under the Truth in Negotiations Act, the certificate is not provided until the parties reach agreement on a price—at the time of the handshake. The current threshold for a Truth in Negotiations certificate is $650,000 for contracts as well as subcontracts under such contracts and modifications to any contract.[161]

---

[155] 41 U.S.C. § 605(c)(7).

[156] *Fischbach & Moore Int'l Corp. v. Christopher,* 987 F.2d 759, 762 (Fed. Cir. 1993).

[157] 10 U.S.C. § 2410(a). *See* **Chapter 1** for a more general review of certifications required of contractors.

[158] $100,000 as of November 12, 2009. *See* FAR § 2.101.

[159] Generally, DFARS § 252.243-7002 (MAR 1998).

[160] 10 U.S.C. § 2306a. This statute applies only to the Department of Defense and the National Aeronautical and Space Administration. By regulation, the requirements of this statute have been extended to the civilian agencies. FAR § 15.804.

[161] 10 U.S.C. § 2306a; 41 U.S.C. § 254b; FAR § 15.403-4; FAC 2005-13 (71 Fed. Reg. 57367) (thresholds are subject to change).

The Act provides a specific remedy when the data does not meet the requirements of accuracy, currency, and completeness. Under the Act and the related regulations, the government is entitled to a price reduction if the data is found to be defective—that is, inaccurate, not current, or not complete.[162]

These certification requirements overlap to some extent. However, there are sufficient differences that can cause confusion. **Table 15.1** compares the current requirements set forth in each of these statutes and their implementing regulations.

E. Pub. L. 85-804 Relief Distinguished from CDA Claims

Although the Contract Disputes Act covers the vast majority of contract claims and disputes, contractors may, in special circumstances, obtain relief even if a CDA claim is not an appropriate or available alternative. Congress has also granted the executive branch extraordinary powers to provide certain contractors relief in the course of procurements related to the national defense.

One of these laws, 50 U.S.C. §§ 1431-1435, permits certain procuring activities to grant relief to contractors that may have no legal right to such relief—for example, a contract amendment without consideration to the government. This avenue for relief is not a substitute for relief under the CDA and will be considered only after it is determined that the CDA does not provide an adequate remedy. The procedures related to extraordinary contractual actions are found in the Federal Acquisition Regulation at Part 50, which implements Public Law (Pub. L.) 85-804.[163]

Pub. L. 85-804 empowers the president to authorize agencies exercising functions in connection with national defense to enter into, amend, and modify contracts, without regard to other provisions of law, whenever that action would facilitate the national defense. By Executive Order No. 10789, certain agencies have been delegated authority to exercise the powers set forth in Pub. L. 85-804.[164] These agencies include:

Department of Agriculture	Department of Transportation
Department of Commerce	Department of Treasury
Department of Defense	Federal Emergency Management
Department of Energy	General Services Administration
Department of Interior	Government Printing Office
Department of the Air Force	National Aeronautics and Space Administration
Department of the Army	Tennessee Valley Authority
Department of the Navy	

The authorized agencies can establish contract adjustment boards to approve, authorize, and direct appropriate action under FAR Part 50. The decisions of such boards are not appealable; however, an agency adjustment board may reconsider, modify, or reverse a prior decision. The boards have the authority to set their own

[162]FAR § 15.407-1.
[163]50 U.S.C. §§ 1431-1435.
[164]FAR § 50.101(b) states that the president may include other agencies.

Table 15.1 Comparison of Claim-Related Certifications

	Contract Disputes Act (CDA)	DOD Contracts	Truth in Negotiations Act
Language:	(1) Claim made in good faith	(1) Claim made in good faith	Data in support of proposal are accurate, current, and complete
	(2) Supporting data accurate and complete to best of contractor's knowledge and belief	(2) Supporting data accurate and complete to the best of contractor's knowledge and belief	
	(3) Amount requested accurately reflects adjustment for which contractor believes government liable	(3) Amount requested accurately reflects adjustment for which contractor believes government liable	
	(4) Certifier duly authorized to certify the claim on behalf of the contractor		
$ Threshold:	Claim > $100,000	Claim or request for adjustment < $100,000	Price or adjustment: $650,000 (subject to adjustment for inflation to nearest $50,000 every five years)
Certified by:	"Any person duly authorized to bind the contractor with respect to the claim"	"Any person duly authorized to bind the contractor with respect to the claim"	"The contractor"— anyone authorized to sign contractual documents
Date Required:	When submitted as a "claim" under CDA	Upon submission of a request for equitable adjustment or request for extraordinary relief	At time of agreement ("handshake") on price

procedures.[165] Generally, contractors seeking relief from an agency adjustment board are represented by counsel; however, the proceedings are not adversarial. For example, cross examination of witnesses is not usually permitted.

A summary of the types of relief available under Pub. L. 85-804 and FAR Part 50 follows.

- *Advance payments.* Such payments are authorized under FAR § 32.405 for sealed bid contracts.
- *Amendments without consideration.* Relief may be available when an actual or threatened loss under a defense contract will impair the productive capacity of a contractor whose continued performance is essential to the national defense. For example, if the government acts in its capacity as the other contracting party and creates a loss, relief may be obtained under Pub. L. 85-804.[166] As a general

---

[165]FAR § 50.202.
[166]*See Technitrol Engrg. Corp.,* ACAB No. 1084 (Feb. 9, 1968), 2 ECR ¶ 53.

rule, relief is not available when the action could be considered to be a sovereign act[167] that is general and public in nature, such as an increase in minimum wage rates.[168]

- *Correction of mistakes.* Relief may be granted if: a mistake or ambiguity resulted in the contract failing to express the agreement as both parties understood it;[169] the contractor's mistake was so obvious that it was or should have been apparent to the contracting officer;[170] or a mutual mistake of fact.[171]

- *Informal commitments.* When an entity or person performed in good-faith reliance on the official's apparent authority to issue the instructions to provide the goods or services.[172]

- *Indemnification.* FAR Subpart 50.4 provides for a process to indemnify a contractor from nuclear risks or unusually hazardous risks.[173]

Relief is not available under Pub. L. 85-804 until all other administrative or agency rights have been exhausted.[174] Finally, the contractor must submit the request for relief before all obligations (including final payment) under the contract have been discharged.[175]

Even though a request for relief under Pub. L. 85-804 is not a CDA claim, the contractor seeking any adjustment under FAR Part 50 that exceeds the simplified acquisition threshold ($100,000)[176] must provide a certification which is similar, in part, to the CDA claim certification. A person authorized to certify the request on behalf of the contractor must state that:

(1) The request is made in good faith.

(2) The supporting data are accurate and complete to the best of that person's knowledge and belief.[177]

In addition to these requirements, FAR § 50.304 details the specific types of information to be provided in support of a request for relief under Pub. L. 85-804. Although not labeled as a claim submission, any request for relief under Pub. L.

---

[167]*See R.E. Lee Elec. Co.,* NASACAB (6 June 1969) 2 ECR ¶ 87; *but see S.W. Electronics & Mfg. Corp.,* NCAB (27 Oct. 1971), 2 ECR ¶ 147.

[168]*See Floors, Inc.,* ACAB No. 1043 (25 May 1962), 1 ECR ¶ 118.

[169]FAR § 50.302-2(a)(1).

[170]FAR § 50.302-2(a)(2).

[171]FAR § 50.302-2(a)(3).

[172]FAR § 50.302-3.

[173]FAR §§ 50.403-1–403-2.

[174]FAR § 50.102(a)(2); *NL Indus.,* AECCAB No. 3-9-72 (26 Apr. 1973) 2 ECR ¶ 189.

[175]FAR § 50.203(c). If some obligation remains under the contract after final payment, relief still may be available. *See York Corp.,* ACAB No. 1009 (29 Feb. 1960); 1 ECR ¶ 30 (obligation to pay liquidated damages).

[176]FAR § 2.101 (the amount may be higher for specific contracts or contracts performed outside of the United States).

[177]FAR § 50.303-2.

85-804 is a form of a legal proceeding, particularly given the requirement for a certification.

IV.   SUBCONTRACTORS AND THE CDA CLAIMS PROCESS

A.   Standing to Pursue a Claim

As discussed in **Section III.C.1**, *privity of contract* between the claimant and the government is an essential element in the CDA claims process. Consequently, there are relatively few situations on which a subcontractor has the right or standing to assert a claim directly against the government even though a large percentage of the work performed on most federal construction projects typically is subcontracted by the prime (general) contractor. In that context, subcontractors, at any tier, and suppliers may incur extra expense due to the actions and directives of the government, differing site conditions, and delays caused by the government.

Even though the subcontractor may not have standing, it is permissible for a prime contractor to assert a claim against the government on behalf of a subcontractor that has been damaged (e.g., incurred additional cost) due to government action or inaction. This *sponsorship* rule applies to appeals before the boards as well as suits in the Court of Federal Claims. The right was confirmed in *United States v. Blair*,[178] where the United States Supreme Court stated that Blair (the prime contractor) "was the only person legally bound to perform his contract with the Government and he had the undoubted right to recover from the Government the contract price for the tile, terrazzo, marble and soapstone work whether that work was performed personally or through another."[179]

Although the decision in *Blair* confirmed the right of a prime contractor to sponsor a subcontractor's claim, the prime contractor may maintain the claim on behalf of its subcontractor only if it has reimbursed the subcontractor for its costs or damages or remains liable for such reimbursement in the future. In federal government contracts, this requirement is known as the *Severin* doctrine.[180]

In *Severin,* the contract between the prime contractor and the subcontractor contained an exculpatory clause totally absolving the prime contractor from liability to the subcontract for any claim caused by actions of the government. The *Severin* doctrine can have harsh results, and later decisions created exceptions and limitations to the basic rule. For example, in 1965, the Court of Claims held that the doctrine had no application when the subcontractor's claims were asserted as claims for equitable adjustments under the provisions of the contract between the government and the prime contractor.[181] Similarly, if the government seeks to invoke the *Severin* doctrine based on a subcontractor's release of a prime contractor, the government

---

[178]321 U.S. 730 (1944).

[179]*Id.* at 737.

[180]*Severin v. United States,* 99 Ct. Cl. 435 (1943), *cert. denied,* 322 U.S. 733 (1944); *see also J.L. Simmons Co. v. United States,* 304 F.2d 886 (Ct. Cl. 1962).

[181]*Blount Bros. Constr. Co. v. United States,* 348 F.2d 471 (Ct. Cl. 1965).

bears the burden of establishing that the subcontractor executed an iron-clad release sufficient to trigger application of that doctrine[182] or that the subcontract completely immunizes the prime contractor from any liability to the subcontractor.[183] Boards demonstrate a similar reluctance to adopt a broad interpretation of this doctrine if the effect is to bar consideration of a subcontractor's claim on its merits.[184]

Although the application of the *Severin* doctrine has been limited by board and court decisions over the last six decades, it has never been expressly rejected or reversed. Consequently, prime contractors and subcontractors at any tier need to be aware of the doctrine's potential application when drafting subcontract or purchase order terms, releases, and claim pass through (cooperation) agreements. Failure to consider the *Severin* doctrine may have unintended consequences.[185]

B. Pass-Through Agreements

Sponsorship of a subcontractor's claim in the Disputes process can present particular risks for the prime (general) contractor due to the possibility of inconsistent results between a decision in the federal claims process and one in the federal/state court proceeding[186] or in an arbitration between the prime contractors and the subcontractor.

Similarly, the courts have disfavored a waiver of a party's Miller Act rights and seek to determine if any such waiver is sufficiently specific.[187] One court has said that to be an effective waiver of Miller Act rights, the Miller Act must be mentioned. Courts do not favor a finding that a subcontractor has contractually waived its rights under the Miller Act.[188]

The Miller Act also addresses the time of any waiver of the right to sue on the payment bond. The Act at 40 U.S.C. § 3133(c) provides:

(c) A waiver of the right to bring a civil action on a payment bond required under this subchapter is void unless the waiver is—

(1) in writing;

---

[182]*Metric Constructors, Inc. v. United States*, 314 F.3d 578 (Fed. Cir. 2002).

[183]*TAS Group v. U.S. Department of Justice*, DOTBCA No. 4535, 06-2 BCA ¶ 33,441.

[184]*801 Market Street Holdings, L. P. v. General Services Administration*, CBCA No. 425, 08-1 BCA ¶ 33,853 (*Severin* doctrine requires the government to demonsrrate an iron-clad release or contract provision exonerating the prime contractor from liability to subcontractor); *see also Acquest Government Holdings, Opp, LLC v. General Services Administration*, CBCA No. 413, 08-1 BCA ¶ 33,720; *Ball, Ball & Brosamer, Inc.*, IBCA No. 2841, 97-1 BCA ¶ 28,897.

[185]*George Hyman Constr. Co. v. United States*, 30 Fed. Cl. 170 (1993). Inclusion of a routine release endorsement on the back of a check negotiated by the subcontractor barred claim under the *Severin* doctrine. Subsequent efforts to revive the claim by the execution of a pass-through agreement or to reform the release to reflect the parties' true intent rejected by the court.

[186]In the context of actions brought by subcontractors/suppliers under the Miller Act, 40 U.S.C. §§ 3131-3134, federal courts have traditionally held that such subcontractors are not subject to the Disputes clause in the contract between the government and the prime contractor. *See, e.g., United States ex rel. Pembroke Steel, Inc. v. Phoenix Gen. Constr. Co.*, 462 F.2d 1098 (4th Cir. 1972); *United States v. R.M. Wells, Inc.*, 497 F. Supp. 541, 544 (S.D. Ga. 1980).

[187]*See* **Chapter 12** for a discussion of Miller Act bonds.

[188]See *W.H. Caldwell & Son, Inc. v. United States ex rel. John H. Moon & Sons, Inc.*, 407 F.2d 21, 23 (5th Cir. 1969); *United States ex rel. DDC Interiors, Inc. v. Dawson Constr. Co., Inc.*, 895 F. Supp. 270 (D. Colo; 1995), *aff'd*, 82 F.3d 427 (10th Cir. 1966).

(2) signed by the person whose right is waived; and

(3) executed after the person whose right is waived has furnished labor or material for use in the performance of the contract.

Given the combination of the provisions of the Miller Act and the court decisions interpreting that Act, many prime contractors seek to address the risk of potentially inconsistent results and the cost of multiple legal actions over the same claim through the mechanism of a *Pass-Through* or *Claims Cooperation Agreement.*[189] The intent of this type of agreement is to provide the framework for a cooperative effort on the claim and to address the potential for inconsistent results.

There is no standard form for a comprehensive Claims Cooperation Agreement, as it must be tailored to reflect applicable provisions of the subcontract or purchase order and the facts of the specific situation. However, in drafting a Claims Cooperation Agreement, these topics need to be considered and possibly addressed in the agreement.

- *Scope of claims covered.* Does the claim submission encompass claims by the prime contractor, as well as one or more subcontractors?

- *Sponsorship of claims.* Confirmation of prime contractor's sponsorship of the subcontractor's claim.

- *Lead party.* Who controls the process and what are the consequences of a decision to drop, settle, or litigate a particular claim?

- *Cooperation by parties.* Cooperation provision and sharing of information.

- *Forum selection.* Board or Court of Federal Claims.

- *Sharing of recovery.* Allocation of recovery among the parties to the agreement.

- *Sharing of costs.* Allocation of costs among the parties to the agreement.

- *Payment.* Payment terms for both recovery and claim prosecution expense.

- *Limitation on remedies.* Exclusivity of remedy in the disputes process.

- *Joint defense privilege.* Necessary to protect certain prime contractor-subcontractor communications related to the claim.[190]

- *Indemnity for claims.* Certifications and related indemnification which reflect that each party must, to some extent, rely on factual data and cost information provided by the other party or parties to the claim.

- *Appeals.* If an appeal of a board or court decision is desired by one party, the agreement should address its effect on the other party or parties to the agreement.

---

[189]Sometimes also termed "liquidation agreements."

[190]*Moniaros Contracting Corp.,* DOTBCA No. 28234, 96-1 BCA ¶ 28,234.

• *Disputes among parties.* Resolution of disagreements or disputes related to Claims Cooperation Agreement need to be addressed.

Although it is difficult to anticipate every possible issue, a reasonably comprehensive Claims Cooperation Agreement can help avoid uncertainty and facilitate a joint effort by the contracting parties.

C. Certification of Subcontractor Claims

Submission and CDA certification of subcontractor claims can present problems for prime contractors. To varying degrees, the prime contractor may not fully agree with the positions asserted by the subcontractor. In addition, it is likely that the prime contractor must, to some extent, rely on factual information and cost data developed outside of its organization. Both circumstances need to be carefully considered in determining if the prime contractor can make a "good faith" submission and certification of the subcontractor's claim. Notwithstanding these practical problems, both the boards and the courts have held that the certification of a subcontractor's claim must be signed by the prime contractor and contain all of the required elements.[191]

The prime contractor is not entitled to qualify its certificate by stating that it is "based on" a certificate provided by the subcontractor[192] or that it was "subject to review."[193] Absolute agreement with the subcontractor's claim is not essential, as the prime contractor may certify a claim with which it does not fully agree, if it concludes the subcontractor's claim is made in good faith and is not frivolous.[194] In *Arnold M. Diamond, Inc. v. Dalton*,[195] the prime contractor, which had previously rejected its subcontractor's claim, certified it on the order of a federal bankruptcy court. The Federal Circuit reversed an earlier board decision that refused to accept the contractor's certification as being in good faith. In the Federal Circuit's opinion, certification of a claim upon the direction of a federal bankruptcy judge satisfied the Act's requirements.

V. GOVERNMENT CLAIMS

The CDA also covers government claims. The Act provides that "[a]ll claims by the Government against a contractor relating to a contract shall be the subject of a decision

---

[191]*United States v. Johnson Controls, Inc.*, 713 F.2d 1541 (Fed. Cir. 1983); *Century Constr. Co. v. United States*, 22 Cl. Ct. 63 (1990) (prime may not substitute subcontractor's name for itself); *Doyon Properties-American, JV*, ASBCA No. 55842, 08-1 ¶ 33,752 (contractor may not adopt subcontractor's certification by reference); *Lockheed Martin Tactical Defense Sys. v. Dept. of Commerce*, GSBCA No. 14450-COM, 98-1 BCA ¶ 29,717; *Harrington Assoc., Inc.*, GSBCA No. 6795, 82-2 BCA ¶ 16,103.

[192]*Cox Constr. Co.*, ASBCA No. 31072, 85-3 BCA ¶ 18,507.

[193]*Alvarado Constr., Inc. v. United States*, 32 Fed. Cl. 184 (1994).

[194]*United States v. Turner Constr. Co.*, 827 F.2d 1554 (Fed. Cir. 1987).

[195]25 F.3d 1006 (Fed. Cir. 1994).

by the contracting officer."[196] Although most government claims are the subject of a final decision, a government withholding of a "sum certain" due a contractor,[197] an assessment of liquidated damages,[198] or a default termination action[199] may constitute an appealable final decision even though no formal final decision is issued. The one clear exception to the requirement for a final decision is the situation where the government asserts a *fraud claim* against a contractor. That type of claim need not be the subject of a contracting officer's final decision.[200] Moreover, such claims generally are beyond the jurisdiction of the boards.[201]

All government demands are not government claims. For example, the ASBCA has refused to consider an appeal of a government demand that a contractor repair defective work. Although the government directed the contractor to perform the work and the contractor disputed that it was required to do so without additional compensation, the board held that there was no government claim that could be appealed until the government either defaulted the contractor or the contractor did the work and submitted a claim.[202]

The question of when a claim is a government claim or a contractor's claim requiring certification to obtain a final decision has not been answered consistently by the boards and the courts. Several boards have held that the government's *withholding of payment* due a contractor is a government claim that does not require contractor certification.[203] Similarly, a demand for repayment of money allegedly paid to the contractor by mistake is a government claim, and no contractor certification is required.[204]

Although the assessment of *liquidated damages* has been characterized as a government claim,[205] the decisions addressing this question are not consistent. Certain board decisions have held that where a contracting officer assesses liquidated damages in a final decision, a government claim exists and no contractor certification

---

[196]41 U.S.C. § 605(a). Some cases indicated that the boards may decline jurisdiction over a government counterclaim where the counterclaim was never presented to the contractor and the contractor had no opportunity to comment on it. *See Osborn Eng'g Co.,* DOT CAB No. 2165, 90-2 BCA ¶ 22,749; *Instruments & Controls Serv. Co.,* ASBCA No. 38332, 89-3 BCA ¶ 22,237. *But see Security Servs., Inc.,* GSBCA No. 11052, 92-1 BCA ¶ 24,704. In *Security Services,* the GSBCA concluded that the contacting officer had the discretion to either first negotiate the claim or issue a final decision. Similarly, a contracting officer's refusal to negotiate a government claim before issuing a final decision did not negate the finality of that decision. *See also Siebe North, Inc. & Norton Co.,* ASBCA No. 34366, 89-1 BCA ¶ 21,487.

[197]*Sprint Communications Co. v. General Services Administration,* GSBCA No. 14263, 97-2 BCA ¶ 29,249. *But see McDonnell Douglas Corp.,* ASBCA No. 50592, 97-2 BCA ¶ 29,199.

[198]*Midwest Properties, LLC v. General Services Administration,* GSBCA Nos. 15822, 15844, 03-2 BCA ¶ 32,344.

[199]*K & S Constr. v. United States,* 35 Fed. Cl. 270 (1996).

[200]*Martin J. Simko Constr., Inc. v. United States,* 852 F.2d 540 (Fed. Cir. 1988).

[201]*Comada Corp.,* ASBCA No. 26599, 83-2 BCA ¶ 16,681; *Warren Beaves d/b/a Commercial Marine Servs.,* DOT CAB No. 1324, 83-1 BCA ¶ 16,232. *But see Martin J. Simko v. United States* 852 F.2d 540 (Fed. Cir. 1988).

[202]*H. B. Zachry Co.,* ASBCA No. 39209, 90-1 BCA ¶ 22,342.

[203]*General Dynamics Corp.,* ASBCA No. 31359, 86-3 BCA ¶ 19,008; *Perkins & Will,* ASBCA No. 28335, 84-1 BCA ¶ 16,953; *TEM Assocs., Inc.,* NASA BCA No. 33-0990, 91-2 BCA ¶ 23,730; *Mutual Maintenance Co., Inc.,* GSBCA No. 7496, 85-2 BCA ¶ 18,098; *Alaska Lumber & Pulp Co., Inc.,* AGBCA No. 83-301-1, *et al.,* 91-2 BCA ¶ 23,890.

[204]*PX Eng'g Co., Inc.,* ASBCA No. 40714, 90-3 BCA ¶ 23,253.

[205]*Midwest Properties, LLC v. General Services Administration,* GSBCA Nos. 15822, 15844, 03-2 BCA ¶ 32,344.

is required.[206] Others treat liquidated damages as a *government claim* but hold that the contractor must submit a CDA claim to receive interest on any remission of liquidated damages.[207] One board has indicated that the burden of proof determines whose claim it is.[208]

In the context of an appeal of a termination for default, the ASBCA has ruled that it did not have jurisdiction over the contractor's claim for a partial remission of liquidated damages because the contractor had never submitted a claim seeking remission of the liquidated damages.[209] The Claims Court, however, required a contractor to certify a claim for the return of liquidated damages that the government withheld from payments due a contractor, apparently because acknowledged delays occurred which the contractor claimed were caused by the government.[210] Thus, although the government generally withholds the liquidated damages, a contractor's claim to recover such amounts withheld as liquidated damages must be certified.[211] Similarly, if a contractor files a claim for an extension of time in response to the assessment of liquidated damages, a board decision resulting in the remission of some or all of the liquidated damages may not entitle the contractor to recover CDA interest on the liquidated damages previously held by the government. The claim for the extension of time is likely to be treated as distinct from the claim for remission of the liquidated damages.[212]

Even if the contractor is not required to submit or certify a claim because it is considered to be the government's claim, the contractor should recognize that the submission of a claim by the contractor probably is needed in order to create the basis to recover CDA interest on the funds held by the government. Many government claims—for example, the assessment of liquidated damages or deductive changes—may result in the government withholding funds that are otherwise due under the contract. Even if the government's position eventually is determined to have no merit, the contractor is not entitled to receive CDA interest on those funds unless it submits a CDA claim.[213] When responding to a government claim and deduction of monies for liquidated damages or a deductive change order, the prudent course of action is to submit a CDA claim.

VI.   CONTRACTING OFFICER'S DECISION

A.   Contracting Officer's Decisions on Claims and Appeals Therefrom

Once a claim meeting all the requirements of the CDA has been submitted, the next step in the dispute resolution process is the issuance of a contracting officer's

---

[206]*Evergreen Int'l. Aviation, Inc.*, PSBCA No. 2468, 89-2 BCA ¶ 21,712.

[207]*Whitesell-Green, Inc.*, ASBCA Nos. 53938, *et al.*, 06-2 BCA ¶ 33,323.

[208]*Equitable Life Assurance Society of the U.S.*, GSBCA No. GS-7699R, 87-2 BCA ¶ 19,733.

[209]*AEC Corp., Inc.*, ASBCA No. 42920, 03-1 BCA ¶ 32,071.

[210]*Warchol Constr. Co. v. United States*, 2 Cl. Ct. 384, 392–93 (1983).

[211]*Sun Eagle Corp. v. United States*, 23 Cl. Ct. 465 (1991).

[212]*Tidewater Contractors, Inc. v. Department of Transportation*, CBCA No.982-C(50)-R, 08-2 BCA ¶ 33,974.

[213]*General Motors Corp.*, ASBCA No. 35634, 92-3 BCA ¶ 25,149.

decision. The issuance of a valid contracting officer's decision, or the failure to issue such a decision within the time allowed by the Act, is a prerequisite to the contractor bringing suit on the claim in the Court of Federal Claims or filing an appeal with either the Armed Services Board of Contract Appeals or the Civilian Board of Contract Appeals.[214]

1. Time Allowed for Issuing the Decision

The CDA provides that, in the case of claims of $100,000 or less, the contracting officer shall issue a decision within 60 days of receipt of a written request from the contractor that a decision be issued within that period.[215] For claims over $100,000, the Act provides that, within 60 days of receipt of a certified claim, the contracting officer shall issue a decision or notify the contractor of the time within which a decision will be issued.[216] If the claim's monetary value requires a contractor certification, the contracting officer has no obligation to render a decision on a claim accompanied by a defective certification so long as the contracting officer notifies the contractor in writing of the basis for the conclusion that the certification is defective within 60 days of the date of receipt of the claim.[217]

The Act states that contracting officer's decisions are to be issued *within a reasonable time* in accordance with agency regulations, taking into account such factors as the size and complexity of the claim and the adequacy of the information in support of the claim.[218] Thus, although the CDA does not require a full evidentiary submission in order to recognize a claim,[219] it is in a contractor's interest to make its claim submission clear, persuasive, and understandable.

The CDA also provides that, in the event of undue delay on the part of the contracting officer in issuing a decision, a contractor may request the appropriate agency board of contract appeals or the Court of Federal Claims to direct that a final decision be issued in a specified period of time.[220] In making such a request, the contractor should be sure that it has provided the contracting officer with all the information reasonably necessary for a proper review of the claim and the issuance of a decision.

A contractor should be aware, however, that even when a claim is properly submitted and the contracting officer fails to issue a decision, the Court of Federal Claims or a board still has the option of staying proceedings for the purpose of obtaining a decision on the claim.[221] It is reasonable to expect, however, that the Court of

---

[214]*See Milmark Servs., Inc. v. United States,* 231 Ct. Cl. 954, 956 (1982).

[215]41 U.S.C. § 605(c)(1).

[216]41 U.S.C. § 605(c)(2); *Cubic Defense Applications, Inc.,* ASBCA No. 56097, 07-2 BCA ¶ 33,695 (contracting officer's statement that the government intended to issue a final decision "by approximately . . ." did not comply with the statute's requirements. A fixed date must be set forth.).

[217]Pub. L. No. 102-572.

[218]41 U.S.C. § 605(c)(3); *Dillingham/ABB-SUSA,* ASBCA Nos. 51195 *et al.,* 98-2 BCA ¶ 29,778; *Suburban Middlesex Insulation, Inc.,* VABCA No. 4896, 96-2 BCA ¶ 28,481; *VECO, Inc.,* DOTBCA No. 2961, 96-1 BCA 28,108. *But see Defense Sys. Co., Inc.,* ASBCA No. 50534, 97-2 BCA ¶ 28,981 (nine months to review $71 million claim not unreasonable).

[219]*Metric Constr. Co., v. United States,* 1 Cl. Ct. 383, 391 (1983).

[220]41 U.S.C. § 605(c)(4).

[221]41 U.S.C. § 605(c)(5).

Federal Claims or a board will not be inclined to exercise this option in the situation where the contracting officer involved has been directed to issue a decision but has failed to do so or in a situation where the contracting officer gave no reason for the failure to issue a decision.

2. Failure to Issue Decision

Notwithstanding the statutory requirement that the contracting officer act on claims within specific time frames, a contractor and its counsel must consider whether to petition the board to set a deadline for a final decision or to commence formal proceedings by filing an appeal or a suit on a *deemed denied* basis if the contracting officer fails to act on the claim.[222] One practical consideration is that a board or the court may stay the proceedings to await the issuance of a final decision,[223] particularly if there is an indication that the agency is attempting to comply with the Act's requirements.

In order to provide the necessary foundation to appeal or file an action from the lack of a final decision or petition a board to set a deadline for the issuance of a final decision, certain basic documentation should be available to clearly establish the key events and their dates. This would include a letter notifying the agency that the matter is in dispute and that a final decision is requested. If the claim is in excess of $100,000, the request for a final decision must be certified. If the proposal in excess of $100,000 has been previously certified and nothing has occurred that would require a new certification,[224] a basic request for a final decision is sufficient.

Any failure by a contracting officer to issue a decision on a claim within the period required by the Act or directed by a board or the Court of Federal Claims is deemed to be a decision by the contracting officer denying the claim (*deemed denied* decision), and such failure authorizes the commencement of suit in the Court of Federal Claims or an appeal to the appropriate board.[225] The fact that the contracting officer fails to issue a decision, however, does not mean that the government is barred from contesting the claim in subsequent proceedings. Failure to issue a decision is deemed a denial and not a default that precludes the government from contesting the merits of the claim at the board or court.[226]

Once it is apparent that no final decision will be received, the contractor must decide whether to petition the board or Court of Federal Claims to set a date by which the contracting officer is required to issue a final decision[227] or alternatively file an appeal with the board or file a suit in the Court of Federal Claims. The latter involves making an election regarding the forum that will eventually decide the matter. Regardless of whether the decision is to file a petition, an appeal, or a suit, it is important to set forth the history of the efforts to obtain a final decision. This

---

[222]*Boeing Co. v. United States,* 26 Cl. Ct. 257 (1992); *Mitcho, Inc.,* ASBCA No. 41847, 91-2 BCA ¶ 23,860.
[223]*Continental Maritime of San Diego,* ASBCA No. 37820, 89-2 BCA ¶ 21,694; *Titan Group, Inc.,* ASBCA No. 28584, 83-2 BCA ¶ 16,803.
[224]*Kunz Constr. Co. v. United States,* 12 Cl. Ct. 74, 79 (1987).
[225]41 U.S.C. § 605(c)(5).
[226]*Maki v. United States,* 13 Cl. Ct. 779, 782 (1987).
[227]41 U.S.C. § 605(c)(4).

approach is more detailed than typical notice pleadings as a detailed event-by-event statement of the facts with supporting documents enables the board or court to quickly evaluate the reasons why an action was instituted prior to the receipt of a final decision.

3. Contents of the Decision

The CDA requires that each contracting officer's decision "state the reasons for the decision reached and . . . inform the contractor of his rights as provided in [the Act]. Specific findings of fact are not required, but, if made, shall not be binding in any subsequent proceeding."[228]

FAR § 33.211 also sets forth the procedure that the contracting officer is to follow if a claim by or against a contractor cannot be settled by mutual agreement. In preparing the final decision, the contracting officer is directed to include:

- Description of the claim or dispute
- Reference to pertinent contract terms
- Statement of the factual areas of agreement and disagreement
- Statement of the final decision, with supporting rationale
- In the case of a final decision asserting a claim that the contractor is indebted to the government, a demand for payment in accordance with FAR § 32.610(b)

As a practical matter, the extent of the findings of fact and the rationale provided in the final decision can vary greatly, depending on the nature of the dispute and the specific contracting officer. Although the degree of detail and explanation may vary, every final decision is required by regulation[229] to contain a paragraph that reads substantially in this way:

This is the final decision of the Contracting Officer. You may appeal this decision to the agency board of contract appeals. If you decide to appeal, you must, within 90 days from the date you receive this decision, mail or otherwise furnish written notice to the agency board of contract appeals and provide a copy to the Contracting Officer from whose decision this appeal is taken. The notice shall indicate that an appeal is intended, reference this decision, and identify the contract by number. With regard to appeals to the agency board of contract appeals, you may, solely at your election, proceed under the board's (1) small claim procedure for claims of $50,000 or less, or in the case of a small business concern (as defined in the Small Business Act and regulations under that Act) $150,000 or less; or (2) its accelerated procedure for claims of $100,000 or less. Instead of appealing to the agency board of contract appeals, you may bring an action directly in the United States Court of Federal Claims (except as provided in the Contract Disputes Act of 1978, 41 U.S.C. 603, regarding Maritime Contracts) within 12 months of the date you receive this decision.[230]

---

[228]41 U.S.C. § 605(a).
[229]FAR § 33.211(a)(4)(v).
[230]*Id.*

A final decision that fails to contain this paragraph is considered defective.[231] However, a contractor may treat a clear written assertion by the contracting officer of the government's right to payment of a sum certain as a final decision even if the contracting officer's letter fails to include the otherwise required final decision and statement of appeal rights.[232]

A defective statement of the contractor's appeal rights does not automatically toll (stop) the period for filing an appeal or suit. To excuse a late appeal, the contractor must demonstrate detrimental reliance on the defective statement of its appeal rights.[233] Since the receipt of the final decision triggers the time periods for an appeal or filing of a suit, the FAR directs the contracting officer to furnish the contractor a copy of the decision by certified mail, return receipt requested, or any other method that provides evidence of receipt.[234]

4. Status of Decision on Appeal

The CDA provides for a *de novo* review of the final decision by the applicable board or the Court of Federal Claims.[235] A *de novo* review means that no finality or presumption of correctness is attached to any of the findings or conclusions set forth in the contracting officer's final decision. To the extent that the final decision reflects a complete rejection of the contractor's claim and disagreement on basic facts such as excusable delays, a *de novo* proceeding provides an opportunity to present the claim in a forum that affords the final decision no deference.

However, some final decisions do not contain a complete denial of the contractor's claim. For example, in a claim involving an alleged differing site condition, the final decision could contain a partial concession of entitlement to time or money or both. When the contractor appeals that decision, any elements of that decision that were favorable to the contractor are also subject to a *de novo* review. In other words, the government is free to contest even those points that were conceded by the contracting officer. As a consequence, time or money allowed in the final decision may be lost at the board or court proceeding. This result was established by the Federal Circuit in *England v. Sherman R. Smoot Corp.*[236]

Beginning with a 1976 decision in *Robert McMullan & Son, Inc.*,[237] the boards held that the issuance of a timely extension by the contracting officer gave rise to a

---

[231] *Pathman Constr. v. United States,* 817 F.2d 1573, 1578 (Fed. Cir. 1987); *Lawrence Harris Constr., Inc.,* VABCA No. 7219, 05-1 BCA ¶ 32,830.

[232] *Lasmer Indus., Inc.,* ASBCA No. 56411, 08-2 BCA ¶ 33,919; *see also Midwest Transport, Inc.,* PSBCA No. 6132, 08-1 BCA ¶ 33,823.

[233] *State of Florida Dept. of Ins. v. United States,* 81 F.3d 1093 (Fed. Cir. 1996); *Decker & Co. v. West,* 76 F.3d 1573 (Fed. Cir. 1996); *TPI Int'l Airways, Inc.,* ASBCA No. 46462, 96-2 BCA ¶ 28,373 (reliance shown); *but see Medina Contracting Co.,* ASBCA No. 53783, 02-2 BCA ¶ 31,979; *and American Renovation & Constr. Co.,* ASBCA No. 54039, 03-2 BCA ¶ 32,296.

[234] FAR § 33.211; *David Grimaldi Co.,* ASBCA No. 49795, 97-2 BCA ¶ 29,201; *Select Contracting, Inc. v. VA Medical Ctr.,* VABCA No. 4541, 95-2 BCA ¶ 27,830; *National Interior Contractors, Inc.,* VABCA No. 4561, 95-2 BCA ¶ 27,695.

[235] 41 U.S.C. § 605(a).

[236] 388 F.3d 844 (Fed. Cir. 2004).

[237] ASBCA No. 19023, 76-1 BCA ¶ 11,728.

rebuttable presumption that the government was responsible for that period of delay. In the *Sherman R. Smoot* decision, the Federal Court overruled the *McMullan* presumption on the basis that it was inconsistent with Section 605(a) of the Contract Disputes Act. Even though the change order issued in the *Smoot* decision was in the form of a unilateral change order, the Federal Circuit stated that the findings of fact reflected by the *unilateral change order* should be given no different status than the findings of fact in the contracting officer's final decision. Accordingly, the government could contest those findings of fact on a "de novo" basis.[238]

Consequently, contractors need to evaluate the potential downside of an appeal or suit from a final decision that allowed some measure of relief (time or money or both).

B. Appeal Deadlines

The CDA provides that a contracting officer's final decision on a claim (whether a contractor or a government claim) is "final and conclusive and not subject to review by any forum, tribunal, or Governmental agency unless an appeal or suit is timely commenced as authorized by [the Act]."[239] Once a contractor receives a contracting officer's final decision, it has two alternatives: (1) take no action on the appeal and it becomes binding on both parties or (2) appeal the decision.

Under the CDA, a contractor has two avenues of appeal from a contracting officer's final decision. Within 90 days of the date of receipt of the decision, the contractor may appeal the decision to the appropriate board of contract appeals.[240] Alternatively, within 12 months of the date of receipt of the decision, the contractor may initiate an action in the Court of Federal Claims.[241]

With respect to either an appeal to a board of contract appeals or an action in the Court of Federal Claims, it is important to consider three basic points.

(1) There can be no appeal or suit unless there has been a valid contracting officer's final decision or the failure to issue such a decision within the period required under the Act.

(2) Once a valid final decision has been issued, it is essential that a board appeal, or Court of Federal Claims suit, whichever the contractor wishes to pursue, be filed within the required time frame.

(3) The contractor should realize that, once it has elected either a board or the Court of Federal Claims as the forum in which to challenge the contracting officer's decision, it may not switch to the other forum.

---

[238] A bilateral change order executed by both the contractor and the contracting officer should avoid the application of this rule. However, it is very likely that the government would insist on a release as part of the bilateral modification process.

[239] 41 U.S.C. § 605(b). The government may not appeal a final decision of its own contracting officer. *Douglass Indus., Inc.,* GSBCA No. 9630, 90-2 BCA ¶ 22,676. However, a final decision favoring a contractor can be rescinded and a new final decision denying the claim may be issued so long as it is done within the CDA appeal period. *Daniels & Shanklin Constr. Co.,* ASBCA No. 37102, 89-3 BCA ¶ 22,060.

[240] 41 U.S.C. § 606. (The contractor does not have a choice of boards, as the applicable board is already designated.)

[241] 41 U.S.C. § 609(a)(1), (a)(2). *See Opalack v. United States,* 5 Cl. Ct. 349, 361 (1984).

As noted previously, the contracting officer is directed to obtain evidence of the date on which the contractor received a final decision.[242] This regulatory directive reflects the requirement for strict compliance with the time limits set forth in the CDA for appealing to a board or the Court of Federal Claims. Neither a board nor the court can consider an appeal that is not timely presented to it,[243] as the periods for challenging a contracting officer's decision set forth in the CDA are jurisdictional and cannot be waived.[244] However, if the agency creates confusion by sending out two versions of a final decision at different times, the contractor may be entitled to rely on the appeal language in the second version of the final decision as triggering the period of time to file an appeal or suit.[245]

The boards' rules typically provide that filing of the appeal occurs when it is mailed or otherwise furnished to the board.[246] "Mailing" has been interpreted as meaning the United States Postal Service. Thus, an appeal that was submitted to a commercial carrier before the expiration of the 90-day appeal period but received after that period expired was untimely.[247]

Sometimes contracting officers send out a copy of the final decision by facsimile followed by a copy sent via certified mail. Unless the facsimile copy clearly indicates that it is an "advance" copy, the 90-day period has been calculated from the date of receipt of the facsimile.[248] If the contractor elects to file its notice of appeal by facsimile, a *filing* on the 90th day has certain risks. For example, the ASBCA has dismissed an appeal as untimely because the facsimile notice of appeal was not received by it until the 91st day. The board ruled that when a notice of appeal is mailed via the U. S. Postal Service, the date of mailing constitutes the filing date. When the notice is sent in a different manner—for example, commercial carrier or facsimile—the date of receipt at the board is the filing date.[249]

---

[242]FAR § 33.211(b). When the government alleges that an appeal is untimely, it bears the burden of proving the date of the contractor's receipt of the final decision. *Alco Mach. Co.,* ASBCA No. 38183, 89-3 BCA ¶ 21,955; *Atlantic Petroleum Corp.,* ASBCA No. 36207, 89-1 BCA ¶ 21,199.

[243]*Cosmic Constr. Co. v. United States,* 697 F.2d 1389 (Fed. Cir. 1982); *Gregory Lumber Co. v. United States,* 229 Ct. Cl. 762 (1982); *L.C. Craft,* ASBCA No. 47351, 94-2 BCA ¶ 26,929; *Contract Servs. Co.,* ASBCA No. 34438, 87-2 BCA ¶ 19,850.

[244]*Cosmic Constr. Co.* 697 F.2d 1389 (Fed. Cir. 1982); *Robert T. Rafferty v. General Services Administration,* CBCA No. 617, 07-1 BCA ¶ 33,577.

[245]*See Fitnet Int'l Corp.,* ASBCA No. 56605, 2009 WL 2359864, 09-2 BCA ¶ 34,210; *Pinnell Brown Constr., Inc., v. Department of Veterans Affairs,* CBCA No. 917, 08-1 BCA ¶ 33,721.

[246]See ASBCA Rule 1(a). Historically, some boards, but not all, accepted facsimile notices of appeal. *See J.C. Equip. Corp.,* IBCA No. 2885-89, 91-3 BCA ¶ 24,322. With the consolidation of the various civilian boards, it is essential to obtain and review the rules of the two remaining boards (ASBCA and CBCA) to determine the applicable rule.

[247]*KAMP Sys., Inc.,* ASBCA No. 55317, 08-1 BCA ¶ 33,748; *Tiger Natural Gas, Inc. v. General Services Administration,* GSBCA No. 16039, 03-2 BCA ¶ 32,321; *C.R. Lewis Co.,* ASBCA No. 37200, 90-3 BCA ¶ 23,152; *North Coast Remanufacturing, Inc.,* ASBCA No. 38599, 89-3 BCA ¶ 22,232; *Associated Eng'g. Co.,* VABCA No. 2673, 88-2 BCA ¶ 20,709.

[248]*Corners and Edges, Inc.,* ASBCA Nos. 55767, 56277, 08-2 BCA ¶ 33,949; *Tyger Constr. Co.,* ASBCA No. 36100 *et al.,* 88-3 BCA ¶ 21,149; *but see AST Anlagen und Sanierungstechnik GmbH,* ASBCA No. 51854, 04-2 BCA ¶ 32,712.

[249]*KAMP Sys., Inc.,* ASBCA No. 55317, 08-1 BCA ¶ 33,748; *Birkart Globistics AG,* ASBCA No. 53458, 06-1 BCA ¶ 33,138.

Often counsel participate directly in the transmission of a claim and may correspond with the agency regarding the claim. If the contracting officer sends the final decision to the contractor's attorney, that attorney may be treated by the boards or the court as the contractor's representative for the purpose of receiving the final decision. Accordingly, the time period for filing an appeal or suit would begin to run upon the attorney's receipt of the final decision.[250] Even if the real party in interest is a subcontractor and the prime contractor is only sponsoring the subcontractor's claim, the period for an appeal or suit begins to run when the prime contractor receives the final decision.[251]

A contracting officer cannot waive the filing deadlines.[252] However, reconsideration of final decision by a contracting officer can have the effect of starting a new appeal period, which would allow the board to assume jurisdiction over a timely appeal of the second final decision[253] or, under certain circumstances, even the lack of a second final decision.[254] However, relying on post–final decision communications to extend the appeal period is very risky, as such communications do not revive appeal rights unless they clearly constitute a reconsideration of the final decision.[255] If both parties are interested in further negotiations after a final decision is issued, a safer course is to file an appeal or suit and then mutually seek a brief stay to explore a negotiated resolution unless there is a clear written record that the decision is being "reconsidered" by the contracting officer.

Although strict compliance is required with the appeal limitation periods set forth in the CDA, the limitations period are not triggered when the contractor's right to proceed to either a board or the Court of Federal Claims arises because the contracting officer has failed to issue a decision on a proper claim within the period of time required by the Act and the claim therefore is deemed denied.[256]

Finally, in the case of a termination for default, the circumstances may be such that the time the contractor has to challenge the termination does not begin to run when the contracting officer issues the final decision terminating the contract, but at a later date. This exception to the general basic requirement for strict compliance with the appeal deadlines reflects the continuing application of the doctrine set forth in *Fulford Manufacturing Co.*[257] The proposition embodied by the *Fulford* doctrine

[250] *Structural Finishing, Inc. v. United States,* 14 Cl. Ct. 447 (1988).

[251] *Colton Constr. Co., Inc.,* ASBCA No. 30313, 85-3 BCA ¶ 18,262.

[252] *Watson Rice & Co.,* AGBCA No. 82-126-3, 82-2 BCA ¶ 16,009 at pg. 79,359.

[253] *Fitnet Int'l Corp.,* ASBCA No. 56605, 09-2 BCA ¶ 34,210; *Summit Contractors v. United States,* 15 Cl. Ct. 806 (1988); *Nash Janitorial Serv., Inc.,* GSBCA No. 7338-R, 89-2 BCA ¶ 21,615.

[254] *West Land Builders,* VABCA No. 1664, 83-1 BCA ¶ 16,235.

[255] Ongoing negotiations without clear evidence of an agreement to reconsider the decision will not prevent the appeal period from running. *Compare Colfax, Inc.,* AGBCA No. 89-159-1, 89-3 BCA ¶ 22,130, *and Birken Mfg. Co.,* ASBCA No. 36587, 89-2 BCA ¶ 21,581, *with Royal Int'l Builders Co.,* ASBCA No. 42637, 92-1 BCA ¶ 24,684.

[256] *Pathman Constr. Co. v. United States,* 817 F.2d 1573 (Fed. Cir. 1987).

[257] ASBCA Nos. 2143, 2144, Cont. Cas. Fed. (CCH) ¶ 61,815 (May 20, 1955) (digest only) (timely appeal of the default action will also preserve right to contest excess cost assessment even though second final decision is not appealed in timely manner). *See also T.E. Deloss Equip. Rentals,* ASBCA No. 35374, 88-1 BCA ¶ 20,497; *El-Tronics, Inc.,* ASBCA No. 5457, 61-1 BCA ¶ 2961. However, one board decision has held that the failure to timely appeal a default termination final decision will preclude a challenge to the default in a subsequent appeal from a government claim for recovery of unliquidated progress payments or property damages. *See Dailing Roofing, Inc.,* ASBCA No. 34739, 89-1 BCA ¶ 21,311. *See also Guidance Sys.,* ASBCA No. 34690, 88-3 BCA ¶ 20,914.

is that when a contractor makes a timely appeal to an assessment of excess reprocurement costs, the propriety of the default termination can be challenged even though the default termination was not appealed.[258] The *Fulford* doctrine has not been altered by the CDA.[259] Thus, in most cases, the limitation periods set forth in the CDA do not "bar a contractor from contesting the propriety of a default termination in an action appealing a contracting officer's decision assessing excess reprocurement costs" if such an action is filed within 90 days (a board appeal) or 12 months (Court of Federal Claims) of the decision assessing excess costs.[260] Failure to seek review of a default termination within the 90-day or 12-month period, however, bars a contractor from challenging the default termination if excess costs are not assessed.[261]

C. Contractor's Choice of Forum

1. Factors

When it becomes apparent that the agency will or is likely to issue an adverse final decision, it is essential that the contractor and its counsel give careful consideration to the election of the forum (board or Court of Federal Claims) in which it will proceed. The CDA gives the contractor the basic right to seek a *de novo,* or complete, review of a final decision in either forum.[262] When considering whether to elect to go to the board or to the court, there are a number of factors to consider, such as:

- *Time and money.* Ordinarily, board proceedings are believed to be less time consuming and costly than court proceedings. Often this perception reflects the fact that the boards' formal rules of procedure are not as extensive as the rules in the Court of Federal Claims.[263] However, some board judges issue extensive prehearing orders that mirror, to a large degree, orders issued by a federal district court or the Court of Federal Claims.

- *Judicial background and experience.* In accordance with the CDA, board judges must have at least five years of experience in government contract law. Typically, they have much more than that minimum level of experience in government contracting.[264] There is no parallel specialized experience requirement for Court of Federal Claims judges. Board judges hear only government contract cases. Judges on the Court of Federal Claims hear a wide range of matters besides contract cases.

[258]*D. Moody & Co. v. United States,* 5 Cl. Ct. 70 (1984).
[259]*D. Moody & Co.* 5 Cl. Ct. at 76; *Southwest Marine, Inc.,* DOTBCA No. 1891, 96-1 BCA ¶ 27,985; *Tom Warr,* IBCA No. 2360, 88-1 BCA ¶ 20,231.
[260]*D. Moody & Co.,* 5 Cl. Ct. 70.
[261]*Id.*
[262]41 U.S.C. §§ 606, 609(a)(1)–609(a)(2).
[263]At least one board has ruled that the CDA gave it subpoena and sanction power over the federal government. *Mountain Valley Lumber, Inc.,* AGBCA No. 2003-171-1, 06-2 BCA ¶ 33,339.
[264]41 U.S.C. § 607(b)(1).

- *Issues.* If the case involves a particular issue or contract provision, it is important to learn how that board or the court has viewed that issue or legal theory for relief in the past or if there are any relevant differences in the jurisdiction of the boards and the Court of Federal Claims. [265] For example, a board or the court may have recently issued a decision reflecting its views on the proof of delay and the use of a CPM to demonstrate delay. If the case warrants the investment, this type of research should be conducted as part of the forum selection process.

- *Agency involvement.* If a case is appealed to a board, the agency will provide the government trial counsel. Accordingly, the counsel representing the government may be the same person who advised the contracting officer when the claim was being denied. When a case is filed in the Court of Federal Claims, the Civil Division of the Justice Department represents the government. Under certain circumstances, the Justice Department can, in theory, settle a case over the procuring agency's objections.

- *Hearing/trial location.* Both the boards and the Court of Federal Claims are located in the metropolitan Washington, D.C., area. In practice, the boards and the court can and often do hold hearings outside of Washington. Location usually depends on the convenience of all of the parties.

- *Representation by legal counsel.* Board practice permits an officer of the corporation to represent it on appeal. At the Court of Federal Claims, a corporation must be represented by an attorney admitted to practice before that court.[266]

2. Election Doctrine

Once a contractor has elected either a board or the Court of Federal Claims as the forum in which to challenge a contracting officer's final decision, it may not switch to the other forum. In this regard, a contractor that is poised to proceed to either a board or the court should be aware of the Election doctrine. The term "Election doctrine" refers to the body of law related to the contractor's right to initially select the forum in which to challenge a contracting officer's decision. However, the Act does not allow the contractor to pursue its claim in both forums.[267] Thus, once a contractor

---

[265]For example, a comparison of the decisions of the ASBCA and the Claims Court (now the Court of Federal Claims) concerning the interpretation of essentially the same specification illustrates the value of this type of research. *Compare Western States Constr. Co. v. United States,* 26 Cl. Ct. 818 (1992), *with Tomahawk Constr. Co.,* ASBCA No. 41717, 93-3 BCA ¶ 26,219. If the contractor's claim theory includes an assertion based on a third-party beneficiary status, the Federal Circuit has ruled that a board may not accept jurisdiction over an appeal to it. *See Winter v. Floorpro, Inc.,* 570 F.3d 1367 (Fed. Cir. 2009). In addition, as discussed in **Chapter 3**, the Court of Federal Claims and the ASBCA have different views regarding whether a direct challenge to a past performance evaluation is a CDA claim.

[266]*Alchemy, Inc. v. United States,* 3 Cl. Ct. 727 (1983).

[267]*Tuttle/White Constructors, Inc. v. United States,* 656 F.2d 644, 649 (Ct. Cl. 1981); *Marshall Associated Contractors, Inc. v. United States,* 31 Fed. Cl. 809 (1994).

makes a binding election to appeal a contracting officer's decision to the appropriate board of contract appeals, the contractor cannot change course and pursue its claim in the court.[268] The converse is also true.

A binding *election* takes place when a contractor files an appeal or initiates a suit in a "forum with jurisdiction over the proceeding."[269] This means that when a contractor initiates proceedings on its claim before a board of contract appeals in a timely manner, it has made a binding election to proceed before the board, and it is barred from initiating suit in the court; any suit it files in the court will be dismissed.[270] However, the filing of an appeal with the appropriate board of contract appeals is not a binding election if it is determined by the board that the contractor's appeal was untimely, and hence the subsequent filing of a claim in the Court of Federal Claims is not barred.[271] The rationale is that a contractor's choice of forums in which to contest the contracting officer's decision is a binding election only if that choice is truly available, which it is not if resort to a board of contract appeals is untimely.[272] In those circumstances, the untimely appeal to the board was an absolute nullity and the Election doctrine is not applicable.[273]

3. Transfer and Consolidation of Cases

The CDA provides that if two or more actions (suits or appeals) arising from one contract are separately filed in the Court of Federal Claims and with a board, the court is authorized to order the consolidation of the suits before it or to transfer the matters to the board involved "for the convenience of parties or witnesses or in the interest of justice."[274] In deciding whether a case should be consolidated or transferred, the court will take into account a number of factors: (1) whether the disputes in the different forums arise out of the same contract, (2) whether the cases present overlapping issues or the same issues, (3) whether the plaintiff initially elected to initiate proceedings at the board, (4) whether substantial effort in the case already has been expended in one forum but not the other, (5) which proceeding involves the most money, and (6) which proceeding presents the more difficult and complex claims.[275]

D. Status of the Contracting Officer's Decision on Appeal

When an appeal is filed with a board or a suit is filed in the Court of Federal Claims, the board or the court will have access to the contracting officer's decision along

---

[268]*Tuttle/White Constructors, Inc.,* 656 F.2d 644. However, a notice of appeal to a board that was retrieved before docketing did not constitute a binding election. *Blake Constr. Co., Inc. v. United States,* 13 Cl. Ct. 250 (1987).

[269]*Tuttle/White Constructors, Inc.,* 656 F.2d 644.

[270]*National Neighbors, Inc. v. United States,* 839 F.2d 1539, 1541–42 (Fed. Cir. 1988).

[271]*Id.*

[272]*Id.*

[273]*Id.*

[274]41 U.S.C. § 609(d). *See Glendale Joint Venture v. United States,* 13 Cl. Ct. 325, 327 (1987); *Multi-Roof Sys. Co. v. United States,* 5 Cl. Ct. 245, 248 (1984); *E.D.S. Fed. Corp. v. United States,* 2 Cl. Ct. 735, 739 (1983).

[275]*Id.*

with other documents as the record is developed. In some legal settings, an administrative decision similar to the contracting officer's decision is given deference by a higher tribunal. However, this is not the case under the CDA. In *Wilner v. United States*,[276] the Federal Circuit ruled that no deference was to be given to the contracting officer's decision. In *Wilner*, the contracting officer found that the contractor was entitled to 260 days of compensable delay and offset the amount of additional compensation due the contractor against a previous allowance and allowed a net adjustment to the contract price in the final decision.[277] The contractor, unsatisfied with the award, filed suit in the Claims Court in order to increase the amount of compensation, but the court found that the contractor had not produced sufficient evidence of the government's liability.[278] On appeal, the Federal Circuit ruled that the contracting officer's decision was not evidence that independently established that the government had caused the delay.[279] The court ruled that the decision of the contracting officer is not "binding upon the parties" and is not entitled to deference, that is, the contractor's claim is to be proven *de novo* at the board or before the Court of Federal Claims.[280] Put simply, once an action is brought before a board or the court, the parties start with a clean slate.[281] The contracting officer's decision is not disregarded at trial or on appeal but rather will be evidence to be considered on an equal basis with the rest of the evidentiary record.[282]

The *Wilner* rule has both a positive impact as well as a negative impact on contractors during the disputes process. For a contractor that is dissatisfied with the decision of the contracting officer on a claim, the *Wilner* rule offers the contractor a second chance to effectively prove the claim. A contractor could develop evidence more thoroughly or potentially introduce new evidence that the contracting officer did not see when the claim was initially submitted. However, utilizing the *Wilner* rule gives the government the same opportunity. The government gets another chance to argue against liability, and there is always the potential, as in *Wilner,* that a board or the court will find that the contractor did not prove its claim even to the extent recognized by the contracting officer in the final decision.

VII. CONTRACT DISPUTES AND ALTERNATIVE DISPUTE RESOLUTION

There has long been a need for alternatives to the traditional manner in which government contract disputes are resolved. The Administrative Conference of the United

---

[276]24 F.3d 1397 (Fed.Cir. 1994).
[277]*Id.* at 1398.
[278]*Id.*
[279]*Id.* at 1400.
[280]*Id.* at 1401.
[281]*Id.* at 1402.
[282]*Id.* at 1403. *See also Flink/Vulcan v. United States*, 63 Fed. Cl. 292, 310 (2004).

States (Conference), whose purpose is to promote efficiency and fairness in federal agency procedures, has been a major proponent of alternative dispute resolution (ADR) in government contracts. The Conference has strongly supported ADR and has several publications that discuss the contract disputes dilemma and various ADR efforts.[283]

FAR § 33.214 states that the objective of using ADR procedures is to increase the opportunity for relatively inexpensive and expeditious resolution of issues in controversy. Essential elements of ADR include: (1) existence of an issue in controversy, (2) a voluntary election by both parties to participate in the ADR process, (3) an agreement on alternative procedures and terms to be used in lieu of formal litigation, (4) participation in the process by officials of both parties who have the authority to resolve the issue in controversy, and (5) contractor certification of claims in excess of $100,000. Consistent with the concept of alternative dispute resolution, the Department of the Navy has issued Secretary of the Navy Instruction 5800.15[284] providing for the voluntary use of binding arbitration to resolve contract controversies, procedures governing the arbitration, and confirmation of an award pursuant to the Federal Arbitration Act.[285]

ADR procedures may be used at any time that the contracting officer has authority to settle the issue in controversy and may be applied to a portion of a claim. When ADR procedures are used subsequent to issuance of a contracting officer's final decision, their use does not alter any of the time limitations or procedural requirements for filing an appeal of the contracting officer's final decision and does not constitute a reconsideration of the final decision. In the event that the contracting officer rejects a request by a small business to use ADR, the contracting officer is required by regulation to set forth a written explanation for that decision and to provide it to the contractor.[286]

The CDA states that the boards shall provide, to the fullest extent practicable, informal, expeditious, and inexpensive resolution of disputes, and this is the authority for their use of ADR.[287] The boards of contract appeals have implemented ADR procedures and issued a Notice Regarding Alternative Methods of Dispute Resolution that strongly endorses the use of ADR and suggests several techniques. Many of the procedures outlined by that notice come from the recommendations made by the Administrative Conference of the United States. The boards routinely provide a notice of the availability of ADR procedures when the docketing notice is sent out to the parties, together with a copy of the board's rules of procedure. In addition, most boards will seek to assist in the resolution of claims prior to the issuance of a final decision by making a judge available to mediate a claim. Similar to ADR after an appeal is filed, this voluntary process requires a joint request by the contractor and the government.

The decision to use ADR is made jointly by the parties, and a board will not accept a unilateral request. However, the board may take the initiative in suggesting ADR as

---

[283]*See generally* 1987 Sourcebook: Federal Agency Use of Alternative Means of Dispute Resolution (Office of the Chairman 1987) and its report on Appealing Government Contract Decisions: "Reducing the Cost and Delay of Procurement Litigation"; DOD Directive 5145.5.

[284]72 Fed. Reg. 13094 (March 20, 2007).

[285]9 U.S.C. §§ 1–9.

[286]FAR § 33.214(b).

[287]41 U.S.C. § 605(d).

an option in dispute resolution. There are a number of ADR methods, and both the parties and the board may agree to the use of any of these methods, such as settlement judge, minitrial, summary trial with binding decision, and other agreed methods.

The Court of Federal Claims has formally approved the voluntary use of ADR. As described in the *Deskbook for Practitioners* published by that court's bar association,[288] the court is sensitive to rising litigation costs and the delay often inherent in the traditional judicial resolution of complex legal claims. Accordingly, General Order 13 established two alternative methods of dispute resolution: the settlement judge and the minitrial.

These techniques are voluntary, and both parties must agree to their use. The court expects the techniques to be invoked in complex cases where the amount in controversy exceeds $100,000, the parties anticipate a lengthy period of discovery, and a trial is expected to consume more than one week.

When both parties agree to utilize one of these alternative methods of dispute resolution, they notify the presiding judge as early as possible in the proceedings or concurrently with the submission of a joint preliminary status report.

If the presiding judge agrees, the case will be referred to the clerk, who will assign the case to another court judge who will preside over the ADR procedure and who will exercise final authority, within the general guidelines adopted by the court, to determine the details of the ADR process in that case. If the ADR method utilized by the parties fails to produce a satisfactory settlement, the case will be returned to the docket of the presiding judge. All representations made in the course of utilizing a method of ADR are confidential and, except as permitted by Federal Rule of Evidence 408, may not be utilized for any reason in subsequent litigation.

VIII. RECOVERY OF ATTORNEYS' FEES IN GOVERNMENT CONTRACT CLAIMS

Generally, the FAR does not allow for the recovery of attorneys' fees and expenses, or claim consultants' fees and expenses, associated with the preparation and prosecution of government contract claims.[289] However, with the passage of the Equal Access to Justice Act (EAJA),[290] the Congress provided a statutory basis for certain eligible contractors to recover some or all of their legal costs and expenses of litigation with the government.

An EAJA application for recovery of legal fees and expenses is filed within 30 days after the conclusion of the primary appeal or suit.[291] To recover its fees and expenses, the claimant must meet these criteria:

---

[288]United States Court of Federal Claims, *Deskbook for Practitioners Fifth Edition* (United States Court of Federal Claims Bar Association) (2008).

[289]FAR § 31.205-33; *Plano Builders Corp. v. United States*, 40 Fed. Cl. 635 (1998). If the contractor can convince the board or court that these costs were incurred in aid of contract administration, rather than claim preparation or prosecution, these costs can be recovered to the extent they are reasonable and allocable. *See Bill Strong Enters., Inc. v. United States*, 49 F.3d 1541 (Fed. Cir. 1995); *Betancourt & Gonzalez, S.E.*, DOTBCA No. 2785 *et al.*, 96-1 BCA ¶ 28,033.

[290]5 U.S.C. § 504.

[291]*Id.; see also Southern Dredging*, ENGBCA No. 6236-F, 97-2 BCA ¶ 29,014; *AIW-Alton, Inc.*, ASBCA No. 47439, 96-2 BCA ¶ 28,399.

- Have a net worth of not more than $7 million
- Have no more than 500 employees[292]
- Be the prevailing party in the litigation with the government[293]

The government will not be held liable for the claimant's legal fees and expenses if it can demonstrate that its position in the litigation was substantially justified.[294] Even though the contractor recovers less than it claimed or prevailed on fewer than all of the issues, the claimant still may be deemed to be the prevailing party.[295]

The size and net worth criteria must be satisfied by the prime contractor. Even though the real party in interest is a subcontractor, it is not in privity of contract with the government and is not eligible to recover EAJA legal fees and expenses.[296]

The EAJA limits the amounts that can be recouped for legal fees to a maximum hourly rate[297] for attorneys plus out-of-pocket expenses. Compensation for paralegals' time is reimbursed at the market rate.[298] Expert witness rates can be no higher than those paid by the government to its expert witness.[299]

---

[292]5 U.S.C. § 504(b)(1)(B). These requirements apply to corporations, partnerships, or unincorporated businesses.

[293]5 U.S.C. § 504(a)(1).

[294]*Scarborough v. Principi*, 541 U.S. 401 (2004) (Court reversed Federal Circuit and held that an EAJA application may be amended after the 30-day filing period to cure an initial failure to assert that the government's litigation position in the underlying litigation lacked substantial justification); *see also Precision Pine & Timber, Inc., v. United States*, 83 Fed. Cl. 544 (2008); *Oneida Constr. Inc./David Boland, Inc., Joint Venture*, ASBCA Nos. 44194 *et al.*, 95-2 BCA ¶ 27,893; *Labco Constr., Inc.*, AGBCA No. 95-104-10, 95-2 BCA ¶ 27,677; *ABC Health Corp.*, VABCA No. 2462E, 94-3 BCA ¶ 27,013; *Sun Eagle Corp.*, ASBCA No. 45985, 94-2 BCA ¶ 26,870. *But see C.H. Hyperbasics, Inc.*, ASBCA Nos. 49375 *et al.*, 05-2 BCA ¶ 33,111 (board reduced EAJA legal fee recovery in part due to contractor's rejection of a significant settlement offer), *and Silver Enters. v. Department of Transportation*, CBCA No. 63-C, 07-1 BCA ¶ 33,496 (board denied EAJA recovery even though appeal was sustained, in part due to the fact that contractor failed to maintain adequate records of its costs and agency's refusal to pay the claim was "substantially justified").

[295]*Midland Maintenance, Inc.*, ENGBCA Nos. 6080-F, 6092-F, 97-1 BCA ¶ 28,849; *Jackson Elec. Co.*, ENGBCA No. 6238-F, 97-1 BCA ¶ 28,848; *Tayag Bros. Enters, Inc.*, ASBCA No. 42097, 96-2 BCA ¶ 28,279.

[296]*SCL Materials & Equip. Co.*, IBCA No. 3866-97F, 98-2 BCA ¶ 30,000.

[297]5 U.S.C. § 504(b)(1)(A). For actions or appeals awarded on or after March 29, 1996, the maximum rate for legal fees is $125 per hour. For actions or appeals commenced prior to that date, the maximum rate was $75 per hour.

[298]*Richlin Security Service Co. v. Chertoff*, 128 S.Ct. 2007 (2008).

[299]*Techplan Corp.*, ASBCA Nos. 41470 *et al.*, 98-2 BCA ¶ 29,954.

➤ LESSONS LEARNED AND ISSUES TO CONSIDER

- While the Contract Disputes Act (CDA) addresses the *procedures* for processing claims on a government contract, it is essential that a contractor understand its obligations and rights under the standard clauses, such as the Changes, Differing Site Conditions, and Suspension of Work clauses.
- Compliance with the contract's *notice provisions* as well as consideration of the CDA's six-year statute of limitations often are essential to preserving the contractor's right to recovery on a claim.
- Contractors need to appreciate the importance of *releases* in contract modifications on their rights to submit subsequent requests for equitable adjustments or claims.
- The concept of a *claim* has *different meanings* under the CDA and under the government's various laws and regulations related to false or overstated claims and proposals.
- Every request for an equitable adjustment or other request for extra contractual relief is not necessarily a *CDA claim*. A *CDA claim* is a nonroutine written submission or demand that seeks, as a matter of right, the payment of money in a sum certain.
- Every CDA claim in excess of $100,000 must be *certified* by an authorized representative of the prime contractor in order to be considered a *claim* and entitle the contractor to recover CDA interest.
- The CDA claim certification is one of *several certifications* that may apply to claims or change order proposals.
- Prime contractors must provide *unqualified certifications of their subcontractors' claims*. In that context, a contractor should consider obtaining an appropriate indemnity agreement.
- A carefully *drafted Claims Cooperation Agreement* may be needed when subcontractor claims are an element in the CDA claims process.
- A *defective claim certification* may delay action by the contracting officer, a board of contract appeals, or the Court of Federal Claims.
- The CDA specifies *time frames* for action by the contracting officer on all claims and provides a means for a contractor to compel consideration of the claim if the contracting officer is unreasonably slow in acting.
- Once a final decision is received, a contractor has *90 days* to file an appeal at board of contract appeals or *one year* to file a suit in the Court of Federal Claims. If these periods are allowed to pass, the final decision is, in almost all cases, final and binding.

- A contractor should carefully consider whether to appeal to the applicable board of contract appeals or file a suit in the Court of Federal Claims. Once an *election* is made, it is, in almost all cases, binding on the parties.
- As a result of an appeal of the final decision to a board or the filing of a suit in the Court of Federal Claims, the claim receives *de novo* consideration. This means that *no presumption of correctness* is attached to the final decision—even those portions favorable to the contractor.

APPENDIX 15A: SAMPLE FREEDOM OF INFORMATION ACT LETTER

To be sent Certified Mail—Return Receipt Requested.

[Name]

[Address]

ATTN:

Agency—FOIA Officer

SUBJECT: Freedom of Information Act Request (FOIA)

RE: Contract No. [Number]

[identify project]

Dear [name of FOIA officer]:

We represent [name of company] and have been requested to file this FOIA request regarding the [insert name of project] in [city, state] ("Project").

Pursuant to the Freedom of Information Act, 5 U.S.C. § 552, and the regulations promulgated thereunder, 40 CFR Part 2 [if applicable, insert specific Agency regulation], we hereby request a copy of the following agency records related to the design and construction of the Project.

All correspondence, documents, investigations, studies, reports, memoranda, etc., including, but not limited to:

(1) Design development and program of requirements;

(2) Selection of the project design firm;

(3) Evaluation of _____ proposal in response to the solicitation leading to the award of Contract _____.

(4) Correspondence to and from [agency] and [contractor];

(5) Correspondence to and from [agency] and the [name of design firm for [project];

(6) Notes of meetings, memoranda, calculations, working papers, letters, phone notes, and reports;

(7) Photographs, maps, diagrams and blueprints.

This request for records includes both documents and nonidentical copies, as well as e-mails and any other records maintained in an electronic format.

We understand there may be search and copy fees incurred as a result of this request and confirm our obligation to reimburse these charges and fees payable in accordance with the FOIA and the applicable regulations. In the event that the estimated cost may exceed $250, please contact us prior to proceeding to fulfill this request.

Sincerely,

16

FEDERAL GRANTS FUNDING CONSTRUCTION CONTRACTS

I. OVERVIEW

Federal government construction projects typically are built using direct contracts with private contractors. Federal assistance or support for construction projects, however, also may be provided to state and local governments, companies, and non-profit organizations through grants and cooperative agreements. Grants and cooperative agreements, collectively referred to as *assistance agreements,* are forms of federal financial assistance to private organizations or to state or local governments to support a public purpose. The purpose of a contract is for the federal government to directly acquire goods or services. The purpose of a grant or an assistance agreement is to provide financial assistance.[1]

The difference between a grant and a cooperative agreement is the degree of the federal agency's involvement in carrying out the contemplated activity.[2] If *substantial involvement* is expected between the federal agency and the recipient, a cooperative agreement is used.[3] When *substantial involvement* is not expected between the federal agency and the recipient, a grant must be used.[4] Direct contracts are used by the government when the principal purpose is acquisition of property or services for the direct benefit or use of the federal government.[5] A grant or cooperative agreement is used when the principal purpose of the transaction at the federal level is to

---

[1]Office of the Gen. Counsel, GAO, Principles of Federal Appropriations Law 5-48 (3d ed. Jan. 2004) (GAO "Red Book").
[2]*See Xcavators, Inc.,* 59 Comp. Gen. 758, B-198297, 80-2 CPD ¶ 229.
[3]31 U.S.C. § 6305(2).
[4]31 U.S.C. § 6304(2).
[5]31 U.S.C. § 6303.

provide financial support or stimulation as authorized by a federal statute.[6] The term "grant," as used in the remainder of this chapter, includes cooperative agreements.

Federal grants provide funding for many types of projects, including highway and bridge construction, airport improvements, water-treatment plants, disaster recovery, low-income housing, and infrastructure enhancing homeland security. Recipients of federal construction grants must comply with the statutes and regulations applicable to their specific program.

II. ROLE OF FEDERAL AGENCIES IN GRANT-FUNDED CONTRACTING

A. Federal Grant and Cooperative Agreement Act

In 1972, the Commission on Government Procurement found that many government agencies failed to distinguish properly between procurement contracts and grants, a failure that led to inappropriate use of grants. In response to these problems, Congress passed the Federal Grant and Cooperative Agreement Act (FGCAA) of 1977.[7] The purpose of the FGCAA was to differentiate grants from procurement contracts and to prevent federal agencies from using the two interchangeably.[8] The FGCAA established a system for federal agencies to utilize in determining whether to use a procurement contract, a grant, or a cooperative agreement (as discussed in **Section I** of this chapter).[9] The FGCAA also authorized the Office of Management and Budget (OMB) to publish supplementary interpretive guidelines, issued through *circulars,* to help agencies establish consistent procedures for the use of grants and procurement contracts. OMB issued Circular A-102 to govern the use of grants provided to state and local governments.[10] Circular A-102 would later be replaced by the Grants Management Common Rule.

B. Grants Management Common Rule and Other Government-wide Grants Requirements

In 1983, a 20-agency task force analyzed grants management and possible ways to streamline grants to state and local governments. In response to the task force's

---

[6]Office of Mgmt. & Budget, OMB Circular A-102, Grants and Cooperative Agreements with State and Local Governments (as revised by 59 Fed. Reg. 52224 (Oct. 14, 1994), as further amended by 62 Fed. Reg. 45934 (Aug. 29, 1997)), available at *www.whitehouse.gov/omb/circulars/a102/a102.html* (accessed Nov. 12, 2009).

[7]31 U.S.C. §§ 6303–6308.

[8]Andreas Baltatzis, *The Changing Relationship Between Federal Grants and Federal Contracts,* 32 Pub. Cont. L.J. 611, 614 (Spring 2003).

[9]*See supra* notes 2–6 and accompanying text.

[10]Office of Mgmt. & Budget, OMB Circular A-102, Grants and Cooperative Agreements with State and Local Governments (as revised by 59 Fed. Reg. 52224 (Oct. 14, 1994), as further amended by 62 Fed. Reg. 45934 (Aug. 29, 1997)), available at *www.whitehouse.gov/omb/circulars/a102/a102.html* (accessed Nov. 12, 2009).

findings, the president, in 1987, directed all affected federal agencies to adopt a common rule in order to implement government-wide terms and conditions for grants to state and local governments. The Grants Management Common Rule (the "Common Rule"),[11] first issued in March 1988,[12] standardized and codified the fiscal and administrative requirements federal agencies impose on state and local governments receiving grants. All federal agencies with grant-making authority "adopted the [C]ommon [R]ule verbatim, except where inconsistent with specific statutory authority."[13] Thus, although there may be some variation among agency rules, the core requirements are consistent with OMB's guidance. The Common Rule generally replaced Circular A-102 regarding an adopting agency's use of grants and setting standards for grantees.

OMB maintains a chart that identifies the parts of the Code of Federal Regulations (C.F.R.) in which federal departments and agencies have codified government-wide grants requirements.[14] That chart, with slight modifications, is reproduced as **Table 16.1.**

In 2004, the OMB, in response to the Federal Financial Assistance Management Improvement Act of 1999,[15] established Title 2 of the C.F.R., "Grants and Agreements," as a central location for grant policies and regulations. Title 2 consists of two subtitles. Subtitle A, "Office of Management and Budget Guidance for Grants and Agreements," sets out OMB government-wide policy guidance to federal agencies. Subtitle B, "Federal Agency Regulations for Grants and Agreements," provides agency-specific regulations implementing OMB guidance.[16] The regulations in Subtitle B, however, are not comprehensive; they simply add policies and procedures for nonprocurement debarment and suspension to each agency's grant regulations codified elsewhere (such as individual implementations of the Grants Management Common Rule). As shown by **Table** 16.1, agencies have not, in fact, centralized their grant policies and regulations.

C. Sources of Grant Authority

Federal agencies have the inherent authority to use procurement contracts to carry on construction activities.[17] By contrast, an agency must have specific statutory authority for the use of grants.[18] Many federal statutes authorize agencies to provide financial

[11]Grants and Cooperative Agreements to State and Local Governments, 60 Fed. Reg. 19638 (Apr. 19, 1995).

[12]Uniform Administrative Requirements for Grants and Cooperative Agreements to State and Local Governments, 53 Fed. Reg. 8034 (Mar. 11, 1988).

[13]Grants and Cooperative Agreements to State and Local Governments, 60 Fed. Reg. 19638 (Apr. 19, 1995), Supplementary Information.

[14]*www.whitehouse.gov/omb/grants_chart/* (accessed Nov. 12, 2009).

[15]Pub. L. No. 106-107, 113 Stat. 1486 (1999) (codified as notes to 31 U.S.C. § 6101).

[16]Governmentwide Guidance for Grants and Agreements; Federal Agency Regulations for Grants and Agreements, 69 Fed. Reg. 26276 (May 11, 2004).

[17]*See* **Chapter 2.**

[18]*See* Federal Facility Contributions to Capital Costs of Sewage Treatment Projects, 59 Comp. Gen. 1, B-194912 (1979).

Table 16.1 Codification of Government-wide Grants Requirements by Department and Agency

Department	Grants Management Common Rule (State & Local Governments)	OMB Circular A-110 (Universities & Non-Profit Organizations)[1]	Nonprocurement Suspension & Debarment[2]	Drug-Free Workplace Act Common Rule[3]	Byrd Anti-Lobbying Amendment Common Rule[4]
Agriculture	7 C.F.R. Part 3016	7 C.F.R. Part 3019	7 C.F.R. Part 3017	7 C.F.R. Part 3021	7 C.F.R. Part 3018
Commerce	15 C.F.R. Part 24	15 C.F.R. Part 14	2 C.F.R. Part 1326	15 C.F.R. Part 29	15 C.F.R. Part 28
Defense	32 C.F.R. Part 33	32 C.F.R. Part 32	2 C.F.R. Part 1125	32 C.F.R. Part 26	32 C.F.R. Part 28
Education	34 C.F.R. Part 80	34 C.F.R. Part 74	34 C.F.R. Part 85	34 C.F.R. Part 84	34 C.F.R. Part 82
Energy	10 C.F.R. Part 600	10 C.F.R. Part 600	2 C.F.R. Part 901	10 C.F.R. Part 607	10 C.F.R. Part 601
Health & Human Services	45 C.F.R. Part 92	45 C.F.R. Part 74	2 C.F.R. Part 376	45 C.F.R. Part 82	45 C.F.R. Part 93
Housing & Urban Development	24 C.F.R. Part 85	24 C.F.R. Part 84	24 C.F.R. Part 24	24 C.F.R. Part 21	24 C.F.R. Part 87
Interior	43 C.F.R. Part 12	43 C.F.R. Part 12	2 C.F.R. Part 1400	43 C.F.R. Part 43	43 C.F.R. Part 18
Justice	28 C.F.R. Part 66	28 C.F.R. Part 70	2 C.F.R. Part 2867	28 C.F.R. Part 83	28 C.F.R. Part 69
Labor	29 C.F.R. Part 97	29 C.F.R. Part 95	29 C.F.R. Part 98	29 C.F.R. Part 94	29 C.F.R. Part 93
State	22 C.F.R. Part 135	22 C.F.R. Part 145	2 C.F.R. Part 601	22 C.F.R. Part 133	22 C.F.R. Part 138
Transportation	49 C.F.R. Part 18	49 C.F.R. Part 19	49 C.F.R. Part 29	49 C.F.R. Part 32	49 C.F.R. Part 20
Treasury	—	—	31 C.F.R. Part 19	31 C.F.R. Part 20	31 C.F.R. Part 21
Veterans Affairs	38 C.F.R. Part 43	—	2 C.F.R. Part 801	38 C.F.R. Part 48	38 C.F.R. Part 45
Agency5	Grants Management Common Rule (State & Local Governments)	OMB Circular A-110 (Universities & Non-Profit Organizations) (see Note 1)	Nonprocurement Suspension & Debarment (see Note 2)	Drug-Free Workplace Act Common Rule (see Note 3)	Byrd Anti-Lobbying Amendment Common Rule (see Note 4)
ADF	—	—	22 C.F.R. Part 1508	22 C.F.R. Part 1509	—
AID	—	22 C.F.R. Part 226	22 C.F.R. Part 208	22 C.F.R. Part 210	22 C.F.R. Part 227
BBG	—	22 C.F.R. Part 518	22 C.F.R. Part 513	—	22 C.F.R. Part 519
CNCS	45 C.F.R. Part 2541	45 C.F.R. Part 2543	45 C.F.R. Part 2200	45 C.F.R. Part 2545	—
EPA	40 C.F.R. Part 31	40 C.F.R. Part 30	2 C.F.R. Part 1532	40 C.F.R. Part 36	40 C.F.R. Part 34
EX-IM	—	—	2 C.F.R. Part 3513	—	12 C.F.R. Part 411

Agency					
FEMA	44 C.F.R. Part 13	—	29 C.F.R. Part 1471	—	44 C.F.R. Part 18
FMCS	29 C.F.R. Part 1470	—	29 C.F.R. Part 1471	29 C.F.R. Part 1472	—
GSA	41 C.F.R. Part 105-71	41 C.F.R. Part 105-72	41 C.F.R. Part 105-68	41 C.F.R. Part 105-74	41 C.F.R. Part 105-69
IMS	45 C.F.R. Part 1183	—	45 C.F.R. Part 1185	45 C.F.R. Part 1186	—
IAF	—	—	22 C.F.R. Part 1006	22 C.F.R. Part 1008	—
NASA	14 C.F.R. Part 1273	14 C.F.R. Part 1260	2 C.F.R. Part 180	14 C.F.R. Part 1267	14 C.F.R. Part 1271
NARA	36 C.F.R. Part 1207	36 C.F.R. Part 1210	2 C.F.R. Part 2600	36 C.F.R. Part 1212	—
NEA	45 C.F.R. Part 1157	—	2 C.F.R. Part 3254	45 C.F.R. Part 1155	45 C.F.R. Part 1158
NEH	45 C.F.R. Part 1174	—	2 C.F.R. Part 3369	45 C.F.R. Part 1173	45 C.F.R. Part 1168
NSF	45 C.F.R. Part 602	—	2 C.F.R. Part 2520	45 C.F.R. Part 630	45 C.F.R. Part 604
ONDCP	21 C.F.R. Part 1403	—	21 C.F.R. Part 1404	21 C.F.R. Part 1404	—
OPM	—	—	5 C.F.R. Part 919	—	—
OPIC	—	—	—	—	22 C.F.R. Part 712
Peace Corps	—	—	22 C.F.R. Part 3700	22 C.F.R. Part 312	22 C.F.R. Part 311
SBA	13 C.F.R. Part 143	—	13 C.F.R. Part 145	13 C.F.R. Part 147	13 C.F.R. Part 146
SSA	—	—	2 C.F.R. Part 2336	20 C.F.R. Part 439	—
TVA	—	—	—	—	18 C.F.R. Part 1315

[1] Additional agencies are expected to codify OMB Circular A-110 (2 C.F.R. Part. 215); in the meantime, the Circular's requirements apply to those agencies and their awards.

[2] Exec. Order No. 12,549, 51 Fed. Reg. 6370 (Feb. 18, 1986) provided that agencies, including those that have not yet codified this common rule, are covered by OMB's government-wide guidelines, 2 C.F.R. Part 180, which are identical to this common rule. See OMB's memorandum to the agencies at 60 Fed. Reg. 33036 (June 26, 1995) and OMB's notice at 53 Fed. Reg. 34474 (Sep. 6, 1988).

[3] For additional information about OMB's government-wide implementation of the Drug-Free Workplace Act, see OMB's notice at 54 Fed. Reg. 4946 (Jan. 31, 1989) and 55 Fed. Reg. 21679 (May 25, 1990).

[4] The law only required major agencies, as identified by OMB, to codify this common rule; all other agencies are covered by OMB's government-wide guidance, 54 Fed. Reg. 52306 (Dec. 20, 1989), which is identical to this common rule. See also OMB's clarification notices at 55 Fed. Reg. 24540 (June 15, 1990) and 57 Fed. Reg. 1772 (Jan. 15, 1992). OMB issued amendments to its government-wide guidance, 61 Fed. Reg. 1412 (Jan. 19, 1996).

[5] Abbreviations used for the following independent agencies: African Development Foundation (ADF), Agency for International Development (AID), Broadcasting Board of Governors (BBG), Corporation for National & Community Service (CNCS), Environmental Protection Agency (EPA), Export-Import Bank of the United States (EX-IM), Federal Emergency Agency (FEMA), Federal Mediation & Conciliation Service (FMCS), General Service Administration (GSA), Institute of Museum Services (IMS), Inter-American Foundation (IAF), National Aeronautics & Space Administration (NASA), National Archives & Records Administration (NARA), National Endowment for the Arts (NEA), National Endowment for the Humanities (NEH), National Science Foundation (NSF), Office of National Drug Control Policy (ONDCP), Office of Personnel Management (OPM), Overseas Private Investment Corporation (OPIC), Small Business Administration (SBA), Social Security Administration (SSA), Tennessee Valley Authority (TVA).

assistance through grants to achieve a public purpose. The American Recovery and Reinvestment Act (ARRA or the Recovery Act) of 2009 is a most recent example.[19] Other examples are the Clean Air Act, the Clean Water Act, and the Safe Drinking Water Act, which authorize grants funded by the Environmental Protection Agency (EPA). The following are but two more examples of federal statutes that are sources of authority for agencies to make grants.

1. Stafford Act

The Robert T. Stafford Disaster Relief and Emergency Assistance Act (Stafford Act)[20] enables the federal government to provide grants to state and local governments in response to major emergencies and natural disasters. The Stafford Act makes the Federal Emergency Management Agency (FEMA) responsible for managing the grants and coordinating government-wide relief efforts. These grants generally require a cost-sharing arrangement, with the federal government providing 75 percent of the disaster assistance costs and the state or local entity contributing the other 25 percent.[21] For selected emergencies, the federal share has been increased to 90 percent (Hurricane Fran) or even 100 percent (Hurricane Andrew and the September 11, 2001, terrorist attacks in New York City and Washington, DC). Grants made under the Stafford Act may attach conditions intended to prevent fraud, such as limiting noncompetitive contracts to a period of not more than 150 days and requiring grantees to maintain integrated databases to collect information on eligible recipients and on disbursements and payments.[22] The Stafford Act also requires, to the extent feasible and practical, preference in awarding contracts to be given to local contractors doing business primarily in the affected area.[23]

2. Homeland Security Act

The Department of Homeland Security (DHS) was created by the Homeland Security Act of 2002.[24] The DHS is responsible for managing federal grants to state and local governments to help in preparing for, preventing, and responding to terrorist attacks and other disasters. DHS grants are for port security, critical infrastructure protection, regional and local mass transit systems, equipment and training for first responders, and homeland security. FEMA oversees the distribution of these grants and administers the grants in accordance with its codification of the Common Rule.[25] Additional required construction-contract provisions for FEMA are shown in **Table 16.2.**

---

[19]*See* **Chapter 17.**

[20]Pub. L. No. 93-288, 88 Stat. 143 (1974), as amended (codified at 42 U.S.C. §§ 5121–5208). *See also* 44 C.F.R. §§ 206.31–206.48 (implementing regulations).

[21]*See* 44 C.F.R. § 206.47 (cost-sharing adjustments). The federal share of assistance is not less than 75 percent of the eligible cost of such assistance. 42 U.S.C. § 5170b(b), (c)(4).

[22]*See* 6 U.S.C. §§ 791–797.

[23]42 U.S.C. § 5150; 44 C.F.R. § 206.10.

[24]Pub. L. No. 107-296, 116 Stat. 2135 (2002).

[25]44 C.F.R. Part 13.

Table 16.2   Required Construction Contract Provisions

Contract Provision	HUD	DOT	EPA	FEMA
Changes	*Permissive**	Required	Required	*Permissive*
Differing Site Conditions	*Permissive*	Required	Required	*Permissive*
Suspension of Work	*Permissive*	Required	Required	*Permissive*
Termination for Default	Required	Required	Required	Required
Termination for Convenience	Required	Required	Required	Required
Davis-Bacon	Required	Required	Required	Required
Work Hours and Safety	Required	Required	Required	Required
EEOC	Required	Required	Required	Required

*The Department of Housing and Urban Development (HUD) and the Federal Emergency Management Agency (FEMA) are allowed, but not required, to include clauses approved by the Office of Federal Procurement Policy, including clauses pertaining to changes, remedies, changed conditions, access to records, retention of records, and suspension of work. *See* 24 C.F.R. § 85.36(i) (HUD); 44 C.F.R. § 13.36(i) (FEMA).

D.   Specific Agency Programs

The EPA awards approximately $4 billion—about half its annual budget—in grants to its state, local, tribal, educational, and nonprofit partners.[26] EPA codified the Grants Management Common Rule at 40 C.F.R. Part 31.[27] Several environmental statutes (i.e., the Clean Air Act,[28] the Clean Water Act,[29] and the Safe Drinking Water Act[30]) provide the authority for funding EPA grants. The management of EPA's grants is a cooperative effort involving its Office of Administration and Resources Management's Office of Grants and Debarment, national program managers, regional program managers, and Grants Management Offices (GMOs).[31] In 1987, the Congress phased out EPA's construction grants program for water-treatment projects and replaced it with the Clean Water State Revolving Fund (CWSRF) program.[32] This program provides low-interest loans to communities for the construction of infrastructure projects to control water pollution.

The U.S. Department of Transportation (DOT) annually provides billions of dollars in grants and cooperative agreements. The DOT codified the Common Rule at 49 C.F.R. Part 18. DOT grants are generally made to state and local governments, with a lesser amount going to tribes, universities, and nonprofit organizations.

---

[26]Office of Grants & Debarment, EPA, *Grants Management Plan 2009–2013* (Oct. 2008), available at *www.epa.gov/ogd/EO/finalreport.pdf* (accessed November 12, 2009).

[27]*See* **Section III.B.1** of this chapter.

[28]42 U.S.C. § 7606.

[29]33 U.S.C. § 1368 (Federal Water Pollution Control Act, more commonly known as the Clean Water Act).

[30]42 U.S.C. § 300h-3(e).

[31]*Office of Grants & Debarment, EPA, Grants Mgmt. Plan 2009–2013* (Oct. 2008), available at *www.epa. gov/ogd/EO/finalreport.pdf* (accessed Nov. 12, 2009).

[32]The CWSRF was created by the Clean Water Act Amendments of 1987 to replace the EPA's Construction Grants Program for funding wastewater treatment projects.

These grants typically are used to assist planning, design, and construction of transportation projects, such as highway, transit, and airport improvements.

The U.S. Department of Housing and Urban Development (HUD) is another agency heavily involved in funding grants for construction projects. HUD codified the Common Rule at 24 C.F.R. Part 85. HUD's Office of Departmental Grants Management and Oversight(ODGMO) coordinates the department's grants management and implements policies and procedures relating to grants. HUD provides grants for the construction of affordable housing and other economic development programs.

III. ADMINISTRATION AND INTERPRETATION OF GRANT-FUNDED CONTRACTS

Unlike direct federal contracts, grants are not governed by the procurement statutes, the Federal Acquisition Regulation (FAR),[33] or the Contract Disputes Act of 1978.[34] Instead, grants are governed by the FGCAA, "the federal statutes that authorize their use,"[35] the federal regulations agencies follow in awarding grant funding, and, if applicable, local procurement laws. However, in the context of the federal government's comprehensive effort to preclude unethical contractors from receiving contracts funded, directly or indirectly, by it, it is very likely that many of the business ethics and compliance programs will be incorporated into grant funded programs.[36]Contractors receiving payments funded by federal grants are also subject federal anti-fraud statues, such as the False Claims Act.

---

[33]*See Trauma Serv. Group, Ltd. v. United States,* 33 Fed. Cl. 426 (1995).

[34]*Rick's Mushroom Serv., Inc. v. United States,* 521 F.3d 1338 (Fed. Cir. 2008) (cooperative agreements are not subject to the Contract Disputes Act). *See also Esco Constr. Co.,* AGBCA No. 95-101-1, 95-1 BCA ¶ 27,324; *MontAna Human Rights Comm'n,* HUDBCA No. 90-5305-C8, 91-2 BCA ¶ 23,993; *Craft Wall of Idaho, Inc.,* HUDBCA No. 83-819-C19, 85-1 BCA ¶ 17,808. *See generally* Contract Disputes Act of 1978, 41 U.S.C. §§ 601–609 (1994).

[35]*See* Jeffrey C. Walker, *Enforcing Grants and Cooperative Agreements as Contracts Under the Tucker Act,* 26 Pub. Cont. L.J. 683, 691 (1997).

[36]For example, Section 872 of the Duncan Hunter National Defense Authorization Act of 2009 (Pub. L. 110-417) requires the Office of Management and Budget and the General Services Administration to establish, within one year of the effective date of that legislation (October 14, 2008), an information database on the integrity and performance of contractors that will be available to all federal agencies and grantees. That legislation also directs the adoption of regulations within one year of the effective date of the legislation requiring contractors with agency awards or grant contracts with a total value in excess of $10,000,000 to provide to the federal government detailed information similar to that currently found in FAR § 52.209-5 with certain critical differences. The new reporting requirements, representations, and database will cover a **five year** period, not three years, and will include disclosure of civil judgments "in connection with" the award or performance of a contract or grant with the federal government, default terminations, and the administrative resolution of suspension or debarment proceedings. The phrase "in connection with" is not defined in the Duncan Hunter Act.

A. State and Local Government Procurement Procedures and the Model Procurement Code

During a five-year period in the mid-1970s, the American Bar Association (ABA) Sections of Public Contract Law and State and Local Government Law drafted the Model Procurement Code for State and Local Governments (MPC). The MPC was adopted by the ABA in 1979. The MPC provided a set of model rules designed to provide a common framework for state and local procurement. Since 1979, the MPC has been adopted in full by 16 states and in part by several others.[37]

The ABA revised the MPC in 2000 in an effort to stay current with new technologies and project delivery methods. The 2000 MPC provides:

(1) the statutory principles and policy guidance for managing and controlling the procurement of supplies, services, and construction for public purposes; (2) administrative and judicial remedies for the resolution of controversies relating to public contracts; and (3) a set of ethical standards governing public and private participants in the procurement process.[38]

In 2002, the ABA published the Model Procurement Regulations, which provide practical guidance to state and local governments for implementing the MPC. The ABA has also published the 2007 Model Code for Public Infrastructure Procurement (2007 MC PIP). The 2007 MC PIP is a condensed version of the 2000 MPC. It is self-described as "suitable for more specific use by subunits of state and local government with long-term responsibility for delivery and operation of infrastructure services and facilities."[39] Its purpose is to enhance procurement capabilities and to improve procurement practices within local governments receiving federal financial assistance for design and construction. Jurisdictions adopting these model procedures streamline the federal review of procurements funded by federal grants because these procurement procedures meet the requirements of the Grants Management Common Rule.

If a state or local government has not adopted a procurement procedure such as the MPC, then it must submit a procedure incorporating the procurement standards of the Grants Management Common Rule and obtain the agency's approval before implementing the procedure. The principles of federal procurement law and policy identified in the Common Rule must be included. Approval by the grantor agency is a prerequisite to eligibility for a grant.

---

[37]John B. Miller, *The 2000 ABA Model Procurement Code,* 36 Procurement Law 4 (Fall 2000).

[38]Section of Pub. Contract Law & Section of State & Local Gov't Law, Am. Bar Ass'n, *The 2000 Model Procurement Code for State and Local Governments xi (2000).*

[39]Section of Pub. Contract Law & Section of State & Local Gov't Law, Am. Bar Ass'n, *2007 Model Code for Public Infrastructure Procurement* ii–iii (2007).

B. Mandatory Contract Clauses

When a state procures construction services under a grant, it must ensure that the construction contract includes any clauses required by federal statutes and executive orders and their implementing regulations.[40] **Table 16.2** summarizes the most common required provisions for four of the major agencies that fund construction projects through grants.

1. Grants Management Common Rule Requirements

The EPA's codification of the Grants Management Common Rule is set forth in 40 C.F.R. Part 31. It will be used in this chapter as the model for discussing the general requirements for grantees' administration and use of federal grant funds. Other federal agencies that fund grants for construction projects, including DOT,[41] FEMA,[42] and HUD,[43] impose nearly identical requirements through implementation of the Common Rule.

Grantees use their own procurement procedures based on state law and local ordinances, but they must ensure that their procedures also comply with federal law. Using EPA's Common Rule as an example, grantees' procedures must include these key elements:

- Grantees must apply the same policies and procedures as used for procurements with non-federal funds.[44]
- State contracts that use grant funds must include any clauses required by federal statutes and executive orders.[45]
- Grantees and subgrantees other than states will use their own procurement procedures reflecting applicable state and local laws and regulations and that comply with applicable federal law and the standards set out in 40 C.F.R. § 31.36(b).

In addition to these key elements, the grantees are obligated to adopt and follow procedures to satisfy these requirements:

- Grantees are required to maintain a contract administration system which will ensure that contractors perform in accordance with the contract plans and specifications.[46]
- Grantees are required to have a written code of standards of conduct for their employees in connection with grant-funded contracts.[47] This code must be implemented to avoid conflicts of interest.

---

[40]*See, e.g.,* 40 C.F.R. § 31.36(a) (EPA); 43 C.F.R. § 12.76(a) (Department of the Interior).
[41]49 C.F.R. Part 18.
[42]44 C.F.R. Part 13.
[43]24 C.F.R. Part 85.
[44]40 C.F.R. § 31.36(a).
[45]*Id.*
[46]40 C.F.R. § 31.36(b)(2).
[47]40 C.F.R. § 31.36(b)(3).

- Grantees are encouraged to enter into intergovernmental agreements for procurement or use of common goods and services when they will enhance efficiency.[48] Grantees also are encouraged to use federal surplus or excess property where feasible rather than purchasing new.[49]
- Grantees may make awards only to responsible contractors, giving consideration to contractor integrity, compliance with public policy, record of past performance, and financial and technical resources.[50]
- Grantees are required to maintain sufficient historical records of a grant-funded procurement to enable a later review of the entire contracting process.[51] The rule strictly limits the use of time and materials contracts to circumstances where no other contract is suitable and where the contract includes a ceiling price.[52]
- Grantees are responsible for settlement of all issues arising out of procurements, including source evaluation, protests, disputes, and claims. With regard to protests related to the award of grant funded contracts, grantees are required to have protest procedures in place to handle and resolve disputes. These procedures for disputes, including bid protests, will provide for resolution by the grantee agency.

Procurement practices are a primary focus of the requirements imposed on grantees. These requirements address both the procurement process and the resolution of protests arising out of the procurement process. The underlying federal policy is to ensure full and fair competition. All procurements involving federal grant funds should be conducted using procedures that will provide full and open competition.[53] Practices that restrict competition are banned. Banned practices include imposing unreasonable qualification requirements, specifying a brand-name product rather than allowing an equivalent product to be offered, requiring unnecessary experience and excessive bonding, and allowing conflicts of interest. The use of state or local geographical preferences in evaluating bids or proposals is prohibited, except where federal statutes require or encourage them.[54] Grantees must have written selection procedures for procurements that ensure all solicitations include a clear and accurate description of the requirements for the goods or services to be procured, that identify all requirements that bidders or offerors must fulfill, and that list all factors to be used in evaluating bids or proposals.

One goal of the bid protest requirements is to resolve protests at the grantee level. A protestor must exhaust the available remedies provided by the grantee before filing a protest with the grantor agency. Federal agencies will not substitute their judgment for that of the grantee unless the matter is primarily a federal concern.[55] Even then,

[48]40 C.F.R. § 31.36(b)(5).
[49]40 C.F.R. § 31.36(b)(6).
[50]40 C.F.R. § 31.36(b)(8).
[51]40 C.F.R. § 31.36(b)(9).
[52]40 C.F.R. § 31.36(b)(10).
[53]40 C.F.R. § 31.36(c)(1).
[54]40 C.F.R. § 31.36(c)(2).
[55]40 C.F.R. § 31.36(b)(11).

federal-agency review of a bid protest is limited to (1) violations of federal law, regulations, and 40 C.F.R. § 31.36 and (2) violations of the grantee's protest procedures for failure to review a protest.[56] A bid protest on other grounds will be referred to the grantee for resolution under its procedures.[57]

2. Other Mandatory Clauses

Federal agencies may impose other agency-specific requirements on grantees and sub-grantees in addition to Common Rule requirements. For example, every EPA construction contract for a treatment project funded by a grant under the Clean Water Act must include the *Supplementary General Conditions* set forth in Appendix C-2 to Subpart E of 40 C.F.R. Part 35.

Construction grants awarded by the EPA under the Clean Water Act are also subject to Buy American Act[58] requirements.[59] Contractors must give preference to the use of domestic materials in the construction of water-treatment facilities funded by EPA grants.[60] Domestic construction materials generally must be used if the price is no more than 6 percent higher than that of nondomestic materials.[61] There are limited exceptions if using domestic materials is not in the public interest or if the cost of the domestic materials is unreasonable.[62]

Recipients of highway construction grants from the Department of Transportation generally are required to use competitive bidding on the construction projects.[63] In emergency situations, highway construction contracts are exempt from the competitive bidding requirement.[64] As noted in **Table 16.2,** highway construction contracts funded through DOT grants are required to have standardized contract clauses concerning differing site conditions, suspension of work, and material changes in the scope of work.[65]

Detailed DOT construction contracting procedures are set out in 23 C.F.R. Part 635, Subpart A. This part of the C.F.R. describes the policies, requirements, and procedures for all federally aided highway projects. The prescribed procedures span from the time of authorization to begin construction to the time of final acceptance by the Federal Highway Administration.[66]

State agencies receiving grant funding for highway and other transit projects must include contract clauses enforcing the Buy America requirements mandated by the Federal Highway Administration (FHWA)[67] and Federal Transit Administration.[68]

---

[56]40 C.F.R. § 31.36(b)(12).

[57]40 C.F.R. § 31.36(b)(12)(ii).

[58]41 U.S.C. §§ 10a–10d. *See also* **Chapter 5** for a more detailed discussion of the Buy American Act.

[59]40 C.F.R. § 31.36(c)(5).

[60]*Id.*

[61]40 C.F.R. § 31.36(c)(5)(i).

[62]40 C.F.R. § 31.36(c)(5)(ii).

[63]49 C.F.R. § 18.36(j); *see also* 23 U.S.C. § 112(a).

[64]49 C.F.R. § 18.36(p); *see also* 23 U.S.C. § 112(b).

[65]49 C.F.R. § 18.36(r); *see also* 23 U.S.C. § 112(e).

[66]23 C.F.R. § 635.101.

[67]23 U.S.C. § 313; 23 C.F.R. § 635.410.

[68]49 U.S.C. § 5323(j); 49 C.F.R. Part 661.

These Buy America requirements are distinct from the Buy American Act.[69] The Buy American Act does not apply to grants by its own terms, but some agencies impose its obligations as a condition of funding a grant.[70] Under the FHWA and Federal Transit Administration Buy America requirements, if steel or iron materials are to be used, then *all* manufacturing processes for those materials must take place in the United States. By contrast, the Buy American Act in Part 25 of the FAR requires only that more than 50 percent of the cost of the components of manufactured construction material (such as steel) be mined, produced, or manufactured in the United States.[71] In addition to these Buy American Act provisions, funds provided in grants under the American Recovery and Reinvestment Act of 2009[72] are also subject to separate Buy American requirements as discussed in **Chapter** 17.

C. Anti-Fraud and Criminal Statutes

One of the EPA's mandatory clauses for construction contracts is *Price Reduction for Defective Cost or Pricing Data*.[73] This clause states, in part:

(11) (a) If the owner or EPA determines that any price, including profit, negotiated in connection with this agreement or any cost reimbursable under this agreement was increased by any significant sums because the engineer or any subcontractor furnished incomplete or inaccurate cost or pricing data or data not current as certified in his certification of current cost or pricing data (EPA form 5700-41), then such price, cost, or profit shall be reduced accordingly and the agreement shall be modified in writing to reflect such reduction.

In addition to the remedy included in this clause, a failure to furnish complete, accurate, and current cost or pricing data may expose a contractor to civil liability and criminal penalties for violation of the False Claims Act[74] or the False Statements Act.[75] If a contractor submits a certification of cost or pricing data that is false, incomplete, or not current, and if the grantee uses the contractor's certification as support when seeking a payment under a cost-reimbursement contract or additional grant funds, then the contractor could be subject to civil liability under the False Claims Act, as provided in 31 U.S.C. § 3729. This statute provides, in part:

[A]ny person who—

(A) knowingly presents, or causes to be presented, a false or fraudulent claim for payment or approval;

---

[69]41 U.S.C. §§ 10a–10d. *See also* **Chapter** 5 for a more detailed discussion of the Buy American Act.
[70]*See, e.g., supra* notes 57-61 and accompanying text.
[71]*See* FAR Part 25.
[72]Pub. L. 111-5.
[73]*See* 40 C.F.R. Part 35, Subpart E App. C-2.
[74]31 U.S.C. §§ 3729-3733; 18 U.S.C. § 287.
[75]18 U.S.C. § 1001.

(B) knowingly makes, uses, or causes to be made or used, a false record or statement material to a false or fraudulent claim;

* * *

is liable to the United States Government for a civil penalty of not less than $5,000 and not more than $10,000 . . . plus 3 times the amount of damages [that] the Government sustains because of the act of that person.[76]

With respect to the False Claims Act, a "claim" is any request or demand for money or property that is made to a contractor, grantee, or other recipient, if the money or property is to be spent or used to advance a federal government program or interest and if the federal government either provides, has provided, or will reimburse any portion of the money or property.[77] "Material" means having a natural tendency to influence or be capable of influencing the payment or receipt of money or property.[78] A contractor or grantee acts "knowingly" if it has actual knowledge that the information is false, acts in deliberate ignorance of the truth or falsity of the information, or acts in reckless disregard of the truth or falsity of the information. No specific intent of proof to defraud is required,[79] but innocent mistakes, mere negligence, or even gross negligence (without more) are insufficient to establish liability.[80] Although the 50 sovereign states and their agencies are immune from liability under the False Claims Act as grantees,[81] municipal corporations are not.[82]

Contractors also face potential criminal liability under the False Statements Act.

Whoever makes or presents to any person or officer in the civil, military, or naval service of the United States, or to any department or agency thereof, any claim upon or against the United States, or any department or agency thereof, knowing such claim to be false, fictitious, or fraudulent, shall be imprisoned no more than five years and shall be subject to a fine in the amount provided in this title.

Contractors on grant-funded projects also face potential criminal liability under 18 U.S.C. § 666, which prohibits theft from programs receiving federal funds or bribery related to them. In *United States v. Vitillo*,[83] a construction manager contracting

---

[76]31 U.S.C. § 3729(a)(1). For a more detailed discussion of the False Claims Act, see **Chapter 1.**
[77]31 U.S.C. § 3729(b)(2).
[78]31 U.S.C. § 3729(b)(4).
[79]31 U.S.C. § 3729(b)(1).
[80]*United States ex rel. Lamers v. City of Green Bay*, 168 F.3d 1013, 1018 (7th Cir. 1999)); *United States ex rel. Ervin & Assocs. v. Hamilton Sec. Group*, 298 F. Supp. 2d 91, 100–01 (D.D.C. 2004) (citing *United States v. Krizek*, 111 F.3d 934, 941 (D.C. Cir. 1997)).
[81]*Vermont Agency of Nat. Resources v. United States ex rel. Stevens*, 529 U.S. 765 (2000).
[82]*Cook County, Ill. v. United States ex rel. Chandler*, 538 U.S. 119 (2003).
[83]490 F.3d 314 (3d Cir. 2007).

with a regional airport authority that received funding from the Federal Aviation Administration (FAA) was convicted of violating this statute, and the principal of the corporation was sentenced to 36 months imprisonment. On appeal, the court held that the construction manager, even if acting as an independent contractor, was nev ertheless an *agent* of the local government agency receiving federal financial assistance. This decision illustrates the significant reach of federal anti-fraud statutes into the administration of local contracts funded by federal dollars.

These provisions clearly affect the administration of state and local public contracts funded by a federal grant. Grantee agencies and their contractors must be aware of the additional requirements, restrictions, and remedies affecting contracts funded by federal money.

D. Interpretation of Contracts

Contracts with state and local governments or other grantees usually are interpreted under state law. Under some circumstances, however, federal law may be applied, particularly when a clause required by the grantor agency is construed.

This was illustrated in *Brinderson Corp. v. Hampton Roads Sanitation District.*[84] The contract at issue was for the construction of a wastewater treatment plant in Hampton Roads, Virginia. Brinderson asserted a claim for an alleged differing site condition.

The project was partially funded by a grant from EPA. EPA's mandatory clauses for construction contracts were included in the contract.[85] The construction contract included the mandatory Differing Site Conditions provision. That clause obligated the contractor to give prompt notice when it encountered an alleged differing site condition.

Brinderson's claim was denied by the grantee because it had failed to give prompt, written notice. Under Virginia law notice provisions are strictly construed. After the rejection of its claim by the public agency, Brinderson filed suit in federal court. The federal court applied Virginia state law and ruled against Brinderson.

Brinderson appealed the decision to the United States Court of Appeals for the Fourth Circuit. This court reversed the district court's decision on the ground that the Differing Site Conditions clause was required by federal regulation, and, therefore, *federal* common law should be applied in construing it. Under federal common law, the owner's actual early notice of the differing site condition was sufficient, and the claim would not be denied for lack of written notice. This principle of constructive notice has evolved over many years of construing the Differing Site Conditions clauses in the FAR and other regulations. Therefore, it would be anomalous not to apply the federal common law to the clause in Brinderson's contract. To hold otherwise would introduce uncertainty and additional risk in the procurement process.

---

[84]825 F.2d 41 (4th Cir. 1987).
[85]*See* 40 C.F.R. Part 35, Subpart E App. C-2.

➤ LESSONS LEARNED AND ISSUES TO CONSIDER

- The purpose of a *procurement contract* is the federal government's direct acquisition of good or services.
- The purpose of a *federally funded assistance agreement* is to provide financial assistance to the recipient to acquire goods or services.
- State and local governments rely on *grants* and other forms of federal assistance to construct many public works programs.
- A grant or cooperative agreement may be used only when the *principal purpose* of a transaction is to accomplish a public purpose of support or stimulation authorized by federal statute.
- An agency must have specific *statutory* authority for the use of grants.
- The Federal Grant and Cooperative Agreement Act differentiates grants and procurement contracts and *prohibits* federal agencies from using the two *interchangeably*.
- The *Grants Management Common Rule* standardized the administrative requirements that federal agencies impose on state and local governments receiving grants.
- The ABA Model Procurement Code lays out a set of model rules designed to provide a *common framework* for state and local procurement that meets the requirements of both the Common Rule and the FGCAA.
- The Grants Management Common Rule allows grantees to use *local procurement procedures*—reflecting state and local laws and regulations—provided that the procurements conform to applicable federal law and have prior approval by the grantor agency.
- Grantees must comply with the laws and regulations applicable to a particular grant or federal assistance program. Application of grant terms that are based on *clauses or policies drawn from federal contracts* may result in an interpretation of a locally awarded and administered contract that reflects both local and federal principles.

17

THE AMERICAN RECOVERY AND REINVESTMENT ACT AND GOVERNMENT CONSTRUCTION PROJECTS

I. OVERVIEW

A. Purpose of Legislation

Congress passed the American Recovery and Reinvestment Act of 2009 (ARRA or Recovery Act)[1] as part of the federal government's financial and economic stimulus program to mitigate the severe economic downturn following the major, worldwide financial crisis in 2008–2009. That crisis and downturn adversely affected every segment of the construction industry in the United States. This complex and controversial legislation seeks to achieve multiple objectives[2] including:

- Creation or preservation of 3.5 million jobs
- Revive and expand the renewable energy industry
- Fund computerization of Americans' health records
- Fund weatherization of federal building spaces and over 1 million homes
- Increase the affordability of college education
- Invest $150 billion in new infrastructure projects
- Provide tax credits for many American households
- Provide $750 million to the Small Business Administration to fund loans to small business concerns

---

[1]Pub. L. No. 111-5. This legislation is sometimes called the Stimulus Bill. In this chapter, it will be referred to as ARRA or "Recovery Act.

[2]Goals and objectives are set forth on the federal government's Web site (*www.recovery.gov*), (accessed November 9, 2009) which was created to explain and promote the benefits of this legislation.

• Provide "unprecedented levels" of transparency, oversight, and accountability

Many of ARRA's elements will have minimal direct effect on government construction contractors. However, the infrastructure and related construction project funding are part of the effort to create and preserve jobs in this country and provide substantial opportunities for new work for construction contractors.[3] In addition, ARRA's transparency, oversight, and accountability objectives will create additional reporting and business compliance obligations for contractors and their first-tier subcontractors performing ARRA-funded projects directly for the federal government or under grants and assistance programs funded by the federal government.

This chapter seeks to identify the major aspects of ARRA that affect construction contractors and describe how the requirements of ARRA differ from and modify the requirements of traditionally funded federal contracts and grant programs.

B. Application to New and Existing Contracts

ARRA mandates that new projects funded, in whole or in part, by Recovery Act funds contain contract provisions addressing:

• Registration and special reporting requirements
• Preferences for products manufactured in America
• Business ethics and compliance
• Whistleblower protections
• Broader Davis-Bacon wage rate application to ARRA-funded projects

In general, the use of Recovery Act funds triggers the application of these contract provisions and will be contained in or referenced in the solicitation issued by the agency (federal, state, or local). For example, under a direct federal government contract FAR § 4.1501 requires the contracting officer to indicate that the contract action is made under the Recovery Act and which item(s) or service(s) are ARRA funded.[4] For contractors working for state and local entities, policy guidance and requirements established by the Office of Management and Budget (OMB) and the federal agencies providing the funds to the state and local public entities will reflect ARRA's requirements. With this notification, a potential contractor can evaluate the effect of these requirements on its operations, its cost of performance and address those requirements in the lower-tier subcontracts or purchase orders if it is awarded that contract.

A critical issue for construction contractors is how the Recovery Act and its various unique requirements apply to contracts awarded prior to March 31, 2009. First, contractors need to appreciate that some of ARRA requirements imposed

---

[3]Although the government's Web site (*www.recovery.gov*) seeks to provide information on such opportunities, many contractors may find a private Web site (*www.recovery.org*) to be a more useful and easily navigated source of information on ARRA-funded construction projects in a specific geographic area.

[4]The various FAR clauses were issued as *interim rules* with a March 31, 2009, effective date. Contracting officers were instructed to include these clauses in solicitations and contracts awarded on or after that date.

by Congress—for example, Buy American requirements, reporting obligations, and so on—are quite different from those found in non–ARRA-funded contracts.[5] Management of ARRA's different requirements and standards may be especially challenging as federal agencies seek to use ARRA funds to implement changes to projects that were not initially funded under ARRA. For example, a contractor may need to address and manage the fact that similar, if not identical, materials or products being incorporated into a construction project would be subject to very different Buy American tests depending on whether ARRA funds are used to pay for that part of the project. Some subcontractors may resist the various reporting requirements.

Second, it is very probable that federal agencies will use Recovery Act monies to fund modifications to existing construction contracts. An initial question for any contractor is whether the federal agency has the authority under the Changes clause[6] to unilaterally modify the contract to add the new FAR clauses implementing the Recovery Act. Under federal procurement law, the scope of permissible contract modifications is limited by the provisions of the FAR Changes clause.[7] As a general rule, the government does not have the power to *unilaterally* modify the general provisions or terms and conditions of the contract that are set forth in the FAR or in the agency supplements to the FAR.[8] Application of this principle means that the agency and the contractor would have to *mutually agree to a bilateral change* to an existing government contract to add the various FAR clauses implementing the Recovery Act.

The FAR Councils expressly recognized this principle when they issued the interim FAR regulations (rules) on March 31, 2009. The guidance comments regarding each of these new FAR clauses contained this statement:

Applicability Date: The rule applies to solicitations issued and contracts awarded on or after the effective date [March 31, 2009] of this rule. Contracting officers shall modify, *on a bilateral basis,* in accordance with FAR 1.108(d)(3) existing contracts to include the FAR clause for future orders, if Recovery Act funds will be used. In the event that the contractor refuses to accept such a modification, the contractor will not be eligible for receipt of Recovery Act funds.[9]

FAR § 1.108(d)(3) authorizes modifications to existing contracts with consideration.[10] In general, consideration is a matter of negotiation and mutual agreement by the parties to the contract. Although many government contractors may elect to

---

[5]*See* **Chapter 5** for a review of the Buy American Act as it applies to non-ARRA construction projects.
[6]FAR § 52.243-4.
[7]*See* **Chapter 8** and FAR § 52.243-4(a).
[8]*See B.F. Carvin Constr. Co.,* VABCA No. 2224, 92-1 BCA ¶ 24,481.
[9]*See* 74 Fed. Reg. 14623, 14633, 14639, 14646 (emphasis added).
[10]The analysis for federal government projects reflects the scope of permissible unilateral modifications under the FAR Changes clause. For state and local contracts, as well as lower-tier subcontracts, the analysis may be very different reflecting the provisions of the applicable Changes clause as well as state law governing public contracts and subcontracts between private entities.

accept a Recovery Act–funded change to the contract, every contractor should first consider:

- The added expense and administrative burden for the contractor associated with the various reporting and compliance obligations
- Consideration of that cost and expense in the pricing of the modification
- The potential for confusion associated with the different Buy American tests for very similar materials and products on a construction project
- The willingness of first-tier subcontractors[11] to accept like modifications, as many of the new FAR provisions contain flow-down requirements
- The value and potential profit in the change order
- The importance of the modification to the federal agency and that project
- The potential effect of an arbitrary rejection of a Recovery Act–funded modification on a contractor's Past Performance Evaluation[12]

As noted, Congress appropriated approximately $150 billion for construction spending and provided ARRA funding for both direct federal contracts and federal grants.[13] This funding reflects one-time supplemental appropriations over and above the government's customary annual appropriations. The ARRA's spending allocations are likely to fund awards for new construction projects and also modifications to existing non–ARRA-funded projects through fiscal year 2013,[14] with actual outlays expected to extend into 2019.[15]

The policy guidance issued by the OMB and government-wide regulations implementing the Recovery Act impose supplementary requirements on awardees of direct contracts and on recipients (and subawardees) of grants funded by the Act. (The term "grant," as used in this chapter, includes cooperative agreements. Federal grants and cooperative agreements for construction projects are addressed in **Chapter 16.**)

Part 176 was added to Title 2 of the Code of Federal Regulations (C.F.R.) to provide guidance and standard award terms to implement selected portions of the ARRA related to *grants* funded under the Act's authority. Federal agencies must continue to use their standard award terms and conditions on award notices for *grants* unless they conflict with the requirements of the ARRA. Recipients and subawardees of ARRA grant funds also must continue to comply with the granting agency's adoption of the Grants Management Common Rule and 2 C.F.R. Part 215.[16]

---

[11]First-tier subcontractors include vendors and suppliers in a direct contractual relationship with the prime contractor. *See* definition of a subcontract at FAR § 44.101.

[12]*See* **Chapter 3** for a discussion of Past Performance Evaluations.

[13]*See www.recovery.gov.*

[14]American Recovery and Reinvestment Act of 2009, Pub. L. No. 111-5, Title X, 123 Stat. 115, 191–93 (making funds available for "Military Construction" until September 30, 2013).

[15]Cong. Budget Office, Estimated Cost of the Conference Agreement for H.R. 1, The American Recovery and Reinvestment Act of 2009 (Feb. 13, 2009), available at *www.cbo.gov/ftpdocs/99xx/doc9989/hr1conference.pdf.*

[16]*See* Office of Mgmt. & Budget, Updated Implementing Guidance for the American Recovery and Reinvestment Act of 2009, M-09-15, §§ 5.5, 5.9, at 50–51 (Apr. 3, 2009), available at *www.whitehouse.gov/omb/assets/memoranda_fy2009/m09-15.pdf.* For a discussion of the Grants Management Common Rule and 2 C.F.R. Part 215, *see* **Chapter 16.**

Contractors working on ARRA-funded projects face special reporting responsibilities, additional Buy American requirements, and new accountability and ethics rules that are not otherwise applicable to other federally funded projects. These issues arise whether the work is Recovery Act funded under a direct federal contract or a grant, but the contractor's specific duties may vary based on that distinction. Where appropriate, those differences are noted in this chapter.

II. REGISTRATION AND REPORTING REQUIREMENTS

ARRA Section 1512,[17] its implementing regulations, and related OMB guidance require extensive reporting of a wide range of information by certain recipients of federal funds appropriated by the Recovery Act. Recipients assume different reporting obligations depending on the type of the award (direct contract or grant), the level at which they receive funds (prime, first tier, etc.), and the amount of annual revenue they receive from federal sources.

The Recovery Act defines "recipient" as "any entity that receives Recovery Act funds directly from the Federal Government (including Recovery Act funds received through grant, cooperative agreement, loan, or contract) other than an individual, and includes a state that receives Recovery Act funds."[18] The OMB guidance indicates that the reporting requirements "apply to the prime non-Federal recipients of Federal funding" and that "[t]he prime recipient (such a state government) is responsible for reporting on [its] use of funds as well as any sub-awards (i.e., any sub-grants, subcontracts, etc.) [it] make[s]."[19] Prime recipients of *grant* funding (states or local governments), however, may choose to delegate certain reporting requirements to subrecipients (i.e., the construction contractor). Prime recipients of grants can require that subrecipients report the Federal Funding Accountability and Transparency Act (FFATA)[20] data elements[21] required by ARRA section 1512(c)(4).[22] By contrast,

---

[17]123 Stat. at 287–88.

[18]ARRA, Pub. L. No. 111-5, § 1512(b), 123 Stat. 115, 287 (2009). *See also* Office of Mgmt. & Budget, Updated Implementing Guidance for the American Recovery and Reinvestment Act of 2009, M-09-15, § 2.10, at 20, available at *www.whitehouse.gov/omb/assets/memoranda_fy2009/m09-15.pdf.*(accessed November 9, 2009)

[19]Office of Mgmt. & Budget, Updated Implementing Guidance for the American Recovery and Reinvestment Act of 2009, M-09-15, § 2.10, at 20–21, available at *www.whitehouse.gov/omb/assets/memoranda_fy2009/m09-15.pdf* ("In limited circumstances, recovery funds will go from a federal agency to a state, and then to a local government or other local organization. In that case, the current reporting model will not track funds to subsequent recipients beyond these local governments or other organizations. OMB plans to expand the reporting model in the future to also obtain this information, once the system capabilities and processes have been established.")

[20]Pub. L. No. 109-282, 120 Stat. 1186 (2006).

[21]FFATA data elements: Dunn & Bradstreet Universal Numbering System (DUNS) number; Central Contractor Registration (CCR) information; type of entity; amount awarded; amount received; subaward date; subaward period; place of performance; area of benefit; and the names and total compensation of each of the subrecipient's five most highly compensated officers.

[22]Office of Mgmt. & Budget, Implementing Guidance for Reports on Use of Funds Pursuant to the American Recovery and Reinvestment Act of 2009, M-09-21, §§ 2.2, 2.3, at 6–11 (June 22, 2009), available at *www.whitehouse.gov/omb/assets/memoranda_fy2009/m09-21.pdf.*

prime contractors with direct federal contracts must require certain first-tier sub-contractors to provide information to the contractor for the contractor's report rather than delegate reporting to subcontractors.[23]

The information reported will be made public on the *recovery.gov* Web site.[24] Most of the required information bears some relationship to the performance of the work and includes:

- Identification of the government contract
- Amount of Recovery Act funds invoiced by the contractor for the reporting period (typically a calendar quarter)
- List of significant services performed and expected contract results
- Progress report on the portion of the project funded by the Recovery Act
- Description of the types of jobs created and retained
- Description of employment impact (number of jobs created or retained). A job cannot be reported as both created and retained.

Reporting requirements also affect first-tier subcontractors.[25] The prime contractor is obligated to obtain and report detailed information on most first-tier subcontractors in excess of $25,000[26] including:

- Dunn & Bradstreet Universal Numbering System (DUNS) number for the sub-contractor and its parent company, if any
- Name of subcontractor, its physical address, and congressional district
- Subcontractor's primary performance address and applicable congressional district
- Subcontractor's North American Industry Classification System (NAICS) code
- Government agency funding work
- Subcontract number assigned by the prime contractor, description of work in the subcontract, dollar value of the subcontract

For infrastructure investments made by state or local governments through federal grants, the recipient government also must report the purpose, total cost, and rationale of the agency for funding the investment. Recipients will be "required to

---

[23]FAR § 52.204-11(d)(10).

[24]*See* ARRA, Pub. L. No. 111-5, § 1512(d), 123 Stat. 115, 288 (2009).

[25]"First-tier subcontractors" includes vendors and suppliers in a direct contractual relationship with the prime contractor. *See* definition of a subcontract at FAR § 44.101.

[26]*See* FAR § 52.204-11(d)(10). Simplified reporting requirements are applicable if the subcontract is less than $25,000 and the subcontractor's gross income for the prior year was less than $300,000. *See* FAR § 52.204-11(d)(9).

report on a number of data elements. Detailed information on the data elements . . . will be provided by Federal agencies to funding recipient[s] in the standard terms and conditions of individual award agreements."[27] To some extent, these economic and progress reporting requirements supplement or duplicate standard reporting (i.e., progress reporting) on a construction project. However, if a Recovery Act–funded change order is performed, it appears that the prime contractor will need to track and report data on that individual modification if the clause at FAR § 52.204-11 is incorporated into the contract.

In addition to the progress and economic data reports, the FAR clause[28] setting forth the Recovery Act data reporting requirements also implement special data collection provisions of the Federal Funding Accountability and Transparency Act of 2006, as amended.[29] If certain thresholds are met, these data collection provisions seek information on the total compensation of the contractor's and its first-tier subcontractors' "five most highly compensated officers" for the calendar year in which the contract was awarded unless the public has access to the same information from certain public filings with the Securities and Exchange Commission.[30] The monetary thresholds triggering these reporting requirements are:

- Receipt by the contractor (first-tier subcontractor) in the fiscal year prior to the award of the contract of $25 million or more in annual gross revenues from federal contracts, loans, grants (and subgrants) and cooperative agreements.
- Receipt by the contractor (first-tier subcontractor) of 80 percent or more of its annual gross revenues in that fiscal year from those sources.

The prime contractor is obligated to obtain similar data from any first-tier subcontractor that also meets these monetary thresholds and receives a subcontract in excess of $25,000 on a project funded, in whole or in part, by ARRA monies.[31] Similarly situated grant recipients and subrecipients must do the same.[32]

---

[27]Office of Mgmt. & Budget, Updated Implementing Guidance for the American Recovery and Reinvestment Act of 2009, M-09-15, § 2.10, at 21; available at *www.whitehouse.gov/omb/assets/memoranda_fy2009/m09-15.pdf*. For detailed lists of the required data elements, *see* FAR § 52.204-11(d) (direct contracts); Office of Mgmt. & Budget, Implementing Guidance for Reports on Use of Funds Pursuant to the American Recovery and Reinvestment Act of 2009, M-09-21, §§ 2.2, 2.3, at 6–11 (June 22, 2009), available at *www.whitehouse.gov/omb/assets/memoranda_fy2009/m09-21.pdf* (grants).

[28]FAR § 52.204-11.

[29]Pub. L. 109-282.

[30]2 C.F.R. § 176.90.

[31]Reporting of compensation data is not required for small firms as defined in FAR § 52.204-11(d)(9).

[32]Office of Mgmt. & Budget, Implementing Guidance for Reports on Use of Funds Pursuant to the American Recovery and Reinvestment Act of 2009, M-09-21, § 2.3, at 8–11 (June 22, 2009), available at *www.whitehouse.gov/omb/assets/memoranda_fy2009/m09-21.pdf*.

Reports are due no later than 10 calendar days after the end of each calendar quarter in which the recipient receives funds.[33] Reporting deadlines are not linked or related to an individual project's schedule. Since the 10-day window to complete reporting for each calendar quarter is narrow, contractors should implement documentation procedures with these obligations in mind.

Contractors will report the required information using the data collection tools available at *www.FederalReporting.gov*.[34] Recipients of grants, their first-tier subrecipients, and direct government contractors must maintain current registrations in the Central Contractor Registration (CCR). The CCR Web site provides a User's Guide that provides the details on registering as a government contractor. A DUNS number is a requirement of that registration.[35]

Recipients must track ARRA funds separately for each project or portion of a project financed by the Act to fulfill public reporting obligations. The contracting officer must structure contract awards to allow for separate tracking. For example, the contracting officer may award separate contracts or establish contract line item number structures to mitigate commingling of funds.[36]

III. BUY AMERICAN REQUIREMENTS

Recipients of ARRA funds must comply with Buy American requirements specific to the statute. ARRA Section 1605 prohibits the use of funds appropriated or otherwise made available by the Recovery Act for "construction, alteration, maintenance, or repair of a public building or public work unless all of the iron, steel, and manufactured goods used in the project are produced in the United States."[37] This prohibition is in addition to, and distinct from, the Buy American Act, which is discussed in **Chapter 5.** The regulations implementing the Recovery Act reiterate the applicability of either the ARRA Buy American provisions or the traditional Buy American Act requirements to *direct federal contracts*.[38] The Buy American Act does not apply to *grants* by its own terms, but some agencies impose its obligations

---

[33]ARRA, Pub. L. No. 111-5, § 1512(c), 123 Stat. 115, 287 (2009); FAR § 52.204-11(c) (direct contracts); 2 C.F.R. § 176.50(c) (grants) (reporting requirements related to compensation may be annual reports). *See* 74 Fed. Reg. 14643.

[34]FAR § 52.204-11(d) (direct contracts); 2 C.F.R. § 176.50(d) (grants). The Federal Reporting.gov website indicates that it will be opened to receive data on October 1, 2009. *See www.federalreporting.gov* (last visited on August 18, 2009).

[35]2 C.F.R. § 176.50(c) (grants).

[36]FAR § 4.1501(b) (direct contracts); 2 C.F.R. § 176.20(b) (grants). *See also* Office of Mgmt. & Budget, Updated Implementing Guidance for the American Recovery and Reinvestment Act of 2009, M-09-15, § 4.3, at 38, available at *www.whitehouse.gov/omb/assets/memoranda_fy2009/m09-15.pdf* ("Agencies in some cases may need to use Recovery Act funds in conjunction with other funds to complete projects. They may do so, but they must separately track and report the use of Recovery Act funds for these projects.").

[37]ARRA, Pub. L. No. 111-5, § 1605, 123 Stat. 115, 303 (2009). *See also* 2 C.F.R. §§ 176.60-176.170 (grants).

[38]Buy American Act requirements concerning preferences for *unmanufactured* construction material in *direct federal contracts* are reiterated in the FAR provisions and mandatory clauses related to the ARRA. *See, e.g.,* FAR §§ 25.602(b), 25.604(c)(2), 25.605(a)(2), 52.225-21(b)(1)(ii), 52.225-22(c)(1)(ii), 52.225-23(b)(1)(ii), 52.225-24(c)(1)(ii).

as a condition of funding a grant.[39] Agencies must include specific provisions in solicitations concerning projects funded by the ARRA to implement its distinct Buy American requirements.[40]

A. International Trade Agreements

The ARRA requires that its Buy American requirements be applied consistent with the U.S. obligations under international agreements.[41] The U.S. trade agreements that require equal treatment of iron, steel, and manufactured goods produced in certain countries *do apply*, therefore, to direct government contracts for construction with an estimated value of $7.433 million or more that are funded by the ARRA.[42] Specific applicability of individual trade agreements is governed by FAR Subpart 25.4, except that Caribbean Basin countries are not included as *designated countries* with respect to projects funded by the ARRA.

Applying international trade agreements to procurements made by state and local governments or other nonfederal entities using *grant* funds received through the ARRA is more complicated. States and other affected nonfederal entities are required to abide by their separate, individual obligations to other countries under trade agreements when making subawards of grants funded by ARRA.[43] Not every state has undertaken these international obligations, and those that have typically reserve certain exclusions on their commitment.[44] For example, many states exclude construction-grade steel. The Buy American requirements for state and local government projects funded by ARRA grants, therefore, vary from state to state and even from project to project within individual states.

B. Defining the Regulated Materials

Unlike the Buy American Act, the ARRA Buy American requirements do not require an analysis of the origin of components or subcomponents of *manufactured goods*. Instead, it simply demands that the manufacturing process that produces the construction material occur in the United States.[45] The prohibition does not apply, therefore, to iron and steel used as *components* of manufactured goods.[46] *Manufactured goods* (also called *manufactured construction material*) are goods brought to the construction site for incorporation into the work that have been either (1) processed into a specific form and shape or (2) combined with other raw material to create a

---

[39]*See, e.g.,* **Chapter 16.**

[40]FAR § 25.1102(e); *see also* FAR §§ 52.225-22, 52.225-24 (direct contracts); 2 C.F.R. §§ 176.150, 176.170 (grants).

[41]ARRA, Pub. L. No. 111-5, § 1605(d), 123 Stat. 115, 303 (2009). *See also* FAR § 52.603(c) (direct contracts); 2 C.F.R. § 176.90 (grants).

[42]FAR § 25.603(c).

[43]2 C.F.R. § 176.90.

[44]*See* 2 C.F.R. Part 176, Subpart B app. (chart titled "U.S. States, Other Sub-Federal Entities, and Other Entities Subject to U.S. Obligations Under International Agreements").

[45]FAR § 25.602(a)(2)(ii) (direct contracts); 2 C.F.R. § 176.70(a)(2)(ii) (grants).

[46]FAR § 25.602(a)(2)(i) (direct contracts); 2 C.F.R. § 176.70(a)(2)(i) (grants).

material that has different properties than the properties of the individual raw materials.[47] With regard to iron and steel used as construction material (as opposed to being used as a component of a manufactured good), all manufacturing processes must take place in the United States except metallurgical processes involving refinement of steel additives.[48]

C. Exceptions

There are three exceptions to the ARRA's Buy American requirements: (1) the iron, steel, or relevant manufactured good is not mined, produced, or manufactured in the United States in sufficient and reasonably available commercial quantities of satisfactory quality; (2) the cost of domestic iron, steel, or manufactured goods will increase the cost *of the overall project* by 25 percent or more; or (3) the restrictions would be inconsistent with the public interest.[49] An offeror for a direct federal contract (i.e., a contractor) or a grant applicant/recipient (*e.g.*, a state or local government) may request a determination that one of these exceptions should be made. This request should typically be made to the contracting officer (for a direct federal contract) or award official (for a grant) before the award.[50] Contractors may submit alternative offers or proposals that include only domestic materials in order to avoid being rejected simply because an exception does not apply.[51] If an exception is made, the federal agency must publish a detailed justification of the waiver in the Federal Register within two weeks of the determination.[52] This duty is not imposed by the Buy American Act.

If agency's authorized representative[53] determines that the cost of domestic iron, steel, and manufactured goods will increase the cost *of the overall project* by 25 percent or more, then that person must apply an evaluation factor to determine whether an offer that includes foreign iron, steel, or manufactured goods will be accepted. The contracting officer or award official must compute a total evaluated price for the offer by increasing the total offered price by 25 percent. For direct contracts, if the offer also includes foreign *unmanufactured* construction materials, then 6 percent of the cost of the foreign unmanufactured materials is added to the previous sum. This total evaluated price is then compared against offers including only domestic materials.[54] Although the Buy American Act does not apply to grants by its own terms, agencies may include its requirements as terms of the grant or as a supplement to the ARRA requirements for grants.[55] These evaluation factors demonstrate the very strong preference for domestic materials in construction projects funded by the ARRA, which is stronger than that of the traditional Buy American Act.

---

[47]FAR § 25.601 (direct contracts); 2 C.F.R. §§ 176.140(a)(1); 176.160(a) (grants).
[48]FAR § 25.602(a)(2)(i) (direct contracts); 2 C.F.R. § 176.70(a)(2)(i) (grants).
[49]FAR §§ 25.603(a), 25.604(c)(1) (direct contracts); 2 C.F.R. 176.80(a) (grants).
[50]FAR §§ 25.604, 25.605, 25.606 (direct contracts); 2 C.F.R. §§ 176.100, 176.110, 176.120 (grants).
[51]FAR § 25.605(c) (direct contracts); 2 C.F.R. § 176.110(b) (grants).
[52]FAR § 25.603(b)(2) (direct contracts); 2 C.F.R. § 176.80(b)(2) (grants).
[53]Typically the contracting officer.
[54]FAR § 25.605.
[55]2 C.F.R. § 176.110.

Agencies have some discretion in the application of ARRA's Buy American requirements. EPA, for example, has published a notice of a nationwide waiver of the ARRA's Buy American requirements for "*de minimis* incidental components" of water infrastructure projects financed through the Clean Water State Revolving Fund (CWSRF) or the Drinking Water State Revolving Fund (DWSRF) using ARRA funds.[56] The EPA issued the waiver under the public-interest exception. This particular exception applies to iron, steel, and manufactured goods that comprise *in total* no more than 5 percent of the total cost of the materials used in and incorporated into a project. The rationale for the waiver is that devoting EPA and recipient resources to analyzing waiver requests for literally thousands of miscellaneous, generally low-cost components (such as nuts, bolts, other fasteners, tubing, gaskets, etc.)—far out of proportion to the percentage of total project materials cost they represent—would create an obstacle to meeting the ARRA's program-specific deadline for CWSRF/DWSRF projects to be under contract within one year of the ARRA's enactment.

D. Noncompliance

If a violation of the ARRA's Buy American requirements or the Buy American Act is alleged and fraud is not suspected, the contracting officer or award official must notify the contractor or the grant recipient of the apparent unauthorized use of foreign construction material and must request a reply.[57] The reply must include proposed corrective action. If a contractor, recipient, or subrecipient has used unauthorized foreign construction material, the contracting officer must take one or more of the these actions:[58]

(1) Determine whether an exception applies and modify the contract to subtract the appropriate price penalty;

(2) Consider requiring removal and replacement of the unauthorized foreign construction material;

(3) Terminate the contract for default, or suspend or terminate the grant;

(4) Prepare and forward a report to the agency suspension or debarment official; or

(5) If the noncompliance appears to be fraudulent, refer the matter to the agency officer responsible for criminal investigation.

E. Modification of Existing Non-ARRA Contracts

As discussed, it is very probable that agencies will use Recovery Act monies to fund modifications to a project that is not otherwise subject to ARRA. In addition to the various ARRA reporting requirements, contractors should consider that a product, especially a steel product, could be considered as a domestic product under one of the Buy American standards and foreign under the other. If an agency proposes to use

[56]74 Fed. Reg. 39,959 (Aug. 10, 2009).
[57]FAR § 25.607(b) (direct contracts); 2 C.F.R. § 176.130(b) (grants).
[58]FAR § 25.607(c) (direct contracts); 2 C.F.R. § 176.130(c) (grants).

Recovery Act monies to fund a change order to a non-ARRA contract, contractors should attempt to identify and resolve any potential Buy American compliance issues during the negotiation of and before accepting that bilateral change.

For example, if the non-ARRA contract involves renovation work, Recovery Act funds might be used to pay for the renovation of additional floors or parts of a floor. Although the basic design and specifications for the added space might not change, the Buy American tests for products manufactured from steel are different with resulting potential of the same product being both domestic and foreign depending on the applicable Buy American standard. Given that potential scenario, a contractor should consider:

- The ability of its subcontractors and vendors at all tiers to readily manage dual requirements
- The ability of the project's quality control staff to readily distinguish materials and goods for the purpose of the varying Buy American standards
- Subcontractor and vendor willingness to accept the application of the two standards
- The willingness of the contracting officer to apply one of the exceptions to the ARRA Buy American requirements to assist in managing the modification without confusion, unnecessary expense, or unintended noncompliance

The time to address these topics or questions is when the modification is initially proposed, not after the discovery of the delivery or installation of a noncompliant product. The combination of two sets of Buy American requirements can be confusing for the contractor, its subcontractors, and vendors at all tiers. Many times the Buy American provisions are incorporated by reference and can be overlooked. Contractors should consider adding an express provision to their subcontracts and purchase orders alerting the lower-tier firms to the existence of varying Buy American standards and tests and requiring first-tier subcontractors and vendors to *verify* that all products furnished by or on behalf of those firms comply with the applicable Buy American provision.

IV.   ACCOUNTABILITY AND ETHICS

A.   Oversight: Government Access to Records and Employees

ARRA Section 1515 authorizes any representative of an agency's inspector general to inspect any records of any contractor, grantee, subcontractor, or subgrantee and to interview any officer or employee of any contractor, grantee, or subgrantee (but not subcontractor) with respect to each contract or grant awarded using ARRA funds.[59] Similarly, ARRA Section 902 permits the Comptroller General and Government Accountability Office (GAO) representatives to inspect any records and to interview

---

[59]ARRA, Pub. L. No. 111-5, § 1515, 123 Stat. 115, 289 (2009). *See also* ARRA, § 1514, 123 Stat. at 289 (requiring agency inspector general review of concerns raised by the public).

any officer or employee of a contractor or subcontractor (but not a grantee or sub-grantee) with respect to each contract or grant awarded using ARRA funds.[60] Neither the Act nor the implementing regulations require the government to provide any advance notice of these interviews or record inspections.

B. Duty to Report Misconduct

Federal agencies must require that recipients and subrecipients of ARRA grants report to an appropriate inspector general any credible evidence of the submission of a false claim (as defined by the False Claims Act) or any other violation of laws related to fraud, conflict of interest, bribery, gratuity, or similar misconduct.[61]

C. Whistleblower Protection

Section 1553 of the ARRA prohibits nonfederal employers receiving funds under the Act from discharging, demoting, or otherwise discriminating against an employee as reprisal for disclosing certain information to certain government officials *or their representatives* or to his or her supervisor.[62] The relevant government officials are the Recovery Accountability and Transparency Board;[63] an inspector general; the Comptroller General; a member of Congress; a state or federal regulatory or law enforcement agency; a court or grand jury; or the head of a federal agency. A *supervisor* is a person with supervisory authority over the employee or another person working for the employer who has the authority to investigate, discover, or terminate misconduct.[64] Employees are protected from reprisal for disclosing information that the employee reasonably believes is evidence of:

(1) gross mismanagement of an agency contract or grant relating to covered funds;
(2) a gross waste of covered funds;
(3) a substantial and specific danger to public health or safety related to the implementation or use of covered funds;
(4) an abuse of authority related to the implementation or use of covered funds; or
(5) a violation of law, rule, or regulation related to an agency contract (including the competition for or negotiation of a contract) or grant, awarded or issued relating to covered funds.[65]

---

[60]ARRA, Pub. L. No. 111-5, § 902, 123 Stat. 115, 191 (2009).

[61]Office of Mgmt. & Budget, Updated Implementing Guidance for the American Recovery and Reinvestment Act of 2009, M-09-15, § 5.9, at 51 (Apr. 3, 2009), available at *www.whitehouse.gov/omb/assets/memoranda_fy2009/m09-15.pdf*.

[62]ARRA, Pub. L. No. 111-5, § 1553, 123 Stat. 115, 297–302 (2009).

[63]The ARRA established the Recovery Accountability and Transparency Board to coordinate and conduct oversight of covered funds to prevent fraud, waste, and abuse. *See* Pub. L. No. 111-5, §§ 1521–1530, 123 Stat. 115, 289–94 (2009).

[64]ARRA, Pub. L. No. 111-5, § 1553(a), 123 Stat. 115, 297 (2009). *See also* FAR § 3.907-2 (direct contracts).

[65]ARRA, Pub. L. No. 111-5, § 1553(a), 123 Stat. 115, 297 (2009). *See also* FAR § 3.907-11 (direct contracts) (defining "covered information").

Claims of reprisal are easily asserted but difficult for an employer to rebut. An employee establishes that reprisal occurred if he or she demonstrates that a covered disclosure was a *contributing factor* to the discharge, demotion, or other discrimination suffered.[66] A whistleblowing employee can demonstrate that a disclosure was a contributing factor to an adverse employment action with only circumstantial evidence. This evidence may include merely a showing that the employer knew of the disclosure or "that the reprisal occurred within a period of time after the disclosure such that a reasonable person could conclude that the disclosure was a contributing factor."[67] To rebut an employee's allegation, an employer must demonstrate *by clear and convincing evidence* that the employer would have taken the action constituting the reprisal in the absence of the disclosure.[68] The clear-and-convincing-evidence standard is a higher burden of proof than the preponderance-of-the-evidence standard typically applicable to civil actions; however, it is not as high as the burden of "beyond a reasonable doubt" that applies to criminal matters.

If the head of the federal agency concerned determines that an employer discharged, demoted, or otherwise discriminated against an employee as reprisal, then the agency must take one or more of these actions:[69]

(1) Order the employer to take affirmative action to abate the reprisal;
(2) Order the employer to reinstate the person to the position that the person held before the reprisal, together with the compensation (including back pay), compensatory damages, employment benefits, and other terms and conditions of employment that would apply to the person in that position if the reprisal had not been taken; or
(3) Order the employer to pay the complainant an amount equal to the aggregate amount of all costs and expenses (including attorneys' fees and expert witnesses' fees) that were reasonably incurred by the complainant in connection with bringing the complaint regarding the reprisal.

If the head of an agency issues an order denying relief in whole or in part, has not issued an order within an appropriate time (210 days after submission of a complaint if there has been no time extension), or decides not to investigate or to discontinue an investigation, then a good-faith complainant may bring an action against the employer in the appropriate district court of the United States.[70] An order issued by the head of an agency can be enforced in the United States district court for the

[66]ARRA, Pub. L. No. 111-5, § 1553(c)(1)(A)(i), 123 Stat. 115, 299 (2009). *See also* FAR § 3.907-6(a)(1)(i) (direct contracts).
[67]ARRA, Pub. L. No. 111-5, § 1553(c)(1)(A)(ii), 123 Stat. 115, 299 (2009). *See also* FAR § 3.907-6(a)(1)(ii) (direct contracts).
[68]ARRA, Pub. L. No. 111-5, § 1553(c)(1)(B), 123 Stat. 115, 299 (2009). *See also* FAR § 3.907-6(a)(2) (direct contracts).
[69]ARRA, Pub. L. No. 111-5, § 1553(c)(2), 123 Stat. 115, 300 (2009). *See also* FAR § 3.907-6(b) (direct contracts).
[70]ARRA, Pub. L. No. 111-5, § 1553(c)(3), 123 Stat. 115, 300 (2009). *See also* FAR § 3.907-6(c) (direct contracts).

district in which the reprisal was found to have occurred.[71] An employer adversely affected or aggrieved by an order issued by the head of an agency may obtain review of the order in the United States court of appeals for the circuit in which the reprisal is alleged in the order to have occurred.[72]

V. WAGE-RATE REQUIREMENTS

ARRA section 1606 applies the wage-rate requirements of the Davis-Bacon Act[73]— already applicable to all direct government contracts for construction—to all construction projects funded *in whole or in part* through grants under the ARRA.[74] All laborers and mechanics employed by contractors and subcontractors with contracts in excess of $2,000 for construction, alteration, or repair (including painting and decorating) must be paid wages at rates not less than those prevailing on projects of a similar character in the locality as determined by the U.S. Secretary of Labor.[75] The contracting agency or grantor agency should include in its solicitation a wage determination setting forth the wage rates and fringe benefits applicable on the project. For a more detailed discussion of the Davis-Bacon Act, see **Chapter 5.**

> ### ➤ LESSONS LEARNED AND ISSUES TO CONSIDER

- The American Recovery and Reinvestment Act (ARRA or Recovery Act) imposes *additional obligations* on government contractors and on recipients and sub-recipients of grants funded by the Act.
- The Recovery Act applies to federal government solicitations issued and contract awarded on or after March 31, 2009. An agency may add the Recovery Act clauses in the FAR by a *bilateral modification*, which typically requires the contractor's agreement and consideration for the change. Contractors should consider both the obligations and benefits associated with a Recovery Act funded change order.
- The ARRA requires contractors and many first-tier subcontractors and vendors to *report* information concerning funds allocated and received as well as the status of projects.

(Continued)

---

[71]ARRA, Pub. L. No. 111-5, § 1553(c)(4), 123 Stat. 115, 300 (2009). *See also* FAR § 3.907-6(d) (direct contracts).
[72]ARRA, Pub. L. No. 111-5, § 1553(c)(5), 123 Stat. 115, 300 (2009). *See also* FAR § 3.907-6(e) (direct contracts).
[73]40 U.S.C. §§ 3141–3148.
[74]ARRA, Pub. L. No. 111-5, § 1606, 123 Stat. 115, 303 (2009).
[75]2 C.F.R. §§ 176.180, 176.190.

- Certain contractors and first-tier subcontractors and vendors must report the total compensation of their five most highly compensated officers.
- Information reported pursuant to the ARRA is available to the public on the *www.recovery.gov* Web site.
- Payments received and work funded by the ARRA must be *tracked separately*.
- The ARRA requires contractors to comply with special additional *Buy American requirements* concerning iron, steel, and manufactured goods.
- The ARRA Buy American tests and requirements *differ* from those in the traditional Buy American Act. Contractors receiving a Recovery Act funded modification to a non-ARRA construction project need to consider how to manage the procurement of similar materials or goods for that project from their subcontractors and vendors, which may be subject to *different Buy American* tests. This management effort involves the contractor's quality control staff as well as specific provisions in subcontracts and purchase orders addressing verification of compliance.
- Government *representatives from the GAO and offices of the agency inspectors general can inspect records and interview employees* with respect to contracts or subgrants awarded to contractors using ARRA funds.
- *Whistleblowers* are protected from retaliation by employers.
- *Davis-Bacon Act wage rates* apply to projects funded by ARRA grants.

APPENDIX A: INTERNET-BASED RESOURCES APPLICABLE TO GOVERNMENT CONTRACTING

I. FED BIZ OPPS

For firms and individuals desiring to do business with the various agencies of the government, most agencies maintain Web sites that will provide information regarding solicitations, both pending and contemplated (i.e., presolicitation notice). The primary source, however, for all contracting is Fed Biz Opps—Federal Business Opportunities: *http://fedbizopps.gov/.*[1]

To access the Fed Biz Opps main page, enter *http://fedbizopps.gov* or *http://fbo .gov* in the address bar of a web browser. Choosing the "Opportunities" button at the top of the page will allow the user to browse all the current opportunities listed on the site. On the main page, it is possible to make a more refined search by selecting the location, agency, type of contract ("y" is the code for construction; "z" encompasses renovation), set-aside program, or type of notice (award, solicitation, etc.).

If, however, a more generalized search is necessary, below the drop-down box there are several hot links to major agencies that procure goods and construction services (e.g., DOD—Department of Defense, and DHS—Department of Homeland Security). If the search is for business opportunities for all agencies, select "all" (or go to the alphabetical listing of agencies).

II. CENTRAL CONTRACTOR REGISTRATION

Currently, in order to participate in most federal agency contracting programs, a contractor must be listed in the Central Contractor Registration (CCR). To access, enter *www.ccr.gov.* This Web site provides information on the procedures for becoming registered in the government's CCR. There is also information on how to become registered on the Fed Biz Opps Web site and on individual agency Web sites.

---

[1] Although verified at time of submission to the publisher (November 2009), Web site addresses are subject to change.

Obtaining a Central Contracting Registration number is not a complex procedure, but if a contractor wants to view an agency's solicitations or submit a bid proposal, the contactor must provide its registered CCR number.

Electronic funds transfer (ETF) is the default payment procedure used by the government on its contracts.[2] All contractors, except for foreign firms working outside of the United States, must provide the data regarding ETF transfers to the contractor's financial institution or bank to enable the government to make payment via an ETF. These sections are mandatory fields in the Central Contractor Registration process.

III. OTHER WEB-BASED RESOURCES—CONTRACTING OPPORTUNITIES

Virtually all agencies have Web sites, both home page sites for the headquarters organization and sites for individual subordinate offices located throughout the United States. Many agencies, such as the United States Department of Defense and its various service branches (e.g., the United States Air Force, the Department of the Army, the United States Corps of Engineers, the United States Navy, etc.), annually issue thousands of solicitations for various types of goods or services.

If a contractor is interested only in federal business opportunities in a specific geographical area, the contractor may search, for example, the United States Army Corps of Engineers. In a search engine (e.g., Google or Yahoo!), enter the name "United States Army Corps of Engineers" (in quotes). The first that comes up is the Corps of Engineers Headquarters Web site. On the first page at the top there is a hot link entitled "Contract with the Corps." The hot link will take a contractor to a Web site that provides information on how to become a registered contractor (a contractor with a CCR number). This page also has a hot link to all Corps of Engineers field offices ("Find a Local Corps Office") in, and outside, the United States

Each Corps field office has its own Web site. By searching for the term "contracting opportunities," information may be obtained for that office's upcoming projects, telephone numbers, e-mail addresses, and procedures for submitting bids.

Other agency Web sites provide contracting opportunities. For example, the Web site www.defenselink.mil/sites lists all DOD agencies and offices. The same information may be obtained for the United States Department of the Interior (*www.doi.gov*), the General Services Administration (*www.gsa.gov*), the Department of Housing and Urban Development (*www.hud.gov*), and so on. On the first page of the Web site of each agency is a search box. A search using "contracting opportunities" will provide substantial information on each agency's programs, procurement procedures, and current and planned projects.

---

[2]*See* FAR Subpart 32.11.

IV. INFORMATION ON REGULATIONS AND PROCEDURES

Various Web sites that provide information on several aspects of federal contracting, including laws and regulations, are listed in the next sections.

A. Regulations

- For research of the basic Federal Acquisition Regulation (FAR), see: *www .acquisition.gov/far*.
- For research in the most current FAR as well as the agency supplements, see: *http://ecfr.gpoaccess.gov*. Select "Title 48—Federal Acquisition Regulation System" from the pull-down menu and hit "Go."
- For research into past regulations that may be applicable to a specific contract, see: *www.gpoaccess.gov/CFR*. Click the link titled "Browse and/or Search" at the left of the page. This will take the user to a page with past versions of the Code of Federal Regulations, including the FAR (Title 48), going back to 1996.
- The United States Air Force maintains a Web site that provides search capability of the Federal Acquisition Regulations as well as the FAR supplementary procurement regulations promulgated by various federal agencies. See: *http:// farsite.hill.af.mil*.
- The Office of Management and Budget has compiled a chart of the parts of the CFR in which departments and agencies have codified grant requirements. See: *www.whitehouse.gov/omb/grants/chart.htm*.

B. Other Federal Agency Programs and Procedures

- The main gateway to all federal Web sites can be found at the official Web site of the United States government at *www.usa.gov*.
- The *Washington Post* maintains a Web site for information on key individuals in the government at *www.whorunsgov.com*.

1. Agency Web Sites

- The United States Army Corps of Engineers site provides information on its various offices, programs, and activities. This site also provides information and access to ongoing and planned procurement opportunities. See: *www.usace .army.mil*.
- For research of the Federal Emergency Management Agency, see: *www.fema .gov/*.
- For research of the General Services Administration (GSA), its programs, offices, and activities, see: *http://gsa.gov/*.

- For research of the Government Accountability Office (GAO), its rules, regulations, and decisions, see: *www.gao.gov/*.
- For research of the Department of Health and Human Services, see: *www.hhs.gov/*.
- For research on the Department of Housing and Urban Development, see: *www.hud.gov*.
- For research of the United States Department of the Interior (DOI) and its programs and activities, see: *www.doi.gov/*.
- For research of Department of Labor (DOL) programs, activities, and forms, see: *www.dol.gov/*.
- For research of the United States Postal Services, its contracting programs and policies, see: *www.usps.com/cpim/manuals/pm/pm.htm*.
- For research of the Small Business Administration, its programs and activities, as well as size standards, see: *www.sba.gov/contractingopportunities/officials/size/index.html*.
- For access to the Small Business Administration's Office of Hearings and Appeals, see: *www.sba.gov/aboutsba/sbaprograms/oha/ohadecisions/index.html*.
- For research of the United States Department of Transportation (DOT), its programs and activities, see: *www.dot.gov/*.
- For research of the Department of Veterans Affairs, its programs and activities, see: *www.va.gov/partners/buspart/index.htm*.
- For information on and access to resources pertaining to the reporting requirements of the American Reinvestment and Recovery Act of 2009 (ARRA), see: *www.FederalReporting.gov*.
- For information the Defense Contract Audit Agency, see: *www.dcaa.mil*.
- For information on the Federal Procurement Data Systems, see: *www.fpds.gov*.
- For the OMB's Circular "Grants and Cooperative Agreements with State and Local Governments," see: *www.whitehose.gov/omb/circulars/a102/a102.html*.
- For the OMB's instructing for federal agencies not to follow a GAO ruling concerning selecting of procurements for set-asides, see: *www.whitehouse.gov/omb/assets/memoranda_fy2009/m09-23.pdf*.
- For research on the guidance given by the OMB for implementing the ARRA, see: *www.whitehouse.gov/omb/assets/memorand_fy2009/m09-15.pdf* and *www.whitehouse.gov/omb/assets/memoranda_fy2009/m09-21.pdf*.
- For the Online Representations and Certifications Application (ORCA), see: *orca.bpn.gov*.

2. *Government Standards, Guidance, and Other Reference Materials*

- For agency reference materials, standards, and design guides, go to the agency Web site. Sites such as those maintained by the Corps of Engineers, Naval

Facilities Engineering Command, the Veterans Administration, and the General Services Administration's Public Buildings Service have links to "publications" or "libraries."

- For information published by the Department of Treasury on interest rates applicable to monies owed by or to the federal government, see: *www.treasurydirect .gov/govt/rates/rates.htm.*

- For Davis-Bacon Act Wage rate determinations, see: *www.access.gpo .gov/davisbacon.*

- For research of the Department of Treasury regulations and procedures affecting prompt payment, see: *www.fms.treas.gov/prompt.*

- For information on the cost of the American Recovery and ReinvestmentAct of 2009, see: *www.cbo.gov/ftpdocs/99xx/doc9989/hr1conference.pdf.*

- For research on the Department of Defense's program Wide Area Work Flow, see: *www.dfas.mil/contractor/pay/electroniccommerce/wideareaworkflow .html.*

- For information on the Department of Defense's budget for audits by the Defense Contract Audit Agency, see: *http://www.defenselink.mil/comptroller/defbudget/ fy2009/budget_justification/pdfs/02_Procurement/Vol_1_Other_Defense_ Agencies/DCAA%20PDW%20PB09.pdf.*

- For research on the Department of Defense's small business programs, go to see *www.acq.osd.mil/osbp* and then choose the link labeled "Program Goals & Statistics."

- For information on the Electronic Subcontracting Reporting System, see: *www .esrs.gov.*

- The Federal Energy Management Program has published a guide entitled "Commissioning for Federal Facilities" available at *www.eere.energy.gov/femp/ pdfs/commissioning_fed_facilities.pdf.*

- For the Department of Energy's publication "Building Commissioning, the Key to Quality Assurance," go to *www.hud.gov/offices/pih/programs/ph/phecc/ resources.cfm* and scroll down to the link for this publication.

- For research on the Small Business Administration's goals, see: *www.sba.gov/ aboutsba/sbaprograms/goals/index.html.*

- For information on the locations of HUBZones, visit: *www.sba.gov/hubzone.*

- For research on the Environmental Protection Agency's management of grants, see: *www.epa.gov/ogd/EO/finalreport.pdf.*

- For information on the President's Council on Sustainable Development, see: *www.usda.gov/oce/sustainable.*

- For access to the Corps of Engineer's Quality Control System (QCS) User Manual & Training Guide, see: *www.rmssupport.com/qcs/guides.aspx.*

- For research on Army Corps of Engineer's "Safety and Health Requirements Manual," see: *www.usace.army.mil/CESO/Pages/EM385-1-1,2008new!.aspx.*

C. Past Performance Evaluations

- For information on the Office of Federal Procurement Policy's (OFPP) guidance on contractor's past performance evaluations, see: *www.whitehouse.gov/omb/procurement/contract_perf/best_practice_re_past_perf.html.*
- For information on the Department of Defense's policy and guidance on contractor's past performance evaluations, see: *www.ogc.doc.gov/ogc/contracts/cld/papers/ppiguide.pdf*
- For the Office of Management and Budget's (OMB) information on past performance information, see: *www.whitehouse.gov/omb/rewrite/procurement/contract_perf/best_practice_re_past_perf.html.*
- For information on and access to the Past Performance Information Retrieval System (PPIRS), see: *www.ppirs.gov.*

D. Nongovernment Reference Sites

- The American Society of Heating, Refrigeration and Air Conditioning Engineers publishes a wide range of materials on construction, including building information modeling and LEED (Leadership in Energy and Environmental Design), at: *www.ashrae.org.*
- The American Subcontractors Association has information on a wide range of issues pertaining to subcontracting and government procurement at: *www.asaonline.com/Web/subcontractor_advocacy/asa_federal_advocacy.aspx.*
- The Associated General Contractors of America has published a paper on reverse auctions in construction procurement available at: *www.necanet.org/pdf/ASC_News/AGC-reverse-auctions.pdf.*
- For information on building commissioning, see the Building Commissioning Association's Web site at: *www.bcxa.org.*
- For information on the National Defense Industrial Association, see: *www.ndia.org.*
- The National Defense Industrial Association has published the "Earned Value Management System Intent Guide," available at: *http://management.energy.gov/documents/NDIA_PMSC_EVMS_IntentGuide_Nov_2006.pdf.*
- For information on the National Institute of Building Sciences, see: *www.nibs.org.*
- For research on green building and the U.S. Green Building Council, see: *www.usgbc.org.*
- For information on the Whole Building Design Guide, visit: *www.wbdg.org.*
- The Project on Government Oversight (POGO) maintains the Federal Contractor Misconduct Database at: *www.contractormisconduct.org.*

APPENDIX B: GLOSSARY OF TERMS AND ACRONYMS

AA	Army Audit Agency
AAO	Army acquisition objective
ACASS	Architect Engineer Contractor Appraisal Support System
ACO	Administrative contracting officer
ADA	Americans with Disabilities Act
ADP	Automatic data processing
ADR	Alternative dispute resolution
AE	Architect engineer
AEC	Atomic Energy Commission
AF	Air Force
AFAC	Air Force Acquisition Circular
AFAFC	Air Force Accounting and Finance Center
AFARS	Army Federal Acquisition Regulation Supplement
AFFARS	Air Force Federal Acquisition Regulation Supplement
AFPC	Air Force Procurement Circular
AFPRO	Air Force Procurement Representative Office
AFRCE	Air Force regional civil engineer
AGBCA	Department of Agriculture Board of Contract Appeals
ANC	Alaska Native Corporation
APA	Administrative Procedures Act
APP	Army Procurement Procedure
AQL	Acceptable quality level
AR	Army regulations
ARD	Army Renegotiation Division, Armed Services Renegotiation Board
ARRA	American Recovery and Reinvestment Act
ASBCA	Armed Services Board of Contract Appeals
ASD/P&I	Assistant Secretary of Defense (Properties and Installations)
ASPM	Armed Services Pricing Manual
ASPR	Armed Services Procurement Regulation
B&P	Bid and proposal
BAA	Buy American Act
BAFO	Best and final offer

BAQ	Basic allowance for quarters
BCA	Board of Contract Appeals
BEQ	Bachelor enlisted quarters
BOAs	Basic Ordering Agreements
BOQ	Bachelor officer's quarters
BPA	Blanket Purchase Agreement
BPN	Business Partner Network
BUSHIPS	Bureau of Ships (Navy)
CAB	Civil Aeronautics Board
CAD	Computer-assisted design
CAFC	Court of Appeals for the Federal Circuit
CAM	DCAA Contract Audit Manual
CAS	Cost Accounting Standards
CASB	Cost Accounting Standards Board
CBCA	Civilian Board of Contract Appeals
CCASS	Construction Contractor Appraisal Support System
CCR	Central Contractor Registration
CDA	Contract Disputes Act
CFR or C.F.R.	Code of Federal Regulations
CICA	Competition in Contracting Act
CLIN	Contract line item number
CM	Construction management
CMc	Construction Manager as constructor
CO	Contracting officer
COB	Close of business
COC	Certificate of Competency (SBA determination affecting responsibility)
COE or C of E	Corps of Engineers
COFC	Court of Federal Claims
CONUS	Contiguous United States
COR	Contracting officer's representative
COTR	Contracting officer's technical representative
COTS	Commercial off-the-shelf
CPAF	Cost plus award fee
CPARS	Contractor Performance Assessment Reporting System
CPD	Comptroller General's Procurement Decisions

CPFF	Cost plus fixed fee
CPIF	Cost plus incentive fee
CPM	Critical path method
CPPC	Cost plus a percentage of cost
CPSR	Contractor Purchasing System Review
CWAS	Contractor weighted average share in cost risk
CWHSSA	Contract Work Hours and Safety Standards Act
CX	Center of expertise
CY	Calendar year
DA	Department of the Army
DAC	Defense Acquisition Circular
DAR	Defense Acquisition Regulation
DBA	Davis Bacon Act (addresses prevailing wages on federal contracts)
DCAA	Defense Contract Audit Agency
DCAAM	Defense Contract Audit Agency Manual
DCAS	Defense Contract Administration Service
DCMA	Defense Contract Management Agency
DFARS	Department of Defense Federal Acquisition Regulation Supplement
DHS	Department of Homeland Security
DLAR	Defense Logistics Acquisition Regulation
DOA	Department of Agriculture
DOD	Department of Defense
DOE	Department of Energy
DOI	Department of the Interior
DOJ	Department of Justice
DOL	Department of Labor
DOS	Department of State
DOT	Department of Transportation
DOTBCA	Department of Transportation Board of Contract Appeals
DOTCAB	Department of Transportation Contract Appeals Board (later called DOTBCA)
DPAS	Defense Priorities and Allocations System
DUNS	Data Universal Numbering System
EAJA	Equal Access to Justice Act
EBCA	Department of Energy Board of Contract Appeals

ECI	Early contractor involvement
ECP	Engineering change proposal
EFT	Electronic funds transfer
ENG	Corps of Engineers (Army)
ENGBCA	Corps of Engineers Board of Contract Appeals
E.O.	Executive Order of the President
EPA	Environmental Protection Agency
EPLS	Excluded Parties List System
F&D	Findings and determinations
FAA	Federal Aviation Administration
FAC	Federal Acquisition Circular
FAQ	Frequently asked question
FAR	Federal Acquisition Regulation
FASA	Federal Acquisition Streamlining Act of 1994
FBI	Federal Bureau of Investigation
FCA	False Claims Act
FEMA	Federal Emergency Management Agency
FERA	Fraud Enforcement and Recovery Act of 2009
FFP	Firm fixed price contract
FGCAA	Federal Grant and Cooperative Agreement Act of 1977
FHWA	Federal Highway Administration
FLSA	Fair Labor Standards Act
FMS	Foreign military sales
FOIA	Freedom of Information Act
FPDS NG	Federal Procurement Data System Next Generation
FPE	Fixed price with escalation
FPIF	Fixed price incentive contracts (firm target incentive)
FPIS	Successive target incentive contract
FPR	Federal Procurement Regulations
FTCA	Federal Tort Claims Act
FY	Fiscal year
G&A	General and administrative
GAO	Government Accountability Office (formerly General Accounting Office)
GC	General contractor
GFE	Government-furnished equipment
GFM	Government-furnished material

GFP	Government-furnished property
GOCO	Government-owned contractor-operated
GPE	Government-wide point of entry
GPO	Government Printing Office
GSA	General Services Administration
GSAR	General Services Acquisition Regulation
GSBCA	General Services Administration Board of Contract Appeals
GTE	Government technical evaluation
GTR	Government technical representative
HCA	Head of Contracting Activity
HHS	Department of Health and Human Services
HPA	Head of Procuring Activity
HUBZone	Historically Underutilized Business Zone
HUD	Department of Housing and Urban Development
HUD BCA	Department of Housing and Urban Development Board of Contract Appeals
IBCA	Department of Interior Board of Contract Appeals
IDBB	Integrated design-bid-build
IDC	Integrated design-construct
IDIQ or ID/IQ	Indefinite delivery indefinite quantity
IFB	Invitation for bids
IFN	Item for negotiation
IG	Inspector general
I&L	Installations and logistics
IR&D	Independent research and development
IRS	Internal Revenue Service
J&A	Justification and authorization for less than full and open competition under CICA
LEED	Leadership in Energy and Environmental Design
LOC	Limitation of cost
LOF	Limitation of funds
MACC	Multiple award construction contract
MBE	Minority business enterprise
MCP	Military construction program

MC PIP	Model code for public infrastructure procurement
MILCON	Military construction
MILSPEC	Military specification
MILSRAP	Military standard contract administration procedures
MIL STD	Military standard
MIRR	Material inspection and receiving report
MOA	Memorandum of agreement
MOU	Memorandum of understanding
MSDS	Material Safety Data Sheets
NAICS	North American Industry Classification System
NAPS	Navy Acquisition Procedures Supplement
NARSUP	Navy Acquisition Regulation Supplement
NASA	National Aeronautics and Space Administration
NASA BCA	National Aeronautics and Space Administration Board of Contract Appeals
NASA PR	National Aeronautics and Space Administration Procurement Regulation
NAVFAC	Naval Facilities Engineering Command
NAVY CAB	Navy Contact Adjustment Board
NFS	NASA FAR Supplement
NLRB	National Labor Relations Board
NPR	NASA Procurement Regulation
O&M	Operations and maintenance
OFCC	Office of Federal Contract Compliance
OFCCP	Office of Federal Contract Compliance Programs
OFPP	Office of Federal Procurement Policy
OICC	Officer in Charge of Construction (Navy)
OMB	Office of Management and Budget
ORCA	Online Representations and Certifications Application
OSHA	Occupation Safety and Health Administration
OSD	Office of the Secretary of Defense
PBS	Public Buildings Service (GSA)
PCO	Procuring contracting officer
PL or Pub. L.	Public law
POC	Point of contact

POD BCA	Post Office Department Board of Contract Appeals (now PSBCA)
PPA	Prompt Payment Act
PPIRS	Past Performance Information Retrieval System
PSBCA	Postal Service Board of Contract Appeals
QA	Quality assurance
QC	Quality control
QPL	Qualified Products List
R&D	Research and development
REA	Request for equitable adjustment
RFI	Request for information
RFP	Request for proposal(s)
RFQ	Request for quotation(s)
ROICC	Resident Officer in Charge of Construction (Navy)
Rule of Two	SBA rule requiring use of small business set-aside if two or more small businesses are available to provide the goods or services
SABER	Simplified Acquisition of Base Engineering Requirements
SBA	Small Business Administration
SBC	Small business concern
SBIR	Small Business Innovation Research Program
SDB	Small disadvantaged business
SDVO	Service-Disabled Veteran-Owned Small Business
Section 8(a)	Section 8(a) of the Small Business Act; Federal Contracting Preference Program for Disadvantaged Small Business Concerns
Set-aside	Procurement reserved for small business concerns
SF	Standard form
SIC	Standard Industrial Classification
SOP	Standard operating procedure
SOW	Statement of work
SSA	Social Security Administration
SSB	Source Selection Board
T&M	Time and materials
T for C or T4C	Termination for convenience
T for D or T4D	Termination for default
TASK ORDER	Order for services under an ID/IQ contract

TBA	To be announced
TBD	To be determined
TCO	Termination contracting officer
TIN	Taxpayer Identification Number
TINA	Truth in Negotiations Act
TVA	Tennessee Valley Authority
UCC	Uniform Commercial Code
UNICOR	Federal Prison Industries
USA	United States Army
USACE	United States Army Corps of Engineers
USAF	United States Air Force
USC or U.S.C.	United States Code
USCIS	United States Citizenship and Immigration Services
USG	United States Government
USPS	United States Postal Service
VA	Department of Veterans Affairs
VAAR	Veterans Administration Acquisition Regulation
VABCA	Veterans Affairs Board of Contract Appeals
VACAB	Veterans Administration Contract Appeals Board
VE	Value engineering
VECP	Value engineering change proposal
VPP	Vendor Past Performance System (module in GSA's contract writing system to collect past performance data; used by some GSA regions)
WOSB	Woman-owned small business
YTD	Year to date

INDEX